Lecture Notes in Physics

For information about Vols. 1–90, please contact your bookseller or Springer-Verlag.

Vol. 91: Computing Methods in Applied Sciences and Engineering, 1977, II. Proceedings, 1977. Edited by R. Glowinski and J. L. Lions. VI, 359 pages. 1979.

Vol. 92: Nuclear Interactions. Proceedings, 1978. Edited by B. A. Robson. XXIV, 507 pages. 1979.

Vol. 93: Stochastic Behavior in Classical and Quantum Hamiltonian Systems. Proceedings, 1977. Edited by G. Casati and J. Ford. VI, 375 pages. 1979.

Vol. 94: Group Theoretical Methods in Physics. Proceedings, 1978. Edited by W. Beiglböck, A. Böhm and E. Takasugi. XIII, 540 pages. 1979.

Vol. 95: Quasi One-Dimensional Conductors I. Proceedings, 1978. Edited by S. Barišić, A. Bjeliš, J. R. Cooper and B. Leontič. X, 371 pages. 1979.

Vol. 96: Quasi One-Dimensional Conductors II. Proceedings 1978. Edited by S. Barišić, A. Bjeliš, J. R. Cooper and B. Leontić. XII, 461 pages. 1979.

Vol. 97: Hughston, Twistors and Particles. VIII, 153 pages. 1979.

Vol. 98: Nonlinear Problems in Theoretical Physics. Proceedings, 1978. Edited by A. F. Rañada. X, 216 pages. 1979.

Vol. 99: M. Drieschner, Voraussage – Wahrscheinlichkeit – Objekt. XI, 308 Seiten. 1979.

Vol. 100: Einstein Symposion Berlin. Proceedings 1979. Edited by H. Nelkowski et al. VIII, 550 pages. 1979.

Vol. 101: A. Martin-Löf, Statistical Mechanics and the Foundations of Thermodynamics. V, 120 pages. 1979.

Vol. 102: H. Hora, Nonlinear Plasma Dynamics at Laser Irradiation. VIII, 242 pages. 1979.

Vol. 103: P. A. Martin, Modèles en Mécanique Statistique des Processus Irréversibles. IV, 134 pages. 1979.

Vol. 104: Dynamical Critical Phenomena and Related Topics. Proceedings, 1979. Edited by Ch. P. Enz. XII, 390 pages. 1979.

Vol. 105: Dynamics and Instability of Fluid Interfaces. Proceedings, 1978. Edited by T. S. Sørensen. V, 315 pages. 1979.

Vol. 106: Feynman Path Integrals, Proceedings, 1978. Edited by S. Albeverio et al. XI, 451 pages. 1979.

Vol. 107: J. Kijowski, W. M. Tulczyjew, A Symplectic Framework for Field Theories. IV, 257 pages. 1979.

Vol. 108: Nuclear Physics with Electromagnetic Interactions. Proceedings, 1979. Edited by H. Arenhövel and D. Drechsel. IX, 509 pages. 1979.

Vol. 109: Physics of the Expanding Universe. Proceedings, 1978. Edited by M. Demiański. V, 210 pages. 1979.

Vol. 110: D. A. Park, Classical Dynamics and Its Quantum Analogues. VIII, 339 pages. 1979.

Vol. 111: H.-J. Schmidt, Axiomatic Characterization of Physical Geometry. V, 163 pages. 1979.

Vol. 112: Imaging Processes and Coherence in Physics. Proceedings, 1979. Edited by M. Schlenker et al. XIX, 577 pages. 1980.

Vol. 113: Recent Advances in the Quantum Theory of Polymers. Proceedings 1979. Edited by J.-M. André et al. V, 306 pages. 1980.

Vol. 114: Stellar Turbulence. Proceedings, 1979. Edited by D. F. Gray and J. L. Linsky. IX, 308 pages. 1980.

Vol. 115: Modern Trends in the Theory of Condensed Matter. Proceedings, 1979. Edited by A. Pekalski and J. A. Przystawa. IX, 597 pages. 1980.

Vol. 116: Mathematical Problems in Theoretical Physics. Proceedings, 1979. Edited by K. Osterwalder. VIII, 412 pages. 1980.

Vol. 117: Deep-Inelastic and Fusion Reactions with Heavy Ions. Proceedings, 1979. Edited by W. von Oertzen. XIII, 394 pages. 1980.

Vol. 118: Quantum Chromodynamics. Proceedings, 1979. Edited by J. L. Alonso and R. Tarrach. IX, 424 pages. 1980.

Vol. 119: Nuclear Spectroscopy. Proceedings, 1979. Edited by G. F. Bertsch and D. Kurath. VII, 250 pages. 1980.

Vol. 120: Nonlinear Evolution Equations and Dynamical Systems. Proceedings, 1979. Edited by M. Boiti, F. Pempinelli and G. Soliani. VI, 368 pages. 1980.

Vol. 121: F. W. Wiegel, Fluid Flow Through Porous Macromolecular Systems. V, 102 pages. 1980.

Vol. 122: New Developments in Semiconductor Physics. Proceedings, 1979. Edited by F. Beleznay et al. V, 276 pages. 1980.

Vol. 123: D. H. Mayer, The Ruelle-Araki Transfer Operator in Classical Statistical Mechanics. VIII, 154 pages. 1980.

Vol. 124: Gravitational Radiation, Collapsed Objects and Exact Solutions. Proceedings, 1979. Edited by C. Edwards. VI, 487 pages. 1980.

Vol. 125: Nonradial and Nonlinear Stellar Pulsation. Proceedings, 1980. Edited by H. A. Hill and W. A. Dziembowski. VIII, 497 pages. 1980.

Vol. 126: Complex Analysis, Microlocal Calculus and Relativistic Quantum Theory. Proceedings, 1979. Edited by D. Iagolnitzer. VIII, 502 pages. 1980.

Vol. 127: E. Sanchez-Palencia, Non-Homogeneous Media and Vibration Theory. IX, 398 pages. 1980.

Vol. 128: Neutron Spin Echo. Proceedings, 1979. Edited by F. Mezei. VI, 253 pages. 1980.

Vol. 129: Geometrical and Topological Methods in Gauge Theories. Proceedings, 1979. Edited by J. Harnad and S. Shnider. VIII, 155 pages. 1980.

Vol. 130: Mathematical Methods and Applications of Scattering Theory. Proceedings, 1979. Edited by J. A. DeSanto, A. W. Sáenz and W. W. Zachary. XIII, 331 pages. 1980.

Vol. 131: H. C. Fogedby, Theoretical Aspects of Mainly Low Dimensional Magnetic Systems. XI, 163 pages. 1980.

Vol. 132: Systems Far from Equilibrium. Proceedings, 1980. Edited by L. Garrido. XV, 403 pages. 1980.

Vol. 133: Narrow Gap Semiconductors Physics and Applications. Proceedings, 1979. Edited by W. Zawadzki. X, 572 pages. 1980.

Vol. 134: $\gamma\gamma$ Collisions. Proceedings, 1980. Edited by G. Cochard and P. Kessler. XIII, 400 pages. 1980.

Vol. 135: Group Theoretical Methods in Physics. Proceedings, 1980. Edited by K. B. Wolf. XXVI, 629 pages. 1980.

Lecture Notes in Physics

Edited by H. Araki, Kyoto, J. Ehlers, München, K. Hepp, Zürich
R. Kippenhahn, München, H. A. Weidenmüller, Heidelberg
and J. Zittartz, Köln

170

Eighth International Conference on Numerical Methods in Fluid Dynamics

Proceedings of the Conference,
Rheinisch-Westfälische Technische Hochschule
Aachen, Germany, June 28 – July 2, 1982

Edited by E. Krause

Springer-Verlag
Berlin Heidelberg GmbH 1982

Editor

Egon Krause
Aerodynamisches Institut
der Rheinisch-Westfälischen Technischen Hochschule Aachen
Wüllnerstr. zw. 5–7, D-5100 Aachen

ISBN 978-3-540-11948-7 ISBN 978-3-540-39532-4 (eBook)
DOI 10.1007/978-3-540-39532-4

Originally published by Springer-Verlag Berlin Heidelberg New York in 1982

2153/3140-543210

CONTENTS

Invited Lectures

Contributed Papers

INTERNATIONAL CONFERENCE ON NUMERICAL METHODS IN FLUID DYNAMICS

First Conference:	Novosibirsk, USSR, 1969
Second Conference:	Berkeley, California, USA, 1970
Third Conference:	Paris, France, 1972
Fourth Conference:	Boulder, Colorado, USA, 1974
Fifth Conference:	Enschede, The Netherlands, 1976
Sixth Conference:	Tbilisi, USSR, 1978
Seventh Conference:	Stanford University and NASA/Ames, USA, 1980
Eighth Conference:	Aachen, West Germany, 1982

Editor's Preface

This volume of Lecture Notes in Physics contains the papers presented at the Eighth International Conference on Numerical Methods in Fluid Dynamics, held at the Rheinisch-Westfälische Technische Hochschule Aachen in West Germany, June 28 to July 2, 1982. The papers presented were selected from abstracts submitted from all over the world by three papers selection committees, one in the U.S.A., another in the U.S.S.R., and the third in Europe. The papers selection committees were headed by M. Holt (U.S.A.), N.N. Yanenko (U.S.S.R.), and the editor. Abbreviated versions of the papers were distributed at the conference. The detailed papers are given in this volume.

Included are the invited lectures by W. F. Ballhaus, P. I. Chushkin and L. V. Shurshalov, K. W. Morton, K. Oshima and Y. Oshima, P. Moin, and K. Gersten, and 65 contributed papers. The invited papers appear first, followed by the contributed papers in alphabetical order by the first author.

The papers were presented in nine sessions, which were entitled: Solution of the Navier-Stokes Equations, Solution of the Euler Equations, Accuracy of Euler Solutions, Gas Dynamics, Transonic Flows, Transitional and Turbulent Flows, Internal Flows, Flows with Strong Rotation, and Boundary Layers.

The conference was attended by over 200 scientists. In addition to the strong representation from West Germany, the participation of delegations from the U.S.A., the U.S.S.R., France, The Netherlands, Great Britain, Japan, Israel, Italy, Belgium, Austria, Canada, Algeria, China, Czechoslovakia, Finland, Ireland, Norway, Sweden, Switzerland, Poland, and Saudi Arabia showed the continuously increasing interest in this conference throughout the world. A list of the participants is given at the end of this volume.

By choosing Aachen as the site of the eighth conference, the permanent members of the Organizing Committee for the International Conferences on Numerical Methods in Fluid Dynamics paid tribute to the early work in flow prediction done here at the Technische Hochschule. Von Kármán and Trefftz completed their analysis of wing sections at the Aerodynamisches Institut as early as 1914. Their paper, published in 1918, introduced a new series of profiles which later became known as the von Kármán-Trefftz profiles. In 1922 Karl Pohlhausen submitted his doctoral dissertation to the Technische Hochschule. In his work he constructed a prediction method by using von Kármán's integral formulation of the boundary-layer equations. For sixty years Pohlhausen's solution and later improvements and extensions thereof proved to be an indispensable tool in the design work of aerodynamicists. Even today integral methods are still being used to study complex boundary-layer problems. In 1927 von

Kármán published a method for calculating the pressure distribution on airships, and shortly thereafter, he and K. O. Friedrichs coauthored a paper on cantilever wings. Friedrichs, at that time a member of the Aerodynamisches Institut, also investigated flow through propellers. During World War II, R. Sauer occupied the chair for practical mathematics here at the Technische Hochschule. His book on gas dynamics, first published in 1943, is a collection of lectures he gave to members of the Aerodynamisches Institut in 1940. The book is now considered one of the fundamental works on gas dynamics.

The editor served as the general conference chairman. He is indebted to many colleagues who helped with the details of the conference, but particularly to Wolfram Limberg of the Aerodynamisches Institut, who supervised all of the local arrangements, and to Hertha Rehfeld, the conference secretary.

Financial support for the conference was provided by the Deutsche Forschungsgemeinschaft, Dornier GmbH, Messerschmitt-Bölkow-Blohm GmbH, the U.S. Department of the Navy, Office of Naval Research, and the European Research Office of the United States Army.

The editor is also indebted to Professor W. Beiglböck and the editorial staff of the Springer Verlag for valuable assistance in preparing these proceedings. Grateful acknowledgement is also due to Prof. M. Holt, who helped in the review of some of the papers published in this volume.

August 10, 1982

Egon Krause

(Editor)

COMPUTATIONAL AERODYNAMICS AND DESIGN

W. F. Ballhaus, Jr.
Ames Research Center, NASA
Moffett Field, California

INTRODUCTION

In the last decade, advances in computer technology and data communications began to drastically change the way we live and the way we work. We bank, shop, make airline reservations, and pay our bills by using computers. Now, a new generation of children is growing up with computers in their homes and in their classrooms. This computer revolution has also had a major effect on the production of new aircraft. With the major investment of the aircraft industry in computer-aided design and computer-aided manufacturing (CAD/CAM), much of the development process from design through manufacturing is computer controlled. Furthermore, great progress is being made in computerizing the aeronautical disciplines that are the elements of design, such as aerodynamics, structures, guidance and control, and propulsion.

Nowhere has this progress been more exciting than in aerodynamics. The availability of modern supercomputers and the ingenuity of computational aerodynamics researchers have resulted in new methods for solving historically intractable nonlinear flow-field problems. Advances in data communications have facilitated remote access to these large computing engines, and advances in computer technology incorporated in mini- and midicomputers now provide sophisticated interactive graphics and data manipulation capability. All of these advances have profoundly affected the aerodynamic design process.

Here we assess the changing role of computational aerodynamics in design and consider the prospects for continued advancement.

THE TOOLS OF THE DESIGNER — THEORY AND EXPERIMENT

From the beginning, aerodynamicists have sought to use a proper combination of theory and experiment to achieve design objectives in a timely and cost-effective manner. The development of theoretical and experimental techniques has been motivated by a desire to assist the aerodynamicist in better understanding the influence of design variables on aerodynamic performance. In the early days of aviation, the Wright brothers built a small wind tunnel and even used some theory to enhance their understanding of aerodynamics (fig. 1).

In the decades since, the quality of wind-tunnel test techniques has continued to improve. Today these techniques are advancing at an <u>evolutionary</u> rate. Theoretical techniques also advanced, but this advancement was impeded by the nonlinearity of

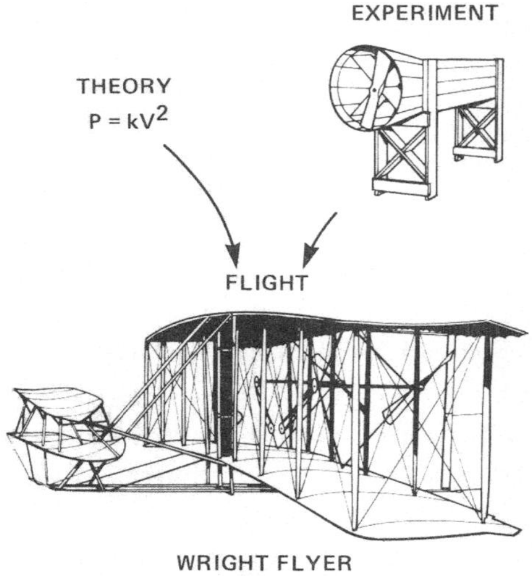

Fig. 1. In the beginning there was a blend of theory and experiment in aerodynamic design.

the simplified form of the partial differential equations governing fluid flows in certain flight regimes, such as the transonic Mach number range. In the late 1960s, computers sufficiently large to permit solution of these equations by finite-difference techniques became available. Computational aerodynamics has since advanced at a revolutionary rate.

In spite of the rapid advancement of computational methods, it is not expected that computational simulations will completely replace wind-tunnel testing in the foreseeable future. Their roles instead are complementary. Computations can provide some of the less intricate flow simulations required in design more quickly and at less cost than wind tunnels. They can also be used to make more effective use of wind tunnels by providing the means to (1) evaluate and improve new design concepts, such as swept forward wings or jet flaps for lift augmentation, before testing; (2) carefully discriminate among candidate configurations, eliminating all but the most promising before testing; (3) assist the aerodynamicist in instrumenting test models to improve resolution of the physical phenomena of interest; and (4) correct wind-tunnel data for scaling and interference errors. Computational simulations can also provide data for conditions that are outside the operating range of existing experimental facilities. An example would be a high-speed planetary probe entry condition, as in the Galileo Probe scheduled to enter the atmosphere of Jupiter in the late 1980s.

Inadequacies in testing (or analog simulations) are associated with limitations in operating range, such as Mach number, Reynolds number, gas composition, and enthalpy level, and the control of boundary conditions, such as flow nonuniformity, wall- and support-interference effects, and model fidelity. When these factors are properly controlled, the physical phenomena in the free-flight condition should be simulated correctly.

The inadequacies in computational (or digital) simulations are primarily associated with poor resolution of physical phenomena, and this is a direct result of insufficient computer power. If one accepts the unsteady compressible Navier-Stokes equations as an adequate system to describe aerodynamic flows, then physical phenomena of interest could be accurately simulated given sufficient computer power. However, current computer power is inadequate to permit numerical solution of these equations with suitable resolution of the wide range of length scales active in high-Reynolds-number turbulent flows. Hence, the aerodynamicist usually resorts to mathematical formulations that are approximations to the Navier-Stokes equations. These approximations introduce phenomenological errors, with the consequence that certain aspects of the flow-field physics are not properly represented. For any given mathematical formulation, the approximating procedures used to solve the governing equations and boundary conditions introduce numerical errors. Specifically, these errors are due to such factors as inadequate grid refinement or incomplete treatment of complex aerodynamic configurations. The consequences are again that physical phenomena are not properly represented. Both phenomenological errors and numerical errors can be reduced by increased computer power.

Because of the inherent differences in the nature of the two types of simulations, wind tunnels and computers are complementary: they have different inherent errors in simulating free-flight conditions. The principal point to be emphasized is that computational and wind-tunnel simulations are merely tools of the designer. Success in design depends very strongly on the judgment and expertise of the designer, his knowledge of aerodynamics, and his ability to use these design tools effectively.

THE ROLE OF COMPUTATIONAL AERODYNAMICS IN DESIGN

The Design Process

In recent years, aircraft design has been complicated substantially by the trade-offs that must be made to accommodate conflicting requirements. For commercial aircraft, these involve performance, cost, noise, and exhaust pollution, all of which are driven by economic and societal pressures. In the case of military aircraft, different mission profiles require trade-offs at multiple design points. For example, an aircraft may be required both to cruise and maneuver efficiently at transonic speeds and to accelerate rapidly to supersonic speeds and perform effectively in that regime. The optimum aerodynamic configurations that correspond to each of these design points are significantly different. This suggests aerodynamic designs with variable geometry, which further complicates the mechanical and structural design of the aircraft.

The process by which these requirements are converted into a production aircraft is illustrated schematically in figure 2 (M. Lores, Lockheed-Georgia, private communication, May 1982). The system requirements, as determined from customer requirements and mission analysis, feed into the conceptual design phase. From simple analyses and parametric variation studies, a conceptual baseline emerges. During the preliminary design phase, the concept is refined by means of more detailed analyses or exploratory tests or both. Design baselines are then allocated to each of the technical specialties. In the detailed design phase, detailed analysis or testing leads to final tests to verify the production baseline.

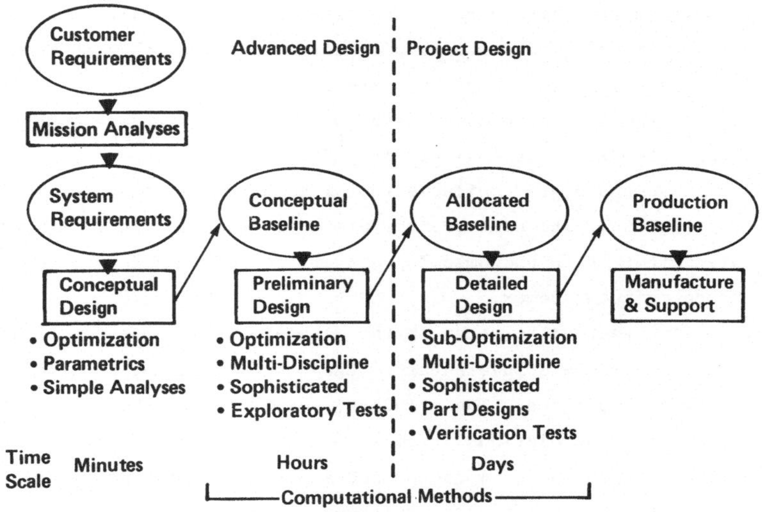

Fig. 2. The aircraft design process.

Levels of Use of Computational Aerodynamics in Design

Pierre Perrier (Dassault, private communication, Feb. 1982) defines four distinct levels at which computational aerodynamics is used in the design process. Level 0 involves no use whatsoever, with the designer relying on analysis, empiricisms, and successive wind-tunnel testing to refine the design. Level 1 involves extensive computation in the preliminary design (PD) phase followed by configuration refinement via wind-tunnel testing in the detailed design (DD) phase. Level 2 involves extensive reliance on computations in PD with a synergistic mixture of computations and wind-tunnel testing in DD. Final performance verification is obtained using wind-tunnel testing. At level 3, computational simulations are the principal design tool in both PD and DD, with some wind-tunnel verification provided at appropriate check-points in both PD and DD.

There is a variation among the broad range of designers of aerospace vehicles in the degree to which computations are relied on in the design process. It is fair to say that all major airframe manufacturers have progressed beyond level 0. That is, they all use computations at least in PD. The result has been a greater degree of management confidence in the designs emerging from the PD phase. Hence, larger performance gains are possible as a result of the consequently greater degree of design flexibility permitted. Furthermore, for the same budget, a greater degree of design refinement can be achieved before encountering the decision point that determines whether the project will continue.

Some manufacturers have not progressed beyond level 1 to the use of computations in DD, either because they do not consider aerodynamic refinements in their products to be highly important or because they lack confidence in their computational simulations. Generally, to reach level 2, a design group must have (1) a turnaround on their computations competitive with their wind-tunnel fabrication and test turnaround and (2) a high degree of confidence in their computational simulations, at least in the operating range near the design point. According to Raimo Hakkinen (McDonnell-Douglas, private communication, Jan. 1982), "The role of wind tunnel testing in transport wing design has evolved from that of almost complete dominance in the 1960s to the current one of verification testing after the final configuration candidates have been selected by computational methods. The total cost of the design process has been correspondingly reduced by an order of magnitude (in a recent project by a factor of 15), and a specific wing design can be evaluated in as little as 24 hours, instead of the several weeks normally required for fabrication and testing of a wind tunnel model." For fighter aircraft designs, which strain the limits of performance and require treatment of complex three-dimensional viscous flow phenomena, computations account for only about 20 to 30 percent of the aerodynamic design effort, according to Richard Bradley (General Dynamics, private communications, Jan. 1982). This percentage is expected to increase as computational treatment of these complex phenomena advances.

Level 3 use of computational aerodynamics is still very rare, although there are some outstanding examples, usually involving only minor to moderate changes to some base configuration that has been extensively tested. The Dassault Falcon 50 is one of these examples. In the case of this transatlantic business tri-jet, a new computationally optimized wing was introduced in fabrication and flight test with only one low-speed and two high-speed wind-tunnel test series (P. Perrier).

The potential payoffs from the extensive use of computational simulations at levels 2 and 3 are high, but there is a significant risk that must be addressed. Generally, testing occurs later in the design cycle in levels 2 and 3 than in levels 0 and 1. Before verification tests are performed, the aerodynamic design cannot be considered frozen, and hence the nonaerodynamics groups cannot freeze their designs. Should difficulties be uncovered late in the design cycle by verification tests, development cost, schedule, and performance can be significantly affected. On the other hand, computations can significantly shorten the time required for design; they also provide the designer a tool with which design deficiencies, once uncovered, can be quickly corrected.

A management technique that has worked effectively at Dassault and some companies in the United States is to pursue design refinements computationally as an off-line project activity. That is, computational simulations are not relied on in DD and hence cannot adversely influence the project critical path in terms of cost, schedule, and performance. These simulations are undertaken by a design team working off-line and seeking improvements that can be made available to the project in a timely fashion. Whether these improvements should be incorporated in the project design is a decision that must be made by the project manager. Frequently this type of off-line effort can reduce design time (and, hence, cost) and improve performance.

If the computational design team is not successful, they do not adversely affect the project cost, schedule, and performance beyond that which was anticipated in the original project plan.

Some Lessons from the Experience of the Last Decade

The current level of integration of computational simulations into the design process has resulted from the learning experience of more than a decade. This experience began with research in numerical solution techniques, followed by the development of computer codes based on the products of that research. Extensive evaluation and experience with these codes has resulted in their gradual acceptance by the designer and his management. And so the process continues, as improved solution techniques are devised.

Many lessons have been learned, including the following: (1) the proper role of computations can be easily misunderstood by the designer and his management; (2) calibration, interpretation, and verification are all essential elements in adapting computational analysis to design; (3) there are definable factors that determine the degree to which a given computational capability is useful; (4) new codes must be marketed carefully — acceptance will be gradual; and (5) aerodynamics is still an art — no entirely satisfactory procedure has yet been devised to assist the designer in defining aerodynamic shapes for specified performance objectives and design constraints. Each of these items is addressed in turn.

Proper role of computations — It is now generally accepted among aerospace designers and managers that computations should be used extensively in vehicle design. The question is no longer whether these methods should be used but how to get them to do more. This current attitude is a result of a lengthy educational process. According to M. George (Northrop, private communication, Feb. 1982), "In the beginning, the codes were expected to provide 'answers,' i.e., actual drag, lift and moment data. Today, with greater acceptance and more extensive use, the codes are used to provide information such as shock formations, for example, and other flow characteristics and phenomena which are the causes of aerodynamic performance. Such analysis was previously impossible without expensive, time-consuming tests. Today, the codes have become an integral part of the design process — that process in fact is dependent on the availability and use of computational methods."

Generally, computations are useful in providing the aerodynamicist with an understanding of the effects of configuration modifications on the features of the flow field. They are normally not reliable in predicting absolute performance values for a given configuration. Failure of the designer or his management to understand the proper use of computational aerodynamics can cause him to abandon its use and, hence, lose the advantages that derive from it.

Importance of calibration, interpretation, and verification — Extensive experience with computations in aerodynamic analysis and design indicates the importance of calibration, interpretation, and verification. The aerodynamicist, because he understands that there are numerical and phenomenological errors associated with computational simulations, must calibrate his computational tools before applying them. Normally, computed results are correlated with available wind-tunnel data for some representative baseline configuration to determine the extent to which the computations represent the essential flow-field physics. Discrepancies are reduced by appropriate "tuning" to empirically model those physical features of the flow that are not adequately treated by the particular computational approach in use.

For the case of an inviscid computational simulation, this could involve modeling the boundary layer, by assuming an appropriate displacement thickness, and the shock/boundary-layer and trailing-edge flow conditions. For the case of wing design using a code that is capable of solving only flow past a wing, some means must be used to model the effects of the fuselage and other aircraft components on the wing flow. Deficiencies in the computations, with appropriate modeling included, must be accounted for by skillful interpretation of computed results. Guidance for interpretation is provided by results of the calibration. Once properly calibrated, the computer code can be used by the designer to solve the aerodynamics problem of interest. In the case of a design problem, the computer code can be used to help define the aerodynamic shape that best meets performance objectives while satisfying design constraints. Because the computation normally is not reliable in establishing the absolute level of performance, experimental verification is required before the design can be accepted.

Usefulness of a given computational capability — For an aeronautical engineer to decide to invest his time to learn how to use a new computer code, there must be clearly evident gains in capability relative to previously existing codes or procedures. Capability is measured in terms of accuracy and reliability, run cost and turnaround time, ease of use, and versatility. A designer will use only those computational tools in which he has confidence. He builds this confidence through extensive experience in comparing computed results with validated data — the calibration process. This gives him a measure of accuracy and reliability, as well as run cost and turnaround time. For any given analysis then, he has the information required to determine which of the tools he has calibrated can provide acceptable accuracy at minimum cost in the time required.

In high-priority design activities, turnaround dominates cost considerations. In design studies constrained by tight computer budgets, the opposite may be true. However, in assessing cost and turnaround time, one must consider the time and cost of people, as well as the time and cost associated with computer facilities. Ease of use, or code "friendliness," is directly associated with these time and cost factors. Friendliness refers to such features as ease of manipulating data; the effort involved in the input of data (especially geometry data) and post-processing of flow-field data; the degree to which input/output is consistent with other codes used by the organization; code transportability and maintainability; and code structure and readability. Versatility refers to the scope of applications to which a code can be applied. It is certainly easier to justify the time investment required to learn the use of a new code if that code can be used in a wide range of applications that are of interest to the organization.

Marketing new codes — New methods and new codes, like any new product, must be marketed very carefully. Premature release can result in reputation assassination, virtually guaranteeing that the new technology will not be accepted by the user community. This discourages early release of new technology to potential users; the code developer seeks to wring all of the "bugs" out of a code before its release. Hence, detailed testing of the software is continued until the code is thoroughly certified, with the consequence that its release is delayed 1 to 2 years. An alternative approach is to risk early release of an incompletely tested code to a carefully selected group of users. These users are then scrupulously serviced by the code development team. For example, the team can keep a record of all cases run, of the input and output, and of other information necessary to determine the causes of failure in any given run. Deficiencies in the use of the code are quickly corrected. Records are kept to document deficiencies in the code itself, which can then be corrected with changes appropriately documented.

The advantage to the code development team is that the select group of users assists in debugging the code in practical applications. Furthermore, this initial group of trained users can then train additional users, relieving the code development team of that burden. The advantage to the users is early access to a new analysis/design capability. Moreover, these initially trained users are recognized as valuable assets within a company — their services are in demand by a number of project managers. Experience indicates that acceptance of any fundamentally new computational technology by the aerodynamicist and his management is a gradual process. It occurs in any company only after an extensive calibration phase during which the capabilities and proper use of the new technology are thoroughly evaluated.

Aerodynamics is still an art — The need to control the vast amounts of information involved in modern design has led to increased use of aircraft synthesis computer codes to combine all of the technical specialties so that an integrated design solution can be achieved quickly and efficiently. In such an environment, the design flexibility available to the aerodynamicist is limited, and the interfaces with the other aeronautical specialties are clearly defined. Within these defined limitations, the aerodynamicist must first formulate a definition of the design problem in terms of aerodynamic performance objectives and then decide on the specific approach to be used to solve the problem.

Two distinct problem solving approaches can be identified. The first, which we might call the conventional approach, begins with analysis and then leads to detailed wind-tunnel testing of a limited number of candidate configurations. From the resulting performance data, a configuration is selected as the solution to the problem, or new configurations are selected for further testing. In this approach, computational simulations may or may not be used in defining candidate configurations for wind-tunnel testing. The second approach is called direct design, or optimization. Here, the aerodynamicist formulates and solves a constrained optimization problem, defining a function to be minimized (e.g., the drag or some combination of performance parameters) for a set of decision variables that represent the configuration shape, subject to a set of imposed constraints. The constraints generally involve limitations on geometrical flexibility and acceptable levels of aerodynamic performance. Computational simulations are a fundamental part of the optimization approach; they are required to compute the objective function for various test design variables used in searching the design space.

Notice that in the conventional approach, the final selection is based on the testing of a limited number of candidate configurations, but over a broad range. Only a narrow design space can be searched, and it may not include the best feasible design. In the optimization approach, the focus is on producing a single configuration that represents the best trade-off of design objectives subject to the imposed constraints, and in principle an infinite number of candidates have been considered in choosing the optimum configuration. However, only a limited number of design points are considered. Consequently, performance difficulties may be encountered at other points in the operating range of the aircraft. The design techniques in current use are combinations and variations of the two approaches outlined above specifically tailored to solve a particular problem.

EXAMPLES OF COMPUTATIONAL AERODYNAMICS IN DESIGN

This first example illustrates how new computational technology is integrated into existing design procedures. The Highly Maneuverable Aircraft Technology (HiMAT) design (fig. 3) took place in the mid-1970s and was the first in which a nonlinear, three-dimensional, transonic finite-difference code was used in an aerodynamic design (ref. 1). The HiMAT project objective was to develop and evaluate high-maneuverability technologies and to synthesize those technologies in a remotely piloted research vehicle (RPRV). The specific aerodynamic objective was to improve fighter maneuver performance without compromising other mission requirements, such as cruise performance and supersonic acceleration capability. The tight budget, typical of this type of technology demonstration project, and the ambitious performance goal required heavy reliance on analysis and computations with a minimum of wind-tunnel testing.

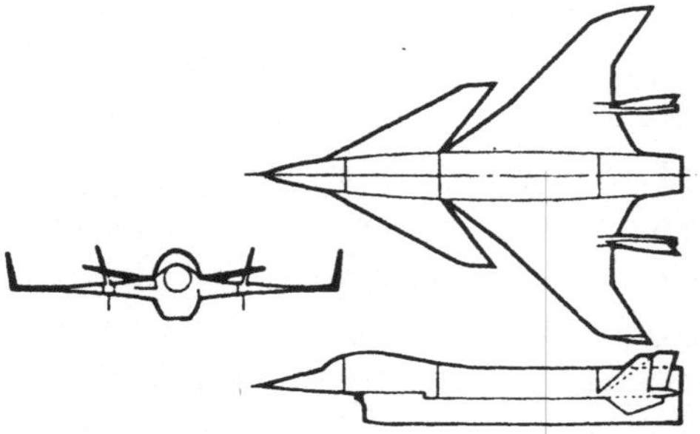

Fig. 3. HiMAT RPRV three-view.

The design-team strategy was to optimize the basic design for high maneuverability and then to determine an acceptable cruise configuration that could be achieved using a variable camber system and preprogrammed twist increment due to structural bending. The preliminary design and initial stages of the detailed design would be conducted using linear theory, a linear inviscid approximation to the Navier-Stokes equations. This computational approach is cost-effective and, in its present mature stage of development, is capable of treating complex aerodynamic configurations. However, it does not properly account for shock waves or other nonlinear phenomena encountered in the transonic regime. Refinement of the configuration to meet transonic performance requirements would subsequently be accomplished using the Bailey-Ballhaus (B^2) nonlinear transonic code (ref. 2), a less approximate formulation for transonic flows than linear theory. Wind-tunnel tests would be conducted at appropriate stages of the design as needed.

The baseline wing-canard design was achieved after several linear theory design iterations. The design exhibited good subsonic drag characteristics but, as expected, was deficient at the transonic design point (M = 0.9). Wind-tunnel oil-flow visualization indicated shock-induced separation on the outboard wing and canard and a strong unswept shock wave near the trailing edge of the inboard wing. These flow features were primarily responsible for the excess drag at $M_\infty = 0.9$ shown in figure 4 for the linear theory design.

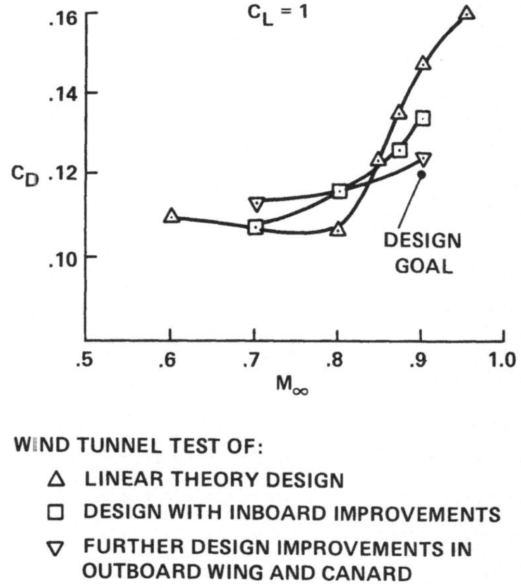

Fig. 4. Reduction of HiMAT maneuver configuration drag rise.

To more thoroughly understand the configuration deficiencies and to provide data to calibrate the B^2 code, a wind-tunnel test was conducted in which upper-surface pressures were measured. Comparisons of these measurements and B^2-computed pressures for two span stations are shown in figure 5a. The B^2 code accuracy was found to be acceptable in the inboard 70 percent of the semispan, where the shock wave was not highly swept. However, it failed to capture the highly swept outboard shock. This deficiency had been anticipated by the code authors (ref. 3) and was eventually corrected (ref. 4), as indicated in figure 5b, but not in time to influence the HiMAT design. Consequently, the B^2 code was applied to improve the design only for the inboard 70 percent semispan region. The B^2 code was also incapable of including the canard in the flow-field computation. During the calibration phase, it was determined that the effect of the canard on the wing flow could be modeled by superimposing an appropriate amount of wing twist on the geometrical twist of the wing in applying the wing boundary condition.

With these improvements, the code could be used effectively to guide the designer in modifying the wing shape to improve the wing flow and thereby reduce the transonic drag. The airfoil sections and wing planform were modified in a trial-and-error manner to minimize the computed strength and extent of the shock without sacrificing lift. This was continued until suitable pressure distributions were obtained. Verification wind-tunnel tests of the resulting configuration indicated that the computationally derived modifications were successful in weakening or eliminating shock waves over much of the inboard 70 percent of the span. This resulted in a major

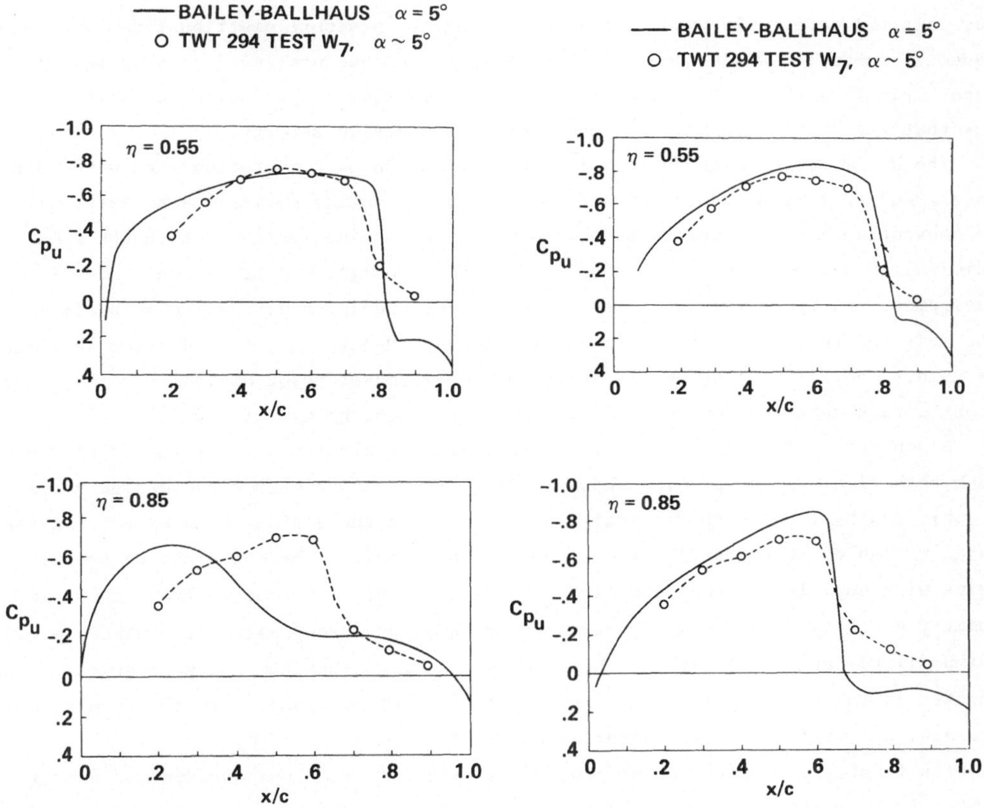

(a) Original B^2 code.

(b) Improved B^2 code.

Fig. 5. HiMAT wing surface pressure comparisons at two span stations.

reduction in drag. In the outboard region, where the flow was nearly two-dimensional in a plane normal to the sweep direction, a two-dimensional nonlinear transonic code and sweep theory were used to improve the configuration. This further reduced the drag so that the final configuration, when tested, indicated a substantial improvement over the linear-theory design (fig. 4).

The HiMAT design illustrates how new computational technology is accepted by the designer. The motivating force behind the use of the B^2 code was the need to solve a particular transonic design problem, in the midst of a design effort, that could not have been solved within schedule and budget constraints by using existing procedures. The Rockwell research staff acquired this code before it had been thoroughly tested and certified, extended it in consultation with the code developers to treat the winglets at the wingtips, calibrated it to understand its deficiencies, and modeled the effect of the canard on the wing by introducing appropriate twist in the wing boundary condition. They applied the code only where calibration studies indicated it could be used reliably (the inboard 70-percent semispan). They used it effectively to find configuration modifications that would improve features of the flow that the code could provide reliably, that is, pressure distributions, and subsequently conducted wind-tunnel tests to verify the absolute level of drag performance. Finally, they interacted with the code developers, providing valuable feedback that was used to improve the code for future design efforts.

The HiMAT design approach involved a combination of computations and wind-tunnel tests. Budget limitations precluded the extensive testing normally associated with the conventional design approach defined in the preceding section. Although some optimization was used to obtain the linear theory design, the primitive stage of development and excessive run time of the B^2 code precluded its use in an optimization mode for refinement of the transonic design. Hence, a trial-and-error approach was used. The following example, design of a helicopter rotor airfoil section, seeks a more direct approach toward configuration refinement by use of optimization.

A schematic flowchart of a typical optimization process is depicted in figure 6. The hypothetical design problem shown is drag minimization with three design variables (h_1, h_2, and h_3). Here the objective is to minimize the airfoil drag by varying the design variables that describe the shape of the airfoil. The optimization process begins with an initial specified airfoil shape. Constraints can be imposed on the geometry of the airfoil (e.g., minimum thickness), on aerodynamic performance (e.g., minimum lift), and on flow-field characteristics (e.g., maximum shock strength). The computer is then free to minimize the objective function, subject to the imposed constraints, by varying the design variables.

The first step in the design process is calculation of the aerodynamic coefficients of the initial airfoil. These coefficients are stored as baseline values for future use in the gradient calculations. The optimization program then perturbs each of the design variables, one by one, returning to the aerodynamics program for evaluation of the aerodynamic coefficients and the partial derivative of drag with respect

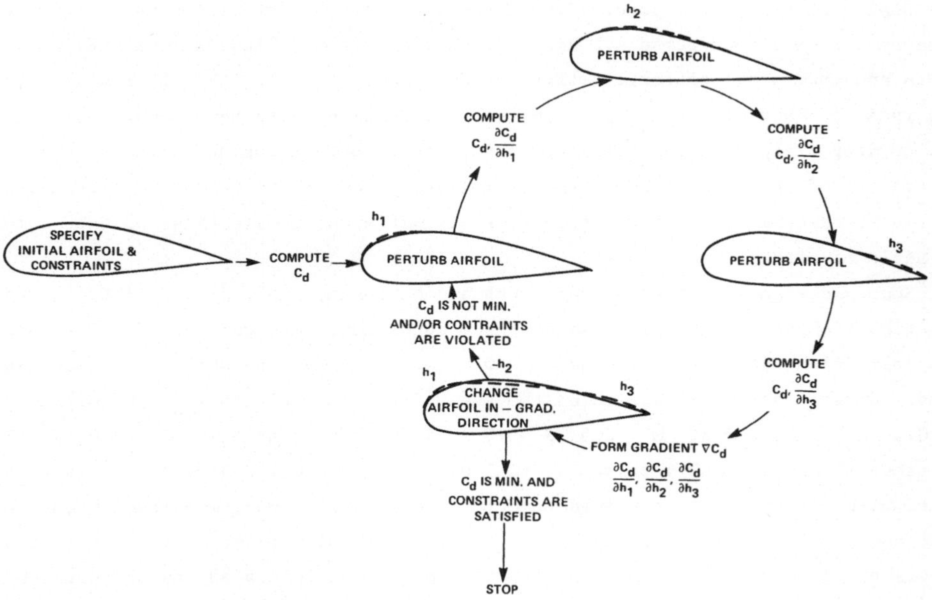

Fig. 6. Drag minimization by numerical optimization, 3 design variables.

to each design variable after each perturbation. The partial derivatives form the gradient of drag (∇C_D). The direction in which (h_1, h_2, and h_3) are displaced to reduce the drag coefficient is $-\nabla C_D$ (the steepest descent direction). The optimization program increments the design variables one to four times in the direction indicated by $-\nabla C_D$. The process continues until the drag begins to increase as a result of nonlinearity in the design space or until a constraint (such as airfoil thickness) is encountered.

The specific choices of objective function, constraints, initial airfoil shape, and design variables describing the airfoil shape relative to the baseline airfoil are all selected by the aerodynamicist. Experience indicates very clearly that the success of the design optimization depends strongly on the judgment of the designer in making these choices.

Design of helicopter rotor airfoil sections requires simultaneous consideration of several design requirements. Example requirements are high maximum-lift coefficients and good stall characteristics for Mach numbers of 0.3 to 0.5, a high lift/drag ratio at $M = 0.6$ and $C_\ell \sim 0.6$, and a drag divergence Mach number of at least 0.8, no drag creep, and low pitching moments over most of the Mach number range. To achieve all of these conditions, the optimization code must monitor and constrain certain aerodynamic parameters at five or six different combinations of Mach number and angle of attack.

Such a design was attempted by Hicks and McCroskey (ref. 5) in which, to reduce computing time and cost, only some of these conditions were considered. The shock drag coefficient at $M = 0.82$ and $\alpha = 0°$ was chosen as the objective to be minimized. Constraints were imposed on the shock drag coefficients ($C_d \leq 0.001$) at $M = 0.4$, $\alpha = 12°$, and at $M = 0.5$, $\alpha = 10°$ to delay retreating-blade shock stall. The standard Jameson FLO-6 transonic full-potential airfoil analysis code (ref. 6), linked with a standard constrained function minimization code, was used in the design. These codes had been extensively calibrated in other airfoil designs. The initial airfoil section from which the design optimization was initiated was the Wortman profile, designed for helicopter applications and known for its desirable rotor characteristics.

The resulting airfoil section, called the A-1 section, was tested in the 2- by 2-foot wind tunnel at Ames. A comparison of the A-1's improved performance relative to the baseline section (Wortman FX69-H-098) and a classical rotor section, the NACA 0012, is shown in figure 7.

Computational simulations have been used effectively in a number of recent major aircraft development programs. The Airbus A-310 and Boeing 757 are examples in which computational simulations have played a significant role in achieving new designs that are substantially more fuel efficient than their predecessors. As these simulations become more useful, aircraft companies are becoming less willing to provide information concerning their application. Consequently, there are many interesting examples that, unfortunately, cannot be described here.

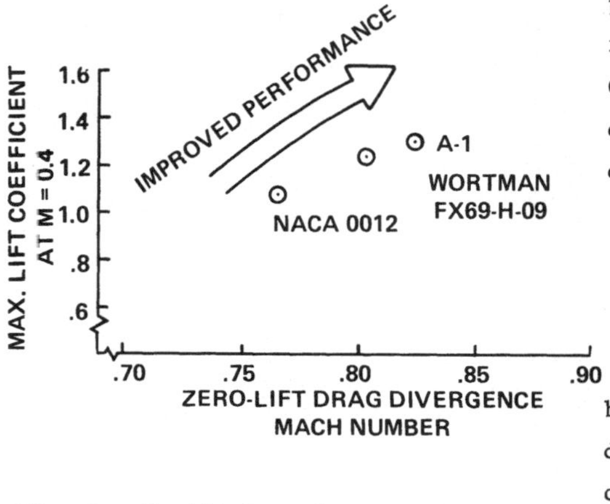

Fig. 7. Verified performance improvement from helicopter airfoil optimization.

PROSPECTS FOR THE FUTURE

Although computational simulations have become considerably more useful in design during the last few years, there is considerable room for improvement. Increasing design cost and complexity, as well as increasing performance objectives of new aircraft, drive the search for more accurate and more cost-effective computational capability. The complexities of the obstacles involved ensure that this search will continue for many years. The major areas of consideration are (1) the availability of advanced computers, (2) advancements in numerical solution techniques, (3) improvements in turbulence models, (4) treatment of complex geometries, and (5) improved computational design procedures.

Advanced Computers

The prospects are good that significant increases in scientific computing capability will become available in this decade. From a technical standpoint, anticipated advances in component technology point to improvements in speed of at least an order of magnitude. Moreover, architectural advances permitting multiple processors to efficiently address a single problem with a common data base offer opportunities for further orders-of-magnitude improvements in speed. There will also be advances in memory size and in data transfer rates to support such improvements in speed. Significant attention will be directed toward developing software to assist the user in taking full advantage of these architectural advances.

The slowdown in advances during the last decade, prompted by marketing considerations rather than technical ones, appears to be reversing. The anticipation of increases in the size of the market for leading-edge scientific computers has led to increased competition within the industry. Moreover, recent government policies reflect a growing realization that the large-scale scientific computer is a fundamental tool affecting the rate of progress in a wide range of technical disciplines (this is especially true in Japan, where advances in scientific computing have become a national objective). Furthermore, large machines are becoming accessible to a broader range of the user community. Until recently, leading-edge scientific computers in the United States have been installed primarily in the national research laboratories, with the airframe manufacturers and universities enjoying only limited access to them. Now, several of these large computing engines have been installed at universities and at least one aircraft company.

Extensive experience with supercomputing, primarily in the national research laboratories, has provided some observations that illustrate both its power and its limitations. The first observation is that <u>it takes a larger computer to develop a new computational capability in a research mode than it does to implement that capability in applications</u>. Once a previously intractable problem is solved, then significant improvements in accuracy and efficiency are subsequently achieved by a combination of analysis and computational experiments. This experimentation requires rapid turnaround which, for problems that strain the limits of existing capability, can only be achieved with the largest and fastest existing computing engines. The algorithm refinements that result frequently reduoe computer run-time to the point where the new capability can be implemented on smaller systems. For example, much of the development work for inviscid transonic aerodynamic methods and codes was done on CDC-7600-class machines. Computer run-times have been reduced by more than an order of magnitude. Now, these codes can be run in minutes on slower, less expensive machines like the VAX 11/780.

The second observation is that people are becoming more expensive than computer hardware. In human-computer interactions, maximum effective use of people is becoming at least as important a measure of efficiency as maximum use of computer cycles. Human effort to construct data bases for processing must be considered along with computer run-time in assessing the total cost of a computational simulation.

The third observation is that large upgrades in computer power are required to produce moderate improvements in simulation capability. Doubling the number of grid points in each direction to improve resolution results in a factor-of-8 increase in the number of total grid points in the field. This usually results in more than a factor-of-8 increase in computer run-time to get only twice the resolution. Hence, minor improvements in computer power usually result in insignificant improvements in accuracy.

Advances in Numerical Solution Techniques

Advances in numerical solution techniques offer prospects for increasing computational simulation capability at a rate comparable to that resulting from increasing computer capability. An excellent illustration of this was provided recently by Holst in a computation of inviscid full-potential transonic flow about the ONERA M-6 wing, $M_\infty = 0.84$, $\alpha = 3°$. This case has been used by a number of researchers to test new methods. The base code used was FLO-28 (ref. 7), which solves the full-potential equation in conservation-law form, using successive-line overrelaxation, a standard solution technique. The computation took 742 sec on the CDC-7600 computer to reach 98 percent convergence on lift coefficient. Holst, using an improved implicit factorization algorithm to solve the same governing equation, achieved the same degree of convergence in 64 sec — an improvement of better than an order of magnitude (ref. 8). By running the code on the faster Cray 1S machine, after first vectorizing the coding to take advantage of the Cray's special architecture, a solution was obtained in 4.8 sec — a further improvement of better than an order of magnitude (T. Holst, NASA Ames, private communication, March 1982). With a run-time less than 5 sec, achieved through advances in both algorithm efficiency and computer power, over 150 wing-flow simulations can be obtained in the same time, and for very nearly the same cost, as for a single flow solution using FLO-28 on the CDC-7600. Furthermore, an aerodynamic analysis code with a 5-sec run-time can be used effectively for design in the optimization mode.

Such dramatic advances in solution techniques for the full-potential formulation provide encouragement for similar advances in techniques for solving the less-approximate Euler and Reynolds-averaged-Navier-Stokes (RANS) formulations. Relative to the full-potential formulation, the Euler formulation is a more accurate simulation of flow-field physics, in that it provides for the treatment of rotational flows, including vortices and strong shock-wave effects. The RANS formulation, combined with a suitable turbulence model, further provides for the treatment of viscous effects, including flow separation.

This capability is fundamentally important for simulating aerodynamic flows near performance boundaries. However, these formulations require considerably more storage and computational work to obtain a solution than the full-potential formulation, and current computer power and solution techniques are inadequate for these formulations to be widely used in design. Nevertheless, major reductions in computer run-time are anticipated during the 1980s by overcoming unnecessarily severe constraints on integration time-steps (frequently a result of stability rather than accuracy considerations).

Further reductions in run-time, and significant reductions in storage, will be achieved by the development of solution-adaptive grid techniques. That is, the distribution of grid points will be determined dynamically, as the solution evolves. Grid-point distribution then will be solution-dependent, with grid-point locations chosen to maximize accuracy for a specified number of points.

Turbulence Models

The dynamics of turbulence in the wide range of scales encountered in practical aerodynamics problems cannot be simulated using either present computer power or that anticipated for many years to come. Hence, the flow variables in the Navier-Stokes equations are averaged over a time period that is long compared with the time scales associated with predominant features of the turbulence (Reynolds-averaged-Navier-Stokes formulation). The number of unknowns then exceeds the number of equations. The process of expressing the unknowns as transport equations or functions in terms of known quantities is called "turbulence modeling." No entirely suitable model for all flow types of engineering interest has yet been discovered.

The current predominant thinking among researchers in the field is that no such universal turbulence model exists. Hence, attention is now focused on developing menus of turbulence models. The process consists of a synergistic use of computation and experiment to develop and test models for various types of flows that are considered building blocks to more complete aerodynamic configurations, such as attached flows with and without curvature and imposed pressure gradients, simple separated and reattached flows, flows with shock waves, and airfoil trailing-edge flows. The state of the art is such that simple attached flows can normally be adequately treated. Three-dimensional cases with large amounts of skewing of flow direction in the boundary layer still present a considerable challenge. Complex flows with massive separation can often be properly simulated qualitatively but not quantitatively. For a more thorough discussion, see reference 9.

Complex Geometries

One of the initial steps in a flow-field computation is the specification of boundary conditions. For aerodynamic flows, digital data describing the configuration shape of interest must be provided. A finite-difference grid system is then generated which encompasses the flow-field domain and serves as the basis for finite-difference approximations to the terms in the governing flow-field partial differential equations. The principal difficulty involved in the analysis of complex aerodynamic configurations is the generation of a suitable grid system.

A number of factors must be considered in assessing the suitability of a grid system. The object is to achieve required accuracy with the minimum number of grid points and with the least effect on the flow-solution algorithm owing to singularities or other special considerations. An example grid for an aircraft nacelle flow field is shown in figure 8 (from ref. 10). The grid is adapted to the surface of the nacelle and the grid points are clustered in regions, such as the leading edge, where large flow gradients in the solution are anticipated. This grid was generated for only a single component — the nacelle.

Major difficulties are encountered in treating multiple component configurations, for which component-adapted grids must be interfaced. These interfaces introduce difficulties in the solution process in that they complicate coding and adversely affect run-time and numerical accuracy. Extensive effort, especially in the aircraft companies, is contributing to a very gradual expansion in the complexity of configurations that can be adequately analyzed. For the time being, however, linear theory will continue to be used for analyzing aerodynamic configurations in detail. Linear theory, as described in the HiMAT example, requires generation of a grid system only on the boundary surface, not for the entire flow field. Because the resulting grid-generation problem is so much simpler, this simple formulation can be used to analyze practical aerodynamic configurations to a much greater degree of detail than can be achieved with the more physically complete nonlinear formulations (full-potential, Euler, and Reynolds-Averaged-Navier-Stokes).

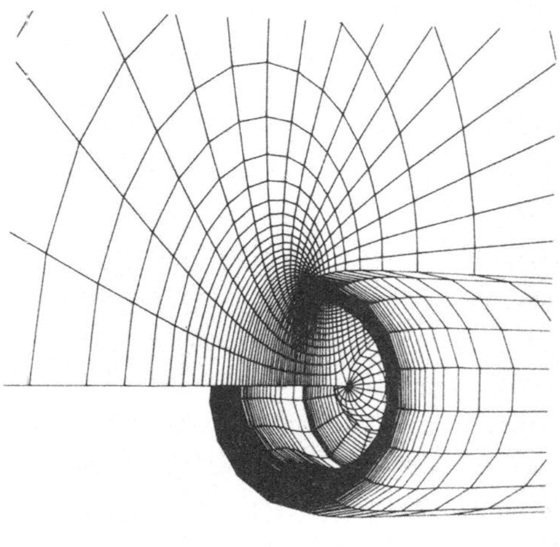

Fig. 8. Perspective view of three-dimensional coordinate system for a nacelle.

Computational Design Procedures

One of the principal advantages of computational simulations relative to testing is the opportunity they offer to compute a vehicle geometry that yields some desirable aerodynamic characteristic.

Optimization, as illustrated by the helicopter airfoil design described in the preceding section, is a promising approach toward taking advantage of this opportunity, providing the designer can supply the judgment required to use it effectively.

To begin with, great care must be taken in selecting some controllable parameter or function of several parameters as the objective to be minimized. Experience indicates that since total drag is usually not predicted reliably, the emphasis must be on obtaining configuration modifications that eliminate or at least weaken shock waves, eliminate embedded supersonic flow on fuselage cabin regions, minimize adverse pressure gradients, etc. Success also depends strongly on the number of design variables that define the configuration shape and the extent to which these design variables permit flexibility in the design (i.e., can a near-optimum shape be described from some suitably chosen values of design variables selected). Finally, since many flow-field computations may be required to achieve an optimum design, only those methods with suitably short run-times for a single flow-field computation can be used in an optimization mode.

Optimization can be used to improve the effectiveness of other design techniques. For example, it can be combined with the inverse approach to permit specification of performance and configuration constraints. In the inverse approach, a configuration shape corresponding to a pressure distribution selected by the aerodynamicist is computed; the pressure distribution selected is one that his experience indicates will provide the desired performance. The combined optimization-inverse approach would seek that configuration that best fits the specified pressure distribution while satisfying the imposed constraints.

Clearly, much remains to be done to provide a more direct and systematic approach to design than trial and error. The usefulness of optimization and other approaches will depend on the degree to which computer run-time for complete three-dimensional flow simulations can be reduced by a combination of more powerful computers and improved solution algorithms. It will also depend on how accurately flow-field physics and geometrical detail can be simulated to relate configuration shape to aerodynamic performance.

CONCLUDING REMARKS

The last decade has been an exciting one for researchers in computational aerodynamics. The next one promises to be just as exciting for aircraft designers. Advances in computer power, solution algorithms, viscous flow simulation, and grid generation will permit much more detailed simulations of complex aerodynamic phenomena and the effects of geometry. These simulations will broaden to include other aeronautical disciplines, such as structures and propulsion. A new generation of managers will be in place, who, unlike their predecessors, have grown up with the computer. Their challenge will be to integrate the major computational advances of the 1980s into the design process in bold and imaginative new ways.

ACKNOWLEDGMENT

The author gratefully acknowledges contributions by several individuals who provided ideas on various aspects of the design process, particularly P. Perrier, R. Bradley, M. George, R. Hakkinen, L. Roberts, N. Malmuth, L. Miranda, M. Lores, and P. Rubbert and his associates at Boeing.

REFERENCES

1. Child, R.; Panageas, G.; and Gingrich, P.: AGARD Conference Proceedings No. 280, 1980.
2. Bailey, F.; and Ballhaus, W.: NASA SP-147, Part 1, 1975.
3. Lomax, H.; Bailey, F.; and Ballhaus, W.: NASA TN D-6933, 1973.
4. Ballhaus, W.; Bailey, F.; and Frick, J.: NASA CP-2001, Vol. 4, 1976.
5. Hicks, R.; and McCroskey, W.: NASA TM-78622, Mar. 1980.
6. Jameson, A.: Commun. Pure & Appl. Math., vol. 27, 1974.
7. Caughey, D.; and Jameson, J.: AIAA J., vol. 17, no. 2, Feb. 1979.
8. Holst, T.; and Thomas, S.: AIAA Paper 82-0105, 1982.
9. Marvin, J.: AIAA Paper 82-0164, 1982.
10. Ives, D.: AIAA Paper 81-0997, 1981.

NUMERICAL COMPUTATIONS OF EXPLOSIONS IN GASES

P.I. Chushkin, L.V. Shurshalov

Computing Centre of the USSR Academy of Sciences, Moscow-117333, USSR

ABSTRACT

The development and the present-day state of the problem on numerical computations of explosions in gases are reviewed. In the first part, different one-dimensional cases are discussed: point explosion with counterpressure, blast-like expansion of volumes filled with a compressed hot gas, blast of charges of condensed explosive, explosion processes in real high-temperature air, in combustible detonating media and under action of other physical-chemical factors. In the second part devoted to two-dimensional flows, we consider explosion in the non-homogeneous atmosphere, blast of asymmetric charges, detonation in gas, explosion modelling of some cosmic phenomena (solar flares, the Tunguska meteorite). The survey includes about 110 works beginning with the first publications on the subject.

INTRODUCTION

The review is devoted to numerical solutions of gasdynamical problems in the theory of explosion in gases. This theory as a separate branch of gasdynamics arose about 40 years ago in the works by L.I.Sedov and G.I.Taylor. For the first decade only analytical approaches were used here. The primary numerical solutions of non-self-similar problems were obtained in the middle of the next decade. Now numerical methods, which enable to compute very difficult and practically interesting cases, play the dominant role in the study of explosion flows. At present, intensive activities concerning numerical solutions of new various and more complicated explosion problems are carrying out.

Explosion flows are characterized by very sharp space-time non-homogeneity of flow parameters, extremely large change of disturbed region extents, presence of a series of interacting gasdynamical discontinuities inside this region. These facts create serious computational difficulties and place exacting demands upon numerical methods used to solve explosion problems.

Explosion flows were frequently computed with the aid of finite-difference net methods. Many various numerical techniques of such type (explicit, implicit, predictor-corrector, Eulerian or Lagrangian, shock fitting or shock capturing) were used. In particular, many authors often applied the explicit second-order method with artificial Neumann-Richtmyer viscosity, developed in H.L.Brode's works, and the explicit first-order method proposed by S.K.Godunov. A number of authors used the flux-corrected transport (FCT) method, worked out by J.P.Boris and D.L.Book, and different variants of the method of particles (FLIC method, method of large particles by O.M.Belotserkovskii and Yu.M.Davydov, EIC method by Ch.L.Mader). Some problems were also solved by the method of characteristics and by the A.A.Dorodnicyn method of integral relations.

In this review, main attention is paid to those works where

concrete problems of explosion theory were solved and certain interesting gasdynamical results were obtained. The contents follow different types of problems. In the first part we discuss one-dimensional computations in such cases as point explosion, blast-like expansion of gaseous volumes and blasts of charges of condensed explosive, consideration of high-temperature effects for blasts in real air, explosion processes in a combustible detonating gas and in gases with other physical-chemical properties. In the second part,we treat two-dimensional unsteady problems: explosion in stratified atmosphere, explosion of charges having non-traditional forms, gaseous detonation and blasts modelling certain phenomena arising in the near cosmos.

These problems are connected with the classical formulation in which rapid concentrated release of large energy in a relatively small volume and subsequent propagation of explosion wave in ambient unbounded gas are considered. We shall not especially discuss the problems on diffraction and reflection of shock waves, their implosion and motion in tubes. Also, we shall not touch upon numerical calculation of explosion processes in astrophysics, in plasma dynamics, in the case of laser beam irradiation of deuterium-tritium targets, since these problems have some specific features and demand special consideration.

1. ONE-DIMENSIONAL EXPLOSION PROBLEMS

1.1. Point explosion in perfect gas

In the classical point explosion problem, the well-known idealized formulation is assumed. It is supposed that a finite amount of energy is instantly released at a point. As a result, the spherical shock wave arises and propagates into the ambient unbounded gas which is taken to be perfect, inviscid and non-heatconducting. In the cylindrical and planar cases the energy is released along a line and a plane correspondingly.

To solve the one-dimensional unsteady non-self-similar problem on point explosion with counterpressure,different authors applied different numerical methods. As a marching variable some variable related to the time was usually chosen and at its small value all the initial conditions were set from the strong blast solution.

In the first works, the calculations were carried out only for the spherical explosion in the diatomic perfect gas. Goldstine and Neumann (1955) used the finite-difference three-level Lagrangian method, Brode (1955) applied the explicit second-order method with artificial viscosity, Okhotsimskii et al. (1957) employed the implicit second-order method with shock front fitting. The last authors brought their solution to the shock front pressure ratio $p_s/p_a = 1.008$ (p_a is the pressure of ambient gas) and published the results as the tables. Further Okhotsimskii and Vlasova (1962) extended this solution to the pressure ratio $p_s/p_a = 1.000011$ using an approximate procedure to calculate the weak shock wave at the late stage.

Korobeinikov and Chushkin (1964, 1966) numerically calculated the point explosion with counterpressure for all three symmetries (planar, cylindrical and spherical cases, $j = 1,2,3$ correspondingly) and for a number of specific heats ratio values γ . Here the Dorodnicyn method of integral relations was worked out. The authors introduced the convenient independent variables $\xi = (R/R_s)^j$, $q = c_a^2/D^2$ ($0 \leqslant \xi \leqslant 1$, $0 \leqslant q \leqslant 1$), where R is the linear coordinate, c_a is the sonic velocity of ambient gas, R_s and D are the shock wave coordinate and velocity. The solution was computed for eight strips and power approximations with respect to ξ . The initial-value problem took place for the approximating system of ordinary differential equations in q . The tables of the obtained solution were published (Korobeinikov, Chushkin, Sharovatova, 1969).

The numerical solution of point explosion problem for various j and γ was computed and then printed as the tables in the book by Kestenboim, Roslyakov and Chudov (1974). These authors used explicit predictor-corrector method. Oppenheim et al. (1978) solved the problem in the phase space generalizing the phase plane concept (applied in the self-similar case) by means of adding the variable q . The equations were solved at several fixed q, while the linear approximation with respect to one phase variable led to the approximating system of ordinary differential equations for which the boundary value problem with three unknown parameters and the singularity point at the explosion centre took place.

Some above mentioned solutions for $\gamma = 1.4$ are compared in Fig.1 where the decay coefficient of the shock wave $\mu = -2d\ln D/d\ln R_s$ is shown as the function of q .

Several other works devoted to numerical computation of point explosion should be noted. Plooster (1970) and Arkhangel'skii (1974) calculated the cylindrical case. Director and Dabora (1977) studied the spherical blast, when energy release was not instant, but followed some power dependence. Andriankin and Myagkov (1981) carried out calculation of double planar explosion when at the centre of the first strong blast after the lapse of certain time, the second one occured.

1.2. Blast-like expansion of volumes of a hot compressed gas

A number of works treat blast-like expansion of finite volumes filled with a gas having high pressure and temperature. Such a model, describing an explosion initiated by a source of finite size, is more realistic than the point explosion model. Besides, the temperature singularity at the blast centre is absent in the former model. However, the flow pattern becomes more complicated here. Additional discontinuity surfaces (interface, secondary shock waves) arise in the disturbed region. They interact between themselves or with the main shock front.

Using the method with pseudo-viscosity, Brode (1956b) analysed blast-like expansion of an air sphere with pressure of 20000 atm and temperature of 386000 K. The ambient air had see-level parameters. Some analitical approximations for real equations of state of air were taken. The calculation revealed complex dynamics of the flow under consideration. The second embedded shock wave formed between the interface and the centre and moved inward. It reflected at the centre and overtook the interface,partly reflecting, partly passing through it and then overtaking the main shock front. Secondary shock waves, repeatedly generated in the course of the process, moved in similar manner and quite weakened at the late stage.

To illustrate the middle stage of the explosion, the relative pressure profiles p/p_a versus the relative radius λ are presented in Fig.2 at several moments τ . Here p_a is the ambient air pressure, λ and τ are refered to the dynamic length and the dynamic time, respectively. Directions of secondary shock waves are indicated by the arrows.

Blast-like expansion of a gaseous sphere in the air was calculated by Chou and Huang (1969) using the method of characteristics with a network of inverse type and by Lutzky and Lehto (1970) using the finite-difference method with pseudo-viscosity. The former authors emphasized the high accuracy of their solution, due to the shock fitting and the consideration of the singularity on the boundary of the hot sphere at the initial moment. The numerical results obtained in both works allowed to study the similarity rules for the main and the second shock waves at the late stage of the blast, when the flow principally depends on the amount rather than the character of energy release. The method of characteristics was also applied by MacLaughlin (1980) who computed blast-like expansion of a hot compressed gaseous sphere into the medium with very low density which varied along the radius.

The shock wave propagation for expansion of a hot spherical cavity with high pressure and prolonged constant energy release inside it was numerically studied by Kovaleva and Nemchinov (1976). In particular, they found pulsations of the cavity after stopping the energy release. Using the FCT method, Oved et al. (1979) solved the problem on simulation of a spherical blast wave produced by electrical discharge in a small cavity in the air. The energy release was supposed to occur in all the volume behind the luminiscence front and to depend on the time in accordance with experimental data. The locations of the blast wave and the luminiscence front were determined as the positions where, correspondingly, the maximum negative gradients of density and temperature were observed. A good agreement of the numerical solution and the results of measurements was established.

It is worth to note a numerical analysis of asymptotic properties of one-dimensional unsteady flows of a perfect gas which occupies all the space and has zero pressure and constant density. In the planar case (Derzhavina, 1976), it was assumed that the internal and the kinetic energies were released instantly within a thin layer (the kinetic energy release was equivalent to the impulse application). As the calculation showed, the asymptotic numerical solution followed first to the short impact regime and then to the strong explosion regime. The time of change of the asymptotics essentially depended on the ratio of these energies. In the analogous spherical and cylindrical cases (Parkhomenko, Popov and Ryzhov, 1977a,b), another intermediate self-similar regime existed. It corresponded to the sink flow and was accompanied by appearance of the second shock wave.

1.3. Blast of charges of condensed explosive

Two stages may be distinguished in this process. The first stage includes the initiation and the development of detonation in condensed explosive. The motion of gaseous products of blast and the shock wave propagation into an ambient media proceed at the second stage. A series of works are devoted to the numerical simulation of the first stage only. Among them the monograph by Mader (1979), which involves also two-dimensional cases, is the most fundamental. We consider here works concerning computations of the second stage.

The first and very complete results of numerical solution of this problem were published by Brode (1959). He calculated blast of a spherical trotyl charge,taking real equations of state for both the explosive and the ambient air. However, he treated only the case of see-level pressure p_a. Flow pattern was similar to the case of blast-like expansion of gaseous volumes.

Fonarev and Chernyavskii (1968) extended this investigation and analysed the influence of the air pressure p_a and specific heat release of explosive E . Their results are given in Fig.3, where the shock excess pressure $\Delta p_s/p_a$ is plotted versus dimensionless shock radius λ . Four calculated variants with different values of p_a and E are depicted by solid lines; here E_t is the specific heat release of trotyl. As it is seen, the initial pressure affects the solution weakly, while the specific heat release of explosive affects significantly at the initial stage of the blast. The comparison with the strong explosion solution (dashed line) demonstrates the influence of finite mass and non-instantaneous energy release in the blast source.

The numerical method with pseudo-viscosity was also employed by other authors who calculated explosion of spherical charges. Widely varying undisturbed parameters of real air, Larson (1970) studied a blast of one-pound pentolite charge. Sternberg and Hurwitz (1976) compared the model of instantaneous detonation and the model of detonation with initiation at the charge centre. The shock wave in the first case was stronger than in the second one. Air blasts of cylindrical charges of various condensed explosives were computed by Vasil'ev and Zhdan (1981), who applied the method with pseudo-viscosity in the central part of the disturbed region and the method of Godunov with a moving network in the remaining part. Kuhl and Seizew (1981) studied explosion of planar or spherical trotyl charges and resulting expansion of blast products into vacuum. Shubin (1981) computed propagation of the shock wave for air blast of a pentolite spherical charge and shock interaction with water surface. He used MacCormack's explicit method in the region of continuous solution and the net-characteristic method to calculate contact discontinuities.

1.4. High-temperature effects for explosion in real gas

As it is well known, various physical-chemical processes (dissociation, ionization, radiation) occur in a real gas at high temperatures accompanying powerful blast. Therefore, it is more reasonable to use description of thermodynamical state of a medium, taking into account its real properties, and not to be contented with the simple model of perfect gas. The first computation of point spherical explosion in the real air was carried out by Brode (1956a) with the aid of his numerical technique. He processed various theoretical and experimental data and constracted analytical approximations for caloric and thermal equations of state of the air. The analysis of the numerical solution showed that consideration of high-temperature processes in the hot air could reduce the blast effect in the sence of the shock excess pressure by 40% in comparison with the perfect gas case. The reason for this fact was that a considerable part of energy was spent on dissociation and ionization. Since the hot zone near the blast centre existed for a rather long time, the reverse transfer of this energy into the shock wave proceeded gradually at the late stage of the blast.

The radiation plays an important role for explosions because the hot zone irradiates a significant part of energy. The solution of gasdynamical problems with radiation encounters serious physical and mathematical difficulties. The former are related with complexity of properties of real substances at high temperatures and with necessity to consider non-equilibrium processes. The latter are due to solution of multi-dimensional system of integro-differential equations, moreover, the radiation transfer equation must be solved for different rays and different wave lengths. Hence, various approximate models and approaches are widespread in the radiation gasdynamics.

A numerical approach to calculate one-dimensional unsteady flows with radiation was developed by Brode (1969). Using the Lagrangian finite-difference method with pseudo-viscosity, the author supposed local thermodynamical equilibrium, diffusion approximation and the gray gas model to describe radiation. As a sample problem, he computed blast-like expansion of an aluminium sphere into the ambient air, caused by a short energy release in the central zone of the sphere.

Nemchinov and Svettsov (1977) numerically studied blast-like expansion and cooling of a spherical air volume heated by a laser beam. To take into account angle directions and spectrum of radiation, the averaging procedure with respect to angles and quantum energies was carried out in the transfer equations. The detailed results were published for the explosion with small energy of 10 kJ. At the initial moment, the sphere had the radius of 2.6 cm, the temperature of 25 eV and the density of 3.7×10^{-5} g/cc. (Later Svettsov (1980) considered the different values of air density). The calculations confirmed the important role of radiation at the early stage of the process. Figure 4 shows the time dependences of maximum temperature T and radiation energy losses W at this stage. As is seen, the latter exceeds 20%.

A number of other works concerning the numerical solution of one-dimensional explosion problems with radiation should be mentioned. Their authors computed cooling of a hot sphere for air blast with small energy release (Markelova et al., 1973), spherical powerful explosion (Zinn, 1975), propagation of planar and spherical blast waves with transparent gas radiation (Erickson and Olfe, 1973), planar laser explosion near a wall (Nemchinov et al., 1980), cylindrical explosion in optically thin or optically thick gases (Higashino, 1981).

1.5. Explosion in combustible gas

A large group of numerical investigations were carried out for blasts in combustible gases, where the shock wave initiated exothermal chemical reactions. In this case,the flow development and its calculation become essentially complicated. Here it is necessary to take into consideration the kinetics of chemical reactions and high-temperature effects. However, some useful simplified models may be introduced in certain cases (see Korobeinikov, 1973).

The simplest model is the model of detonation wave where all the reactions and all the heat release are assumed to proceed on the wave front. In the case of point explosion in a combustible gas,an overdriven detonation wave is initiated. A numerical study, realized with the aid of the method of integral relations (Korobeinikov, 1969), demonstrated that such spherical or cylindrical waves propagating in a detonating gas approached the Chapman-Jouguet regime rather rapidly. The planar detonation wave still remained overdriven within the calculated finite distances, though its velocity approached the Chapman-Jouguet velocity.

Further, another complicated and more realistic model was used. Here direct and inverse reactions, subdued to the Arrhenius law, started to run after an induction period. A combustible gas was supposed to be perfect and the chemical kinetics was described by a single variable - the concentration of unreacted molecules. At first, such a two-front model was realized by Korobeinikov and Levin (1969) for strong explosion when it was possible to neglect the influence of combustion on the flow. Here decoupling of the overdriven detonation wave was numerically discovered.

Subsequent works (Levin and Markov, 1974, 1975; Korobeinikov and Markov, 1977), in which explosion in a combustible gas was computed by the method of Godunov, demonstrated two possible regimes of combustion. If the blast energy was relatively small, the propagating overdriven detonation wave splited up in time into a usual shock wave and a front of slow burning. When the blast energy increased and reached some critical value, the self-sustaining detonation combustion arose and it was of a pulsation nature for large activation energy of the induction period.

In several works, devoted to calculation of blast process in various gaseous mixtures, real rather than model equations of chemical kinetics were considered. Levin et al. (1981) predicted how the value of the critical energy, which was required for blast initiation of detonation combustion, depended on the parameters of chlorine-hydrogen mixture. Using the MacCormack technique, Sanders and Dwyer (1976) studied planar detonation wave in stoichiometric methane-air mixture.

Propagation of shock wave in the air, caused by blast of a spherical gaseous charge,was calculated by Zhdan (1975) for acetylene-air or propane-air mixtures and by Fishburn (1976) for oxygen-hydrogen mixture. The second author considered various cases of detonation initiation: central, surface and instantaneous in all the volume of the charge. Applying the Godunov method, Zhdan and Fedenok (1981) studied planar shock wave for air blast of gaseous charges consisting of five variants of detonating mixtures.

At present, special attention is paid to numerical computations of detonation and explosion in two-phase media. Heterogeneous detonation in combustible gas-droplet media was treated in several works. With the aid of an implicit technique, Zhdan (1976, 1977) calculated point spherical blast in kerosene-air mixture with various sizes of droplets. His results for the blast energy of 335 kJ are presented in Fig.5 where the pressure distributions p/p_a along the dimensionless radius λ are given for several successive moments. At first, the pressure on the initiating shock wave drops faster than for equivalent point blast in perfect gas (dash-dotted line). Then the Chapman-Jouguet surface is formed (its positions are marked by circles on the curves), the detonation starts and gradually tends to stationary self-sustaining regime.

Detonation, initiated by blast in two-phase gas-droplet media, was also studied in other works (Eidelman and Burcat, 1980, 1981; Burcat and Eidelman, 1980), where the cases of heptane-oxygen and kerosene-air mixtures were calculated by the FCT method. Here the influence of fuel droplet sizes and parameters of initiating point spherical explosion was considered. Zhdan (1981) carried out numerical investigation of heterogeneous detonation initiated by blast of spherical or cylindrical trotyl charges in kerosene-oxygen aerosol. The author determined the values of critical energy of the charge required to initiate detonation for different radii of droplets.

Eidelman, Oved et al. (1981) computed spherical blast waves in various gases (argon, air, oxygen) containing small solid particles of coal or aluminium. They numerically integrated the equations of two-phase flow and the equations describing solid particles evaporation.

1.6. Explosion in gases with specific properties

In this section, we discuss explosion in electrically conducting media in presence of external electromagnetic field in the case, when the magnetogasdynamical equations are valid. To maintain the one-dimensional nature of the problem, one has to be restricted with planar or cylindrical symmetry and special types of external field.

The method of integral relations was used by Korobeinikov(1965) to solve the problem on point explosion in a gas with infinite electrical conductivity provided a constant external magnetic field. A planar detonation wave was computed by Grigorenko and Levin (1975) who considered constant intensity of applied electromagnetic field, small magnetic Reynolds numbers and constant electrical conductivity of the medium.

Using the numerical method with pseudo-viscosity, Korobeinikov, Markov and Putjatin (1977) studied a very complicated case of blasts of cylindrical charges in real radiating electrically conducting gases. The same approach was employed in another work to investigate the blast process in laser pulse propulsion device (Ageev et al., 1981). A calculated example from the former work is presented in Fig.6 illustrating explosion of an RDX charge with radius of 10 cm in argon (p_a = 100 mm Hg, T_a = 300 K) at constant magnetic field with intensity H = 50 kGs. Here the trajectories of the shock wave $R = R_s(t)$ and of the interface $R = R_i(t)$, drawn by dashed lines, are compared with the case when the magnetic field is absent, i.e. H = 0 (solid lines). The magnetic field affecting the ionized argon slows down the detonation products and later on even forces them to turn back.

Blast waves may serve as generators of inversion media in gasdynamical lasers. Particles of a working gaseous mixture heated by a shock wave then get cool rapidly in a rarefaction wave. Here non-equilibrium relaxation processes can create population inversion of vibrational-rotational levels. That kind of the population inversion was studied in some papers (Lee, 1973; Levin and Tunik, 1977), where the cylindrical shock wave was caused by a line source explosion in the working mixture composed of carbon dioxide, nitrogen and helium. These authors concluded that it was possible under certain conditions to obtain sufficient population inversion and optical gain coefficient behind the blast wave. Podduev (1979) also carried out calculations for the analogous working mixture, but the cylindrical shock wave was caused by blast of a charge of the real explosive RDX. In this case to provide a sufficient value of the optical gain coefficient for intensive laser radiation, an additional expansion of the working mixture into vacuum was necessary. The similar results for explosion of an acetylene-oxygen charge were obtained by Korobeinikov and Podduev (1981).

Expansion and cooling of a hot volume of hydrogen was computed by Popov and Romashkevich (1977) taking into account non-equilibrium high-temperature processes. They worked out a splitting algorithm successively solving the gasdynamical and relaxation equations. One should also mention numerical calculations of point explosion in viscous heat-conducting perfect gas with zero temperature (Shidlovskii, 1977) and calculations of strong planar explosion in rarefied gas (Aristov and Shakhov, 1979).

2. TWO-DIMENSIONAL EXPLOSION PROBLEMS

2.1. Explosion in non-homogeneous atmosphere

The investigation of explosion in non-homogeneous media, particularly in the Earth's atmosphere, where the pressure and the density distributions along the height are stratified, is of significant interest. In general case, this problem is multi-dimensional, for this reason its solution becomes very difficult. However, if the properties of the medium depend only on the distance from the point of the blast, then the gas motion remains one-dimensional. This situation takes place for the planar explosion in the Earth's atmosphere, if the explosion plane is horizontal. That kind of numerical solutions were obtained by Enstrom and Brode (1971), who used the method with pseudo-viscosity, and by Kestenboim et al. (1974, 1976), who used both the implicit technique and the predictor-corrector method. Here the simple formulation of the problem enables to bring the calculation to the late stage of the blast and to study in detail the development of the rising and the descending shock fronts.

There exist also some approximate approaches where the solution is reduced to consideration of one-dimensional problems along separate rays passing through the blast centre (the so-called sector, or quasi-one-dimensional approximation). Thus, the flow is supposed to depend on the radius only, while the angle variable appears as a parameter. All the characteristics of the undisturbed air along the rays are taken in accordance with their actual values in the atmosphere. This model does not allow for interactions of gas motions in each sector and energy redistribution between them. Therefore, the model appears to be reliable only to a certain extent.

In the framework of this approximation, Lutzky and Lehto (1968) computed the propagation of descending shock waves caused by spherical explosions in exponential atmospheres with various non-homogeneity scales. The authors employed the Lagrangian method with artificial viscosity. Their numerical results proved the validity of the known empirical modified Sachs rule for prediction of the blast wave parameters in a stratified atmosphere.

The sector approximation was successfully applied by Korobeinikov, Chushkin and Shurshalov (1972, 1977) and by Putjatin (1980) to compute the early stage of propagation of shock waves modelling the flight and the blast of the Tunguska meteorite. This approach was also used by Kestenboim et al. (1974) who calculated spherical explosion in the exponential atmosphere.

The problem on blast in a non-homogeneous atmosphere was just the first unsteady axisymmetrical explosion problem which was numerically treated. As early as in 1955 K.I.Babenko, A.M.Molchanov and V.V.Rusanov computed the initial stage of blast in the exponential atmosphere of the Earth.

In the late sixties, using the explicit predictor-corrector method, Kestenboim and Chudov (see the book by Kestenboim, Roslyakov and Chudov, 1974) carried out numerical study of strong point blast in the exponential (isothermal) atmosphere. The obtained results quantitatively demonstrated the distortion of the shock wave, which became egg-like at late times, and non-uniformity of its intensity. The process was characterized by the acceleration and the relative strengthening of the top portion of the shock wave after some moment and by lifting of the hot gaseous mass in the central region of the blast.

A series of the works by Shurshalov (1976, 1978, 1980) and also the work by Korobeinikov, Chushkin and Shurshalov (1977) were devoted to a detailed numerical investigation of explosion in the non-homogeneous atmosphere with the purpose of modelling the flight and the blast-like decomposition of the Tunguska meteorite. Here the model of expansion of a hot gaseous volume was employed. The results were obtained both in spherical and in cylindrical cases for various heights h and energies E at all blast stages up to the late one, allowing for such factors as counterpressure, gravity, real properties of air, non-isothermal character of the Earth's atmosphere and radiation, as well as reflection of the shock wave from the ground. These two-dimensional calculations confirmed the effectiveness of the approximate modified Sachs rule. The computations were carried out by the Godunov method with the moving adapting non-uniform network and the shock wave fitting.

Numerical experiment gives an unique opportunity to consider the above-mentioned factors both independently and jointly and to reveal the role of each factor. For instance, it was proved that consideration of gravity was quite necessary even for a large initial energy release and that the effect of this factor manifested itself rather early. It was found that for moderate heights of blast, the shock front shape differed little from spherical (or cylindrical) while the shock wave intensity changed significantly along the front. The formation of a large ring-shaped vortex was observed (see Fig.7 where the velocity field for an air spherical explosion with $h = 6.5$ km, $E = 6.5 \times 10^{22}$ erg is given at the moment $t = 19$ sec). The comparison of the results for the exponential and for the non-isothermal atmospheres showed that the latter factor had rather weak influence within the distances of 10 - 20 km, however this influence gradually increased with the distance. On the contrary, the high-temperature properties of real air were mostly pronounced at the beginning of the blast process, when the temperatures were great enough. Owing to dissociation and ionization, the temperature in the hot region considerably decreased and the intensity of the shock wave was reduced by 30-50%.

Since calculation of radiation in two-dimensional problems on explosion is very laborious work, here various simplified approaches and models are usually introduced. In particular, Shurshalov (1980) used the diffusion approximation to describe the spatial distribution of radiation and the gray gas model or the more accurate multi-groups model to allow for the frequency dependence of the air absorption coefficient. The calculated dynamics of the radiation losses W in the disturbed region for a spherical blast in the Earth's atmosphere (h = 6.5 km, E = 10^{23} erg) is presented in Fig.8 as a function of the radius R_s corresponding to the lowest point of the shock wave. Here the dashed line is related to the gray gas model, the solid line corresponds to the ten-groups model, while two dotted lines give the contributions of the visible and the infrared parts of the spectrum (the lower and the upper lines accordingly). The effective action of radiation lasted for about 1.5 sec and the total energy losses amounted to 10-30% of the blast energy in different variants.

An interesting but intricate approach based on the Monte-Carlo method was developed by Sandford and Anderson (1973) to compute radiation processes. However, only an illustrative example was given in their work.

An important feature of explosion in a stratified atmosphere is the spatial redistribution of energy. The principal conclusion about energy transfer upwards was made in the very first solutions of the problem. A more detailed behaviour of this process was described by Shurshalov (1978). His results for a spherical blast (h = 8 km, E = 10^{23} erg) are presented in Fig.9. Here $E(\theta)$ is the energy contained inside each solid angle minus the energy of undisturbed gas in the same volume; $E_0(\theta)$ is the initial blast energy, contained in the same solid angle (the ray $\theta = 0°$ is directed vertically downwards). As it is seen, the energy flow from lower sectors to upper ones proceeds so intensively that not only the energy, which was put there by the blast, transfers upwards,but also a part of the internal and potential energy, which the gas had in these zones before the explosion. On the other hand, locally near the horizontal plane passing through the blast centre ($\theta = 90°$), there exists not only the energy transfer upwards but also slight opposite energy transfer is observed because of the complex circular character of the flow in the ring-shaped vortex.

Among other publications concerning the numerical solution of the problem on explosion in the Earth's non-homogeneous atmosphere, it is worth mentioning the work by Brode (1970) where some computational difficulties are discussed.

2.2. Explosion of charges having non-standard form

Interesting gasdynamical problems are set,when one treats blasts of charges having more general form than standard spherical, cylindrical or planar. For explosions of such charges, energy redistribution occurs, the flow becomes non-one-dimensional and the effect of the shock wave is different along different directions.

A number of generalizations of cylindrical explosion lead to the problems of that kind. The works by Shurshalov (1973) and also by Korobeinikov, Chushkin and Shurshalov (1972, 1973) are devoted to computation of blasts of cylindrical gaseous charges having finite or infinite length and constant or variable energy release. Here propagation of the main and the second shock waves was studied, approach of the solution to the limiting one-dimensional regime was traced. The influence of the form of the charge end was investigated. Peculiar motion of the interface and its connection with the second shock wave were emphasized. For the case of concentrated energy release at the end of the charge, the interaction between the quasi-spherical and the cylindrical portions of the blast wave was ascertained and the effect of appreciable strengthening of the blast wave at its intermediate portion was found.

As another generalization of cylindrical explosion, Shurshalov (1975b) considered infinitely long charges of condensed explosive, having triangular, square or star-like cross-sections. The motions of the main and the second shock wave as well as the interface were treated in detail. The author analysed features of the flow due to the sharp corners of cross-sections and, in particular, disclosed the corresponding increase or decrease of energy concentration. Two patterns of the flows (in two half-quadrants) are shown in Fig.10 concerning the blasts of trotyl charges having square and star-like cross-sections and located in the air under the see-level conditions. The main shock waves are drawn by solid lines, the interfaces are depicted by dashed lines and the initial boundaries of the charges are marked by shading. Shurshalov (1981) also treated a cylindrical explosion of an infinite charge with energy release, which was variable with respect to the time and the radial direction, and studied the influence of an outer stream ("the wind") on the flow.

Special attention was paid to the effects on the blast flow, caused by sharp-pointed and sharp-edged parts of the charge surface. Shurshalov (1974, 1975a), Arkhangel'skii and Shurshalov (1977), Mozzhilkin and Shishkin (1981) numerically investigated this question. Here the classical problem on decomposition of arbitrary gasdynamical discontinuity (the Riemann problem) was extended to two-dimensional planar and axisymmetrical cases. We discuss the features of such flows, namely, planar expanding (Fig.11a) and converging (Fig.11b) flows of a gas, initially compressed to the pressure of 100 atm and filling the regions inside and outside the right angle, accordingly. Since this problem is self-similar, the graphs are given in dimensionless variables x/At and y/At (A is the constant dimensional quantity). The main shock wave is depicted by the solid line, the interface is drawn by the dashed line, the second shock wave and the shock wave reflected from the symmetry plane are shown by the circles. It was found that if the flow near the sharp-pointed or sharp-edged parts of the charge was of expansion type, the second shock wave always occured and led to deceleration of expanding gas and to formation of a bend on the interface.If the flow near the point or the edge was of converging type, the other regimes arose and their basic features were the Mach or the regular reflection of the shock wave and the presence or the absence of a cumulative gaseous jet. That sort of effects were also observed by Shurshalov (1973, 1975a,b) in non-self-similar flows and by Arkhangel'skii (1975) who considered the spark discharge in the air for laser beam focusing.

Recently,Shurshalov treated another class of axisymmetrical blast flows generated by explosion of ellipsoidal gaseous charges of small density. Figure 12 gives the deviation δp_s of the shock excess pressure from its mean value for the charges having the form of an elongated ellipsoid of rotation and different energies. The approach of the flow to the one-dimensional regime was found to be realized within considerable distances. As to the shock wave intensity and the flow parameters in the central zone of the blast, the corresponding distances were equal to 30-60 and more the characteristic size of the charge and essentially exceeded that distances (6-8 characteristic lengths) at which the non-point-blast spherical solution approached the point-blast regime in the appropriate one-dimensional case. Thus, it is quite reasonable to formulate and to solve the problems on point explosion with variable energy release depending on the angle.

All the computations mentioned in this section were carried out by the Godunov method, as a rule, with treatment of the blast wave and the interface as boundary lines of the moving network.

2.3. Some problems on detonation

Numerical solutions of non-one-dimensional problems of blast type for non-equilibrium reacting gaseous mixtures have not achieved as yet extensive development, particularly in the case allowing for more or less strict chemical kinetics. There are not so many publications where two-dimensional detonation and combustion are numerically investigated. Usually, the simplest models of chemical reactions and the simplest flows are considered.

A number of works dealt with calculation of motion of a metal tube filled with explosive along which a detonation wave propagated. Probably, the first unsteady solution of this problem was obtained by Wilkins (1964) who applied his finite-difference code with artificial viscosity. This solution is the most complete,since elastic-plastic properties of material of the tube walls were taken into account. This factor noticeably affected the motion, specifically, the shape of the deformable tube and its wall thickness.

Later, extensive numerical investigation of this problem,though without considering the strength of the tube material,was carried out in the works by Kashirskii et al. (1974a,b). Using an explicit two-step second-order method, the authors calculated the flow with various charge/shell mass ratio, different ways of detonation initiation and real equations of state for detonation products. The detonation wave was assumed to have the Chapman-Jouguet velocity. The motions of the shell (heavy lines) and of the detonation products boundary (thin lines) at three moments 1-3 are illustrated in Fig.13. The detonation wave was initiated at the central point of the right open tube end ($z = 2$) and its front at moment 1 is marked by figure 4.

To compute detonation in a cooper tube containing an explosive mixture, Gol'din et al. (1973) proposed another approach, where a model heat release in each mesh was considered. The strength of the metal material was not allowed for. The finite-difference schemes with triangular and quadrangular meshes were constructed. In the first scheme, it was easier to overcome some instability of the numerical technique.

One of the first non-one-dimensional computations of detonation was fulfiled by Elliott (1962) who applied the method of characteristics. He treated detonation of a spherical charge of explosive, initiated at a point on its surface. The one-front model of detonation wave and the real equation of state for the explosion products were assumed. The flow was computed in the region between the detonation wave (which on the diverging portions followed the Chapman-Jouguet regime, while on the converging portions was overdriven) and the free surface of the gas expanding into vacuum.

Many computations of homogeneous and heterogeneous detonations were carried out by Mader (1974, 1979). For instance, he studied detonation of an explosive plate enveloped in an aluminium shell (or without it). A two-dimensional Lagrangian finite-difference code was used for various models of detonation combustion. The detonation front was found to remain practically planar with small distortions near the edges of the plate.

The Mader method was employed by Tanaka and Hikita (1976) to solve the analogous problem on detonation. The special attention was paid to charges of small cross-sections slightly exceeding the critical size for which detonation still proceeded. The authors revealed a periodical self-sustaining process of interaction between the detonation wave and the rarefaction waves from the side surface of the charge.

The development of two-dimensional perturbations accompanying propagation of a detonation wave in a planar channel containing a combustible mixture was studied in several papers within the framework of the two-front model with direct and inverse reactions. Levin and Markov (1975), Markov (1981) used the Godunov method, while Taki and Fujiwara (1976) employed another finite-difference first-order method. The numerical results reproduced experimentally observed features of such flows, in particular, their cellular structure, dynamics of triple configurations of shock waves, variation of extents of induction zones, pulsations of the longitudinal velocity etc. Some of that features are seen in Fig.14 taken from the work by Markov (1981), where the periodical pattern of propagation of the shock wave (solid lines) and the flame front (dashed lines) in the hydrogen-oxygen mixture is presented.

Ivanov et al. (1981) numerically modelled development of detonation in an unconfined finite-thickness layer of the fire-damp,when a lateral expansion of detonation products took place. The gasdynamical equations were integrated by the predictor-corrector method,while the kinetic equations were solved by an implicit technique.

Using the method with artificial viscosity, Podduev (1980) calculated propagation of a blast wave in a planar channel of constant cross-section, filled with the relaxing gaseous mixture utilized in gasdynamical lasers. He especially interested in two-dimensional outflow of the mixture from the channel into vacuum in order to obtain the inversion population.

2.4. Numerical simulation of some cosmic phenomena

Some cosmic phenomena have the blast-like nature. They can be successfully modelled upon the basis of the gasdynamical theory of explosion. We discuss two types of such phenomena, namely, solar flares and entry of large meteorite bodies into dense layers of the atmosphere.

Solar flares are characterized by extremely great energy release in a relatively small volume and during a relatively short time interval. Despite the fact that the perturbations from solar flares propagate in a very rarefied space, the methods of mechanics of continuous medium turn out to be suitable for their analysis. However, the investigation becomes especially complicated owing to such specific factors as the solar wind, the solar gravity and rotation, the interplanetary magnetic field and so on.

So far available numerical solutions do not describe this phenomenon with all the completeness. One-dimensional approximation was supposed in the early publications (Hundhausen and Gentry, 1969; Shidlovskaya, 1975; Steinolfson and Dryer, 1978; etc.). Later, axisymmetrical model of plasma motion accompanying a solar flare was worked out (Korobeinikov and Shidlovskaya, 1975; Shidlovskaya, 1977). The solar gravity and the radial component of the interplanetary magnetic field were taken into account. The computations were fulfiled with the aid of the Belotserkovskii and Davydov method of large particles. Using the numerical method with pseudo-viscosity, De Young and Hundhausen (1971) calculated two-dimensional propagation of solar flare-associated perturbations without consideration of the interplanetary magnetic field. Wu et al. (1979) and Han et al. (1979) confined themselves to the two-dimensional planar formulation of the problem, in which the calculations were limited by the equatorial plane of the Sun. Here the MHD equations were numerically integrated by the Lax-Wendroff method or by the method with artificial viscosity.

Now we discuss the interaction between the atmosphere and large cosmic bodies entering it with a hypersonic speed. Such bodies can undergo the strong destruction and decomposition, sometimes involving the process of explosion character. That kind of phenomenon occured for the Tunguska cosmic body.

In the case of the Tunguska meteorite, the problem consists in determination of its unknown trajectory and energy parameters, taking as a basis the registered picture of the ground surface destructions at the fall area. With that end in view, it is necessary to calculate propagation of shock waves in the Earth's stratified atmosphere, which were caused by the flight and the blast-like expansion of the body. Here high-temperature effects, including radiation, and reflection of the shock waves from the ground must be allowed for. The formulation of this complex problem and its successive solution involving at each step new physical factors were carried out by Korobeinikov, Chushkin and Shurshalov (1972, 1974, 1976, 1977) in a series of their investigations continued in the works by Korobeinikov et al. (1979, 1981) and by Putjatin (1980).

The first authors worked out the model of the phenomenon, in which the ballistic and the blast waves were simulated by the shock waves arisen from an equivalent explosion of a semi-infinite cylindrical charge with variable specific energy along its axis coinsiding with the trajectory. As a result, the trajectory and energy parameters of the meteorite were determined, satisfying the condition that the calculated and the actual pictures of the flattened forest at the Tunguska catastrophe region should accord.

The actual picture is presented in Fig.15a where the fallen trees field is indicated by the dashes, while the radial and the closed curves describe the shock front propagation on the ground. The calculated picture is given in Fig.15b where the shock waves isochrones are drawn by the dashed lines, while the fallen trees directions are shown by the arrows. It is seen the good correspondence between these two pictures with respect to the outer form and the inner structure. The computations resulted in the following values of parameters of the Tunguska meteorite: the inclination angle of the final active part of the trajectory $a = 40^\circ$, the height of the blast $h = 6.5$ km, the blast energy $E_0 = 10^{23}$ erg, the specific energy of the ballistic wave $E_1 = 1.4 \times 10^{17}$ erg/cm. The total energy release is in a good agreement with available barograms and seismograms of the Tunguska phenomenon.

The observed and the calculated zones of the forest radiant burn at the Tunguska meteorite fall area are shown in Fig.16. Here the circles of different sort designate different burn degree measured at the fall area. The dashed and the solid curves, corresponding to the gray gas model and to the ten-groups model, are the calculated isolines of the normal radiation thermal impulse $J_n = 16$ cal/cm^2. That isoline may be chosen as the theoretical boundary of the radiant burn zone in the case under consideration. It is evident that these computational results are also in good accordance with the observed data.

CONCLUSION

Computational methods have found wide application to solution of various explosion problems. Finite-difference methods of explicit type are the most popular for calculation of flows caused by blast. High-accurate numerical investigation of one-dimensional explosion flows with various physical-chemical processes taken into account has become quite common. The number of available computations of two-dimensional (planar and axisymmetrical) blast problems is still rather limited. In the most complicated cases, the obtained solutions have model character. As to the numerical computations of three-dimensional explosion problems, they only begin to be carried out. Here very few results which quite often have an illustrating character are published (Pracht, 1975; Stein, Gentry and Hirt, 1977; Mader and Kershner, 1980). Three-dimensional computations demand the use of powerful computers with very high speed and very large memory capacity.

REFERENCES

Ageev V.P., Barchukov A.I., Bunkin F.V., Konov V.I., Korobeinikov V.P., Prokhorov A.M., Putiatin B.V., Khudiakov V.M. (1981) Progress

in Astronaut. and Aeronaut. 76, 33-45.
Andriankin E.I., Myagkov N.N. (1981) Zh. Prikl. Mekh. i Tekhn. Fiz. No.4, 119-125.
Aristov V.V., Shakhov E.M. (1979) Zh. Vychisl. Mat. i Mat. Fiz. 19, 1276-1287.
Arkhangel'skii N.A. (1974) Zh. Vychisl. Mat. i Mat. Fiz. 14, 1281-1291.
Arkhangel'skii N.A. (1975) Izv. AN SSSR. Mekh. Zhidk. i Gaza No.3, 161-164.
Arkhangel'skii N.A., Shurshalov L.V. (1977) Izv. AN SSSR. Mekh. Zhidk. i Gaza No.1, 83-88.
Brode H.L. (1955) J. Appl. Phys. 26, 766-775.
Brode H.L. (1956a) Research Mem. RM-1824-AEC. Rand Corp., Santa Monica.
Brode H.L. (1956b) Research Mem. RM-1825-AEC. Rand Corp., Santa Monica.
Brode H.L. (1959) Phys. Fluids 2, 217-229.
Brode H.L. (1969) Astronaut. Acta 14, 433-444.
Brode H.L. (1970) Problems Associated with Air Blast Interaction Calculation. Rand Corp., Santa Monica.
Burcat A., Eidelman S. (1980) AIAA Journal 18, 1233-1236.
Chou P.Ch., Huang L. (1969) J. Appl. Phys. 40, 752-759.
Derzhavina A.I. (1976) Prikl. Mat. i Mekh. 40, 185-189.
De Young D.S., Hundhausen A.J. (1971) J. Geophys. Res. 76, 2245-2253.
Director M.N., Dabora E.K. (1977) AIAA Journal 15, 1315-1321.
Eidelman S., Burcat A. (1980) AIAA Journal 18, 1103-1109.
Eidelman S., Burcat A. (1981) J. Comp. Phys. 39, 456-472.
Eidelman S., Oved Y., Hasson A., Burcat A. (1981) Combust. Sci. and Technol. 25, 21-30.
Elliott L.A. (1962) Proc. Roy. Soc. London A267, 558-565.
Enstrom J.E., Brode H.L. (1971) Rep. RDA-TR-042-DNA. Rand Corp., Santa Monica.
Erickson G.G., Olfe D.B. (1973) Phys. Fluids 16, 2121-2131.
Fishburn B.D. (1976) Acta Astronaut. 3, 1049-1065.
Fonarev A.S., Chernyavskii S.Yu. (1968) Izv. AN SSSR. Mekh. Zhidk. i Gaza No.5, 169-174.
Gol'din V.Ya., Kalitkin N.N., Levitan Yu.L., Rozhdestvenskii B.L. (1973) Chisl. Metody Mekh. Splosh. Sredy 4, No.3, 62-70.
Goldstine H., Neumann J. (1955) Communs Pure and Appl. Math. 8, 327-354.
Grigorenko V.L., Levin V.A. (1975) Izv. AN SSSR. Mekh. Zhidk. i Gaza No.5, 116-120.
Han S.M., Wu S.T., Nakagawa J. (1979) Comput. and Fluids 7, 97-108.
Higashino F. (1981) 8th Int. Colloq. Gasdyn. Explos. and React. Syst., Minsk. Book Abstracts, 179.
Hundhausen A.J., Gentry R.A. (1969) J. Geophys. Res. 74, 2908-2919.
Ivanov M.F., Fortov V.E., Borisov A.A. (1981) Fiz. Goreniya i Vzryva 17, 108-116.
Kashirskii A.V., Korovin Yu.V., Odintsov V.A., Chudov L.A. (1974a) Zh. Prikl. Mekh. i Tekhn. Fiz. No.2, 167-168.
Kashirskii A.V., Korovin Yu.V., Chudov L.A. (1974b) Zh. Prikl. Mekh. i Tekhn. Fiz. No.6, 170-172.
Kestenboim Kh.S., Kuzina Z.N., Markov A.A. (1976) Izv. AN SSSR. Mekh. Zhidk. i Gaza No.3, 124-131.
Kestenboim Kh.S., Roslyakov G.S., Chudov L.A. (1974). Point Explosion. Methods of Calculation. Tables. Nauka, Moscow.
Kovaleva I.N., Nemchinov I.V. (1976) Fiz. Goreniya i Vzryva 12, 113-116.
Korobeinikov V.P. (1965) Dokl. AN SSSR 165, 1019-1022.
Korobeinikov V.P. (1969) Astronaut. Acta 14, 411-419.

Korobeinikov V.P. (1973) Problems of Theory of Explosions in Gases. Trudy Matem. Inst. AN SSSR 119.
Korobeinikov V.P., Chushkin P.I. (1964) Dokl. AN SSSR 154, 549-552.
Korobeinikov V.P., Chushkin P.I. (1966) Trudy Matem. Inst. AN SSSR 87, 4-34.
Korobeinikov V.P., Chushkin P.I., Sharovatova K.V. (1969) Gasdynamical Functions of Point Explosion. Vychisl. Tsentr AN SSSR, Moscow.
Korobeinikov V.P., Chushkin P.I., Shurshalov L.V. (1972) Astronaut. Acta 17, 339-348.
Korobeinikov V.P., Chushkin P.I., Shurshalov L.V. (1973) Arch. Mech. Stosow. 25, 993-1006.
Korobeinikov V.P., Chushkin P.I., Shurshalov L.V. (1974) Izv. AN SSSR. Mekh. Zhidk. i Gaza No.3, 94-100.
Korobeinikov V.P., Chushkin P.I., Shurshalov L.V. (1976) Acta Astronaut. 3, 615-622.
Korobeinikov V.P., Chushkin P.I., Shurshalov L.V. (1977) Zh. Vychisl. Mat. i Mat. Fiz. 17, 737-752.
Korobeinikov V.P., Chushkin P.I., Shurshalov L.V. (1981) Preprint IAF-81-407. XXXIII Congr. Inter. Astronaut. Feder., Rome.
Korobeinikov V.P., Levin V.A. (1969) Izv. AN SSSR. Mekh. Zhidk. i Gaza No.6, 48-51.
Korobeinikov V.P., Markov V.V. (1977) Arch. Termodyn. i Splania 8, 101-120.
Korobeinikov V.P., Markov V.V., Putjatin B.V. (1977) Izv. AN SSSR. Mekh. Zhidk. i Gaza No.4, 133-138.
Korobeinikov V.P., Podduev M.I. (1981) Progress in Astronaut. and Aeronaut. 76, 89-105.
Korobeinikov V.P., Putjatin B.V., Chushkin P.I., Shurshalov L.V. (1979) Lect. Notes Phys. 90, 325-332.
Korobeinikov V.P., Shidlovskaya L.V.(1975) Chisl.Metody Mekh.Splosh. Sredy 6, No.4, 56-68.
Kuhl A.L., Seizew M.R. (1981) Progress in Astronaut. and Aeronaut. 75, 226-241.
Larson R.A.L. (1970) J. Appl. Phys. 41, 2747-2748.
Lee J.H., Bui T.D., Knystautas R.(1973) Appl.Phys.Let. 22, 434-436.
Levin V.A., Markov V.V. (1974) Izv. AN SSSR. Mekh. Zhidk. i Gaza No.5, 89-93.
Levin V.A., Markov V.V. (1975) Fiz. Goreniya i Vzryva 11, 623-633.
Levin V.A., Markov V.V., Osinkin S.F.(1981) Dokl.AN SSSR 261, 50-52.
Levin V.A., Tunik Yu.V. (1977) Fiz. Goreniya i Vzryva 13, 447-454.
Lutzky M., Lehto D.L. (1968) Phys. Fluids 11, 1466-1472.
Lutzky M., Lehto D.L. (1970) J. Appl. Phys. 41, 844-846.
MacLaughlin R. (1980) J. Comput. Phys. 35, 77-89.
Mader Ch.L. (1974) Acta Astronaut. 1, 373-384.
Mader Ch.L. (1979) Numerical Modeling of Detonations. Univ. Calif. Press, Berkley.
Mader Ch.L., Kershner J.D. (1980) Rep. LA-8206. Los Alamos Scient. Lab., Los Alamos.
Markelova L.P., Nemchinov I.V., Shubadeeva L.P. (1973) Zh. Prikl. Mekh. i Tekhn. Fiz. No.2, 54-63.
Markov V.V. (1981) Dokl. AN SSSR 258, 314-317.
Mozzhilkin V.V., Shishkin Yu.V. (1981) Vychisl. Metody i Programmir., 68-74. Saratov.
Nemchinov I.V., Polozova I.A., Svettsov V.V., Shuvalov V.V. (1980) Dinamika Izluch. Gaza 3, 33-45. Vychisl. Tsentr AN SSSR, Moscow.
Nemchinov I.V., Svettsov V.V. (1977) Zh. Prikl. Mekh. i Tekhn. Fiz. No.4, 24-32.
Okhotsimskii D.E., Kondrasheva I.L., Vlasova Z.P., Kazakova R.K. (1957) Trudy Matem. Inst. AN SSSR 50.
Okhotsimskii D.E., Vlasova Z.P. (1962) Zh. Vychisl. Mat. i Mat. Fiz.

2, 107-124.
Oppenheim A.K., Kuhl A.L., Kamel M.M. (1978) Arch. Mech. Stosow. 30, 553-571.
Oved Y., Millinazzo F., Clements R.M., Smy P.R. (1979) AIAA Journal 17, 601-605.
Parkhomenko V.P., Popov S.P., Ryzhov O.S. (1977a) Uch. Zap. Tsentr. Aero-Gidrodinam. Inst. 8, No.3, 32-38.
Parkhomenko V.P., Popov S.P., Ryzhov O.S. (1977b) Zh. Vychisl. Mat. i Mat. Fiz. 17, 1325-1329.
Plooster M.N. (1970) Phys. Fluids 13, 2665-2675.
Podduev M.I. (1979) Kvant. Electronika 6, 379-381.
Podduev M.I. (1980) Dokl. AN BSSR 24, 702-705.
Popov S.P., Romashkevich Yu.I. (1977) Zh. Vychisl. Mat. i Mat. Fiz. 17, 1602-1607.
Pracht W.E. (1975) J. Comput. Phys. 17, 132-159.
Putjatin B.V. (1980) Dokl. AN SSSR 252, 318-322.
Sanders B.R., Dwyer H.A. (1976) Lect. Notes Phys. 59, 384-390.
Sandford M.T.II, Anderson R.S. (1973) J. Comput. Phys. 13, 130-157.
Shidlovskaya L.V. (1975) Dokl. AN SSSR 225, 272-275.
Shidlovskaya L.V. (1977) Zh. Vychisl. Mat. i Mat. Fiz. 17, 196-208.
Shidlovsky V.P. (1977) AIAA Journal 15, 33-38.
Shubin G.R. (1981) Comput. and Fluids 9, 299-312.
Shurshalov L.V. (1973) Zh. Vychisl. Mat. i Mat. Fiz. 13, 971-983.
Shurshalov L.V. (1974) Izv. AN SSSR. Mekh. Zhidk. i Gaza No.2, 69-74.
Shurshalov L.V. (1975a) Izv. AN SSSR. Mekh. Zhidk. i Gaza No.4, 116-122.
Shurshalov L.V. (1975b) Izv. AN SSSR. Mekh. Zhidk. i Gaza No.5, 130-135.
Shurshalov L.V. (1976) Dokl. AN SSSR 230, 803-806.
Shurshalov L.V. (1978) Arch. Mech. Stosow. 30, 629-643.
Shurshalov L.V. (1980) Izv. AN SSSR. Mekh. Zhidk. i Gaza No.3, 105-112.
Shurshalov L.V. (1981) 8th Int. Colloq. Gasdyn. Explos. and React. Syst., Minsk. Book Abstracts, 162.
Stein L.R., Gentry R.A., Hirt C.W. (1977) Comput. Methods in Appl. Mech. and Eng. (Netherlands) 11, 57.
Steinolfson R.S., Dryer M. (1978) J. Geophys. Res. A83, 1576-1582.
Sternberg H.M., Hurwitz H. (1976) Prep. Pap. 6th Symp. (Int.) Detonation, San Diego, 2, 557-568.
Svettsov V.V. (1980) Dinamika Izluch. Gaza 3, 46-57. Vychisl. Tsentr. AN SSSR, Moscow.
Taki S., Fujiwara (1976) AIAA Pap. 404.
Tanaka K., Hikita Ts. (1976) Acta Astronaut. 3, 1005-1013.
Vasil'ev A.A., Zhdan S.A. (1981) Fiz. Goreniya i Vzryva 17, 99-105.
Wilkins M.L. (1964) Methods in Comput. Physics 3, 211-262. Acad. Press, New York - London.
Wu S.T.,Han S.M., Dryer M. (1979) Planet. and Space Sci. 27, 255-264.
Zhdan S.A. (1975) Zh. Prikl. Mekh. i Tekhn. Fiz. No.6, 69-74.
Zhdan S.A. (1976) Fiz. Goreniya i Vzryva 12, 586-594.
Zhdan S.A. (1977) Dinamika Splosh. Sredy 32, 36-46. Novosibirsk.
Zhdan S.A. (1981) Fiz. Goreniya i Vzryva 17, 105-111.
Zhdan S.A., Fedenok V.I. (1981) Dinamika Splosh. Sredy 51, 42-52. Novosibirsk.
Zinn J. (1975) J. Comput. Phys. 13, 569-590.

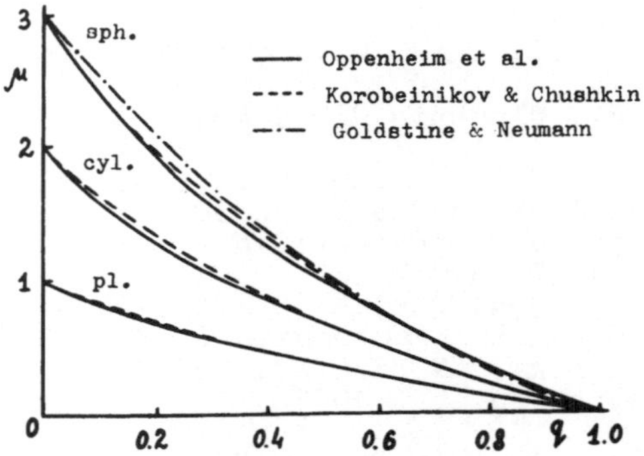

Fig.1. Shock wave decay coefficient for point explosion.

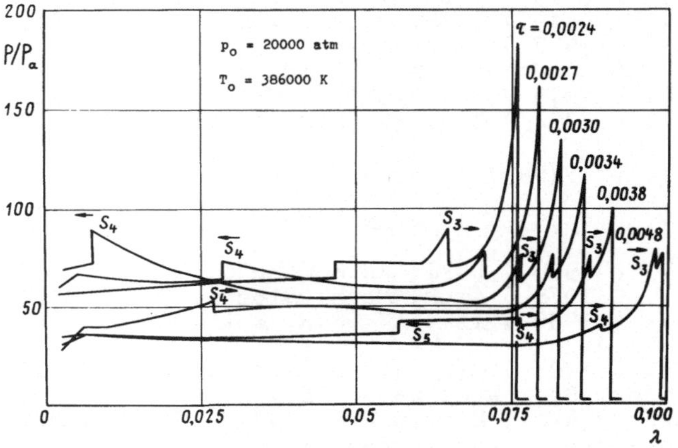

Fig.2. Pressure versus relative radius at different moments for blast-like expansion.

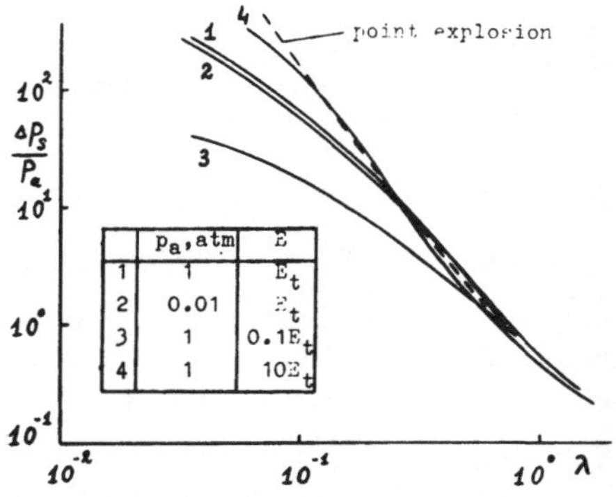

Fig.3. Shock excess pressure for blasts of condensed explosives.

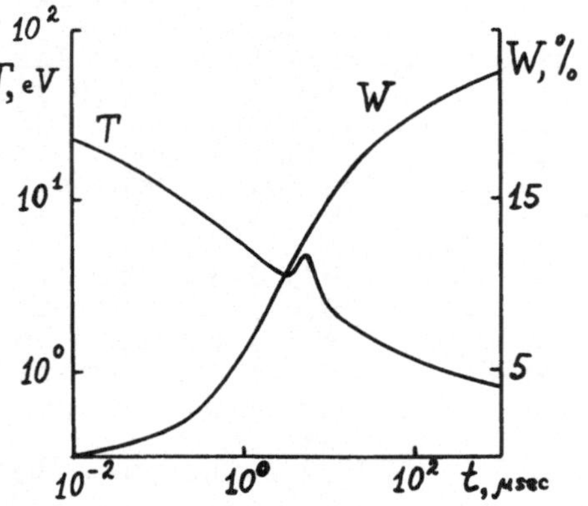

Fig.4. Maximum temperature and radiation energy losses for blast-like expansion.

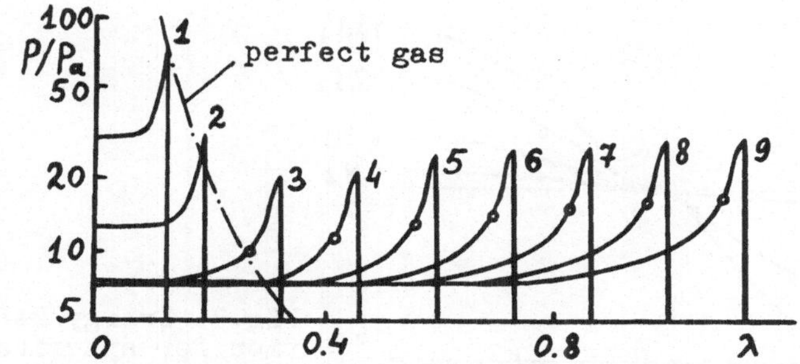

Fig.5. Pressure distribution at different moments for blast in air-kerosene mixture.

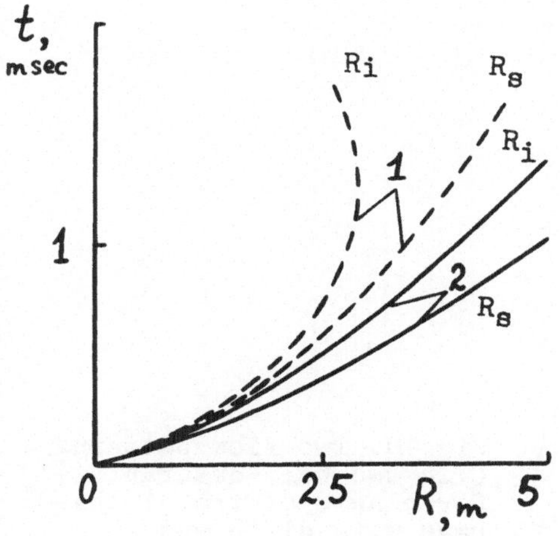

Fig.6. Shock front and interface for RDX charge blast.

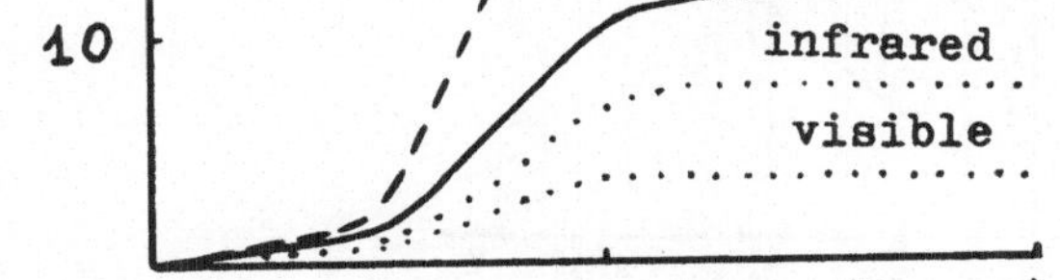

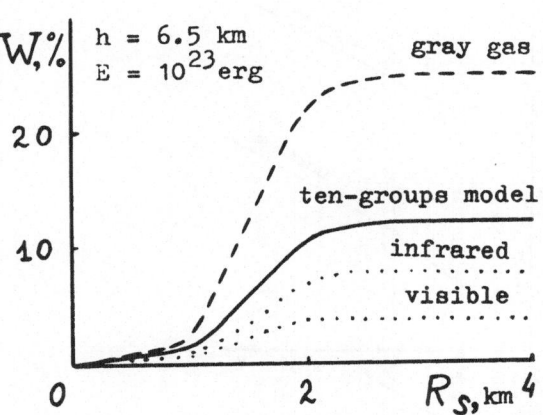

Fig.8. Radiation energy losses for explosion in the atmosphere.

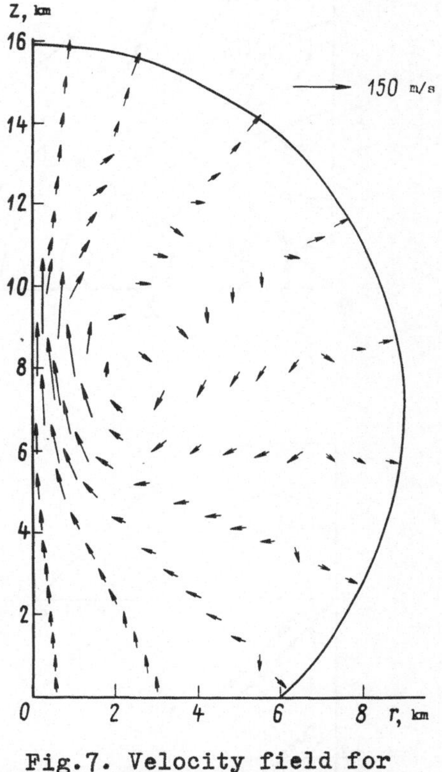

Fig.7. Velocity field for spherical explosion in the atmosphere.

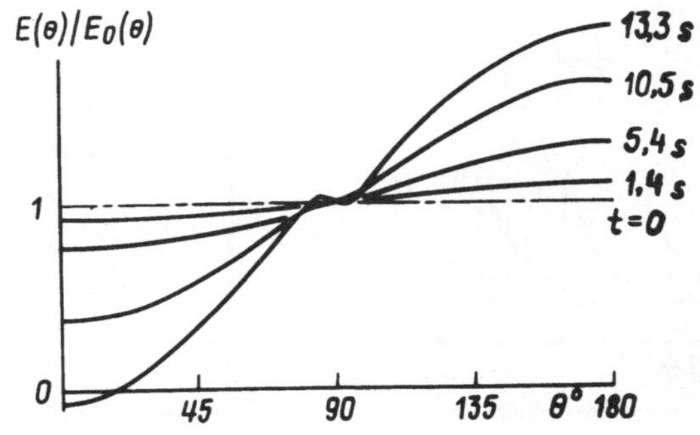

Fig.9. Energy redistribution for spherical explosion in the atmosphere.

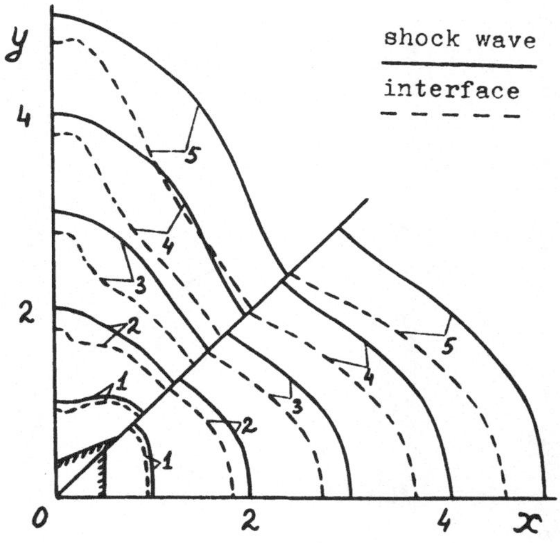

Fig.10. Two flow patterns (in two half-quadrants) for blasts of trotyl charges with different cross-sections.

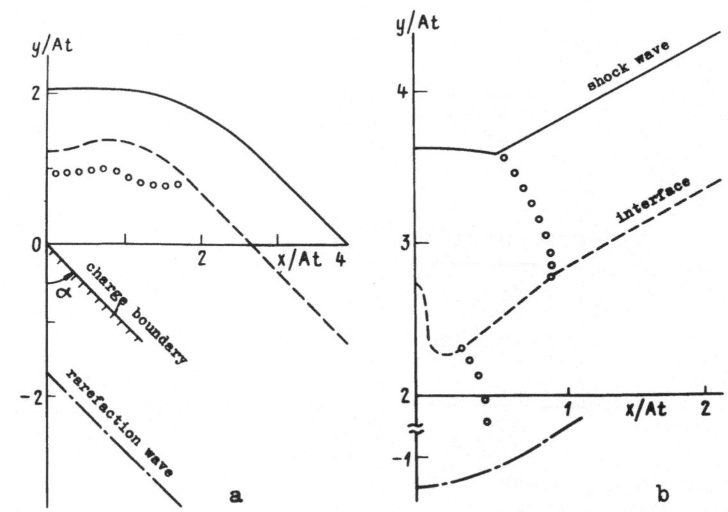

Fig.11. Flow patterns for blast-like expansion of plane regions inside (a) and outside (b) right angle.

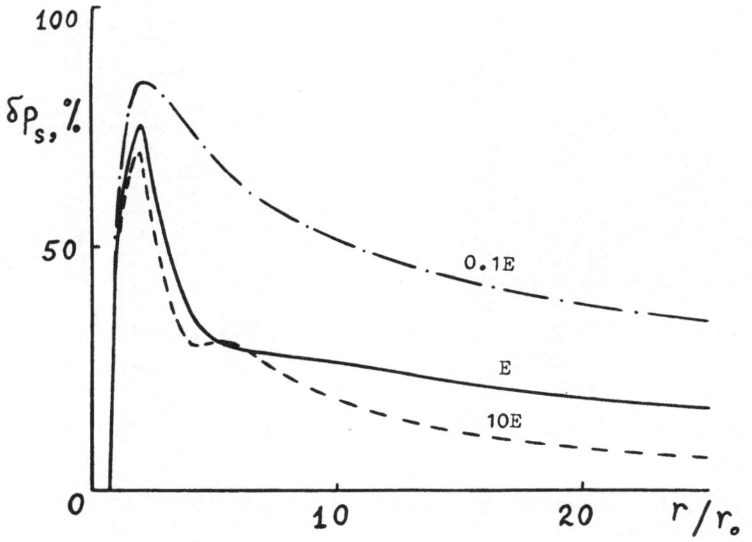

Fig.12. Deviation of shock excess pressure for blasts of ellipsoidal charges with different energies.

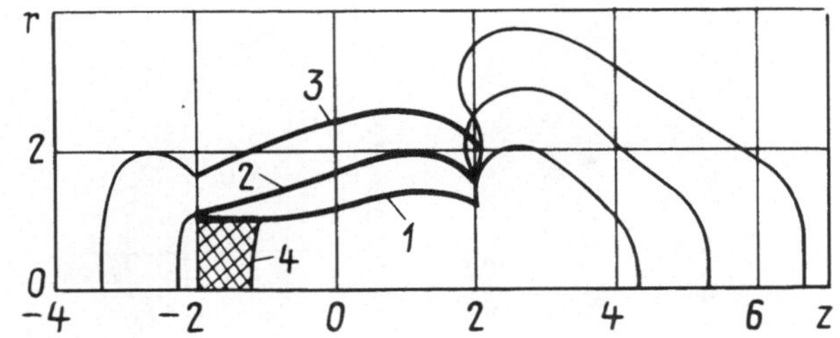

Fig.13. Flow patterns at three moments 1-3 for detonation of charge contained in metal tube.

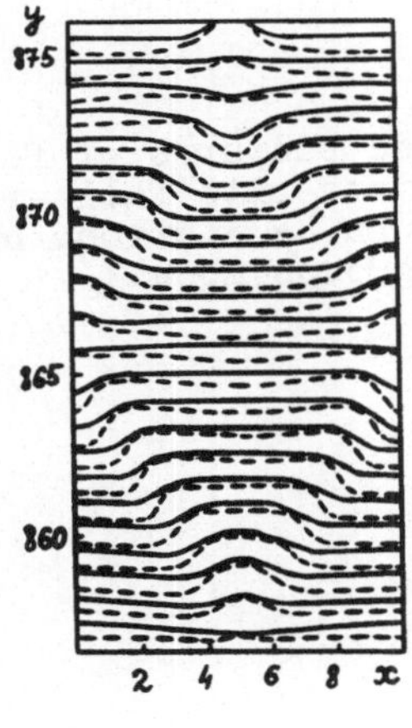

Fig.14. Propagation of detonation shock wave (solid lines) and flame front (dashed lines) in plane channel filled with O_2-H_2 mixture.

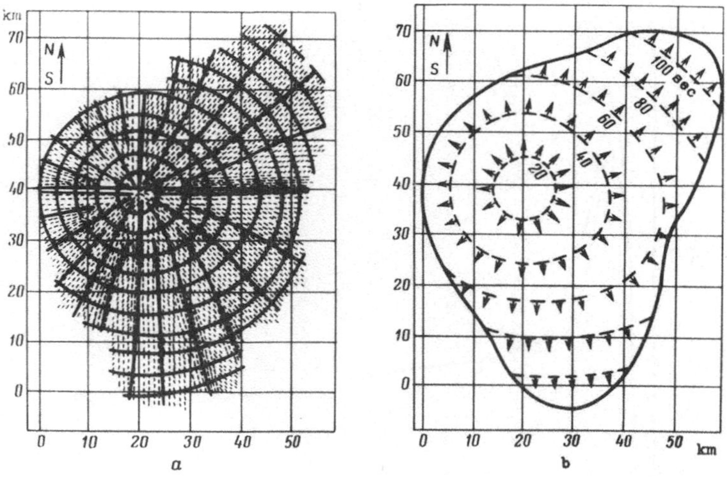

Fig.15. Observed and calculated pictures of forest flattening at the Tunguska meteorite fall area.

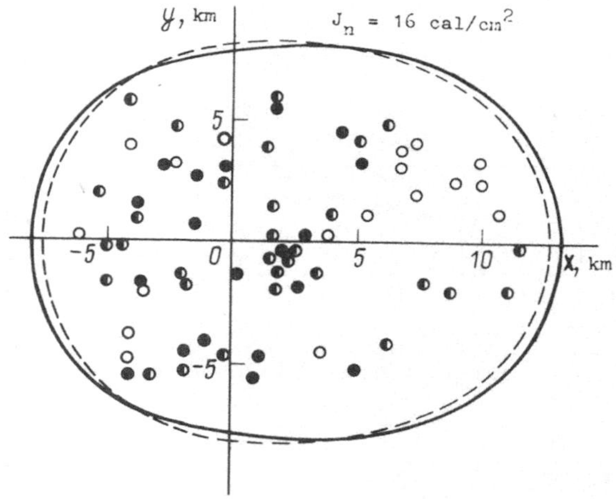

Fig.16. Observed and calculated pictures of forest radiant burn at the Tunguska meteorite fall area. o - weak burn, ◐ - moderate burn, ● - strong burn. - - - gray gas, ——— ten-groups model.

TWO-DIMENSIONAL SEPARATED FLOWS

K. Gersten
Institut für Thermo- und Fluiddynamik
Ruhr-Universität
D-4630 Bochum, F.R.G.

1. Introduction

The phenomenon of flow separation and the analysis of separated flows are considered from the view point of asymptotic theory, i.e. from theory of Navier-Stokes equations at high Reynolds numbers. The purpose of this paper is to discuss a few examples that show the state of the art of asymptotic theory with respect to separated flows. In order to simplify the matter, the discussion will be restricted to incompressible, two-dimensional flows. For further details see review papers by Stewartson (1974, 1977, 1980, 1981), Shen (1968), Brown and Stewartson (1969), Williams (1977), Messiter (1978), Herwig and Brown (1982).

A powerful tool to predict flows at high Reynolds numbers is Prandtl's boundary layer theory, see Schlichting (1979). It can be considered as the first-order approximation of the asymptotic theory for the Navier-Stokes equations for the limit $Re \to \infty$. But the basic concept of Prandtl's theory, i.e. the two parts of the flow field - the outer flow and the boundary layer flow - can be determined one after the other, fails when flow separation occurs. In case of flow separation both the outer flow due to Prandtl as well as the boundary layer flow are not correct. Within Prandtl's theory the outer flow is determined by assuming attached flow. On the other hand, Prandtl's boundary layer equations lead to a singularity at the separation point, the so-called Goldstein singularity, so that boundary layer calculations cannot be continued beyond the point of separation.

Hence, the following questions arise: Is flow separation an insurmountable barrier for the asymptotic theory? Is the Goldstein singularity a reality, i.e. do the solutions of the Navier-Stokes equations have a singularity in the limit $Re \to \infty$, or is the occurence of the singularity a hint that Prandtl's theory in its original form is not able to cover separated flows?

The answer is: The solutions of the Navier-Stokes equations have usually no singularity. The barrier of the Goldstein singularity can be overcome by an appropriate modification of Prandtl's boundary-layer concept. These modifications and some aspects of asymptotic theory in connection with separated flows will be discussed in what follows.

2. Definition

In the book by Telionis (1981), p. 285, the following statement can be found: "If the full Navier-Stokes equations are adopted as a model, no specific criterion (for separation) is available, but then presumably the intire flow field can be predicted and there is no special need for identifying the specific location of separation. In fact, in this case the concept of separation itself becomes academic." In other words: Flow separation is a phenomenon that is understandable only within the framework of asymptotic theory.

By Shen and Nenni (1975), the specific proposal was made to associate separation directly with the condition that the boundary layer should become "unmatchable", in the sense that the induced normal velocity at the outer edge of the boundary layer attains such a magnitude as to invalidate the basic assumption of $v \cdot Re^{1/2} \sim O(1)$, Re being the Reynolds number. Separation occurs if the unmatchability condition

$$v' \equiv \frac{\partial}{\partial x} (U \, \delta^*) > O(1) \tag{1}$$

is satisfied, v' denoting the perturbation normal velocity on the inviscid flow. Separation can happen, consequently, if the displacement thickness

$$\delta^* = \int_0^\infty (1 - \frac{u}{U}) dy \tag{2}$$

or its x-derivative becomes unbounded. This definition has the advantage that it is valid for steady as well as unsteady flows.

3. Scales of Separated Region

As a typical example for separation the flow over an indented flat plate is considered. Two limiting cases have to be distinguished according to the geometrical scales of the dent: a small separation bubble within the lower part of the boundary layer and the "catastrophic" separation of the whole boundary layer with a large recirculating eddy, see Figure 1. In the first case (Figure 1a) the depth H of the dent is much smaller than the thickness δ of the oncoming boundary layer, whereas in the second case the depth H is much larger than δ. The asymptotic theories are completely different for these two limiting cases. The case $H \ll \delta$ can be treated by triple deck theory, whereas the case $H \gg \delta$ leads to the so-called Kirchhoff free-streamline theory. For other geometries an equivalent distinction between these two limiting cases has to be made.

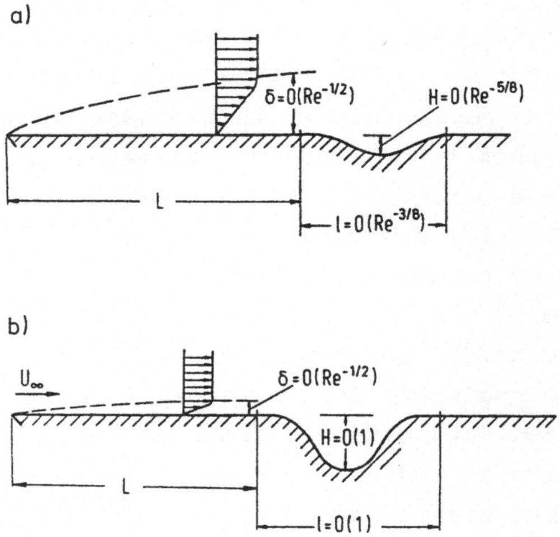

Figure 1:

Flow over an indented flat plate. Scales of the dent. $Re = U_\infty L/\nu$

a) $H \ll \delta$: $\delta/H = O(Re^{1/8})$
Triple-deck theory

b) $H \gg \delta$: $\delta/H = O(Re^{-1/2})$
Free-streamline theory

4. Triple-Deck Theory

The failure of Prandtl's boundary layer theory near separation can be overcome by the concept of multistructured boundary layers or in particular by the so-called "Triple-Deck-Concept", see Stewartson (1974). In Figure 2 the geometry of a small dent in a flat plate is shown, where separation might occur. In the neighbourhood of the dent the boundary layer along the flat plate has a three-layer structure. Each of the three layers (called "decks") has its own scaling shown in Figure 2 and

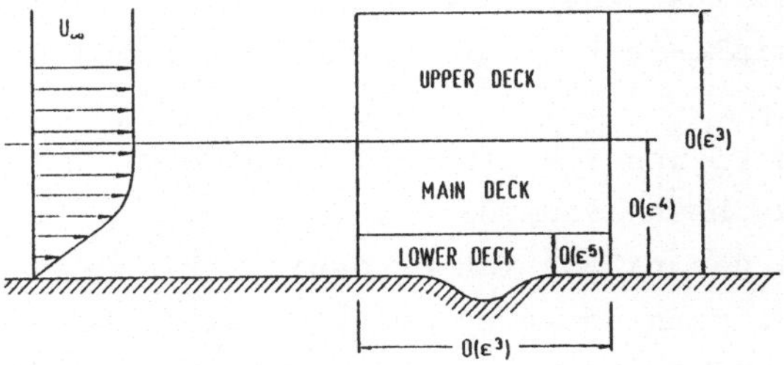

Figure 2:
Triple-deck structure, $\varepsilon = Re^{-1/8}$

its characteristic function in the process of viscous-inviscid interaction between outer flow and viscous region. For further details of the flow equations see Stewartson (1974). The scale of the dent depth is $H/L = O(Re^{-5/8})$. Figure 3 shows a typical example of a calculation of a recirculation in the dent according to triple-deck theory. Due to this theory the originally three-parameter problem (Re, H/L, L/l) has been reduced to a two-parameter problem ($Re^{1/4} \cdot H/L$, $Re^{-3/8}$ L/l). The border between flows with recirculating bubble (separated) and attached flows is shown in Figure 3. For one particular solution of the triple-deck flow equation representing an infinite number of flows (see Table in Fig. 3) the distributions of pressure and shear stress at the wall are depicted, after Herwig (1981).

Most of the dent geometries, that lead to attached flows according to triple-deck theory would have led to separation by Prandtl's theory. In this connection it should be noted that sometimes one can read the following statement: If Prandtl's boundary layer equations lead to separation, the real flow will reach zero shear stress in the neighbourhood of a few boundary-layer thickness tion point. This, however, is generally not true. There are dent geometries, where Prandtl's theory would lead to separation, whereas the triple-deck solution can correspond to attached flows.

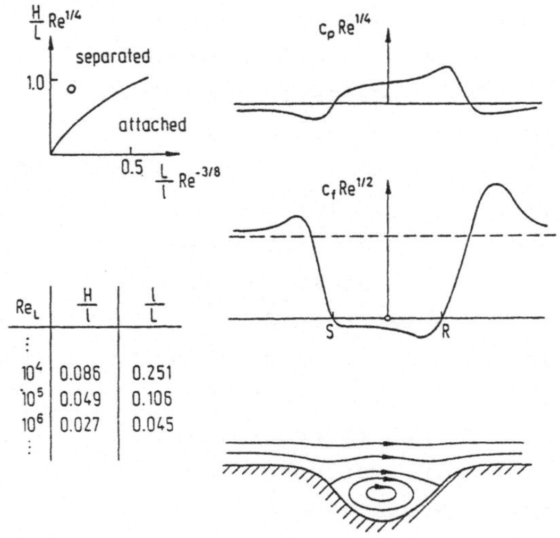

Re_L	$\frac{H}{l}$	$\frac{l}{L}$
⋮		
10^4	0.086	0.251
10^5	0.049	0.106
10^6	0.027	0.045
⋮		

Figure 3:
Triple-deck solution for an indented flat plate, after Herwig (1981). $Re^{1/4}$ $H/L = 0.86$, $Re^{-3/8}$ $L/l = 0.125$

In Figure 3 a particular ratio of the depth and the length of the dent has been assumed: $H/l = O(Re^{-1/4})$. For smaller ratios (long dents) and larger ratios (short dents) triple-deck theory can still be applied, as has been shown by Smith et al. (1981). The concept of multistructured boundary layers has also been applied to the free convection flow along a vertical flat plate with a hump leading to a double-deck theory. Recirculating flow on such a configuration has been calculated by Merkin,

see Herwig and Brown (1982).
It is worth mentioning that triple-deck theory requires to solve again Prandtl's boundary-layer equations. Only the boundary conditions are different, in particular at the outer edge of the viscous region. Since triple-deck theory takes into account the mutual interaction between outer flow and viscous region, the boundary layer equations are usually solved numerically by an inverse method, i.e. the displacement thickness distribution is given and the induced outer pressure has to be found. In practice so-called inverse boundary layer methods are applied to calculate flows with recirculating regions, see Cebeci and Bradshaw (1977). Strictly speaking, these are not asymptotic methods any more, because for each Reynolds number a new calculation has to be made. Obviously, triple-deck theory is the only justification for inverse boundary layer methods.

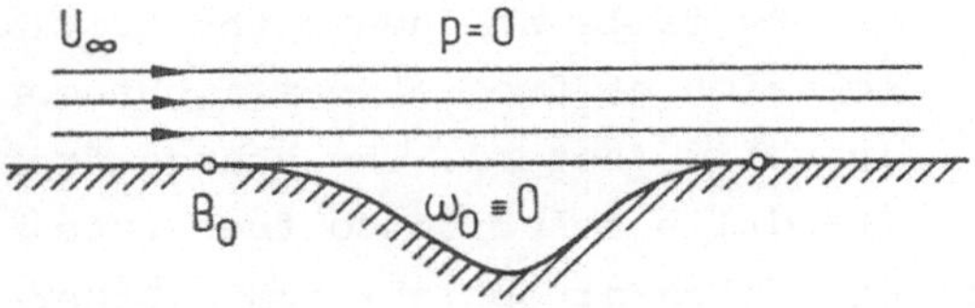

Figure 4:
Limiting solution for Re → ∞ for an indented flat plate with recirculating flow, after Herwig (1981)

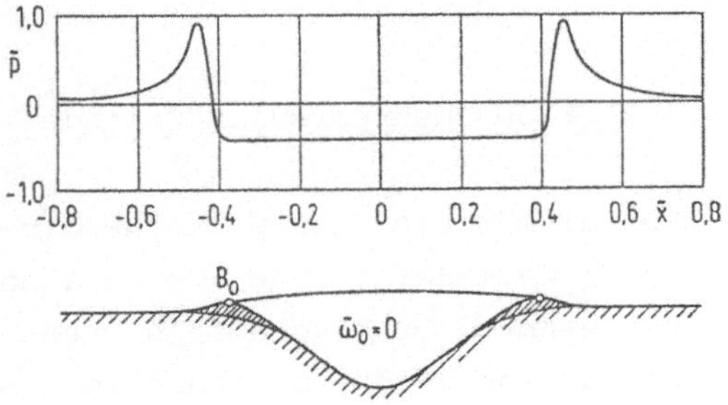

Figure 5:
Limiting solution for Re → ∞ for an indented flat plate with humps, after Herwig (1981).
$\tilde{p} = (p - p_\infty)/\kappa \rho U_\infty^2$
Hump height h = 0.1 H, $\kappa = H/l$

5. Free-Streamline Theory

5.1. Indented Flat Plate

The limiting case shown in Figure 1b (H >> δ) has been investigated in detail by Herwig (1981). Dealing with the asymptotic structure of the flow he found that the only possible limiting solution for infinite Reynolds number (Limiting solution of the Euler equations) is the Kirchhoff free streamline solution. This solution can be considered as a special case of Batchelor's separation-bubble model for bubble vorticity equal to zero, see Batchelor (1955). For the simple dent this limiting solution is the trivial solution shown in Figure 4. Non-trivial solutions occur for dents with humps as shown in Figure 5.

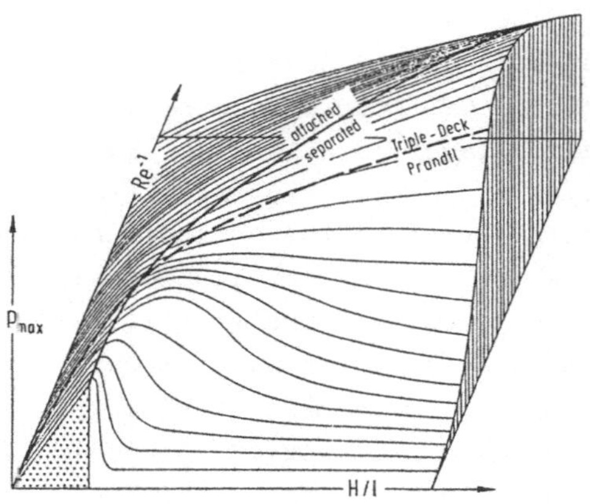

Figure 6:

Maximum pressure for indented flat plate at L/l = const.

The connection of the two limiting cases ($H << \delta$ and $H >> \delta$) can be demonstrated by considering a characteristic parameter of the dent flow. In Figure 6 the pressure maximum on the dent wall is plotted as function of Reynolds number Re and relative dent depth H/l , whereas the third parameter L/l is kept fixed. The limiting solution ($Re \to \infty$) is given by Prandtl's theory for small values H/l as long as no separation occurs. At the occurrence of separation, Prandtl's boundary layer theory fails and hence the trivial solution of free-streamline theory is valid instead. The change from Prandtl's solution to free-streamline solution is discontinuous in form of a ("catatrophic") jump. Otherwise the limiting solutions of the Navier-Stokes equations do not experience any singularities. The diagram shows clearly that triple-deck solutions refer to flows of finite Reynolds number which reduce to the simple flat-plate flow (Blasius solution) without dent. It can be seen from the diagram, that at a given ratio H/l the flow may be attached for moderate Reynolds numbers, whereas it may eventually show recirculation for increasing Reynolds number.

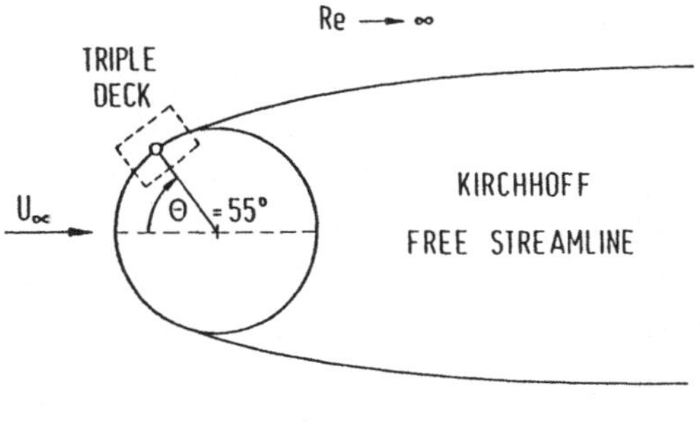

Figure 7:

Asymptotic solution ($Re \to \infty$) for the flow past a circular cylinder

5.2. Flow Past Circular Cylinder

The asymptotic flow past the circular cylinder for $Re \to \infty$ is also represented by Kirchhoff's free-streamline theory as shown in Figure 7. The local detail near the point where the free streamline leaves the cylinder contour are again described by triple-deck theory, after Sychev (1972) and Smith (1977, 1979). The drag

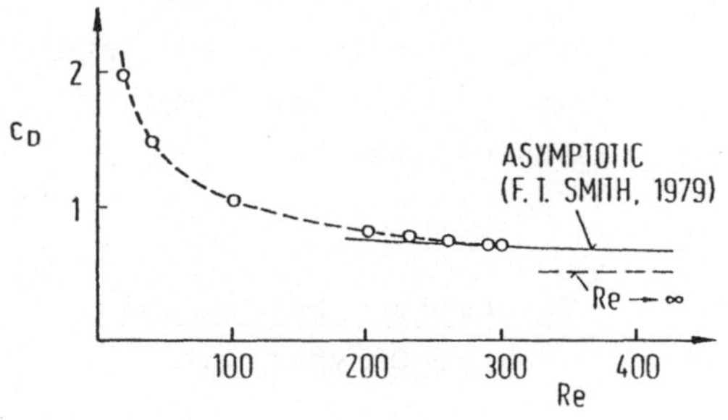

Figure 8:

Drag coefficient of circular cylinder

———— Asymptotic theory after Smith (1979)

- - o - - Numerical solution after Fornberg (1980)

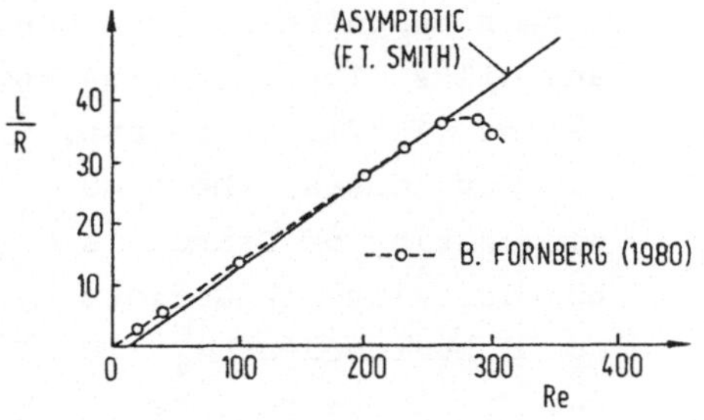

Figure 9:

Length of wake bubble past a circular cylinder

———— Asymptotic theory after Smith (1979)

- - o - - Numerical solution after Fornberg (1980)

coefficient of the cylinder according to Smith (1979) is in excellent agreement with numerical solutions of the Navier-Stokes equations, presented by Fornberg (1980) and shown in Figure 8. Unfortunately, this agreement does not refer to the length of the wake bubble, that according to Fornberg's numerical solution stops growing at about Re = 280 in contrast to asymptotic theory; see Figure 9. It is hoped that this serious inconsistency will be settled in the near future.

6. Flow Control by Continuous Injection

If continuous injection proportional to $1/\sqrt{x}$ is applied the flow past a flat plate is described by self-similar solutions of Prandtl's boundary-layer equations, see Figure 10. For the critical value f_{wc} of the injection intensity the wall shear stress becomes zero. The asymptotic theory of full Navier-Stokes equations, however, will not lead to zero shear stress as has been shown by Kassoy (1974). Figure 10 shows the wall shear stress in the neighbourhood of the critical injection intensity. Prandtl's theory leads to a singularity at $f_w = f_{wc}$ and hence to separation, whereas the correct asymptotic theory shows a completely regular flow with non-zero wall shear stress at $f = f_{wc}$.

Equivalent investigations for constant injection velocity along the flat plate have been carried out by Kassoy (1973).

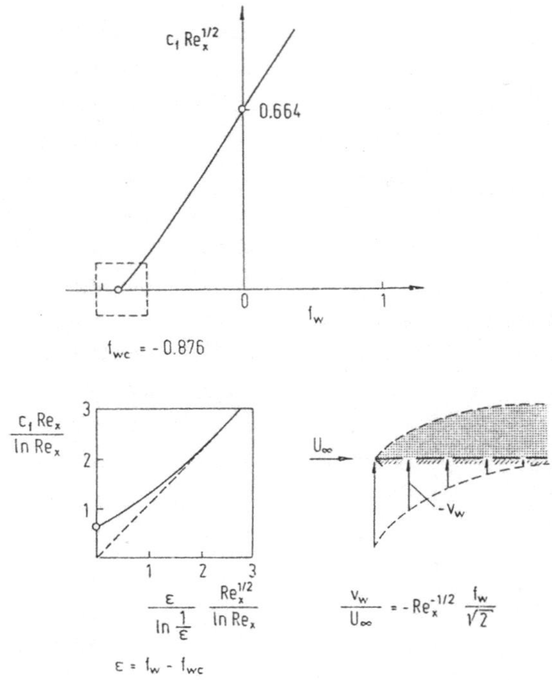

Figure 10:
Flow past a flat plate with injection $v_w \sim 1/\sqrt{x}$.
$c_f = 2\tau_w / U_\infty^2$,
$Re_x = U_\infty x/\nu$.

7. Combined Forced and Free Convection

In flow problems where forced and free convection are opposing each other zero shear stress may occur, see Wilks (1974), Buckmaster (1970).
The point of zero shear stress is usually connected with a singularity, see Hunt and Wilks (1980), Davies and Walker (1977). There are, however, cases, where no singularity of Prandtl's boundary-layer equations could be found at $\tau_w = 0$, see for instance Jones (1973). Such cases would not be called separation by our definition. But in this connection it should be emphasized that the results of a numerical calculation could never be considered as a proof that a specific solution is free of singularities since with a coarse mesh the solution may appear perfectly normal in the entire domain of integration.

8. Unsteady Laminar Flows

In unsteady flow it is clear that usually points of $\tau_w = 0$ are not singular. But the definition of separation, Eqn.(1), is still valid. An important question is whether a singularity will appear in a finite time in unsteady boundary layers with no initial singular behavior. The answer is 'yes.' As a matter of fact, F.T. Smith (London) and S. Cowley investigated the emergence of the singularity for the flow past the impulsively started circular cylinder, see Herwig and Brown (1982).

9. Turbulent Flows

There are investigations on the asymptotic theory for turbulent flows, connected with the names Yajnik (1970), Mellor (1972), Bush and Fendell (1972) and Afzal (1973). The triple-deck concept has also been applied to turbulent flows by Melnik and Chow (1975), Adamson (1976), Messiter (1980) and Sykes (1980). The theory is still in a preliminary status as far as separation is concerned. But further work on asymptotic theory can be expected along the same lines as for laminar flows. Turbulence modelling is of course an additional difficulty in connection with flow separation, see Simpson (1981).

10. Conclusions

a) Separation is a phenomenon of asymptotic theory and can be defined as a point where Prandtl's boundary layer equations lead to a singularity.

b) By this definition, $\tau_w = 0$ is not a sufficient condition for separation. Particularly in unsteady boundary layers, but obviously also in combined forced and free convection cases exist where the point $\tau_w = 0$ is not a singularity. These zero skin friction points - by the above definition - are not called separation points.

c) Separation means failure of Prandtl's boundary layer theory. Cases with separation can be treated either by triple-deck theory or by free streamline theory without appearance of singularities. Triple-deck theory may even lead to flows without zero skin friction.

d) Asymptotic theory for flows with recirculation is not complete with respect to reattachment, wakes and turbulence modelling.

Adamson, Jr. T.C.(1976): The Structure of Shock Wave-Boundary Layer Interactions in Transonic Flow, in Symposium Transonicum II, ed. K. Oswatitsch, D. Ruess, Springer, Berlin, pp. 244-251.

Afzal, N.(1973): A Higher Order Theory for Compressible Turbulent Boundary Layers at Moderately Large Reynolds Number, J. Fluid Mech., vol. 57, pp. 1-25.

Banthiya, N.K., and Afzal, N.(1980): Mixed convection over a semi-infinite horizontal plate. ZAMP, vol. 31, pp. 646-652.

Batchelor, G.K.(1955): A proposal concerning laminar wakes behind bluff bodies at large Reynolds number. J. Fluid Mech., vol. 1, pp. 388-398.

Brown, S.N., and Stewartson, K.(1969): Laminar Separation. Annual Reviews of Fluid Mechanics, vol. 1, pp. 45-72.

Buckmaster, J.(1970): The behaviour of a laminar compressible boundary layer on a cold wall near a point of zero skin friction. J. Fluid Mech., vol. 44, pp. 237-247.

Bush, W.B., and Fendell, F.E.(1972): Asymptotic Analysis of Turbulent Channel and Boundary-Layer Flow. J. Fluid Mech., vol. 56, pp. 657-681.

Cebeci, T., and Bradshaw, P.(1977): Momentum Transfer in Boundary Layers.

Cebeci, T.(1978): An unsteady laminar boundary layer with separation and reattachment. AIAA Journal, vol. 16, pp. 1305-1306.

Davies, T., and Walker, G.(1977): On solutions of the compressible laminar boundary-layer equations and their behaviour near separation. J. Fluid Mech., vol. 80, pp. 279-292.

Dean, W.R.(1950): Note on the motion of a liquid near a position of separation. Proc. Camb. Phil. Soc., vol. 46, pp. 293-306.

Fornberg, B.(1980): A numerical study of steady viscous flow past a circular cylinder. J. Fluid Mech., vol. 98, pp. 819-855.

Gersten, K., and Schilawa, S.(1978): Buoyancy effects on forced-convection heat transfer in horizontal boundary layers. Proc. 6th Internat. Heat Transfer Conference, Toronto MC-13, pp. 73-78.

Herwig, H.(1981): Die Anwendung der Methode der angepaßten asymptotischen Entwicklungen auf laminare, zweidimensionale Strömungen mit endlichen Ablösegebieten. Doctoral Thesis, University Bochum.

Herwig, H., and Brown, S.(1982): Report on EUROMECH 148: Two-Dimensional Separated Flows. Z. Flugwiss. Weltraumforsch., vol. 6(1982), pp. 202-205.

Hunt, R., and Wilks, G.(1980): On the behaviour of the laminar boundary layer equations of mixed convection near a point of zero skin friction. J. Fluid Mech., vol. 101, pp. 377-391.

Jones, D.R.(1973): Free convection from a semi-infinite flat plate inclined at a small angle to the horizontal. Quart. Journ. Mech. and Appl. Math., vol. 26, pp. 77-97.

Kassoy, D.R.(1970): On laminar boundary-layer blow-off. SIAM J. Appl. Math., vol. 18, pp. 29-40.

Kassoy, D.R.(1970): On laminar boundary-layer blow-off. Part 2. J. Fluid Mech., vol. 48, pp. 209-228.

Kassoy, D.R.(1973): The singularity at boundary layer separation due to mass injection. SIAM J. Appl. Math., vol. 25, pp. 105-123.

Kassoy, D.R.(1974): A resolution of the blow-off singularity for similarity flow on a flat plate. J. Fluid Mech., vol. 62, pp. 145-161.

Lagerstrom, P.A.(1977): Solutions of the Navier-Stokes equation at large Reynolds number. In. J. Rom (Edr.) Proceedings of the International Symposium on Modern Developments in Fluid Dynamics. Society of Industrial and Applied Mathematics, Philadelphia,pp. 364-376.

Mellor, G.L.(1972): Mellor, G.L.(1972): The large Reynolds number asymptotic theory of turbulent boundary layers. Int. J. Engng. Sci., vol. 10, pp. 851-873.

Melnik, R.E., and Chow, R.(1975): Asymptotic theory of two-dimensional trailing edge flows. NASA SP-347, pp. 177-250.

Melnik, R.E., and Grossmann, B.(1981): On the turbulent viscid-inviscid interaction at a wedge shaped trailing edge. In: T. Cebeci (Ed.), Numerical and Physical Aspects of Aerodynamics Flows, Springer, Berlin.

Merkin, J.H.(1969): The effect of buoyancy forces on the boundary-layer flow over a semi-infinite vertical flat plate in a uniform free stream. J. Fluid Mech., vol. 35, pp. 439-450.

Messiter, A.F.(1978): Boundary-layer separation. Proceedings 8th U.S. National Congress of Applied Mechanics. June 26-30, 1978. University of California, Los Angeles.

Messiter, A.F.(1980): Interaction between a normal shock wave and a turbulent boundary layer at high transonic speeds Part I: Pressure distribution, ZAMP, vol. 31, pp. 204-226.

Schilawa, S.(1981): Auftriebseffekte in laminaren Grenzschichten an horizontalen Wänden. Doctoral Thesis, University Bochum, 1981.

Schlichting, H.(1979): Boundary-Layer Theory. 7th Edition. McGraw-Hill, New York.

Sears, W.R., and Telionis, D.P.(1975): Boundary layer separation in unsteady flow. SIAM J. Appl. Math., vol. 28, pp. 215-235.

Sears, W.R., and Telionis, D.P.(1977): Boundary-layer separation in unsteady flow. In: J. Rom (Edr.), Proceedings of the International Symposium on Modern Developments in Fluid Dynamics. Society of Industrial and Applied Mathematics, Philadelphia, pp. 323-343.

Shen, S.F.(1968): Unsteady separations according to the boundary-layer equations. Adv. Appl. Mech., vol. 18, pp. 177-220.

Shen, S.F., and Nenni, J.P.(1975): Asymptotic solution of the unsteady two-dimensional incompressible boundary layer and its implications on separation. In "Unsteady Aerodynamics", ed. R.B. Kinney, Univ. of Arizona, Tucson, vol. 1, pp. 245-259.

Simpson, R.L.(1981): A review of some phenomena in turbulent flow separation. J. Fluids Engng., vol. 103, pp. 520-533.

Smith, F.T.(1977): The laminar separation of an incompressible fluid streaming past a smooth surface. Proc. R. Soc. London. A 356, pp. 443-463.

Smith, F.T., and Duck, P.W.(1977): Separation of jets and thermal boundary layers. Quart. J. Mech. Appl. Math., vol. 30, pp. 143-156.

Smith, F.T.(1979): Laminar flow of an incompressible fluid past a bluff body: the separation, reattachment, eddy properties and drag. J. Fluid Mech., vol. 92, pp. 171-205.

Smith, F.T., Brighton, P.W.M., Jackson, P.S. and Hunt, J.C.R.(1981): On boundary-layer flow past two-dimensional obstacles. J. Fluid Mech., vol. 113, pp. 123-152.

Stewartson, K.(1974): Multistructured boundary layers on flat plates and related bodies, Advances in Applied Mechanics, vol. 14, Academic Press, pp. 145-239.

Stewartson, K.(1977): On the asymptotic theory of separated and unseparated fluid motions. In: J. Rom.(Edr.), Proceedings of the International Symposium on Modern Developments in Fluid Dynamics. Society of Industrial and Applied Mathematics, Philadelphia, pp. 305-322. See also: SIAM J. Appl. Math. vol. 28 (1975), pp. 501-518.

Stewartson, K.(1980): High Reynolds-number flows. In: R. Rautmann (Edr.), Approcimates Methods for Navier-Stokes Problems. Lecture Notes in Mathematics No. 771, Springer-Verlag, Berlin, pp. 505-518.

Stewartson, K.(1981): D'Alembert's paradox. SIAM Review, vol. 23, pp. 308-343.

Stewartson, K., and Williams, P.G.(1969): Self-induced separation. Proc. Roy. Soc. A 312, pp. 181-206.

Sykes, R.I.(1980): An asymptotic theory of incompressible turbulent boundary-layer flow over a small hump. J. Fluid Mech., vol. 101, pp. 647-670.

Telionis, D.P.(1981): Unsteady Viscous Flows. Springer-Verlag, New York.

Wilks, G.(1974): A separated flow in mixed convection. J. Fluid Mech., vol. 62, pp. 359-368.

Williams, J.C.(1977): Incompressible boundary-layer separation. Annual Reviews of Fluid Mechanics, vol. 9, pp. 113-144.

Yajnik, K.S.(1970): Asymptotic theory of turbulent shear flows. J. Fluid Mech., vol. 42, pp. 411-427.

NUMERICAL SIMULATION OF WALL-BOUNDED TURBULENT SHEAR FLOWS

Parviz Moin

NASA Ames Research Center
Moffett Field, California 94035 U.S.A.

1. Introduction

Advances in computer hardware and numerical methods in conjunction with carefully designed computer programs have made meaningful numerical simulation of wall-bounded turbulent flows possible. The physical realism of the resultant computer-generated data has been validated by detailed structural and statistical comparisons with experimental measurements. These calculations have proven to be a very useful complement to the laboratory experiments.

This paper reviews some recent developments in three-dimensional, time-dependent numerical simulation of turbulent flows bounded by a wall. We shall be considering both direct and large-eddy simulation techniques within the same computational framework. In the following section, we have outlined the governing equations. In Section 3, the computational spatial-grid requirements as dictated by the known structure of turbulent boundary layers are presented. Next, we review the numerical methods currently in use. Some of the features of these algorithms, including spatial differencing, time advancement, and data management, will be discussed in some detail. In Section 5 we provide a selection of the results of recent calculations of turbulent channel flow, including the effects of system rotation and transpiration on the flow structure. Finally, in Section 6 we shall make our concluding remarks.

2. Dynamical Equations

To date, attention has been largely focused on the incompressible flows governed by the Navier-Stokes equations:

$$\frac{\partial u_i}{\partial t} - \varepsilon_{ijk} u_j \omega_k = - \frac{\partial P}{\partial x_i} + \frac{1}{Re} \frac{\partial^2 u_i}{\partial x_j \partial x_j} \qquad (1a)$$

$$\frac{\partial u_i}{\partial x_i} = 0 \qquad (1b)$$

where Re is the Reynolds number. To satisfy the incompressibility constraint (1b), certain numerical techniques use the Poisson equation for the dynamic pressure:

$$\frac{\partial^2 p}{\partial x_i \partial x_i} = \frac{\partial}{\partial x_i}\left(\varepsilon_{ijk} u_j \omega_k\right) \tag{1c}$$

obtained from application of the divergence operator to Eq. (1a).

In the direct simulation (DS) approach, the above equations are solved numerically with appropriate boundary conditions. Aside from errors due to numerical implementation, no further approximations are required. In the large-eddy simulation (LES) technique, the dependent variables are the resolvable portion of the velocity and pressure field. Every flow variable f is decomposed to large-scale and subrid-scale components.

$$f = \bar{f} + f' \tag{2}$$

The large-scale component is defined by

$$\bar{f}(\underline{x}) = \int G(\underline{x},\underline{x}')\, f(\underline{x}')\, d\underline{x}' \tag{3}$$

where G is a filter function with a characteristic length Δ, which is a function of the computational grid resolution. Applying the filtering operation (3) to Eqs. (1a), (1b), or (1c) leads to the exact equations for the large-scale field. The major difference between the filtered and unfiltered equations of the direct simulation is the inclusion of the additional terms associated with the subgrid-scale stresses (SGS) in the governing equations for the large eddies. These terms are modeled to close the system of equations. Different eddy-viscosity models [1, 2, 3] as well as multi-equation models [4] have been successfully used to relate SGS stresses to the resolvable turbulence. These models should display an important feature: as the grid resolution is refined, the characteristic length of the SGS eddies becomes smaller.

In this paper, most of the discussion will be in reference to a Cartesian coordinate system. The x and z (x_1, x_3) axes are parallel to the wall, with x increasing in the mean-flow direction. The y axis is normal to the wall. We shall primarily discuss numerical simulation of flows that are homogeneous or nearly homogeneous in the x and z directions. This is the area where most of the current effort has been concentrated. Some computations of unidirectional flows in cylindrical geometries have been performed [2, 5]. Currently, however, numerical simulation of the turbulent flows in complex geometries involving generalized coordinate systems has not been undertaken. Other notions used in this paper include: δ, the channel half-width or boundary-layer thickness; U_o, free-stream velocity or centerline velocity in channel; u_τ, shear velocity; U_m, average mean velocity; Re, the Reynolds number based on U_o and δ; Re_τ, the Reynolds number based on u_τ and δ; ν, the kinematic viscosity; y_w, the distance to the wall; and N_i, number of grid points in the x_i-direction.

3. Spatial Resolution Requirements

Generally, numerical simulation of wall-bounded turbulent shear flows requires a large number of grid points in all spatial directions. This requirements is much more stringent than the corresponding one for free-turbulent shear flows (e.g., jets and wakes). The difference stems from physical observations that locally large eddies near the wall are much smaller than those away from the wall. Moreover, in free-turbulent flows, large-scale structures exhibit an appreciable degree of Reynolds number independence, whereas, near the walls, the important large-scale structures decrease in size with increasing Reynolds number.

In the direction normal to the solid boundary, one should and can easily distribute grid points with variable spacings. A sufficient number of grid points should be placed near the wall to resolve the viscous sublayer and the buffer layer. As the Reynolds number increases, more points are required in this region. The grid size can be increased in the regions away from the walls; however, it should be bounded by the Prandtl mixing length ($\sim 0.1\ \delta$).

In the lateral (spanwise) direction, the required number of grid points can be prohibitively large. The difficulty arises due to the fine spacing of the streaky structures [6] in the vicinity of the wall. Kline and his co-workers have shown that streaks play a significant role in the production and dynamics of turbulence in the entire flow field. Therefore, it is important that the numerical simulation of wall-bounded turbulent shear flows resolve these structures or account for their effects. For a limited range of Reynolds numbers, laboratory observations, as well as some quantitative measurements indicate that the mean spacing of the streaks, λ_m, is about 100 wall units, i.e., $\lambda_m^+ = \lambda_m u_\tau / \nu \simeq 100$, and their most probable spacing is about 80 wall units. The mean width of the smaller of the high- and low-speed streaks can be at most 50 ν/u_τ. In fact, from the measurements of Blackwelder and Kaplan [7], one can deduce that the mean width of the high-speed wall-layer structures is about 20-40 wall units. These values are based on an ensemble of measurements, and, at a given instant, structures with smaller (as well as larger) widths are formed. Therefore, in order to capture the wall structures at their proper scale, it is not unreasonable [8] to require that the computational grid resolution in the spanwise direction be fine enough to resolve eddies with a spanwise extent of 20 wall units. In the numerical integration of the governing nonlinear equations, if we assume that at least four grid points are required to represent an eddy and its evolution for a short period of time, the computational grid size in the z-direction should be about 5 wall units, i.e., $h_3^+ \simeq 5$. It should be pointed out that this estimate is based on experimental data for moderately low Reynolds number turbulent flows and may not apply at very high Reynolds numbers.

In the absence of physical boundaries in the spanwise direction, the extent of the computational domain in this direction, L_z, should be large enough that artificialities introduced by the use of periodic boundary conditions do not seriously affect the statistics of the flow. Based on two-point correlation measurements [9], L_z should be at least three times the boundary-layer thickness [3]. With these two estimates, the required number of grid points in the z-direction, N_z, is

$$N_z \simeq \frac{3}{5/Re_\tau} = 0.6\ Re_\tau$$

Using the universal velocity-distribution law [10] , we may relate Re to Re_τ:

$$Re = Re_\tau \left(\frac{1}{\kappa}\ \ell n\ Re_\tau + 5.0 + E\right)$$

where $\kappa \simeq 0.4$, $E \simeq 0$ for channel and pipe flows, and $E \simeq 2.8$ for the boundary layer. Figure 1 shows the required number of grid points in the spanwise direction vs. the Reynolds number for channel flows.

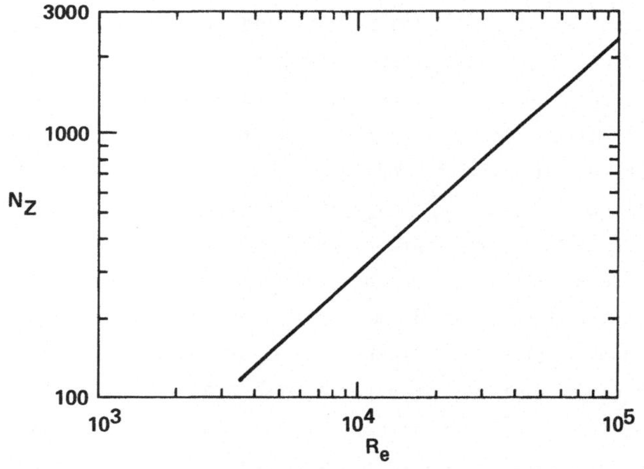

Figure 1. Grid-point requirements in the spanwise direction.

The computational grid requirement is not as stringent in the streamwise direction as in the spanwise direction. Similar considerations of the physics of turbulent boundary layers tend to indicate [3] that, in the streamwise direction, the required number of grid points is about half that for the spanwise direction.*

As an example, for the moderate Reynolds number $Re = 10^4$, 64 nonuniformly spaced grid points appear to be sufficient in the direction normal to the wall(s). For this case, therefore, about 4×10^6 mesh points are necessary to resolve the energetic turbulent structures. The computational effort required for this calculation overly taxes the capabilities of the presently available supercomputers. On the other hand, low Reynolds number flows, such as the channel flow experiment of Eckelmann [11] with $Re = 2800$, is definitely within reach of such computers. For this flow, less than half a million grid points are required.

It is emphasized that the spatial resolution requirements just given are based on the experimentally determined "large" eddy sizes. In the above calculations, one must use subgrid scale (SGS) models to represent the small dissipative eddies. It is difficult to characterize the size of these very small eddies. As a reference, however, we may consider the Kolmogoroff length scale $\eta = (\nu^3/\varepsilon)^{1/4}$ as their typical length scale. For channel flows based on the mean dissipation rate per unit mass for the entire flow field, η can be expressed in the wall units as $\eta^+ = \left[Re_\tau(u_\tau/U_m)\right]^{1/4}$. For the low Reynolds number, $Re = 2800$, that was just considered, η^+ is approximately 2 wall units. Based on the wall value of ε, the limiting value of η^+ is slightly less than 1 (η^+ is exactly equal to 1, if only the dissipation due to mean motion is considered). Clearly, if eddies whose extent is about η^+ in <u>all</u> the spatial directions are to be resolved, the required number of grid points is prohibitively large. Thus, it appears that, with the present computers, <u>direct</u> numerical simulation of wall-bounded turbulent shear flows is not feasible, and calculations should incorporate subgrid scale models to represent the dissipative eddies. However, this conclusion has been based on using η^+ as the characteristic size of the dissipative eddies that must be resolved. The question arises as to whether eddies of this size make significant contributions to the local dissipation rate in turbulent boundary layers. With the available experimental data, it is difficult to answer this question conclusively. A rough estimate of the dissipation spectra obtained from the one-dimensional energy spectra measurements of Bakewell and Lumley [13] seems to indicate that these eddies do make appreciable contributions to the local dissipation rate. Perhaps, the easiest way to answer this question is to attempt a direct simulation of a very low Reynolds number

---

*This estimate is based on the typical streamwise extent of the wall-layer structrues [12]. In visual studies [6], the "lifted" sublayer streaks were observed to oscillate. Depending on the wavelength of these oscillations, more points in the streamwise direction may be necessary.

turbulent channel flow and validate it by detail structural and statistical comparison with the available experimental data.

The demand for a large number of grid points for resolving the wall-layer structures can be significantly reduced by the grid-imbedding technique [14]. One can place a large number of grid points in the x and z directions only in the vicinity of the walls. Since implicit time advancing will generally be used in conjunction with derivatives in the normal direction, this approach is not as convenient to implement as one where the same number of grid points in the x and z directions are used at all the y-locations. However, careful grid imbedding can lead to enormous savings on computer time and storage. If at moderate Reynolds numbers the wall-layer structures are to be resolved at their proper scale, grid imbedding appears to be the only course of action at present.

Another method for alleviating the need for a large number of grid points in the simulation of wall-bounded shear flows is that of Deardorff [1] and Schumann [2]. In this method the flow in the vicinity of the walls is ignored. The calculations are carried out to a point in the logarithmic layer where boundary conditions consistent with the logarithmic velocity distributions are applied. For high Reynolds number flows and certain practical problems, this approach is very promising. With considerably less effort than is required to extend the calculations to the wall, successful comparison of the mean velocity and turbulent stresses with experimental data has been obtained. However, the applicability of these empirical boundary conditions to other flow situations has not been established. Moreover, the effect of perturbations to these boundary conditions on the computed flow field is not yet known. If a two-dimensional, time-dependent "wall function" can indeed be constructed with a sufficient degree of generality, then this type of calculation can be of considerable practical value for the numerical simulation of high Reynolds number, wall-bounded turbulent flows. The calculations that do extend to the wall can serve as a viable testing ground to validate the proposed wall conditions. A novel and inexpensive method for evaluating these conditions is described by Chapman and Kuhn [15]. They calculated the inner region of a turbulent boundary layer by specifying space- and time-dependent boundary conditions at the outer edge of the viscous sublayer. Considerable care was exercised in assuring that these conditions were consistent with the known dynamics of the near-wall turbulence. The appropriate "outer" boundary conditions used in this work can be used as wall conditions in the large-eddy simulation of the outer region of turbulent boundary layers.

4. Numerical Methods

In this section we shall discuss the numerical methods used to solve the three-dimensional, time-dependent, Navier-Stokes equations for wall-bounded turbulent flows. As a result of modeling the subgrid scale stresses, the dynamical equations in the large-eddy simulation approach are somewhat more complicated than Eqs. (1). However, when eddy-viscosity models are used, these equations can be recast into a form for which virtually identical numerical methods can be used. Equation (1a) can be written as:

$$\frac{\partial u_i}{\partial t} = H_i - \frac{\partial P}{\partial x_i} + \frac{1}{R}\frac{\partial^2 u_i}{\partial x_j \partial x_j} \tag{4}$$

where H_i represents the nonlinear terms (including the subgrid scale terms) and the dependent variables (u_i,P) are to be interpreted either as the full velocity and pressure field in DS, or their resolvable portions in LES.

4.1 Spatial Derivatives

Finite-difference and spectral (pseudospectral) methods have been used to approximate the partial derivatives $\partial/\partial x_i$. Deardorff [1] and Schumann [2] used second-order finite differences in conjunction with staggered grids [16] in all spatial directions. Moin and Kim [3] evaluated partial derivatives in two of the spatial directions (x_1, x_3) pseudospectrally, whereas, derivatives in the direction normal to the walls were evaluated by second-order central-difference formulae. Orszag and co-workers [17, 18], Moin and Kim [19], Kleiser [20] and Taylor and Murdock [21] computed all the spatial derivatives by pseudospectral methods. When using pseudospectral methods, the dependent variables are expressed as a linear combination of a set of orthogonal functions. Fourier series is the appropriate representation of the flow field in the directions for which periodic boundary conditions are used. In other dierctions, orthogonal polynomial decompositions generally lead to a high convergence rate, irrespective of the nature of boundary conditions [22]. For smooth functions, pseudospectral methods are much more accurate numerical differentiators than the conventional second- and fourth-order finite-difference approximations. However, this superiority of the spectral methods may not be very pronounced when turbulent flows which often involve small-scale fluctuations are calculated. In numerical simulation of two-dimensional, Navier-Stokes equations in a periodic box, Herring et al. [23] have systematically compared their results obtained with second-order finite-difference and spectral methods. They showed that spectral calculations are approximately equivalent in accuracy to finite-difference calculations with only <u>twice</u> the resolution in each space dimension. It is quite likely that a more favorable comparison can be obtained if fourth-order finite-difference methods are used. Another important conclusion from their study was that the accuracy or inaccuracy of spectral methods can be deduced from the computed energy spectra, whereas, the spectra obtained from the corresponding finite-difference calculations tend to hide their inaccuracy. To illustrate this property of spectral methods, we shall consider two numerical simulations of turbulent channel flow, one inadequately resolved and the other with sufficient grid resolution.

Figure 2a shows the one-dimensional lateral energy spectra in the vicinity of the wall ($y_w/\delta = 0.025$) from a turbulent channel-flow simulation at $Re = 13800$, Ref. [3]. In this calculation, the pseudospectral method with 128 grid points was used in the lateral direction. However, for this Reynolds number the resulting computational resolution was not adequate to resolve the wall-layer streaks at their proper scale. The energy accumulation at the high wave-number portions of $E_i(k_3, y_w/\delta = 0.025)$ signals this inadequacy. Note that the longitudinal spectra obtained at the same vertical location (Fig. 2b) do not have energy buildup at high wave numbers. In this calculation, 64 grid points were used in the longitudinal direction. These appear to be sufficient to resolve the streamwise variation of turbulent structures (which, incidentally, suggests that the streamwise grid resolution estimate given in section 3 may be too stringent (see below)). Figure 2c shows the one-dimensional, lateral energy spectra $E_i(k_3, y_w/\delta = 0.389)$ at a distance away from the wall. In this region, the finely spaced near-wall structures are absent, and no resolution problems are expected. This is reflected in the behavior of E_i and the absence of excessive energy buildup at high values of k_3. Figure 3 shows the corresponding near-wall ($y_w/\delta = 0.025$) one-dimensional spectra from a channel flow simulation at the relatively low Reynolds number $Re = 3850$. In this calculation, 128 grid points were also used in the spanwise direction and were apparently sufficient to resolve the wall-layer structures. For this case, no excessive energy accumulation is evident at the high-wave-number end of $E_i(k_3, y_w/\delta = 0.025)$. One should be cautious in using energy spectra as the sole indicator of grid resolution adequacy or inadequacy. Insufficient computational resolution may totally suppress certain instability mechanisms and the subsequent formation and growth of the corresponding turbulent structures. This phenomenon may be concealed from energy spectra.

If the spatial grid resolution is sufficient to resolve all the important scales of motion, spectral or pseudospectral methods certainly are the best possible numerical differentiators. However, in some cases the accuracy afforded by spectral methods should be balanced against inherent inefficiencies in the data-management process and difficulties encountered with application of these methods to complex geometries.

4.2 Explicit Time Advancement

The three momentum Eqs. (4) must be integrated in time, subject to the incompressibility constraint. Explicit methods offer the advantages of low cost per step and ease of formulation and computer programming. In calculations where the wall-layer dynamics have been excluded [1, 2], only explicit time advancement has been

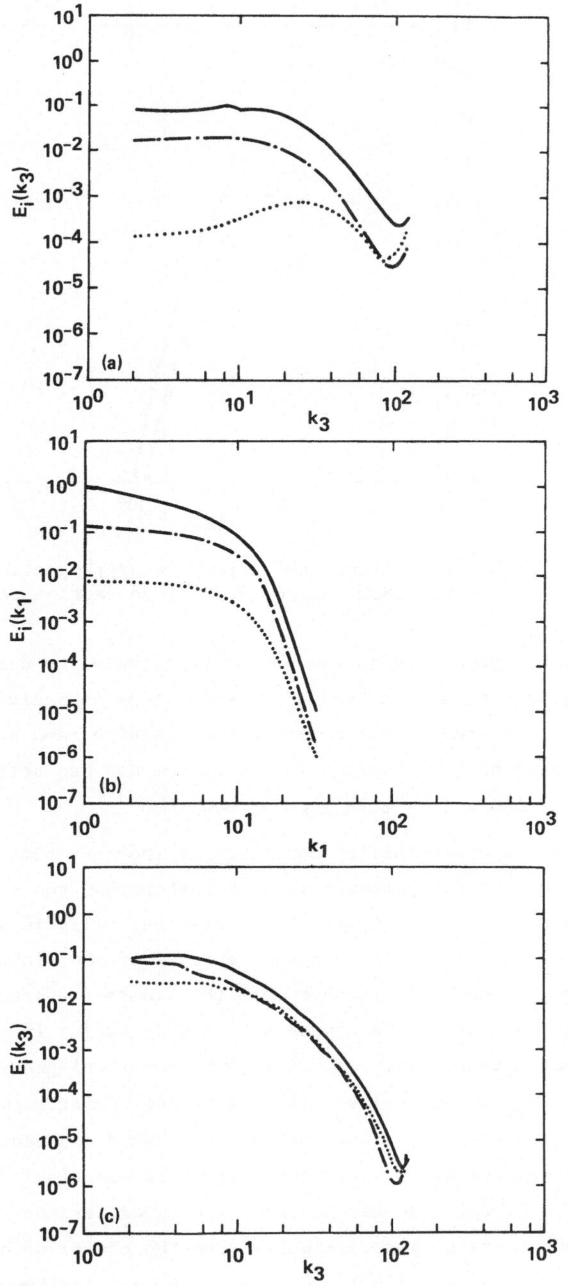

Figure 2. One-dimensional energy spectra at Re 13800. — E_1 spectrum of the streamwise fluctuating velocity; —·—, E_2 spectrum of spanwise fluctuating velocity; ----, E_3 spectrum of normal fluctuating velocity. (a) lateral spectra, $y_w/\delta = 0.025$; (b) longitudinal spectra, $y_w/\delta = 0.025$; (c) lateral spectra, $y_w/\delta = 0.389$. Note that in (a) the ratio of the peak value of E_2 to its value at the highest resolved wave number is only 4.2.

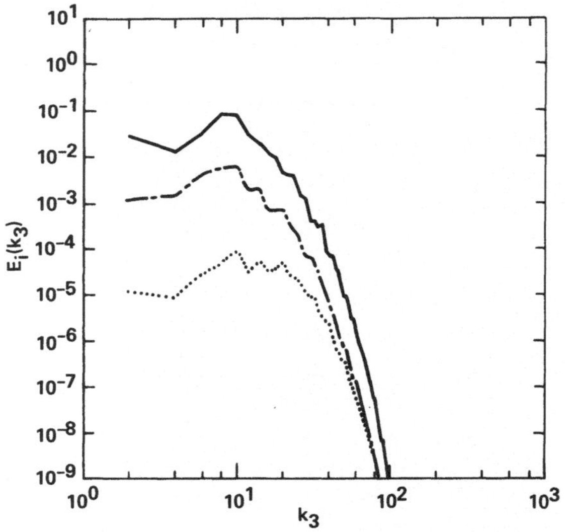

Figure 3. One-dimensional lateral energy spectra at Re = 3850. $y_w/\delta = 0.025$ (see caption of Fig. 2).

used. In these simulations, due to the use of relatively few uniformly or nearly uniformly spaced grid points, the stability restriction (especially those of the diffusion type) on time step is not severe. Both leapfrog and second-order Adams-Bashforth methods have been employed. The latter method has better overall accuracy and stability properties and is more popular now.

To enforce the incompressibility condition at the next time step, usually, the Poisson equation (1c) for pressure is used rather than the continuity equation. For several flow geometries of interest, noniterative, elliptic solvers are available [24] for exact solution (to within round-off errors) of the discretized Poisson equation. It is important, however, to note that the finite-differenced ∇^2 operator for the Poisson equation cannot be chosen arbitrarily [25]. In order to ensure compliance with the incompressibility condition, the numerical gradient operator used to approximate $\partial P/\partial x_i$ in Eq. (4) and the divergence operator must be the same. (When a staggered grid is used, a combination of forward and backward difference schemes for the divergence and the gradient operators also leads to the incompressibility condition.). Except when second-order finite-difference methods on a staggered grid are used, there is some ambiguity with the choice of boundary condition for the Poisson equation (Ref. [19]). Usually, one uses the Neumann boundary condition for pressure (which is obtained from the normal momentum equation). It is also possible to obtain a Dirichlet condition from the boundary evaluation of the tangential momentum equations. In general, the Neumann and Dirichlet problems for pressure may not have the same solution. It can be shown [19] that, when pseudospectral methods are used in the direction normal to the boundary, in order to satisfy the boundary conditions and the equation of continuity at the wall, it is imperative that both Neumann and Dirichlet conditions for pressure be satisfied. However, only one of them can be used to solve the Poisson equation. This inconsistency leads to serious numerical difficulties if pseudospectral methods are used. With standard finite-difference techniques, although the above ambiguity is still present, the corresponding numerical difficulties can be avoided [26, 27]. If second-order finite-difference methods are used in conjunction with staggered grids, the incompressibility condition at the interior cell adjacent to the boundary provides the additional boundary relation needed to solve the system of algebraic equations for pressure. In this case it is not necessary to extract pressure boundary conditions from the momentum equations.

4.3 Partially Implicit Time Advancement

The calculations that have extended to the wall and applied the no-slip boundary conditions have used semi-implicit numerical methods. These numerical schemes circumvent the prohibitive time-step restrictions arising from the viscous term and the necessity of using very fine mesh spacing in the vicinity of the wall. Moreover, when spectral methods are used to approximate the derivatives in the direction normal to the boundaries, implicit methods provide the means for convenient imposition of boundary conditions. This is because, in contrast to the fully explicit methods, at each time step the problem is treated as a boundary value rather than an initial value problem.

The flows considered to date have been restricted to those that are homogeneous in two space dimensions. The direction of inhomogeneity is normal to the wall(s). Specifically, plane channel flow, pipe flow, axial flow between two concentric cylinders, plane and circular Couette flows have been simulated numerically. For these cases, the use of periodic boundary conditions in the homogeneous directions allows the application of Fourier transforms, which alleviate the need for split or factored-type algorithms. Perhaps, the most direct approach for the solution of Eq. (4) and the continuity equation is to solve them simultaneously. For simulation of turbulent channel flow, Moin and Kim [3, 19] used the second-order Adams-Bashforth method for H_i and the Crank-Nicolson method for $\partial P/\partial x_i$ and $\partial^2 u_i/\partial x_j \partial x_j$ in Eq. (4). This, together with the continuity equation at the new time step, $n + 1$, led to a system of four coupled, linear, partial-differential equations for u_i and P of the form

$$L(u_i^{n+1}, P^{n+1}) = F(u_i^n, u_i^{n-1}, P^n) \qquad (5)$$

where the superscripts denote the time step. Fourier transforming Eq. (5) in the x_1 and x_3 directions produces linear, ordinary differential equations of the form

$$\mathscr{L}(\hat{u}_i^{n+1}, \hat{P}^{n+1}) = \hat{F}$$

where $\mathscr{L}$ denotes Fourier transform. To solve these ordinary differential equations, both finite-difference operators on a modified staggered grid [3] and Chebyshev polynomial expansions [19] were used to approximate $\partial/\partial x_2$ and $\partial^2/\partial x_2^2$. The result is a system of algebraic equations for the Fourier transform of the dependent variables at the new time step. For the case where finite-difference operators were used, this system is of block-tridiagonal form; and for the case where Chebyshev polynomials were used, it is nearly block-tridiagonal. Both systems can be solved with $0(N_2)$ operations, where N_2 is the number of mesh points in the x_2-direction.

For their study of transition to turbulence in plane channels and Couette flows, Orszag and Kells [17] used a three-step fractional step method. Chebyshev polynomials and Fourier decomposition were used to represent the dependent variables spatially. In the first step, the Adams-Bashforth method is used for the nonlinear terms, H_i. The result is a set of intermediate velocity field, $\tilde{u}_i^{n+1}$. Next, the pressure correction (incompressibility condition) is applied, leading to another set of intermediate velocity field $\tilde{\tilde{u}}_i^{n+1}$ that satisfies the continuity equation $\partial\tilde{\tilde{u}}_i^{n+1}/\partial x_i = 0$. This step involves solving a Poisson equation for $\tilde{u}_2^{n+1}$ with the boundary conditions $\tilde{\tilde{u}}_2^{n+1} = 0$ at the walls. The velocity field at the new time step is obtained by applying the viscous correction which involves the solution of three Helmholtz equations for the velocity field. The velocity boundary conditions are applied at this stage. This method has a global error of order $0[\Delta t^2 + (1/Re)\,\Delta t]$. Thus, strictly speaking, it is a first-order method. In addition, the velocity field at the new time step does not satisfy the continuity equation. Only the intermediate velocity field, $\tilde{u}_i$, is solenoidal. If the last two steps were interchanged, the velocity field would be divergence-free, but the boundary conditions could not be enforced. Apparently, having the exact boundary conditions was preferred over the continuity equation. Another characteristics of note is that, to solve the Helmholtz equations in the final step using Chebyshev-Fourier expansions, one must transfer the forcing terms in these equations to the wave (Chebyshev-Fourier) space. These terms include $\tilde{\tilde{u}}_i^{n+1}$. Thus, to carry out the transformation, $\tilde{\tilde{u}}_i^{n+1}$ must be defined at the boundaries as well as in the interior of domain. It is necessary, therefore, to concoct boundary conditions for the intermediate velocities, $\tilde{\tilde{u}}_i^{n+1}$.

Recently, Leonard [5] developed a partially implicit method based on a special vector-function decomposition. An important feature of these vector functions is that each vector is solenoidal and satisfies the boundary conditions. When this series expansion for the velocity vector is used in Eq. (4) and the inner product of the result with a set of adjoint vectors is formed, the pressure terms are eliminated. In fact, since the velocity bases functions satisfy the continuity equation, only two dependent variables per mesh point remain. This leads to considerable savings in computer-memory requirements. Like the previous methods, the nonlinear terms are treated by an explicit method, whereas, the Crank-Nicolson scheme is used for the viscous terms. Leonard and Wray [28] have applied this method to the pipe-flow problem. They rigorously treat the behavior of the flow variables near the computational singularity at the pipe centerline which leads to the use of the Jacobi polynomial expansions in the radial direction. Moser et al. [27] applied this method to plane and curved channel flow problems. In the direction normal to the walls, a particular combination of Chebyshev polynomials was used as bases functions. With this choice, for each Fourier mode, the resulting system of algebraic equations was solved with $O(N_2)$ operations.

4.4 Data Management

Limitations of the high-speed central memory of presently available computers and the large number of grid points required for numerical simulation of turbulent flows necessitate the use of secondary memory. Generally, the entire data base must reside on secondary memory (SM) and only portions of it are successively transferred to the core memory (CM) for processing. In order to minimize the data transfer (I/O) time, an efficient data-management algorithm should be an essential part of each computer program.

The particular choice of data-management technique depends on the numerical method and the computer used for the calculations. However, all the algorithms have the objective of overlapping the data transfer from SM to CM with arithmetic operations. It is also important to minimize the number of passes over the data base. In general, one or two passes are required at each time step.

As an example of a data-management algorithm, we consider the scheme employed in Ref. [3]. In order to solve Eq. (6), a two-pass, double-buffer, data-management algorithm was used. In the first pass over the data base, the required pressure-velocity data from previous time steps are transferred to CM, and $\hat{F}$ (Eq. (6)) is computed and transferred to SM. Since second-order finite-difference formulas were used to approximate the derivatives in the x_2-direction, each time only three $(x_1 - x_3)$ planes of data are required to compute $\hat{F}$ in one plane. In the second pass, $(x_2 - x_3)$ planes of $\hat{F}$ were transferred to CM, and the block-tridiagonal system was solved. Note that, with this algorithm, the data were accessed in two different ways — first in the $x_1 - x_3$ planes and then in the $x_2 - x_3$ planes. Since the data are stored sequentially in, say, the $x_1 - x_3$ planes, the second access ($x_2 - x_3$ planes) is nonsequential; therefore, it is not as efficient. Alternatively, another data-management scheme can be used for solving the same set of equations. With this algorithm, in the first pass, in addition to computing $\hat{F}$, the forward sweep portion of the Gauss-elimination process can also be performed. In the second pass, the backward sweep portion is performed again, using $x_1 - x_3$ planes. Thus, the data base is accessed sequentially. In fact, the entire process can usually be accomplished with only one pass through the data base per time step [30]. It should be pointed out that, if pseudospectral rather than finite-difference methods are used in the x_2-direction, two passes through the data base (one of them nonsequential) are required.

We emphasize that the numerical methods discussed here have been largely designed for and applied to the simplest wall-bounded, turbulent shear flows, namely, those that are homogeneous in two spatial directions. The flow homogeneity has led to the use of periodic boundary conditions as a reasonable approximation to the unknown flow conditions at the "open" boundaries. The use of Fourier transforms converts the task of solving a partial differential equation to that of solving a set of uncoupled ordinary differential equations. This is a very significant fringe benefit associated with periodic boundary conditions. In order to calculate flows with two or more directions of inhomogeneity, in addition to having to specify the unknown (turbulent) inflow and outflow conditions, one should use split- or factored-type algorithms. An adaptation of the fractional step method appears to be an attractive candidate for this purpose.

5. Results

To illustrate the versatility and usefulness of the aforementioned calculations, we briefly present here a selection of recent results from the large-eddy simulation of wall-bounded turbulent flows. The main calculations were performed on the ILLIAC IV computer with $64\times63\times128$ grid points in the x, y, and z directions, respectively. The total computational time ranged from 20 hours to 92 hours.

Figure 4 shows the mean velocity profile, $\langle\bar{u}\rangle$, from numerical simulation of turbulent channel flow at $Re = 13800$, Ref. 3. Different symbols represent calculations with different grid resolutions and different sizes of the computational box in the x- and z-directions, where periodic boundary conditions are used. The calculations have predicted the logarithmic region with the proper slope. The agreement with the experimental data [31] is good. The distributions of the Reynolds stresses (Fig. 5) and higher-order statistics are also in good agreement with measurements. The contribution of subgrid scale turbulence to second- and higher-order statistical correlations is appreciable only in the vicinity of the walls. This is a consequence of the grid-resolution inadequacy in this region to represent the wall-layer turbulence structures at their proper scale. Figure 6 shows a contour plot of the instantaneous normal component of vorticity fluctuations, $\omega_2 = (\partial w/\partial x - \partial u/\partial z)$ in an $x - z$ plane close to the wall ($y^+ \simeq 6$). It is clear that, in accordance

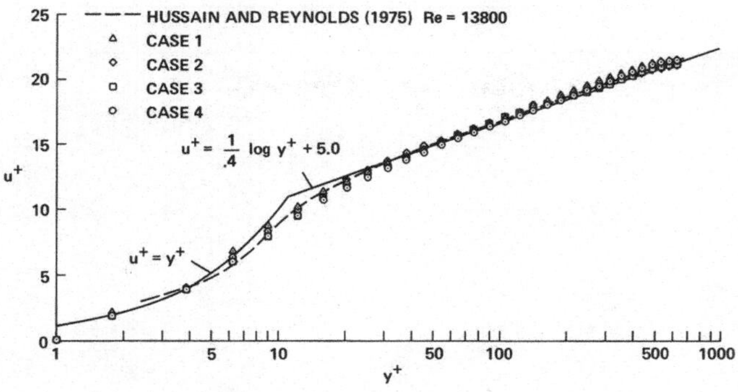

Figure 4. Mean velocity profile from four computed cases and comparison with experimental data. Four results are obtained from using different computational grids.

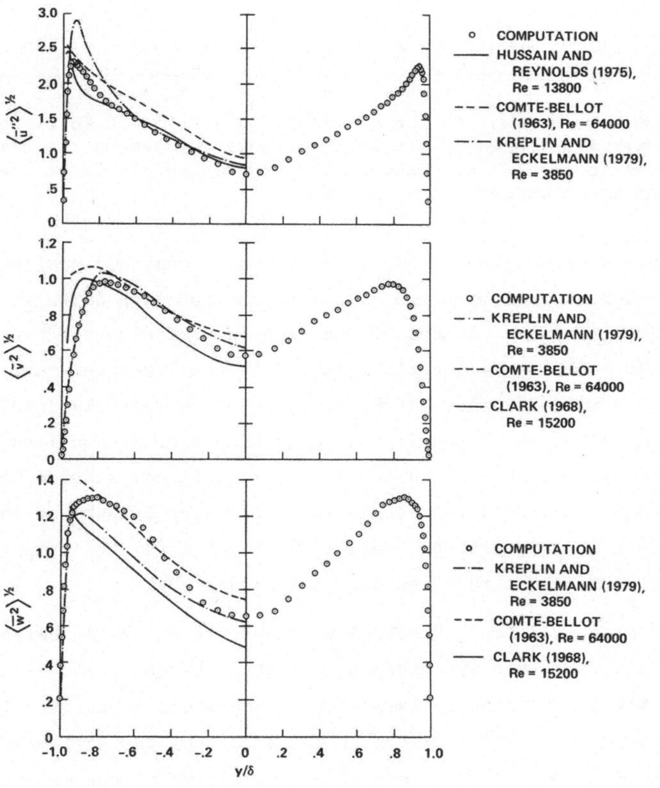

Fig. 5 Resolvable turbulence intensities and comparison with experimental data. The intensities are non-dimensionalized with the shear velocity, u_τ.

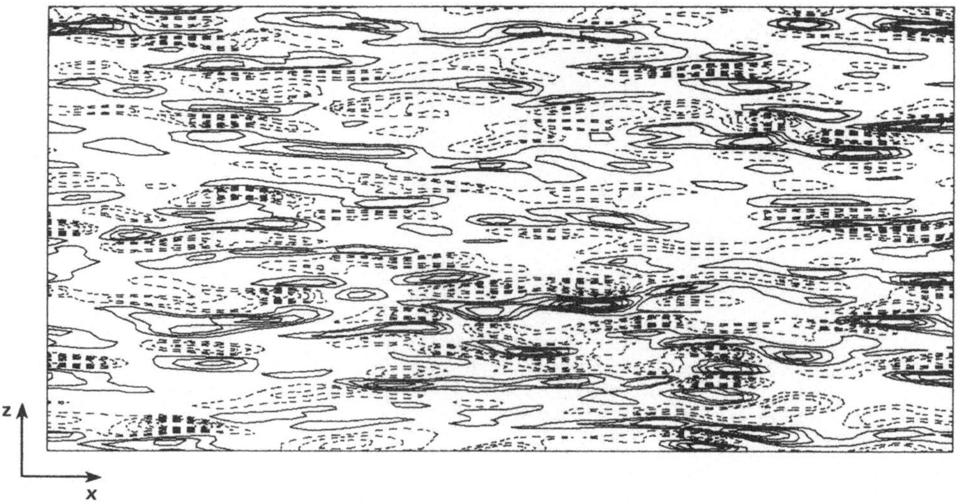

Figure 6. Contours of ω_2 in the (x, z)-plane at $y^+ = 6.26$. The rectangle on the lower right-hand corner of the figure represents the computational grid cell in the (x, z) planes. The streamwise extent of the figure is $2\pi\delta$ and its spanwise extent is $\pi\delta$.

with experimental measurements [6], this region is composed of flow structures which are long in the flow direction and narrow in the spanwise direction. The rapid spanwise variation of ω_2 is due to the existence of elongated regions of high-speed fluid $((\bar{u} - \langle\bar{u}\rangle) > 0)$ located adjacent to the low-speed regions [6, 3]. This figure is a vivid display of that particular characteristic of wall-bounded turbulent shear flows which requires a large number of grid points in the lateral direction. As was pointed out in section 3, in this calculation, the spanwise grid resolution was not adequate to resolve the wall-layer streaks at their proper scale. However, it is quite significant that, in spite of this, the computed flow field did display the streaky structures but at a larger scale.

The data generated from these calculations are currently being used to study the physical structure and dynamics of turbulent channel flow. In one study, for example, our aim is to identify large-scale, energetic structures in the flow field. In particular, we wanted to investigate the frequency and dominance of horseshoe or hairpin vortices that have been observed to originate at the wall and extend to outer regions with the characteristic inclination angle 40° - 50° (see, e.g., Ref. [32].) The vorticity field at several points in time was computed. At each grid point in various x - z planes, the angle $\theta = \tan^{-1}(\omega_2/\omega_1)$, and

$|\omega_{12}| = (\omega_1^2 + \omega_2^2)^{1/2}$ was calculated. Figure 7 shows the resultant distribution at $y_w/\delta = 0.2$. The contribution from each grid point was weighted by $|\omega_{12}|$. Indeed, the distribution attains its maximum at $\theta = 45°$. However, the probability of finding vorticity vectors with inclination angle in the range $0 \leqslant \theta \leqslant 90°$ is also appreciable. A detailed description of the results of this study will be presented elsewhere. In another investigation, Kim [33] has used the computed velocity-pressure field to examine the structure of the flow by conditional sampling techniques. Figure 8 shows the conditionally averaged, streamwise velocity obtained by using the VITA technique [7]. It is remarkably similar to the experimental results of Blackwelder and Kaplan [7]. The figure displays the burst and sweep events and their vertical extent. Kim has extended the experimental findings utilizing conditionally averaged pressure, vorticity, and the spanwise velocity component.

By simple modifications of the channel flow code described above, the effects of transpiration and spanwise rotation on the flow were computed. In the former case, uniform blowing through one wall and uniform suction at the same rate was applied through the other wall. Figure 9 shows that, in agreement with experimental measurements, the calculations predict the wall-shear-stress diminution caused by blowing

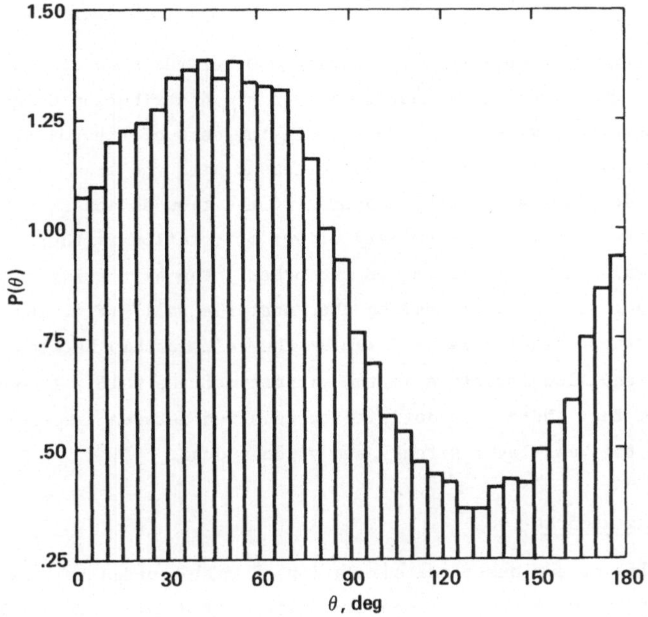

Figure 7. Distribution of the inclination angle of vorticity vectors (weighted by $|\omega_{12}|$) at $y_w/\delta = 0.2$.

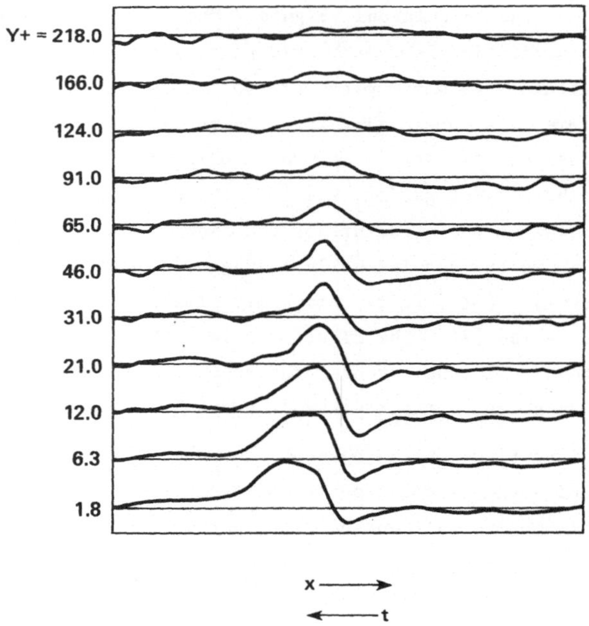

Figure 8. Conditionally averaged streamwise velocity fluctuations at various distances from the wall.

and its augmentation resulting from suction. Other features of the computed flow field, such as the effect of transpiration on the distribution of Reynolds stresses, are also in agreement with measurements. In the case of the rotating channel, the computational results [35] are in good agreement with the experimental data [36] and reproduce the detailed structural features of the flow as observed by flow-visualization techniques. The primary effect of rotation is that the flow is stabilized on one wall and destabilized on the other. The skin friction is reduced on the "stable wall" and is increased on the "unstable wall" (Fig. 10). One of the objectives of these studies was to examine the relationship between the changes in skin friction and flow structure in the vicinity of the wall. In both flows, it has been shown that there is a definite correlation between the characteristic dimensions of the wall-layer streaks and viscous drag.

6. Concluding Remarks

The results of the numerical simulation of wall-bounded turbulent shear flows have been most encouraging. For geometrically simple cases, it has been possible to predict many of the statistical and time-dependent features of the flows considered. The potential of these calculations for increasing our understanding of the physics of turbulent boundary layers is just beginning to be tapped.

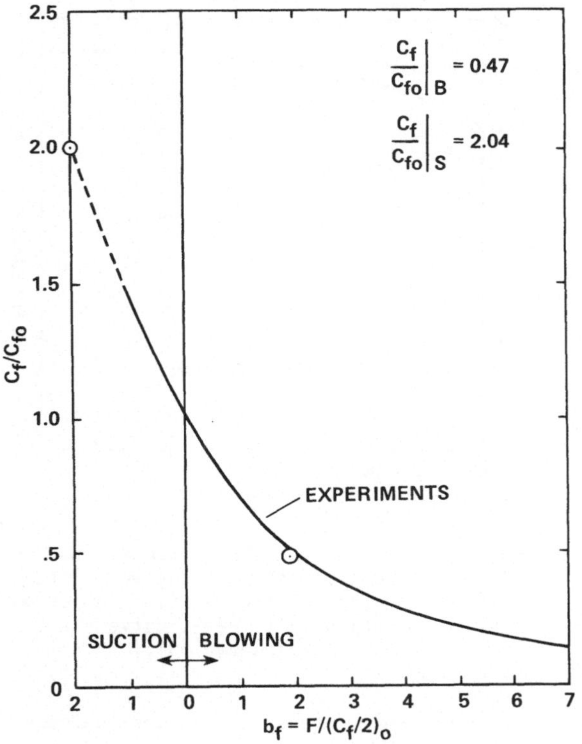

Figure 9. Ratio of friction coefficient to friction coefficient without transpiration, plotted as a function of a modified blowing parameter b_f [34]. ⊙, computation; ——— correlation curve of the experimental data; ----, linear extrapolation of the experimental correlation curve.

The large number of grid points required to resolve the wall-layer structures is a formidable hindrance to numerical simulation of high Reynolds number flows. The grid-imbedding technique can significantly ease this burden. Consideration of the bursting frequency and the frequency spectra of streamwise velocity fluctuations [37, 13] indicates that, for moderate Reynolds numbers, it may be possible to perform accurate LES calculations with larger time steps than is currently permitted by numerical stability restrictions. Thus, further improvements in numerical methods are likely to yield high dividends.

Extension to more complex geometries requires progress in our ability to prescribe turbulent inflow and outflow boundary conditions. Another practical difficulty in calculating these inhomogeneous flows is acquiring adequate ensemble averages. In contrast to homogeneous flows, one cannot integrate turbulence quantities over spatial grid points to secure a better statistical sample. To obtain adequate statistics, the only resort is time-averaging or, in the case of

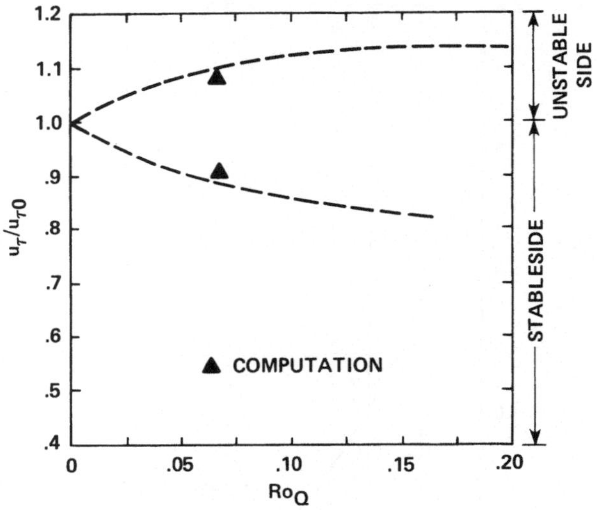

Figure 10. Ratio of the shear velocity to the shear velocity without rotation, plotted as a function of the rotation number. ▲, computation [35]; ----, curve fit to the experimental data [36].

nonstationary flows, averaging over several independent calculations. In both cases, the computational cost is significantly higher than for flows with one or two directions of homogeneity.

Acknowledgments

I am grateful to J. Kim for helpful discussions and for generously providing me with some results of his unpublished work. Useful discussions with A. Leonard are also appreciated.

References

1. Deardorff, J. W., 1970. A Numerical Study of Three-Dimensional Turbulent Channel Flow at Large Reynolds Numbers. J. Fluid Mech., 41, 453-480.
2. Schumann, U., 1975. Subgrid Scale Model for Finite Difference Simulations of Turbulent Flows in Plane Channels and Annuli. J. Comp. Phys., 18, 376-404.
3. Moin, P., and J. Kim, 1982. Numerical Investigation of Turbulent Channel Flow. J. Fluid Mech., 118 , 341.
4. Deardorff, J. W., 1973. The Use of Subgrid Transport Equations in a Three-Dimensional Model of Atmospheric Turbulence. J. Fluids Engr., 95, 429.
5. Leonard, A., 1981. Divergence-Free Vector Expansions for 3D Flow Simulations. Bull. of Amer. Phys. Soc., 26, 9, 1247.
6. Kline, S. J., W. C. Reynolds, F. A. Schraub, and P. W. Rundstadler, 1967. The Structure of Turbulent Boundary Layers. J. Fluid Mech., 30, 741-773.
7. Blackwelder, R. F., and R. E. Kaplan, 1976. On the Wall Structure of Turbulent Boundary Layer. J. Fluid Mech., 76, 89.
8. Gupta, A. K., J. Laufer, and R. E. Kaplan, 1971. Spatial Structure in the Viscous Sublayer. J. Fluid Mech., 50, 493.
9. Comte-Bellot, G., 1963. Contribution a l'Etude de la Turbulence de Conduite.
10. Hinze, J. O., 1975. Turbulence, McGraw-Hill, Inc., 2nd ed.
11. Eckelmann, H., 1974. The Structure of Viscous Sublayer and The Adjacent Wall Region in a Turbulent Channel Flow. J. Fluid Mech., 65, 439.
12. Clark, J. A., and E. Markland, 1971. Flow Visualization in Turbulent Boundary-Layers. Proc. ASCE, J. Hydraulics Div., 97, 10, 1653-1664.
13. Bakewell, H. P., and J. L. Lumley, 1967. Viscous Sublayer and Adjacent Wall Region in Turbulent Pipe Flow. Phys. Fluids, 10, 1880.
14. Chapman, D. R., 1979. Computational Aerodynamics Development and Outlook. AIAA J., 17, 1293-1313.
15. Chapman, D. R., and G. D. Kuhn, 1981. Two Component Navier-Stokes Computational Model of Viscous Sublayer Turbulence. AIAA Paper 81-1024, AIAA 5th CFD Conf., Palo Alto, Calif.
16. Harlow, F. H., and J. E. Welch, 1965. Numerical Calculation of Time-Dependent Viscous Incompressible Flow. Phys. Fluids, 8, 2182.
17. Orszag, S. A., and L. C. Kells, 1980. Transition to Turbulence in Plane Poiseulle and Plane Couette Flow. J. Fluid Mech., 96, 159.
18. Patera, A. T., and S. A. Orszag, 1980. Transition and Turbulence in Planar Channel Flows. Lecture Notes in Physics (W. C. Reynolds, and R. W. MacCormac, eds.) 114, 329, Springer-Verlag.
19. Moin, P., and J. Kim, 1980. On the Numerical Solution of Time-Dependent Viscous Incompressible Fluid Flows Involving Solid Boundaries. J. Comp. Physics, 35, 381-392.
20. Kleiser, L., 1979. Solution of Coupled Velócity-Pressure Equations in the Fourier-Chebyshev Spectral Method for Incompressible Flows. Private communication.
21. Taylor, T. D., and J. W. Murdock, 1981. Application of Spectral Methods to the Solution of Navier-Stokes Equations. Comp. and Fluids, 9, 255.
22. Orszag, S. A., 1971. Galerkin Approximations to Flows Within Slabs, Spheres, and Cylinders. Phys. Rev. Lett., 26, 1100.

23. Herring, J. R., S. A. Orszag, R. H. Kraichnan, and D. G. Fox, 1974. Decay of Two-Dimensional Homogeneous Turbulence. J. Fluid Mech., 66, 417.

24. Buzbee, B., G. Goulub, and C. Nielsen, 1970. On the Direct Methods for Solving Poisson's Equation. SIAM J. Numer. Anal., 7, 627.

25. Shaanan, S., J. H. Ferziger, and W. C. Reynolds, 1975. Numerical Simulation of Turbulence in the Presence of Shear. Report No. TF-6, Dept. of Mech., Engrg., Stanford University.

26. Moin, P., W. C. Reynolds, and J. H. Ferziger, 1978. Large Eddy Simulation of Incompressible Turbulent Channel Flow. Report No. TF-12, Dept. of Mech. Engrg., Stanford University.

27. Kim, J., and P. Moin, 1979. Large Eddy Simulation of Turbulent Channel Flow --ILLIAC IV Calculation. In Turbulent Boundary Layers--Experiments, Theory, and Modeling, The Hague, Netherlands. AGARD Conf. Proc. no. 271.

28. Leonard, A., and A. Wray, 1982. Numerical Solution of Three-Dimensional Transitional Flow in a Pipe. Proc. of this conference.

29. Moser, R. D., A. Leonard, and P. Moin, 1982. To be published.

30. Wray, A., and M. Y. Hussaini, 1980. Numerical Experiments in Boundary-Layer Stability. AIAA Paper 80-0275, AIAA 18th Aerospace Science Meeting, Pasadena, Calif.

31. Hussain, A.K.M.F., and W. C. Reynolds, 1975. Measurements in Fully Developed Turbulent Channel Flow. J. Fluids Engrg., 97, 568-578.

32. Head, M. R., and P. Bandyopadhyay, 1981. New Aspects of Turbulent Boundary-Layer Structure. J. Fluid Mech., 107, 297.

33. Kim, J., 1982. Stanford University, to be published.

34. Kays, W. M., 1972. Heat Transfer to the Transpired Turbulent Boundary-Layer. Int. J. Heat and Mass Transfer, 15, 1023.

35. Kim, J., 1982. Stanford University, to be published.

36. Johnston, J. P., R. M. Halleen, and D. K. Lezius, 1971. Effects of Spanwise Rotation on the Structure of Two Dimensional Fully Developed Turbulent Channel Flow. J. Fluid Mech., 56, 533.

37. Rao, K. N., R. Narasimha, and M. A. Badri Narayanan, 1971. The Bursting Phenomenon in a Turbulent Boundary Layer. J. Fluid Mech., 48, 339.

SHOCK CAPTURING, FITTING AND RECOVERY

K. W. Morton

University of Reading, Reading, England

1. INTRODUCTION

During the 1950's two very different approaches to numerical shock modelling were developed - shock fitting, in which the position and evolution of shocks is approximated explicitly, and shock capturing by finite difference methods operating on a fixed mesh. There was a prolonged lull in developments from the late 1950's to the early 1970's, as the two methods seemed to have reached a stalemate, but the last ten years have seen rapid progress on many fronts - theory of conservation laws, numerical methods and analysis of convergence. An indication of the situation is provided by the excellent survey of Gary Sod (1978). This stimulated widespread interest and even the best results given there now look very poor by comparison with those from current methods.

Yet still these mainly shock capturing methods lack the precision achievable with shock fitting techniques on the coarse meshes that must often be used in two and three dimensions. One way of achieving this and bridging the gap between the two viewpoints is to use shock recovery: the results obtained on a fixed grid by a shock capturing method are scanned for the presence of shocks whose positions and strengths are found and this information used in the subsequent evolution. The mesh may or may not be adjusted.

Such a viewpoint is natural to those working with finite elements and we shall consider here some recent developments of these methods for evolutionary problems which are pertinent to shock recovery.

In contrast to finite difference methods, the objective with finite elements is to create at each time level the best least squares fit to the solution from an appropriate space of trial functions. Typical choices are piecewise constants or piecewise linears. Obtaining intermediate values in order to model advection accurately or to calculate flux functions requires formulae comparable with those for polynomial interpolation. Recovery of sub-gridscale information, such as shocks or boundary layers, requires more specialised techniques which take maximum advantage of any available knowledge about the solution.

Let us consider first the problem of obtaining intermediate results from those given on a uniform mesh. Typically a finite difference method will provide grid-point values u_j at $x_j = jh$, j an integer; then, for example, mid-point values are given by truncating the interpolation formula

$$u_{j-\frac{1}{2}} \approx (1 - \tfrac{1}{8}\delta^2 + \tfrac{3}{128}\delta^4 - \dots)\frac{(u_{j-1} + u_j)}{2}, \qquad (1.1)$$

where $\delta^2 u_j := u_{j-1} - 2u_j + u_{j+1}$, in the standard notation. On the other hand suppose U_j are the nodal values defining a continuous piecewise linear approximation which is the best L_2 fit to $u(x)$: that is, the U_j are given by the Galerkin equations

$$\int [\textstyle\sum_{(j)} U_j \phi_j(x) - u(x)]\phi_i(x)dx = 0, \qquad \forall i; \tag{1.2}$$

where $\phi_j(x)$ are the hat-shaped linear basis functions. Then intuitively we can see that the U_j will overshoot the true nodal values u_j of $u(x)$ and indeed we have the well-known recovery formula

$$u_j \sim (1 + \tfrac{1}{12}\delta^2 - \tfrac{1}{360}\delta^4 + \ldots)U_j \,. \tag{1.3}$$

For reasonably smooth $u(x)$, the three-term formula $12u_j \sim U_{j-1} + 10U_j + U_{j+1}$ is remarkably accurate. To compare with (1.1), suppose now we need the nodal values for a similar approximation on a mesh shifted by $\frac{1}{2}h$: then the natural formula is

$$(1 + \tfrac{1}{6}\delta^2)U_{j-\frac{1}{2}} \sim (1 + \tfrac{1}{24}\delta^2 + \tfrac{1}{384}\delta^4 \ldots)(\frac{U_{j-1} + U_j}{2}). \tag{1.4}$$

The operator on the left is natural since it corresponds to the mass matrix in the tridiagonal system of equations given by (1.2). We see that the implicit four point formula given by truncating (1.4) is nine times more accurate than the four point explicit formula given by (1.1). This is one of the bases of the greater accuracy that is often attainable with finite element methods - at the cost of more work because of the implicitness.

One of the overlap areas between finite differences and finite elements is provided by the use of piecewise constant approximations, as for instance in some finite volume methods. We shall use a notation in which the value U_j is understood to extend from $(j-\frac{1}{2})h$ to $(j+\frac{1}{2})h$ on a uniform mesh, or from $x_{j-\frac{1}{2}}$ to $x_{j+\frac{1}{2}}$ on a non-uniform mesh: the basis function, which we shall also denote by ϕ_j when there is no confusion, has unit value over this interval and is zero elsewhere. The Galerkin equations (1.2) for an L_2 best fit have a diagonal matrix in this case and U_j is the average of $u(x)$ over the interval, an interpretation which is shared by most finite difference schemes. In contrast to the piecewise linears, U_j undershoots the point values and the recovery formula corresponding to (1.3) becomes

$$u_j \sim (1 - \tfrac{1}{24}\delta^2 + \tfrac{3}{640}\delta^4 - \ldots)U_j. \tag{1.5}$$

For the shifted projection, we now have exactly the same formula for the $U_{j-\frac{1}{2}}$ as (1.1) for the point values. Thus the only important point about the interpretation of U_j as element averages is to use (1.5) to obtain u_j before, for instance, calculating flux functions.

All the above formulae apply to smooth underlying functions $u(x)$. In the presence of boundary layers or shocks, the recovery process is the same in principle but must use different recovery functions: while formulae like (1.3) may be derived by replacing u in (1.2) by an interpolating spline, Barrett & Morton (1980) in their work on diffusion-convection problems used recovery formulae of the general form

$$< U - \tilde{u}, \phi_i > = 0, \tag{1.6}$$

in which the recovery function $\tilde{u}$ was exponential in form and the recovery was performed very locally. (Here and below we use the notation $<\cdot,\cdot>$ to denote the L_2 inner product in the space variables). The main point is to use whatever information is known about the approximated function $u(x)$ and to exploit the fact that $U(x)$ is its L_2 best fit - or nearly so in evolutionary problems.

In the next and main section of the paper, we shall describe Characteristic Galerkin methods for approximating unsteady conservation laws and their application to shock capturing and recovery. Finite element methods are strongly based on the Galerkin formulation but as their development for problems other than those which are steady, linear, elliptic and self-adjoint progresses one finds that the formulation has to be generalised and the Characteristic Galerkin methods result from such a generalisation. This part of the paper is not meant as a review nor is it primarily intended to promulgate new methods which have been widely tested on practical problems: it is rather my aim to present a new viewpoint and to show how this naturally links together important ideas and algorithms developed by Godunov (1959), Boris & Book (1973), van Leer (1979), Engquist & Osher (1980), Roe (1981) and several others. In the final section of the paper the moving finite element method due to Gelinas, Doss & Miller (1981) will be described from the same viewpoint and developments of it to capture shocks presented briefly.

2. CHARACTERISTIC GALERKIN SCHEMES

2.1 Basic ECG scheme

Consider the scalar conservation law in one dimension

$$\partial_t u + \partial_x f(u) = 0 \tag{2.1a}$$

or

$$\partial_t u + a(u)\partial_x u = 0, \tag{2.1b}$$

where $a(u) = \partial f/\partial u$: and suppose $u(x,t)$ is approximated at time level $n\Delta t$, in terms of basis functions $\phi_j(x)$, by

$$U^n(x) = \textstyle\sum_{(j)} U_j^n \phi_j(x). \tag{2.2}$$

Then the characteristic Galerkin method based on Euler time-stepping, and hence called the Euler Characteristic Galerkin or ECG method by Morton & Stokes (1981) and Morton (1982), takes the form

$$\langle U^{n+1} - U^n, \phi_j \rangle + \Delta t \langle \partial_x f(U^n), \Phi_j^n \rangle = 0. \tag{2.3}$$

Here Φ_j^n is the upwind-averaged test function,

$$\Phi_j^n(x) := \frac{1}{a^n(x)\Delta t} \int_x^{x+a^n(x)\Delta t} \phi_j(z)dz, \tag{2.4}$$

where $a^n(x) := a(U^n(x))$. The method was introduced in order to model accurately the advection of continuous profiles. It is equivalent to exactly tracing the evolution of $U^n(x)$ along characteristics by the equation

$$u(y,t + \Delta t) = u(x,t) \quad \text{where} \quad y = x + a(u(x,t))\Delta t \tag{2.5}$$

and following this by L_2 projection onto $\text{span}\{\phi_j\}$.

A continuous piecewise linear approximation is appropriate for accurate advection and the corresponding test function, when the characteristic speed a^n is constant and for various positive values of the CFL number $\mu = a\Delta t/h$, is shown in Fig. 1. The effectiveness of such a scheme in advecting a steep ramp function over a coarse grid is shown by Fig. 2: the small scale oscillation shown there is typical of least squares best fits by piecewise linears and it is seen that there is little change in it from that produced by the projection of the initial data. In this constant coefficient case on a uniform mesh, the scheme (2.3) reduces, in terms of nodal values, to

$$(1 + \tfrac{1}{6}\delta^2)(U_j^{n+1} - U_j^n) + \mu(\Delta_0 - \tfrac{\mu}{2}\delta^2 + \tfrac{\mu^2}{6}\delta^2\Delta_-)U_j^n = 0 \tag{2.6}$$

and corresponds to shifting the projection by the distance $a\Delta t$ that the characteristic has travelled in one time step. Clearly for $\mu = \frac{1}{2}$ the formula (2.6) reduces to that obtained from the first two terms of (1.4) and the ninefold improvement of this over (1.1) is consistent with the advantage of (2.6) over any of the usual explicit four point difference schemes for advection. As it stands (2.3) may be quite expensive to evaluate exactly but several efficient approximations of the ideal test function Φ_j^n which reproduce (2.6) are given by Morton (1982).

2.2 Shock modelling with piecewise constants

In shock modelling, however, discontinuous approximations are often favoured for either finite difference or finite element schemes. Let us consider piecewise constants first. Then (2.3) needs careful interpretation even for rarefaction waves because not only is $f(U^n)$ discontinuous but so is Φ_j^n, through the dependence on $a(U^n)$, and these discontinuities coincide. Several finite difference methods have used the idea of resolving the jumps in $U^n(x)$ by the correct physical rarefaction fans. This could be done here but is unnecessary if we remember that, even if the objectives of our calculation are perfectly achieved, $U^n(x)$ is only the L_2 projection of the exact solution at time $n\Delta t$: thus the exact evolution of this approximate solution is hardly justified. Suppose then that $a(U_k^n) =: a_k^n > a_{k-1}^n$ so that (2.5) leaves a gap in the definition of the mapping $y \to x$ and hence (2.3) is ill-defined. We need use our knowledge of the exact solutions of the differential equation only to the extent of recognising that the jump in $U^n(x)$ would have been resolved. So let us suppose, for instance, that $U^n(x)$ is the projection of a function $\tilde{u}^n(x)$ which is linear between $(k-\frac{1}{2}-\frac{1}{2}\theta)h$ and $(k-\frac{1}{2}+\frac{1}{2}\theta)h$, with $0 < \theta < 1$, and takes the constant value $\tilde{u}_k^n$ in $[(k-\frac{1}{2}+\frac{1}{2}\theta)h, kh]$ and the value $\tilde{u}_{k-1}^n$ in $[(k-1)h, (k-\frac{1}{2}-\frac{1}{2}\theta)h]$. Then $\partial_x f(\tilde{u}^n)$ is well defined and a little algebra shows that, if $a(\tilde{u}^n) > 0$, $\frac{1}{2}\theta < \min a(\tilde{u}^n)\Delta t/h$ and $\max a(\tilde{u}^n)\Delta t/h \le 1$ for $x \in [(k-1)h, kh]$,

$$\int_{(k-1)h}^{kh} \partial_x f(\tilde{u}^n)\tilde{\Phi}_j^n \, dx = \begin{cases} \Delta_- \tilde{f}_k^n - \frac{\theta h}{8\Delta t}\Delta_- \tilde{u}_k^n & \text{for } j = k \\ \frac{\theta h}{8\Delta t}\Delta_- \tilde{u}_k^n & \text{for } j = k-1, \end{cases} \tag{2.7}$$

where $\tilde{f}$ and $\tilde{\Phi}$ are defined using $\tilde{u}$ rather than U. Suppose also that a similar jump exists at $(k+\frac{1}{2})h$ and is similarly resolved. Then combining the two results we obtain from (2.3)

$$h(U_k^{n+1} - U_k^n) + \Delta t(\Delta_- \tilde{f}_k^n + \frac{\theta h}{8\Delta t}\delta^2 \tilde{u}_k^n) = 0 \quad . \tag{2.8}$$

Furthermore, from the fact that U^n is the projection of $\tilde{u}^n$, we easily deduce that

$$U_k^n = [1 + (\theta/8)\delta^2]\tilde{u}_k^n, \tag{2.9}$$

so that (2.8) becomes

$$U_k^{n+1} = \tilde{u}_k^n - (\Delta t/h)\Delta_- \tilde{f}_k^n. \tag{2.10}$$

In other words, U^{n+1} depends only on the constant sections of $\tilde{u}^n$ and the dependence on the parameter θ appears only in the recovery formula (2.9). We will below generally use the limiting case $\theta \to 0$ so that (2.10) becomes just the familiar first order upwind scheme: however, it should be noted that increasing θ will reduce the false diffusion that ruins this scheme for smooth flows and, indeed, taking $\theta = 1$ to make $\tilde{u}^n$ piecewise linear with knots at jh gives a scheme similar to that of Fromm (1968).

For a general rarefaction case, suppose $f(u)$ has just a single sonic point $\bar{u}$; that is $a(\bar{u}) = 0$ and otherwise $a(u) \neq 0$. Then from (2.4), we see that

$$\tilde{u}^n(x) = \bar{u}, \quad \phi_j(x) = 1 \quad \Rightarrow \quad \tilde{\Phi}_j^n(x) = 1. \tag{2.11}$$

It is also easy to see that if the CFL condition

$$a(u)\Delta t/h < 1 \quad \text{for } u \in [\tilde{u}_{k-1}^n, \tilde{u}_k^n] \tag{2.12}$$

is satisfied, then

$$\tilde{\Phi}^n_{k-1}(x) + \tilde{\Phi}^n_k(x) = 1 \quad \text{for} \quad x \in [(k-1)h, kh]. \tag{2.13}$$

Thus we can deduce that, in the limit $\theta \to 0$, (2.3) yields the following algorithm:

$$a^n_{k-1}, a^n_k \geq 0 \ : \ \text{use} \ \Delta_- f^n_k \ \text{to update} \ U^n_k \to U^{n+1}_k \tag{2.14a}$$

$$a^n_{k-1}, a^n_k \leq 0 \ : \ \text{use} \ \Delta_- f^n_k \ \text{"} \quad \text{"} \ U^n_{k-1} \to U^{n+1}_{k-1} \tag{2.14b}$$

$$\begin{aligned} a^n_{k-1}\, a^n_k < 0 \ : \ &\text{use} \ f^n_k - f(\bar{u}) \text{"} \quad \text{"} \ U^n_k \to U^{n+1}_k \\ &\text{use} \ f(\bar{u}) - f^n_{k-1} \ \text{to update} \ U^n_{k-1} \to U^{n+1}_{k-1}. \end{aligned} \tag{2.14c}$$

It is clear that this algorithm is exactly equivalent in this case to the key flux-splitting idea of Engquist & Osher (1980), in which U_j is updated using $\Delta_- f^+_j + \Delta_+ f^-_j$. On the other hand, the derived fluxes f^-, f^+ are not introduced and the algorithm has the form of those due to Roe (1981), in which the total flux difference $\Delta_- f_j$ between each pair of intervals is in turn broken up into contributions to update U in these (and possibly neighbouring) intervals. We shall see that this feature of sharing the properties of these two finite difference schemes occurs naturally in all our ECG schemes. Note that if $f(u)$ has several sonic points then the contribution from each which is correctly given by the Engquist-Osher scheme is picked up by the ECG scheme (2.3) only if the correct physical waves are used in the recovery process rather than piecewise linears for $\tilde{u}$.

Suppose now that we have a shock at $(k-\frac{1}{2})h$, that is that $a^n_{k-1} > a^n_k$. Then the characteristics drawn from points just to the left and just to the right of $(k-\frac{1}{2})h$ overlap, and the mapping (2.5) gives a multivalued $y(x)$ and hence a multivalued $u(x,t+\Delta t)$. However, (2.3) can still be properly defined and we have purposely delayed the derivation of this formula until we came to this case. We suppose that we have recovered from $U^n(x)$ a continuous function $\tilde{u}^n(x)$, such as that used in obtaining (2.7), which is determined by the projection relation

$$\langle U^n - \tilde{u}^n, \phi_j \rangle = 0 \qquad \forall j. \tag{2.15}$$

Then $y(x) = x + a(\tilde{u}^n(x))\Delta t$ defines a continuous (x,y) path which gives a possibly multivalued mapping $y \to x$. We define $U^{n+1} \in \text{span}\{\phi_j\}$ by

$$\langle U^{n+1}, \phi_j \rangle = \int_{-\infty}^{\infty} \tilde{u}^n(x(y))\phi_j(y)dy, \tag{2.16}$$

the integral being defined along the (x,y) path. Then, introducing $\tilde{\Phi}^n_j(x)$ by (2.4), we have

$$\begin{aligned} \langle U^{n+1} - U^n, \phi_j \rangle &= \int_{-\infty}^{\infty} \tilde{u}^n(x) \, d\Big[\int_x^y \phi_j(z)dz\Big] \\ &= -\int_{-\infty}^{\infty} d\,[\tilde{u}^n(x)] \int_x^y \phi_j(z)dz = -\Delta t \int_{-\infty}^{\infty} f_x(\tilde{u}^n(x))\tilde{\Phi}^n_j(x)dx, \end{aligned}$$

i.e. $\langle U^{n+1} - U^n, \phi_j\rangle + \Delta t \langle \partial_x f(\tilde{u}^n), \tilde{\Phi}_j^n\rangle = 0.$ (2.17)

This is now the general formula with which we shall work. For the case of a shock at $(k-\frac{1}{2})h$ with piecewise constant elements, it is clear that where $f_x \neq 0$ we have $dy < 0$ and the calculation is the same as for the rarefaction case but with $\tilde{u}_{k-1}^n$ and $\tilde{u}_k^n$ interchanged. The net effect is to yield the same algorithm as given in (2.14) which is now shown to cover all possible cases.

2.3 Piecewise linear basis functions

For continuous piecewise linear functions no recovery is necessary and (2.3) may be used directly even in the presence of shocks, though there may be considerable loss of accuracy then,which we shall discuss below in 2.4. It is convenient again to arrange the algorithm to deal in turn with the contributions from $\partial_x f(U^n)$ arising from each interval. For $x \in [x_{k-1}, x_k]$, we have $\phi_{k-1}(x) + \phi_k(x) = 1$ and if we assume the CFL condition holds locally then

$$\Phi_{k-2}^n(x) + \Phi_{k-1}^n(x) + \Phi_k^n(x) + \Phi_{k+1}^n(x) = 1 \quad \text{for} \quad x \in [x_{k-1}, x_k] \qquad (2.18)$$

with only the first three non-zero if $a(x) > 0$ and the last three if $a(x) < 0$. Thus the total contribution to updating up to four nodal values of $U^n(x)$ is again $\Delta_- f_k^n$. In general the integrals will need to be approximated by quadrature rules but always this property should be retained. Also, of course, as in (2.6), a tri-diagonal system has to be solved for the nodal values once all contributions to the updating have been accumulated.

For shock modelling, van Leer (1979) has suggested that discontinuous piecewise linear approximations should be used and has developed in his MUSCL code an approximate Riemann solver for them. Usually, basis functions consisting of piecewise constants and of linear functions varying from -1 to $+1$ over an interval have been used for such schemes. However, for an ECG scheme it is more convenient to use the two parts of the continuous linear basis functions, namely $\phi_{j+}(x)$, which varies from 1 at x_j to 0 at x_{j+1}, and $\phi_{j-}(x)$, which varies from 1 at x_j to 0 at x_{j-1}. Then we can write

$$U^n(x) = \sum_{(j)} [U_{j+}^n \phi_{j+}(x) + U_{j-}^n \phi_{j-}(x)] \qquad (2.19)$$

with the jump at x_j equal to $U_{j+}^n - U_{j-}^n$. The sum of test functions in (2.18) splits into eight parts, but each is non-negative: the two outermost terms Φ_{k-2-}^n and Φ_{k+1+}^n can be discarded and for each particular x only four terms are non-zero. Thus contributions from $\partial_x f(U^n)$ for $x \in (x_{k-1}, x_k)$ are distributed at worst to six nodal values of U^n, and usually only to four. As in the continuous case above, these contributions need to be approximated by quadrature rules.

The contributions from the jump in $f(U^n)$ at x_k can be calculated however as for the piecewise constant case. There is a little more dependence on the form used to resolve the discontinuity but we note that

$$\Phi_{k+}(x) + \Phi_{k-}(x) + \Phi_{k+1-}(x) + \Phi_{k-1+}(x) = 1 \quad \text{for} \quad |x-x_k| \le \varepsilon \quad (2.20)$$

for sufficiently small ε so that only these four corresponding nodal values can be affected. The only uncertainty in the calculation results from the fact that each term in (2.20) may be discontinuous at x_k because of the discontinuity in $a(U^n)$: the simplest resolution of the uncertainty is to define $\Phi_j(x_k)$ as the mean of the limits from above and below. Then, dropping unnecessary superscripts, the allocation of updates corresponding to (2.14) can be set out as follows:

$$a_{k-}, a_{k+} \ge 0: \quad [f_{k+}-f_{k-}]\Phi_{k+}(x_k) \text{ to } \tfrac{1}{3}U_{k+} + \tfrac{1}{6}U_{k+1-}$$
$$[f_{k+}-f_{k-}]\Phi_{k+1-}(x_k) \text{ to } \tfrac{1}{6}U_{k+} + \tfrac{1}{3}U_{k+1-} \qquad (2.21a)$$

$$a_{k-}, a_{k+} \le 0: \quad [f_{k+}-f_{k-}]\Phi_{k-}(x_k) \text{ to } \tfrac{1}{6}U_{k-1+} + \tfrac{1}{3}U_{k-}$$
$$[f_{k+}-f_{k-}]\Phi_{k-1+}(x_k) \text{ to } \tfrac{1}{3}U_{k-1+} + \tfrac{1}{6}U_{k-} \qquad (2.21b)$$

$$a_{k+}a_{k-} < 0: \quad [f_{k+}-f(\bar{u})]\Phi_{k+}(x_k) \text{ to } \tfrac{1}{3}U_{k+} + \tfrac{1}{6}U_{k+1-}$$
$$[f_{k+}-f(\bar{u})]\Phi_{k+1-}(x_k) \text{ to } \tfrac{1}{6}U_{k+} + \tfrac{1}{3}U_{k+1-}$$
$$[f(\bar{u})-f_{k-}]\Phi_{k-}(x_k) \text{ to } \tfrac{1}{6}U_{k-1+} + \tfrac{1}{3}U_{k-} \qquad (2.21c)$$
$$[f(\bar{u})-f_{k-}]\Phi_{k-1+}(x_k) \text{ to } \tfrac{1}{3}U_{k-1+} + \tfrac{1}{6}U_{k-}$$

Choosing the basis functions $\Phi_{j\pm}$ to be non-negative, and therefore ensuring that the corresponding $\Phi_{j\pm}$ have the same property, makes it very much easier to introduce approximations to the latter to be used in these formulae and in the quadrature over each open interval (x_{k-1},x_k). The penalty for not having an orthogonal basis is very minor: for each interval is independent of the others and the two nodal parameters at its ends, U_{j+} and U_{j+1-} say, are given by a simple pair of equations.

2.4 Shock recovery

It was shown by Cullen & Morton (1980) with the shallow water equations how for smooth flows the accuracy of a purely Galerkin procedure could be greatly improved on by a two stage procedure for approximating the non-linear terms $\underline{u}\cdot\nabla\underline{u}$: from a piecewise linear approximation to $\underline{u}$ they formed a best least squares fit to $\nabla\,\underline{u}$ by piecewise linears before doing the same for the product $\underline{u}\cdot\nabla\,\underline{u}$. This can be regarded as a simplified recovery procedure: for in one dimension the first step corresponds to forming a quadratic spline approximation to $\underline{u}$. More general use of recovery procedures was envisaged by Barrett & Morton (1980) for diffusion-convection problems and a number of results in one dimension are collected together by Barrett, Moore & Morton (1982).

In the context of modelling conservation laws recovery of u from a best fit in any integral norm is clearly very important before evaluating the flux $f(u)$. When a piecewise constant approximation is used, peaks are cut off and need to be restored where possible: and for a continuous piecewise linear approximation, overshoots occur which need to be smoothed out. In the former case we have already seen, in the course of interpreting the basic ECG scheme (2.3), that recovery by piecewise linears to give (2.17) can restore some of the information lost by false diffusion. There is a similarity here with the philosophy of SHASTA codes (Boris & Book, 1973) and an appropriate algorithm for the choice of the parameter θ in (2.7) would seem to put this scheme into the more general class of flux-corrected transport algorithms described by Zalesak (1979). For the continuous piecewise linears, we have also already given the spline-derived recovery formula (1.5) which can be used very simply to improve accuracy in smooth parts of the flow.

But it is in the neighbourhood of shocks that enhancement of accuracy is most important. Loss of information clearly results primarily from the use of a fixed set of mesh points if discontinuous approximations are used: even for continuous linear elements, fixed nodes together with the resulting fixed spacing is the prime cause of inaccuracy. Thus for each approximation U, we define a recovery function $\tilde{u}$ which has discontinuities just where we deduce that the exact solution u has shocks. There are two steps involved:

(i) recognition of the presence and position of the shock;

(ii) estimation of the shock parameters.

As Rusanov has pointed out (see Rusanov (1981) and references therein) every difference scheme has its own limiting shock profile, and these can be used to deal with (i). Generally speaking this will mean scanning for a local maximum in $-\Delta_- U_j$, that is checking a criterion such as

$$\delta^2 \Delta_- U_j > \text{tol}. \tag{2.22}$$

and this was what was used in the work reported in (Morton, 1980). However, it is more natural with ECG schemes to monitor the crossing of characteristics in each interval and to use a criterion (with ν a free parameter) like

$$a(U_{j-1}) - a(U_j) > (x_j - x_{j-1})/\nu\Delta t \tag{2.23}$$

to detect a shock. This can in fact be used for either piecewise constants or piecewise linear elements: note that the criterion does not have to be completely foolproof since our schemes are valid even in the presence of a shock and we are merely seeking to enhance their accuracy. The estimation of the shock parameters will however be dependent on the approximation used.

For the piecewise constant approximation, the simplest configuration of a steady shock between $x_{k-\frac{1}{2}}$ and $x_{k+\frac{1}{2}}$ linking two constant states clearly leads to a projection with one intermediate value U_k. If the shock is strong enough this situation will be recognised by (2.23) being satisfied for the two isolated values $j = k, k+1$: alternatively it may be desirable to replace the left-hand side of (2.23) by $a(U_{j-1}) - a(U_{j+1})$ which would be recognised for $j = k$. In either case from the three nodal values involved we can recover

$$\tilde{u}_L = U_{k-1}, \; \tilde{u}_R = U_{k+1}, \; \eta = \Delta_+ U_k/(\Delta_+ U_k + \Delta_- U_k) \tag{2.24}$$

with $x_S = (1-\eta)x_{k-\frac{1}{2}} + \eta x_{k+\frac{1}{2}}$ giving the shock position. In contrast to shock fitting, the idea is now not to attempt to follow the shock movement but to incorporate this recovered information into the general ECG scheme (2.17): there is an independent choice as to whether (2.7) with $\theta > 0$ should be used for recovery between the shocks that have been identified. Taking $\theta = 0$ for simplicity, the effect of introducing the recovered shock defined by (2.24) into the ECG formula (2.17) is just to change the allocation of contributions to the updating process arising from $\Delta_+ f_k^n$ and $\Delta_- f_k^n$. The shock jumps from U_{k-1}^n to U_{k+1}^n so compared with (2.14) the allocation process is applied to $\Delta_- f_k^n + \Delta_+ f_k^n$ and is dependent on a_{k-1}^n and a_{k+1}^n. The other difference from (2.14) is the dependence on the shock position x_S: if for instance $a_{k-1}^n, a_{k+1}^n \geq 0$ and η is sufficiently small then $\tilde{\Phi}_{k+1}^n(x_S^n)=0$ and an allocation as in (2.14a) takes place; but for larger η, $\tilde{\Phi}_{k+1}^n(x_S^n) \neq 0$ and a contribution to U_{k+1}^n results. Thus the general formulae will be like (2.21) for the piecewise linear discontinuous case: as there, Φ_j^n is generally discontinuous at x_S^n and needs to be defined as an average of limits from the left and right. To sum up we give the formula for the simplest case, dropping the superscripts:

$$\begin{aligned} a_{k-1}, a_{k+1} \geq 0 : [f_{k+1}-f_{k-1}]\Phi_k(x_S) \quad &\text{to} \quad U_k \\ [f_{k+1}-f_{k-1}]\Phi_{k+1}(x_S) \quad &\text{to} \quad U_{k+1}. \end{aligned} \tag{2.25}$$

This is sufficient to show that in this case, in the situation envisaged in the derivation of (2.24) and for small time steps, the recovered shock moves with the correct speed as given by the Rankine-Hugoniot condition

$$\frac{x_S^{n+1} - x_S^n}{\Delta t} = \frac{f_{k+1} - f_{k-1}}{U_{k+1} - U_{k-1}}. \tag{2.26}$$

The superscripts to f and U have been omitted here as $U_{k\pm 1}$ are not changed in this simple case. If the CFL condition is satisfied with this shock speed one can show that the monotonicity of U is preserved.

We conclude this section with a case in which shock recovery is clearly necessary, the continuous piecewise linear approximation on a uniform mesh. It is relatively easy to recognise shocks which are at least four mesh widths apart by either (2.22) or (2.23): in contrast to the piecewise constant case, a shock between x_{k-1} and

x_k typically leads to an overshoot of U_{k-1} and an undershoot of U_k so that (2.23) will be satisfied just for $j = k$. One may then obtain the shock parameters as part of a global recovery process as given by (1.6) with $\tilde{u}$ consisting of shocks joined by cubic splines with knots at the original nodes, and this works well. However, an alternative is to use a simple explicit local recovery formula of the form

$$\tilde{u}_L = \langle U, \phi_{k-2} + \phi_{k-1}\rangle/2h, \quad \tilde{u}_R = \langle U, \phi_k + \phi_{k+1}\rangle/2h \tag{2.27a}$$

$$\eta = [h^{-1}\langle U, \phi_{k-1} + \phi_k\rangle - \tfrac{1}{2}\tilde{u}_L - \tfrac{3}{2}\tilde{u}_R]/[\tilde{u}_L - \tilde{u}_R]. \tag{2.27b}$$

To incorporate the recovered shock into (2.17) one can either make use of the results obtained with the discontinuous linear elements or, more straightforwardly, introduce two extra nodes at $x_S \pm \frac{1}{2}\varepsilon$, $\varepsilon \ll h$, and give u a linear variation between them. Then unless recovery is used in the smooth flow, which is hardly necessary, the formula (2.17) is little changed from the standard treatment of continuous piecewise linears: one has only to note that, in the definition of $\tilde{\Phi}_j$ for instance, $\tilde{a}$ has extra nodes compared with ϕ_j.

2.5 Extensions to two dimensions, systems, etc.

It is a simple matter in principle to extend (2.3) and (2.17) into more dimensions with Φ_j defined as in (2.4) by an upwind average of the basis function ϕ_j:

$$\Phi_j^n(\underline{x}) = \frac{1}{|\underline{a}^n(\underline{x})|\Delta t} \int_{\underline{x}}^{\underline{x}+\underline{a}\Delta t} \phi_j(\underline{y})\,d\ell \,. \tag{2.28}$$

For continuous linear elements on triangles the details together with a practical approximate scheme are given in (Morton, 1982). The derivation of formulae for piecewise constants corresponding to (2.14) is perhaps more interesting. From a flux vector $\underline{f}$ one obtains contributions from $\underline{\nabla}\cdot\underline{f}$ just along the edges of each triangle and clearly it is only $\partial_n f_n$, the normal derivative of the normal component of $\underline{f}$, which plays a rôle. Just as in one dimension, one can spread the discontinuity in u and $\underline{f}$ across a thin strip either side of the edge and take limits as the strip shrinks to zero. However, the more complicated geometry now means that several Φ_j may be non-zero along the edge and correspondingly several U_j be affected in the update. A typical situation is shown in Fig. 3, and the update formulae take more the form of those in (2.21) and (2.25). A typical contribution to $U_j^{n+1} - U_j^n$ will be

$$-\Delta t V_j^{-1}[f_n]_e \int_e \tfrac{1}{2}\,[\Phi_j^n(\underline{x}_+) + \Phi_j^n(\underline{x}_-)]\,d\ell, \tag{2.29}$$

where V_j is the area of element j, $[f_n]_e$ is the jump in f_n across the edge and the integral of Φ_j^n along either side of the edge, using corresponding characteristic speeds $\underline{a}_+^n$ and $\underline{a}_-^n$, can be calculated as indicated in Fig. 3: a trapezium is constructed along the edge using the vector $\underline{a}_{\pm}\Delta t$ and the length of the edge is allocated to each element in proportion to the area of the trapezium lying in each element. Recalling that, in one dimension and for convex f, (2.14) is identical with the Engquist-Osher algorithm, it is interesting to compare the formulae resulting from (2.29) with those obtained from their scheme in two dimensions: thus in Osher (1981) only the edge lengths are used rather than the trapezia which reduce to them when $\Delta t \to 0$; and also the normal component a_n of $\underline{a}$ rather than $|\underline{a}|$ is used and zeros in this lead to a splitting of the flux difference as in (2.14c).

Extensions of the ECG schemes to systems of equations is of course both more important and more difficult. From the derivation leading to (2.17), one sees that in effect one is constructing the exact solution at time level $n + 1$ corresponding to the approximate solution at level n using the characteristic relation (2.5) and then projecting it. Thus for piecewise constants the correct generalisation is provided by the method of Godunov (1959) in which the Riemann problem is solved for each discontinuity in $\underline{U}^n$ and the result projected back onto the piecewise constants: or, alternatively and more in line with the use of (2.5), one could say that the Engquist-Osher algorithm, which resolves each discontinuity by using the full set of rarefaction and compression waves for the system on the overturned manifolds created by the crossing characteristics, is the most appropriate generalisation. Furthermore, work is in progress on developing these approaches to piecewise linear discontinuous elements.

However, early experiments indicate that more direct generalisations of (2.17) could form practical and effective alternatives to these established approaches. From considering the characteristic normal form for the system, it is clear that $\Phi_j(\lambda_i)$ must in principle be constructed for each characteristic speed λ_i: these then form a diagonal matrix so that transforming back to the original variables suggests approximation schemes of the form

$$\langle \underline{U}^{n+1} - \underline{U}^n, \phi_j \rangle + \Delta t \langle \partial_x \underline{f}(\underline{U}^n), \underline{\underline{\Phi}}_j^n \rangle = 0 \qquad (2.30a)$$

where

$$S^{-1} \underline{\underline{\Phi}}_j^T S = \text{diag.}\ \{\Phi_j(\lambda_i)\},\ S^{-1}AS = \text{diag.}\ \{\lambda_i\} \qquad (2.30b)$$

and A is the Jacobian matrix of the system. To calculate this exactly is clearly expensive: but recall that in (2.14) with piecewise constants only the signs of λ_i are important and that we have ensured that all the Φ_j are necessarily positive. Thus there is considerable scope for effective approximation, just as the Engquist-Osher and the Roe schemes owe much of their success to approximate solution of the Riemann problem.

Finally, it should be mentioned that though Euler time-stepping is perfectly adequate for the scalar problem (2.1) because the characteristics are straight, for systems it may be desirable to use other time-stepping algorithms: for in (2.30b) not only are the λ_i changing but so is the matrix S. Thus unlike (2.3), (2.30) is only an approximation to evolution over one time step followed by projection, both because of the x-variation of S and the t-variation of both S and λ_i. Some of this might be improved by alternative time-stepping and it is not difficult to calculate what the corresponding Φ_j should be for some of the common schemes.

2.6 Numerical tests

Each of the schemes has been tested with the inviscid Burger's equation and tests continue with the Sod (1978) problem. Fig. 4 shows initial data and its exact evolution for the Burger test: the $\cos^2\frac{1}{2}\pi x$ data forms a shock at $t = 2/\pi$; the ramp tests resolution of a rarefaction shock. Fig. 5 shows that for piecewise constants simple shock recovery is extremely effective in improving accuracy. For continuous linears it works well (Fig. 6) but the shock recognition test used needs improving: Fig. 7 shows that recovery is essential with discontinuous linears to control the overshoots and get useful results.

3. MOVING FINITE ELEMENT SCHEMES

Having explored in the previous section to what extent one can carry forward the best L_2 fit to the true solution on a fixed mesh, and having found it helpful to introduce extra nodes, it is a natural next step to include the node positions in the L_2 fitting. This is the starting point for the development of moving finite element methods such as that of Gelinas, Doss & Miller (1981). With continuous piecewise linear elements and using their notation, one seeks approximate solutions in the form

$$v(x,t) = \sum_{(j)} a_j(t)\ \alpha_j(x,\underset{\sim}{s}(t)), \tag{3.1}$$

where α_j are the usual hat-shaped linear basis functions but based on the set of nodes denoted by the vector $\underset{\sim}{s}$ One then has

$$\partial_t v = \sum_{(j)} [\ \dot{a}_j \alpha_j(x,\underset{\sim}{s}) + \dot{s}_j \beta_j(x,\underset{\sim}{a},\underset{\sim}{s})], \tag{3.2}$$

where the β_j are discontinuous basis functions depending on the amplitudes $\underset{\sim}{a}$ as well as the node positions $\underset{\sim}{s}$. Unlike U^n in (2.2), v in (3.1) lies not in a linear but in a non-linear manifold determined by the parameters $\underset{\sim}{a}$ and $\underset{\sim}{s}$, while $\partial_t v$ lies in its linear tangent space. Equations for $\underset{\sim}{a}$ and $\underset{\sim}{s}$ are obtained by taking an L_2 best fit for $\partial_t v$ in this tangent space: thus for the conservation law (2.1) one obtains

$$\langle \partial_t v + \partial_x f(v), \alpha_j \rangle = 0 = \langle \partial_t v + \partial_x f(v), \beta_j \rangle\ . \tag{3.3}$$

This gives a set of ordinary differential equations for $\underset{\sim}{a}$ and $\underset{\sim}{s}$ to be integrated by an appropriate ODE solver.

In the scalar case this last task is greatly simplified by the observation that if the flux function is quadratic then the nodes exactly follow the characteristics so that $\dot{\underset{\sim}{a}} = 0$ and $\dot{\underset{\sim}{s}} = \text{const.}$ Thus even Euler's method is exact for any time step in this case and quite good enough for many problems.

To deal with shocks, Gelinas et al. introduced spring functions into the objective function, rather than just the L_2 norm, so as to prevent node overtaking. However with moving nodes it does seem natural to capture shocks explicitly. This has been done by Wathen (1982) in work on oil recovery problems. A shock is recognised by neighbouring nodes overtaking one another: and when this happens, two coincident nodes with differing amplitudes are followed by satisfying the Rankine-Hugoniot conditions. Results for a standard model problem using the Buckley-Leverett equations in which $f(v) = v^2/[v^2 + \frac{1}{2}(1-v)^2]$ are given in Fig. 8a.

For a system of equations, corresponding to more than two phases in the oil recovery problem, only one set of nodes is used so that clearly some compromise has to be struck as to which characteristics are followed most closely and so as to be able to recognise shocks by the phenomenon of node-crossing. Such a compromise is introduced by using a matrix weighting function W in the L_2 norm for $\partial_t \underline{v}$. Preliminary results obtained by Wathen for a three phase problem are shown in Fig. 8b.

ACKNOWLEDGEMENTS

I am greatly indebted to Stanley Osher, Phil Roe and Bram van Leer for lengthy discussions on the latest developments in finite difference methods for shock modelling, and for stimulating our application of ECG methods to these problems. Thanks are due to Alan Stokes for the calculations in Section 2 and to Andy Wathen for those in Section 3.

REFERENCES

Barrett, J.W., Moore, G. & Morton, K.W., 1982. Optimal recovery and defect correction in the finite element method. Univ. of Reading, Num. Anal. Report 7/82.

Barrett, J.W., & Morton, K.W., 1980. Optimal finite element solutions to diffusion-convection problems in one dimension. Int. J. Num. Meth. Engng. 15, 1457-1474.

Boris, J.P., & Book, D.L., 1973. Flux corrected transport. I SHASTA a fluid transport algorithm that works. J. Comp. Phys. 11, pp38-69.

Cullen, M.J.P., & Morton, K.W., 1980. Analysis of evolutionary error in finite element and other methods. J. Comp. Phys. 34, 245-268.

Engquist, B., & Osher, S., 1980. Stable and entropy satisfying approximations for transonic flow calculations. Math. Comp. 34, 45-75.

Engquist, B., & Osher, S., 1981. One sided difference equations for non-linear conservation laws. Math. Comp. 36, 321-352.

Fromm, J.E., 1968. A method for reducing dispersion in convective difference schemes. J. Comp. Phys. 3, 176-189.

Gelinas, R.J., Doss, S.K., & Miller, K., 1981. The moving finite element method: applications to general partial differential equations with multiple large gradients. J. Comp. Phys. 40, 202-249.

Godunov, S.K., 1959. A finite difference method for the numerical computation of discontinuous solutions of the equations of fluid dynamics. Mat. Sb. 47, 271-290.

Morton, K.W., 1982. Generalised Galerkin methods for steady and unsteady problems. Proc. IMA Conf. on Num. Meth. for Fluid Dynamics (eds. K.W. Morton & M.J. Baines), Academic Press,(to appear).

Morton, K.W., 1980. Petrov-Galerkin methods for non-self-adjoint problems. Proc. Dundee Conf. on Numerical Analysis, (ed. G.A. Watson), Lect. Notes Math. 773, Springer-Verlag. 110-118.

Morton, K.W., & Stokes, A. Generalised Galerkin methods for hyperbolic equations. Proc. MAFELAP 1981 Conf. (ed. J.R. Whiteman), (to appear).

Osher, S., 1981. Numerical solution of singular perturbation problems and hyperbolic systems of conservation laws. Conf. Proc., North-Holland Math. Studies 47, 179-205.

Roe, P.L., 1981. The use of the Riemann problem in finite difference schemes. Proc. VIIth Int. Conf. on Num. Meth. in Fluid Dynamics, Lect. Notes Phys. 141, Springer-Verlag, 354-9.

Roe, P.L., 1981. Approximate Riemann solvers, parameter vectors and difference schemes. J. Comp. Phys. 43, 357-372.

Rusanov, V.V., 1981. On the computation of discontinuous multi-dimensional gas flows. Proc. VIIth Int. Conf. Num. Meth. in Fluid Dynamics, Lect. Notes Phys. 141, Springer-Verlag, 31-43.

Sod, G.A., 1978. A survey of several finite difference methods for systems of nonlinear hyperbolic conservation laws. J. Comp. Phys. 27, 1-31.

Van Leer, B., 1979. Towards the ultimate conservative differencing scheme V. A second order sequel to Godunov's method. J. Comp. Phys. 32, 101-136.

Wathen, A., 1982. Moving finite elements and applications to some problems in oil reservoir modelling. Univ. of Reading, Num. Anal. Report 4/82.

Zalesak, S., 1979. Fully multidimensional flux-corrected transport algorithms for fluids. J. Comp. Phys. 31, 335-362.

Fig. 1 Upwind averaged test functions for $\mu = 0, \frac{1}{4}, \frac{1}{2}, \frac{3}{4}, 1$.

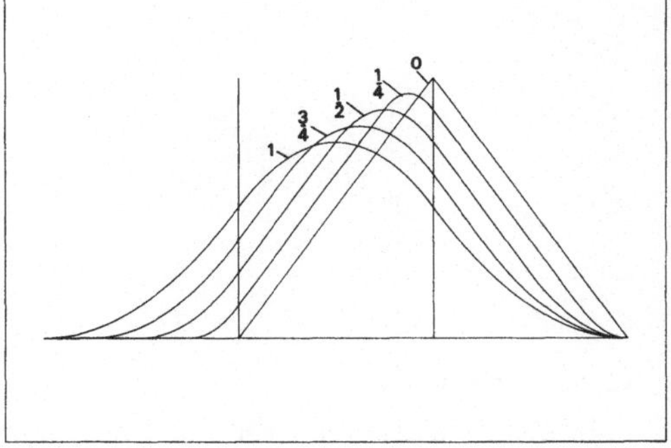

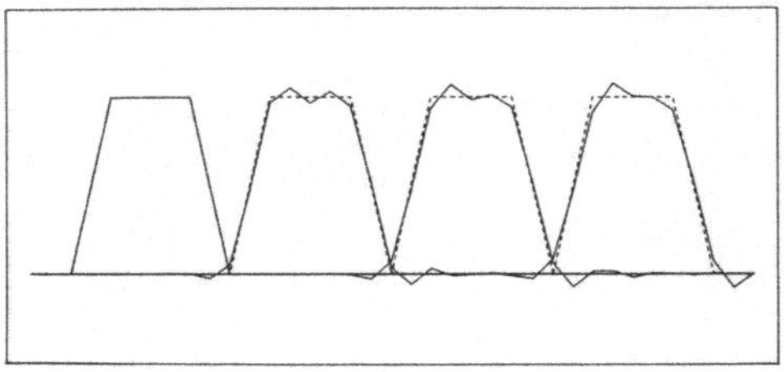

Fig. 2 Linear advection by ECG through 0, 20, 40 time steps with $\mu = 0.8$.

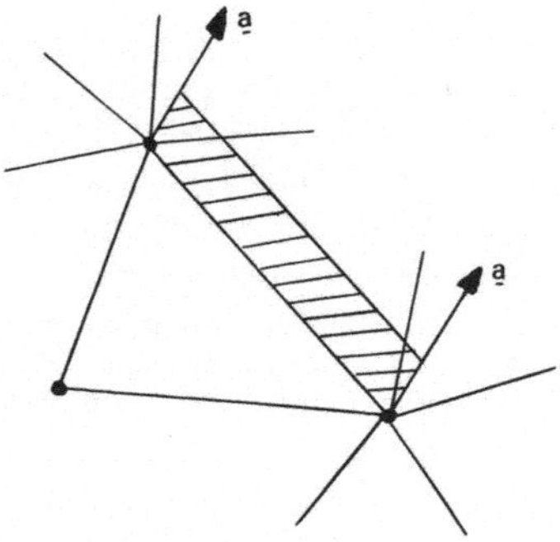

Fig. 3 Allocation of flux differences in 2D.

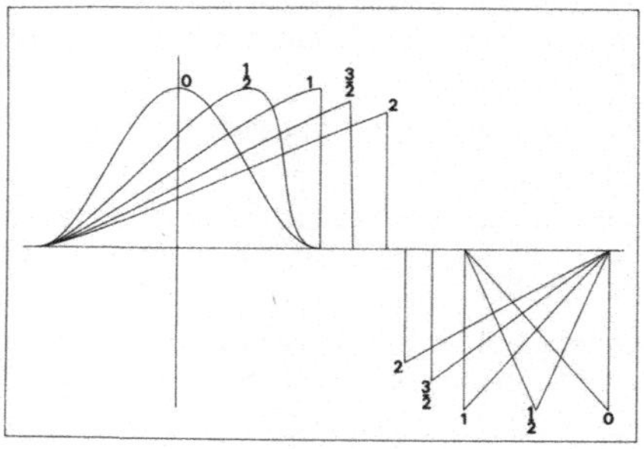

Fig. 4 Exact solution of model problem at $t = 0, \frac{1}{2}, 1, 1\frac{1}{2}, 2$.

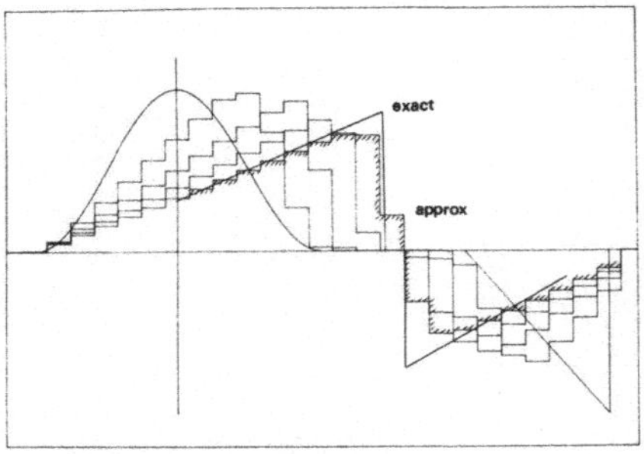

(a) without recovery

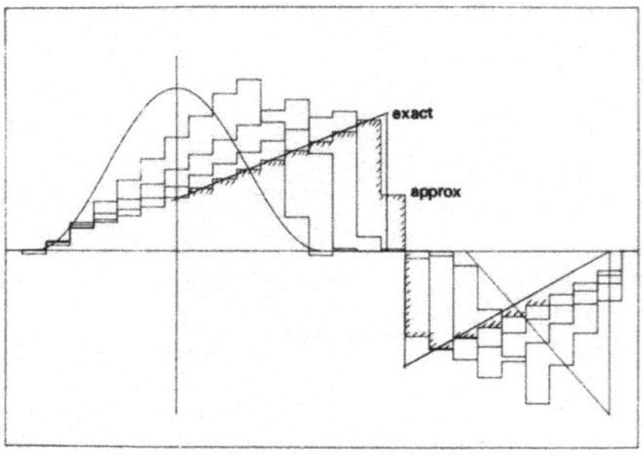

(b) with shock recovery

Fig. 5 Piecewise constant ECG approximation to model problem

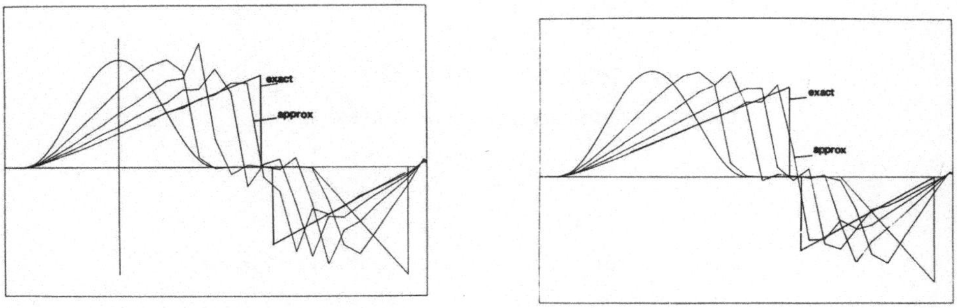

(a) without recovery (b) with shock recovery

Fig. 6 Continuous piecewise linear ECG approximation to model problem.

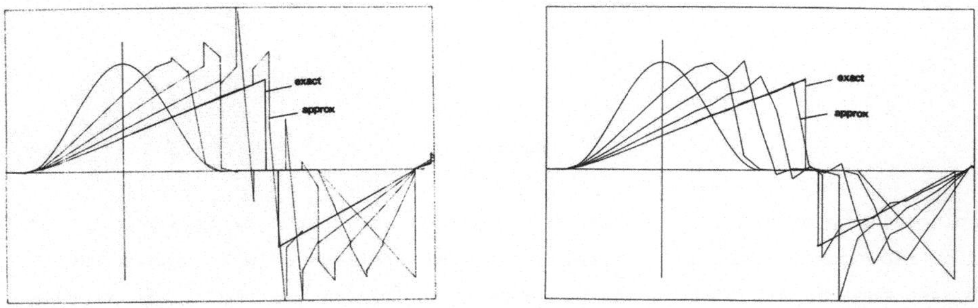

(a) without recovery (b) with shock recovery

Fig. 7 Discontinuous piecewise linear ECG approximation to model problem.

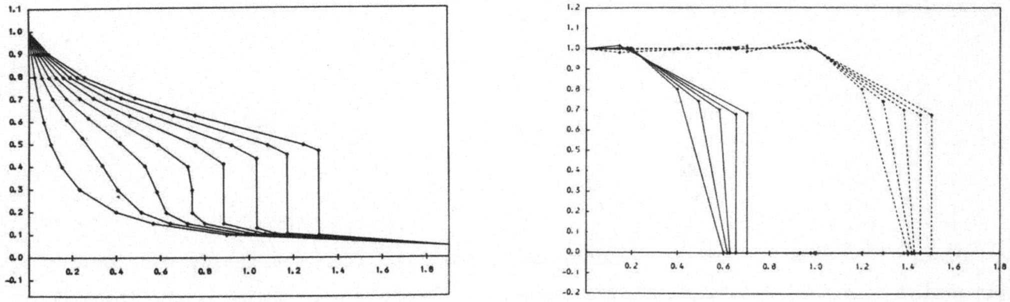

(a) two phase (b) three phase

Fig. 8 Moving finite element approximation to model oil recovery problems.

FLOW SIMULATION BY DISCRETE VORTEX METHOD

by

Koichi OSHIMA

The Institute of Space and Astronautical Science

Komaba, Meguro-ku, Tokyo 153

and

Yuko OSHIMA

Dept. Physics, Ochanomizu University

Ohtsuka, Bunkyo-ku, Tokyo 112

Summary

Applicability of discrete vortex approximation was tested experimentally for four types of flow conditions; an oscillating airfoil, roll-up of wake vortex layer originated from an oscillating plate, an impulsively started flat plate with an angle-of-attack and a two-dimensional rotating elliptic airfoil. Detailed flow visualization reveals the mechanism of creation, growth and migration of vortices and the comparison with those predicted by discrete vortex method has been done. It is concluded that this numerical simulation method is most usefull to predict global feature of the flow fields and care must be taken not to excessively increase the spacial and time resolution.

1. Introduction

Discrete vortex method is to numerically simulate the flow field, considering the potential flow imbedded with finite number of discrete vortices and tracing kinematically their convective motion due to mutual interaction of them in time. In the potential flow field, there is neither creation nor decay of vorticity, then some kinds of modelling concerning the vorticity creation and its decay are inevitable, and the applicability of this modellings has to be checked based on experimental facts, which is the purpose of this paper.

Mathematical foundation and detailed derivation of discrete vortex method have been well doccumented and quite extensive reviews were given, for example, by

P.G.Saffman and G.R.Baker (1979) and by A.Leonard (1980). Therefore, the detailed description of the mathematical manipulation will not be included in this report, rather we concentrate our effort to discuss experimental evidence of such numerical modelling, considering some particular examples of unsteady flow fields such as; an oscillating airfoil, roll-up of wake vortex sheets, an impulsively started flat plate and a rotating elliptic airfoil. Some of them are the quite successful samples of this numerical simulation method. For some other cases, extreme care must be taken to select the suitable values of the computational parameters in order to obtain physically reasonable solution.

Generally say, this method of flow simulation is most suitable to simulate a potential flow field containing large vortical zone in it, where other numerical or analytical methods frequently experience some difficulty. On the other hand, detailed simulation of viscous-inviscid interaction about flow separation points may not be well fitted to this method, which remains to be useful, though. One of the interesting results obtained is simulation of the vortex trapping on a large angle of attack airfoil. Transient trapping of such vortex on the rotating airfoil was observed by this method and it was also visualized in the experiments.

2. Discrete Vortex Method

The first attempt to simulate the flow by a vortex method made by Rosenhead (1931) is to approximate the motion of a two-dimensional vortex sheet by following the movement in time of a system of point vortices. With the advent of computers, many attempts to achieve the better results have been done, but in many cases, the vortices eventually achieved a chaotic state of motion. It now appears that using an increased number of point vortices of decreased strength will not yield a converged solution. Some discussions concerning the convergence of this method have been given by P.T.Fink and W.K.Soh (1978) and later by G.R.Baker (1980). Point vortices are simply too singular, and,instead, K.Kuwahara and H.Takami (1973) proposed to use viscous vortex cores. More recently, many investigators (A.J.Chorin and P.S.Bernard (1973), D.W.Moore (1974), etc.)have used vortices with finite cores or vortex blobs. The structure of them may be constructed accounting for the viscous dissipation in various ways. Thus, almost in any cases, one can obtain a converged flow pattern using some sort of adjustment for the vortex structure, and also claim that the decay of vortices due to the viscous effect is effectively accounted for. However, this is only qualitative argument, and the exact correspondence to real flow has to be proved by experiments.

The flow around a moving body immersed in a uniform flow at infinity creates vortices along its surface due to the viscous effect, which eventually separate into the flow field as free vortices. They are determined using various kind of modelling. For the flows past a sharp corner, the Kutta condition is most conveniently applied, and for those around a smooth body, K.Kuwahara proposed the so-called no-slip condition, which represents the viscous boundary layer along the body surface. Many papers concerning such type of flow have appeared (for example, T.Sarpkaya and R.L.Schoaff (1977), etc.). It is our finding that the former condition sometimes is too strong to give a realistic solution, and the latter, if used with less number of vortices, is able to give most realistic flow.

In principle, the shed vorticity is determined by solving the boundary layer equations and finding out the separation point, as proposed by W.R.Sears (1976). However, this viscous-inviscid interaction problem about separation points is quite difficult one, as reviewed by J.C.Williams, III (1977), even for totally laminar flow (C.A.Koromilas and D.P.Telionis (1980), etc.). Some reports (A.Sugavanam and J.C.Wu (1982), T.J.Mueller and S.M.Bastill (1982), etc.) have appeared, though. Therefore, further efforts to include this viscous-inviscid interaction effect into discrete vortex algorism, it seems to us, are not worthwhile.

The location where the nascent vortices create, the number of them and the time step of integration are the parameters which have to be chosen based only on experience of researchers. In most cases, fortunately, choice of these parameters is not so influencial to the final results. In fact, a parameter set of only two vortex shedding points chosen near the airfoil surface was most successful in our study. Of course, there are cases where choice of the parameters is critical. Examples of such cases will be discussed in this report.

The vortices thus created interact each other and travel through the flow field. They can simulate large dispersed vortical flow region, such as separation blobs or large wake region. The computational procedures are simple and the computer capacity need to solve ordinary fluiddynamical problems is well within reasonable range. This is the most distinguished feature of this method.

3. Flow Around an Oscillating Airfoil

With relations to such phenomena as flutter of airplanes, buffeting of helicopter blades, cascade flow and circulation control of aircraft wing, the flow around an oscillating airfoil has been extensively studied in aerospace society. Complete surveys of these investigations are found in two review papers by W.J.McCroskey (1977) and (1982). For weakly oscillating airfoil, the linear theory is applicable and some higher order theories were also proposed. Nonlinear problem of this type of flow fields have been numerically analysed using discrete vortex method by B.C.Basu and G.J.Hancock (1978), W.Geissler (1978) and J.Katz (1981). Here we consider such flow under large angle-of-attack condition, where the linearized solution fails to be useful. The numerical simulation using discrete vortex method has been carried out by K.Ono, K.Kuwahara and K.Oshima (1980), and the flow pattern around an NACA 0012 airfoil under pitching oscillation with the reduced frequency of 2.2 are reproduced in Fig.1. The experimental study using a towing water tank was also carried out and the flow patterns were visualized by aluminum dust and electrolysis methods, one of which are shown in Fig.2. Apparent similarity between them is evident.

In this numerical simulation, the nascent vortices were created near the leading and trailing edges using the no-slip condition between them, that is, there is no net flow in average along each of the upper and the lower surfaces. The flow separates near the trailing edge at low angle-of-attack position, but its influence to the shed vortex pattern is not critical. The vortices shed at the leading edge clearly show flow separation and following flowing down of the separated blob along the airfoil surface, which exactly corresponds to the experimental observation. The trace of the separted blob were taken from the observed photographs and shown in Fig.3, on which, it is seen, the flowing down of the separated blob along the wing surface, its eventual flowing out into the wake zone, and interaction with the shed roll-up vortices. Thus the discrete vortex approximation is proved to be useful to simulate the flow field containing the large separation region. Its creation, growth, convection and interaction with the other vortical regions were properly simulated.

4. Roll-up of Wake Vortex Sheets

As another example, the motion of vortex sheet originated at the sharp corner under unsteady motion is considered. Rolled-up of the vortex sheet thus produced is widely studied by vortex method, called as the Kaden's problem; Kaden (1931). In fact, this was the first problem solved by discrete vortex method and, comparing with the self-similar solutions available for this problem (D.I.Pullin (1978), D.I.Pullin and W.R.C.Phillips (1981), etc.) and also with many experimental studies carried out (D.I.Pullin and A.E.Perry (1980), Y.Oshima and K.Oshima (1980), etc.), the convergence of this method in this type of flow was well confirmed.

Rolled-up of the wake pattern due to an oscillating airfoil is also similar phenomenon, and widely studied, since J.B.Bratt (1953) reported a series of beautiful photographs. In Fig.4, some examples of such flow patterns were reproduced. (Y.Oshima and K.Oshima (1980)). The wake patterns computed by discrete vortex method in this problem are one of the most successful examples and they agreed quite well with the visualized flow pattern and available analytical solutions.

A discrete vortex calculation was carried out for a flow caused by a sharp corner in oscillating uniform flow. In this case, a flow around a sharp trailing edge of a wedge fixed in the oscillating uniform flow is considered, and the stagnation point is assumed to swing around the trailing edge, and also the Kutta condition was applied at the trailing edge. This is the modelling of an heaving airfoil at the trailing edge flow. The obtained flow pattern shown in Fig.5 is quite reasonable, though, the thrust coefficient obtained shows no drag effect, which is contrary to the experimental observation. In Fig.6, the measured and the calculated thrust coefficients are plotted against the reduced frequency and it is seen that the measured thrust coefficient gives strong negative thrust under low reduced frequency condition. Then, however reasonable is the flow field predicted by the discrete vortex simulation near the trailing edge, the gloval flow character obtained fails to be useful.

5. An Impulsively Started Flat Plate

The flow field past a flat plate with an angle-of-attack impulsively started is one of the most frequently discussed problems in discrete vortex method, for example, by K.Kuwahara (1973) and M.Kiya and M.Arie (1977). For the cases with large angle-of-attack, choice of the computational parameters such as the location and the number of the nascent vortices and the integration time step is not critical, and the resulted flow pattern shows large separation zone and periodic wake structure, and they are also observed experimentally. However, the flow field with an angle-of-attack of about 15 degrees is quite sensitive to these parameter values. Fig.7 gives the flow pattern of such cases calculated using a set of the parameters, and Fig.8 shows the experimentally visualized flow pattern. A large stagnant, vortex region trapped on the whole plate surface obtained computationally was not seen experimentally. Contrary that, the experiment shows continuous vortex sheet shedding, especially the one originated from the leading edge shows a wavy pattern and eventually develops into roll-up pattern, and the vortex sheet originated from the trailing edge flows upstream along the airfoil surface and eventually merges into the trapped vortex region on the surface.

Fig.9 is the results computed using a set of the parameters slightly different from the previous case. Here, the location of the nascent vortex is selected a little further from the leading edge, and the integration time interval is taken a little longer. Then, as seen in the figure, the resulted flow pattern becomes quite similar to the experimental observation. In order to take account of the leading edge sharpness, the local Kutta condition has to be used in this problem, but the solution using the parameters with higher spacial and time resolution gives unrealistic solution. In this case, therefore, accuracy concerning the leading edge sharpness has to be discarded.

6. Rotating Elliptic Airfoil

Autorotation of an airfoil is rather common phenomenon observed in nature, such as falling tree leaf or playing card, and it has also some applications to wind mills. Especially, the autorotation of a two-dimensional symmetrical airfoil has been extensively investigated by E.H.Smith (1971), H.J.Lugt and S.Ohring (1977), J.D.Iversen (1979) and H.J.Lugt (1980). In this report, the flow around a rotating elliptic airfoil has been studied both experimentally and numerically using discrete vortex method, in which the vortex blob model including the viscous dissipation effect was applied. The separation point of the flow is not fixed at particular point on the body and then the two nascent vortices were created near the both edges and the no-slip condition was applied to the region between them.

The experimental investigations were carried out using a wind tunnel and a towing water tank, and the flow fields were visualized by smoke wire method in the former case and electrolysis method in the latter case. In both cases, the model was driven to a constant rotation by digitally controlled stepping motor. The rotating frequency was set from quite low frequency up to slightly over the autorotating frequency without load. Especially, in the wind tunnel test, the aerodynamic forces were measured by the tunnel balance and the torq to drive the airfoil was also measured by a torq meter. All these measurements were carried out synchronizing with the rotation. In the smoke wire visualization experiment, the photographs were taken by strobolight which synchronized with the rotation at various phase angles.

In Figs.10 and 11, the flow pattern calculated by the discrete vortex method and those visualized by the wind tunnel experiment were presented, respectively. The apparent similarity are evident. The characteristic features of these patterns are that the vortex shedding into the wake zone delays, which means the trapping of a free vortex onto the airfoil, and that the large-scale vortex structures are insensitive to the local flow condition, that is, the fine structure of the boundary layer is not important in these flow field. Based on these flow patterns observed, it is found that the driving force of the autorotation is the lift force due to vortex trapping on the downstream side of the airfoil surface during receding cycle. This driving mechanism is also suggested on Fig.12, which shows the calculated moment coefficient during one cycle and the time-averaged force and moment coefficients. In Fig.13, the experimentally measured torq, lift and drag coefficients time-averaged over one rotation cycle were plotted against the rotating frequency. The zero crossing point of this torq curve with positive gradient corresponds to the autorotation frequency without load, which agreed with the experimentally observed autorotation frequency. This type of vortex trapping on a large angle-of-attack airfoil was originally suggested by P.G.Saffman and J.S.Sheffield (1977), and recently discussed by M-K.Huang and C-Y.Chow, keeping in mind its close connection with lift generating mechanism of insect flight and circulation control wings.

Another remarkable feature of these flow patterns is the biharmonic structure of the flow field. In the other words, the flow structure with a half frequency of the rotation is observed experimentally as well as in numerical simulation, as the double trace of the streak lines. The flow survey using hot-wire anemometery also confirms this feature. This corresponds to the lock-in phenomena of an oscillating wake of the bluff body into the rotating frequency. Such phenomenon has been discussed by J.M.R.Graham (1980), L.E.Ericsson (1979), etc. This means that the discrete vortex simulation can reproduce real flow field which was not explicitly or apriori assumed. The flow field around a vibrating bluff body, which was reviewed by P.W.Bearman and J.M.R.Graham (1980), then, may be quantitatively simulated by this method.

7. Conclusions

Experimental studies of four typical flow conditions; an oscillating airfoil, roll-up of wake vortex sheet, an impulsively started flat plate and a rotating airfoil have been carried out using the wind tunnel and the towing water tank, and the applicability of discrete vortex method has been tested, comparing with the corresponding experiments. In the computing process by the discrete vortex method, a set of parameters such as the location of the nascent vortices and the time step of integration has to be chosen on cut-and-try bases. Generally say, these parameters should be selected so as not to improve excessively the spacial and time resolution of the problem domain. Accordingly, the strength of the nascent vortices is better to be determined by the global character of the flow field, not by the local condition. Viscous effect during migration of the discrete vortices can be simulated considering viscous decay of the vortex core.

In conclusion, discrete vortex method can give physically realistic flow pattern, best in global sense. However, its physical reality always has to be checked experimentally. Probably, mathematicians may hate this indefinitness of this method, but fluiddynamicists know that mathematically rigorous solutions of the Navier-Stokes equations are not necessarily physical reality.

Acknowledgements. All the computational and experimental data presented in this report were taken in our laboratory of The Institute of Space and Astronautical Science. The authors are extremely grateful to all the menbers of the laboratory for their devotion. Their names are Drs.K.Kuwahara, K.Ono, A.Natsume, Ho Phung, Messrs.Y.Tokunaga, J.Kato and N.Izutsu. Some data were taken from part of their thesis, and some are not published yet.

References

Baker,G.R. (1980) A test of the method of Fink and Soh for following vortex-sheet motion, J.Fluid Mech. vol.100 pp.209-220

Basu,B.C. and Hancock,G.J. (1978) The unsteady motion of a two-dimensional aerofoil in incompressible inviscid flow, J Fluid Mech. vol.87 pp.159-178

Bearman,P.W. and Graham,J.M.R. (1980) Vortex shedding from bluff bodies in oscillatory flow: A report on Euromech 119, J.Fluid Mech. vol.99 pp.225-245

Bratt,B.A. (1953) Flow patterns in the wake of an oscillating aerofoil, R and M 2773

Chorin,A.J. and Bernard,P.S. (1973) Discretization of a vortex sheet with an example of roll-up, J.Comp.Phys. vol.13 pp.423-429

Ericsson,L.E. (1979) Karman vortex shedding and the effect of body motion, AIAA J. vol.18 pp.935-944

Fink,P.T. and Soh,W.K. (1978) A new approach to roll-up calculation of vortex sheets, Proc.R.Soc.Lond.A. vol.362 pp.195-208

Geissler,W. (1978) Nonlinear unsteady potential flow calculation for three-dimensional oscillating wings, AIAA J. vol.16 pp.1168-1174

Graham,J.M.R. (1980) The forces on sharp-edged cylinders in oscillatory flow at low Keulegan-Carpenter numbers, J.Fluid Mech. vol.97 pp.331-346

Huang,M-K and Chow,C-Y (1982) Trapping of a free vortex by Joukowski airfoil, AIAA J. vol.20 pp.292-298

Iversen,J.D. (1979) Autorotating flat-plate wings: the effect of the moment of inertia, geometry and Reynolds number, J.Fluid Mech. vol.92 pp.327-348

Kaden,H. (1931) Aufwicklung einer unstabilen Unsteigkeitsflach, Ing.Arch. vol.2 pp.149-239

Katz,J. (1981) A discrete vortex method for the non-steady separated flow over an airfoil, J.Fluid Mech. vol.102 pp.315-328

Kiya,M and Arie,M (1977) A contribution to an inviscid vortex-shedding model for an inclined flat plate in uniform flow, J Fluid M. vol.82 pp.223-240

Koromilas,C.A. and Telionis,D.P. (1980) Unsteady laminar separation: an experimental study, J.Fluid Mech. vol.97 pp.347-384

Kuwahara,K. and H.Takami (1973) Numerical studies of two-dimensional vortex motion by a system of point vortices, J.Phys.Soc.Japan vol.34 pp.247-253

Kuwahara,K. (1973) Numerical study of flow past an inclined flat plate by an inviscid model, J.Phys.Soc.Japan vol.35 pp.1545-1551

Kuwahara,K. (1978) Study of flow past a circular cylinder by an inviscid model, J.Phys.Soc.Japan vol.45 pp.292-297

Leonard,A. (1980) Vortex methods for flow simulation, J.Comp.Phys. vol.37 pp.289-335

Lugt,H.J. and Ohring,S. (1977) Rotating elliptic cylinders in a viscous fluid at rest or in a parallel stream, J.Fluid Mech. vol.79 pp.127-156

Lugt,H.J. (1980) Autorotation of an elliptic cylinder about an axis perpendicular to the flow, J.Fluid Mech. vol.99 pp.817-840

McCroskey,W.J. (1977) Some current research in unsteady fluid dynamics-the 1976 Freeman scholar lecture, J.Fluids Engrg. vol.99 pp8-38

McCroskey,W.J. (1982) Unsteady airfoils, Ann.Rev.Fluid Mech. vol.14 pp.285-311

McCroskey,W.J. and Pucci,S.L. (1982) Viscous-inviscid interaction on oscillating airfoil in subsonic flow, AIAA J. vol.20 pp.167-174

Moore,D.W. (1974) A numerical study of the rolled-up of a finite vortex sheet, J.Fluid Mech. vol.63 pp.225-2335

Mueller,T.J. and Batill,S.F (1982) Experimental studies of separtion on a two-dimensional airfoil at low Reynolds numbers, AIAA J. vol.20 pp.457-463

Ono,K.,Kuwahara,K.and Oshima,K. (1980) Numerical analysis of dynamic stall phenomena of an oscillating airfoil by the discrete vortex approximation, 7 Int.Conf.Numeri.Method.Fluid Dyn. Springer-Verlag pp.310-315

Oshima,Y. and Oshima,K. (1980) Vortical flow behind an oscillating airfoil, Theo.Appl.Mech. 15 ICTAM pp357-368

Pullin,D.I. (1978) The large:scale structure of unsteady self-similar rolled-up vortex sheets, J. Fluid Mech. vol.88 pp.401-430

Pullin,D.I. and Perry,A.E. (1980) Some flow visualization experiments on the starting vortex, J.Fluid Mech. vol.97 pp.239-255

Pullin,D.I. and Phillips,R.C. (1981) On a generalization of Kaden's problem, J.Fluid Mech. vol.104 pp.45-53

Rosenhead,L. (1931) The formation of vortices from a surface of discontinuity, Proc.R.Soc.A. vol.134 pp.170-192

Saffman,P.G. and Sheffield,J.S. (1977) Flow over a wing with an attached free vortex, Studies Appl.Math. vol.57 pp.107-117

Saffman,P.G. and Baker,G.R. (1979) Vortex interactions, Ann.Rev.Fluid Mech. vol.11 pp.95-122

Sarpkaya,T. and Schoaff,R.L. (1979) Inviscid model of two-dimensional vortex shedding by a circular cylinder, AIAA J. vol.17 pp.1193-1200

Sears,W.R. (1976) Unsteady motion of airfoils with boundary-layer separation, AIAA J. vol.14 pp.216-220

Smith,E.H. (1971) Autorotating wings: an experimental investigation, J.Fluid Mech. vol.50 pp.513-534

Sugavanam,A. and Wu,J.C. (1982) Numerical study of separated turbulent flow over airfoils, AIAA J. vol.20 pp.464-470

Williams,III,J.C. (1977) Incompressible boundary-layer separation, Ann.Rev.Fluid Mech. vol.9 pp.113-144

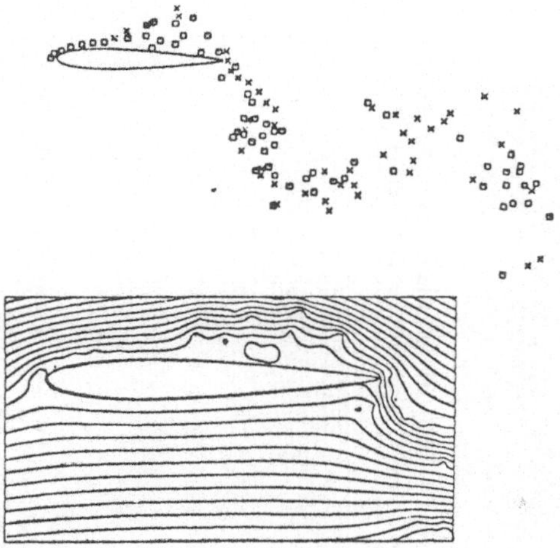

Fig.1 Calculated flow pattern around an oscillating airfoil

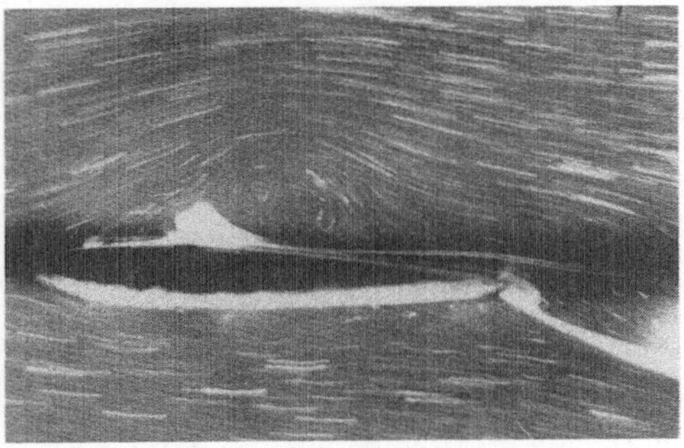

Fig.2 Visualized flow pattern around an oscillating airfoil

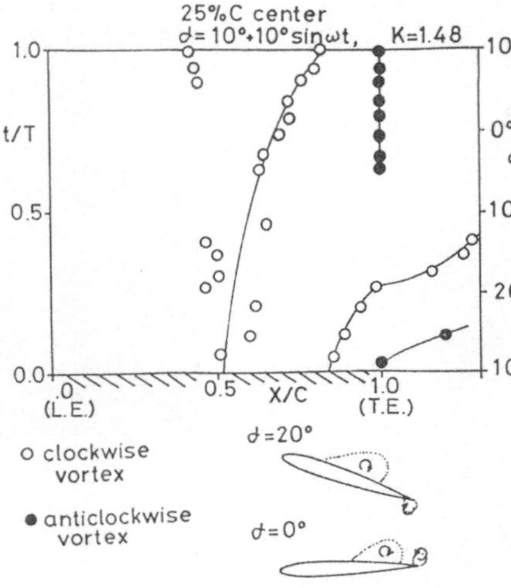

Fig.3 Trace of the separated blob on the airfoil surface

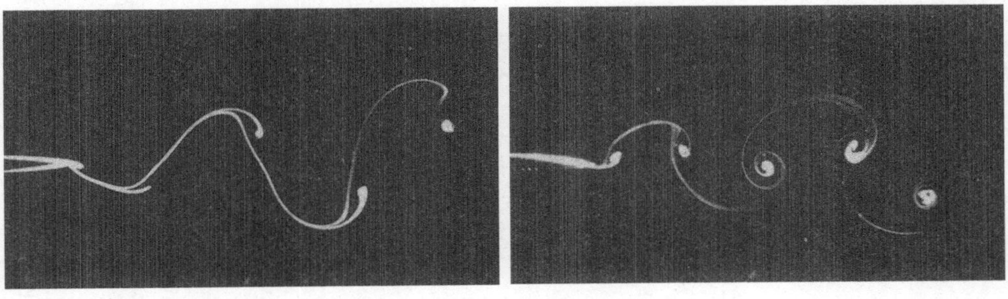

Fig.4 Roll-up of the wake vortex sheet behind an oscillating airfoil

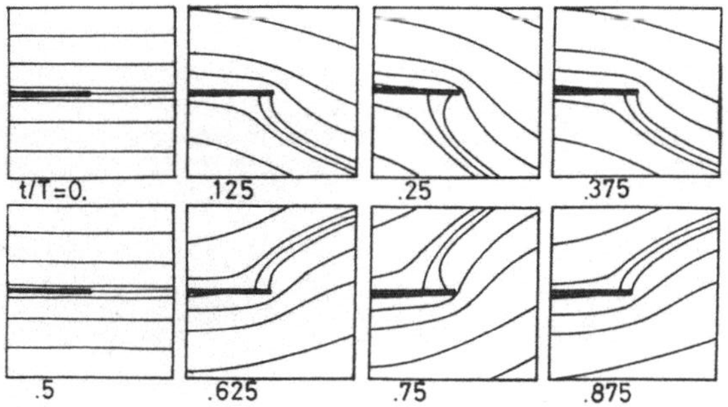

Fig.5 Flow patterns behind an oscillating flat plate

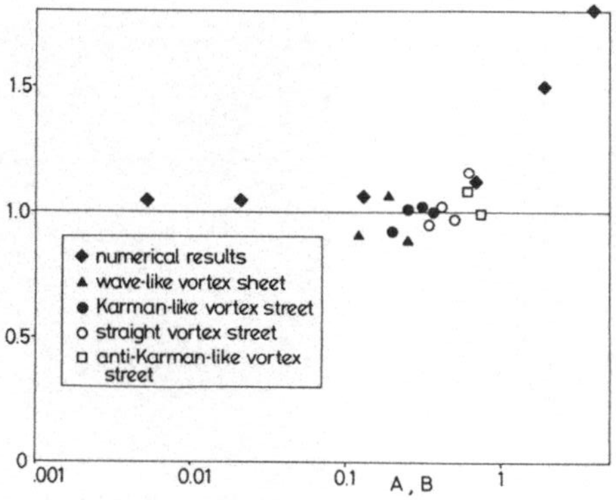

Fig.6 Thrust coefficients calculated and measured

Fig.7 Impulsively started flat plate - Unsuccessful numerical simulation

Fig.8 Flow pattern of impulsively started flat plate observed experimentally

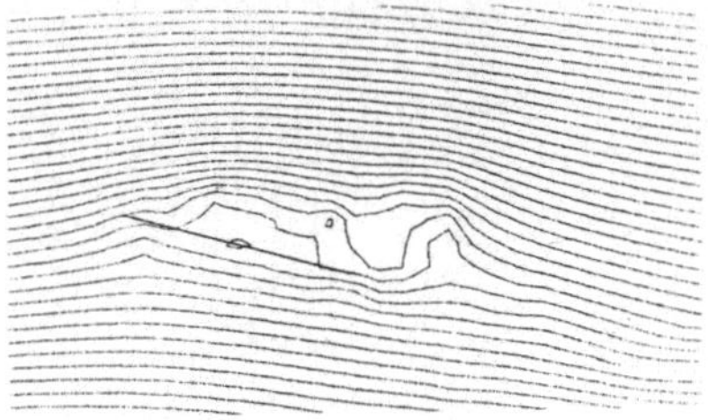

Fig.9 Successfully simulated flow pattern around a flat plate

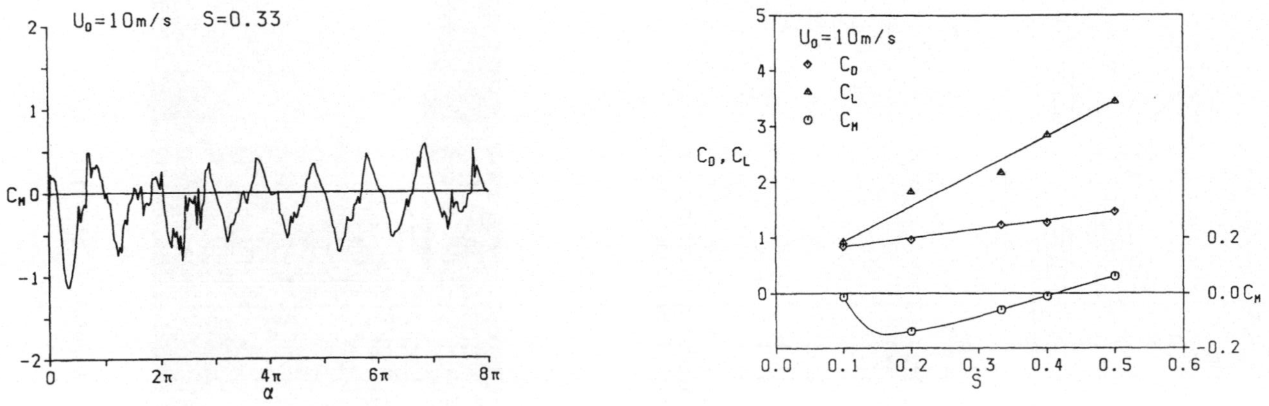

Fig.12 Variation of the moment coefficient during rotation, and the time-averaged force coefficients

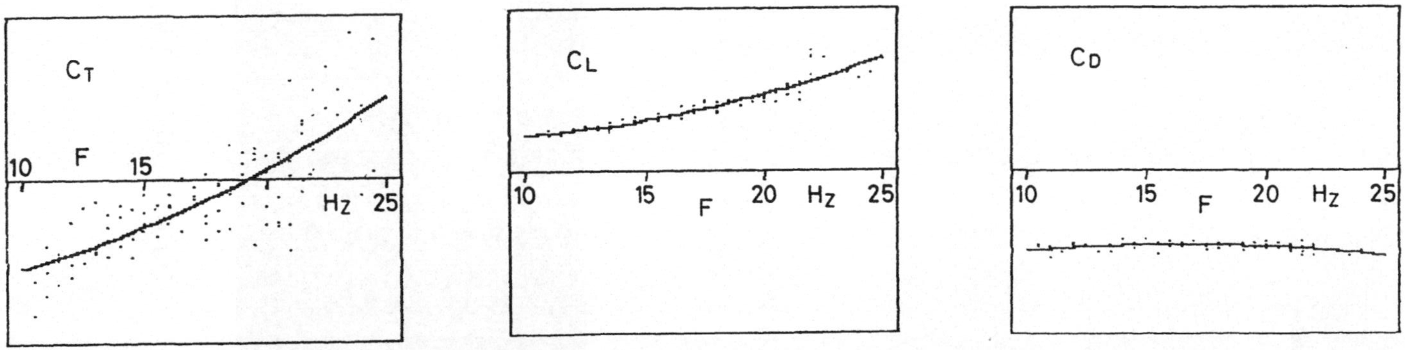

Fig.13 The measured torq, lift and drag coefficients against the rotating frequency

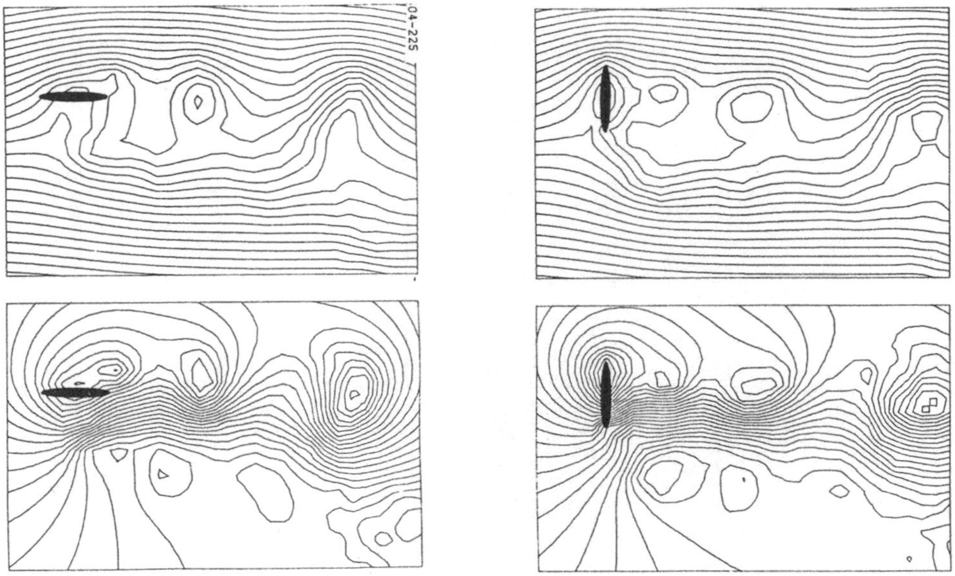

Fig.10 Numerically simulated flow patterns around a rotating airfoil: Instantaneous stream lines with uniform flow (upper figures) and without it (lower figures)

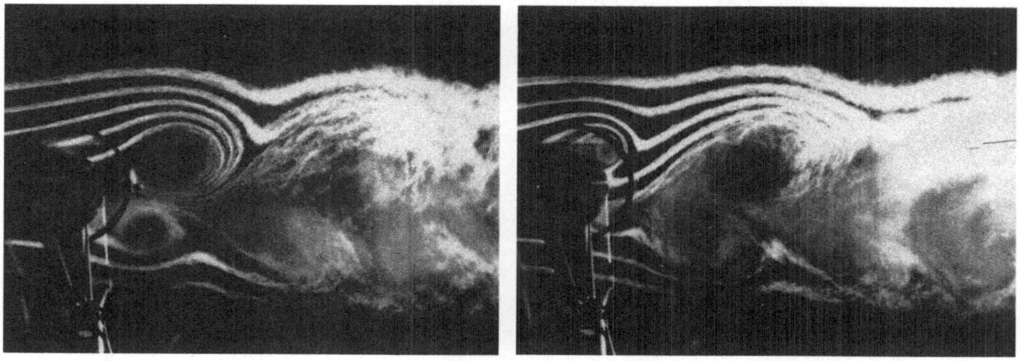

Fig.11 Visualized flow pattern around a rotating airfoil

A GRID INTERFACING ZONAL ALGORITHM FOR THREE-DIMENSIONAL TRANSONIC FLOWS ABOUT AIRCRAFT CONFIGURATIONS

E.H. Atta and J. Vadyak

Lockheed-Georgia Company, Marietta, Ga., U.S.A.

Abstract

An efficient grid interfacing zonal algorithm has been developed for computing the transonic flow field about three-dimensional multicomponent configurations. The algorithm uses the full-potential formulation and the fully-implicit approximate factorization scheme (AF2). The flow field solution is computed using a component adaptive grid approach in which separate grids are employed for the individual components in the multicomponent configuration, where each component grid is optimized for a particular geometry. The component grids are allowed to overlap, and flow field information is transmitted from one grid to another through the overlap region. An overlapped-grid scheme is implemented for a wing and a wing/pylon/nacelle configuration. Numerical results show that the present algorithm is stable, accurate, and can be used effectively to compute the flow field about complex configurations.

Introduction

Reliable and efficient three-dimensional transonic analysis methods are needed to make realistic and cost-effective predictions of aircraft aerodynamics. Early efforts to predict the transonic flow fields about aircraft multiple-component configurations are based on the transonic small disturbance formulation.[1-3] This allows the geometry of the configuration to be greatly simplified and the surface boundary condition to be applied on a mean approximate surface. Accurate prediction of such flow fields requires the use of the full-potential formulation and the generation of a suitable surface-fitted grid. Because each aircraft component (wing, nacelle, fuselage) requires, in general, a grid system that is usually incompatible with the grid systems of the other components, the generation of a single surface fitted grid for the entire configuration is a difficult task. In such a global grid, control of grid point distribution, skewness, and clustering will be difficult to achieve. Efforts to predict the flow field about a complete aircraft configuration using a single grid approach have been made recently by Yu.[4]

In the present paper the alternate approach of using component adaptive grids is investigated. The basic idea of this approach is to develop separate grid systems for the individual components of a complex configuration where each grid is optimized for a particular component. The flow solver for each component employs the fully-implicit approximate factorization scheme AF2.[5] Earlier studies in two dimensions showed that by allowing the component grids to overlap, fast convergence can be achieved.[6] This approach is generalized here to three dimensions, where an overlapped-grid scheme is implemented for an isolated wing and a wing/pylon/nacelle configuration.

Grid Generation

The basic idea underlying the present method is to employ an optimum grid for each component in the multicomponent configuration and then to interface the grids in a manner which allows for the efficient and accurate solution of the governing equations. Use of a separate body-fitted grid for each component greatly facilitates obtaining an accurate numerical solution, and circumvents the need for generating a single grid suitable for a complex multicomponent geometry.

A two-overlapped grid scheme is used to implement the present method for an isolated wing and for a wing/pylon/nacelle configuration. The grid topologies for these two cases are shown in Figs. 1 through 7.

The cartesian grid shown in isolation in Fig. 3 is a sheared grid which employs stretched spacings using a geometric progression in the X, Y, and Z directions. The wing grid in Fig. 2 is essentially identical to the one used in the NASA-AMES TWING code[5]. This grid is obtained by generating a series of two-dimensional O-type grids for a number of spanwise wing stations. Each O-grid is obtained by solving a system of two coupled Laplace equations. For the wing/pylon/nacelle case, an arbitrary region is excluded from the wing grid to accommodate the nacelle grid (see Fig. 7). The boundary of this region represents the overlap region inner boundary. The nacelle grid (see Fig. 6) is obtained using a two-dimensional numerical grid generation technique for a series of meridional planes, each plane containing the longitudinal axis of the nacelle. The grid generation procedure is applied only once for axisymmetric geometries, but is applied for each meridional plane in the case of asymmetric nacelle geometries. The NASA-AMES GRAPE code[7] is used to generate the grid for a given meridional plane. The code generates a two-dimensional grid by solving two coupled Poisson equations. Each meridional plane grid is of the C-type and wraps around the nacelle hilite as shown in Fig. 6.

Governing Equations

The governing equations for steady three-dimensional potential flow in an arbitrary curvilinear coordinate system (ξ, η, ζ) are given by

$$\left(\frac{\rho U}{J}\right)_\xi + \left(\frac{\rho V}{J}\right)_\eta + \left(\frac{\rho W}{J}\right)_\zeta = 0 \tag{1}$$

$$\rho = [1 - \frac{\gamma-1}{\gamma+1} (U\phi_\xi + V\phi_\eta + W\phi_\zeta)]^{\frac{1}{\gamma-1}} \tag{2}$$

where U , V , W are the contravariant velocity components in the ξ, η, and ζ curvilinear coordinate directions, respectively, ρ is the density, J is the Jacobian of the transformation from a cartesian coordinate system to the general curvilinear coordinate system (ξ, η, ζ), and γ is the specific heat ratio. The density and the contravariant velocity components are normalized by the stagnation density and the critical speed of sound, respectively. Equation (1) is the full-potential equation in strong conservation form and expresses mass continuity for three-dimensional steady flows. Equation (2) expresses entropy conservation and is used to determine the density ρ in terms of the velocity potential ϕ.

The contravariant velocity components U, V, and W can be expressed in terms of ϕ as

$$U = A_1\phi_\xi + A_4\phi_\eta + A_5\phi_\zeta \tag{3}$$

$$V = A_4\phi_\xi + A_2\phi_\eta + A_6\phi_\zeta \tag{4}$$

$$W = A_5\phi_\xi + A_6\phi_\eta + A_3\phi_\zeta \tag{5}$$

Expressions for the metric parameters A and the Jacobian J are given by

$$\begin{aligned} A_1 &= \xi_x{}^2 + \xi_y{}^2 + \xi_z{}^2 \quad & A_2 &= \eta_x{}^2 + \eta_y{}^2 + \eta_z{}^2 \quad & A_3 &= \zeta_x{}^2 + \zeta_y{}^2 + \zeta_z{}^2 \\ A_4 &= \xi_x\eta_x + \xi_y\eta_y + \xi_z\eta_z \quad & A_4 &= \xi_x\zeta_x + \xi_y\zeta_y + \xi_z\zeta_z \quad & A_6 &= \eta_x\zeta_x + \eta_y\zeta_y + \eta_z\zeta_z \end{aligned} \tag{6}$$

$$J = \xi_x\eta_y\zeta_z + \xi_y\eta_z\zeta_x + \xi_z\eta_x\zeta_y - \xi_z\eta_y\zeta_x - \xi_y\eta_x\zeta_z - \xi_x\eta_z\zeta_y \tag{7}$$

Numerical Algorithm

The present numerical algorithm uses the finite-difference formulation developed in Ref. 5. The full-potential equation is approximated by the finite-difference expression

$$\overleftarrow{\delta}_\xi \left(\frac{\tilde{\rho}U}{J}\right)_{i+\frac{1}{2},j,k} + \overleftarrow{\delta}_\eta \left(\frac{\bar{\rho}V}{J}\right)_{i,j+\frac{1}{2},k} + \overleftarrow{\delta}_\zeta \left(\frac{\hat{\rho}W}{J}\right)_{i,j,k+\frac{1}{2}} = 0 \tag{8}$$

where i,j,k are the grid point indices in the ξ (wraparound), η (spanwise or circumferential), and ζ (normal) directions, respectively. The operators

$$\overleftarrow{\delta}_\xi(\ ),\ \overleftarrow{\delta}_\eta(\ ),\ \overleftarrow{\delta}_\zeta(\ )$$

are first-order backward difference operators in the ξ, η, and ζ directions, respectively. The terms $\tilde{\rho}$, $\bar{\rho}$, and $\hat{\rho}$ are upwind-biased density coefficients given by expressions of the form

$$\tilde{\rho}_{i+\frac{1}{2},j,k} = (1-\nu)\rho_{i+\frac{1}{2},j,k} + \nu_{i+\frac{1}{2},j,k}\,\rho_{i+\frac{1}{2}+r,j,k} \tag{9}$$

where r denotes an upwind point, and ν is an artificial viscosity coefficient given by

$$\begin{aligned} \nu &= 0 \text{ IF } M_{i,j,k} < 1 \\ \nu &= C\,(M^2_{i,j,k} - 1) \text{ IF } M_{i,j,k} > 1 \end{aligned} \tag{10}$$

where M is the local Mach number, and C is a user-specified constant. Similar expressions hold for $\bar{\rho}$ and $\hat{\rho}$ which effect upwinding in the η and ζ directions, respectively.

The finite-difference equations are solved using the AF2 approximate factorization scheme which has proved to be significantly more efficient than successive-line-overrelaxation schemes. The AF2 algorithm is written in a three-step form as

$$\left(\alpha - \frac{1}{A_k}\overleftarrow{\delta}_\eta A_j \overrightarrow{\delta}_\eta\right) g^n_{i,j} = \alpha\omega L\phi^n_{i,j,k} + A_{k+1}\, f^n_{i,j,k+1} \tag{11}$$

$$\left(A_k \mp \beta_\xi \overleftrightarrow{\delta}_\xi - \frac{1}{\alpha}\overleftarrow{\delta}_\xi A_i \overrightarrow{\delta}_\xi\right) f^n_{i,j,k} = g^n_{i,j} \tag{12}$$

$$(\alpha + \overrightarrow{\delta}_\zeta) C^n_{i,j,k} = f^n_{i,j,k} \tag{13}$$

In equations (11) to (13), α is a free parameter chosen to maintain stability and attain fast convergence, β_ξ is a factor that controls the amount of dissipation required in regions of supersonic flow, ω is a relaxation factor, n is the iteration number, L_ϕ is the mass residual, f and g are intermediate functions which are obtained during the solution process,

and C is the potential function correction given by

$$C^n_{i,j,k} = \phi^{n+1}_{i,j,k} - \phi^n_{i,j,k} \tag{14}$$

The terms A_i, A_j, and A_k are defined by

$$A_i = (\tilde{\rho}A_1/J)^n_{i-\frac{1}{2},j,k} \qquad A_j = (\bar{\rho}A_2/J)^n_{i,j-\frac{1}{2},k} \qquad A_k = (\hat{\rho}A_3/J)^n_{i,j,k-\frac{1}{2}} \tag{15-17}$$

In Steps 1 and 2 the g and f functions are obtained by solving a tridiagonal system of equations while in Step 3 the correction C is obtained by solving a bidiagonal system of equations.

Boundary conditions for the f and g functions on the overlap boundaries were selected such that if ϕ is specified on these boundaries f and g are set to zero while if ϕ_n (normal derivative) is specified then g_η and f_ξ are set to zero.

Boundary Conditions

For solid boundaries such as the wing or nacelle surfaces, the tangency condition is implemented by setting the contravariant velocity component that is normal to the surface equal to zero. For outflow boundaries, such as the compressor face inside the nacelle (see Fig. 6), the contravariant velocity component normal to the outflow surface is specified in terms of the nacelle mass flow ratio.

For the pylon surface, small disturbance boundary conditions were used. Free-stream conditions are imposed at infinity. For the overlap region inner and outer boundaries, the velocity potential ϕ or the velocity potential normal derivative ϕ_n can be specified. Both options were tried and no significant difference in the final results was observed. However, specifying the velocity potential which is a scalar quantity is easier to implement and requires less computational effort compared to specifying the velocity potential normal derivative.

Iteration Procedure

As illustrated in Figures 1 and 5, the component grids are allowed to overlap. For example, in the case of the wing/pylon/nacelle configuration, the inner bound for the wing component grid lies within the nacelle/inlet computational domain. Likewise, the outer bound for the nacelle/inlet component grid lies within the wing computational domain.

The computation is started by initializing the entire velocity potential field. Then the wing algorithm is executed for a specified number of iterations (typically 10) holding ϕ constant on the overlap inner boundary. The nacelle algorithm is then executed for a specified number of iterations using overlap outer boundary conditions as determined from the wing solution. At this stage, the wing algorithm is again executed using updated overlap inner boundary conditions obtained from the nacelle solution. This process is repeated until overall convergence is achieved. Typically 10 to 15 cycles are required for convergence.

To transfer information between the component grids, a trivariate interpolation polynomial based on a linear Taylor series expansion is employed.

Computed Results

The first step to implement the present scheme was to develop a code capable of solving the full-potential equation in a domain containing more than one grid system. Fig. 1 shows such a composite grid around a wing. The inner grid is a surface-fitted grid, while the outer grid is a sheared cartesian grid. Figs. 2 and 3 show both grids in isolation. The velocity potential is specified on the outer boundary of the inner surface-fitted grid and on the inner boundary of the outer cartesian grid. The NASA-AMES full-potential wing code TWING is modified and used as the flow solver for the inner grid. For the outer cartesian grid a full-potential code that solves the flow around box type geometries was developed.

Second-order two-dimensional interpolation was used to feed the velocity potential values between the two grids during the iterative solution process. Performance of the present scheme was evaluated by comparing with results obtained by using the standard TWING code and the Lockheed-SIP code[8]. Both of these codes use a single grid to solve for the transonic flow on a wing. Fig. 8 gives a sample of the results obtained for a nonlifting NACA0012 swept wing. The results are in good agreement. Because the inner curvilinear grid which uses the major portion of the total computer time is now localized in a small region of the flow field,a 15% reduction in computer run time was obtained compared with the single grid code TWING. In the overlapped grid scheme, the inner boundary of the cartesian grid is located one half chord around the wing and the results were obtained using (101x17x11) grid points for the curvilinear grid and (31x17x26) for the cartesian grid. The TWING code solution was computed using (101x17x25) grid points.

A wing/pylon/nacelle code (TWPN) that employs the present scheme was developed. This required the developmednt of a three-dimensional nacelle code and the modification of AMES wing (TWING) code to allow for the inclusion of the nacelle.

The overlapped grid topology for such a configuration is shown in Fig. 5. Presently, the pylon is treated as a vertical slit in the wing grid with small disturbance boundary conditions imposed on it. A separate grid for the pylon is currently under evaluation. Figs. 6 and 7 show the wing and nacelle grids in isolation. The wing grid contains the overlap region inner boundary, while the nacelle grid outer boundary constitutes the overlap region outer boundary. The code is structured to allow for noninteracting or mutual interference run modes, that is, the velocity potential values that are imposed on the overlap region inner boundary in the TWPN code can be fixed or allowed to be updated during the solution process.

To check the code, a converged wing solution was obtained from the AMES wing code TWING. The converged velocity potential values were used as boundary conditions for the overlap region inner boundary in the TWPN code. The code was then executed with different overlap boundary shapes, (see Fig. 9) and in each case the converged wing flow field solution was reproduced.

The three-dimensional nacelle code is currently under development. However, computations were performed for the two axisymmetric inlet configurations shown in Figs. 10 and 11 on relatively coarse grids. These calculations were performed using isolated nacelle grids similar to the one shown in Fig. 6 but which have their outer computational boundaries extended far enough to represent free-stream conditions.

Computed nacelle Mach number distributions for both internal and external surfaces are illustrated in Figs. 10 and 11. Also shown on these figures are the computed results of the Jameson FL049[9] axisymmetric nacelle code. The FL049 code is a two-dimensional, full-potential, finite volume transonic nacelle code which uses the successive-line over-relaxation scheme with multi-grid convergence acceleration. The Mach number distributions produced by the two analyses agree very well. These results were performed on a (128X32) grid for the FL049 code and a (61X13X13) grid for the three-dimensional nacelle code.

To evaluate the performance of the TWPN code for non-interacting wing/nacelle flow fields, a converged nacelle solution was computed using the FL049 code and imposed as boundary conditions for the overlap inner boundary. A 3-D Taylor expansion series was used to interpolate the velocity potential from the FL049 code grid system to the TWPN code grid system. The wing grid in the TWPN code employed (101X26X25) grid points, while the FL049 code used (128X32) grid points.

Figs. 12 to 13 show the interference effect of the nacelle on two transport wing models. The pressure distributions are computed at 5% span inboard and outboard of the nacelle. The nacelle flow field produced a marked change on the wing pressure distribution especially on the wing lower surface. This resulted in a loss of lift. These predictions agree qualitatively with the experimentally observed nacelle interference effect. A mutual interference solution should provide more accurate prediction. The computations were performed on the VAX11/780 mini computer and a typical noninteracting case requires about 2.5-3 hours of CPU time.

Conclusions

An efficient grid interfacing zonal algorithm has been developed to compute the transonic flow field about complex configurations. In the present method the difficult task of generating one global grid for a multicomponent configuration is replaced by the easier task of generating separate grids for each component. The optimized grids produce the desired balance between convergence speed and accuracy of the flow field solution. Numerical results using an overlapped-grid scheme show that the present algorithm is stable and the use of interpolation to transfer flow field information from one grid to another did not degrade the accuracy of the solution.

Acknowledgement

The authors would like to thank Kathleen Hall of Southern Technical Institute for developing the grid plotting routines and George Shrewsbury of the Advanced Flight Sciences Department at Lockheed-Georgia Company for providing the FL049 code test runs.

References

1. Boppe, C. W., and Stern, M. A.: "Simulated Transonic Flows for Aircraft with Nacelles, Pylons, and Winglets," AIAA Paper 80-0130,1980.
2. Shankar, V., and Malmuth, N. D.: "Computational and Simplified Analytical Treatment of Transonic Wing-Fuselage-Pylon-Store Interactions," AIAA Paper 80-0127,1980
3. Srokowski, A. J., Shrewsbury, G. A., and Lores, M. E.: "A Transonic Mutual Interference Program for Computing the Flow about Wing-Pylon/Nacelle Combinations", AIAA Paper 80-1333,1980.
4. Yu, N. J.: " Transonic Flow Simulation for Complex Configurations with Surface Fitted Grids," AIAA Paper 81-1258, 1981.
5. Holst, T., and Thomas, S.: "Numerical Solution of Transonic Wing Flow Fields," AIAA Paper 82-0105, 1982.
6. Atta, E. H.: "Component-Adaptive Grid Interfacing," AIAA Paper 81-0382, 1981.
7. Sorensen, R. L.: "A Computer Program to Generate Two-Dimensional Grids About Airfoils and Other Shapes by Use of Poisson's Equation," NASA TM-81198, 1980.
8. Sankar,N.L.,Malone,J.,and Tassa,Y.:"A Strongly Implicit Procedure For Steady Three-Dimensional Transonic Potential Flows",AIAA Journal,Vol.20, No.5,May 1982.
9. Jameson,A.:"Transonic Flow Analysis For Axially Symmetric Inlets With Center Bodies",Report NO.2 Antony Jameson and Associates,INC.,1981.

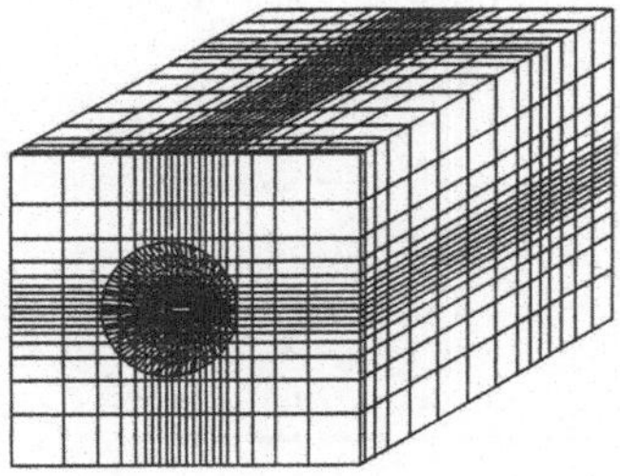

Figure 1. Overlapped Grid Topology For An Isolated Wing

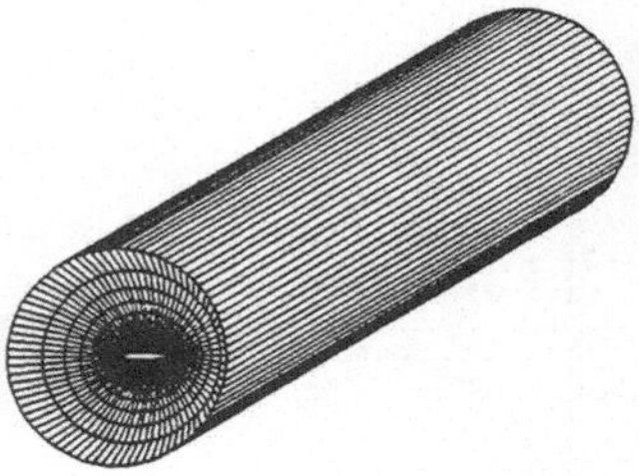

Figure 2. Wing Curvilinear Inner Grid

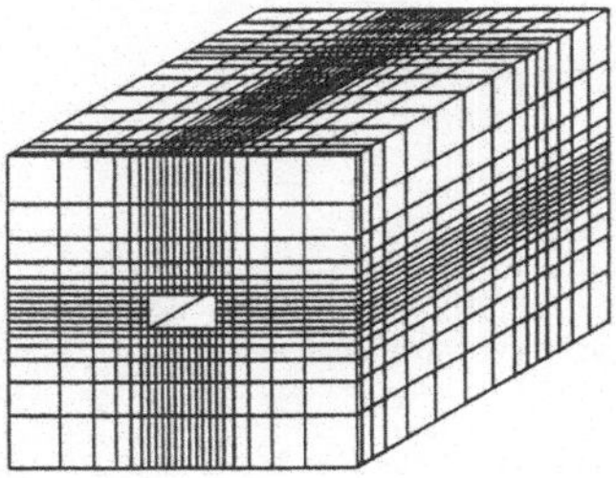

Figure 3. Cartesian Outer Grid

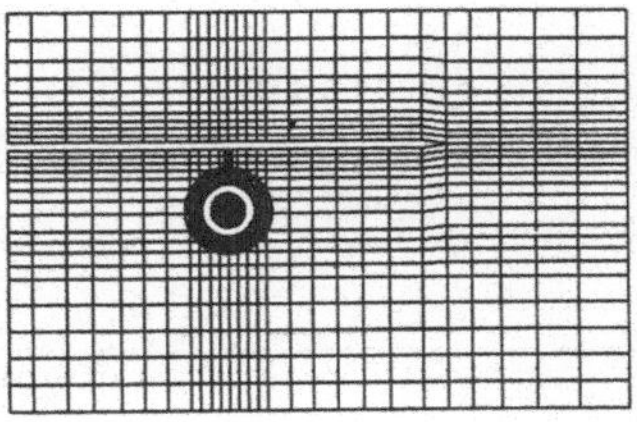

Figure 4. Front View Of Overlapped Wing/Pylon/Nacelle Grid

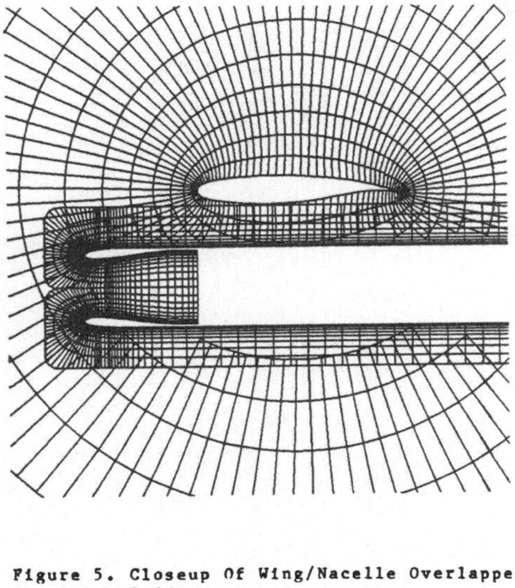

Figure 5. Closeup Of Wing/Nacelle Overlapped Grid

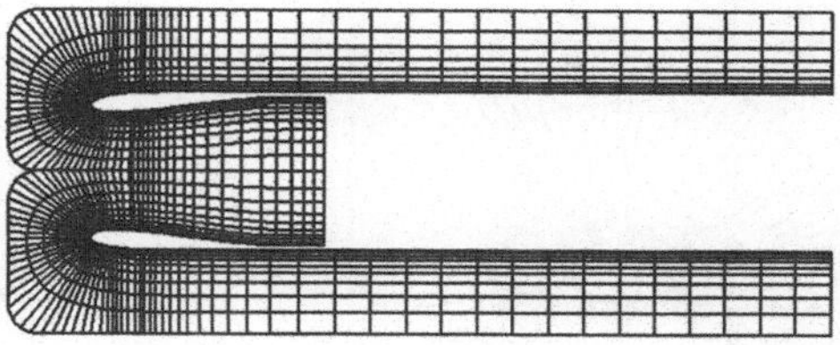

Figure 6. Nacelle Component Grid

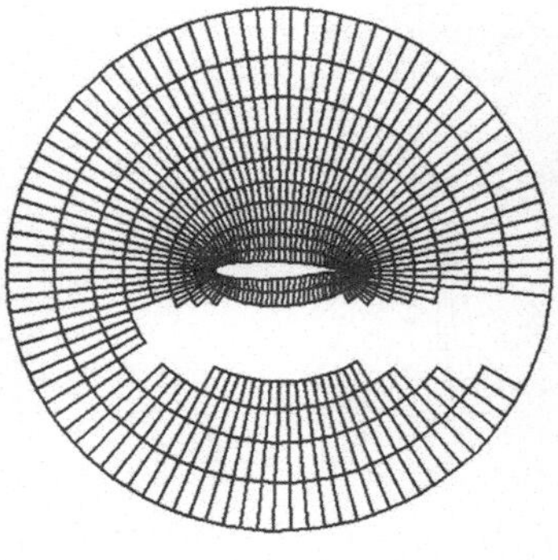

Figure 7. Wing Component Grid

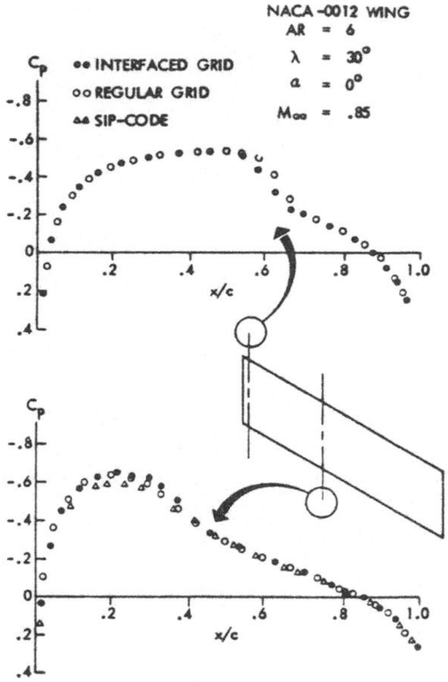

Figure 8. Comparison Of Overlapped Grid Results With Single Grid Results

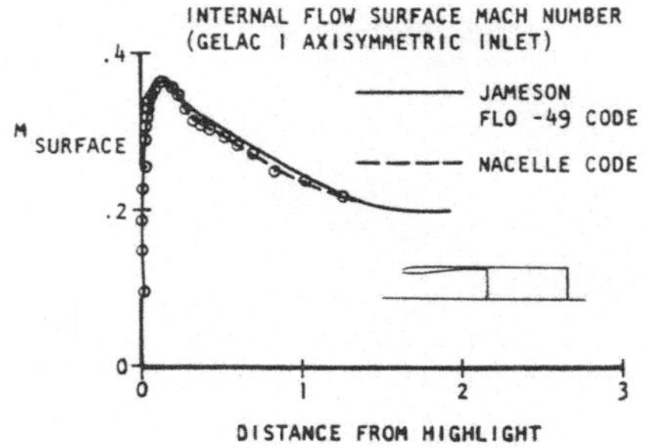

Figure 10. Isolated Nacelle Results

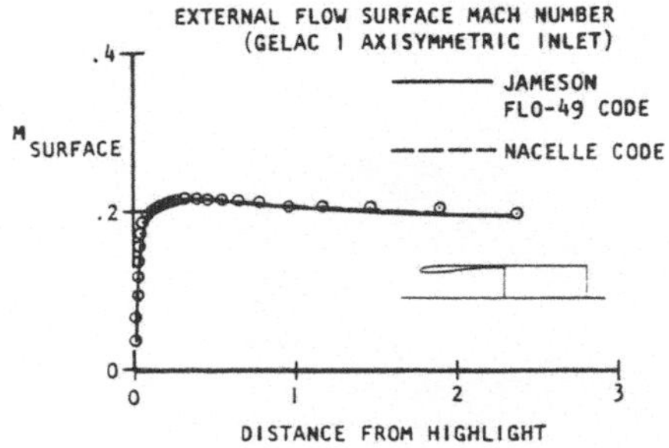

Figure 11. Isolated Nacelle Results

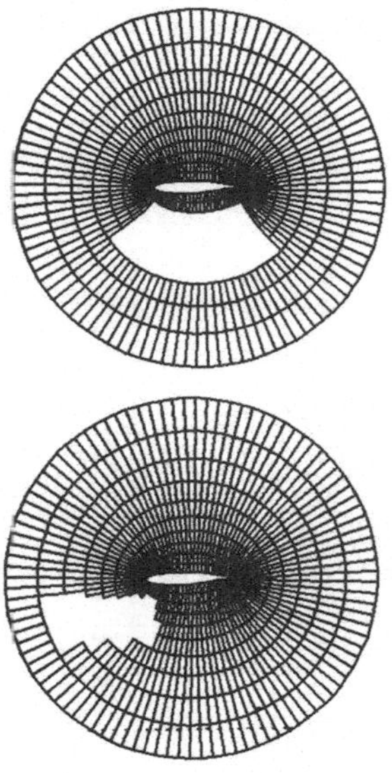

Fig.9 Grid Topology For Different Overlap Boundary

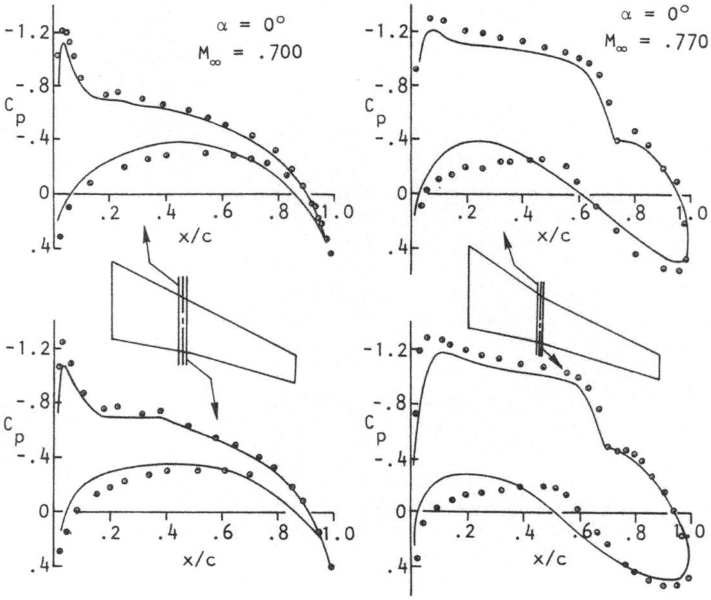

Figs.12,13 Nacelle Interference Effect On Wing Pressure Distribution
(o wing alone, — wing/nacelle)

AN ACCURATE AND EFFICIENT
FINITE ELEMENT EULER EQUATION ALGORITHM

A. J. Baker and M. O. Soliman
Department of Engineering Science and Mechanics
University of Tennessee
Knoxville, TN 37996 USA

SUMMARY

An implicit finite element algorithm is established for numerical solution of the multi-dimensional Euler equations. The theoretical construction employs a multi-pole expansion of the Galerkin-Weighted Residuals concept requiring the error in the semi-discrete (finite element) approximation to be rendered orthogonal to the finite dimensional subspace of H^1 selected to generate the solution. Identification of the tensor matrix product resolution of the Newton iteration algorithm Jacobian, valid for arbitrary degree k of the approximation subspace, and a generalized-coordinates framework amenable to use with any regularizing coordinate transformation, renders the algorithm economically competitive. Numerical results document accurate solutions for mixed subsonic-supersonic flows, including robust resolution of shocks on rather coarse discretizations of R^n, as well as convergence estimates in H^1 and E. The algorithm concept is equally applicable to the complete Navier-Stokes equations.

INTRODUCTION

The past two decades have witnessed a rapidly expanding interest in characterization of high speed flows, principally in aerodynamics. This has fostered formulation of numerical algorithm constructions to solve various forms of the governing partial differential equation system, which in turn has yielded development of computational fluid dynamics (CFD). The true "workhorse" of the CFD aerodynamics community over the past decade has been the "MacCormack algorithm" [1] and variations thereof. This is due principally to the ultimate simplicity of this explicit, predictor-corrector algorithm, and its proven track record for prediction of compressible, supersonic inviscid flowfields. In the split-operator construction, the programming requirements are elementary, and the resultant code runs quite economically. The basic theoretical formulation enjoys continuing refinement, including application to the viscous Navier-Stokes equations, cf. [2,3].

During the past few years, CFD research attention has focused on development of implicit Navier-Stokes and Euler algorithms. Here, in simplest terms, the small time-step restriction of explicit integration is replaced by the requirement to solve matrix equation systems. As a result, principal attention has focused on matrix factorization procedures that reduce large sparse matrices to block-banded forms. The approximate-factorization, implicit finite difference (AFFD) algorithms of Beam and Warming

[4], and Briley and McDonald [5], exemplify the concept. An integral ingredient of the AFFD algorithm construction is definition and use of the "generalized coordinates" description. Basically, the divergence operator in the equation set is transformed into scalar components parallel to principal coordinates of a coordinate transformation "regularizing" the boundary ∂R of the solution domain R^n to the unit square (or cube, n=3). Thompson and coworkers, cf [6], pioneered the concept, the subject of which has become rather popular, cf. [7]. Steger and Pulliam [8] first reported a generalized coordinate AFFD algorithm for a two-dimensional cascade flow, and topical results are summarized by Steger [9].

There are four basic requirements to be met in the construction of <u>any</u> CFD algorithm. Denoting the representative partial differential equation set as $L(q(\vec{x},t))$, the first and obvious requirement is specification of an approximation. The options available include (at least) finite element (FE), finite difference (FD) and/or finite volume (FV). Since the approximation cannot be exact, the second requirement is a statement regarding constraints placed on the generated approximation error. The third requirement is definition of a method for controlling phase dispersion error, and the fourth relates to construction of an efficient matrix equation solution algorithm. The FD and FV constructions differ fundamentally from FE concepts principally in regards to categories 1 and 2, which are lumped together in replacing $L(q(\vec{x},t))$ with difference quotients yielding $L(Q(\Delta\vec{x}_j,\Delta t)) \equiv 0$. In the FE construction, the semi-discrete approximation is made to the solution rather than $L(q(\vec{x},t))$. A vast reservoir exists for selection of members that endow the solution approximation $q^h(\vec{x},t)$ with specific properties, eg., completeness. Then, $L(q^h(\vec{x},t)) \neq 0$ is the error in the governing differential equation. The category 2 specification simply requires this error to be orthogonal to the function space from which q^h was extracted. Calculus and vector field operations permit an exact evaluation of this constraint, yielding the matrix equivalent statement $L(\{Q(t)\}) \equiv \{0\}$. Category 3 is enforced in the FE construction using a multi-pole expansion on the basic error orthogonalization statement, and category 4 is addressed using tensor matrix product approximations. This paper presents the derivation and evaluation of this FE algorithm for the Euler equations.

Regarding category 3, the principal error mechanism in discrete approximations to hyperbolic equations is dispersion error resulting from the distinct phase velocities of the Fourier components. The algorithm requirement is to control (diffuse) this error and the standard FD(FV) practice is to add an artificial diffusion term to $L(q(\vec{x},t))$, cf. [10]. For the FE algorithm, and interpreting $L(q^h(\vec{x},t))$ as a hyper-surface, regions of large phase error are characterized by large changes in surface amplitude and direction of the normal. Noting that $\nabla L(q^h(\vec{x},t))$ characterizes both magnitude and direction of the non-smoothness, the multi-pole FE constraint requires this error also be orthogonal to the approximation sub-space, subject to a constraint set $\vec{\beta}$. This of course yields an "artificial viscosity" term, as well as some additional terms, and it is informative to compare results of the FD and FE formalism distinctions.

The definition of categories 1-3 yields a matrix equation system {F}, that is non-linear unless explicit time-integration has been specified. In all other instances, including a direct steady-state algorithm, and as a function of the dimension of R^n and the number of dependent variables defined in L(q), the Jacobian of {F} is a large albeit sparse matrix. The direct matrix solution is usually ill-advised, and as noted, one FD approach is to construct an approximate factorization. One FE approach, presented and discussed herein, is to construct the tensor matrix product approximation to the Jacobian. Since category 4 is strictly a problem in linear algebra, there exists a wide range of alternative procedures including SOR, SLOR, and ADI. The specific selection can be the prime determining factor in the cost of obtaining the solution. This paper presents and documents the FE algorithm construction for the Euler equations solution on R^n.

FINITE ELEMENT ALGORITHM

The Euler equation set, governing inviscid, rotational compressible flows on R^n, in conservative form and using Cartesian index notation, is

$$L(\rho) = \frac{\partial \rho}{\partial t} + \frac{\partial}{\partial x_j}[m_j] = 0 \quad (1) \qquad L(g) = \frac{\partial g}{\partial t} + \frac{\partial}{\partial x_j}[u_j g + u_j p] = 0 \quad (3)$$

$$L(m_i) = \frac{\partial m_i}{\partial t} + \frac{\partial}{\partial x_j}[u_j m_i + p\delta_{ij}] = 0 \quad (2) \qquad L(p) = p - (\gamma-1)[g - \tfrac{1}{2}u_j m_j] = 0 \quad (4)$$

In equations 1-4, ρ is density, $m_j = \rho u_j$ is the momentum vector, where u_j is the convection velocity, $g = \rho e$ is the specific total energy, and γ is the ratio of specific heats for a polytropic perfect gas. Identifying the dependent variable set $q_\alpha(\vec{x},t) \equiv \{q(\vec{x},t)\} = \{\rho, m_i, g, p\}^T$, equations 1-4 belong to the description

$$L(q) = \frac{\partial q_\alpha}{\partial t} + \frac{\partial}{\partial x_j}\left[u_j q_\alpha + f_j^\alpha\right] + s_\alpha = 0 \qquad (5)$$

where $f_j^\alpha(q_\beta)$ and $s_\alpha(q_\beta)$ are specific non-linear functions of their argument. The partial differential equation system 5 is defined on the n-dimensional Euclidean space R^n spanned by the $\vec{x}$ coordinate system with scalar components x_i, $1 \leq i \leq n$. The solution domain is $\Omega \equiv R^n \times t = \{(\vec{x},t): \vec{x}\varepsilon R^n \text{ and } t\varepsilon[t_0,t)\}$ and the solution domain boundary is $\partial\Omega = \partial R \times t$. Thereupon, the generalized differential boundary condition admitting flow tangency and energy flux constraints is

$$\ell(q_\alpha) = a_1^\alpha q_\alpha + a_2^\alpha \frac{\partial}{\partial x_j} q_\alpha \hat{n}_j + a_3^\alpha(q_\beta) = 0 \qquad (6)$$

where the a_i^α are specified coefficients and $\hat{n}_j$ is the outwards pointing unit normal vector. Finally, q_α on $\Omega_0 = R^n \times t_0$ is $q_\alpha(\vec{x},t_0) = q_\alpha^0(\vec{x})$.

For the finite element algorithm for equations 5-6, the semi-discrete approximation $q_\alpha^h(\vec{x},t)$ to the exact solution $q_\alpha(\vec{x},t)$ is constructed from members of a finite-dimensional subspace of $H_0^1(R)$, the Hilbert space of all functions possessing square integrable first derivatives and satisfying the boundary conditions, equation 6. The typical selection is polynomials truncated at degree k, and defined on disjoint interior subdomains R_e^n forming the discretization of $R^n \equiv \cup R_e^n$. Hence,

$$q_\alpha(\vec{x},t) \approx q_\alpha^h(\vec{x},t) = \sum_{e=1}^{M} q_\alpha^e(\vec{x},t) \equiv \sum_{e=1}^{M} \{N_k(\vec{x})\}^T\{QI(t)\}_e \qquad (7)$$

where the semi-discrete free index "I" denotes q_α^h evaluated at the nodal coordiantes of the discretization UR_e^n at anytime t, and sub- or super-script e indicates pertaining to the $e^{\underline{th}}$ finite element domain $\Omega_e = R_e^n xt$. The elements of the row matrix $\{N_k(\vec{x})\}^T$ are polynomials written on $\vec{x}$, complete to degree k and forming a cardinal basis.

The theoretical finite element algorithm constraint for equation 5-6 requires the semi-discrete approximation error in equation 5 and 6 ie., $L(q_\alpha^h)$ and $\ell(q_\alpha^h)$, as well as the vector gradient $\nabla L^c(q^h)$ to be orthogonal to the function space used to construct q_α^h, ie.,

$$\int_{R^3}\{N_k\}L(q_\alpha^h)d\vec{x} + \vec{\beta}_1\cdot\int_{R^3}\{N_k\}\nabla L(q_\alpha^h)d\vec{x} + \beta_2\int_{\partial R}\{N_k\}\ell(q_\alpha^h)d\vec{\sigma} \equiv \{0\} \qquad (8)$$

Equation 8 defines a system of ordinary differential equations. Integrating using the Θ-implicit algorithm, where $\Theta \equiv \frac{1}{2}$ defines the trapezoidal rule, yields

$$\{FI\} = \{QI\}_{j+1} - \{QI\}_j - \Delta t\left[\Theta\{QI\}'_{j+1} + (1-\Theta)\{QI\}'_j\right] \qquad (9)$$

Equation 8 provides the definition of the derivatives $\{QI\}'$. Hence, equation 9 is a non-linear algebraic equation system, solved using the Newton iteration algorithm.

$$[J(FI)]_{j+1}^p\{\delta QI\}_{j+1}^{p+1} = -\{FI\}_{j+1}^p \qquad (10)$$

The solution set $\{QI\}_{j+1}$ and the Newton Jacobian are defined as

$$\{QI\}_{j+1}^{p+1} \equiv \{QI\}_{j+1}^p + \{\delta QI\}_{j+1}^{p+1} \quad (11) \qquad [J]_{j+1}^p \equiv \frac{\partial\{FI\}_{j+1}^p}{\partial\{QJ\}_{j+1}^p} \qquad (12)$$

For solution domains R^n exhibiting boundaries ∂R not coincident with coordinate surfaces of $\vec{x}$, it is convenient to cast the finite element algorithm statement, equations 8-12, into generalized coordinates. Assuming a dual space $\bar{R}^n$ exists, with boundaries $\partial\bar{R}$ which are coordinate surfaces of a rectangular Cartesian coordinate system $\vec{\eta}$, assures existence of the transformation $x_i = x_i(\eta_j)$. On the discretizatton UR_e^n, and in terms of the finite element tensor product cardinal basis $\{N_k^+(\vec{\eta})\}$, the interpolation of the local transformation is, $x_i^e = \{N_k(\vec{\eta})\}^T\{XI\}_e$ where elements of $\{XI\}_e$, are the nodal coordinates in $\vec{x}$ defining the (vertices of the) domain R_e^n, hence the discretization UR_e^n. Letting J_e denote the Jacobian of the transformation, and employing the chain rule, the first term in equation 8 becomes

$$\int_{R^n}\{N_k\}L(q_\alpha^h)d\vec{x} = S_e\int_{R_e^n}\{N_k\}L(q_\alpha^e)d\vec{x} = S_e\left[\int_{R_e^n}\{N\}\left(\frac{\partial q_\alpha^e}{\partial t} + s_\alpha^e\right)\det J_e\, d\vec{\eta}\right.$$
$$\left. - \int_{R_e^n}\frac{\partial\{N\}}{\partial\eta_k}\left(\frac{\partial\eta_k}{\partial x_j}\right)_e\left(u_j^e q_\alpha^e + f_{\alpha j}^e\right)\det J_e\, d\vec{\eta} + \oint_{\partial\bar{R}_e\cap\partial\bar{R}}\{N\}\left(u_j^e q_\alpha^e + f_{\alpha j}^e\right)\cdot\hat{n}_j d\vec{\sigma}\right] \qquad (13)$$

In equation 13, S_e is the matrix assembly operator projecting contributions to equation 8, computed on R_e^n, into the global matrix statement, equation 9. Define the elemental contravariant convection velocity.

$$\bar{u}_k^e \equiv \det J_e\left(\frac{\partial\eta_k}{\partial x_j}\right)_e (m_j^e/\rho^e) \qquad (14)$$

Substituting for q_α^e using equation 7, and rearranging scalar terms in equation 13, yields definitions of integrals of products and derivatives with respect to η_k of elements of $\{N_k\}$, which are easy to evaluate, cf. [11]. Name those matrices by the convection [M $\alpha\beta\beta$..], where α equals the number of cardinal bases $\{N_k\}$ defined in the integral, and $\beta \equiv (0,K)$ is a Boolean index denoting undifferentiated elements in $\{N_k\}$, or differentiated with respect to η_k. The resultant FORTRAN statement equivalent to equation 13, upon noting that $f^e_{\alpha j} = p^e_\alpha \delta_{ij}$, is

$$\int_{R^n} \{N_k\} L(q^h_\alpha) d\vec{x} = S_e\Big[DET_e[M200]\big(\{QI\}'_e + \{SI\}_e\big) - \{UBARK\}^T\big([M30K0] - \hat{n}_k[B3000]\big)\{QI\}_e - \big(ETAKI_e[M2K0] - \hat{n}_I DET_e[B200]\big)\{P\}_e\Big] \quad (15)$$

In equation 15, DET_e is the scalar measure of R^n_e, $ETAKI_e$ represents the scalar components of $\partial\eta_k/\partial x_i$ on R^n_e, and $\hat{n}_I$ is the unit outward normal and [B··] denotes an integral evaluation on $\partial R_e \cap \partial R$. The matrices [·3···] are hypermatrices of degree one, with elements that are themselves column matrices of rank equal to that of $\{UBARK\}_e$, the contravariant convection velocity nodal distribution on R^n_e. The remaining terms in equation 10 are evaluated in this manner, cf. [11].

Equation 12 defines the Jacobian formulation for the Newton matrix iteration solution algorithm, equation 10. Formation of exact Jacobian is ill-advised, especially for problems defined on R^3. The finite element algorithm employs the tensor matrix product decomposition of [J] as

$$[J(FI)] \equiv S_e\Big[[J(FI)]_e\Big] = S_e\Big[[J_1]_e \otimes [J_2]_e \otimes [J_3]_e\Big] \quad (16)$$

Defining two intermediate solutions {PI}, the Newton algorithm statement 10 becomes replaced by the sequence of three solutions,

$$\begin{aligned} S_e\Big[[J_1]_e\{P1\}^{p+1}_e &= -\{FI\}^p_e\Big] \\ S_e\Big[[J_2]_e\{P2\}^{p+1}_e &\equiv \{P1\}^{p+1}_e\Big] \\ S_e\Big[[J_3]_e\{\delta QI\}^{p+1}_e &\equiv \{P2\}^{p+1}_e\Big] \end{aligned} \quad (17)$$

Construction of the matrix elements of each $[J_\eta]$ is easy, since equation 9 is an analytically-differential function of {QJ}. For example, noting in equation 15 that $\{QI\} \rightarrow \{QI\}^{p+1}_{j+1} - \{QI\}_j$ when inserted into equation 9, the lead term in the Newton Jacobian is

$$\frac{\partial\{FI\}}{\partial\{QJ\}} = S_e\Big[DET_e[M200]\delta_{IJ} + \cdots\Big] \quad (18)$$

where δ_{IJ} is the Kronecker delta. The element matrix $[M200]_e$, as defined on R^3_e, is replaced in terms of the elemental tensor product matrices $[A200_n]$ as

$$DET_e[M200]_e = DET_1[A200_1] \otimes DET_2[A200_2] \otimes DET_e[A200_e] \quad (19)$$

For the rectangular mesh on $\bar{R}^3$, it is an elementary operation to prove that the assembly of equation 19 yields an identity. Most importantly, from the solution economy standpoint, each tensor matrix Jacobian $[J_\eta] = S_e[J_\eta]_e$ is block tridiagonal (pentadiagonal) for k=1 (2) in equation 7, in distinction the sparse (but very large) structure of [J].

ACCURACY, CONVERGENCE AND EFFICIENCY

A linearized theoretical analysis [11] yields an estimation of the multi-pole expansion term $\vec{\beta}_1$ in equation 8. The generalized coordinates form is,

$$\vec{\beta}_{1e}^{\alpha} \equiv \det J_e[\vec{\nu}_\alpha^1\, \delta t + \vec{\nu}_\alpha^2\, \delta x]$$

where $\det J_e$ is the measure of R_e^n, the parameter vectors $\vec{\nu}_\alpha^i$ have magnitudes of the order $\nu \equiv (15)^{-\frac{1}{2}}$, and δt and δx are limit operators for the temporal and spatial terms in D/Dt. The refinement of this elementary estimate is accomplished by numerical solution of the one-dimensional Riemann shock tube specification of [12]. Figure 1 summarizes the k=1, M=200 discretization solution for ρ^h, m_1^h and p^h for the theoretical definitions $\nu_\alpha^1 = (15)^{-\frac{1}{2}} = \nu_\alpha^2$ for each q_α^h. Excessive overshoot at the shock is indicated, which is effectively removed without inducing excessive artificial diffusion by the refined definitions $\nu_\alpha^1 = \nu\{3/8,0,1/4\}$ and $\nu_\alpha^2 = \nu\{3/4,2,1\}$. Figure 2 summarizes this solution, which is crisp in every detail regarding resolution of the shock, contact discontinuity, and rarefaction wave. Using this definition for $\vec{\nu}_\alpha^i$ in the k=1 algorithm, and defining $\nu_\alpha^1 = \nu\{0,0,0\}$ and $\nu_\alpha^2 = \nu\{1/4,3/4,1/2\}$ for the k=2 algorithm, the results of a numerical study measuring semi-discrete approximation error convergence in H^1 and energy (E) with discretization refinement (M) are summarized in Figure 3. The data for the k=2 algorithm solution, interpolated by the dashed lines, lies uniformly above the corresponding k=1 algorithm data for each q_α^h, confirming the superior resolution of solution gradients evident in an "eyeball norm," eg., Figure 2. Importantly, these solution data adhere to the convergence curves even for the extremely coarse discretizations $12 \leq M \leq 25$. A modest convergence slope distinction exists among the various q_α^h, but none exists among the $1 \leq k \leq 2$ solutions. The common nominal slope of 2/3 is in qualitative agreement with that for a $p^{\underline{th}}$ order accurate finite difference scheme, ie.,

$$\|q^h\| \leq C\, \Delta^{p/p+1}$$

where C is a constant independent of the mesh measure Δ and p = 2.

Reverting to the "eyeball norm," significant distinctions exist between the k=1 finite element algorithm solution, and the equivalent complexity, second-order accurate AFFD algorithm, which on one-dimensional space is the familiar Crank-Nicolson form. For an M=100 discretization, Figure 4 compares the FE and FD solution predictions of convection velocity u_1^h and temperature T^h. The light solid line is the theoretical solution, to which the k=1 FE algorithm results are a close interpolation. In distinction, the FD solution for u_1^h shows excessive overshoot, as well as interpolation of the shock over five domains. The FD solution for T^h also exhibits excessive overshoot and a poor high temperature plateau behind the shock. Increasing the level of $\vec{\nu}_\alpha^2$ would smooth the overshoot, at the expense of additional artificial diffusion of the shock.

Discretization of this refinement on R^n, $n < 1$, are impractical in terms of solution costs on present computers. The FE algorithm appears to retain its sharp resolution capacity on the progression to coarser discretizations. For example, Figure 5 summarizes the k=1 FE solution for u_1^h for M=50 and M=25 elements. A modest increase in overshoot is exhibited, but the shock remains crisply resolved. The Riemann shock tube specification has been extended, to permit evaluation of the FE algorithm on R^2 and R^3. For example, Figure 6 summarizes the k=1, M = 32 x 6 solution for the shock tube aligned at an arbitrary angle ϕ on R^2. Since the resolution of m_i^h is on $\vec{x}$, two components are computed, neither of which distinctly exhibits the shock, Figures 6a-b. However, computation from m_i^h of the convection velocity component u^h parallel to the tube confirms the shock location, Figure 6c. A similar prediction occurs on R^3, as well as excellent resolution of the shock oblique to the mesh, cf. [11]. The total number of node point dependent variable degrees of freedom for the three-dimensional case was 3200. The Newton algorithm tensor product Jacobian formulation exhibited almost quadratic convergence, and the total central memory requirement was 64 K words.

REFERENCES

1 MacCormack, R.W., "The Effect of Viscosity in Hypervelocity Impact Cratering," Technical Paper AIAA-69-354, 1969.

2 MacCormack, R.W., "An Efficient Explicit-Implicit-Characteristic Method for Solving the Compressible Navier-Stokes Equations," Proc. SIAM-AMS Symp. Comp. Fluid Dyn., New York, 1977.

3 MacCormack, R.W., "A Numercical Method For Solving the Equations of Compressible Viscous Flow," Technical Paper AIAA-81-0110, 1981

4 Beam, R.M. and Warming, R.F., "An Implicit Factored Scheme for the Compressible Navier-Stokes Equations," AIAA J., V. 16, 1978, pp. 393-402.

5 Briley, W.R. and McDonald, H., "Solution of the Multi-Dimensional Compressible Navier-Stokes Equations by a Generalized Implicit Method," J. Comp. Phys., V. 24, 1977, pp. 372.

6 Thames, F.G., Thompson, J. F., Mastin, C. W., and Walker, R.L., "Numerical Solutions for Viscous and Potential Flow About Arbitrary Two-Dimensional Bodies Using Body-Fitted Coordinate Systems," J. Comp. Phys., V. 24, No. 1, 1977, pp. 245-273.

7 Proceedings, NASA Workshop On Numerical Grid Generation Techniques for Partial Differential Equations, NASA Report CP-2166, 1980.

8 Steger, J.L. and Pulliam, T.H., "An Implicit Finite Difference Code for Inviscid and Viscous Cascade Flow," Technical Paper AIAA 80-1427, 1980.

9 Steger, J.L., "Finite Difference Simulation of Compressible Flows," Presented at ASME-AIAA Symposium on Computers In Flow Predictions and Fluid Dynamics Experiments, ASME Winter Annual Meeting, Washington, DC, Nov. 1981.

10 von Neumann, J. and Richtmyer, R.D., "A Method For the Numerical Calculation of Hydrodynamic Shocks," J. Appl. Phys., V. 21, 1950, pp. 232-237.

11 Baker, A.J., "Research On A Finite Element Numerical Algorithm for the Three-Dimensional Navier-Stokes Equations," USAF Report AFWAL-TR-82-3012, 1982.

12 Sod, G.A., "A Survey of Several Finite Difference Methods for Systems of Non-Linear Hyperbolic Conservation Laws," J. Comp. Phys., V. 27, pp. 1-31, 1978.

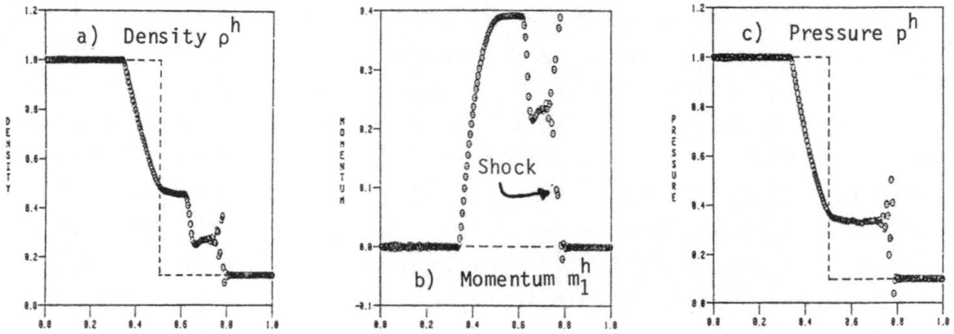

Figure 1. Finite Element Algorithm Solution, Riemann Shock Tube, k=1, M=200, t = 0.14154, $\nu_\alpha^1 \equiv \nu \equiv \nu_\alpha^2$, $\nu_m^1 \equiv 0$, (---) Denotes Initial Conditions.

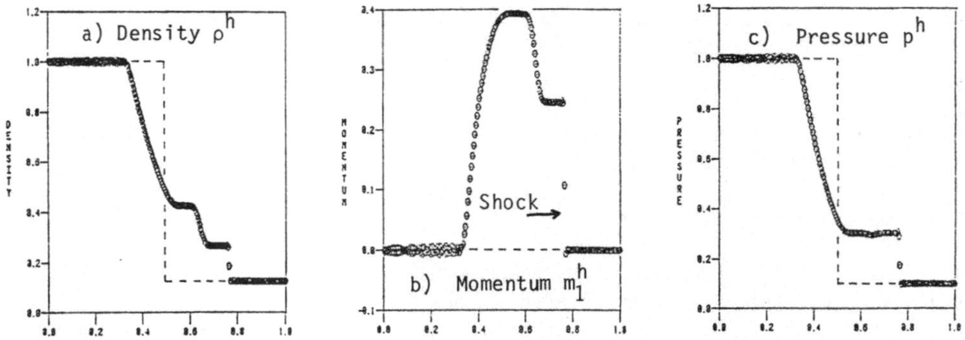

Figure 2. Finite Element Solution, Riemann Shock Tube, k=1, M=200, $\nu_\alpha^1 = \nu\{3/8,0,1/4\}$, $\nu_\alpha^2 = \nu\{3/4,2,1\}$, t = 0.14154s, (---) Denotes Initial Condition.

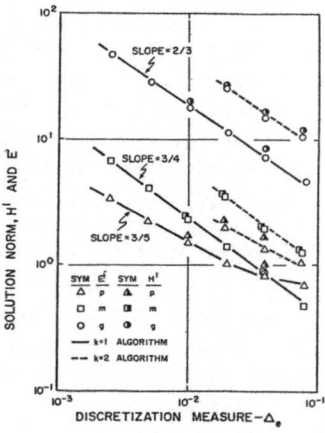

Figure 3. Semi-Discrete Approximation Accuracy And Convergence in $\|q_\alpha^h\|^2_{H^1}$ and $\|q_\alpha^h\|_E$, Finite Element Algorithm Solution For Riemann Shock Tube.

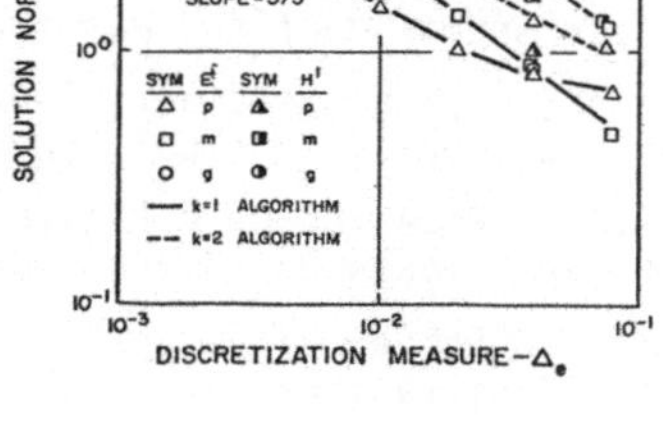

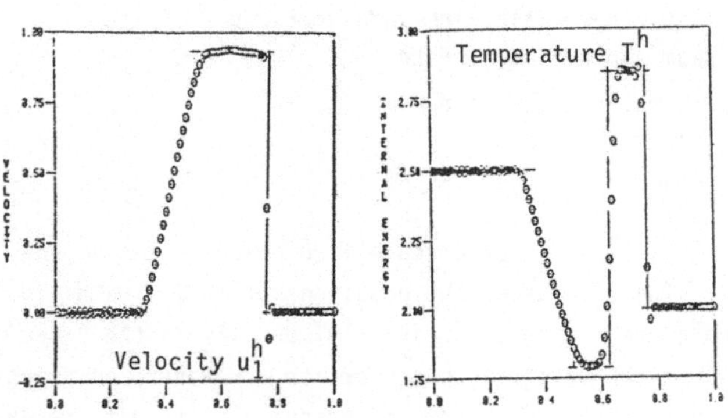

a) M=100, k=1 Finite Element Solution, $\vec{\nu}_{opt}$

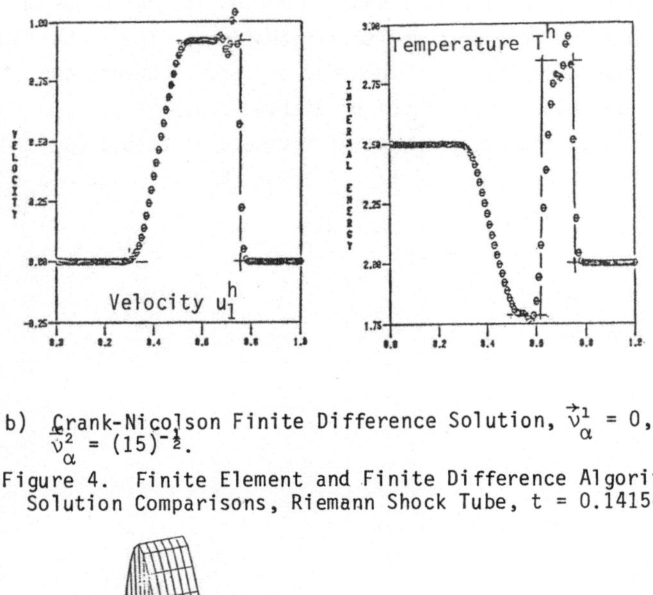

b) Crank-Nicolson Finite Difference Solution, $\vec{\nu}_\alpha^1 = 0$, $\vec{\nu}_\alpha^2 = (15)^{-\frac{1}{2}}$.

Figure 4. Finite Element and Finite Difference Algorithm Solution Comparisons, Riemann Shock Tube, t = 0.14154s.

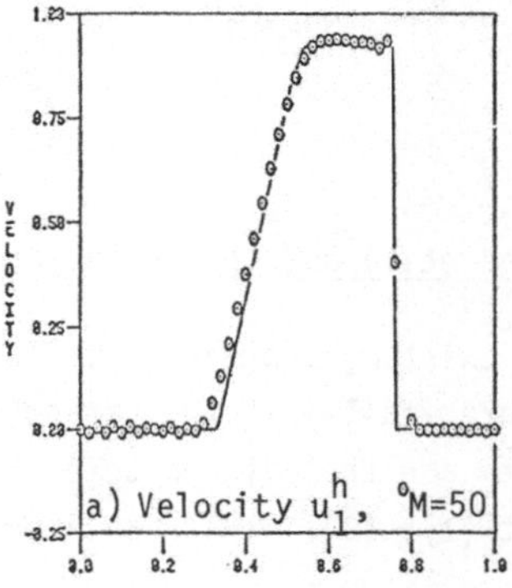

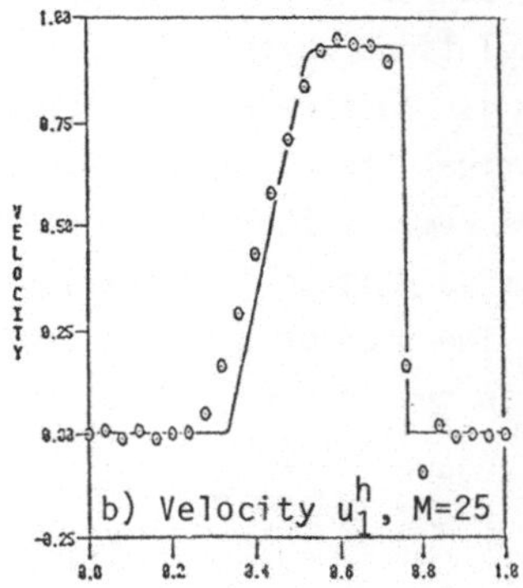

Figure 5. Finite Element Solution Velocity Distribution For Riemann Shock Tube, k=1, $\nu_\alpha^1 = \nu_{opt}$.

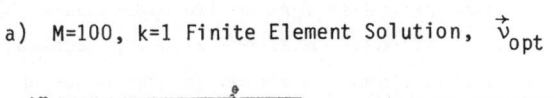

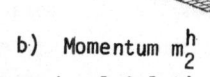

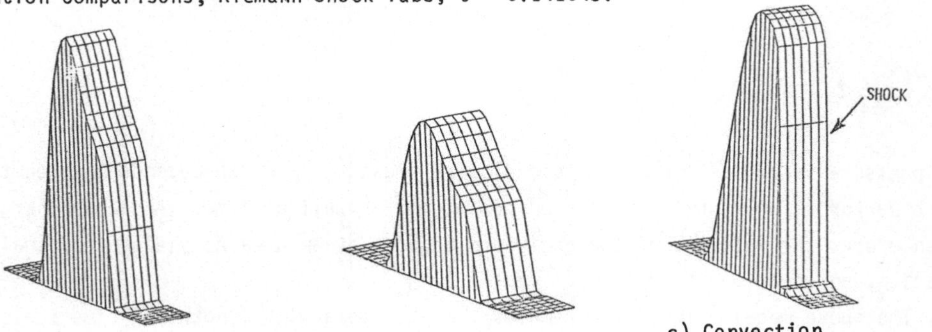

a) Momentum m_1^h b) Momentum m_2^h c) Convection Velocity u^h

Figure 6. Finite Element Two-Dimensional Solution For Inviscid Reimann Shock Tube, M = 32 x 6, k=1, ϕ = 26°, $\nu_\alpha^1 = 0$, $\nu_\alpha^2 = \nu\{3/4,2,1\}$, t = 0.14154s.

Numerical Computation of Large-Scale Fire-Induced Flows

Howard R. Baum and Ronald G. Rehm
National Bureau of Standards

Introduction and the Hydrodynamic Model

Existing approaches to the calculation of large scale fire driven flows in enclosures fall into two categories. The time dependent equations of motion with viscosity and thermal conductivity have been integrated directly, usually in the Boussinesq limit (1), (2). This is in principle the most rigorous approach. However, in practice it is difficult to achieve, because the Grashof numbers of practical interest are extremely high. The simultaneous calculation of all the dynamically active length and time scales is then beyond the power of most computers, even for the two dimensional calculations which are the subject of most investigations. An alternative approach (3), (4) is to perform an averaging operation on the equations of motion and introduce an empirical "turbulence model" to represent transport due to fluctuations. The turbulence model removes the problem of computing over the wide ranges of scales. However, the price paid is an inherent loss, to an unknown extent, of predictive capability.

The present approach uses the following filtered set of inviscid equations (5), driven by a prescribed volumetric heat source, Q(x,t) to represent the fluid motion.

$$\frac{\partial \rho}{\partial t} + \nabla\cdot(\rho\vec{u}) = 0$$

$$\rho\left\{\frac{\partial \vec{u}}{\partial t} + (\vec{u}\cdot\nabla)\vec{u}\right\} + \nabla\tilde{P} - \rho\vec{g} = 0$$

$$\rho C_p\left\{\frac{\partial T}{\partial t} + (\vec{u}\cdot\nabla)T\right\} - \frac{dP_0}{dt} = Q(\vec{x},t) \qquad (1)$$

$$P_0(t) = \rho R T$$

$$\left\{\frac{\gamma}{\gamma-1}\oint \vec{U}\cdot\vec{n}dA\right\} P_0(t) + \frac{V}{\gamma-1}\frac{dP_0}{dt} = \int Q dv$$

The pressure is divided into a mean enclosure pressure P_0 which depends only on time t and a perturbation $\tilde{P}$ which depends on position $\vec{x}$ as well as time. A global energy balance over the volume V of the enclosure bounded by an area A, yields the final equation, an equation for $P_0(t)$.

The above model effectively removes all phenomena with length and time scales too small to be resolved by direct computation on medium performance computers. Rapid time scales associated with acoustic and shock waves are removed by the analytic filtering process (5). All combustion phenomena are replaced by the heat source $Q(\vec{x},t)$ which is required to be resolvable on the finite difference grid used to solve the problem.

Grashof number limitations, together with boundary layer information, are removed by the invisid approximation. The resulting system of equations contains both large scale fluctuations as well as the "mean" flow. Hence no "turbulence model" is needed.

Eqs. (1) are recast for computation into evolution equations for density and velocity, together with an elliptic partial differential equation for a modified perturbation pressure P*, defined as $\tilde{P}$ minus the hydrostatic pressure at ambient density ρ_0.

$$\frac{\partial\rho}{\partial t} = -\left\{(\vec{u}\cdot\nabla)\,\rho + \rho D(\vec{x},t)\right\}$$

$$\frac{\partial\vec{u}}{\partial t} = \left(\frac{\rho-\rho_0}{\rho}\right)\vec{g} - \left\{\frac{1}{2}\nabla\,[(u)^2] + \vec{\omega}\times\vec{u}\right\} + \frac{1}{\rho}\nabla P^*$$

$$\nabla\cdot\left(\frac{1}{\rho}\nabla P^*\right) = (\vec{g}\cdot\nabla)\left(\frac{\rho-\rho_0}{\rho}\right) - \frac{\partial D}{\partial t} - \left\{\nabla^2(\frac{1}{2}u^2) + \nabla\cdot(\vec{\omega}\times\vec{u})\right\} \qquad (2)$$

$$P^* = \tilde{P} + \rho_0(\vec{g}\cdot\vec{x})$$

$$\nabla\times\vec{u} = \vec{\omega}$$

$$\nabla\cdot\vec{u} = D(\vec{x},t)$$

$$D(\vec{x},t) = \frac{1}{\gamma P_0(t)}\left[(\gamma-1)Q(\vec{x},t) - \frac{dP_0}{dt}\right]$$

Eqs. (2) are to be solved subject to initial conditions corresponding to the fluid at rest in the enclosure at ambient density,

$$\vec{u}(\vec{x},0) = 0;\ \rho(\vec{x},0) = \rho_0 \qquad (3)$$

At solid boundaries the normal component of velocity vanishes, while at an opening either the normal component of velocity or the pressure is specified (5). The density is specified at an inflow boundary; otherwise no boundary conditions are necessary in an inviscid model.

All variables are made non-dimensional using a simple Froude number scaling based on enclosure height H, ambient pressure P_∞ and maximum total rate of heat release Q_0.

Numerical Computations

To actually carry out the numerical computations, considerable care in the choice of methodology is needed to preserve the advantages inherent in eqs. (1). Analytical studies of the ability of several candidate finite difference schemes to calculate internal gravity waves (6) led to the conclusion that methods of second order accuracy in space and time would be necessary. The scheme chosen is dispersive rather than diffusive to avoid numerical viscosity damping an inviscid model. The finite difference scheme can be considered in two parts, the time marching procedure and the spatial discretization. All time derivaties are replaced by central differences over twice the time step δ. Thus, for any function f(t):

$$\frac{\partial f}{\partial t}(t) \tilde{=} \frac{f(t+\delta)-f(t-\delta)}{2\delta} \tag{5}$$

All terms on the right hand side of the evolution equations, (the first two of eqs. (2)) are evaluated at time t except for the density ρ in the last term in the evolution equation for ρ. In this term, stability requirements dictate that

$$\rho(t)\ D(t) \tilde{=} \frac{D(t)}{2}\ [\rho(t+\delta) + \rho(t-\delta)] \tag{6}$$

A staggered spatial grid is used. Vector components are evaluated at the faces and scalar quantities at the center of the fundamental grid. The staggered grid permits central differences to second order accuracy for all linear operations. The non-linear terms must be considered separately. The density evolution equation in continuous form is the mass conservation equation minus the expression for the velocity divergence. Each of these two equations are approximated by central differences and then subtracted. The density at all faces is approximated by the mean of the density at the centers of adjacent cells. This procedure ensures global mass conservation as well as second order accuracy.

The momentum equation is actually differenced in the vector invariant form shown as the second of eqs. (2). This ensures non-linear stability and complete compatability between the present "primitive variable" (i.e. $\vec{u}$, P, ρ) formulation and vorticity-stream function formulations (7). The pressure gradient and gradient of $(\vec{u})^2$ are linear operations, with $(\vec{u})^2$ obtained by averaging each component of $\vec{u}$ at the faces to get the value at the cell center. The density multiplying each component of the pressure gradient is the value at the location of the velocity component being updated. The $\vec{u} \times \vec{\omega}$ terms are handled by averaging adjacent values of each velocity component to obtain a second order accurate estimate at the location of the appropriate component of $\vec{\omega}$. Then each component of $\vec{u} \times \vec{\omega}$ is formed by multiplication and averaged to evaluate the product at the location of the velocity component being updated. Thus, a typical component of the **x** momentum eq. is:

$$v(x,y+h_y/2)\ \omega_z(x,y+h_y/2) \equiv K$$

$$K = \frac{1}{2}\left\{v(x-h_x/2,y+h_y) + v(x+h_x/2,y+h_y)\right\}\ \omega_z(x,y+h_y)$$
$$+ \frac{1}{2}\left\{v(x-h_x/2,y) + v(x+h_x/2,y)\right\}\ \omega_z(x,y) \tag{7}$$

The pressure equation is discretized by taking the central difference approximation to the divergence of the momentum equation, with the velocity divergence replaced by the last two of eqs. (2). The solution to this equation constitutes the bulk of the numerical computation, since the density and velocity can be updated once the pressure gradients are known. Mathematically, the calculation of the pressure consistent with prescribed density and velocity fields requires the solution of a non-separable, elliptic, self-adjoint partial differential equation. The boundary conditions are obtained from the momentum equation using the requirement that $\vec{u}.\vec{n} = 0$ on the enclosure boundaries. Thus, for the closed systems investigated to date the pressure boundary condition is:

$$\vec{n}\cdot\nabla P^* = \left(\frac{\rho-\rho_o}{\rho}\right)\vec{g}\cdot\vec{n} \tag{8}$$

The solution procedure combines the use of a fast direct (non-iterative) Poisson equation solver with a conjugate gradient iterative scheme to account for variable density effects (8). The result is a very efficient solver, which routinely produces solutions accurate to six significant figures in three to eight iterations, with four or five being typical.

The time and space differencing are linked by a stability requirement that limits the time step δ according to the inequality (9)

$$\delta < \mathrm{Max}\left\{D^2 + \left(\frac{|u|}{h_x} + \frac{|v|}{h_y} + \frac{|w|}{h_z}\right)^2\right\}^{-1/2} \tag{9}$$

The calculations are started with a first order time step in density. The pressure consistent with this density field is then computed and used in conjuction with the density field to obtain a first order time step in velocity. Once velocity and density at two time levels and pressure at one time step are known, the velocity and density can be updated. The stability criterion given by eq. (9) is then applied at each point in the grid. If eq. (9) is satisfied, the pressure is computed and the calculation proceeds. If δ is too large, it is halved and the calculation restarted with the results of the previous time step as initial conditions. Further details on all aspects of the algorithm may be found in (9).

All computations to date have been carried out in two space dimensions. Temperature (density) contours at four different times during the heating process in a square room simulation are shown in figure (1). The temperature contours at time t=2 are almost exact replicas of the heat input $Q(\vec{x},t)$ contours, since almost no motion has yet occurred. At t=8.5 the buoyant plume has begun to form from the heated bubble of gas. The plume has begun to spread along the ceiling at t=11.5, accompanied by the development of considerable large scale structure. The enclosure filling process is well advanced at t=14.5. The process is dominated by momentum driven descending wall layers, a factor also seen in experiments. Late in the computation a considerable amount of structure has begun to appear at the grid scale. In reality, energy at this scale should cascade to smaller, sub grid scales and eventually be dissipated. Since this energy is lost to the resolvable motion, it is desirable to remove the energy at the grid scale before too much buildup occurs. The computer program has an optional Lanczos smoothing algorithm

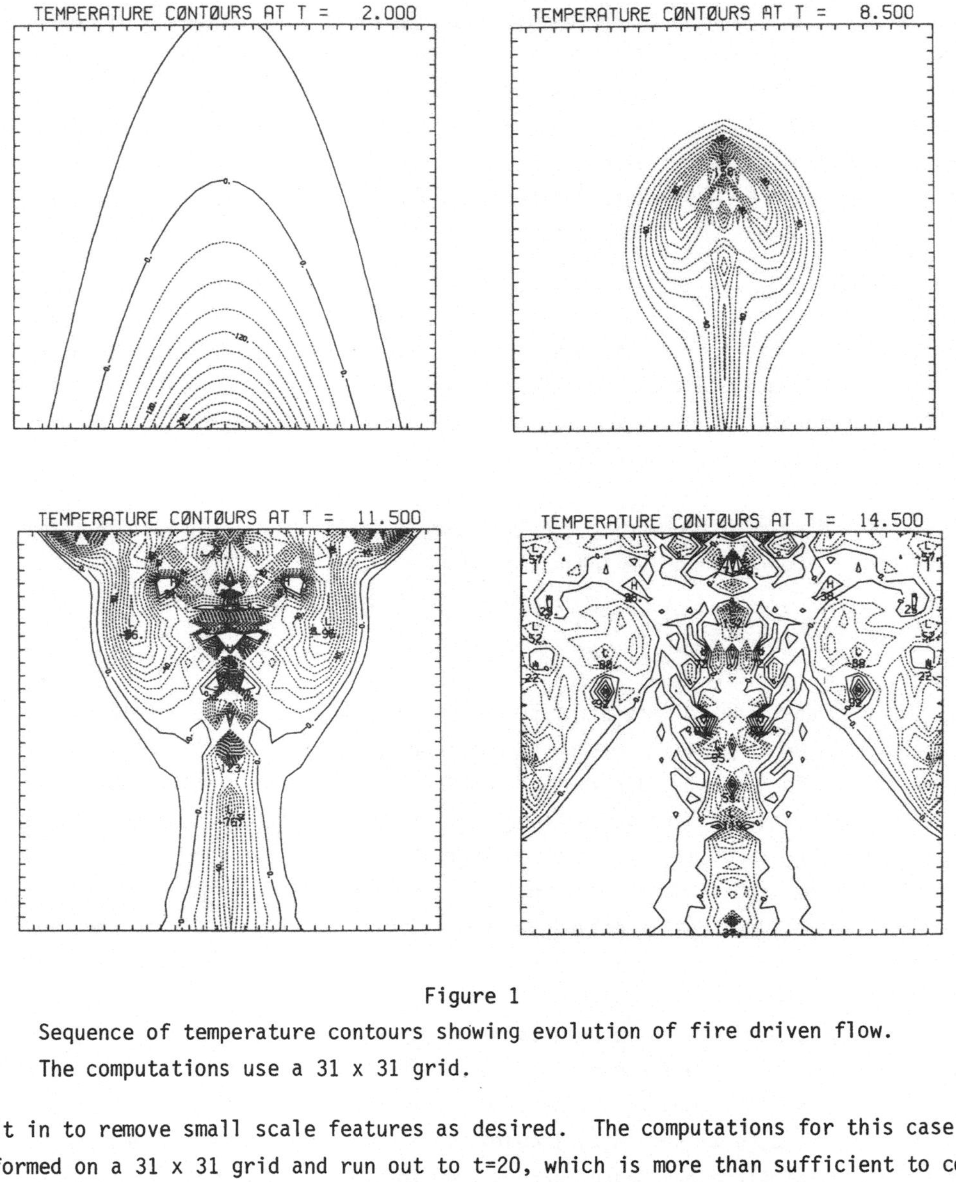

Figure 1

Sequence of temperature contours showing evolution of fire driven flow. The computations use a 31 x 31 grid.

built in to remove small scale features as desired. The computations for this case were performed on a 31 x 31 grid and run out to t=20, which is more than sufficient to complete the filling process. The calculation required 245 seconds to execute on a CDC Cyber 170/750.

Experimental Verification

Experiments were performed to test the ability to predict the two dimensional buoyant plume front velocity by releasing dyed salt water into a fresh water tank. Two sample frames are reproduced in Figure 2. Similar experiments were performed by Tsang

(10). Buoyancy induced flows with small density differences can be described by the Bussinesq equations. Moreover, the hydrodynamic model described earlier contains the Boussinesq equations as a special case. Thus, given a salt water volume flux per unit length $\dot{A}$, the trajectory of the plume front as it descends in the tank should be the same as that calculated from the hydrodynamic model if the dimensionless height y/H is plotted against the appropriate reduced time t/τ. The relevant time scales are:

$$\begin{aligned} \tau\ (\text{salt}) &= H \left\{ g(\Delta\rho/\rho)\,\dot{A} \right\}^{-1/3} \\ t\ (\text{heat}) &= H \left\{ gQ_o/\rho_o C_p T \right\}^{-1/3} \end{aligned} \qquad (10)$$

Experiments and computations both indicate that the dimensionless plume front velocity is approximately constant, until the front reaches the enclosure boundary. Moreover, qualitative features, like the descending wall layers in the last plot of Figure (1) are observed experimentally. Comparison of theory with experiment is shown in the table.

COMPARISON OF THEORY AND EXPERIMENT

QUANTITY	DIMENSIONLESS PLUME SPEED
Average of all computed points	1.15
63 x 64 Grid	1.09
Richardson Extrapolation to continuous 2-D limit	1.06
Tsang Experiment	0.96
Present Experiment	0.65 - 0.90

References

1. Torrance, K.E. and Rockett, J.A., "Numerical Study of Natural Convection in an Enclosure with Localized Heating from Below - Creeping Flow to the Onset of Laminar Instability," J. Fluid Mech., 36, p.33, (1969).
2. Knight, C., "Numerical Studies of Natural Convection in an Enclosure," Tech. Rept. 15, Div. Engrg. and Applied Phys., Harvard Univ. (1976).
3. Yang, K.T. and Liu, V.K., "UNDSAFE-II A Computer Code for Buoyant Turbulent Flow in an Enclosure with Radiation," Tech. Rept. TR-79002-78-3, Dept. Aero. and Mech. Engrg., Univ. of Notre Dame, (1978).
4. Ku, A.C., Doria, M.L., and Lloyd, J.R., "Numerical Modeling of Unsteady Buoyant Flows Generated by Fire in a Corridor," Proc. 16th Intl. Symposium on Combustion, p. 1373, (1977).
5. Rehm, R.G. and Baum, H.R., "The Equations of Motion for Thermally Driven, Buoyant Flows," J. Res. Nat. Bur. Stds., 83, p. 297, (1978).
6. Baum, H.R. and Rehm, R.G., "Finite Difference Solutions for Internal Waves in Enclosures," Nat. Bur. Stds. Rept. (in preparation).

7. Arakawa, A., "Computational Design for Long-Term Numerical Integration of the Equations of Fluid Motion: Two Dimensional Incompressible Flow." Part I, J. Comp. Phys., 1, p. 119, (1966).

8. Lewis, J.G. and Rehm, R.G., "The Numerical Solution of a Nonseparable Elliptic Partial Differential Equation by Preconditioned Conjugate Gradients," J. Res. Nat. Bur. Stands., 85, p. 367, (1980).

9. Baum, H.R., Rehm, R.G., Barnett, P.D. and Corley, D.G., "Finite Difference Calculations of Buoyant Convections in an Enclosure, Part I, the Basic Algorithm," Nat. Bur. Stds. Rept. NBSIR 81-2385, 1981.

10. Tsang, G., "Laboratory Study of Two-Dimensional Starting Plans," Atmos. Environment, 4, p. 519, (1970).

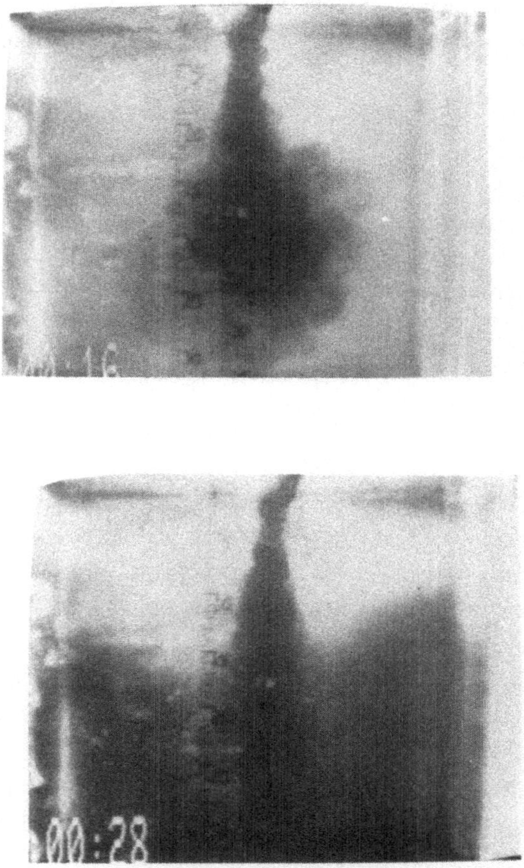

Figure 2
Photographs of descending plume (top) and filling process (bottom) in salt experiment. Note descending wall layers in the last frames of Figures 1 and 2.

SIMULATION OF THE FLUCTUATING FIELD OF A FORCED JET

Alvin Bayliss
Courant Institute of Mathematical Sciences

Lucio Maestrello
NASA Langley Research Center

Eli Turkel
Tel-Aviv University

A numerical method for simulating the axisymmetric fluctuating field of a forced circular jet has been developed. The method is based on a solution of the Euler equations for homentropic flow in cylindrical coordinates. The jet is forced by transient point mass injection. The source strength is scaled by a parameter ε and the state vector $W = (\bar{p},\bar{\rho},\bar{u},\bar{v})$ is expanded as a mean state plus a fluctuating state of order ε. Here $\bar{p}$ is the pressure, $\bar{\rho}$ the density and $\bar{u}$ and $\bar{v}$ are the axial and radial velocity components respectively.

The solution is expanded in the form $\bar{p} = p_0 + \varepsilon p$, etc. where the mean state (with subscript "0") is assumed to satisfy the unforced Euler equations. The variables (p,ρ,u,v) will represent the fluctuating field in response to the jet forcing. New variables $e = (\rho_0+\varepsilon\rho)u$ and $f = (\rho_0+\varepsilon\rho)v$ are introduced for simplicity. With cylindrical spatial coordinates z(axial) and r(radial) the Euler equations for the fluctuating field become

$$\rho_t + \left(\rho U_0+\rho_0 u\right)_z + \left(\rho V_0+\rho_0 v\right)_r + \frac{(\rho V_0+\rho_0 v)}{r} = m$$

$$e_t + \left(e(U_0+\varepsilon u)\right)_z + \left(e(V_0+\varepsilon v)\right)_r + p_z + fU_{0r} - eV_{0r} - \frac{\varepsilon fu}{r} = \varepsilon mu \qquad (1)$$

$$f_t + \left(f(U_0+\varepsilon u)\right)_z + \left(f(V_0+\varepsilon v)\right)_r + p_r + eU_{0z} - fV_{0z} - \frac{\varepsilon fv}{r} = \varepsilon mv.$$

In the development of system (1) it is assumed that the derivatives of the mean pressure and density in space can be neglected. It is an exact consequence of the full Euler equations with a source of mass injection, and of the expansion into mean and fluctuating states. The source is εm (units density/time). The mean velocities U_0 and V_0 are taken from measurements of a spreading jet. The system is solved in a cylindrical rectangle including a semi-infinite pipe from which the jet exits. The computational domain is illustrated in Fig. 1.

This work was partially supported by NASA Contract Nos. NAS1-14472 and NAS1-16394 for the first and third authors while they were in residence at ICASE, NASA Langley Research Center, Hampton, VA. Additional support for the first author was provided by the U. S. Air Force Contract No. AFOSR-76-2881 and U. S. Department of Energy Contract No. DE-AC02-76ER03077.

The use of the quasi-momentum variables e and f gives a system from which the fluctuating field can be computed directly rather than as a small part of the total field. In addition the linear limit can be recovered by simply setting $\varepsilon = 0$ in (1). The nonlinear terms are explicitly exhibited. The fluctuating pressure p is obtained from the density ρ by the homentropic relation

$$\overline{p} = A\overline{\rho}^{\gamma} = p_0\left(1 + \varepsilon \frac{\rho}{\rho_0}\right)^{\gamma}, \tag{2}$$

where $\gamma = 1.4$ in air.

The fundamental assumption entering into the derivation of (1) is that the mean state is a solution to the unforced Euler equations. This is not true for a state determined from experimental measurements. Numerical experiments have verified however, that the qualitative features of the fluctuating solution are insensitive to small changes in the mean state and thus we believe that the solution to (1) qualitatively represents the fluctuating field in response to a a given source.

The source is assumed to be a delta function in space (modelled by a sharp Gaussian) with a pulse-like time dependence. The source location, z_s, is on the jet center-line approximately 1.2 jet diameters downstream of the nozzle exit. Specifically

$$m(t,z,r) = f(t)\delta\left(r^2+(z-z_s)^2\right), \tag{3}$$

where the function f(t) is

$$f(t) = e^{-(at^2+bt^{-2})}, \tag{4}$$

for suitable constants a and b. The use of (4) permits the investigation of a broad band spectrum.

The fluctuating field described by (1) reduces to the acoustic field for large distances. The near field and flow field are dominated by instability waves which are generated because the mean flow is linearly unstable. The pulse is assumed to dominate the natural sources of jet noise. These natural sources are both the linear and nonlinear terms in (1). In real jet these terms are determined from the turbulent fluctuations whereas in the numerical model these natural sources are excited by the pulse. The important physical effect is the generation of packets of instability waves in the flow. Large scale structures which are related to mean flow instabilities have been observed experimentally in both forced and unforced jets [1,2]. These structures interact with and modify the resulting acoustic fields. The model permits this interaction to be studied in both the linear ($\varepsilon = 0$) and nonlinear ($\varepsilon \neq 0$) regimes. Calculations with the linear model and comparison with experiments are described elsewhere [3,4]. In the rest of this paper we describe the numerical requirements in order to compute with this model and the effects of the nonlinear terms on the fluctuating field.

Numerical Model

In order to numerically integrate (1) it is necessary to resolve the solution over large length scales (far field, near field, and flow field). This necessitates the use of a higher order difference scheme. The system can be written in the form

$$w_t + F_z + G_r = H, \tag{5}$$

where w is the vector (ρ,e,f) and F, G, and H are appropriate functions. Equation (5) is split into two one-dimensional operators in z and r. Each 1-d system is integrated by using a fourth order version of the MacCormack scheme [5]

$$\begin{aligned} \tilde{w}_i^{n+1} &= w_i^n + \frac{\Delta t}{6\Delta x}\left(-7F_i+8F_{i+1}-F_{i-2}\right) + \Delta t H_1 \\ w_i^{n+1} &= \tfrac{1}{2}\left(\tilde{w}_i^{n+1} + w_i^n + \frac{\Delta t}{6\Delta x}\left(7\tilde{F}_i-8\tilde{F}_{i-1}+\tilde{F}_{i-2}\right) + \Delta t\tilde{H}_1\right), \end{aligned} \tag{6}$$

together with a symmetric variant (H_1 is obtained from some splitting of H). Typical grids require of the order of 40,000 grid points over distances of the order of 50 jet diameters in all directions. Our experience has been that on problems of this size, second order schemes are not sufficient to obtain accurate solutions with the amount of resolution that is feasible. The explicit scheme (6) naturally lends itself to vectorization and has been implemented on the CDC CYBER-203 with great efficiencies.

In addition it is necessary to impose boundary conditions which accurately simulate outgoing radiation at the far field boundaries. A family of radiation boundary conditions has been developed which provide increasingly accurate approximations to outgoing radiation. The leading member of this family is

$$p_t + \rho_\infty c_\infty \tilde{u}_t + \frac{p}{d} = 0, \tag{7}$$

where c_∞ is the ambient sound speed and ρ_∞ the ambient density. Here $d^2 = r^2 + z^2$ and $\tilde{u}$ is the outgoing radial velocity based on a spherical coordinate system near the source m.

System (1) includes terms proportional to r^{-1}. This singularity at the axis is resolved by including these terms in the flux vector G when r=0. In addition it is necessary to modify the difference formula (6) at boundaries. This is accomplished by introducing fictitious grid points outside of the computational domain and using a third order extrapolation of the flux function (F or G). This approach was found to be the most readily vectorizable. It has been verified that the resulting scheme is fourth order accurate.

Nonlinear Results

We next describe results illustrating the effect of the nonlinear terms on the fluctuating field. In Figs. 2 and 3 the fluctuating pressure is shown as a function of axial location z/D (D is the jet diameter) and non-dimensional time tc_∞/D for two different radial positions and for $\varepsilon = 0.00$ and $\varepsilon = 0.05$. All figures show an acoustic wave (speed of sound normalized to unity) in the downstream direction trailed by several much larger waves. These are instability waves which travel with a speed of approximately $.7U_j$ where $U_j = .66c_\infty$ is the jet exit velocity. A series of acoustic ripples can also be seen propagating upstream. These are due to diffraction of the upstream acoustic wave by the nozzle lip.

The figures indicate that the nonlinear terms have little effect on the primary acoustic pulse and on the acoustic diffraction from the nozzle lip. The nonlinearity has a pronounced effect on the instability waves. Increasing the nonlinearity causes these predominantly large scale structures to break up into smaller scale structures. This can be seen in both the additional ripples which trail the instability waves and a sharpening of the individual pulses indicating an enhanced high frequency content. It is also evident that for increasing r/D these smaller scale structures are comparable in amplitude to the primary instability waves.

In Figs. 4 and 5 the fluctuating vorticity is shown for two fixed times and for $\varepsilon = 0.00$ and $\varepsilon = 0.05$. The intense vortices in Fig. 4 correspond to the instability waves in Figs. 2 and 3 while the vortices at the later time in Fig. 5 represent a residual shedding of vorticity from the nozzle lip. It is apparent from the figures that nonlinear effects tend to slow down the vortices as they propagate downstream. Thus the trailing vortices catch up with the leading vortices and a possible pairing of vortices can be observed. True vortex merging, which has been observed experimentally [6], depends heavily on viscosity as well as nonlinearity and is not simulated here.

In Figs. 6a and 6b the normalized power spectral densities (PSD) for the fluctuating axial velocity and pressure are plotted for the linear and nonlinear cases as a function of the Helmholtz number fD/c_∞ (f is the frequency). These figures clearly indicate the shift into higher frequencies and the overall broadening of the spectra caused by the nonlinear effects. Then results are typical for the fluctuating field at all points.

Discussion

The fundamental difference between the nonlinear and linear computations is the breakdown of the large scale structures into smaller scale structures. This is associated with a transfer of energy into higher frequencies or equivalently a broadening of the spectral content of the fluctuating field. It is also illustrated by the increased interaction between the different vortices. In real jets this is a fundamental step in the transition to fully developed turbulence. The results indicate that at least the initial stages of this energy cascade into smaller scales can be simulated just by the nonlinear terms of an axisymmetric and inviscid calculation.

The generation of smaller scale fluctuations fully justifies the use of fourth order spatial discretizations. The numerical scheme is accurate and is in general stable. Higher values of ε can be readily computed although the Gaussian approximation to the δ function source will have to be smoothed out.

References

[1] Crow, S. C. and Champagne, F. H.: Orderly Structure in Jet Turbulence, J. Fluid Mech., Vol. 48, Part 3, 1971, pp. 547-591.

[2] Maestrello, L. and Fung, Y. T.: Quasi-Periodic Structure of a Turbulent Jet, J. Sound & Vib., Vol. 64, 1979, pp. 107-122.

[3] Maestrello, L., Bayliss, A., and Turkel, E.: On the Interaction of a Sound Pulse with the Shear Layer of an Axisymmetric Jet, J. Sound & Vib., Vol. 74, 1974, pp. 281-301.

[4] Bayliss, A. and Maestrello, L.: Simulation of Instabilities and Sound Radiation in a Jet, AIAA J., Vol. 19, No. 7, 1981, pp. 835-841.

[5] Gottlieb, D. and Turkel, E.: Dissipative Two-Four Methods for Time-Dependent Problems, Math. Comput., Vol. 30, 1976, pp. 703-723.

[6] Ho, Chi-ming and Huang, Lein-saing: Subharmonics and Vortex Merging in Mixing Layers, to appear in J. Fluid. Mech., 1982.

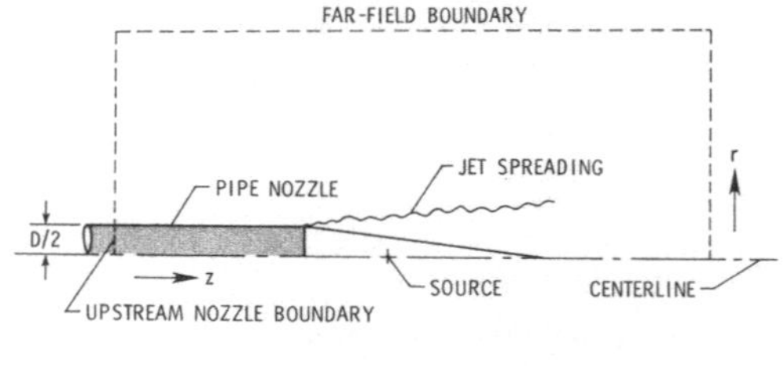

Figure 1. Computational Domain

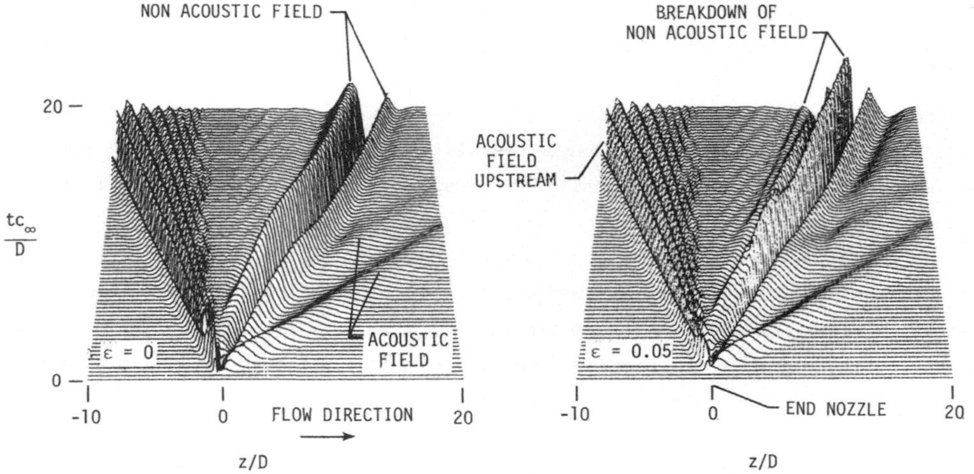

Figure 2. Three-dimensional plots of the fluctuating pressure r/D = .29.

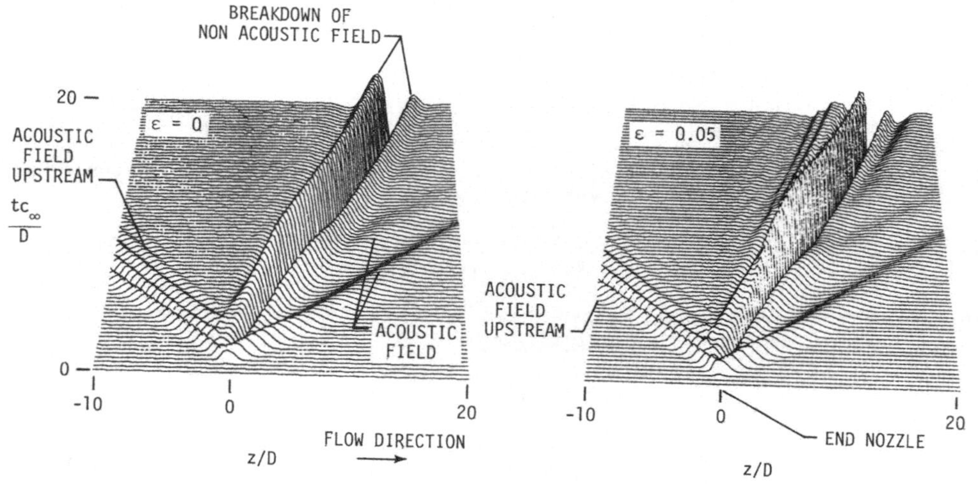

Figure 3. Three-dimensional plots of the fluctuating pressure r/D = .61.

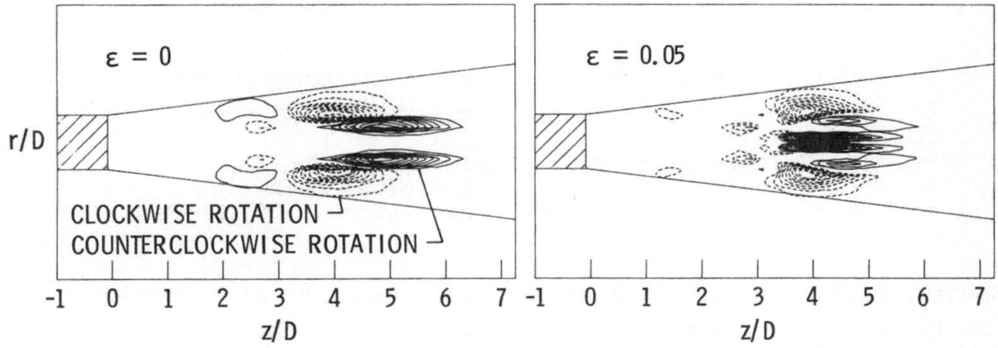

Figure 4. Fluctuating vorticity at $tc_\infty/D = 10$.

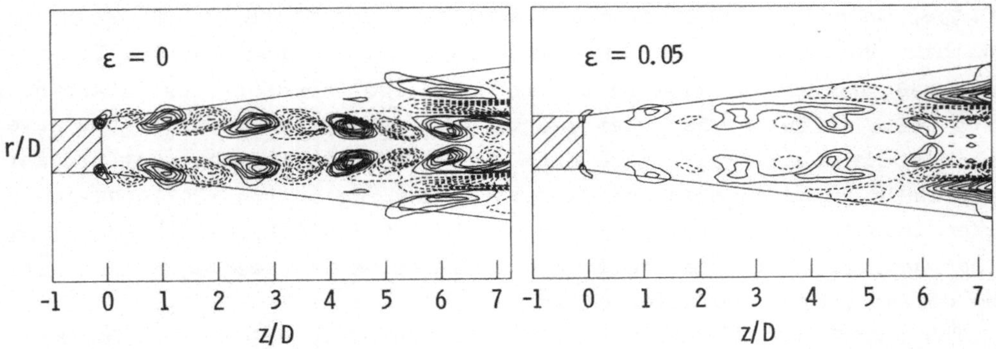

Figure 5. Fluctuating vorticity at $tc_\infty/D = 30$.

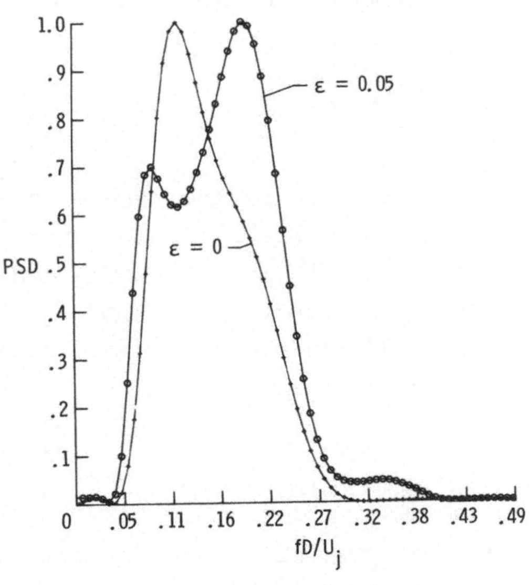

Figure 6a. Flow field velocity spectra $z/D = 7.3$, $r/D = 1.5$

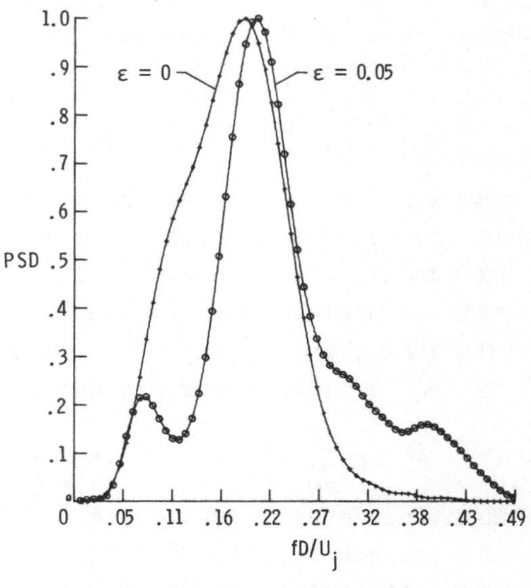

Figure 6b. Acoustic far-field spectra $z/D = 38$, $r/D = 20$.

MODELING OF GASDYNAMIC PROCESSES IN NEUTRON STARS WITH PHASE TRANSITIONS

Yu.A.Berezin, O.E.Dmitrieva, N.N.Yanenko
Institute of Theoretical & Applied Mechanics
USSR Academy of Sciences, Novosibirsk 630090 USSR

Neutron stars are the remnants of supernova stars with mass ~I M⊙ (M⊙ is the mass of the Sun, M=2 10^{33}g), radius ~10km, density in the center $2\cdot 10^{14} \div 4\cdot 10^{15}$ g/cc.

We consider the simplest model of a neutron star with the mass 0.45÷2 M⊙ on the basis of one-dimensional equations of gas dynamics within the framework of the Newtonian gravitational theory. It is assumed that the stellar matter is cold, and therefore, as is known, its state is fully characterized by density and composition. We have chosen the equation of state from the most reliable tables at present available as an equation of state: for the region $1.05\cdot 10^4 \div 0.77\cdot 10^{14}$ g/cc from the paper [1], for the region $0.84\cdot 10^{14} \div 1.26\cdot 10^{16}$ g/cc from the paper [2]. In what follows a combination of these equations will be called a basic equation of state.

The phase transitions with condensation of π-mesons and formation of pion condensates are possible in a superdense matter [3]. If n_* is the critical density, at which the condensation starts developing, then in this case the interpolation formula

$$\mathcal{E}(n)=\begin{cases}\mathcal{E}_N(n)-\frac{1}{2}\beta(n)(n-n_*)^2, & n>n_* \\ \mathcal{E}_N(n) & , n<n_*\end{cases} \qquad (1)$$

applies for the energy density of the neutron matter, where n is the numbers of nucleons per volume unit, $\mathcal{E}_N(n)$ is the energy density of nucleons, $\mathcal{E}_\pi(n)=-\frac{1}{2}\beta(n)\cdot(n-n_*)^2$ is the energy density connected with the formation of pion condensates, $\beta(n)= a+b(n_*/n)+c(n_*/n)^2$. The pressure P and the compressibility K of the matter are defined by expressions:

$$P=n^2\frac{d}{dn}\left(\frac{\mathcal{E}}{n}\right) \quad , \quad K=n\frac{d^2\mathcal{E}}{dn^2} \qquad (2)$$

When the density approaches n_* from above, the compressibility $K\to K_N-(a+b+c)\equiv K_N-\beta_0$; if $\beta_0>K_N$, then the compressibility is negative at densities higher than the critical one and an unstable state can take place. At $n=n_*$ there exists a compressibility discontinuity, hence, a phase transition can occur. The pressure defined from (1) is a nonmonotonic function of the density

$$P = n^2 \frac{d}{dn}\left(\frac{\mathcal{E}}{n}\right) = n\frac{d\mathcal{E}}{dn} - \mathcal{E} = P_0(n) - \tfrac{1}{2}n^2\{(a + b(n_*/n) + c(n_*/n)^2)(1 - n_*^2/n^2) - (n_*/n)(1-n_*/n)^2\cdot(b + 2c(n_*/n))\}\cdot\theta(n-n_*)\,,$$
$$\theta(n-n_*) = \begin{cases} 1, & n > n_* \\ 0, & n < n_* \end{cases} \qquad (3)$$

The relations (1)-(3) are written in pion units.

The pion condensation in a superdense matter leads to a "softening" of the equation of state, and at some sets of parameters n_*, a, b, c the equation of state can become nonmonotonic. A region where $dp/d(1/n)>0$ is unstable.

The governing equations have the form:

$$\begin{aligned} &\rho_t + \frac{1}{r^2}(r^2\rho u)_r = 0 \\ &u_t + uu_r = -\frac{1}{\rho}P_r - \frac{Gq(r)}{r^2} \\ &q(r) = \int_0^r 4\pi\rho(r')r'^2 dr' \\ &P = P(\rho)\,. \end{aligned} \qquad (4)$$

Here G is the gravitational constant. The equations were solved numerically with the help of the well-known von Neuman-Richtmeyer scheme with an artificial viscosity.

The evolution of the following initial configurations of the stars was studied: 1) equilibrium corresponding to the case of degenerate neutron gas ($P(\rho)$ has an analytical expression [3]): 2) approximate equilibrium corresponding to the basic monotonic equation of state; 3) nonequilibrium configuration with fixed distribution $\rho = \rho_0(1-(r/R_0)^K)$; ρ_0 is the density in the center, K is determined by expressions

$$K = \frac{3}{M_0/M - 1}\,, \quad M_0 = \frac{4}{3}\pi\rho_0 R_0^3\,, \quad M = 4\pi\int_0^{R_0}\rho r^2 dr\,.$$

The initial density in the center was set equal to $\rho_0 = 3\cdot10^{14}, 6\cdot10^{14}, 10^{15}$ g/cc for the masses within the range 0.45-2M☉.

The rest initial and boundary conditions: $u(0,t)=0$, $u(q,0)=0$ (or $u(q,0) = -0.7\sqrt{GM/r(q,0)}$ that corresponds to the half of a local parabolic velocity), $P(1,t)=0$, $r(0,t)=0$, $r(q,0)=r^0(q)$, $r^0(q)$ is calculated from the initial configuration of the radius distribution.

Most calculations were carried out with the following values of parameters:

n_*=0.69, a =0.9, b=-0.23, c =0.14 (the values of parameters are given in pion units).

As time increases in all the runs the shock wave formation which moves towards the star periphery is observable.

Within the framework of the present model at the time $\sim 1ms$ there exists a stationary contact discontinuity whose location corresponds to the boundary of a dense interior core and a less dense envelope. The density jump occurs at one step h of the grid in the Lagrangean coordinate independently on the value of this step (h =1/60 in the most calculations). Fig.1 illustrates the $\rho(q)$ distributions at the different moments of time for M=1M⊙, R_o=14.3km, ρ_o=10^{15} g/cc, Fig.2 shows the $\rho(q_i)$ distributions depending on time for the same case.

The study of the question about a possibility of the blowing off the envelope or part of it was carried out with the use of a refining grid in Lagrangian coordinate: $h = 5\cdot 10^{-2} - 10^{-5}$M. It is found that the blowoff of the portion in this model is not associated with the formation of density discontinuity. For example, for M =0.45M⊙, R_0=26km we have M_{core} =0.14M⊙, the blowoff ≈ 0.068M⊙.

For modeling the unsteady processes with regard to stellar cooling we have carried out the calculations with a time dependent equation of state $P(\rho) = P_o(\rho) - (1-e^{-\gamma t})P_1(\rho)$ where $P_o(\rho)$ is the basic equation of state, $P_1(\rho)$ is the nonmonotonic part of the pressure defined in accordance with the formula (2). Depending on the parameter γ a "rapid cooling" occurs (a fast transition from a monotonic equation of state to a nonmonotonic one) or a "slow cooling". Here as before the onset and the following smoothing of density oscillations and the discontinuity formation takes place. In the case of a "slow cooling" the oscillations of the density profile are fully absent which are available on other cases.

Thus, the study of the neutron star dynamics carried out within the framework of the simplest model with a nonmonotonic equation of state has shown the existence of some stable quasistationary regime at which a sphere consists of a dense core and a less dense envelope. The establishment of such a regime does not depend of the initial configurations of neutron stars and on the means of "incorporation" of the monotonic equation of state and is fully determined by the star mass and the equation of state.

In conclusion we give a table combining some results of the numerical experiments (here all the parameters are given at the stage of quasirelaxation: R is the star radius, R_* is the core radius, ρ_c

is the density in the center, ρ_1 and ρ_2 are discontinuity densities

$M/M_\odot$	R(km)	R_*(km)	ρ_c(g/cc)	ρ_1(g/cc)	ρ_2(g/cc)
0.45	8.4	5.1	$1.3\cdot10^{15}$	$7.9\cdot10^{14}$	$3.7\cdot10^{14}$
1	8.6	6.9	$1.6\cdot10^{15}$	$7.9\cdot10^{14}$	$3.5\cdot10^{14}$
1.4	9.1	7.7	$1.8\cdot10^{15}$	$8.1\cdot10^{14}$	$3.5\cdot10^{14}$
1.85	9.5	8.3	$2\cdot10^{15}$	$8.4\cdot10^{14}$	$3.6\cdot10^{14}$
2	9.6	8.5	$2\cdot10^{15}$	$7.8\cdot10^{14}$	$3.2\cdot10^{14}$

References

1. Baym G., Pethick C., Sutherland P., 1971, Astrophys.J.,170,229.
2. Pandharipande V.R., 1971, Nucl.Phys., A178,123.
3. Zel'dovich Yu.B., Novikov I.D. Relyativistskaya Astrophizika, 1967, Moscow, "Nauka".

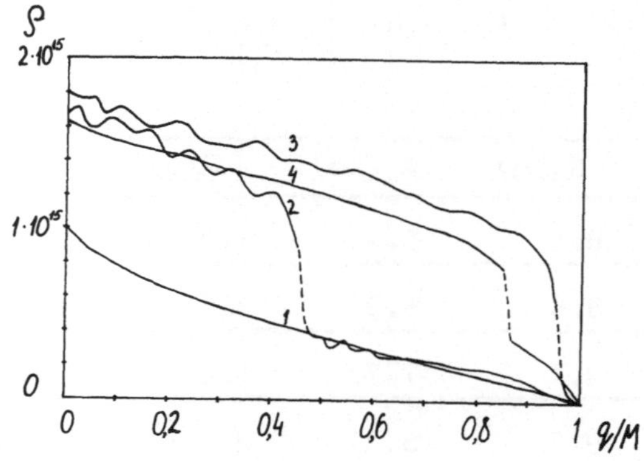
ρ
2·10^15
1·10^15
0
3
4
2
1
0
0,2
0,4
0,6
0,8
1
q/M

Fig. 1

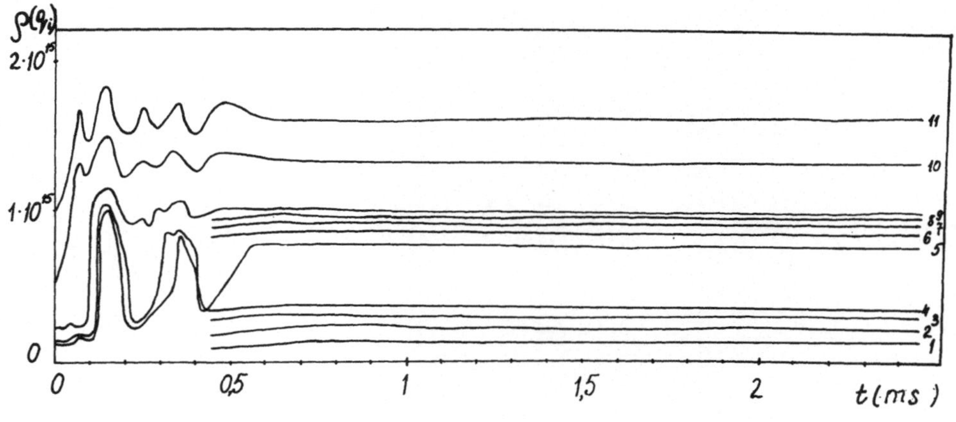
ρ(q_i)
2·10^15
1·10^15
0
11
10
8 9
6 7
5
4 3
2
1
0
0,5
1
1,5
2
t(ms)

Fig. 2

Adaptation of Flux-Corrected Transport Algorithms
for Modeling Blast Waves

D. L. Book and J. P. Boris, Laboratory for Computational
Physics, Naval Research Laboratory, Washington, DC
M. A. Fry, Science Applications, Inc., McLean, VA
R. H. Guirguis, JAYCOR, Alexandria, VA
A. L. Kuhl, R & D Associates, Marina del Rey, CA

Blast wave phenomena include reactive and two-phase flows associated with the motion of chemical explosion products; the propagation of shocks, rarefaction waves, and contact discontinuities through a nonideal medium (real air, possibly thermally stratified and containing dust and water vapor); and the interaction of the blast waves (including boundary layer effects) with structural surfaces. Flux-Corrected Transport (FCT) represents an accurate and flexible class of methods for solving such nonsteady compressible flow problems (Boris and Book, 1976). Coupled with a nondiffusive adaptive gridding scheme (Book, et al., 1980; Fry, et al., 1981), it enables complex time-dependent shocks to be efficiently "captured."

In models which treat all the physical effects required for blast wave simulation, truncation errors inherent in the underlying finite-difference scheme are exacerbated by nonlinear coupling between the fluid equations and by the greater complexity of the phenomena being simulated. Typical of these errors are the "terraces" which develop under some circumstances on the flanks of sloping profiles when the growth of ripples due to phase errors at short length scales is terminated by the action of the flux limiter. Two approaches are possible toward eliminating them: improving the short-wavelength phase and amplitude properties of the underlying algorithm, and switching on additional diffusion locally. The latter approach folds information about the shape of the profile and the nature of the physical process taking place (e.g., rarefaction) into the switch criterion, thus changing the FCT technique from a "convective equation solver" to a "fluid system solver." In doing this, care must be taken to avoid losing the accuracy, robustness and problem-independence which constitute valuable attributes of FCT algorithms (Book, et al., 1981).

Tests carried out on scalar advection of simple density profiles by a uniform flow field show that terracing does not require either diverging velocities or discontinuities in the profile, but appears typically (for $v > 0$) where the first and second derivatives of density have the same sign (Fig.1). In order to improve the properties of the basic difference scheme, we propose a new algorithm for integrating generalized continuity equations over a timestep δt. Consider the following three-point transport scheme:

$$\tilde{\rho}_j = \rho_j^o - \eta(\rho_{j+1}^o - \rho_{j-1}^o) + \kappa(\rho_{j+1}^o - 2\rho_j^o + \rho_{j-1}^o);$$

$$\bar{\rho}_j = \tilde{\rho}_j - \theta(\rho_{j+1}^o - \rho_{j-1}^o) + \lambda(\rho_{j+1}^o - 2\rho_j^o + \rho_{j-1}^o);$$

$$\rho_j^n = \bar{\rho}_j - \mu(\phi_{j+1/2} - \phi_{j-1/2}),$$

where

$$\phi_{j+1/2} = \tilde{\rho}_{j+1} - \tilde{\rho}_j.$$

The arrays $\{\rho_j^o\}$ and $\{\rho_j^n\}$ are the old and new densities, $\tilde{\rho}_j$ and $\bar{\rho}_j$ are temporary intermediate densities, and η, θ, κ, λ, and μ are velocity-dependent coefficients. Here κ and λ are diffusion coefficients, and μ is the antidiffusion coefficient. In the actual algorithm, $\phi_{j+1/2}^c$ is corrected (hence the name FCT) to a value $\phi_{j+1/2}^c$ chosen so no extrema in $\bar{\rho}_j$ can be enhanced or new ones introduced in ρ_j. Previous FCT algorithms had $\theta = 0$; the widely used ETBFCT and related algorithms (Boris, 1976) have in addition $\kappa = 0$. If we define ρ_j to be sinusoidal with wave number k on a mesh with uniform spacing δx, so that $\rho_j^o = \exp\,(ij\beta)$ where $\beta = k\delta x$, then the new density array satisfies

$$\rho_j^n/\rho_j^o \equiv A = 1-2i(\eta+\theta)\sin\beta + 2(\kappa+\lambda)(\cos\beta-1) - 2\mu(\cos\beta-1)\,[1-2i\eta\sin\beta+2\kappa(\cos\beta-1)].$$

From A we can determine the amplification $\alpha = A$ and relative phase error $R = (1/\varepsilon\beta)\tan^{-1}(-\mathrm{Im}A/\mathrm{Re}A)-1$, where $\varepsilon = v\delta t/\delta x$ is the Courant number. Expanding in powers of β we find

$$\alpha = 1 + \alpha_2\beta^2 + \alpha_4\beta^4 + \alpha_6\beta^6 + \ldots \;;$$

$$R = R_0 + R_2\beta^2 + R_4\beta^4 + R_6\beta^6 + \ldots \;.$$

First-order accuracy entails making R_0 vanish, which requires that $\eta + \theta = \varepsilon/2$. Second-order accuracy ($\alpha_2 = 0$) implies that $\mu = \kappa + \lambda - \varepsilon^2/2$. Analogously, the "reduced-phase-error" property $R_2 = 0$ (Boris and Book, 1976) determines $\mu = (1-\varepsilon^2)/6$, thus leaving two free parameters. One of these can be used to make R_4 vanish also. The resulting phase error $R(\beta)$ is small not only as $\beta \to 0$, but also for larger values of β, corresponding to the short wavelengths responsible for terraces (Fig. 2). The remaining parameter η can be chosen to relax the Courant number restriction needed to ensure positivity from $\varepsilon < 1/2$ to $\varepsilon < 1$. When coded, these changes necessitate a small increase in the operation count of ETBFCT along with a small increase in overhead to precalculate the two new arrays of velocity-dependent transport coefficients. On advection tests, the new algorithm completely eliminated terraces (Fig. 3). When applied to the coupled systems of gas dynamic equations, it produced profiles which closely approximate the Riemann solution of the exploding diaphragm problem (Fig. 4).

The second approach uses a rarefaction flux limiter (RFL) to eliminate numerical ripples in strong rarefaction waves. This approach is physically motivated. Raw anti-diffusive fluxes $\phi_{j+1/2}$ are limited so that the slope of local flow field profiles decays with time in a rarefaction wave. In effect, additional diffusion is left in the field to maintain monotonicity of local slopes. For multi-material calculations a "contact surface sensor" is needed to detect physical discontinuities and shut off the RFL locally.

In addition we found that some care was required when applying generalized continuity equation solvers to a system of equations. Truncation errors of the various equations can interact, causing undershoots or overshoots in nonconvective quantities such as pressure. We found that it was necessary to monotonize derived quantities (pressure, velocity) before using them in minimal-diffusion transport algorithms.

The above methodology has been applied to a series of test problems initiated by a spherical high-explosive (HE) detonation in air. An ideal Chapman-Jouguet detonation was used to specify the initial conditions; afterburning was neglected. In the absence of reflecting surfaces, spherical symmetry is maintained and the calculation remains one-dimensional (Fig. 5). A nonuniform radial grid was used with extremely fine zoning near the shock front. The grid was moved so that the shock remained approximately fixed with respect to the mesh (Fig. 6). The original version of the FCT algorithm gave rise to pronounced terraces in the rarefaction region. This would have rendered any two-dimensional calculation involving shock diffraction or nonideal effects dubious. Introduction of the techniquesdescribed here improved the blast wave results considerably. The decrease in phase error reduced terracing dramatically.

Next, a two-dimensional (2D) numerical calculation was performed to simulate one of Carpenter's (1974) height-of-burst experiments which used spherical 8-lb. charges of PBX 9404 at 51.6 cm. The previous fine-zoned 1D calculation was used to initialize the problem. It was mapped onto the 2D grid just prior to the onset of reflection. The solution was then advanced in time, with pressure being calculated from a real-air equation of state and a JWL equation of state for the combustion products. The front of the blast wave was captured in a finely gridded region which moved outward horizontally. Special care was taken to ensure that the grid moved smoothly. The resulting solution, particularly the curve of peak overpressure vs. range, was consistent with Carpenter's experimental data (Fig. 8). Although this calculation represents a reasonable accurate simulation of the double-Mach-stem region, no doubt improvements can and will be made to numerically model such phenomena.

References

Book, D., Boris, J., Kuhl, A., Oran, E., Picone, M., and Zalesak, S., Seventh International Conf. on Num. Methods in Fluid Dynamics, Stanford (1980).

Book, D., Boris, J., and Zalesak, S., in Finite-Difference Techniques for Vectorized Fluid Dynamics Calculations, D. Book, Ed. (Springer-Verlag, New York, 1981).

Boris, J., "Flux-Corrected Transport Modules for Generalized Continuity Equations," NRL Memo Report 3327 (1976).

Boris, J., and Book, D., in Methods in Computational Physics, J. Killeen, Ed., (Academic Press, New York, 1976) Vol. 16, p. 85.

Carpenter, H. J., Proc. Fourth International Symp. on Military Appliations of Blast Simulation (1974).

Fry, M., Tittsworth, J., Kuhl, A., Book, D., Boris, J., and Picone, M., "Shock-Capturing Using FCT Algorithms with Adaptive Gridding," NRL Memo Report 4629 (1981).

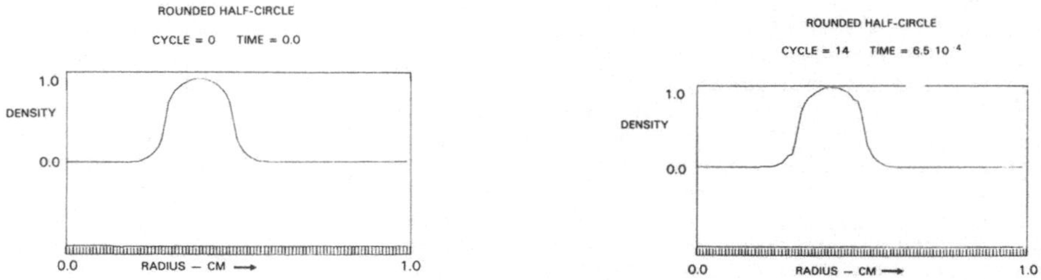

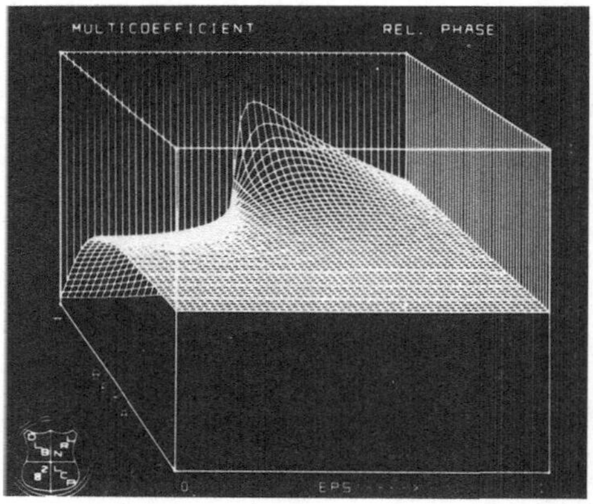

1. Rounded half circle used in passive scalar advection tests (a) initially, and (b) after propagation for 14 cycles using JPBFCT. Note that terraces form even, as here, in the absence of corners in the profile. Tick marks indicate computational zones (N = 100).

2. Contour plot of $R(\beta,\varepsilon)$ for new multicoefficient FCT algorithm. Note $R\approx 0$ except for $\beta \gtrsim 3\ \pi/2$. The relative phase error vanishes exactly for $\varepsilon=1/2$ and $\varepsilon=1$.

3. (a) Blowup of Fig. 1a (dashed line) compared with (b) same profile as computed using new sixth order-phase-accurate FCT algorithm. Solid traces are exact solutions.

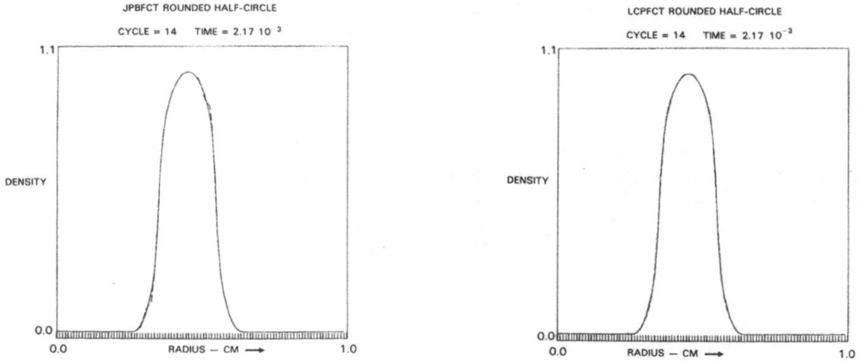

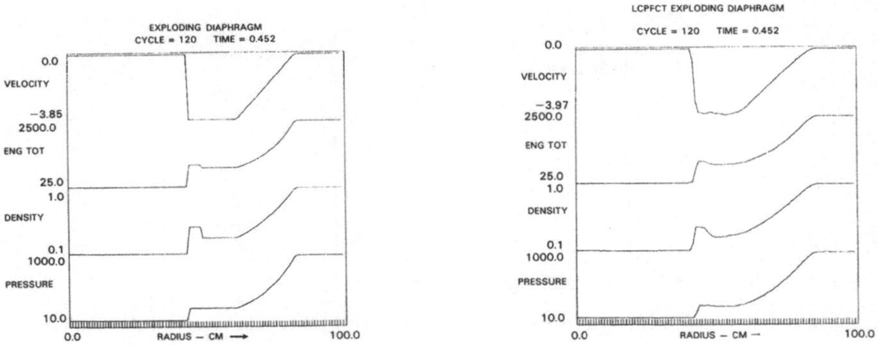

4. (a) Exact and (b) computed solution of exploding diagragm problem (10-to-1 initial density jump, 100-to-1 initial pressure jump).

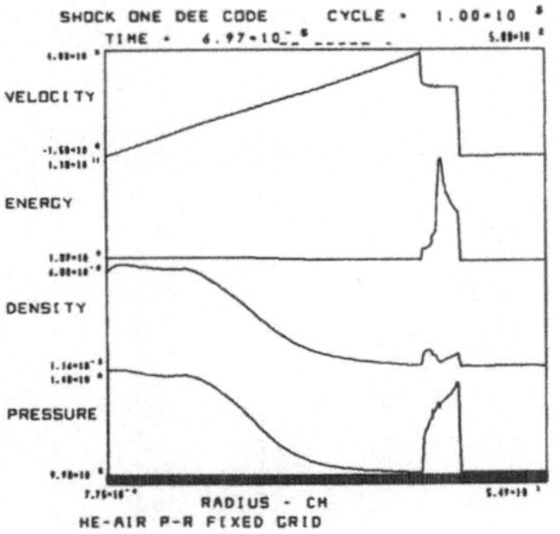

5. One-dimensional solution of expanding HE products and air calculated with the new algorithm using 500 equally spaced zones. Note contact surface separating He products from air.

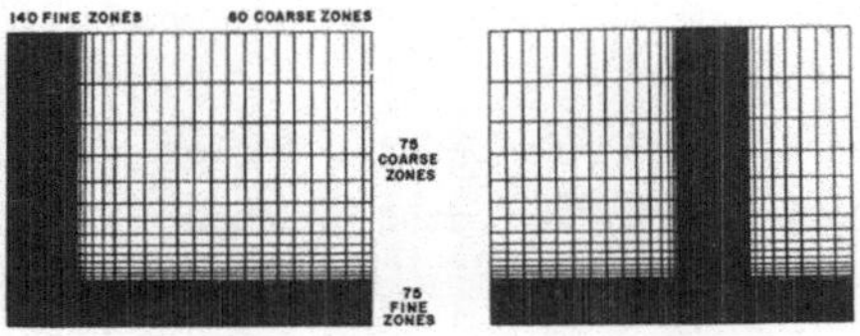

6. Adaptive grid for height-of-burst problem shown (a) initally and (b) at time when transiton to Mach reflection occurs.

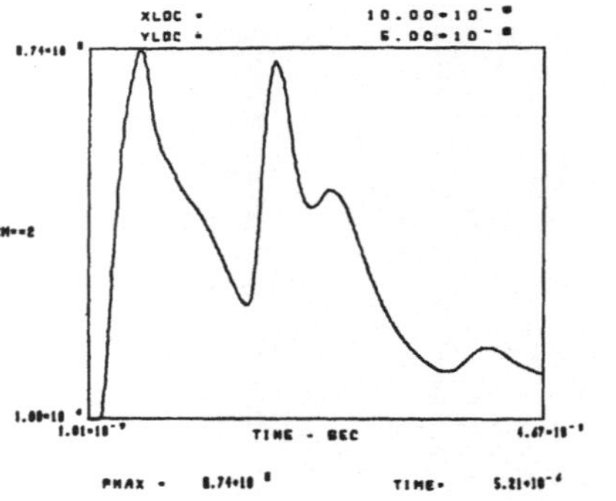

7. Pressure-time histories directly beneath burst site. Note second peak, associated with interaction between shock reflected from ground and following contact surface.

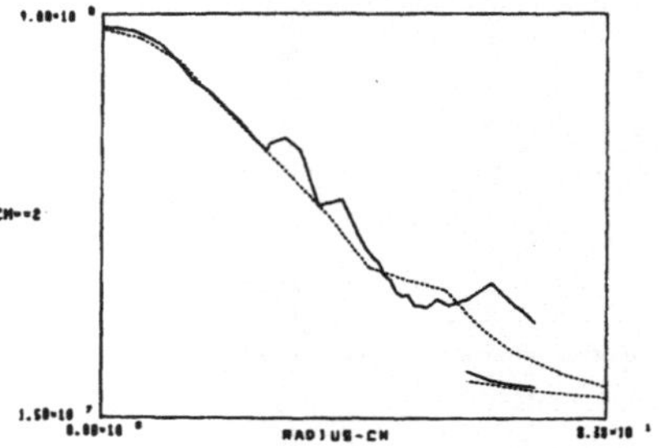

8. Computed peak overpressure vs distance along ground surface. Broken curve represents Carpenter's (1974) data.

This work was supported by the U. S. Defense Nuclear Agency under Subtask Y99QAXSG, Work Unit 00027, "Flux-Corrected Transport Code."

A LOCAL-RELAXATION METHOD FOR SOLVING CONVECTION-DIFFUSION EQUATIONS

E.F.F. Botta and A.E.P. Veldman

Department of Mathematics, University of Groningen

National Aerospace Laboratory NLR, Amsterdam

The Netherlands

INTRODUCTION

As is well-known, iterative schemes for solving second-order central-difference approximations of convection-diffusion equations exhibit convergence difficulties when the diffusion coefficient (the inverse of the Reynolds number or Péclet number) is small. For this reason frequently a trade-off is made in favour of upwind schemes or intermediates between upwind differencing and central differencing. Leading to diagonally-dominant matrices these schemes can be handled with standard techniques like Gauss-Seidel. This numerical convenience is paid for in the form of a loss of accuracy as compared with the central scheme. For a comprehensive discussion hereof see [1].

The trade-off seems to reflect a certain fear of working with non-diagonally-dominant matrices. This fear can be understood partly since, with the usual trial-and-error procedures, it may be difficult to find suitable relaxation factors. It is not fully justified, however, since Young's SOR theory [2] can be of appreciable aid. For matrices with constant coefficients an explicit expression for the optimum relaxation factor can be derived in many situations.

To exploit the latter idea for matrices with varying coefficients we have been inspired by the defect-correction methods which have been designed to transform upwind-difference results into central-difference results [3-6]. When rewritten, the better of these methods appear to be equivalent with SOR methods in which the relaxation factor needs no longer be constant; the relaxation factor in a certain grid point becomes a function of the coefficients of the corresponding local equation. We will call methods of this type local-relaxation (LR) methods. The LR idea is not new - two decades ago Russell [7] already used a very fine LR method - but it has not obtained widespread recognition.

In the paper we will treat an LR method which is based on the optimum SOR method. From a large number of numerical experiments it turns out that for equations with varying coefficients the optimum SOR method can be outperformed by several orders of magnitude. Also for nonlinear equations an LR-method can be very profitable since the relaxation factor is adapted in each relaxation sweep to the momentary magnitude of the coefficients.

SOR THEORY

Let A = D (I-L-U) be a consistently ordered matrix, and let L_ω denote the corresponding SOR matrix for a relaxation factor ω. Then the following relation exists between the eigenvalues μ of the Jacobi matrix B = L + U and the eigenvalues λ of L_ω ($\lambda \neq 0$, $\omega \neq 0$)

$$(\lambda + \omega - 1)^2 = \omega^2 \mu^2 \lambda \quad . \qquad (1)$$

Using this relation Young [2] showed that the SOR method converges for some ω if and only if all eigenvalues μ of B satisfy

$$|\text{Real } \mu| < 1 \quad . \qquad (2)$$

Frequently occurring is the case in which $\mu = \pm \mu_R \pm i \mu_I$ are eigenvalues of B, whereas all other eigenvalues lie in the rectangle $|\text{Real } \mu| \leqq \mu_R$, $|\text{Im } \mu| \leqq \mu_I$. For this case the optimum relaxation factor ω_{opt}, which minimizes the spectral radius ρ (ω) of L_ω, can be derived from (1) using only elementary algebra [8]. The resulting expression is a bit complicated, but a very good approximation has been found which is exact when $\mu_R = 0$ or $\mu_I = 0$, and which fits to the asymptotic behaviour of ω_{opt} as $\mu_I \to \infty$:

$$\tilde{\omega}_{opt} = 2/\{1 + [1-\mu_R^2 + \mu_I^2 \, (1-\mu_R^{2/3})^{-1}]^{1/2}\} \quad . \qquad (3)$$

An impression of the behaviour of the spectral radius ρ (ω) as a function of the relaxation factor ω and the Jacobi spectrum can be obtained from figure 1. It is noted that when μ_I is large (occurring when diagonal dominance is heavily violated) the optimum relaxation factor is very close to ω_{max}, where ω_{max} is the minimum relaxation factor for which convergence occurs.

LR STRATEGY

The LR method we propose uses the following strategy for selecting the relaxation factor ω_i in the i-th equation:

i) consider the Jacobi matrix based on the coefficients of the i-th equation and compute μ_R and μ_I;

ii) find the relaxation factor ω_i by substituting μ_R and μ_I into (3); if desired the resulting expression may be simplified.

APPLICATION TO CONVECTION DIFFUSION EQUATIONS

Convection-diffusion equations of the type

$$\Delta u - f(x,y)\,\frac{\partial u}{\partial x} - g(x,y)\,\frac{\partial u}{\partial y} = 0 \tag{4}$$

will be solved with Dirichlet boundary conditions on a rectangle $0 \leqq x \leqq l_1$, $0 \leqq y \leqq l_2$. The domain is covered with a uniform grid consisting of (N+1) X (M+1) grid points. Applying central differences a discrete equation is generated of the form

$$C_W\, u_{i-1,j} + C_S\, u_{i,j-1} - u_{i,j} + C_E\, u_{i+1,j} + C_N\, u_{i,j+1} = 0 \quad ,$$

in which the coefficients satisfy

$$C_E + C_W + C_N + C_S = 1,\ C_E + C_W \geqq 0,\ C_N + C_S \geqq 0 \quad . \tag{5}$$

It can readily be derived that

$$\mu_R + i\,\mu_I = 2\,(C_E\,C_W)^{1/2}\cos(\pi/N) + 2\,(C_N\,C_S)^{1/2}\cos(\pi/M) \quad . \tag{6}$$

Substituting (6) into (3) the relaxation factor is expressed in the local coefficients. Under conditions (4) the resulting expression is approximated by

$$\omega^* = \min\left\{\omega_o,\ \frac{2}{1 + |C_E - C_W| + |C_N - C_S|}\right\} \quad \text{when } C_E\,C_W\,C_N\,C_S \geqq 0\ ;$$

$$\omega^* = \frac{2}{1 + \gamma_1 |C_N - C_S|} \quad \text{when } C_W\,C_E > 0 \text{ and } C_N\,C_S < 0\ ; \tag{7}$$

$$\omega^* = \frac{2}{1 + \gamma_2 |C_E - C_W|} \quad \text{when } C_W\,C_E < 0 \text{ and } C_N\,C_S > 0\ ;$$

where $\gamma_1 = [1 - (C_E + C_W)^{2/3}]^{-1/2}$ and $\gamma_2 = [1 - (C_N + C_S)^{2/3}]^{-1/2}$. ω_o is the optimum SOR factor for the second-order part in (3) (i.e. the Laplace equation). It is noted that for equations of the type (4) γ_1 and γ_2 depend only on the mesh sizes; in the special case of equal mesh sizes we have $\gamma_1 = \gamma_2 = 1.644$. It can be proved that $\omega^* < \omega_{max}$ when (5) is fulfilled [9].

EXAMPLES

The performance of the LR method (7) will now be compared with the performance of the optimum SOR method. Due to the construction of the LR method, for equations with constant coefficients both methods behave about the same. For equations with variable coefficients, however, the LR method can be much more efficient, as can be seen from the examples below.

Table I corresponds to some one-dimensional equations (set $C_N = C_S = 0$ in (7)). The linear equation $u_{xx} - 10^4 x^2 u_x = 0$ has been solved on $0 \leqq x \leqq 1$ with homogeneous Dirichlet boundary conditions. Starting from $u = x(1-x)$ the number of iterations to reach max $|u_i| < 10^{-6}$ is tabulated for various mesh sizes h. The optimum SOR results (obtained by means of a discrete scan of the ω-axis with small steps Δω) are clearly outperformed by the LR method.

Also a nonlinear equation is presented in Table I: $u_{xx} - Re\, u^2 u_x = 0$, again with homogeneous Dirichlet conditions on $0 \leqq x \leqq 1$. A mesh size $h = 1/20$ has been used for Reynolds numbers Re = 1, 10^2, 10^4. For the largest value of Re a significant gain over the optimum SOR method is obtained. This can be explained since the LR method uses overrelaxation in the end of the iterative process when the iterates u_i approach zero, whereas in the SOR method the underrelaxation which is required in the first sweeps is maintained throughout.

Table II gives some two-dimensional comparisons on the unit square for homogeneous Dirichlet conditions. A mesh size 1/20 has been used in both directions. In the first example ($f = g = Re\, x^2$) the Jacobi eigenvalues lie on either the real axis (small Re) or the imaginary axis (large Re). In the second example ($f = Re\, x^2$, $g = 0$) for large Re both the real and imaginary part of the eigenvalues is nonzero.

A large number of numerical experiments has been performed [9]. The above examples are believed to be representative. The LR strategy (7) can compute with the optimum SOR method; for equations with large first-order terms of which the coefficients are strongly varying the optimum SOR method is clearly outperformed. Moreover it is emphasized that the LR strategy provides an explicit formula for the relaxation factor, whereas in the optimum SOR method usually a trial-and-error procedure is required.

In [9] a comparison has been made with several other LR methods presented in the literature [4,5,7,10,11,12]. As a main conclusion the (remarkable) observation is made that the relative performance of the LR methods for variable-coefficient equations can be predicted by SOR theory (which is derived for constant-coefficient equations).

A DRIVEN-CAVITY EXAMPLE

The present LR method has also been tested on a driven-cavity problem. We have recomputed the case with the largest Reynolds number for which central-difference results have been presented. It is a calculation on a 17x17 grid for Re = 5.10^4 by Kurtz, et al. [13]. They used the method-of-lines and required more than

8000 CPU seconds on a CDC 6600. We have used the LR strategy to find the relaxation factors for the vorticity-transport equation. In the stream-function equation and in the boundary condition for the vorticity the relaxation factor has been chosen by trial and error. The best choice for the latter two relaxation factors resulted in a number of 494 iterations to reach a truncation level believed to be comparable to the one obtained in [13]. This required about 2 CPU seconds on a CDC Cyber 170-760 (which is about 3 times faster than a CDC 6600).

CONCLUSIONS

For constant-coefficient equations Young's SOR theory can be used to find an explicit formula for the optimum relaxation factor, even in cases when diagonal dominance is heavily violated. When desired simple approximations of ω_{opt} can be derived. Combining these with the LR strategy leads to a simple method which for varying-coefficient equations, and for nonlinear equations, is able to outperform the optimum SOR method. The LR method can easily solve central-difference approximations of convection-diffusion equations, thus eliminating the need for the trade-off in favour of the upwind-type schemes mentioned in the introduction.

REFERENCES

[1] P.M. Gresho and R.L. Lee, Comp. Fluids 9 (1981) 223-253.
[2] D.M. Young, Iterative solution of large linear systems, Academic Press, New York, 1971.
[3] S.C.R. Dennis and G.Z. Chang, Phys-Fluids 12, Suppl. II (1969) 88-93.
[4] A.E.P. Veldman, Comp. Fluids 1 (1973) 251-271.
[5] D. Dijkstra, Ph. D. thesis, University of Groningen, 1974.
[6] C.W. Richards and C.M. Crane, Appl. Math. Modelling 2 (1978) 59-61.
[7] D.B. Russell, ARC Report R & M 3331, Oxford, 1963.
[8] A. Rigal, J. Comput. Phys. 32 (1979) 10-23.
[9] E.F.F. Botta and A.E.P. Veldman, to be published in J. Comput. Phys.; also NLR MP 81052 U (1981).
[10] N. Takemitsu, J. Comput. Phys. 36 (1980) 236-248.
[11] J. Strikwerda, SIAM J. Sci. Stat. Comput. 1 (1980) 119-130.
[12] L.W. Ehrlich, J. Comput. Phys. 44 (1981) 31-45.
[13] L.A. Kurtz, R.E. Smith, C.L. Parks and L.R. Boney, Comp. Fluids 6 (1978) 49-70.

TABLE I

Method	$u_{xx} - 10^4 x^2 u_x = 0$			$u_{xx} - \mathrm{Re}\, u^2 u_x = 0$, $h = 1/20$		
	h= 1/10	h= 1/40	h= 1/160	Re= 1	Re= 10^2	Re= 10^4
optimum SOR	1525	3409	15595	46	43	5050
present method	433	227	109	51	48	44

TABLE II

Method	$\Delta u - \mathrm{Re}\, x^2 u_x - \mathrm{Re}\, x^2 u_y = 0$			$\Delta u - \mathrm{Re}\, x^2 u_x = 0$		
	Re= 1	Re= 10^2	Re= 10^4	Re= 1	Re= 10^2	Re= 10^4
optimum SOR	46	310	2053	46	202	1328
present method	50	26	300	50	36	366

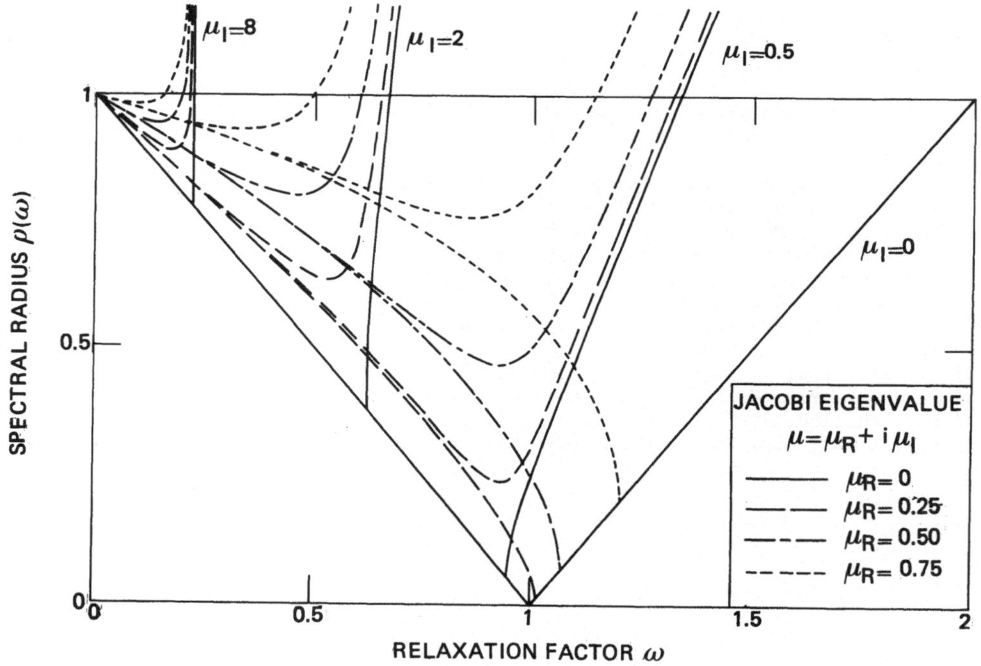

Fig. 1 Relation between spectral radius, relaxation factor and Jacobi spectrum

A NUMERICAL TREATMENT OF TWO-DIMENSIONAL FLOW IN A BRANCHING CHANNEL

J. S. Bramley, and S. C. R. Dennis

Department of Mathematics,
University of Strathclyde,
Glasgow, United Kingdom

Department of Applied Mathematics,
University of Western Ontario,
London, Ontario, Canada

ABSTRACT

A numerical method for treating the steady two-dimensional flow of a viscous incompressible fluid in a branching channel is given. The upstream and downstream boundary conditions are discussed and a logarithmic transformation is applied to the coordinate measuring distance downstream in order to extend the numerical solution far enough downstream. Two methods are presented for dealing with the singularity in the vorticity at the sharp corners associated with the geometrical division of the flow. The Navier-Stokes equations are written in terms of the stream function and vorticity giving the usual two coupled nonlinear partial differential equations. These equations are solved using the method of Dennis and Hudson (1978). The effect of the relative widths of the channels upstream and downstream of the branch on the separation of the flow is discussed using results obtained from three separate grid sizes.

INTRODUCTION

The problem of determining the flow in branching tubes is of considerable importance in the fluid mechanics of blood vessels; many interesting problems have been described by Pedley (1980). In most of the practical applications the geometrical configurations are complicated and the motion is three-dimensional, but at least some information is obtainable from simplified two-dimensional models. Blood flow is also pulsatile, but again information obtained from simplified models in which the flow is assumed to be steady can provide a useful start. The assumption of steady two-dimensional channel flow is made in this paper and one type of branched configuration which has been studied is indicated in figure 1. The tube (or channel) into which the fluid flows after branching is termed the daughter tube. The situation shown in figure 1 can therefore be referred to as the case of a wide daughter tube, in which $d/c = \sqrt{2}$. The angle of the geometrical bifurcation, or the branching angle, is defined to be the total angle between the channel walls at O. In the present paper this is taken to be 90°; this is a particularly suitable angle which enables a very simple grid to be fitted to the computational domain. There is also a second simple geometrical configuration in which the point of bifurcation is at O' rather than O in figure 1. In this case $c/d = \sqrt{2}$, which may be referred to as the case of a narrow daughter tube.

One of the main interests in this problem is to determine under what conditions the fluid will separate near the point A of figure 1. Several previous investigations have considered this question. Kandorpa and Davids (1976) used a finite element method to solve the two-dimensional branching problem with a branching angle of 60° and $d/c = 1.5$ for a Reynolds number of 1000. They found that the flow separates from the wall at A and reattaches downstream of A. Lynn, Fox and Ross (1972) have considered a very similar problem with $d/c = 1$ using the upwind difference method proposed by Greenspan (1968). Calculations were carried out for Reynolds numbers 100, 500 and 1000 and it was found that the flow does not separate at any stage. Gokhate, Tanner and Bischoff (1978) used a finite element method to consider a model of the canine aorta and compare the numerical results with those available from experimental data.

It is clear from previous work that the question of whether separation takes place or not depends in some way on the geometry of the branching. This is found to be the case in the present study. Two cases are considered, in each of which the branching angle is 90°, but distinguished by the cases of the wide daughter tube, with branching at the point O of figure 1 and the narrow daughter tube, with branching at O'. The results with regard to the separation of the flow are considerably different in the two cases. The numerical method is essentially the same in the two cases and incorporates several features, notably in the treatment of the boundary condition upstream, the flow in the downstream portion of the channel, and the singularities which exist at the points O and A of figure 1. The vorticity is infinite at each of these points and two separate methods of dealing with the problem are considered which give completely consistent results. The boundary conditions upstream are dealt with by the use of the theory of the asymptotic decay to Poiseuille flow; the flow downstream is calculated by employing a logarithmic transformation in this region of the domain.

BASIC EQUATIONS

A typical configuration for the branching channel is shown in figure 1, where only the upper half of the branch is shown. The axis of symmetry is the dashed line EO and the walls of the channel are shown hatched. The angle of bifurcation at O is 90°, this being a realistic branching angle for blood vessels. The coordinate axes used are also defined in figure 1 where the flow is from left to right. We introduce the dimensionless streamfunction ψ and the dimensionless vorticity ζ in the Navier-Stokes equations

$$\frac{\partial^2\psi}{\partial x^2} + \frac{\partial^2\psi}{\partial y^2} = -\zeta \,, \tag{1}$$

$$\frac{\partial^2\zeta}{\partial x^2} + \frac{\partial^2\zeta}{\partial y^2} = R\left(\frac{\partial\psi}{\partial y}\frac{\partial\zeta}{\partial x} - \frac{\partial\psi}{\partial x}\frac{\partial\zeta}{\partial y}\right), \tag{2}$$

where R is the Reynolds number defined as Uc/ν, where 2Uc is the flux through the upstream channel of width 2c. Downstream of the bifurcation, the variables x' and y' replace x and y in equations (1) and (2) and this formulation is used in the downstream part of the flow. Upstream the channel is situated between $y = \pm 1$ and we obtain a numerical solution between $y = 0$ and $y = 1$. Downstream of the branch we solve for the flow in the region between $y = 0$ and $y = \sqrt{2}$ for the wide daughter tube and between $y = 0$ and $y = 1/\sqrt{2}$ for the narrow daughter tube. The boundary conditions are: $\psi = 0$ along EO and OF, $\zeta = 0$ along EO, $\psi = 1$ along CA and AB, and the Woods (1954) boundary condition

$$\zeta_b = 3(\psi_b - \psi_i)/h^2 - \frac{1}{2}\zeta_i \tag{3}$$

is used along OF, CA and AB. Here the subscript b refers to a value at the appropriate boundary point, the subscript i refers to the internal grid point most immediate to b and h is the mesh spacing. At a distance equal to one half of the channel width upstream we apply the boundary condition,

$$\frac{\partial\psi}{\partial x} = \alpha_1\left(\frac{3}{2}y - \frac{1}{2}y^3 - \psi\right), \tag{4}$$

described by Bramley and Dennis (1982). Here α_1 is the dominant eigenvalue for the stationary perturbation of Poiseuille flow. The values of α_1 for the range of Reynolds numbers used in this paper are given by Bramley and Dennis (1982). At a distance equal to d/2 downstream of the bifurcation we apply the transformation $r = \ell n(1+\varepsilon x')$ to the x' coordinate, where ε is a constant for a particular mesh such that $h = \ell n(1+\varepsilon h)$ and h is the spacing between mesh points. The use of the constant ε avoids the necessity of interpolation at the station where the change of variable is made. At the downstream boundary we apply the condition of Poiseuille flow

$$\psi = \frac{1}{2} y^2(3-\sqrt{2}y). \tag{5}$$

There are several reasons for not using a condition similar to (4) downstream. The logarithmic transformation means that with relatively few extra mesh points it is possible to apply the Poiseuille flow condition a long way further downstream. One problem associated with using a condition similar to (4) downstream is that the flow is not symmetrical at small distances from A and so we would need at least two eigenvalues in obtaining a result comparable with (4). After putting $\eta = \ln(1+\varepsilon x')$ equations (1) and (2) become

$$E^2\left(\frac{\partial^2\psi}{\partial\eta^2} - \frac{\partial\psi}{\partial\eta}\right) + \frac{\partial^2\psi}{\partial y'^2} = -\zeta, \text{ and} \tag{6}$$

$$E^2\frac{\partial^2\zeta}{\partial\eta^2} + \frac{\partial^2\zeta}{\partial y'^2} = \frac{\partial\zeta}{\partial\eta}\left(E^2+ER\,\frac{\partial\psi}{\partial y'}\right) - RE\,\frac{\partial\zeta}{\partial y'}\frac{\partial\psi}{\partial\eta}, \tag{7}$$

where $E \equiv \varepsilon/e^{\eta}$.

Before explaining how we deal with the problem caused by the singular vorticity at the sharp corners we give some details of the mesh near A for the wide daughter tube in figure 2. The mesh will be different for the narrow daughter tube but similar ideas are used. Figure 2 shows how the 90° bifurcation enables the upstream mesh and the downstream mesh to be joined together. A similar mesh is used at O but due to lack of space we will only give details of the mesh at A. Two different methods are used for dealing with the singularities in vorticity. The stream function is not singular anywhere so an alternative form is only needed for deriving the vorticity at points where the usual difference equations employ values of ζ at the singular points themselves. We use an expansion derived by Moffatt (1964) for small Reynolds numbers. In the present case the Reynolds numbers are not small but near enough to the singular point the local Reynolds number will be small enough for the Moffatt expansion to give a reasonable result. Using this expansion it can be shown that near A

$$\zeta = C_1 r^{\lambda_1-2}(\lambda_1-1)\cos(\lambda_1-2)\theta + C_2 r^{\lambda_2-2}(\lambda_2-1)\sin(\lambda_2-2)\theta, \tag{8}$$

where r is the radial distance from A and the angle θ is as defined in figure 2 with $\theta = 0$ bisecting the channel wall at A. The values of λ_1 and λ_2 in (8) are given by $\lambda_1 = 1.67358$ and $\lambda_2 = 2.30209$. In equation (8) the term containing C_1 corresponds to the antisymmetric part of the flow and the term containing C_2 the symmetric part of the flow about $\theta = 0$. To use equation (8) we need to derive the constants C_1 and C_2 and this is done by using the vorticity at points p_1 and p_4. The corresponding expression for the flow about O is

$$\zeta = C_3 r^{\lambda_3-2}(\lambda_3-1)\sin(\lambda_3-2)\theta \tag{9}$$

where $\lambda_3 = 1.90853$. After obtaining the constants C_1 and C_2 the vorticity at points p_2 and p_3 of figure 2 can be calculated.

The alternative approach to using the Moffatt expansion at A is to introduce the five extra points, denoted by a circle in figure 2, and use various five point meshes to calculate the vorticity at p_2 and p_3 without using the singular vorticity at A. A similar method at O only requires one extra point.

The elliptic partial differential equations are solved by using the method derived by Dennis and Hudson (1978) and described by Dennis and Smith (1980). If we use in the usual manner the subscripts 0, 1, 2, 3, 4 to denote quantities at the grid points (x_o,y_o), (x_o+h,y), (x_o,y_o+h), (x_o-h,y_o) and (x_o,y_o-h), the finite difference equations can be written as

$$\psi_1 + \psi_2 + \psi_3 + \psi_4 - 4\psi_o + h^2\zeta_o + C_o = 0$$

$$(1-\frac{1}{2}Rhu_o+\frac{1}{8}R^2h^2u_o^2)\zeta_1 + (1-\frac{1}{2}Rhv_o+\frac{1}{8}R^2h^2v_o^2)\zeta_2$$
$$+ (1+\frac{1}{2}Rhu_o+\frac{1}{8}R^2h^2u_o^2)\zeta_3 + (1+\frac{1}{2}Rhv_o+\frac{1}{8}R^2h^2v_o^2)\zeta_4$$
$$- \{4+\frac{1}{4}R^2h^2(u_o^2+v_o^2)\}\zeta_o + C_o^* = 0$$

where (u_o,v_o) are the velocity components at (x_o,y_o). Here C_o and $C_o{}^*$ are correction terms of order h^4 which may be used to improve the accuracy of the solutions and are not used in this paper but will be the subject of future work. The equations are solved systematically using the Gauss-Seidel iterative procedure.

The details of the above method all pertain to the wide daughter tube but similar methods can be applied in the case of the narrow daughter tube. The most drastic difference will be that the extra points in figure 2 will now be on the upstream side A. The details of the analysis are virtually identical.

RESULTS

We first comment upon the results for the wide daughter tube. The calculations were performed for Reynolds numbers R = 50, 100, 500 and 1000. Each solution was carried out with M_o = 10, 20 and 40 meshes across the upper half of the upstream channel with a corresponding number of meshes across the downstream channel. The boundary condition given by (4) was applied at a distance c upstream of A. To test this boundary condition the domain of computation was then extended to a distance 2c upstream of A and Poiseuille flow was used as an upstream boundary condition. The eigenvalue α_1 was then calculated from the numerical solution using (4) at a position approximately halfway between C and A. Table I compares the theoretical eigenvalue $\alpha_1(R)$ with the eigenvalues calculated from the numerical solution for each value of R. It will be seen that the agreement is good for the fine mesh but accuracy is lost for the coarse mesh at the larger Reynolds numbers. This shows that the application of equation (4) as an upstream boundary condition gives accurate results; it has therefore been used throughout the remainder of this paper.

For the wide daughter tube the flow does not separate for Reynolds number 50 but there is separation for Reynolds number 100, 500 and 1000. Figure 3 gives the stream lines for a typical case of Reynolds number 500. The results obtained using the coarse mesh do not give separation. The corresponding result for the medium mesh gives separation but the region of recirculation cannot be traced. On the finest mesh the recirculation can clearly be seen. While the branching angle used in this paper is 90° and that used by Kandarpa and Davids (1976) is 60°, both results give separation downstream of A for Reynolds numbers near 500 and 1000. For the narrow daughter tube the flow does not separate for the range of Reynolds numbers (50, 100, 500 and 1000) currently calculated.

Figure 4 compares the wall vorticity downstream of A for the three different meshes in the case of the wide daughter tube for R = 500; the results shown were those obtained using the extra points around A. There are considerable differences between the results for the three mesh sizes. This is to be expected because the coarse mesh will not be able to model the sharp variations in vorticity near the walls. The results for the finest mesh M_o = 40 could be checked using an even finer grid but the amount of computer time required is prohibitive. For these results the logarithmic transformation was applied to the variable x' after a distance 0.5d downstream of A and then $3M_o$ mesh points were employed in the η direction before applying the downstream boundary condition at a station at which $x' \approx 19.3$.

Figure 5 compares on the finest grid, the wall vorticity downstream of A using the Moffatt expansion (crosses) with the wall vorticity obtained using the extra points (continuous line) for Reynolds number 100. The Moffatt expansion is, of course, in theory only strictly applicable in the limit of zero Reynolds number since it is derived from consideration of the equations obtained by neglecting the convective terms in (2). Nevertheless, the expansion must hold over a small region close enough to the appropriate singular point for any value of the Reynolds number because the local Reynolds number is zero here. Thus the expansion at least indicates the nature of the leading terms in the singularity at a corner point for all values of the Reynolds number and to this extent deals effectively with the situation at the corner. The comparison of the results in figure 5 indicate that the Moffatt expansion method does in fact give a good comparison with calculations which are completely independent of the use of this technique. The good agreement of the results obtained from two quite different methods of dealing with the singular points gives a measure of confidence that each method is satisfactory.

In future work we hope to obtain more results for further comparisons, particularly in the case of the narrow daughter tube. It is hoped to apply a more accurate method when solving the partial differential equations. The present results do, however, indicate that the various techniques discussed in this paper can be used satisfactorily to obtain reasonable numerical results. It is quite clear that small grids must be used if satisfactory results near the corner A are to be obtained This is obvious from figure 4, where the use of smaller grids would undoubtedly bring the point of separation nearer the corner. It is impossible to say at the moment whether in the limit of very small grid size the fluid would actually separate at the corner itself. It is, however, verified that from the evidence of the present and previous investigations of this problem, the geometry of the branching has an important effect on the question of separation.

Part of this work was carried out while one of us (J.S.B.) was a visitor to the University of Western Ontario. Financial support by the Natural Sciences and Engineering Research Council of Canada is acknowledged.

REFERENCES

Bramley, J.S. and Dennis S.C.R., 1982, J.Comp. Phys. 47.

Dennis, S.C.R. and Hudson, J.D., 1978, Proceedings of the First International Conference on Numerical Methods in Laminar and Turbulent Flow. p.69, Pentech Press, London, U.K., 1978.

Dennis, S.C.R. and Smith, F.T., 1980, Proc.Roy.Soc.Lond. A372, 393.

Gokhale, V.V., Tanner, R.I. and Bischoff, K.B., 1978, J.Biomechanics 11, 241.

Greenspan, D., Lectures on the Numerical Solution of Linear, Singular and Nonlinear Differential Equations. Prentice-Hall, 1968.

Kandarpa, K. and Davids, N., 1976, J.Biomechanics, 9, 735.

Lynn, N.S., Fox, V.G. and Ross, L.W., 1972, Biorheology, 9, 61.

Moffatt, H.K., 1964, J.Fluid Mech., 18, 1.

Pedley, T.J., The Fluid Mechanics of Large Blood Vessels. Cambridge University Press, 1980.

Woods, L.C., 1954, Aeronaut.Q. 5, 176.

Reynolds number	Number of meshes 10	20	40	$\alpha_1(R)$
50	3.04	3.09	3.106	3.107
100	2.8	2.97	2.995	2.999
500	1.7	2.56	2.79	2.831
1000	1.4	2.12	2.68	2.782

Table I
Comparison of dominant eigenvalue calculated from upstream flow compared with the theoretical eigenvalue $\alpha_1(R)$

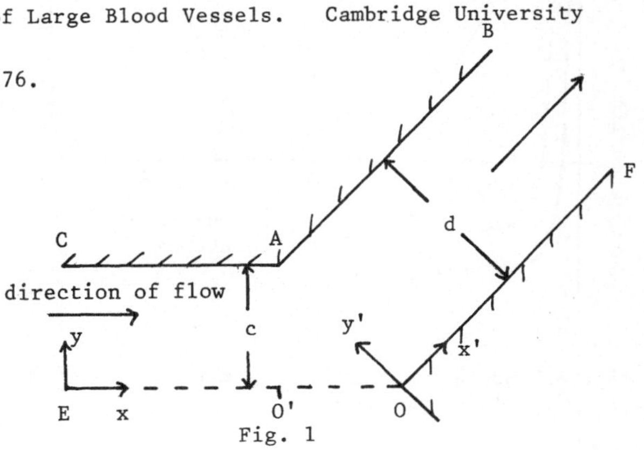

Fig. 1

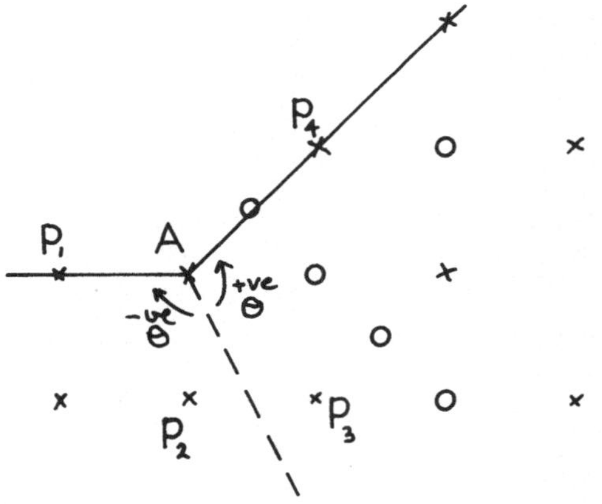

Fig. 2.

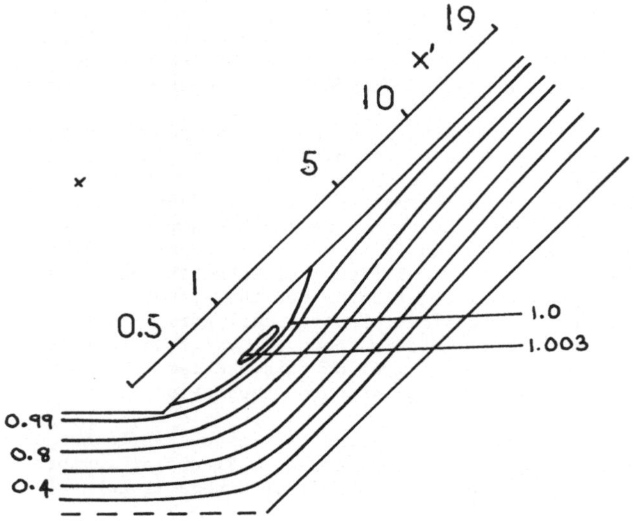

Fig. 3. Streamlines for R = 500

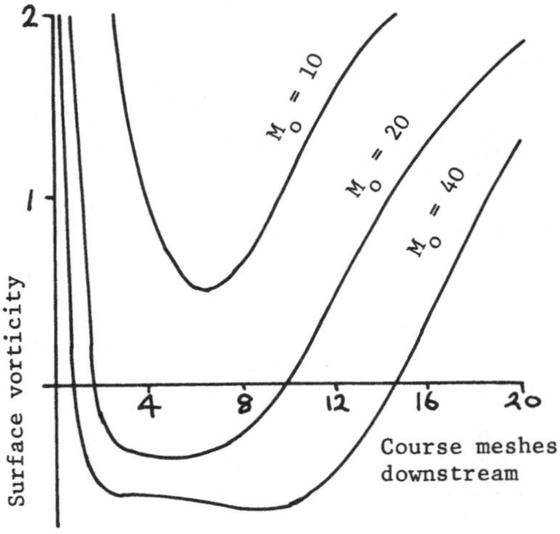

Fig. 4. Surface vorticity against x' for R = 500

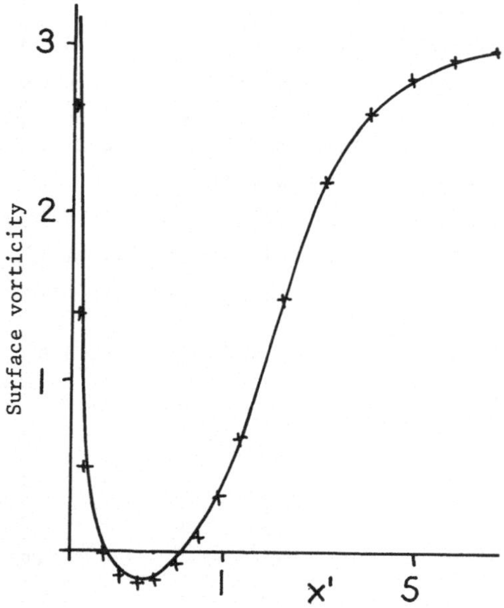

Fig. 5. Surface vorticity against x' for R = 100, - extra points, + Moffatt.

FINITE ELEMENT LEAST SQUARE METHOD FOR SOLVING FULL STEADY EULER EQUATIONS IN A PLANE NOZZLE†

C.H. Bruneau*, J.J. Chattot**, J. Laminie***, J. Guiu-Roux****

* Université de Paris-Sud, Orsay, France, et Office National d'Etudes et de Recherches Aérospatiales (O.N.E.R.A.), Châtillon, France.
** MATRA, Vélizy, France.
*** Université de Paris-Sud, Orsay, et C.N.R.S. E.R.A. 297, Orsay, France.
**** Université de Paris-Sud, Orsay, France.

SUMMARY

A finite element least square method is applied to the steady Euler equations in a nozzle or a channel. For the capture of shock waves an artificial density formula is used. Fast convergence is achieved with I.C.C.G. algorithm.

INTRODUCTION

Progress has been accomplished towards the solution by relaxation of the full Euler equations modelling steady transonic flows of perfect fluid [1-3]. The approach is based on a least square formulation which transforms hyperbolic problems into elliptic type problems, and first order systems into second order systems with positive-definite associated matrix. This insures a proper domain of dependance of the numerical scheme, regardless of the local type of the flow, as well as favourable convergence properties.

In this paper we present the first results obtained with this method concerning the full Euler equations. In the first part, the equations, the boundary conditions and the least square formulation are described. The choice of the minimization variables, the finite element approximation and the fixed-point algorithm on ρ are discussed. For the capture of shock waves an artificial compressibility formula [4] is implemented in the numerical code, and our experience is reported in the second part. The third part is devoted to the presentation of some numerical solutions of transonic flows in nozzles and channels.

I. Variational formulation of the Euler equations.

The conservation of mass, momentum and energy (Bernoulli's equation) in divergence form are written with the usual notations :

$$(1)\quad \begin{cases} \dfrac{\partial \rho u}{\partial x} + \dfrac{\partial \rho v}{\partial y} = 0 \\[2mm] \dfrac{\partial \rho u^2 + p}{\partial x} + \dfrac{\partial \rho uv}{\partial y} = 0 \\[2mm] \dfrac{\partial \rho uv}{\partial x} + \dfrac{\partial \rho v^2 + p}{\partial y} = 0 \end{cases}$$

$$(2)\quad H = \frac{\gamma}{(\gamma-1)}\frac{p}{\rho} + \frac{u^2+v^2}{2} \quad ; \quad (\gamma=1.4)$$

This system is closed by the following boundary conditions : the entropy ($s = c_v \operatorname{Log} \frac{\gamma p}{\rho^\gamma}$) and the flow direction are given in the inlet section, the tangency condition is imposed on the walls and the pressure is specified downstream (subsonic

---

† Work performed with financial support of D.R.E.T.

exit condition assumed) (figure 1). In the system (1) and (2) of three P.D.E. and one algebraic equation the four unknowns are ρ, u, v, p. It was chosen to get the solution of this system through a fixed point algorithm (Eq.(2)) and to solve (1) by a minimization method.

u, v and p are the minimization variables. This choice is based on the following arguments :

i) the minimization with respect to p yields a well-behaved elliptic equation regardless of the local flow type ;
ii) the pressure is specified at the exit of the nozzle or channel which gives a Dirichlet boundary condition for pressure ;
iii) ρ is computed from Bernoulli's equation and the formulation (u,v,p) implies a fixed-point algorithm on ρ which can be conveniently modified to include an artificial density formula ;
iv) this formulation allows computation of low speed flows (incompressible flows) and stagnation points are not singular points as in the unsteady Euler equations.

The fixed point algorithm reads :

$$(3) \qquad \begin{cases} \rho_1 \text{ given} \\ \rho_{n+1} = \dfrac{\gamma p_n}{(\gamma-1)\left(H - \dfrac{u_n^2 + v_n^2}{2}\right)} \end{cases}$$

where $Q_n = (u_n, v_n, p_n)$ is solution for fixed ρ_n of equations (1) linearized by Newton's method as :

$$Q_{n-1+j/q} = Q_{n-1+(j-1)/q} + \tilde{Q}_{j/q} \quad \text{where } j = 1,\ldots,q$$

and $\tilde{Q}_{j/q} = (\tilde{u}_{j/q}, \tilde{v}_{j/q}, \tilde{p}_{j/q})$ solution of

(we note $r_j = n-1+(j-1)/q$) :

$$(4) \qquad \begin{cases} A = \dfrac{\partial}{\partial x}(\rho_n \tilde{u}_{j/q}) + \dfrac{\partial}{\partial y}(\rho_n \tilde{v}_{j/q}) - \dfrac{\partial}{\partial x}(\rho_n u_{r_j}) - \dfrac{\partial}{\partial y}(\rho_n v_{r_j}) = 0 \\[2ex] B = \dfrac{\partial}{\partial x}(2\,\rho_n u_{r_j}\tilde{u}_{j/q} + \tilde{p}_{j/q}) + \dfrac{\partial}{\partial y}(\rho_n v_{r_j}\tilde{u}_{j/q} + \rho_n u_{r_j}\tilde{v}_{j/q}) \\ \qquad\qquad - \dfrac{\partial}{\partial x}(\rho_n u_{r_j}^2 + p_{r_j}) - \dfrac{\partial}{\partial y}(\rho_n u_{r_j} v_{r_j}) = 0 \\[2ex] C = \dfrac{\partial}{\partial x}(\rho_n v_{r_j}\tilde{u}_{j/q} + \rho_n u_{r_j}\tilde{v}_{j/q}) + \dfrac{\partial}{\partial y}(2\,\rho_n v_{r_j}\tilde{v}_{j/q} + \tilde{p}_{j/q}) \\ \qquad\qquad - \dfrac{\partial}{\partial x}(\rho_n u_{r_j} v_{r_j}) - \dfrac{\partial}{\partial y}(\rho_n v_{r_j}^2 + p_{r_j}) = 0 \end{cases}$$

together with homogeneous boundary conditions.

At each step j the boundary conditions for Q_{r_j} are updated according to :

. Entrance section $u_{r_j} = \sqrt{\frac{2}{1+g^2}\left[H - \frac{1}{(\gamma-1)}(\gamma p_{r_j})^{(\gamma-1)/\gamma}\right]}$; (s=0)

$v_{r_j} = g\, u_{r_j}$; (g is the streamline slope)

. Wall condition $v_{r_j} = f' u_{r_j}$; (f' is the wall slope)

. Exit section $p_{r_j} = p_{exit}$.

The system (4) is not readily solvable (first order equations) and is thus embedded into least square formulation which yields a symmetric positive definite matrix with better conditioning.

The following quadratic functional

$$J_{\rho_n, r_j}(\tilde{Q}_{j/q}) = \frac{1}{2}\int_\Omega (A^2 + B^2 + C^2)\,dxdy$$

is minimized with respect to its 3N variables (N is the number of nodes). This is done with the use of the incomplete Cholesky conjugate gradient method (I.C.C.G.).

The discretization is performed by linear quadrilateral finite element approximation dealing with the groups of variables ($\rho u, \rho v, \rho u^2+p, \ldots$) to enforce the conservation properties [3] ; for instance $(\rho u)_h = \sum_{i=1}^{N} (\rho u)_i \phi_i = \sum_{i=1}^{N} \rho_i u_i \phi_i$.

II. <u>Artificial compressibility</u>.

The fixed point algorithm on ρ is stable for subsonic flows ; however for supersonic flows an artificial viscosity term must be introduced. In a finite element formulation the most convenient way consists in adding an upwind term to the density [4] . ρ_n is replaced everywhere in (4) by $\overline{\rho_n}$ according to :

(5) $$\overline{\rho_{n,i}} = \rho_{n,i} - \alpha\mu_i \Delta s_i \frac{\partial \rho_{n,i}}{\partial s_i}$$

where i denotes the index along the streamlines,

α is a coefficient,

$\mu_i = \text{Max}\ \frac{1}{M_c^2}(0, M_c^2 - \frac{M_c^2}{M_i^2})$

s_i is the curvilinear abscissa at point i,

(in numerical experiments the critical Mach number $M_c = 1$).

With $\alpha = 1$, this method does not allow the shock to move correctly and the flow overshoots before the shock.

In order to decrease the speed of the flow before the shock we take $1<\alpha<M_i^2$ so that $\alpha\mu_i \leq 1$. Besides this, we use a method similar to that proposed by Habashi and Hafez [4] to solve the potential equation ($\mu_i = \text{Max}(\mu_{i-1}, \mu_i)$). Here we define μ_i from a mean Mach number value $M_i^2 = \frac{1}{2}(M_{i-1}^2 + M_{i+1}^2)$. In this way we add

an upwind term to the density at these subsonic points for which the new M_i^2 is greater than 1. In this case the formula (5) is applied with $\alpha \leq 1$.

III. Numerical results.

The numerical tests concern either a convergent divergent nozzle with parabolic walls of ratio 0.9 or a parallel channel having a 4.2 % thick circular arc bump on the upper wall. This second problem is a test from the GAMM workshop [5].

The calculation of the flow in the nozzle has been done for several values of the exit pressure. We get either a subsonic or slightly supercritical shockless flow, or a chocked flow with a shock wave. Figures 2 and 3 show this latter case for an exit pressure $p_{exit} = 0.983$. On figure 4 we can see ρ and $\overline{\rho}$ on the upper wall ; the difference between these two quantities gives an estimation of the error introduced by the upwind term. This difference increases with the intensity of the shock when we lower the exit pressure. It seems that the method of artificial compressibility is not well adapted for strong shock for which the density gets close to zero.

For the parallel channel we impose an exit pressure $p_{exit} = 0.843$ corresponding to a Mach number $M = 0.85$ in isentropic flow [5]. The mesh and the results are shown in figures 5, 6 and 7. Notice the difference in shape of the iso-Mach curves before and after the profile due to the jump of entropy through the shock. The use of a mesh size much cruder than that given for the workshop explains the more upstream position of the shock.

CONCLUSION.

First results for the full steady Euler equations show the interest of the least square formulation and of the finite element discretization for the groups of variables. The solution of the algebraïc systems by the incomplete Choleski conjugate gradient algorithm (I.C.C.G.) yields an implicit scheme having good properties of convergence (figure 8). Further progress is necessary for the capture of strong shocks.

REFERENCES.

[1] . Chattot,J.J., Guiu-Roux,J., and Laminie, J., Résolution numérique d'une équation de conservation par une approche variationnelle, Proc. 6th ICNMFD, Tbilissi, USSR, June 20-25, 1978. Lecture Notes in Physics, 90, Springer-Verlag, 1979.

[2] . Chattot,J.J., Guiu-Roux,J., and Laminie,J., Finite element calculation of steady transonic flow in nozzles using primary variables, Proc. 7th ICNMFD, Stanford and NASA/Ames, USA, June 23-27, 1980, Lecture Notes in Physics, 141, Springer-Verlag, 1981.

[3] . Chattot,J.J., Guiu-Roux,J., and Laminie,J., Numerical solution of a first-order conservation equation by a least square method, Int. Journal for Num. Meth. in Fluids, vol.2, 209-219, 1982.

[4] . Habashi,W.G., and Hafez,M.M., Finite element method for transonic cascade flows, AIAA/SAE/ASME 17th Joint Propulsion Conference, Colorado Spring, USA, July 27-29, 1981, AIAA paper n° 81-1472.

[5] . Rizzi,A., and Viviand,H., Numerical methods for the computation of inviscid transonic flows with shock waves, a GAMM Workshop, (Notes on numerical fluid Mechanics, vol. 3), Vieweg, 1981.

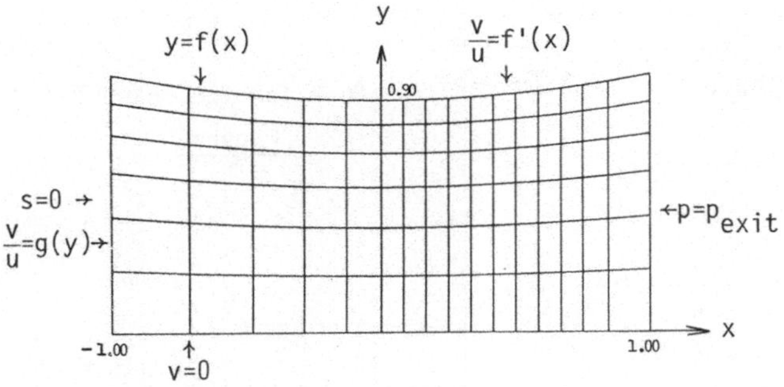

Boundary conditions and mesh for the nozzle

Figure 1

Mach number along the mesh lines

Figure 2

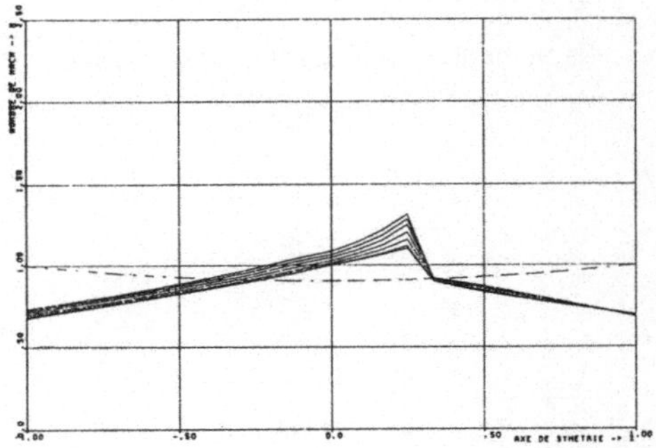

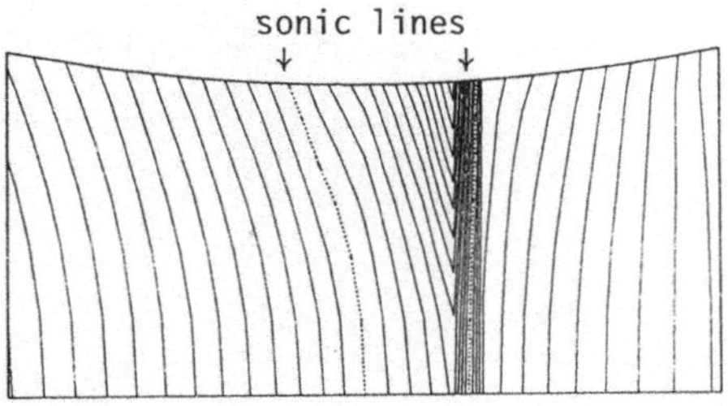

Iso-Mach curves

Figure 3

ρ and $\overline{\rho}$ on the upper wall

Figure 4

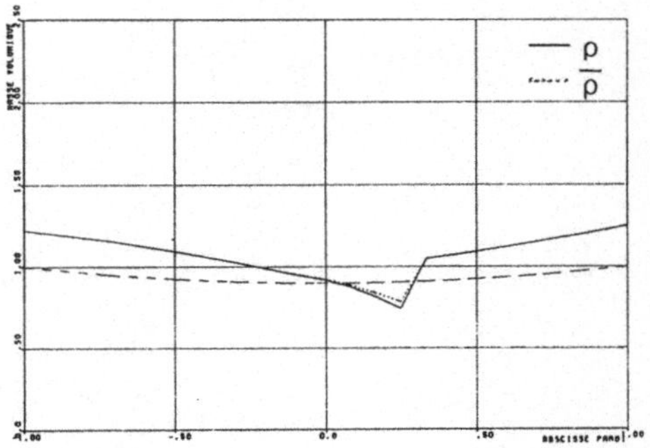

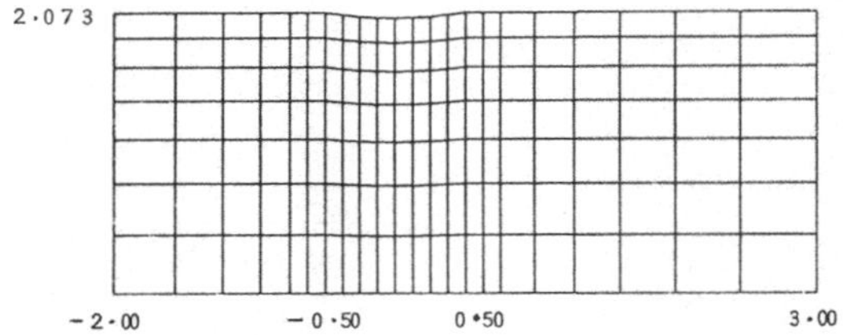

Mesh for the channel

Figure 5

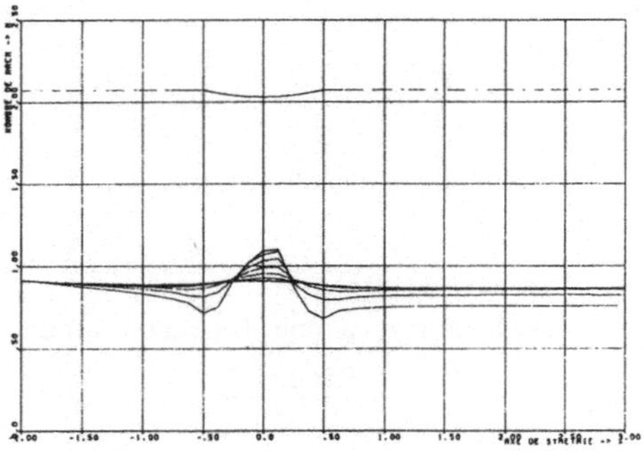

Mach number along the mesh lines

Figure 6

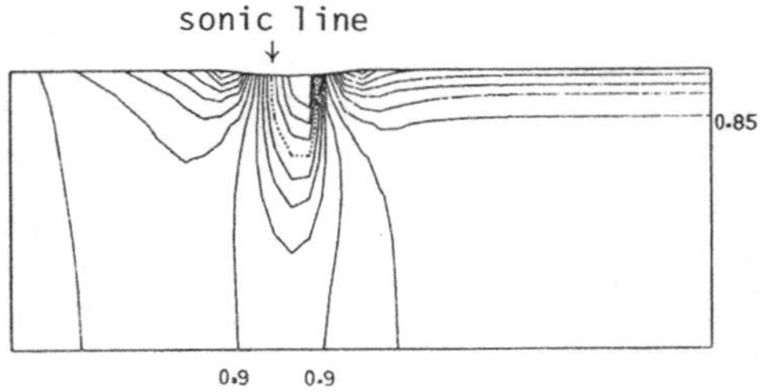

Iso-Mach curves

Figure 7

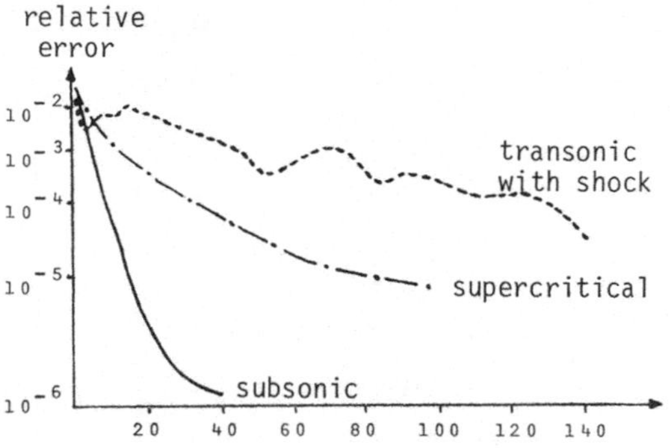

Convergence curves

Figure 8

ANALYSIS OF SEPARATED BOUNDARY-LAYER FLOWS

J. E. Carter and V. N. Vatsa
United Technologies Research Center
East Hartford, CT 06108/USA

INTRODUCTION

The development of computational procedures for the analysis of separated boundary-layer flows continues to be an active research area since flow separation plays an important role in determining the upper limit of performance of aerodynamic configurations in both external and internal flow. The objective of this paper is to present an overview of a method for predicting the strong interaction between the viscous and inviscid flows which occurs when flow separation exists. In the first part of this paper, a new inverse boundary-layer procedure is briefly presented which approximately accounts for normal pressure gradients that can be important in strongly interacting flows. The second part of this paper focuses on viscous-inviscid interacting flows where the viscous formulation is first-order since the pressure is assumed constant across the boundary layer. Results obtained with this procedure for three separated flow problems are presented: 1) transitional separation bubbles near airfoil leading edges, 2) subsonic boattail separated turbulent flow, and 3) transonic turbulent shock wave boundary-layer interaction on an axisymmetric bump configuration. Comparisons with experimental data and solutions of the Navier-Stokes equations, where available, are also shown.

VISCOUS ANALYSIS INCLUDING NORMAL PRESSURE GRADIENTS

Although the occurrence of flow separation does not lead in all cases to significant normal pressure gradients in the viscous region, there are strongly interacting flows in which this effect can be important. For example, it has been observed experimentally that significant normal pressure gradients are induced in the viscous region when a transonic normal shock wave, of sufficient strength to result in flow separation, impinges on an incoming flat plate turbulent boundary layer. As a first step toward the inclusion of normal pressure gradients in a viscous-inviscid interaction analysis, a generalization of the previous inverse boundary-layer formulation developed by Carter[1] is presented in which the pressure $p(x,y)$ is set equal to the inviscid pressure $p_i(x,y)$, which is deduced from the inviscid flow over the generalized displacement thickness given by

$$\int_0^\infty \rho u \, dy = \int_{\delta^*}^\infty \rho_i u_i \, dy \qquad (1)$$

where the i-subscripted variables denote inviscid flow quantities. In this interaction model, which is similar to that proposed by LeBalleur[2], the inviscid flow is permitted to vary over the scale of the boundary layer in contrast with the usual first order assumption of constant pressure, $p(x,y) = p_i(x,\delta^*)$, where $p_i(x,\delta^*)$ is the inviscid pressure at the displacement thickness.

In order to facilitate the finite-difference solution of this new formulation, the x-momentum, continuity and energy equations are transformed with a modified form of the Levy-Lees transformation given by:

$$\xi = \int_0^x \rho_{\delta^*} u_{\delta^*} \mu_{\delta^*} r_0^{2j} \, dx \qquad \eta = \frac{y}{\delta^*} \qquad (2)$$

where x and y are the coordinates along and normal to the surface, respectively, and the δ^*- subscripted variables denote inviscid quantities on the displacement body. Based on previous work for separated boundary layers, the governing equations and boundary conditions are cast into an inverse formulation in which the pressure on the displacement body is deduced from the solution of the viscous equations for a prescribed streamwise distribution of boundary-layer perturbation mass flow, $m = \rho_{\delta^*} u_{\delta^*} r_o^j \delta^*$. The key to this inverse formulation is the introduction of a perturbation stream function, $\tilde{f}$, which is related to the usual stream function ψ by:

$$\psi = m(\eta - \tilde{f}) + \psi_\Delta \tag{3}$$

where ψ_Δ is the deviation of the inviscid stream function ψ_i from the usual first-order linear variation and is given by $\psi_\Delta \equiv \psi_i - m(\eta - 1)$. With this formulation the surface boundary condition, $\psi = 0$ becomes $\tilde{f} = 0$ since $\psi_\Delta = 0$ at the wall; the outer boundary condition, $\psi \to \psi_i$ as $\eta \to \infty$, becomes $\tilde{f} \to 1$. The use of the stream function automatically guarantees that Eq. (1) is satisfied.

The transformed boundary-layer equations are written as follows in which the pressure gradient $\partial p/\partial x$ has been set equal to the inviscid pressure gradient expressed in the present boundary-layer variables:

$$\frac{\partial \tilde{f}}{\partial \eta} = 1 - \bar{\rho}F + \frac{1}{m}\frac{\partial \psi_\Delta}{\partial \eta} \tag{4}$$

$$m^2 \bar{\rho} F \frac{\partial F}{\partial \xi} - m \frac{\partial}{\partial \xi}\left[m(\eta - \tilde{f}) + \psi_\Delta\right]\frac{\partial F}{\partial \eta} = m^2 \beta (\bar{\rho}_i F_i^2 - \bar{\rho} F^2)$$
$$+ m^2 \bar{\rho}_i F_i \frac{\partial F_i}{\partial \xi} - m \frac{\partial \psi_i}{\partial \xi}\frac{\partial F_i}{\partial \eta} + \frac{\partial}{\partial \eta}\left(\bar{\mu}_t \frac{\partial F}{\partial \eta}\right) \tag{5}$$

$$m^2 \bar{\rho} F \frac{\partial g}{\partial \xi} - m \frac{\partial}{\partial \xi}\left[m(\eta - \tilde{f}) + \psi_\Delta\right]\frac{\partial g}{\partial \eta} =$$
$$\frac{1}{Pr}\frac{\partial}{\partial \eta}\left[\bar{\mu}\left(1 + \frac{\epsilon}{\mu}\frac{Pr}{Pr_t}\right)\frac{\partial g}{\partial \eta}\right] + \alpha \frac{\partial}{\partial \eta}\left[\bar{\mu}\left(1 - \frac{1}{Pr}\right)\frac{\partial F^2/2}{\partial \eta}\right] \tag{6}$$

in which the following definitions have been used

$$F = \frac{u}{u_{\delta^*}} \quad \bar{\rho} = \frac{\rho}{\rho_{\delta^*}} \quad g = \frac{H}{H_{\delta^*}} \quad \bar{T} = \frac{T}{T_{\delta^*}} \quad \bar{\mu} = \frac{\mu}{\mu_{\delta^*}} \quad \bar{\mu}_t = \frac{\epsilon + \mu}{\mu_{\delta^*}} \quad \alpha = \frac{u_{\delta^*}^2}{H_{\delta^*}} \quad \beta = \frac{1}{u_{\delta^*}}\frac{du_{\delta^*}}{d\xi} \tag{7}$$

In these equations ε is the eddy viscosity coefficient for turbulent flows for which the algebraic turbulence model of Cebeci-Smith[3] has been used. The laminar and turbulent Prandtl numbers are denoted by Pr and Pr_t, respectively. Equations (4)-(6) are solved with an implicit finite-difference scheme for $\tilde{f}$, F, g and the unknown pressure gradient parameter β for a prescribed m-distribution and with the inviscid flow solution over the generalized displacement body assumed known. The boundary conditions are given by

$$\eta = 0 \quad F = \tilde{f} = 0 \quad g = g_w$$
$$\eta \to \infty \quad F \to F_i \quad \tilde{f} \to 1 \quad g \to 1 \tag{8}$$

The density is deduced from the state equation $\bar{\rho} = (\bar{\rho}_i \bar{T}_i)/\bar{T}$ after the normalized static temperature $\bar{T}$ is deduced from the definition of total enthalpy, H. An important feature of this formulation is that as the boundary-layer edge is approached and the viscous shear and heat conduction terms vanish, the viscous flow solution is required to asymptotically approach the inviscid solution over the generalized displacement body. Further details of this generalized inverse boundary-layer formulation and computed examples are given by Carter and Hafez.[4]

The computed results presented in this paper were obtained with a first-order viscous formulation which is deduced by setting $\psi_\Delta = 0$ and $\bar{\rho}_i = F_i = 1$ in Eqs. (4)-(6). In the actual calculations the first-order inverse boundary-layer formulation presented by Carter[1] was used; however, these two inverse boundary-layer formulations are quite similar and one can easily be derived from the other by a slight change in the definition of the perturbation stream function, transformed normal coordinate, and the pressure gradient parameter.

INTERACTION ITERATION PROCEDURE

The present analysis is based on a global viscous-inviscid iteration technique, previously presented by Carter,[1] in which the inverse boundary-layer solution is solved iteratively with a direct analysis of the inviscid flow including injection to represent the displacement thickness effects. The key feature of this iteration procedure is the simple update formula

$$m^{i+1} = m^i \left[1 + \omega\left(\frac{u_{e_V}}{u_{e_I}} - 1\right)\right] \tag{9}$$

where ω is the relaxation factor, and u_{e_V} and u_{e_I} denote the viscous and inviscid predictions, respectively, of the velocity tangent to the displacement body for the m-distribution at the ith global iteration. This update procedure, which is similar to that used by LeBalleur,[2] is general as it has been used with a variety of inverse boundary-layer and direct inviscid solution procedures as pointed out in Ref. 1.

RESULTS AND DISCUSSION

In this section three applications of the first-order viscous-inviscid interaction analysis are presented. In each case a brief discussion is given of the particular features of each analysis and the inviscid solution procedure which was utilized. References are given for each of these applications in which a more detailed discussion is presented.

Airfoil Leading-Edge Separation Bubbles

The first application of this interaction theory is for the transitional separation bubble which occurs near the leading edge of an airfoil. If the Reynolds number is sufficiently low such that the boundary layer remains laminar up to the point of minimum pressure, then the onset of the adverse pressure gradient generally results in separation of the laminar boundary layer; subsequently, transition from laminar to turbulent flow occurs in the separated shear layer, and if the bubble does not burst, then turbulent reattachment occurs further downstream. Since the streamwise length scale of this viscous-inviscid interaction is typically only a few percent of the airfoil chord, this localized interaction is treated as a perturbation to a global viscous airfoil analysis, which in the present effort was the GRUMFOIL code developed by Melnik, et. al[5]. The deviation of the edge velocity, u_e, due to the transitional bubble, from the reference global airfoil solution is represented by a Cauchy integral given by

$$u_{e_I} = u_{e_{ref}} + \frac{1}{\pi}\int_{s_1}^{s_2} \frac{\frac{d}{d\xi}\left[u_e\left(\delta^* - \delta^*_{ref}\right)\right]d\xi}{s - \xi} \tag{10}$$

in which the source strength is proportional to the streamwise derivative of the corresponding deviation in the displacement thickness.

The local interaction analysis was applied to the transitional separation bubble, measured experimentally by Gault[6], which occurs at the leading edge of a NACA 66_3-018 airfoil at a chord Reynolds number of 1.5×10^6 and $\alpha = 12$ deg. At this angle of attack, the reference solution, obtained from the GRUMFOIL code, was found to significantly overpredict the lift coefficient for this airfoil. Correspondingly the pressure in the strong acceleration region was overpredicted thereby providing an inaccurate reference solution for this case. These errors are probably due to the inability of the GRUMFOIL code to correctly model the lift decrement due to the massive trailing edge separation which occurred in this case at 0.75 chord on the upper surface. In order to compensate for this problem, the GRUMFOIL code was run at a reduced angle of attack at 11.25° which provided a much better match of the experimental lift coefficient and the pressure distribution between the stagnation point and the peak suction region as shown in Fig. 1(a).

The transition model used in the present calculation is the streamwise intermittency function of Dhawan and Narasimha[7] which requires the specification of the onset and length of the transition region. In the present case the onset of transition was assumed to occur midway between the experimental separation point and the "break" point in the experimental pressure distribution shown in Fig. 1. The length of the transition region was established so that the intermittency function had a value of $\gamma = 0.5$ at the experimental pressure "break" point.

The computed results for this case are shown in Figs. 1(a) and 1(b) for the pressure, and skin friction and displacement thickness, respectively. The good agreement shown between the predicted results and Gault's pressure data and measured separation point verifies the use of a local interaction model for the airfoil leading edge transitional bubble problem. Figure 1(a) also shows the inviscid airfoil pressure distribution which was obtained from the GRUMFOIL code with $\alpha = 11.25°$. The large difference between this solution and the viscous airfoil solution obtained with the GRUMFOIL code shows the importance of including the viscous effects in the reference solution which is input to the present perturbation interaction analysis. Further details of this interaction analysis and other computed examples are presented by Carter and Vatsa.[8]

Subsonic Boattail Separated Turbulent Flow

The second application presented in this paper is the subsonic turbulent separated flow in the juncture region of an axisymmetric boattail-sting configuration. The conservative full potential analysis of Green[9] was used to represent the outer inviscid flow. In order to reduce the injection velocity associated with the large displacement thickness growth which occurs in the boattail-sting juncture region, the inviscid flow was solved over a shear layer coordinate which was assumed a priori to approximate the displacement body position. Injection was then used along this shear layer coordinate, but the injection magnitude was significantly reduced since it was now proportional only to the difference between the assumed and the actual displacement body locations. The boundary-layer equations are also expressed in terms of this shear layer oriented coordinate system, but the usual form of these equations is recovered with the use of the Prandtl transposition theorem.

Figure 2 shows the computed pressure and skin friction distributions from both the present interaction analysis and the numerical solution of the Navier-Stokes equations[10] for the $M_\infty = 0.7$ flow over the circular-arc boattail-sting configuration measured experimentally by Reubush.[11] Figure 2 shows excellent agreement between the two theoretical analyses, thereby confirming that the neglected terms in the approximate interaction analysis are of secondary importance in this separated flow case. Nonetheless it is observed that both analyses overpredict the experimental pressure in the separated region and predict the separation point too far downstream. This discrepancy was thought to be due to the use of the Cebeci-Smith algebraic turbulence model as it has been observed in other analyses of separated flow to result in an underprediction of the displacement thickness and a corresponding overprediction of the pressure in regions of strong adverse pressure gradient. This observation led Shang and Hankey[12] to modify the algebraic turbulence model through a global streamwise relaxation model which approximately accounts for the so-called "history of the flow." Figure 3 shows the computed results which were obtained with the Shang-Hankey model implemented into both the interaction and the Navier-Stokes analyses. The overall agreement with the experimental data is significantly improved with the use of this model, although it is observed that the interacting boundary-layer results show greater sensitivity to this modification than the Navier-Stokes analysis thereby giving better agreement with the experimental data.

Transonic Shock Wave Boundary Layer Interaction

The third application of this viscous-inviscid interaction approach is the transonic shock induced separated flow over an axisymmetric circular arc-bump configuration for which experimental data and a solution of the Navier-Stokes equations were given by Johnson, et. al.[13] Calculations using the present approach for this configuration have previously been presented by Carter;[1] however, several numerical improvements have been made in the interim and these are briefly discussed here with more details given in Ref. 14.

In the previous calculations it was found necessary to use numerical smoothing to eliminate oscillations which occurred when the fully conservative potential analysis of Green[9] was used with the inverse boundary-layer analysis. It was found in the present work that these oscillations were eliminated, and hence the need for smoothing, by placing a grid point on the shear layer coordinate to precisely correspond to the location of the body-sting corner, and by using central differencing of the displacement thickness in the inviscid transformed plane to numerically compute the injection velocity. In addition, it was found that using the same x-grid in both the viscous and inviscid calculations, which eliminates interpolation between the two solutions, enhanced the overall interaction convergence rate. Figure 4 shows the computed pressure and displacement thickness distributions in comparison with the experimental data and Navier-Stokes calculation presented by Johnson, et. al.[13] The interaction calculations were made with both the nonconservative potential flow analysis of South[15] and the conservative analysis of Green.[9] In contrast with the Navier-Stokes solution both analyses show good agreement with the data for the shock wave position. It is well known that in an inviscid analysis a conservative potential flow calculation predicts a stronger shock wave located downstream of that given by a nonconservative computation. The present results show that this difference is significantly reduced when viscous interaction effects are included since the stronger shock given by the conservative analysis produces a larger displacement thickness, as shown in Fig. 4(b), thereby weakening the shock and moving it forward to place it in better agreement with that predicted by the nonconservative analysis.

As was found in the subsonic boattail calculation, this interaction calculation overpredicts the pressure and underpredicts the displacement thickness in comparison with the data in the body-sting juncture region. This difference is probably due to the algebraic turbulence model and can be substantially reduced by use of the Shang-Hankey[12] model or by reducing the Clauser constant, as was shown by Carter[1], in the outer region eddy viscosity model.

CONCLUDING REMARKS

The applicability of the viscous-inviscid interaction analysis presented herein to various separated flow problems has been demonstrated. Overall, good agreement with experimental data has been observed with this analysis, although it is concluded from these calculations that the use of an algebric turbulence model is inadequate for flows with significant turbulent separated flow. In addition, the generalization of the first-order inverse boundary-layer analysis to approximately account for normal pressure gradients has been presented. It now remains to combine this new viscous analysis with an inviscid solution procedure and demonstrate its use in viscous-inviscid interacting flows.

ACKNOWLEDGEMENTS

The authors express their graditude for the support of this work to the following technical monitors and agencies: James D. Wilson, Air Force Office of Scientific Research; Joel L. Everhart, NASA-Langley Research Center; and Robert E. Whitehead, Office of Naval Research.

REFERENCES

1. Carter, J. E.: AIAA Paper No. 81-1241, 1981.
2. Leballeur, J. C.: La Recherche Aerospatiale, English Edition, pp. 21-45, 1981-3.
3. Cebeci, T. and A. M. O. Smith: Analysis of Turbulent Boundary Layers, Academic Press, 1974.
4. Carter, J. E. and M. Hafez: AFOSR Report to be published, 1982.
5. Melnik, R. D., R. Chow, and H. R. Mead: AIAA Paper No. 77-680, 1977.
6. Gault, D. E.: NASA TN 3505, 1955.
7. Dhawan, S. and R. Narasimha: J. Fluid Mechanics, Vol. 3, 1958.
8. Carter, J. E. and V. N. Vatsa: NASA CR-165935, 1982.
9. Green, L. L.: AIAA Paper No. 81-1204, 1981.
10. Vatsa, V. N., J. E. Carter, and R. D. Swanson: ISCME Conference, Washington, D. C., June 30-July 2, 1982.
11. Reubush, D. E.: NASA TN D-7795, 1974.
12. Shang, J. S. and W. L. Hankey, Jr.: AIAA Journal, Vol. 13, pp. 1368-1374, 1975.
13. Johnson, D. A., C. C. Horstmann, and W. D. Bachalo: AIAA Journal, Vol. 20, No. 6, pp. 737-744, 1982.
14. Carter, J. E. and V. N. Vatsa: ONR Report to be published, 1982.
15. South, J. C. and A. Jameson: AIAA CFD Conference, 1973.

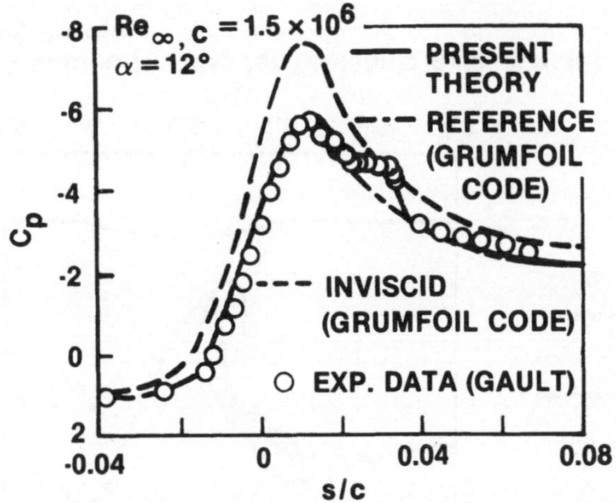

Fig. 1 Predicted Results for NACA 66_3-018 Airfoil
a) Pressure

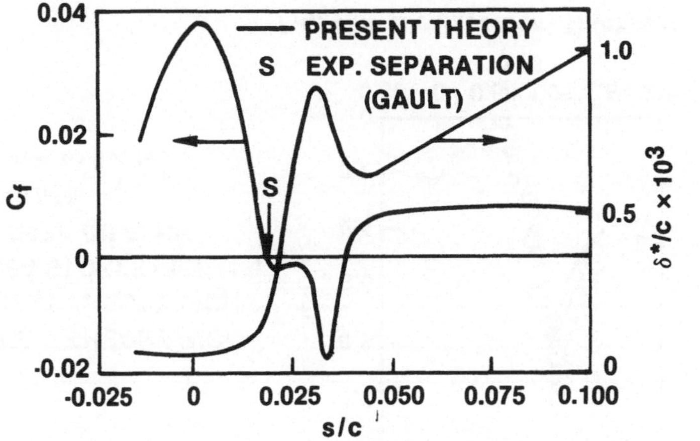

Fig. 1(b) Skin Friction and Displacement Thickness

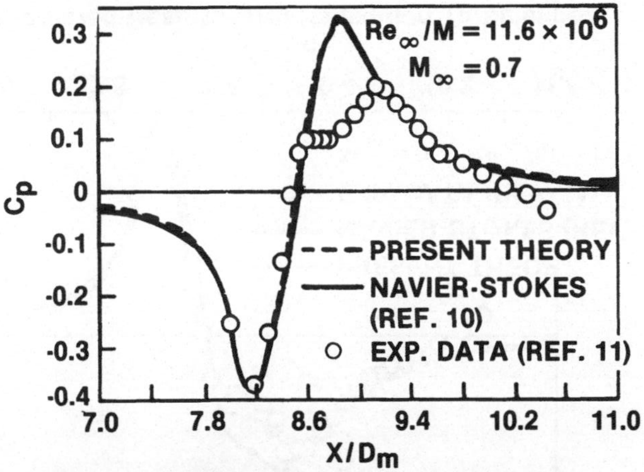

Fig. 2 Predicted Results for Boattail Flow
a) Pressure

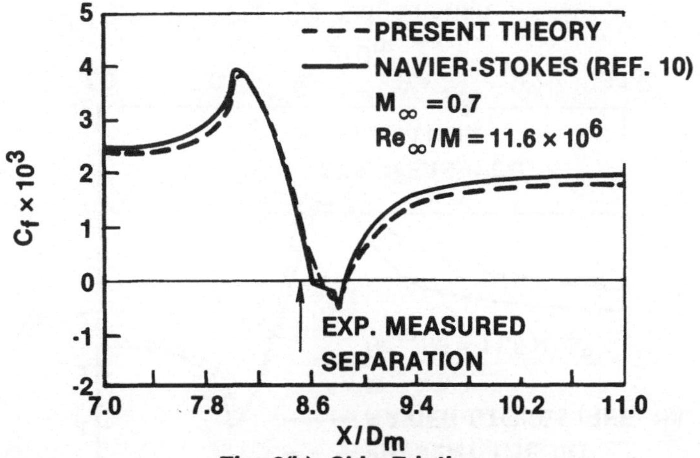

Fig. 2(b) Skin Friction

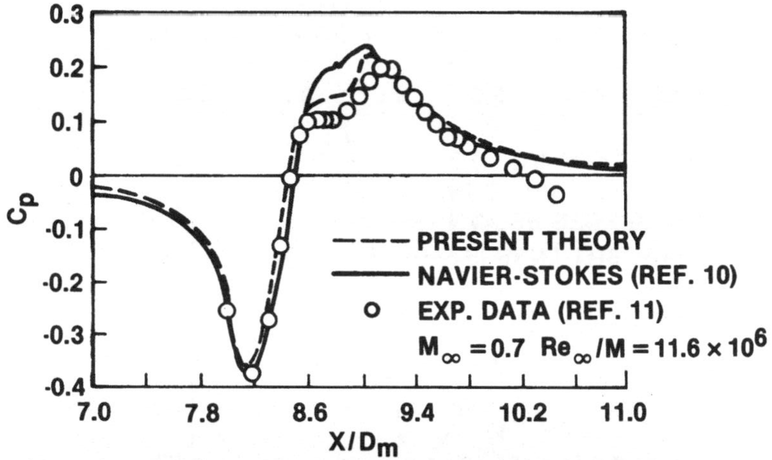

Fig. 3 Predicted Results with Relaxation Turbulence Model for Boattail Flow. a) Pressure

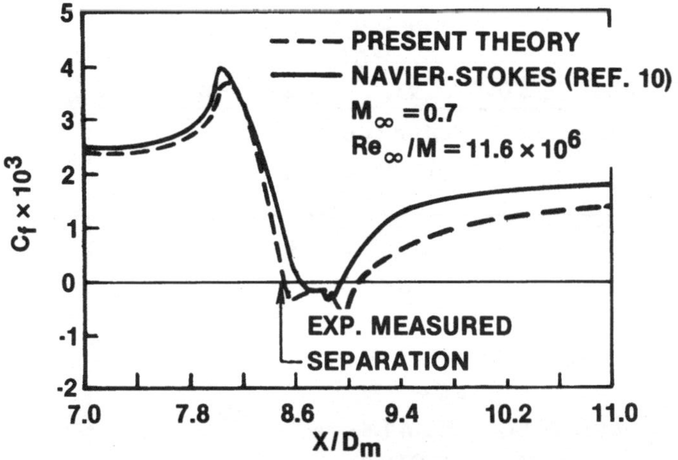

Fig. 3(b) Skin Friction

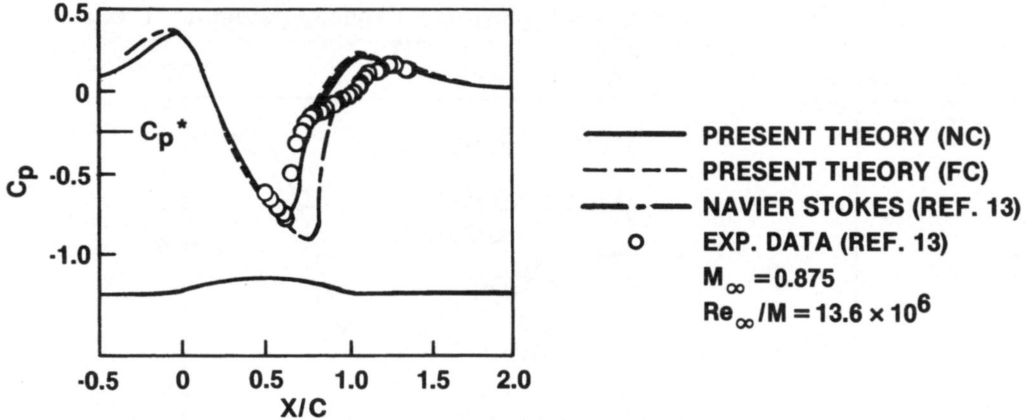

Fig. 4 Predicted Results for Transonic Axisymmetric Flow (a) Pressure

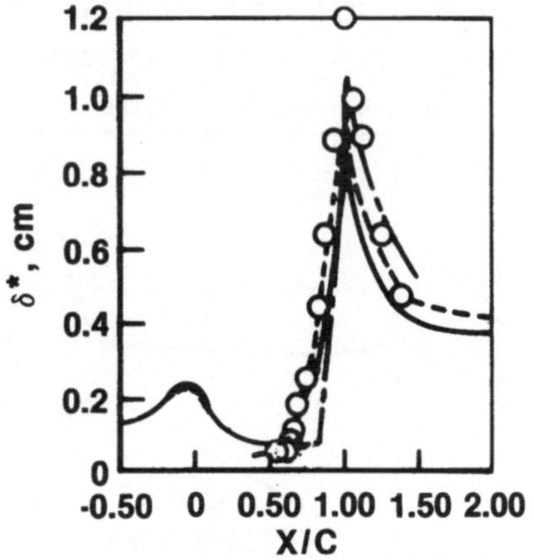

Fig. 4(b) Displacement Thickness

Numerical Modelling of Inviscid Shocked Flows of Real Gases*

Phillip Colella
Lawrence Berkeley Laboratory
University of California
Berkeley, California 94720

Harland M. Glaz
Naval Surface Weapons Center
White Oak Laboratory
Silver Spring, Maryland 20910

Introduction

In the past several years, there has been substantial development of various upstream-centered schemes for solving numerically the equations of inviscid compressible flow. A large subclass of these schemes rely on the existence of efficient algorithms for solving the Riemann problem. Such algorithms exist for fluids with a polytropic or isothermal equation of state, with the possible inclusion of simplified models for chemical reactions. In the following, we give a procedure for constructing solutions to the Riemann problem for fluids with a general convex equation of state. In the case where the Riemann problem is to be used to calculate fluxes for a conservative finite difference scheme, we approximate this exact Riemann solver by introducing a local parameterization of the equation of state, treating the parameters which describe the equation of state as separate dependent variables. This enables us to reduce the number of calls of the equation of state to one per zone per sweep and yields a set of algorithms for which the increase in the cost of performing a real gas calculation over that of the corresponding polytropic gas scheme, other than the single equation of state call per zone, is only a few percent. Furthermore, the real gas algorithms retain the resolution of the polytropic gas algorithms.

We consider the inviscid compressible flow equations in one space variable in conservation form

$$\frac{\partial U}{\partial t} + \frac{\partial F(U)}{\partial x} = 0$$

$$U = \begin{bmatrix} \rho \\ \rho u \\ \rho E \end{bmatrix} \qquad F(U) = \begin{bmatrix} \rho u \\ \rho u^2 + p \\ \rho u E + up \end{bmatrix}. \tag{1}$$

Here ρ is the density, u the velocity, and E the total energy per unit mass. E is the sum of the internal energy e, and the kinetic energy: $E = e + \frac{u^2}{2}$. The pressure p is derived from these quantities via an equation of state: $p = p(\tau, e)$, $\tau = \frac{1}{\rho}$.

We also need the equations (1), expressed in nonconservation form:

$$\frac{\partial V}{\partial t} + A \frac{\partial V}{\partial x} = 0$$

$$V = \begin{bmatrix} \tau \\ u \\ p \end{bmatrix} \qquad A(V) = \begin{bmatrix} u & -\tau & 0 \\ 0 & u & \tau \\ 0 & \rho c^2 & u \end{bmatrix}.$$

The adiabatic speed of sound is expressed in terms of p as $C^2 = pp_e - p_\tau$, $c = \tau C$. In the following, we will assume $C^2 > 0$ and that the equation of state is convex, i.e., $pC_e - C_\tau > 0$. The matrix $A(V)$ has left and right eigenvectors $(l_\#(V), r_\#(V))$, $l_\# \cdot r_{\#'} = \delta_{\#\#'}$, $\# = 0,+,-$, associated with the eigenvalues $\lambda_+ = u + c$, $\lambda_0 = u$,

*This research was supported by the U.S. Department of Energy, the U.S. Defense Nuclear Agency, and the Naval Surface Weapons Center Independent Research Fund.

$\lambda_- = u - c$.

The Riemann Problem

The Riemann problem is the initial value problem for (1) for which the initial data consists of two constant states separated by a single jump discontinuity.

$$U(x,0) = \begin{cases} U_L, & x < 0 \\ U_R, & x > 0 \end{cases}$$

We look for non-entropy-decreasing solutions which depend on (x,t) only in the similarity variable $\frac{x}{t} = \xi$. Such solutions consist of four constant states separated by three waves (figure 1), each of which associated with one of the characteristic speeds. The 0-wave is a jump discontinuity in the density, across which the pressure and velocity are continuous, and which propagates at a velocity u^*. The $\pm$ waves are shock discontinuities, or centered rarefaction waves, depending on the sign of the pressure jump:

$$\begin{array}{ll} p^* > p_S & p^* < p_S \\ e(p^*, \tau_S^*) - e_S + \dfrac{(p_S + p^*)}{2}(\tau_S^* - \tau_S) = 0 & \dfrac{d\tau}{dp} = - C^2 \\ (\tau_S^* - \tau_S) + \dfrac{(p^* - p_S)}{W_S^2} = 0 & \dfrac{du}{dp} = \pm C \\ W_S(u^* - u_S) = \pm(p^* - p_S) & p_S \le p \le p^* \\ & u \pm c = \xi \end{array} \tag{2}$$

$$(S, \pm) = (L, -), (R, +) .$$

For either a shock or a rarefaction wave, the post-wave state is uniquely determined by the pre-wave state and p^*. In particular, we can define u_S^*, the post-wave velocity, as a function of p^* and U_S. The condition that the left and right waves match up to give a solution to the Riemann problem is that $u_L^* = u_R^* = u^*$ We can also define the mean Lagrangian wave speeds

$$\begin{aligned} W_S &= \frac{p^* - p_S}{u^* - u_S} & u^* \neq u_S \\ &= C_S & u^* = u_S \end{aligned}$$

In the case of a shock, W_S is the quantity which appears in the shock jump relations (2). In either case, W_S is uniquely determined by U_S and p^*

We obtain a solution to the Riemann problem by a double iteration scheme: we alternate between iterating to obtain values for W_L, W_R, given p^*, and iterating on p^* so that $u_L^* = u_R^*$. If $p^{*,l}$ is the value of p^* at the l-th iteration, then $p^{*,l+1}$ is given by

$$W_S^l = W_S(p^{*,l}, U_S)$$

$$u_S^{*,l} = u_S \pm \frac{p^{*,l} - p_S}{W_S^l}$$

$$p^{*,l+1} = p^{*,l} - (u_R^{*,l} - u_L^{*,l}) \left[\frac{p^{*,l} - p^{*,l-1}}{u_L^{*,l} - u_L^{*,l-1} + u_R^{*,l} - u_R^{*,l-1}} \right].$$

This is a secant iteration corresponding to the Newton iteration in [4]. The first two guesses used to start the iteration are obtained using Godunov's iteration scheme. The calculation of W_s is performed by an iterative scheme to solve the shock jump equations if $p^* > p_s$ and by solving the ODE's using finite differences if $p^* < p_s$. The calculation of the solution as a function of $(\frac{x}{t})$ is an immediate generalization of the procedure used in the polytropic case except when $(\frac{x}{t})$ is inside a rarefaction fan. In that case, one applies inverse interpolation to the solution of the ODE's which define the solution inside the rarefaction fan as a function of the pressure.

Local Parameterization of the Equation of State

We parameterize the equation of state in terms of a function $\gamma(\rho,e)$, defined to be

$$\gamma = \frac{p(\tau,e)}{\rho e} + 1 .$$

In the case of smooth flow, the behavior of γ as a function of the solution is most naturally specified along the streamline:

$$d\gamma = \left[1 - \frac{\gamma}{\Gamma} \right] (\gamma - 1) \frac{1}{p}\, dp \qquad \text{along } dx = u\,dt. \tag{3}$$

Here Γ is derived from the sound speed to be $\Gamma(\rho,e) = \frac{\rho c^2}{p}$.

In the case of discontinuities across which the Rankine-Hugoniot conditions hold, it is not possible in general to specify how γ behaves as a function of the change in the state variables without solving the equations (2). However, if the jump is not too large, then the jump relations for γ are well-approximated by an integrated form of the characteristic equations (3):

$$\gamma_S^* - \gamma_S = \left[1 - \frac{\hat{\gamma}}{\hat{\Gamma}} \right] (\hat{\gamma} - 1) \frac{1}{\hat{p}} (p^* - p_S) + O(p^* - p_S)^3 \tag{4}$$

where $\hat{\gamma}$, $\hat{\Gamma}$, and $\hat{p}$ are some suitably centered values for those variables.

In order to apply this model in a specific case, we will describe how to extend the single-step Eulerian scheme given by Colella and Woodward [2] for a polytropic gas to the general equation of state case. In outline, this scheme consists of four steps:

1) the interpolation of values for the dependent variables (not necessarily the conserved quantities);
2) the construction of effective left and right states $\overline{V}_{j+½,L}$, $\overline{V}_{j+½,R}$ at $x_{j+½}$;
3) the solution of the Riemann problem at $x_{j+½}$ with left and right states constructed as in 2), to give $\overline{V}_{j+½}$;

4) the conservative differencing of the fluxes $F_{j+1/2} = F(U(\bar{V}_{j+1/2}))$.

$$U_j^{n+1} = U_j^n - \frac{\Delta t}{\Delta x}(F_{j+1/2} - F_{j-1/2}).$$

The states $\bar{V}_{j+1/2,L}$, $\bar{V}_{j+1/2,R}$ are constructed such that $\bar{V}_{j+1/2}$ is, in smooth regions, an approximation to a solution to the characteristic form of the equations at $(x_{j+1/2}, t^n + \frac{\Delta t}{2})$ up to terms of second order, so that the scheme is second order accurate in space and time. It is principally in steps 2) and 3) where the modifications are made to accommodate the general equation of state.

In the first step, we interpolate the variables $q = q(U_j^n)$ $q = p, \rho, u, \gamma$ using a monotonized interpolation scheme. We do not interpolate Γ but treat it as piecewise constant.

The second step is performed in two parts. First, we calculate $\tilde{V}_{j+1/2,L}$, $(\tilde{V}_{j+1/2,R})$, a first guess for the effective left (right) state, by calculating the average of U to the left (right) between the zone edge and the + (–) characteristic (figure 2). In the case where the flow is supersonic, so that there is no characteristic reaching $x_{j+1/2}$ from one side, we use the appropriate limiting value at $x_{j+1/2}$. We then make corrections to $\tilde{V}_{j+1/2,S}$ using the characteristic projection operators (figure 3) for example, for the left state

$$\bar{V}_{j+1/2,L} = \tilde{V}_{j+1/2,L} - \sum_{\lambda_\#(U_j^n)>0} P^\#(\tilde{V}_{j+1/2,L} - V^\#_{j+1/2,L})$$

$$P^\# v = (l_\#(\tilde{V}_{j+1/2,L})\cdot v)\, r_\#(\tilde{V}_{j+1/2,L}) \qquad \# = +,-,0$$

$$\bar{\gamma}_{j+1/2,L} = \gamma^0_{j+1/2,L} - \left[1 - \frac{\gamma_j}{\Gamma_j}\right](\gamma_j - 1)\frac{(\bar{p}_{j+1/2,L} - p^0_{j+1/2,L})}{\frac{1}{2}(\bar{p}_{j+1/2,L} + p^0_{j+1/2,L})} \qquad \text{if } u_j^n > 0$$

$$= \tilde{\gamma}_{j+1/2,L} \qquad \text{otherwise.}$$

We calculate $\bar{V}_{j+1/2,R}$ similarly, replacing L by R and $\lambda_\#(U_j^n)>0$ by $\lambda_\#(U_{j+1}^n)<0$.

In step 3) we solve the Riemann problem with the secant method for the general Riemann solver described above, using the model equation (4) to provide a non-iterative method for obtaining W_S^2. Given p^*, $\bar{V}_{j+1/2,S}$, $\bar{\gamma}_{j+1/2,S}$, we calculate γ_S^* to be

$$\gamma_S^* = \bar{\gamma}_{j+1/2,S} + (1 - \frac{\hat{\gamma}}{\hat{\Gamma}})(\hat{\gamma} - 1)\frac{p^* - \bar{p}_{j+1/2,S}}{\frac{1}{2}(p^* + \bar{p}_{j+1/2,S})}$$

$$\hat{\gamma} = \tfrac{1}{2}(\gamma_j^n + \gamma_{j+1}^n)$$

$$\hat{\Gamma} = \tfrac{1}{2}(\Gamma_j^n + \Gamma_{j+1}^n).$$

Given γ_S^*, it is easy to solve (2) to obtain W_S^2:

$$W_S^2 = \frac{(p^* - \bar{p}_{j+1/2,S})(p^* + \frac{1}{2}(\gamma_S^* - 1)(p^* + \bar{p}_{j+1/2,S}))}{p^*\bar{\tau}_{j+1/2,S} - \dfrac{(\gamma_S^* - 1)}{(\bar{\gamma}_{j+1/2,S} - 1)}\bar{p}_{j+1/2,S}\bar{\tau}_{j+1/2,S}}$$

The evaluation of the Riemann problem at $x_{j+1/2}$ proceeds as before, except that we use linear interpolation between the pre- and post-wave states to evaluate the solution inside a rarefaction fan.

Numerical Results

We have implemented the method described above, and tested it for a variety of equations of state. In figure 4, we show the results of a Cartesian shock tube calculation. In these results, and the figures that follow, all quantities are displayed in cgs units. The dotted line is the computed solution, and the solid line the exact solution, which was obtained using the exact Riemann solver described above. There are 180 zones in the computational domain, with the initial discontinuity located between zones 60 and 61.

The material on the left of the initial discontinuity is the product of a completely burned explosive; the material on right is air, initially at atmospheric conditions. The boundary between the two materials is tracked by solving an additional advection equation for the fraction of air in a zone. If the fraction is not 0 or 1, we take γ_j ,Γ_j in a zone to be a weighted average of those quantities for each of the two materials.

The computed solution is in good agreement with the exact solution. The slight undershoot in the internal energy to the left of the contact is a starting error, which occupies a fixed number of zones as the mesh is refined. The apparent overshoot in γ comes from evaluating the equation of state for the unphysical values of density and energy inside the shock. However, the density, energy, and pressure profiles are all monotone across the shock.

In figure 5, we show the results of the same shock tube but in spherical coordinates with the initial jump located at 3.78 cm. In this case, we see in the solution a shock and a contact discontinuity from the initial jump, as well as a second backward-facing shock which forms due to the effect of the geometry. Since this problem lacks an exact solution, we compare solutions obtained using 800 mesh points (solid line) and 400 mesh points (dotted line). The pressure is converged in the 400 zone results, but the density is not. This is not surprising, since we have only a tenth of the zones between the primary and secondary shocks to resolve the strongly varying density profile. Calculations performed previously with a constant γ EOS indicate that density profiles converge between 800 and 1600 zones.

Finally, we present in figure 6 a two-dimensional Cartesian calculation of a shock in N_2 reflecting off an oblique surface. The Mach number, shock angle, and ambient state were chosen to coincide with a shock tube experiment performed by Ben-Dor and Glass ([1], case g). We obtain the correct shock reflection pattern, that of a double Mach reflection. The wall densities are in good agreement with the experiment, except in the region just behind the shock. In the latter case, the experimental data did not resolve the detailed structure. More highly resolved experiments (Glass [3]) confirm that, at least qualitatively, the behavior seen in the present calculations is correct.

References

[1] Ben-Dor, G. and Glass, I.I., J. Fluid Mech. *92* (1979), p. 459.

[2] Colella, P. and Woodward, P.R., "The Piecewise-Parabolic Method for Gas Dynamical Simulations", in preparation.

[3] Glass, I.I., "Beyond Three Decades of Continuous Research at UTIAS on Shock Tubes and Waves", UTIAS Review, No. 45, July 1981.

[4] van Leer, B., J. Comp. Phys. *32* (1979), p. 101.

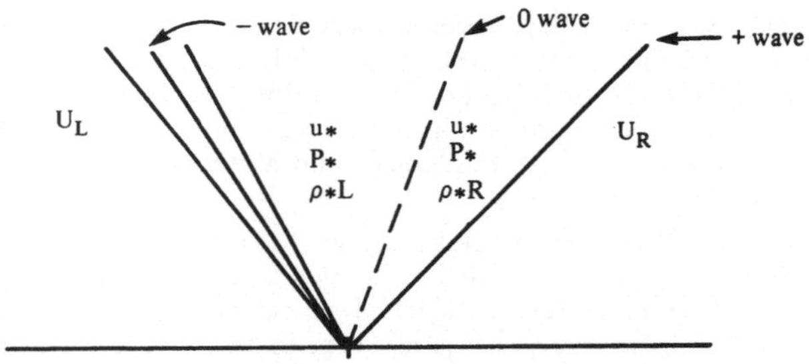

Figure 1

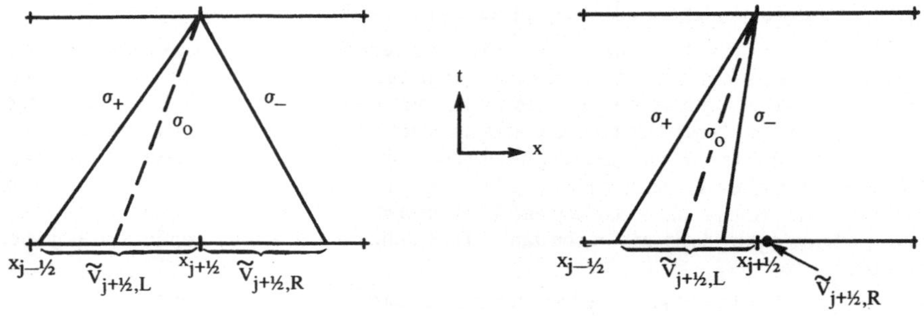

Figure 2

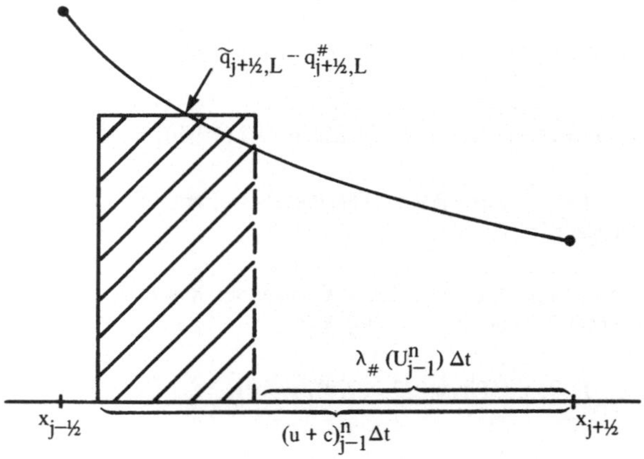

Figure 3

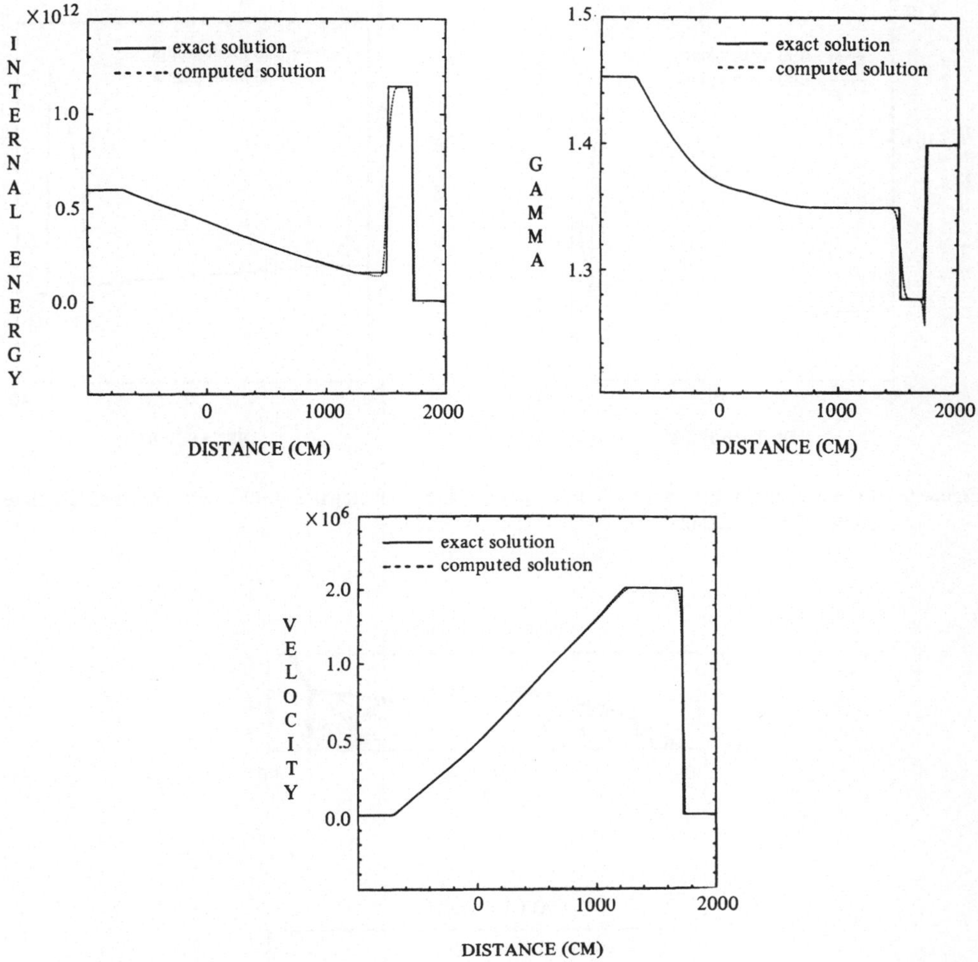

Figure 4. Comparison of exact solution with a calculation (180 mesh points, 155 time steps, T = 1.0059E-03); Cartesian Riemann problem

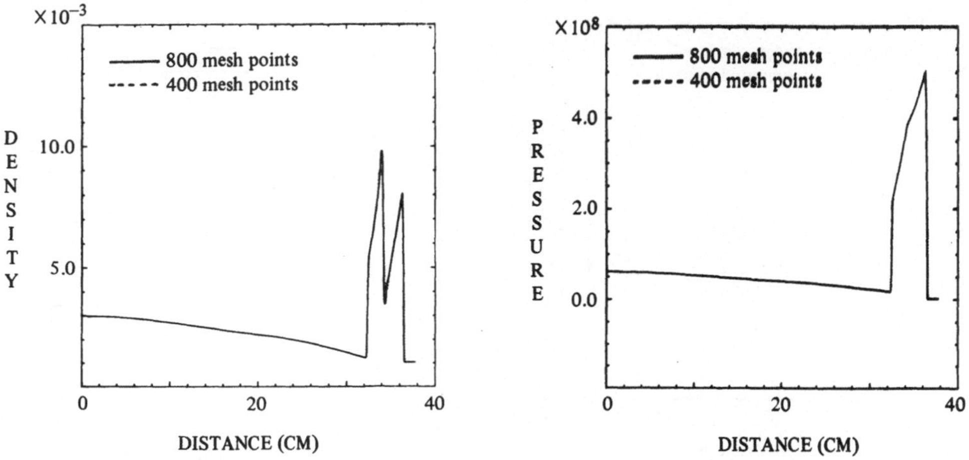

Figure 5 – Spherical shock tube with 400 mesh points (548 time steps) and 800 mesh points (1102 time steps); T = 3.203E-05

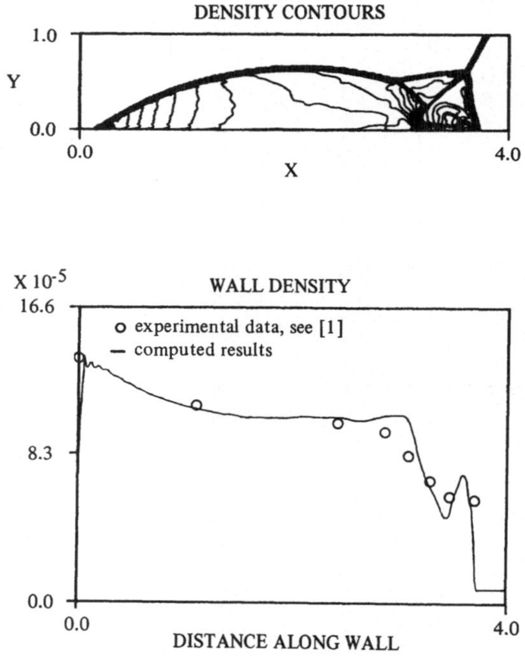

Figure 6 – Density contours and wall density plot for an N_2 oblique shock wave, see [1]. Ramp angle = 26.56°, ambient density = 0.769 x 10^{-5}, ambient pressure = 6730.0 (cgs units), and shock Mach number M_s = 8.06. There are 30 equispaced contour levels with increment = 4.344 x 10^{-6}.

two previous schemes but which avoids the artificial state variables.

A TWO-STEP OVERRELAXATION SCHEME

The following relaxation scheme is suggested for (1) :

downwind step :

$$\xi(x,t+\tilde{\Delta}t) - \xi(x,t) + \lambda\left(\xi(x+\Delta x,t) + \alpha\xi(x,t+\tilde{\Delta}t) - (1+\alpha)\xi(x,t)\right) = 0 \qquad (6)$$

upwind step :

$$\xi(x,t+\tilde{\tilde{\Delta}}t) - \xi(x,t) + \lambda\left(\alpha\xi(x,t+\tilde{\tilde{\Delta}}t) + \xi(x,t+\tilde{\Delta}t) - \alpha\xi(x,t) - \xi(x-\Delta x,t+\Delta t)\right) = 0 \quad (7)$$

relaxation step :

$$\xi(x,t+\Delta t) = .5r\left(\xi(x,t+\tilde{\Delta}t) + \xi(x,t+\tilde{\tilde{\Delta}}t)\right) + (1-r)\xi(x,t) \qquad (8)$$

This gives :

$$(1+\alpha\lambda)\xi(x,t+\tilde{\Delta}t) = (1+\alpha\lambda)\xi(x,t) - \lambda\left(\xi(x+\Delta x,t) - \xi(x,t)\right)$$

$$(1+\alpha\lambda)\xi(x,t+\tilde{\tilde{\Delta}}t) = (1+\alpha\lambda)\xi(x,t) - \lambda\left(\xi(x,t+\tilde{\Delta}t) - \xi(x-\Delta x,t+\Delta t)\right)$$

and :

$$\xi(x,t+\Delta t) = RX\,\xi(x-\Delta x,t+\Delta t) + (1-RX^2)\xi(x,t) + (-RX+RX^2)\xi(x+\Delta x,t) \qquad (9)$$

with $R = .5r$ and $X = \frac{\lambda}{1+\alpha\lambda}$

In (8) $r > 1$ corresponds to overrelaxation.

In this scheme, no artificial state variables are used. Instead, the relaxation proces is based on preliminary values of the natural state variables. As soon as a new value of the state variables is calculated, the old value and the preliminary values are no longer used. In this sence, this relaxation scheme is similar to the classic relaxation scheme for the potential equation, when the analogy between iteration levels and time levels is used. The preliminary values need not to be stored and do not introduce extra dynamics.

The scheme can easily be analysed by the Fourier-method.

Substitution of a perturbation with respect to steady state :

$$\xi(x,t) = \phi(t)\, e^{j\omega x}$$

gives the amplification factor :

$$G = \frac{\phi(t+\Delta t)}{\phi(t)} = \frac{1 - RX^2 + (-RX+RX^2)e^{j\omega\Delta x}}{1 - RX\, e^{-j\omega\Delta x}} \qquad (10)$$

The amplification factor is identically zero for : $\begin{cases} R = 1 & (r=2) \\ X = 1 \end{cases}$

Then, there is complete annihilation of perturbations with respect to steady state for all frequency components. Hence, with the model equation (1), steady state can be reached. in one relaxation sweep.

AN OVERRELAXATION METHOD FOR TRANSONIC FLOW CALCULATIONS BY EULER EQUATIONS

DICK Erik
Department of machinery
State University of Ghent
Sint Pietersnieuwstraat 41
9000 GENT
BELGIUM

INTRODUCTION

In time marching methods for solving steady transonic Euler equations, the unsteady equations are integrated until steady state is reached. This procedure is suggested by the hyperbolic character of the system of equations with respect to time. The convergence rate of these methods is however extremely small. By defining convergence rate as the inverse of the number of iterations or time steps necessary to damp a perturbation of the steady state a fixed number of magnitudes, this convergence rate is found to be of the order of Δx^2. Relaxation methods for the steady potential equation have by the same definition a convergence rate of the order of Δx. Relaxation techniques are however not applicable to steady Euler equations since in the subsonic parts of the flow field the system of equations has simultaniously hyperbolic and elliptic features.

Overrelaxation to accelerate the convergence of unsteady Euler equations is possible since these are uniformally hyperbolic. First attempts to use overrelaxation are independently due to Wirz [1] and Désidéri and Tannehill [2].

The model equation :

$$\frac{\partial \xi}{\partial t} + V \frac{\partial \xi}{\partial x} = 0 \tag{1}$$

is discretised by Wirz as :

$$\bar{\xi}(x,t+\Delta t) - \bar{\xi}(x,t) + \frac{\lambda}{2}\left(\xi(x+\Delta x,t) - \xi(x-\Delta x,t)\right) = 0 \tag{2}$$

$$\xi(x,t+\Delta t) = r_1\bar{\xi}(x,t+\Delta t) + r_2\bar{\xi}(x,t) + (1 - r_1 - r_2)\xi(x,t) \tag{3}$$

with $\lambda = V\Delta t/\Delta x$.

The discretisation of Désidéri and Tannehill on (1) has two steps :

downwind :

$$\xi(x,t+\widetilde{\Delta t}) = \xi(x,t) - \lambda\left(\xi(x+\Delta x,t) - \xi(x,t)\right)$$

$$\bar{\xi}(x,t+\Delta t) = (1-r_1)\bar{\xi}(x,t) + r_1\xi(x,t+\widetilde{\Delta t}) \tag{4}$$

upwind :

$$\bar{\xi}(x,t+\widetilde{\Delta t}) = \bar{\xi}(x,t+\Delta t) - \lambda\left(\bar{\xi}(x,t+\Delta t) - \bar{\xi}(x-\Delta x,t+\Delta t)\right)$$

$$\xi(x,t+\Delta t) = (1-r_2)\xi(x,t) + r_2\bar{\xi}(x,t+\widetilde{\Delta t}) \tag{5}$$

In both methods, artificial variables $\bar{\xi}$ are introduced.

It has been shown by Wirz and by Désidéri and Tannehill that the relaxation factors r_1 and r_2 can be chosen to optimise the convergence rate of the methods.

For non-optimal relaxation factors, the convergence rate is of the order of Δx^2.
For optimal relaxation factors, it is of the order of Δx.

A drawback of both schemes is that the number of dependent variables is doubled by the use of artificial state variables. This complicates the calculations since these artificial variables have dynamics. They need boundary conditions and they interact with the natural dependent variables.

In this paper a scheme is presented which is constructed in the same spirit as the

AN OVERRELAXATION SCHEME FOR ONE DIMENSIONAL EULER EQUATIONS

The one-dimensional Euler equations, written in quasi-linear form are :

$$\frac{\partial}{\partial t}\xi + A\frac{\partial \xi}{\partial x} = 0$$

$$\text{with} \qquad \xi = \begin{pmatrix} \rho \\ u \\ p \end{pmatrix} \qquad A = \begin{pmatrix} u & \rho & 0 \\ 0 & u & \frac{1}{\rho} \\ 0 & \gamma p & u \end{pmatrix} \tag{11}$$

The eigenvalues of the system matrix A are u, u+c and u-c; with $c = \sqrt{\gamma p/\rho}$

The scheme (6)(7)(8) can be applied to (11) by replacing the scalar λ by $A\frac{\Delta t}{\Delta x}$. This means that the stability condition which results from (10) is to be fulfilled by the X-values corresponding to u, u+c and u-c. For R = 1, this condition is :

$$(1-\sqrt{5})/2 \leqq X \leqq (1+\sqrt{5})/2 \tag{12}$$

The X corresponding to u+c, can be chosen to be 1.

$$X^{+} = \frac{(u+c)\frac{\Delta t}{\Delta x}}{1 + \alpha(u+c)\frac{\Delta t}{\Delta x}} = 1 \tag{13}$$

In supersonic flow (u>c), the X-values corresponding to u and u-c are then in the range 0<X<1 and automatically statisfy (12). In subsonic flow (u<c), the X-value corresponding to u-c is negative and has to fulfil :

$$(1-\sqrt{5})/2 \leqq X^{-} \tag{14}$$

Combination of

$$X^{-} = \frac{(u-c)\frac{\Delta t}{\Delta x}}{1 + \alpha(u-c)\frac{\Delta t}{\Delta x}} \tag{15}$$

and (13) gives :

$$\alpha = \frac{(X^{-}-1)M + X^{-}-1}{2X^{-}} \tag{16}$$

$$C_o = \frac{2X^{-}}{1-X^{-}}\;\frac{1}{1-M^2} \tag{17}$$

with $M = \frac{u}{c}$ and $C_o = \frac{c\,\Delta t}{\Delta x}$

By (16) and (17) perturbations transferred with the physical velocity u+c are completely damped. Complete damping is not possible for the transfer along u and u-c. The best scheme is thus reached when the numerical transfervelocity corresponding to u and u-c is maximised. This happens when Δt is as large as possible. This implies that in subsonic flow, X^{-} has to be as close as possible to the stability limit (14) and, in supersonic flow, as close as possible to 1.

A practical choise for X^{-} is given in figure 1.

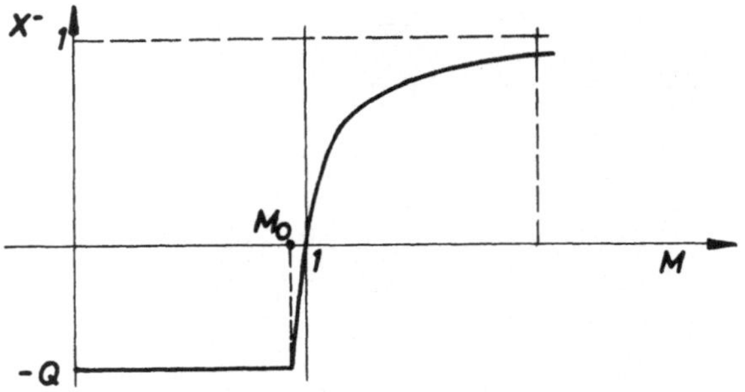

Figure 1

For $M < M_o$:

$$\left.\begin{array}{l} X^+ = 1 \\ X^- = -Q \end{array}\right\} \rightarrow \left\{\begin{array}{l} \alpha = (M-1+Q_o)/Q_o \\ C_o = Q_o/(1-M^2) \end{array}\right. \tag{18}$$

For $M > M_o$:

$$\left.\begin{array}{l} X^+ = 1 \\ X^- = (M-1)/(M-M_o + \frac{1-M_o}{Q}) \end{array}\right\} \rightarrow \left\{\begin{array}{l} \alpha = (M_o - 1+Q_o)/Q_o \\ C_o = Q_o/\big((1-M_o)(1+M)\big) \end{array}\right. \tag{19}$$

with $Q_o = 2Q/1+Q$.

Possible values are : $Q = .615 \quad M_o = .95$.

The attainment of the steady state is thus not determined by the internal damping but by the expulsion of perturbations along u and u-c. The complete damping along u-c is however absolutely necessary to avoid reflection along u-c at the outflow boundary.

Since there is only higher order coupling for small perturbations between the acoustic movement (u+c,u-c) and the convective movement (u), there is asymptotically no interaction between u and u-c. Hence, for sufficiently long time a perturbation along u and u-c is expulsed in a number of iterations which is proportional to the number of elements in the flow field.

The asymptotic rate of convergence is thus $O(\Delta x)$ since increasing the number of elements in the flow field, only lineary increases the number of iterations necessary to reach a certain level of convergence when starting from the same initial state.

To illustrate the convergence rate, the flow is calculated in a nozzle divided in 28 constant Δx segments.

The section is S_o between nodes 1 and 3 and between nodes 21 and 29. Between nodes 3 and 21 the section is :

$$S(i) = S_o\left\{0.9 + 0.1 \times \left[2\left(\frac{i-12}{9}\right)^2 - \left(\frac{i-12}{9}\right)^4\right]\right\}$$

The outlet pressure is $p_2 = .718025 \times p_{01}$.

The exact solution has a shock on node 16.

For each nodal calculation, the conservative one-dimensional Euler equations are linearised to a form (11). The matrix A is iteratively corrected untill the change in Mach number becomes less than some value C. In the shock region, the usual artificial viscosity is necessary to damp the post-shock oscillations. In the momentum equation, a term is added of the form :

$$D_X \, \Delta x \, \frac{\partial^2 u}{\partial x^2} \qquad \text{with} \qquad D_X = O(1)$$

The calculation is done with a relaxation factor r=2 (R=1), a damping term in the artificial viscosity $D_X=.01$, a transition Mach number $M_o=.95$, maximum negative value of $X^-=-Q=-.615$ and convergence factor C=.015.

Since the formulae (18) are very sensitive to Mach number, M has to be multiplied by a safety factor S = .995 before it is used in (18). Overestimation of M causes X^- to exceed the stability limit X^- = -.618.

A convergence norm is calculated as the mean absolute deviation in the nodal points between the calculated Mach number and the exact Mach number. The initial state is a uniform flow with M = .6.

The convergence history for the norm is depicted in figure 2.

The steady state is reached very abruptly. This is due to the expulsion mechanism on which the convergence is based.

The calculations are repeated for the same geometry but subdivided in 56, 84 and 112 elements. The norm is still calculated in the nodal points of the first case. The accuracy is defined as the value of the norm.

The results are shown in table I.

Table I : One-dimensional calculations

number of elements	28	56	84	112
optimal relaxation factor	2.000	2.000	1.996	1.992
field iterations for an accuracy .01	55	104	153	201
idem for .001	-	119	168	218
nodal iterations for an accuracy .01	1828	6428	13382	23173
idem for .001	-	7262	14637	25073
final accuracy	.00201	.00053	.00044	.00027

These results show clearly the linear convergence rate.

Figure 3 shows the Mach number distribution for the coarsest grid.

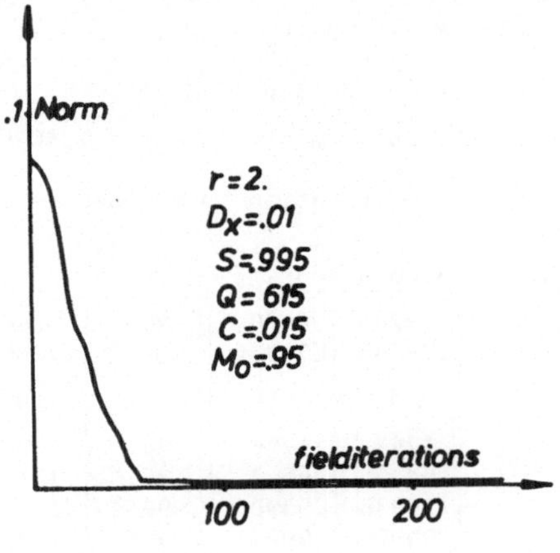

Figure 2

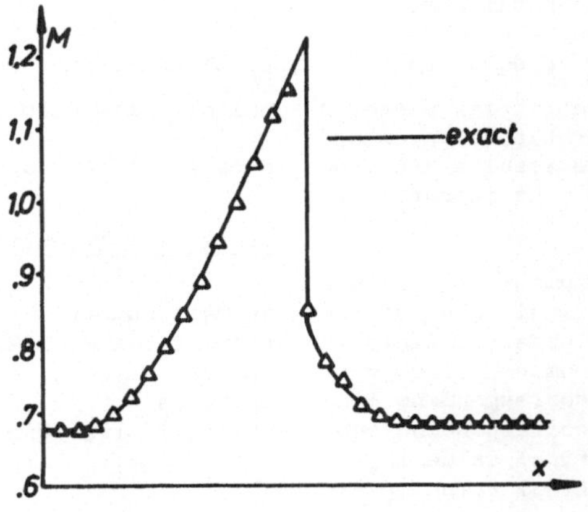

Figure 3

TWO-DIMENSIONAL APPLICATIONS

The one-dimensional scheme can immediately be extended to two-dimensional applications for line relaxation on a mesh in which the transversal lines are straight and parallel. In the formulae (18)(19) the Mach number is to be replaced by the component of the Mach number in the direction perpendicular to the transversal lines.

The discretisation can easily be done by the finite volume technique. The upwind and the downwind volume corresponding to the mesh point (i,j) are depicted in figure 4. In order to have formal similarity with the one-dimensional algorithm, the time derivative term has to be distributed on the points (i+1,j)(i,j)(i-1,j) proportional to the coefficients of these points in the flux through AB.

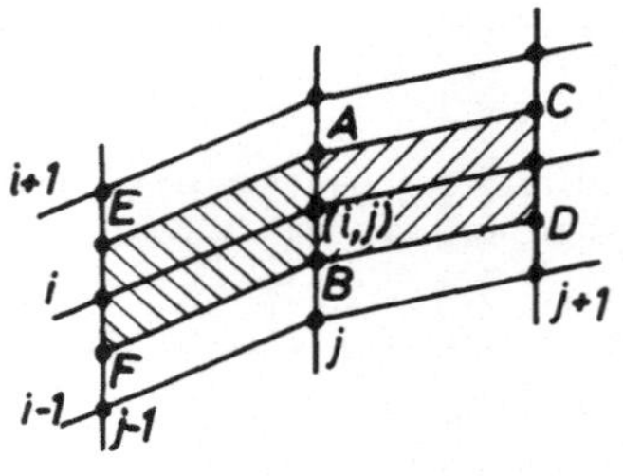

Figure 4

The stability analysis on this two-dimensional discretisation reveals that there is a slight instability in the transversal direction. This instability can easily be compensated by a small artificial viscosity in this direction.
It can for instance be done by introducing in the momentum equations a corrected viscosity of the form : for the X-momentum :

$$D_{y1}\ \Delta y\ \frac{\partial^2 u}{\partial y^2}\ (t^*)\ -\ D_{y2}\ \Delta y\ \frac{\partial^2 u}{\partial y^2}\ (t_o)$$

t^* is taken to be $t+\Delta t$ in the downwind step and $t+\Delta t$ in the upwind step. t_o is a time level that is fixed for a cycle of N_v iterations and renewed after that cycle.
$D_{y1} - D_{y2}$ is sufficiently small.
Typical values are : $D_{y1} = 1$, $D_{y2} = .99$.
With these values the cycle should have approximately as many iterations as there are nodes in the transversal direction of the flow field, in order to ensure stability.

The same nozzle as was used in the one-dimensional example is subdivided into 4x28, 8x28, 4x56 and 8x56 elements with constant height on a transversal line.

The parameters are : $r = 1.99$, $S = .995$, $Q = .615$, $M_o = .95$, $D_x = .01$, $D_{y1} = 1$, $D_{y2} = .99$, $C = .013$.
N_v is equal to the number of nodes in the transversal direction (5 or 9).

The initial state is a uniform flow with $M = .6$, the outflow pressure is .718025 x inlet total pressure.
The results are shown in table II. An equivalent field iteration is defined as 27 or 55 line iterations.

Table II : Two-dimensional calculations

number of elements	4x28	8x28	4x56	8x56
final value of the mean Mach number	.78444	.78375	.78860	.78868
number of field iterations for a mean Mach number = .999 x final value	57	58	132	135
corresponding line iterations	1742	1764	7483	7207
corresponding equivalent field iterations	64.5	65.3	136.0	131.0
final value T_{02}/T_{01}	.99938	.99987	.99954	.99958
final value $\dot{m}_2/\dot{m}_1$	.99950	.99945	.99953	.99993

The final values of the total temperature ratio and of the mass flow ratio are included in the table to show the conservativity of the used finite volume method. The results show clearly the linear convergence rate in Δx. They proove that the number of nodes in the transversal direction has no influence at all on the convergence rate.

REFERENCES

1. WIRZ H.J. : Relaxation methods for time dependent conservation equations in fluid mechanics. VKI-LS-97, Computational Fluid Dynamics, 1977.

2. DESIDERI J.A., TANNEHILL J.C. : Over-relaxation applied to the Mac-Cormack finite-difference scheme. J. Comp. Phys., Vol. 23, 1977, p 313-326.

A DORODNITSYN FINITE ELEMENT BOUNDARY LAYER FORMULATION

C.A.J. Fletcher and R.W. Fleet
University of Sydney, Sydney, Australia

1. INTRODUCTION

The Dorodnitsyn formulation offers a number of *significant advantages* for attached boundary layer computations. In two dimensions the independent variables are x and u/u_e. Consequently an infinite domain in the y direction is replaced by a finite domain in the u direction ($0 \leqslant u/u_e \leqslant 1$). A uniform grid in the u direction automatically captures downstream boundary layer growth and provides high resolution (in physical space) close to the wall. This is particularly relevant for turbulent boundary layer flows.

For two dimensional, incompressible boundary layers only one equation is solved. The normal velocity, v, does not appear explicitly but can be recovered subsequently, if required. The dependent variable, in the Dorodnitsyn formulation, is the nondimensional velocity gradient, $\tau = \partial u/\partial \eta$. Consequently the skin friction is computed very accurately.

In the last twenty years the Dorodnitsyn formulation has been applied to two and three dimensional, incompressible and compressible, laminar and turbulent boundary layer flows (Holt, 1977). The Dorodnitsyn formulation is particularly effective when given a *spectral* interpretation (Fletcher and Holt, 1975). The spectral or orthonormal variant of the Dorodnitsyn formulation has been applied to compressible laminar boundary layer flow (Fletcher and Holt, 1976) and to incompressible turbulent boundary layer flow (Yeung and Yang, 1981).

Here we give the Dorodnitsyn formulation a *finite element* interpretation which permits a very economical algorithm to be constructed. The Dorodnitsyn finite element method is applied to three representative *turbulent* boundary layer flows taken from the 1968 Stanford conference: Wieghardt (zero pressure gradient), Bradshaw, Flow C (adverse pressure gradient) and Ludwieg and Tillmann (favourable pressure gradient).

Solutions obtained with the Dorodnitsyn finite element formulation (DOROD-FEM) are compared with those obtained with a Dorodnitsyn spectral formulation (DOROD-SPEC) and a representative finite difference package, STAN5 (Reynolds, 1976). DOROD-FEM produces results that are as accurate as those produced by STAN5 but, typically, *are* 10 *times more economical.*

2. DORODNITSYN FINITE ELEMENT BOUNDARY LAYER FORMULATION

The Dorodnitsyn formulation, applied to the equations governing two-dimensional, turbulent boundary layer flow, can be written

$$\frac{\partial}{\partial x}\int_0^1 f(u)\; u \text{Ⓗ}\, du = \frac{u_{ex}}{u_e}\int_0^1 f_u\;(1-u^2)\,\text{Ⓗ}\, du + u_e\int_0^1 f_u\left\{\left(1+\frac{\nu_T}{\nu}\right)T\right\}_u du = 0, \qquad (1)$$

where ν_T is an *eddy viscosity* introduced to account for the Reynolds shear stress. The subscript u denotes differentiation with respect to u and f(u) is a general weight function. u is nondimensionalised with respect to u_e and u_e with respect to u_∞. In eq.(1) u and x are independent variables and Ⓗ and T are dependent variables defined by

$$T = 1/\text{Ⓗ} = \partial u/\partial\eta \qquad \text{where} \qquad \eta = Re^{\frac{1}{2}} u_e\, y/L\ . \qquad (2)$$

To apply the finite element method to eq.(1) we introduce the following *trial solutions*,

$$\text{Ⓗ} = \sum_{j=1}^{M} N_j(u)/(1-u)\;\theta_j(x) \qquad (3)$$

and $$\left(1+\nu_T/\nu\right)T = \sum_{j=1}^{M}(1-u)N_j(u)\,(1+\nu_T/\nu)_j\,\tau_j(x)\ , \qquad (4)$$

and let $$f_k(u) = (1-u)N_k(u). \qquad (5)$$

The functions, $N_j(u)$ and $N_k(u)$, are one-dimensional *shape functions*, either linear or quadratic. The simultaneous prescription of the analytic form (in u) of Ⓗ and $(1 + \nu_T/\nu)T$ prevents eq.(2) being satisfied everywhere, except in the limit $M \to \infty$. However eq.(2) is satisfied for the nodal values, i.e. $\theta_j = 1/\tau_j$.

The *group representation* of terms, as in eq.(4), avoids the computationally expensive evaluation of integrals at every x step associated with the analytic expression for the eddy viscosity (section 3). The group representation of terms has been used previously with the spectral Dorodnitsyn formulation (Fletcher and Holt, 1976) and with the finite element method (Fletcher, 1979). The same idea, called the product approximation, has been discussed theoretically (Christie et al., 1981).

Substitution of eqs.(3), (4) and (5) into eq.(1) indicates that a *modified Galerkin* finite element formulation is generated (Fletcher, 1982). Evaluation of the various integrals produces the following system of ordinary differential equations,

$$\sum_j CC_{kj}\frac{d\theta j}{dx} = \frac{u_{ex}}{u_e}\sum_j EF_{kj}\,\theta_j + u_e\sum_j AA_{kj}(1+\nu_T/\nu)_j\,\tau_j \qquad (6)$$

where $CC_{kj} = \int_0^1 N_k N_j\; u\, du$, $EF_{kj} = \int_0^1 N_j\left(\frac{dN_k}{du}(1-u)-N_k\right)(1+u)\; du$

and $$AA_{kj} = \int_0^1 \left(\frac{dN_j}{du}(1-u)-N_j\right)\left(\frac{dN_k}{du}(1-u)-N_k\right) du .$$

A generalised implicit algorithm for marching eq.(6) downstream can be written

$$\sum_j CCC^n_{kj} \; \Delta\theta^{n+1}_j = P^n_k \qquad (7)$$

where $\Delta\theta^{n+1}_j = \theta^{n+1}_j - \theta^n_j$, $P^n_k = \Delta x\left[\frac{u_{ex}}{u_e} \sum_j EF_{kj}\theta_j + u_e \sum_j AA_{kj}(1 + \nu_T/\nu)_j \tau_j\right]^n$

and $$CCC^n_{kj} = CC_{kj} - w\,\Delta x\left[\frac{u_{ex}}{u_e} EF_{kj} - u_e AA_{kj}\left\{(1+\nu_T/\nu)_j \tau^2_j - \frac{\partial(\nu_T/\nu)}{\partial\theta_j}{}_j \tau_j\right\}\right]^n$$

The parameter, w, controls the degree of implicitness; here w = 0.6. Eq.(7) is *tridiagonal* for linear elements and *pentadiagonal* for quadratic elements. For both types of element eq.(1) can be solved efficiently using a generalised Thomas algorithm at each x^n step, *without the need for iteration.*

The Dorodnitsyn spectral formulation, DOROD-SPEC, is described elsewhere (Fleet and Fletcher, 1982)

3. TURBULENCE MODEL

The eddy viscosity, ν_T, is evaluated via the *van Driest mixing length* formulation in the inner region. In the current notation this becomes

$$(\nu_T/\nu)_j = (1 - u)\,\ell^{+2}\tau_j/\tau_1 \qquad \text{where} \qquad \ell^+_j = \kappa y^+_j\,(1-\exp(-y^+_j/A^+)) \qquad (8)$$

$$A^+ = A^+_o/(1 + 10\,p^+) \quad , \; p^+ = -(u_{ex}/u^2_e)/(Re\,u^3_\tau) \; , \quad u_\tau = \tau_1^{0.5}\,R_e^{0.25} \; ,$$

$$y^+_j = u_\tau R_e \int_0^{u_j} du' / T(u') \qquad \text{and} \qquad u^+_j = u_j / u_\tau \; .$$

In eqs.(8) τ_1 is the wall value of τ. For the results presented in section 4, $\kappa = 0.41$ and $A^+_o = 25$. A slight disadvantage using the van Driest formula with the Dorodnitsyn formulation is the explicit appearance of y^+. For DOROD-SPEC the prescription of $\nu_T(u)$, due to Spalding (Yeung and Yang, 1981), avoids this problem.

In the outer region the *Clauser* eddy viscosity formulation is utilised i.e.

$$\nu_T/\nu = 0.0168\, u_e R_e\, \delta^*/L$$

where δ^* is the displacement thickness. No intermittency factor is used in DOROD-FEM.

4. RESULTS AND DISCUSSION

The number of gridpoints across the boundary layer and the size and number of steps in the downstream direction for DOROD-FEM and STAN5 are shown in Table 1. For DOROD-SPEC six unknown coefficients were used, in the equivalent of eq. (3), to represent the variation of Ⓗ across the boundary layer. Downstream stepsizes varied between 0.000015 and 0.4, typically.

	zero press. grad.		adverse press. grad.		fav. press. grad.	
	DOROD-FEM	STAN5	DOROD-FEM	STAN5	DOROD-FEM	STAN5
Gridpoints across boundary layer	11	33-39	11	47-48	11	32-35
No. of steps, $\Delta x/L$	205	401	294	660	124	301
$\Delta x/L$	0.0001 -0.071	0.0004 -0.031	0.0001 -0.049	0.0002 -0.0039	0.0001 -0.044	0.0017 -0.0126
Relative execution time	1	8.99	1.55	17.70	0.60	6.83

Table 1 Comparison of DOROD-FEM and STAN5

All the DOROD-FEM solutions presented in Figs 1 to 8 have used 11 *equally-spaced* grid points across the boundary layer. Using 21 equally-spaced grid points gave slightly more accurate predictions of displacement thickness, δ^*/L, and momentum thickness, θ/L, but made little difference to the accuracy of the skin friction prediction. The execution time increases linearly with the number of grid points across the boundary layer.

Using *quadratic* elements produces a slight improvement in accuracy for all predicted quantities. However the improvement is not as substantial as for laminar boundary layers. The execution time per downstream step is about 20% greater when quadratic elements are used instead of linear elements; but there is some indication that the overall algorithm is less robust. The lack of robustness manifests itself in a more frequent reduction in the downstream stepsize, $\Delta x/L$. Consequently more steps are required to cover a specific downstream interval and overall execution times are typically 30-40% greater when quadratic elements are used.

For the Wieghardt flat plate flow (Coles and Hirst, 1968, flow no. 1400), the skin friction variation is shown in Fig.1. It is apparent that all methods predict the skin friction quite closely. Corresponding results for displacement and momentum thickness are shown in Fig. 2. Whereas DOROD-FEM and DOROD-SPEC produce good agreement with the experimental results, STAN5 generally predicts too high a value for large values of x/L.

A typical velocity distribution (x/L = 3.49) for the zero pressure gradient case is shown in Fig. 3. The solution obtained with STAN5 and DOROD-FEM are almost interchangeable. Only sufficient STAN5 data points have been included to define the profile. Both computational results agree closely with the experimental velocity distribution except at the innermost data point ($\log_{10}y^+ \approx 1.5$).

The downstream development of the skin friction for Bradshaw Flow C (Coles and Hirst, 1968, flow no. 3300) is shown in Fig. 4. Two 'experimental' results are given. The results due to Bradshaw come from the original published results. The experimental results due to Coles and Hirst are based on a log-law interpretation of the velocity profiles.

For this (adverse pressure gradient) case STAN5 gives a more accurate prediction of skin friction particularly at small values of x/L. DOROD-FEM tends to underpredict at small values of x/L and DOROD-SPEC tends to overpredict for large values of x/L. Corresponding predictions of displacement and momentum thickness (Fig. 5) bracket the experimental values. STAN5 is more accurate for displacement thickness and DOROD-FEM and DOROD-SPEC are more accurate for momentum thickness.

However velocity profiles (at x/L = 2.50) generated by DOROD-FEM and STAN5 agree closely (Fig. 6). But both methods slightly overpredict the velocity at a given y^+ in the outer part of the boundary layer. The velocity profiles define quite sharply the three regions of the turbulent boundary layer - inner region, log-law region and outer region.

Results for a favourable pressure gradient (Coles and Hirst, 1968, flow no. 1300) are shown in Figs 7 and 8. Although all three methods provide a reasonably accurate prediction of skin friction variation (Fig. 7), all methods tend to overpredict the displacement and momentum thicknesses (Fig. 8), particularly for large x/L.

It is clear from the results for the three cases considered that the Dorodnitsyn finite element method, DOROD-FEM, is producing solution of *comparable accuracy* to those produced by a typical finite difference method, STAN5. In Table 1 the execution times for the two methods are compared. It can be seen that DOROD-FEM is typically 10 times *more economical* than STAN5.

The greater economy comes from two main sources. The Dorodnitsyn formulation permits an accurate representation of the variation across the boundary layer with approximately *a quarter to a third of the number of grid points* required by STAN5. This provides roughly a factor of four improvement in execution time.

The other major economy arises from the number of step sizes in the downstream direction (Table 1). DOROD-FEM increases $\Delta x/L$ by a factor of 50% if the change in τ_1 for the current step is less than 0.2%. If the change is more than 2% the step size is halved. STAN5 appears to use a considerably more conservative step change criterion that is dependent on the boundary layer thickness. Both methods demonstrate comparable execution times per grid point per downstream step.

In conclusion a Dorodnitsyn finite element formulation for attached boundary layers has been described and tested for three representative turbulent boundary layer flows. The method is *as accurate* as a representative finite difference method *but an order-of-magnitude more economical.*

REFERENCES

Bradshaw, P. (1967) ARC R&M 3575

Christie, I., Griffiths, D.F., Mitchell, A.R. and Sanz-Serna, J.M., (1981) Inst. Maths. Applic. Num. Anal., Vol. 1, pp 253-266.

Coles, P. and Hirst, E. (eds) (1968), "Computation of Turbulent Boundary Layers - 1968 AFOSR-IFP Stanford Conference", Vol. 2.

Fleet, R.W. and Fletcher, C.A.J. (1982), "A Comparison of the Finite Element and Spectral Methods for the Dorodnitsyn Boundary Layer Formulation", Fourth International Conference in Australia on Finite Element Methods, Melbourne, August 1982.

Fletcher, C.A.J. (1979), J. Comp. Phys., Vol. 33, pp 301-312.

Fletcher, C.A.J. (1982), "Computational Galerkin Methods", Springer-Verlag, Heidelburg.

Fletcher, C.A.J. and Holt, M (1975), J. Comp. Phys., Vol 18, pp 154-164.

Fletcher, C.A.J. and Holt, M. (1976), J. Fluid Mech., Vol. 74, pp 561-591.

Holt, M., (1977) "Numerical Methods in Fluid Dynamics", Springer-Verlag, Heidelburg.

Reynolds, W.C. (1976), Ann. Rev. of Fluid Mech., Vol. 8, pp 183-208.

Yeung, W.S. and Yang, R.J. (1981), J. Appl. Mech., Vol. 48, pp 701-706.

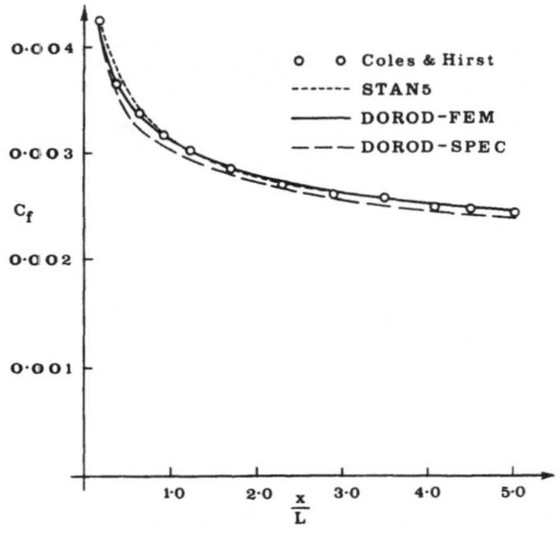

Fig. 1 Skin friction variation: zero pressure gradient

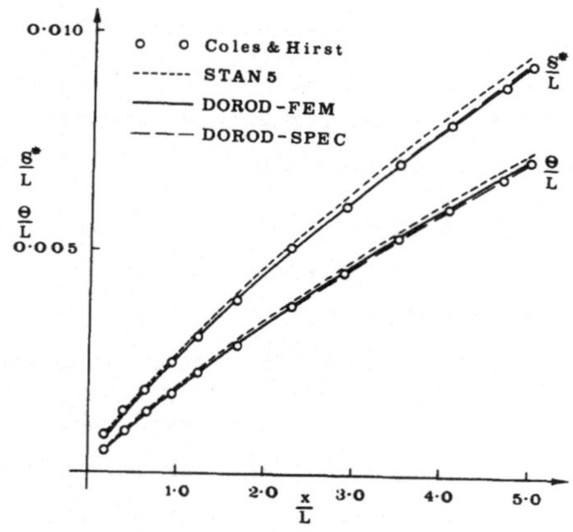

Fig. 2 Displacement and momentum thickness variation: zero pressure gradient

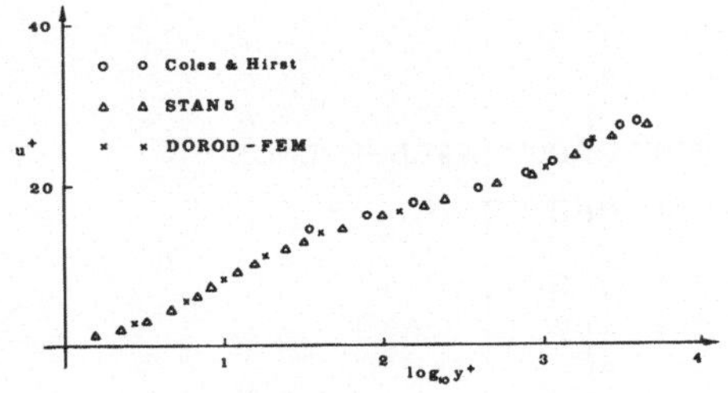

Fig. 3 Velocity distribution in a zero pressure gradient (x/L = 3.49)

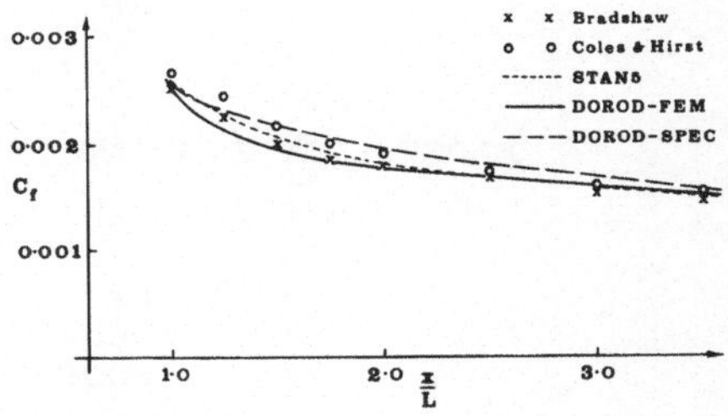

Fig. 4 Skin friction variation: adverse pressure gradient

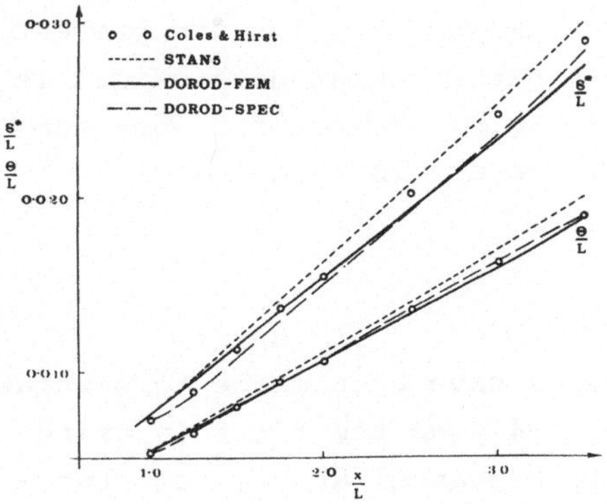

Fig. 5 Displacement and momentum thickness variation: adverse pressure gradient

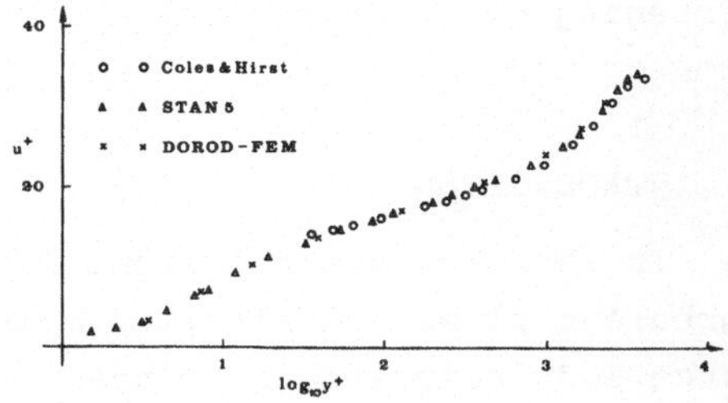

Fig. 6 Velocity distribution in an adverse pressure gradient (x/L = 2.50)

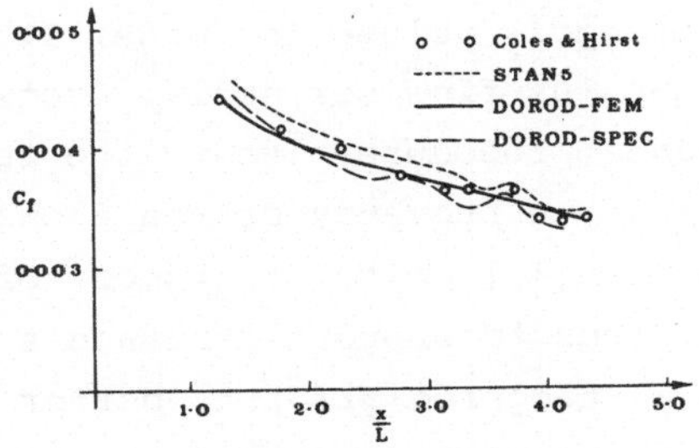

Fig. 7 Skin friction variation: favourable pressure gradient

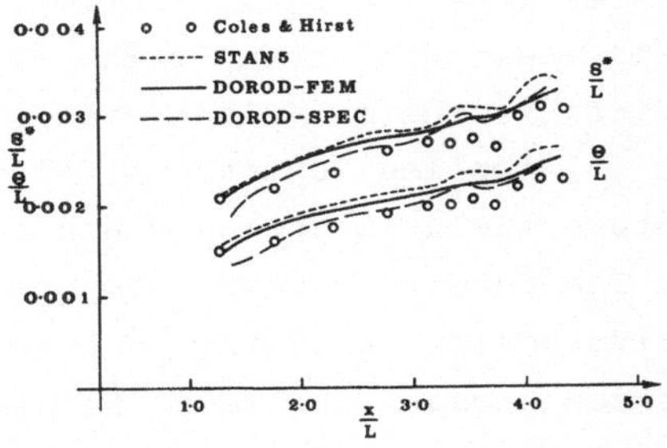

Fig. 8 Displacement and momentum thickness variation: favourable pressure gradient

LARGE EDDY SIMULATION OF A TURBULENT WALL-BOUNDED SHEAR LAYER WITH LONGITUDINAL CURVATURE

R. Friedrich and M.D.Su
Technical University
8000 Munich 2, FRG

ABSTRACT

The statistically steady and two-dimensional turbulent flow over a longitudinally curved convex wall for an incompressible Newtonian fluid is investigated using the large eddy simulation method based on Schumann's volume balance procedure. Preliminary results are presented for a 32x32x64 grid. The comparison with experimental data of Gillis et al. shows fairly good agreement. The global effect of downstream dampening of turbulence due to convex curvature is worked out.

1. INTRODUCTION

In the past several promising attempts have been meade to simulate turbulent flows numerically. Deardorff [1-4] was the first in using large eddy simulations to predict straight channel as well as atmospheric boundary layer flow. While his approach corresponds to some sort of 'filtering' of the basic equations, as shown by Leonard [5], Schumann [6,7] avoids filtering and defines the subgrid scale stresses as grid surface mean values of the deviations of velocity components from their surface mean values. Since Schumann splits each of these stresses into a locally isotropic and an inhomogeneous part, his model is meaningful even if the size of the grid volume is large. It predicts isothermal plane channel and annulus flow correctly. Grötzbach [8] applies Schumann's volume balance procedure also to the temperature field in channels and simulates even buoyancy driven flows [9]. The recent work of Antonopoulos-Domis [10] on the transport of a passive scalar in isotropic turbulence seems to support Schumann's approach because it leads (at least on a staggered grid) to better results than the prefiltering procedure of Leonard [5]. The latter has been applied to channel flow in one of the most impressive works in this field, that of Moin and Kim [11]. With their high resolution of 64^3 grid volumes they succeed in predicting the logarithmic law of the wall with its precise gradient. Prefiltering of the flow equations is also used by Love and Leslie [12] and in Cain, Reynolds and Ferziger's study of the time development of a free shear layer [13]. Orszag and Pao [14] have simulated the turbulent wake of a self-propelled body by means of Fourier series expansions of the velocity field applying periodic boundary conditions in two directions and free-slip conditions in the remaining direction. Their concept of interpreting the downstream variation of the wake as time evolution of the flow with the help of a suitable Galilean transformation did not appear us applicable to wall-bounded shear layers. Therefore we have resolved the whole flow field on a longitudinally curved convex wall extending Schumann's method to the simulation of a statistically steady and two-dimensional turbulent boundary layer.

2. VOLUME AVERAGED BASIC EQUATIONS

In an orthogonal (s,n,z)-coordinate system in which s is measured along the single curved wall, n along straight lines normal to the wall and z normal to the (s,n)-plane any quantity y integrated over a grid volume V (fixed in space) is denoted by ${}^{V}\overline{y}$:

$$^{V}\overline{y} = \frac{1}{{}^{V}\overline{h}\Delta s\Delta n\Delta z} \int\int\int_{\Delta s\Delta n\Delta z} yhdsdndz \; ,$$

where $h = 1 + n/R$, ${}^{V}\overline{h} \approx 1 + \frac{n}{2}\left(\frac{1}{R_{s-\Delta s/2}} + \frac{1}{R_{s+\Delta s/2}}\right)$, R = radius of wall curvature. Surface-averaged quantities are defined through

$$^{j}\overline{y} = \frac{1}{{}^{j}F} \int\int y \, d^{j}F \; , \qquad d^{s}F = dndz \; , \qquad d^{n}F = hdsdz \; , \qquad d^{z}F = hdsdn$$

and appear if divergence-type terms are integrated using Gauss' theorem. The volume-averaged *continuity equation* in its approximated form

$$\frac{1}{{}^{V}\overline{h}} \delta_s {}^{s}\overline{u} + \delta_n {}^{n}\overline{v} + \delta_z {}^{z}\overline{w} + {}^{V}\overline{v}/({}^{V}\overline{h}\,{}^{V}\overline{R}) = 0 \tag{1}$$

contains the resolvable variables ${}^{s}\overline{u}$, ${}^{n}\overline{v}$, ${}^{z}\overline{w}$ in a suitable finite difference form (${}^{V}\overline{v}$ is linearly related to ${}^{n}\overline{v}$). δ_j stands for the usual finite difference operator, e.g. $\delta_s {}^{s}\overline{u} = ({}^{s}\overline{u}(s+\Delta s/2) - {}^{s}\overline{u}(s-\Delta s/2))/\Delta s$. In order to obtain conservation equations for the resolvable quantities, the Navier-Stokes equations are averaged over volumes V_i equal in size, but shifted so as to surround the position of the corresponding velocity component within the staggered grid. Thus, ${}^{s}\overline{u}$ (e.g. follows from the following *momentum equation* :

$$\frac{\partial\, {}^{V_s}\overline{u}}{\partial t} + \frac{1}{{}^{V_s}\overline{h}}\, \delta_s\, {}^{s}\overline{u^2} + \delta_n\, {}^{n}\overline{uv} + \delta_z\, {}^{z}\overline{uw} + \frac{2\, {}^{V_s}\overline{uv}}{{}^{V_s}\overline{h}\, {}^{V_s}\overline{R}} = \tag{2}$$

$$- \frac{1}{\rho\, {}^{V_s}\overline{h}}\, \delta_s\, {}^{s}\overline{p} - \nu\left[\delta_n\, {}^{n}\overline{\frac{1}{h}\left(\frac{\partial v}{\partial s} - \frac{\partial (hu)}{\partial n}\right)} + \delta_z\, {}^{z}\overline{\frac{1}{h}\left(\frac{\partial w}{\partial s} - \frac{\partial (hu)}{\partial z}\right)}\right] .$$

In this equation not all the differentials could be converted into finite difference operators. Especially the diffusive terms must be approximated in the finite difference scheme. This is of minor importance because these terms do not contribute much in a high Reynolds number flow. Averaging of the convective terms, however, has severe consequences, because it leads to new unknown quantities like ${}^{s}\overline{u^2}$, ${}^{n}\overline{uv}$, ${}^{z}\overline{uw}$, ${}^{V_s}\overline{uv}$ which must be expressed in terms of the resolvable variables. If we decompose each flow quantity y into a resolvable (surface averaged) and an unresolvable part (deviation from this average):

$$y = {}^{j}\overline{y} + y' , \tag{3}$$

then the nonlinear terms in (2) can be written as

$${}^{s}\overline{u^2} = ({}^{s}\overline{u})^2 + {}^{s}\overline{u'^2} , \quad {}^{n}\overline{uv} = {}^{n}\overline{u}\,{}^{n}\overline{v} + {}^{n}\overline{u'v'} , \quad \text{etc.} \tag{4}$$

and we face a new closure problem in which the unresolvable *subgrid scale stresses* ${}^{s}\overline{u'^2}$, ${}^{n}\overline{u'v'}$,... have to be modeled. Quantities like ${}^{n}\overline{u}$ can always be calculated from neighbouring staggered grid values (${}^{s}\overline{u}$). Similarly ${}^{V_s}\overline{uv}$ will follow from ${}^{n}\overline{uv}$ in a simple way.

3. SGS EDDY VISCOSITY MODEL

We follow the suggestion of Schumann [6,7] and split each of the *subgrid scale* (SGS) *stresses* into a locally isotropic and an inhomogeneous (statistically averaged) part:

$${}^{j}\overline{u_i'u_j'} = ({}^{j}\overline{u_i'u_j'})'' + \langle {}^{j}\overline{u_i'u_j'} \rangle . \tag{5}$$

The first, locally isotropic contribution is assumed proportional to the fluctuating part of the deformation tensor in order to get zero time mean values of the SGS stresses for $i \neq j$, viz:

$$- ({}^{j}\overline{u_i'u_j'})'' = {}^{j}\mu\overline{D_{ij}}'' - \frac{1}{3}\,\delta_{ij}\,({}^{k}\overline{u_k'u_k'})'' , \tag{6}$$

where $\overline{D_{ss}}'' = 2\left(\frac{1}{{}^{V}\overline{h}}\,\delta_s\, {}^{s}\overline{u''} + \frac{{}^{s}\overline{v''}}{{}^{V}\overline{hR}}\right)$, $\overline{D_{sn}}'' = \frac{1}{{}^{V}\overline{h}}\,\delta_s\, {}^{n}\overline{v''} + \delta_n\, {}^{s}\overline{u''} - \frac{{}^{n}\overline{u''}}{{}^{V}\overline{hR}}$

$$u'' = u - \langle u\rangle \ , \quad v'' = v - \langle v\rangle \quad \text{etc.} \tag{7}$$

In $\overline{D_{ss}''}$ the term containing dR/ds has been neglected. The eddy viscosity ${}^j\mu$ is modeled according to Schumann as

$${}^j\mu = c_2({}^jF\,{}^jc_5\ {}^V\overline{E'})^{1/2} \tag{8}$$

with the only difference that no care has been taken of the anisotropy of the grid spacings, $h\Delta s$, Δn, Δz. The coefficients c_2, ${}^jc_5$ are at present taken from Schumann's simulations. The SGS *kinetic energy* ${}^V\overline{E'}$, however, is determined from a separate *transport equation* containing the full curvature effect.

The second part in (5) plays a predominant role in the near-wall region and, of course, in the whole turbulent field if only a small number of grid volumes is taken to simulate the flow. If the integral length scale of turbulence becomes much smaller than the root value of ${}^jF$ (which is true in our simulation for the near-wall region) all momentum is transported by the SGS motion. The inhomogeneous part of the SGS stress should therefore correspond to common Reynolds stress models. We use the mixing-length concept in writing

$$\langle {}^j\overline{u_i'u_j'}\rangle = -\,\delta_{is}\,\delta_{jn}\,\ell^2\,\left|\,\delta_n\langle {}^s\overline{u}\rangle - \frac{\langle {}^s\overline{u}\rangle}{R+n}\,\right|\,\left(\delta_n\langle {}^s\overline{u}\rangle - \frac{\langle {}^s\overline{u}\rangle}{R+n}\right) , \tag{9}$$

$$\ell^2 = \min\,(L^2,\ 0.01\ {}^V\overline{h}\Delta s\Delta z) .$$

The length scale L equals Bradshaw's mixing length [15]. The linear dimension $(0.01\ {}^V\overline{h}\Delta s\Delta z)^{1/2}$ has been formulated with reference to [6] until a more detailed investigation is possible.

4. NUMERICAL INTEGRATION IN TIME

We apply a mixed integration procedure consisting of an explicit first order Euler scheme with respect to diffusive terms and a second order leapfrog scheme for the convective terms. Averaging steps after 10 time steps avoid $2\Delta t$-oscillations which are typical for such schemes [7]. At each time step a tentative velocity field (denoted by a tilde) is first evaluated, neglecting the pressure gradient. This field does not fulfil the continuity condition (1). The correct velocity components at time step n+1 follow, symbolically written, from:

$$\vec{u}^{(n+1)} = \tilde{\vec{u}}^{(n+1)} - f\,\Delta t\,\frac{1}{\rho}\,\text{grad}\,p^{(n+1)} , \qquad f = \begin{cases} 1 & \text{Euler} \\ 2 & \text{leapfrog} \\ 3/2 & \text{average.} \end{cases} \tag{10}$$

Taking the divergence of (10), we find that the pressure has to satisfy the 'Poisson' equation:

$$\frac{1}{\rho}\,\Delta p^{(n+1)} = \mathrm{div}\,\tilde{\vec{u}}^{(n+1)}/(f\Delta t). \tag{11}$$

Instead of solving the continuity equation (1), we solve equation (11) iteratively by means of the block iteration or semi-direct scheme [16]:

$$\Delta_o p^{(\ell+1)} = \Delta_o p^{(\ell)} - \lambda(\Delta p^{(\ell)} - \frac{\rho}{f\Delta t}\,\mathrm{div}\,\tilde{\vec{u}}). \tag{12}$$

Δ, Δ_o are non-separable and separable Laplacian operators, respectively. Δ_o is made separable introducing a proper average of the Lamé coefficient h over the whole flow field. The parameter λ is taken equal to 1. The solution of (12) 'converges' after 3-4 iterations, i.e. when the bracket on the right-hand side is smaller than 10^{-5}. An Euler scheme serves to integrate the balance equation for the SGS kinetic energy in time. Upwind difference formulas are used to approximate the convection terms.

Initial and *boundary conditions* are crucial points of such simulations. The initial values for ${}^{s}\bar{u}$, ${}^{n}\bar{v}$, ${}^{z}\bar{w}$, ${}^{V}\overline{E'}$ are calculated from statistical (experimental) and fluctuating quantities, e.g.

$${}^{s}\bar{u} = {}^{s}\overline{u''} + {}^{s}\overline{\langle u\rangle} ,$$

where u"-distributions in z-direction are established using random numbers in such a way that the one-dimensional averages in z-direction of u", u''^2 correspond to the experimental values $\langle u''\rangle = 0$, $\langle u''^2\rangle$ at s,n. In a similar way the experimental data of Gillis et al. [17] for $\langle v''^2\rangle$, $\langle u''v''\rangle$ serve to determine v". w" follows from continuity. The geometry of the problem allows for periodic boundary conditions only in z-direction. At the *wall* the no-slip condition can be satisfied only for ${}^{n}\bar{v}$. Linear interpolation is used for ${}^{z}\bar{w}(\Delta n/2)$ and instead of specifying $u(n=0)$, the wall shear stress ${}^{n}\overline{\tau_{sn}}(n=0)$ is taken from experiment. The pressure satisfies a Neumann-type condition. At the *outer edge* of the flow field we assume potential flow, zero ${}^{V}\overline{E'}$ and vanishing gradients of ${}^{n}\bar{v}$, ${}^{z}\bar{w}$. Velocity components at the *upstream* edge coincide with experimental values and the pressure gradient (in s-direction) is zero. Vanishing second derivations of velocity and SGS kinetic energy form the *downstream* boundary conditions. The pressure distribution reflects the centrifugal effect.

5. RESULTS

The experiment of Gillis et al. [17] to determine the effect of convex longitudinal curvature ($\delta/R = 0.1$) on boundary layer turbulence, has been simulated on a CDC Cyber 175 with a grid of 32x32x64 volumes, equally spaced in n- and z-directions. The simulated flow field had an extension of 2δ in n- and z-directions and of 18δ in s-direction. 200 time steps were performed, consuming 19660 seconds of CPU-time. With 5 variables this corresponds to a grind time of 0.3 msec. Preliminary results are shown for a flow with U_∞ = 16 m/s and a Reynolds number of $RU_\infty/\nu = 5.2 \times 10^5$. In figures 1 and 2 profiles of mean velocity and turbulence intensity at different streamwise stations are compared with experiment. Typical flow patterns of the instantaneous fluctuating pressure p" in (n,z)-planes are presented in figure 3 for two stations showing the inhibiting effect of convex curvature.

REFERENCES

1 DEARDORFF, J.W., J. Fluid Mech. 41, pp 453-480 (1970)
2 DEARDORFF, J.W., J. Comp. Phys. 7, pp 120-133 (1971)
3 DEARDORFF, J.W., J. Atmos. Sci. 29, pp 91-115 (1972)
4 DEARDORFF, J.W., J. Fluids Eng., pp 429-438 (1973)
5 LEONARD, A., Advan. Geophysics 18A, pp 237-248 (1974)
6 SCHUMANN, U., J. Comp. Phys. 18, 405-420 (1975)
7 SCHUMANN, GRÖTZBACH, G. and KLEISER, L., VKI-Lecture Series 1979-2
8 GRÖTZBACH, G., Dissertation Univ. Karlsruhe (1977)
9 GRÖTZBACH, G., to appear in J. Fluid Mech.
10 ANTONOPOULOS-DOMIS, M., J. Fluid Mech. 104, pp. 55 (1981)
11 MOIN, P. and KIM, J., to appear in J. Fluid Mech.
12 LOVE, M.D. and LESLIE, D.C., Symp. Turb. Shear Flows, pp 14.1-14.10 (1977)
13 CAIN, A.B., REYNOLDS, W.C. and FERZIGER, J.H., Report TF-14, Stanford University (1981)
14 ORSZAG, S.A. and PAO, Y.H., Advan. Geophysics 18A, pp 225-236 (1974)
15 BRADSHAW, P., J. Fluid Mech. 36, pp 177-191 (1969)
16 CONCUS, P. and GOLUB, G.H., SIAM J. Num.Anal. 10, pp 1103-1120 (1973)
17 GILLIS, J.C., JOHNSTON, J.P., KAYS, W.M. and MOFFAT, R.J., Report HMT-31, Stanford University (1980)

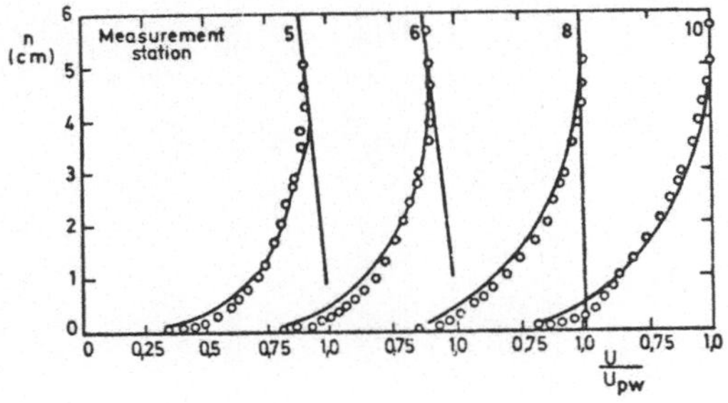

Fig.1 Mean velocity profile at different stations downstream

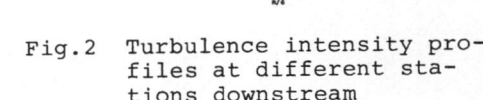

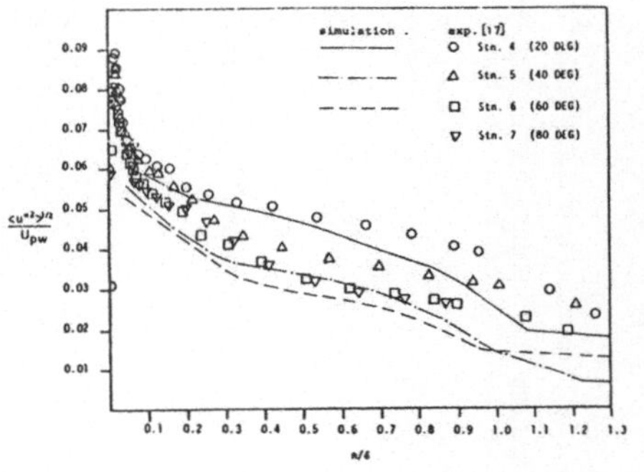

Fig.2 Turbulence intensity profiles at different stations downstream

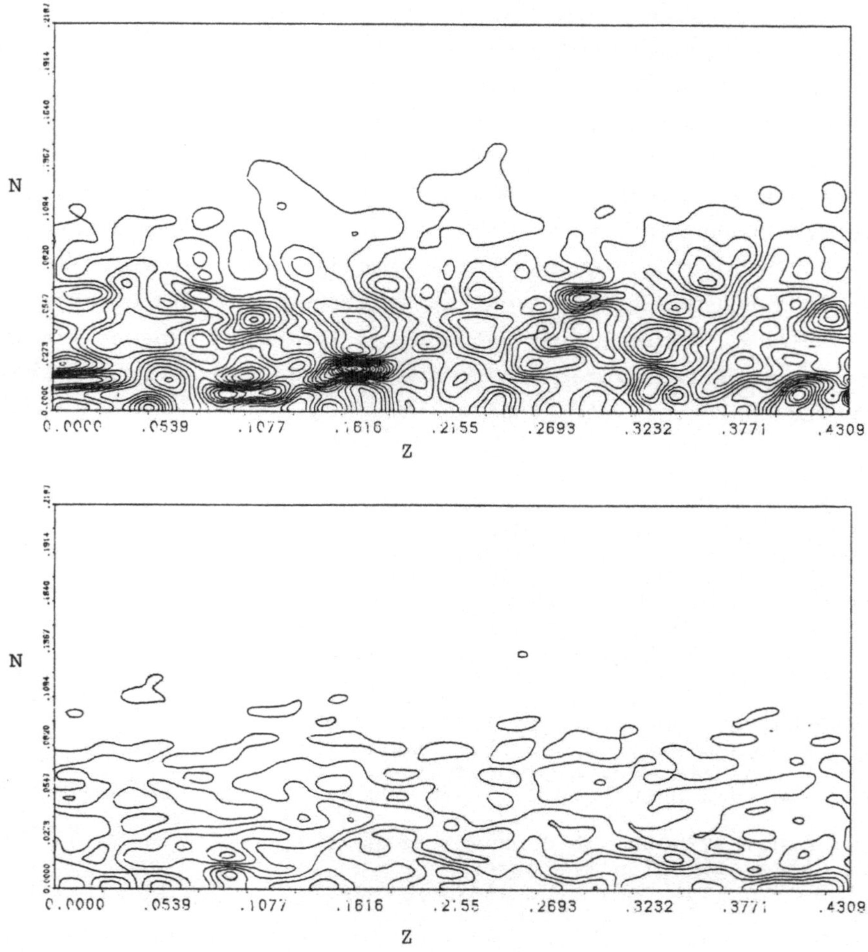

Fig.3 Instantaneous fluctuating pressure p" in two different (n,z)-planes (above: upstream, below: downstream)

MIXED INITIAL AND BOUNDARY VALUE PROBLEMS: UPWIND SCHEMES AND THEIR APPLICATION

by B. GABUTTI and L. ZANNETTI
Associate Professors,
Università di Torino and Politecnico di Torino, ITALY

INTRODUCTION

In this paper we deal with the numerical solution of hyperbolic mixed initial and boundary value problems as those arising when studying inviscid, unsteady flow fields governed by the Euler equations. We refer here to the numerical solution obtained by finite differences approximations.

The peculiar aspect of this matter we are interested on is that the Finite Difference Equations (FDE) may need "additional" boundary conditions, beside the "physical" boundary conditions needed by the corresponding Partial Differential Equations (PDE). In Ref. [6] it is pointed out that additional boundary conditions are introduced when the FDE are inconsistent with the physical phenomena described by the PDE, which are essentially wave propagation phenomena. In fact the use of additional boundary conditions can be generally avoided when schemes of the λ-family ([4],[7]) are used. These schemes are based on one-sided differences, according to the direction of propagation of the signals whose interaction defines the flow properties at each grid point during a transient.

In our opinion the shortcomings a numerical scheme causes at the boundaries represent the main point to be discussed. The quality of a numerical solution depends completely on the procedure at the boundary, provided the scheme is stable and accurate at the interior points. Any time "additional" boundary conditions not defined by the actual physical problem, are needed, the quality of the numerical solution is jeopardized. In fact the most we can state about additional boundary conditions is that they are stable from a numerical point of view, but we can not avoid the, by definition, spurious effects they can introduce in the computation.

Our choice of using schemes of the λ-family is due to the great advantages we get on treating boundaries, beside the good qualities these schemes show when applied to transonic flow fields.

Such schemes ask for three points formulas, at least, to evaluate one-sided differences, when accurate at the second order [2], as consequence they need some special treatment at the point next to the boundary, where only two points are available to describe the signals coming from the boundary. The difficulty can be overcome in many different ways. For instance a suitable extrapolation from inner points could be used to simulate an extra point beyond the boundary. If the computation wants to be maintained second-order accurate, the procedure would involve at least one point outside the domain of dependence. Therefore it would be inconsistent with the general principle of the λ-scheme. In Ref. [8] an algorithm for computing points next to the boundaries has been proposed. It is based on physical considerations, it uses the two available points on the domain of dependence and explicitely the boundary condition in order to abide by the general principle of the λ-scheme and to maintain the accuracy.

In the following it is shown that the proposed algorithm is stable, according to the normal mode analysis (Ref.[3]). A numerical example is discussed.

FORMULATION OF THE PROBLEM

Let us consider a strictly hyperbolic system of the first order,

$$\partial u/\partial t = A\,\partial u/\partial x + F \quad , \quad 0 \le x \le 1 \quad , \quad t \ge 0 \qquad (1)$$

where $u = (u_1(x,t),\dots,u_N(x,t))^T$ is the transposed vector of unknowns of the real variables x,t; $F = (f_1(x,t),\dots,f_N(x,t))^T$ is a known vector function and

$$A = \begin{bmatrix} A^I & 0 \\ 0 & A^{II} \end{bmatrix} \quad , \quad A^I > 0 \quad , \quad A^{II} < 0 \qquad (2)$$

is a $N \times N$ constant matrix. It is also convenient to introduce $\tilde{A}^I$, $\tilde{A}^{II}$ such that

$$A = \tilde{A}^I + \tilde{A}^{II} \quad , \quad \tilde{A}^I \ge 0 \quad , \quad \tilde{A}^{II} \le 0$$

Without restriction we can assume that A is a diagonal matrix.
Together with (1), initial values

$$u(x,0) = f(x) \quad , \quad 0 \le x \le 1 \qquad (3)$$

are given for $t = 0$, which are compatible with the boundary conditions

$$\begin{aligned} u^I(0,t) &= S_I\, u^{II}(0,t) + b_I(t) \qquad \text{for } x = 0 \\ u^{II}(1,t) &= S_{II}\, u^I(0,t) + b_{II}(t) \qquad \text{for } x = 1 \end{aligned} \qquad (4)$$

Here the splitting $u^I = (u_1,\dots,u_\ell)^T$, $u^{II} = (u_{\ell+1},\dots,u_N)^T$ reflects the splitting of (2) and S_I, S_{II} are $\ell \times (N-\ell)$ and $(N-\ell)\times\ell$ matrices, respectively.

Following Gustafsson et al. [3], we know that the stability of the problem (1),(3),(4) is equivalent to the stability of two related quarter-plane problems. We will analyze the right quarter-plane problem on $0 \le x < \infty$, $t \ge 0$,

$$\begin{aligned} &\partial u/\partial t + A\,\partial u/\partial x = 0 \\ &u(x,0) = f(x) \\ &u^I(0,t) = S_I\, u^{II}(0,t) + b_I(t) \end{aligned} \qquad (5)$$

and the left quarter-plane problem on $-\infty < x \le 1$, $t \ge 0$

$$\partial u/\partial t + A\,\partial u/\partial x = 0$$
$$u(x,0) = f(x) \tag{6}$$
$$u^{II}(1,t) = S_{II}\,u^{I}(1,t) + b_{II}(t)$$

separately, for the sake of simplicity, instead of (1), (3), (4).

THE DIFFERENCE APPROXIMATIONS

We consider only approximations of the right quarter-plane problem; approximations of the left quarter-plane problem are analogous and will be omitted. To approximate the initial-boundary value problem (5) we introduce a set of uniform mesh points (x_j, t_n) where

$$x_j = j\Delta x \quad , \quad j = 0,1,\ldots,J \quad ; \quad \Delta x = 1/J \tag{7}$$

and

$$t_n = n\Delta t \quad , \quad n = 0,1,\ldots. \tag{8}$$

The mesh-sizes $\Delta x > 0$, $\Delta t > 0$ are such that $r = \Delta t/\Delta x$ is constant. We denote, as usual,

$$v_j(t) = v_j(j\Delta x, t) \quad , \quad v_j^I(t) = v_j^I(j\Delta x, t) \qquad \text{etc.}$$

$$E^k v_j(t) = v_{j+k}(t) \quad , \quad k = 0,1,\ldots \qquad \text{etc.}$$

By applying upwind finite differences schemes to the first equation of (5) we obtain finite-difference approximations of the form

$$v_j(t+\Delta t) = v_j(t) + P^I v_j(t) + P^{II} v_j(t) \quad , \quad j \; M, M+1, \ldots$$
$$P^I = \sum_{m=0}^{M} Q_m^I E^{m-M} \quad , \qquad P^{II} = \sum_{m=0}^{M} Q_m^{II} E^{M-m} \tag{9}$$

where Q_m^I , Q_m^{II} are fixed diagonal $N \times N$ matrices which depend on r and on $\tilde{A}^I$, $\tilde{A}^{II}$, respectively, and M is the number of grid points involved in the computation of a single point.

A typical example is supplied by the λ-scheme (see [4]); we have $M=3$ and

$$Q_0^I = -\frac{r^2}{2}(\tilde{A}^I)^2 \quad , \qquad Q_1^I = \frac{r}{2}\tilde{A}^I + 2r^2(\tilde{A}^I)^2$$
$$Q_2^I = -2r\tilde{A}^I - \frac{5}{2}r^2(\tilde{A}^I)^2 \quad , \qquad Q_3^I = \frac{3}{2}rA + r^2(\tilde{A}^I)^2$$
$$Q_0^{II} = -\frac{r^2}{2}(\tilde{A}^{II})^2 \qquad Q_1^{II} = -\frac{r}{2}\tilde{A}^{II} + 2r^2(\tilde{A}^{II})^2$$
$$Q_2^{II} = 2r\tilde{A}^{II} - \frac{5}{2}r^2(\tilde{A}^{II})^2 \qquad Q_3^{II} = -\frac{3}{2}r\tilde{A}^{II} + r^2(\tilde{A}^{II})^2$$

In order to use (9), we prescribe initial values

$$v_j(0) = f(j\Delta x) \quad , \quad j = 0,1,2,\ldots \tag{10}$$

and boundary values in the following way. From [6],[7] we know that the outflow unknowns of problem (5) do not require any upwind approximation. For the inflow part of this problem we adopt the physical boundary conditions

$$v_0^I(t) = S_I v_0^{II}(t) + b_I(t) \tag{11}$$

If $M > 1$, schemes of the form (9) require suitable additional equations at points (x_j, t_n) , $j = 1,2,\ldots,M-1$; $n = 0,1,\ldots$.

We assume

$$v_m^I(t+k) = \sum_{j=0}^{m} B_{m,j} E^{m-j} v_j^I(t) + g_m(t) \quad , \quad m = 1,2,..,M-1 \qquad (12)$$

where $B_{m,j}$ are $\ell \times (N-\ell)$ fixed diagonal matrices depending on r , A^I and the $g_m(t)$ are vectors involving Δt, A^I , $b_I(t)$. Formulae of this type are easily found as explained in [8] in detail. For an example of these formulae see next section.

STABILITY

Our stability result is referred to Kreiss's criterion for stability: ([3] definition 3.3.).

Definition 1 - An approximation of (1) with homogeneous initial value $f(x) \equiv 0$ is stable for the right quarter-plane $0 \leq x < \infty$, $t \geq 0$, if

$$\frac{\alpha-\alpha_0}{\alpha\Delta t+1} \sum_{m=0}^{M-1} \| e^{-\alpha t} v_m \|_t^2 + \left(\frac{\alpha-\alpha_0}{\alpha\Delta t+1}\right)^2 \| e^{-\alpha t} v \|_{x,t}^2$$

$$\leq K_0^2 \left(\frac{\alpha-\alpha_0}{\alpha\Delta t+1} \sum_{m=0}^{M-1} \| e^{-\alpha(t+\Delta t)} b_I(t) \|_t^2 + \| e^{-\alpha(t+\Delta t)} F \|_{x,t}^2 \right)$$

holds for constants α_0 , K_0 and for any given $\alpha > \alpha_0$.

Here

$$\| v_m \|_t^2 = \Delta t \sum_{n=0}^{\infty} | v_m(n\Delta t) |^2$$

$$\| v \|_{x,t}^2 = \Delta t \, \Delta x \sum_{n=0}^{\infty} \sum_{m=0}^{\infty} | v_m(n\Delta t) |^2$$

We also need:

Definition 2 - An approximation is said to be Cauchy stable if it is stable for the related Cauchy problem.

We state:

Theorem 1 - If the approximation of interior points (9) is Cauchy stable and if the boundary approximation (12) is Cauchy stable and dissipative then the upwind schemes given by (9),(10),(11),(12) are stable for the right quarter-plane problem.

Proof - Since all Q_m^I , Q_m^{II} , $B_{m,j}$ are diagonal matrices the inflow problem splits into ℓ independent approximations. Analogously the outflow problem splits into $(N-\ell)$ independent approximations. Thus it will be sufficient to consider the two scalar problems on the right quarter-plane $0 \leq x < \infty$, $t \geq 0$:

$$\begin{aligned} \partial w/\partial t + a\,\partial w/\partial x &= 0 \\ w(0,t) &= b(t) \\ w(x,0) &= f(x) \end{aligned} \qquad (13)$$

if $a > 0$ and

$$\begin{aligned} \partial w/\partial t + a\,\partial w/\partial x &= 0 \\ w(x,0) &= f(x) \end{aligned} \qquad (14)$$

if $a < 0$

Only interior approximations of the form (9) are required to solve the pure initial value problem (14). These finite-difference schemes, which are one-sided approximations, are Cauchy stable by assumption. Then the stability of (9) for the right quarter-plane problem (14) is a consequence of Lemma 3.1 of [5].

For problem (13) we use the approximations (11) and (12), if $M > 1$. The stability of these schemes can be proved by arguments similar to those used in the proof of theorem 3.1 of [5]. The proof rests on the theory of matching of stable schemes ([1] as it is used in [5]. We omit the details.

AN APPLICATION

In this section we investigate the stability of boundary approximations considered in [8] associated with a commonly used upwind finite-difference scheme ([8], [2]). Let us consider the system (1),(3),(4) with $F \equiv 0$ and the finite difference approximations

$$v_j(t+\Delta t) = v_j(t) - r\tilde{A}^{I}(I + \frac{1}{2}D_-)D_-v_j(t) - r\tilde{A}^{II}(I - \frac{1}{2}D_+)D_+v_j(t) \qquad (15)$$
$$+ \frac{r^2}{2}(\tilde{A}_j^{I})^2 D_-^2 v_j(t) + \frac{r^2}{2}(\tilde{A}_j^{II})^2 D_+^2 v_j(t) \quad ,$$
$$j = 2, 3, \dots, J-2$$

where I denotes the identity matrix and the discretization defined by (7),(8) and the standard notations $D_- v_j(t) = v_j(t) - v_{j-1}(t)$, $D_+ v_j(t) = v_{j+1}(t) - v_j(t)$ are adopted.

Together with (15) we use the initial values (10), the physical boundary conditions (11) and the following boundary approximations

$$v_j^{I}(t+\Delta t) = v_j^{I}(t) - \frac{3}{2} r A_j^{I} D_- v_j^{I}(t) + \frac{r^2}{2}(A_j^{I})^2 D_- v_j^{I}(t) + h_I(t) \qquad (16)$$
$$v_j^{II}(t+\Delta t) = v_j^{II}(t) - r A_j^{II}(I^{II} - \frac{1}{2}D_+)D_+ v_j^{II} + \frac{r^2}{2}(A_j^{II})^2 D_+^2 v_j^{II}(t)$$

for $j = 1$, and

$$v_j^{I}(t+\Delta t) = v_j^{I}(t) - r A_j^{I}(I^{I} + \frac{1}{2}D_-)D_- v_j^{I} + \frac{r^2}{2}(A_j^{I})^2 D_-^2 v_j^{I}(t) \qquad (17)$$
$$v_j^{II}(t+\Delta t) = v_j^{II}(t) + \frac{3}{2} r A_j^{II} D_+ v_j^{II}(t) + \frac{r^2}{2}(A_j^{II})^2 D_+ v_j^{II}(t) + h_{II}(t)$$

for $j = J-1$. The inhomogeneous terms $h_I(t)$, $h_{II}(t)$ are given by

$$h_I(t) = -\frac{\Delta t}{2}[I^{I} + \Delta t\, A_0^{I}(t)](\partial b_I/\partial t) \quad ,$$
$$h_{II}(t) = -\frac{\Delta t}{2}[I^{II} + \Delta t\, A_J^{II}(t)](\partial b_{II}/\partial t)$$

where I^{I}, I^{II} are $(\ell \times \ell)$ and $(N-\ell)\times(N-\ell)$ identity matrices, respectively.

We can prove:

Theorem 2 - The finite-difference schemes (15),(16),(17) are a stable approximation (in the sense of Definition 1) of the initial boundary value problem (1),(2),(4).

Proof - Firstly we consider the right quarter-plane problem (5). The approximations of interior points (15) are proved to be Cauchy stable in [2]. Thus only the stability of the inflow boundary approximations (16) are to be considered.

Since the stability Definition 1 gives bounds for inhomogeneous equations it is sufficient, for the inflow approximation (16), to consider homogeneous boundary values. Therefore, we may restric the stability discussion to the single difference-equation

$$v_j(t+\Delta t) = v_j(t) - \frac{3}{2} r a D_- v_j(t) + \frac{r^2}{2} a^2 D_- v_j(t) \tag{18}$$

where $a > 0$.

To study the stability of this equation we replace $v_j(n\Delta t)$ by $\hat{v} \exp(i j \eta \Delta x) \varrho^n$ where $\hat{v}$ is a constant amplitude factor, η is the wavenumber and ϱ is the growth factor. The stability condition is $|\varrho| < 1$ for any given η in the interval $[0, \pi/\Delta x]$. By applying this procedure to equation (18) we obtain

$$\varrho = 1 + \frac{1}{2} s(s-3) [1 - \exp(-ik)]$$

where $s = ra$, $k = \eta \Delta x$. Hence

$$|\varrho| = 1 + s(s-1)(s-2)(s-3) \sin^2(k/2) \tag{19}$$

The required condition on ϱ for all significant η gives the stability condition $s \leq 1$. Moreover from (19) it follows that the scheme (18) is second order dissipative, if $0 < s < 1$.

This proves that the inflow boundary approximations (16) are Cauchy stable.

All the hypotheses of theorem 1 are satisfied for the schemes (15),(16) ; thus the stability of these approximations for the right quarter-plane problem is proved. An analogous procedure applies to the approximations (15),(17) of the left quarter-plane problem.

By using a well-known theorem of Gustafsson et al. [3] it follows that the overall stability of the schemes (15),(16),(17) is assured. This completes the proof of the theorem.

NUMERICAL EXAMPLES

An application of the proposed algorithm is here shown. It refers to the time-dependent computation of the steady supercritical flow field in a convergent-divergent nozzle. For the sake of simplicity the flow is considered isentropic, hence the discontinuity of the flow properties in the divergent portion of the nozzle has to be regarded as an isentropic jump.

As boundary condition the total pressure is prescribed at the inlet boundary and the static pressure at the outlet boundary.

A first computation has been performed by using a linear extrapolation to get an extra point beyond the boundary to be used at the points next to the boundaries. In this way the domain of dependence is saved, but the algorithm is first-order accurate as regards the signals coming from the boundary. A second computation uses the above described procedure to compute the points next to the boundaries. In Fig. 1 and Fig. 2 the results of the first computation (squares) and of the second one (circles) are compared with the theoretical solution (solid lines). In the first computation the lack of accuracy at the points next to the boundaries produces a displacement of the solution along the whole flow field, while the second computation appears to be more accurate.

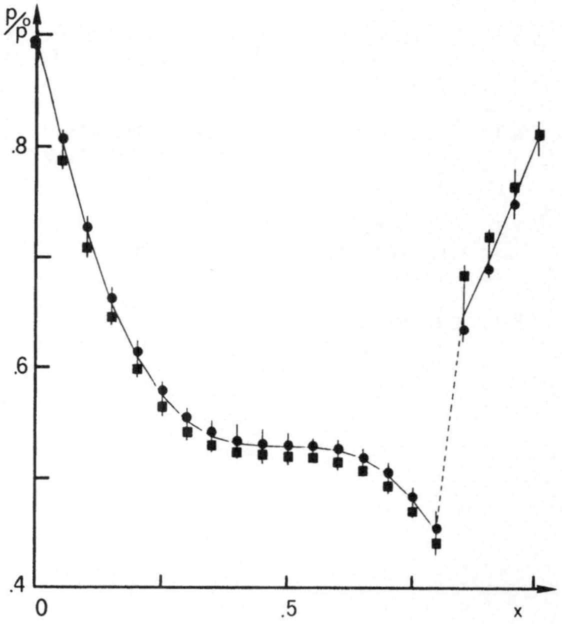

Fig. 1

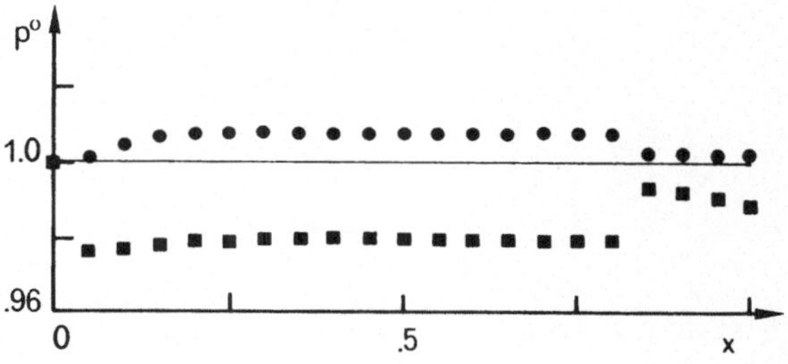

Fig. 2

As stated in Section 1, a parabolic extrapolation to get an extra-point beyond the boundary would use one point outside the domain of dependence, but, beside that, such a procedure applied to the numerical scheme here used, would be unstable, as it can be easily shown by the theoreme 1.

REFERENCES

[1] Ciment, M., "Stable matching of difference schemes", SIAM J. Numer. Anal., 9 (1972), 695-701.

[2] Gabutti, B., "On two upwind finite-difference schemes for hyperbolic equations in non-conservative form", to appear on 'Computer and Fluids'.

[3] Gustafsson, B., Kreiss, H.O. and Sundström A., "Stability theory of difference approximations for mixed initial boundary-value problem - II", Math. Comp. 26 (1972), 649-686.

[4] Moretti, G., "The λ-scheme", Computer and Fluids, 7 (1979), 191-224.

[5] Oliger, J., "Constructing stable difference methods for hyperbolic equations, Numerical Methods for Partial Differential Equations", Seymour Parter, ed., Academic Press, 1980.

[6] Zannetti, L., "Numerical treatment of boundaries in compressible flow problem", 4th GAMM Conference on Numerical Methods in Fluid Mechanics, Paris, 1981.

[7] Zannetti, L., G. Colasurdo, "Unsteady compressible flow: A computational method consistent with the physical phenomena", AIAA Journal, 19 (1981), 852-856.

[8] Zannetti, L., Moretti G., "Numerical experiments on the leading edge flow field", Proceedings of the AIAA 5th Computational Fluid Dynamic Conference, 1981, 149-156.

AN EULERIAN METHOD FOR CALCULATION OF GAS MOTION IN A VARYING REGION

B.P. Gerasimov, S.A. Semushin
The Keldysh Institute of Applied Mathematics
USSR Academy of Sciences
Moscow, USSR

1. Introduction

A method is presented for computation of compressible gas flows in a region with a boundary moving relative to a fixed orthogonal Eulerian grid. Minimal restrictions are imposed on the shape and behavior of the boundary, the resulting calculation formulae being very simple and conservative. In the case of a fixed curved boundary they can be easily extended to nonuniform grids. The difference scheme is constructed by using a split approximation technique [1] with successive involvement of different physical processes. An control volume approach provides the conservativeness of the scheme. Essentially internal points are treated by the FLIC method [2]. Note that the proposed treatment of boundary motion is applicable to other explicit computational techniques as well as to the case of movable difference grids not adjusted with the boundary motion.

Homogeneous calculation through the entire solution domain including the boundaries combined with explicit integration in time provides a rather simple program implementation [3]. The gas is supposed compressible, inviscous and nonconductive; the effect of dissipative processes, however, can be easily taken into account. The case of Cartesian and cylindrical coordinates is considered.

2. General description of the method

Consider the equations for a two-dimensional compressible flow

$$\rho_t' + \mathrm{div}(\rho\vec{W}) = 0 \qquad (1)$$

$$\rho\vec{W}_t' + \mathrm{Div}(\rho\vec{W}\vec{W}) + \mathrm{grad}\, P = 0 \qquad (2)$$

$$\rho E_t' + \mathrm{div}\,(\rho E\vec{W}) + \mathrm{div}(P\vec{W}) = 0 \qquad (3)$$

$$E = \mathcal{E} + (u^2 + v^2)/2, \quad P = P(\rho, \mathcal{E}) \qquad (4)$$

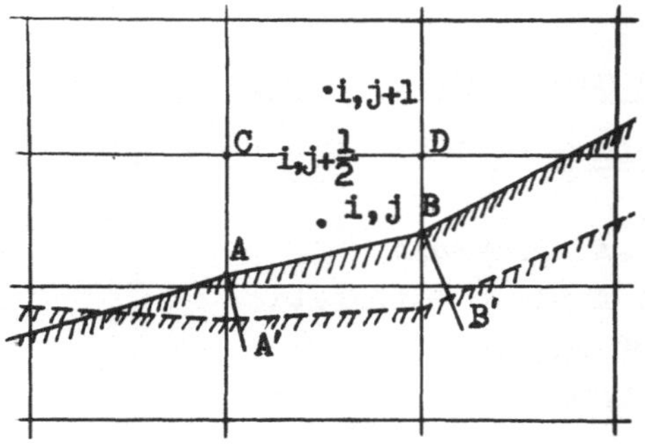

Fig. I. Computational mesh near the moving boundary. ABDC and AA'B'BDC are two configurations of a control volume

By introducing an uniform rectangular grid in complex regions some cells called partial are cut by the boundary. The sections of the boundary not coinciding with the grid lines are approximated by a broken line, the ends of its segments lying on the grid lines. The boundary motion may be either predetermined, i.e. given explicitly, or calculated depending on the current gas parameters (Fig. 1).

A numerical algorithm is constructed by using the split approximation technique based on the FLIC method [2] . During the first stage of calculation the convective transport is excluded, the density being unvaried. Using an explicit central-difference scheme the velocity and energy variations under the pressure gradients are calculated as well as the work performed by the moving wall. At the second stage transport effects are considered. The boundary first is assumed fixed. Applying the donor cell method the mass, momentum and total energy flows across the cell interfaces are calculated making the contribution to corresponding cells. Then the boundary is advanced in time. Due to the explicit nature of transport and the advancement of

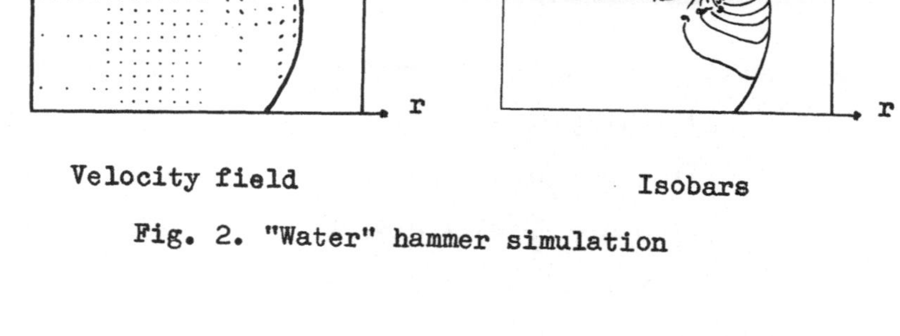

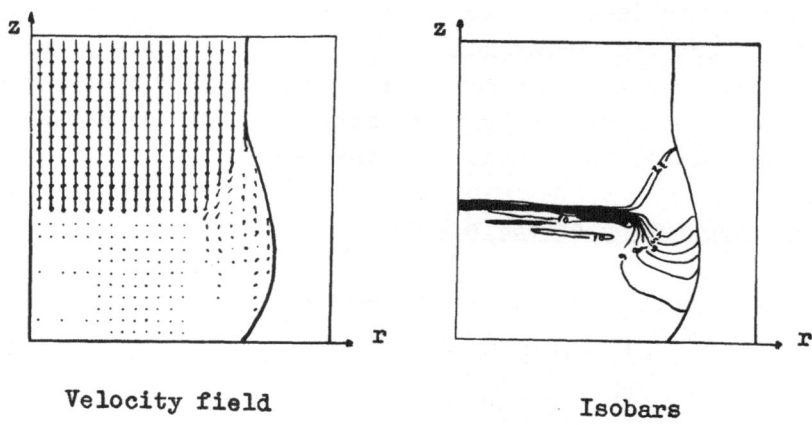

Fig. 2. "Water" hammer simulation

the boundary such an approach is algebraically an equivalent to simultaneous occurrence of these processes. The partial cells of small dimensions impose heavy restrictions on the time step and to avoid this one may join them with the adjacent cell. For simplification the mergence of the cells is replaced by averaging the values in them since the respective difference expressions are equivalent.

Having an intrinsic viscosity proportional to velocity the method described needs an artificial viscosity only in large stagnation regions. Similar to [2] a linear artificial viscosity was introduced only for the compression waves and small velocities.

3. The pressure effects near the boundary

Let us derive equations for calculation of values in the partial cells at the first stage when the transport effects are not taken into account. It is assumed that the condition

$$\frac{\partial P}{\partial n} = 0 \tag{5}$$

is fulfilled at the impenetrable wall with the normal n. Thus the centrifugal force effects are ignored and the wall acceleration is assumed small. Let V_{ij} be the volume of the partial cell, occupied by fluid, and $S_{i+1/2,j}$, $S_{i-1/2,j}$, $S_{i,j+1/2}$, $S_{i,j-1/2}$ be the areas of the cell sides open to gas flows. Now we integrate (2) and the equation for internal energy with omitted convective terms over the cell volume. In the case of the impenetrable wall under the condition (5) the equations for various types of the partial cells (including those in cylindrical coordinates) prove to be

$$\tilde{u}_{ij} = u_{ij} - \frac{\tau}{2V_{ij}\rho_{ij}} \left(S_{i+1/2,j}(P_{i+1,j}-P_{ij}) + S_{i-1/2,j}(P_{ij}-P_{i-1,j})\right) \tag{6}$$

$$\tilde{v}_{ij} = v_{ij} - \frac{\tau}{2V_{ij}\rho_{ij}} \left(S_{i,j+1/2}(P_{i,j+1}-P_{ij}) + S_{i,j-1/2}(P_{ij}-P_{i,j-1})\right) \tag{7}$$

$$\varepsilon_{ij} = \varepsilon_{ij} - \frac{\tau P_{ij}}{V_{ij}\rho_{ij}} \left(S_{i+1/2,j}\,\bar{u}_{i+1/2,j} - S_{i-1/2,j}\,\bar{u}_{i-1/2,j} + S_{i,j+1/2}\,\bar{v}_{i,j+1/2} - S_{i,j-1/2}\bar{v}_{i,j-1/2}\right) - P_{ij}\,\Delta V_{ij}/V_{ij}\rho_{ij}; \quad \bar{\vec{W}} = (\tilde{\vec{W}}+\vec{W})/2 \tag{8}$$

For simplicity the artificial viscosity terms are omitted here. Equations (6)-(8) for various types of the partial cells are homogeneous because of the introduction of dummy pressures outside the boundary of calculation domain (in fictitions cells) where they generally need not be specified. The pressure values in these fictitious cells are involved in calculations only formally since they are multiplied by the zero areas. Complete boundary cells unlike the partial ones are calculated by using a common fictitious cell procedure [2] .

4. Transport calculation

The transport effect is computed in two steps. First, the boundary is assumed fixed and the density, velocity and energy values are calculated. For the cells not adjacent to the moving boundary these values are final at the time t^{k+1}. The transport effect calculation for the partial cells with fixed boundaries meets no difficulties and is given in detail in [3] . Then a new position of the boundary, i.e. at the next time step, is determined and the values are recalculated in the cells adjacent to the new boundary, thus completing the advancement to a new time layer. To simplify notations the intermediate values will be used without superscripts.

In order to follow the conservative principle we define subsidiary control volumes near the moving boundary. Note that the partial cells cannot be used as the control volumes since the former may appear and disappear during the boundary motion. After construction of a set of the control volumes the mass, momentum and energy are calculated in the control volume prior the boundary motion by assuming the distribution of basic flow parameters to be uniform within a cell. These amounts will be the same after the boundary advancement into the position at the next time step. Then by preserving the uniform distribution within the deformed control volume we share the mass, energy and momentum among the cells overlapping with it. Thus the problem is reduced to determination of volumes formed by overlapping the old and new cells with the control volumes. We denote a common part of the control volume α and the cell $\beta = (i,j)$ prior the boundary advancement by $\Delta V_{\alpha\beta}$ and after that by $\Delta V_{\alpha\beta}^{k+1}$. Then for $\varphi = \{\rho, \rho u, \rho v, \rho E\}$ we have

$$\varphi'_\alpha \sum_\beta \Delta V_{\alpha\beta}^{K+1} = \sum_\beta \varphi_\beta \Delta V_{\alpha\beta} \tag{9}$$

where φ'_α are the parameters of the control volume α . The values at the new time step are obtained from

$$\varphi_\beta^{K+1} V_\beta^{K+1} = \sum_\alpha \varphi'_\alpha \Delta V_{\alpha\beta}^{K+1} \tag{IO}$$

The following definition of the control volumes is proposed. The part of the old or new broken boundary line is called external if it is not located within the body at any of the two positions of the boundary; otherwise it is called internal. As a base for the control volumes we take the cells containing the part of the external boundary. The fraction of the cell occupied by fluid at the both positions of the boundary is called a free volume. From the external apices of the broken line we draw bisectrices of the angles between the segments up to their intersection with the internal boundary. Thus the entire zone between the two boundary positions is partitioned in the volumes enveloped by the boundary and bisectrix fragments. (Fig. I). We join each of these volumes to the cell that contains the common with the volume segment of the broken boundary line. Thus we obtain a set of the control volumes after the boundary advancement. Obviously the control volumes may be designated by the indices of the corresponding cells. Provided the Courant condition is satisfied for the boundary velocity and the time step the summation in (9) and (IO) involves nine cells - the basic one and its eight neighbours having common points with it.

The method described has the first order approximation in time and space. The stability of calculations is practically ensured by the Courant condition.

5. Numerical results

Various test problems were solved to check the method. One of these was a highly nonstationary problem of reflection of a strong plane shock wave from a rigid wall. In order to form the partial cells both the wave front and the parallel wall cross the grid lines at an angle 45°. Also problems considering a piston pushed into a gas were solved. In these cases the gas parameters differed from those in the Hugoniot relation by a fraction of percent. Experience shows that the partial cells whose volume at current step is less than 0.2 to 0.5 from that of a fullsized cell should provisionally joined to the adjacent ones.

The method proposed was used to compute flows around curvilinear bodies including invariable ones. The results obtained for the flow around a cylinder were compared to the tabulated ones [4] accurately computed by the method with isolated peculiarities. The difference was a few percent and only at some points it reached IO%.

The test problem concerned with an expanding cylindrical piston was also considered. In order to form a curvilinear moving boundary non-adjusted with the grid the problem was solved in Cartesian coordinates on a square grid of 50x50 cells. As a result the one-dimensional problem was transformed into the essentially two-dimensional one. Comparison was made with the one-dimensional solution of the same problem in Lagrangian variables on a very fine grid. The difference between solutions did not exceed 5%, the shock wave smeared over 3-4 spaces of the grid. A similar accuracy was obtained for a spherical piston in cylindrical coordinates.

The method considered was also applied to problem on flows around bodies with variable configuration. Diffraction of a shock wave was computed for a deforming cone and wedge. The gas motion was studied under "water" hammer in a channel with elastic wall the profile of which R(z,t) was governed by the equation

$$\frac{\partial^2 R}{\partial t^2} + a^2\frac{\partial^2 R}{\partial Z^2} + \frac{P(R(Z,t),\ Z) - P_0}{m} = 0 \qquad (11)$$

with the initial and boundary conditions $R(Z,0)=R_0$; $R_t'(Z,0)=0$; $R(0,t)=R_0$ where R_0 is the initial radius of the channel, P_0 is the pressure outside the channel, m is the specific mass of the channel wall, a is the velocity of the disturbance propagation along the channel wall. At the initial time a rigid diaphragm blocks the channel with an uniformly moving gas. A shock wave arises and begins to deform the channel wall. The results of the combined solution of equations (I)-(4), (II) are shown in Fig.2. Differences between solutions obtained on various space-time grids did not exceed a few percent.

Thus the proposed method is concluded to be applicable to a variaty of gasdynamics problems in complex domains. It is simple in algorithmic implementation, efficient and requires minor a priori information about the solution.

References

1. Samarsky, A.A. Theory of difference schemes, Moscow, "Nauka", I977 (in Russian).
2. Gentry, R.A., Martin, R.E. and Daly, B.J. J.Computational Physics, v.I, I966, p.87.
3. Gerasimov, B.P., and Semushin, S.A. Differential equations, v.I7, I98I, p.I2I4 (in Russian).
4. Lyubimov, A.N., and Rusanov, V.V. Gas flows around blunt bodies, Pt.2, Moscow, "Nauka", I970 (in Russian).

COMPUTATION OF INVISCID TRANSONIC INTERNAL FLOW

U. Giese
Aerodynamisches Institut, RWTH Aachen
Aachen, Germany

Abstract

Compressible internal flows are studied with an implicit finite difference solution of the full potential equation in conservative form. For numerical stability in supersonic regions an artificial compressibility formulation is introduced. Boundary fitted curvilinear coordinates are used which are stretched in order to cope with regions of strong gradients. The equations are solved by an approximate factorization technique. Results are presented for nozzle flow and for flows through valve inlets with different wall contours. Some results are compared with Mach-Zehnder Interferograms.

Formulation of the Problem

Inlets of safety valves are often rounded in order to prevent flow separation as it would occur at a sharp corner. However, if the curvature of the wall is not appropriately chosen, the flow may still separate, as the boundary layer experiences too strong a deceleration downstream of the point of minimum pressure. The goal of this investigation is to study the influence of the wall curvature on the inviscid transonic flow for several mass flux rates.

Irrotational, inviscid, transonic, plane flow is described by the full potential equation

$$(\rho J U)_\xi + (\rho J V)_\eta = 0$$

with
$$U = A_1 \Phi_\xi + A_2 \Phi_\eta \qquad (1)$$
$$V = A_2 \Phi_\xi + A_3 \Phi_\eta$$

In Eqs. (1) Φ is the velocity potential, ρ the density, U and V are the contravariant velocity components. A_1, A_2, and A_3, and the Jacobian J depend on the curvilinear coordinates ξ and η. For inlet flow, the mesh is generated numerically by solving two Poisson equations for the physical coordinates (x,y) [1] . They are stretched by a suitable distribution of mesh points along the wall, such that a fine spacing is achieved in regions where the velocities and their gradients are expected to be large (Fig. 1).

The boundary conditions for Eqs. (1) are

(i) inflow boundary $\Phi = 1$

(ii) outflow boundary $\Phi_x = u_e$ (2)

(iii) axis and wall $\Phi_n = 0 \rightarrow \Phi_\eta = -A_2 / A_3 \cdot \Phi_\xi$

The main differences between calculations of external flows, e.g. flows around profiles, and internal flow calculations are due to the different boundary conditions. For external flows the gradient of the velocity potential on the outer boundary is given by a far field solution; hence the computation may be regarded as an adjustment of the correct distribution of the potential from a known set of values. In internal flow, however, the potential can be prescribed at one cross section only. The difference between the values of the potential at the inflow and the outflow boundary has to be determined during the calculation. In all problems considered in this paper the velocity is assumed to be known at the computational outflow cross section.

Eqs. (1) were discretized by central differences. For supersonic regions the domain of dependence condition was satisfied by using the artificial compressibility method [2]. This leads to a retarded density of the form [3]

$$\tilde{\varrho}_{I+1/2,J} = \varrho_{I+1/2,J} - \mu \cdot \Delta\xi \cdot \delta_\xi \, \varrho_{I+1/2,J}$$

$$\tilde{\varrho}_{I,J+1/2} = \varrho_{I,J+1/2} - \mu \cdot \Delta\eta \cdot \delta_\eta \, \varrho_{I,J+1/2} \tag{3}$$

with $\mu = \max \left\{ 0;\ 2(Ma^2 - 1) \right\}$

where ϱ is the density calculated from isentropic relations and the difference quotients δ_ξ and δ_η are taken in the upwind direction.

Method of Solution

The system of difference equations

$$L(\Phi) = 0 \tag{4}$$

is solved by iteration (iteration index n)

$$N(\Phi^{n+1} - \Phi^n) = -\omega L(\Phi^n) \tag{5}$$

Approximate factorization methods which have succesfully been applied in calculation of external flows [3, 4, 5] are also used here. For subsonic flow and transonic flow with a small supersonic region the AF1 scheme

$$[\alpha - \overleftarrow{\delta}_\eta (\varrho A_3 J)_{I,J+1/2} \overrightarrow{\delta}_\eta] \, C^*_{I,J} = \alpha\omega L(\Phi^n)_{I,J}$$

$$[\alpha - \overleftarrow{\delta}_\xi (\varrho A_1 J)_{I+1/2,J} \overrightarrow{\delta}_\xi] (\Phi^{n+1} - \Phi^n)_{I,J} = C^*_{I,J} \tag{6}$$

gives a very fast convergent solution. In cases in which the supersonic flow occupies a large part of the cross section, e.g. nearly choked channel flow, the destabilizing effect of the Φ_t-term introduced by (6) may cause divergence [3, 4]. If the ξ-coordinate is aligned with the main flow direction, addition of a $\Phi_{\xi t}$-term is known to stabilize the calculation [7]. This term can be introduced by a factorization, similar to the AF2 schemes used in [3, 4, 5, 6].

$$\left[\alpha - \overleftarrow{\delta}_\xi (\rho A_1 J)_{I-1/2,\, J}\right] C^*_{I,\, J} = \alpha\omega L(\Phi^n)_{I,\, J}$$

$$\left[\alpha\overleftarrow{\delta}_\xi - \overleftarrow{\delta}_\eta (\rho A_3 J)_{I,\, J+1/2}\, \overrightarrow{\delta}_\eta\right](\Phi^{n+1} - \Phi^n)_{I,\, J} = C^*_{I,\, J} \qquad (7)$$

This splitting of the ξ-direction leads to a marching procedure, for which a boundary value for $C^*_{I,J}$ at the outflow section is required; the second equation together with the boundary condition (2) gives exactly

$$C^*_{IMAX,\, J} = 0 \qquad (8)$$

Unlike Holst's experiences in airfoil flow calculations [5, 6] experiments with formulations differing from those given in (7), e.g. an interchange of the forward and the backward differencing in ξ, failed, because an exact condition for the intermediate variable C^* could not be derived. The algorithm did not produce the appropriate inflow-outflow difference of Φ in an acceptable number of iterations and in some cases did not even converge. The overall rate of convergence is very sensitive to the choice of α. Small values speed up the correct determination of the potential difference between inflow and outflow boundary; for too small values of α, however, the solution diverges, as the choking condition is violated during the iteration. This behaviour is specific for internal flow calculations, in which the mass flux has to be determined correctly. Large values of α are necessary to reduce the high-frequency error components. Therefore a sequence α_i is used repetitively during the course of the computation, the lower and upper limit of which were estimated by a linearized analysis [4] and adjusted by experiment. For AF1 $\alpha_i = 0.1, \ldots, 10^5$ and for AF2 $\alpha_i = 1, \ldots, 100$ were found to give fast convergence, when a geometric sequence of eight values is used.

Results

As a test case the transonic flow in a channel with a circular arc bump (Fig. 2), as presented in a GAMM workshop [8], was used. Fig. 3 and Fig. 4 show the lines of constant Mach number and the pressure coefficient, respectively, for an inflow Mach number of 0.8435, where the flow is nearly choked. Agreement with the calculation in [8] is good.

The main part of this investigation is concerned with transonic flow through plane inlets. The geometry (Fig. 1) is characterized by straight walls connected with a circular arc. Computations were performed for radius of curvature R_k = 0.5, 0.3 and 0.15 and different outflow Mach numbers. The calculations were performed with grids of (141x21) points. The convergence criterium was defined by the reduction of the residue by four orders of magnitude. This condition was met after e.g. 100 iterations (Fig. 5a) whereas the potential reached its correct value at the outflow section already after 40 iterations (Fig. 5b). The results for a radius of 0.5 height of the outflow channel at an outflow Mach number of 0.5 (Fig. 6) show that the flow is accelerated up to Ma >0.5 in the region near the wall. For outflow Mach number $Ma_e = 0.7$ and the same geometry the flow becomes supersonic near the wall (Fig. 7). The supersonic region extends to about the point where the curved inlet contour meets the straight wall of the channel. A shock wave has developed there. Decreasing the radius of curvature from 0.5 to 0.15 has a similar effect, but the region of low pressure becomes very concentrated to the wall (Fig. 8). A comparison of the effects of a variation of the Mach number at constant wall contour and of the curvature at constant mass flux is presented in Fig. 9-12. If the pressure coefficient c_p is normalized with c_{po}, the curves (Fig. 11) are nearly independent of the outflow Mach number as long as the flow remains subsonic throughout. In Fig. 12 the arc length is stretched by the inverse radius of curvature. Fig. 13-15 show a comparison of experimental and computed results at approximately the same outflow conditions. The lines in Fig. 13 should correspond to the black stripes in Fig. 14. They are shown together in Fig. 15. Generally both show the same features, but the computational result indicates higher Mach numbers near the wall than the Mach-Zehnder Interferogram. The difference could be due to the boundary layers on the wall contour and on the side walls of the windtunnel in the experiment.

References

[1] Thompson, J. F., Thames, F. C., Mastin, C.W.: Boundary-fitted curvilinear coordinate system for solution of partial differential equations on fields containing any number of arbitrary two-dimensional bodies. NASA CR-2729 (1977).

[2] Hafez, M., South, J., Murman, E.: Artificial compressibility methods for numerical solutions of transonic full potential equation. AIAA-J., Vol. 17 No. 18, pp. 838-844 (1979).

[3] Holst, T. L., Ballhaus, W. F.: Fast, conservative schemes for the full potential equation applied to transonic flows. AIAA-J., Vol. 17 No. 2, pp. 145-152, (Feb. 79).

[4] Ballhaus, W. F., Jameson, A., Albert, J.: Implicit approximate-factorization schemes for steady transonic flow problems. AIAA-J., Vol. 16 No. 6, pp. 573-579, (June 78).

[5] Holst, T. L.: Fast, conservative algorithm for solving the transonic full-potential equation. AIAA-J., Vol. 18 No. 12, pp. 1431-1439, (Dec. 80).

[6] Holst, T. L.: Implicit algorithm for the conservative transonic full-potential equation using an arbitrary mesh. AIAA-J., Vol. 17 No. 10, pp. 1038-1045, (Oct. 1979).

[7] Jameson, A.: Iterative solution of transonic flows over airfoils and wings, including flows at Mach 1. Communications on Pure and Applied Mathematics, Vol. 27, pp. 283-309 (1974).

[8] Rizzi, A., Viviand, H. (Eds.): Numerical methods for the computation of inviscid transonic flows with shock waves. Friedr. Vieweg u. Sohn, Braunschweig/Wiesbaden (1981).

[9] Föllmer, B. Aerodynamisches Institut, RWTH Aachen (1981), private communication.

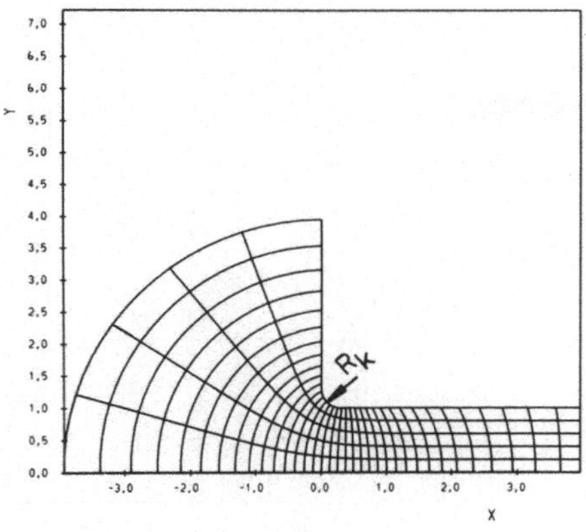

Fig. 1: Inlet Mesh (141x21)

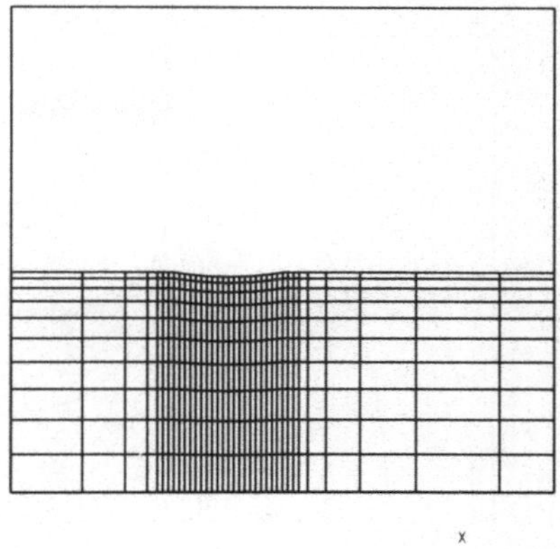

Fig. 2: Channel Mesh (72x21)

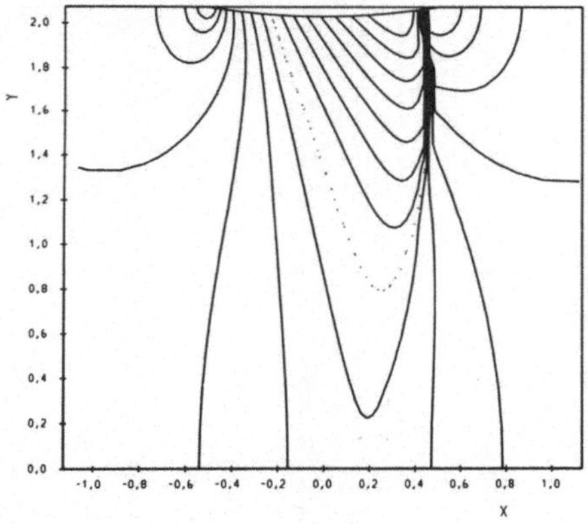

Fig. 3: Channel Flow, Isomachs
$Ma_\infty = 0.8435$, $\Delta Ma = 0.04$

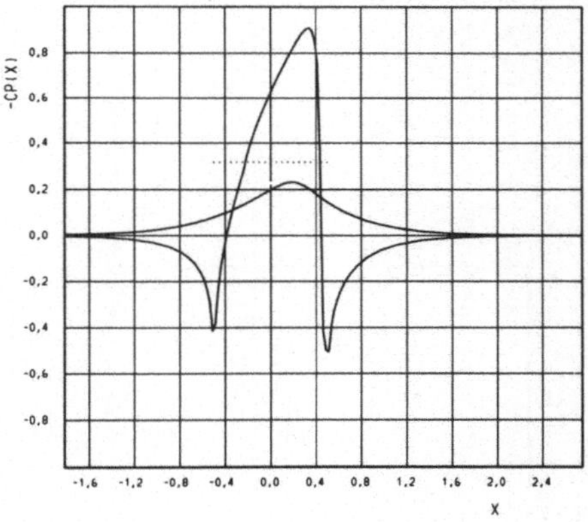

Fig. 4: Channel Flow, Pressure Coefficient

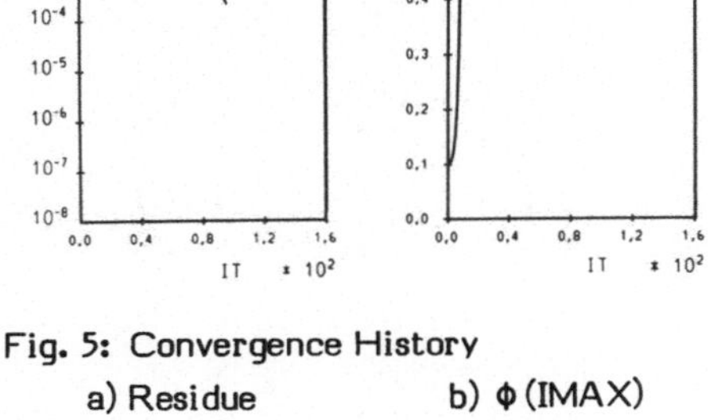

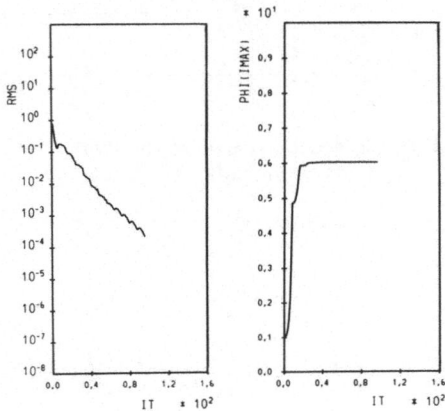

Fig. 5: Convergence History
a) Residue b) ϕ (IMAX)
$R_k = 0.5$, $Ma_e = 0,7$, AF2

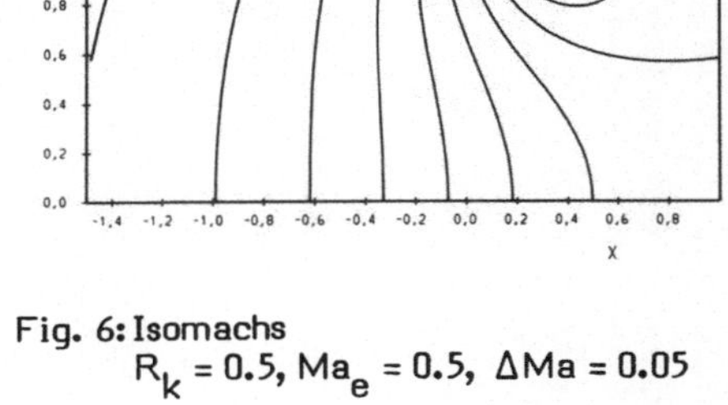

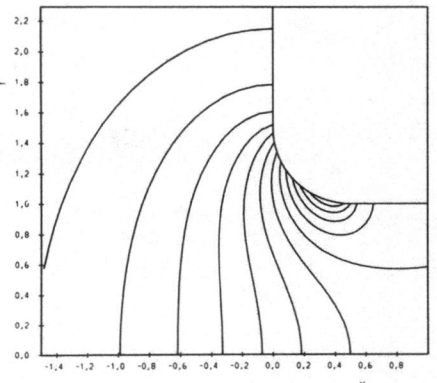

Fig. 6: Isomachs
$R_k = 0.5$, $Ma_e = 0.5$, $\Delta Ma = 0.05$

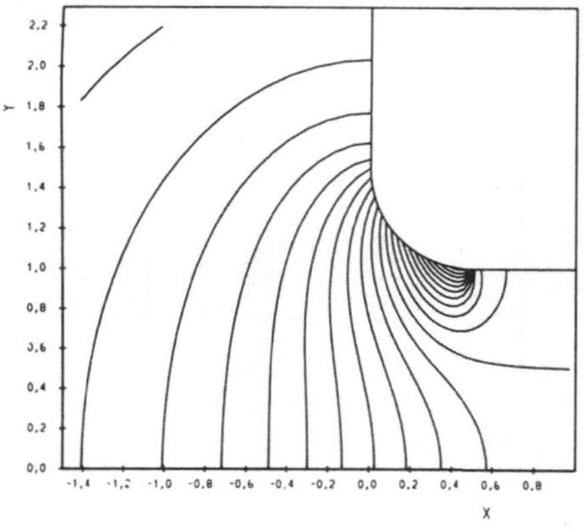

Fig. 7: Isomachs
$R_k = 0.5$, $Ma_e = 0.7$, $\Delta Ma = 0.05$

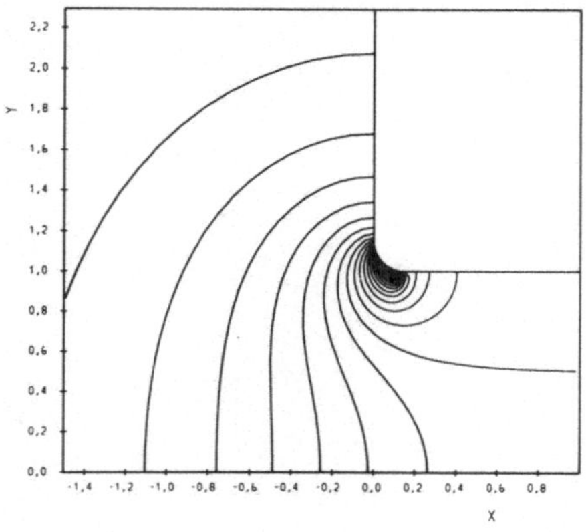

Fig. 8: Isomachs
$R_k = 0.15$, $Ma_e = 0.5$, $\Delta Ma = 0.05$

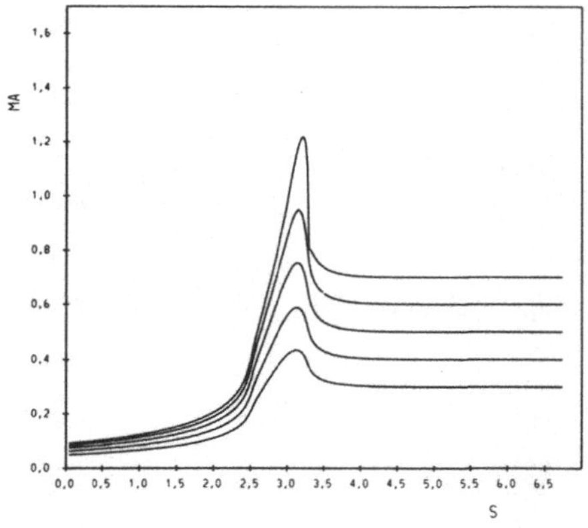

Fig. 9: Mach Number at the Wall
$R_k = 0.5$, $Ma_e = 0.3, 0.4, 0.5, 0.6, 0.7$

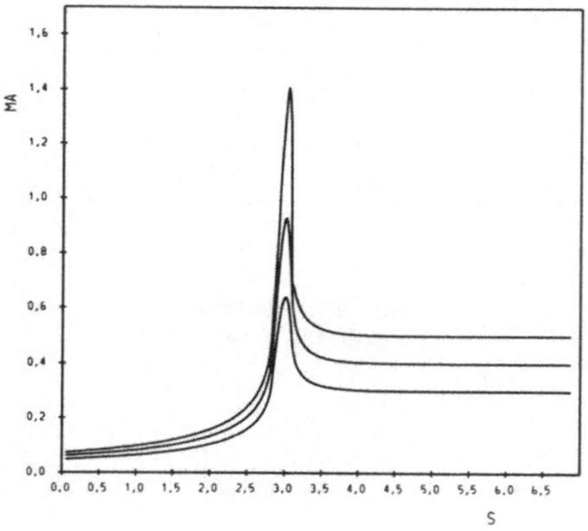

Fig. 10: Mach Number at the Wall
$R_k = 0.15$, $Ma_e = 0.3, 0.4, 0.5$

RANDOM ELEMENT METHOD FOR NUMERICAL MODELING OF DIFFUSIONAL PROCESSES

A.F. Ghoniem and A.K. Oppenheim
University of California
Berkeley, CA 94720 U.S.A.

Abstract

The random element method is a generalization of the random vortex method that was developed for the numerical modeling of momentum transport processes as expressed in terms of the Navier-Stokes equations. The method is based on the concept that random walk, as exemplified by Brownian motion, is the stochastic manifestation of diffusional processes. The algorithm based on this method is grid-free and does not require the diffusion equation to be discritized over a mesh, it is thus devoid of numerical diffusion associated with finite difference methods. Moreover, the algrithm is self-adaptive in space and explicit in time, resulting in an improved numerical resolution of gradients as well as a simple and efficient computational procedure.

The method is applied here to an assortment of problems of diffusion of momentum and energy in one-dimension as well as heat conduction in two-dimensions in order to assess its validity and accuracy. The numerical solutions obtained are found to be in good agreement with exact solution except for a statistical error introduced by using a finite number of elements, the error can be reduced by increasing the number of elements or by using ensemble averaging over a number of solutions.

Introduction

Numerical solutions of transport systems which include diffusion of an active component, such as diffusion of vorticity in momentum transport or diffusion of energy in natural convection systems, have had difficulties obtaining accurate results when the diffusivity is small, e.g. flow at high Reynolds numbers. The difficulty stems from the tendency of these solutions to spread out sharp gradients by introducing extra diffusion into the system that is inversley proportional to the actual molecular diffusivity. Other problems associated with solving diffusion transport systems using conventional numerical methods include accurately resolving moving sharp gradients such as boundary layers and flame fronts, rendering these methods inefficient except if they are equipped with special techniques to accommodate for them like adaptive griding and moving elements.

The random vortex method [1,2,3], developed by Chorin as a solution to the Navier-Stokes equations, provides solutions to some of these problems. The crux of this method, in which vortex transport is split into transport by convection and transport by diffusion, is the notion that transport by diffusion can be appropriately modeled

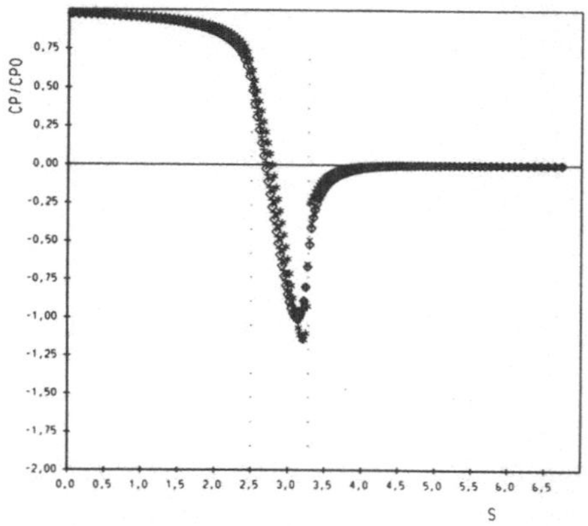

Fig. 11: Wall Pressure
$R_k = 0.5$, $Ma_e = 0.3, 0.4, 0.5, 0.6, 0.7$

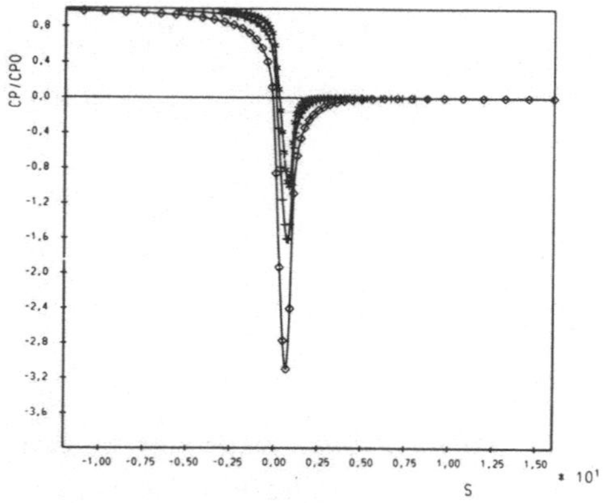

Fig. 12: Wall Pressure
$Ma_e = 0.4$, $R_k = 0.5, 0.3, 0.15$

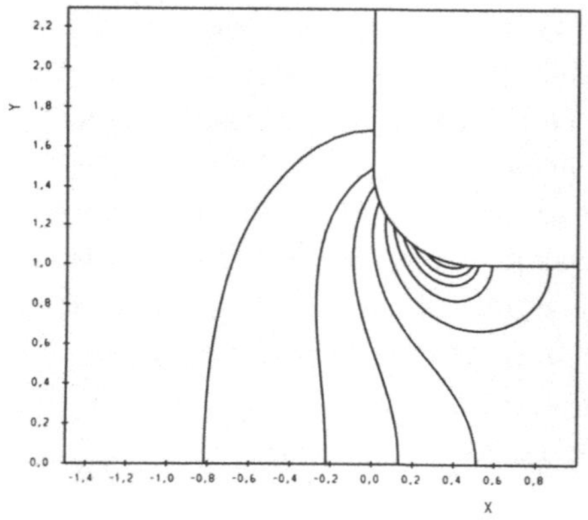

Fig. 13: Lines of Constant Density
$R_k = 0.5$, $Ma = 0.55$

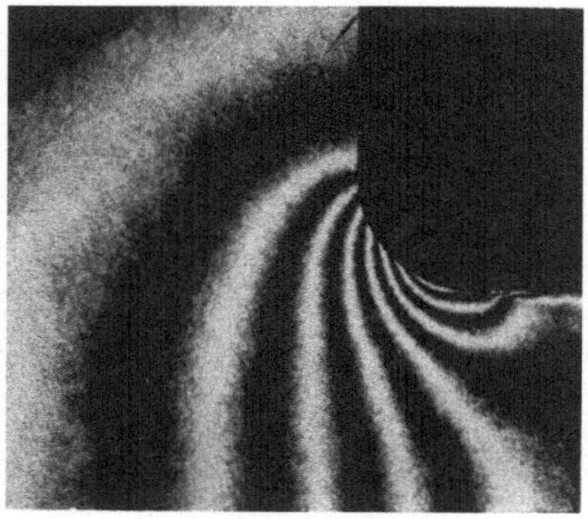

Fig. 14: Mach-Zehnder Interferogram
$R_k = 0.5$, $Ma_e \approx 0.55$ [9]

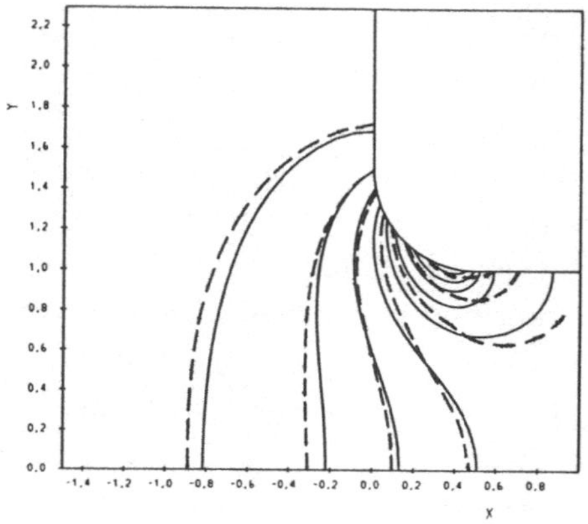

Fig. 15: Lines of Constant Density
—— Computation
- - - Experiment

by the random walk of vortex elements. This observation was first made by Einstein in his studies on Brownian motion, in which he found that "the probable distribution of the resulting displacement (of the diffusing element) in a given time is the same as that of fortiutous error (Gaussian distribution)" [4]. Courant, Friedrichs, and Levy [5] suggested using this concept in solving partial differential equations of the Poisson type, which later developed into the Monte Carlo technique.

The random vortex method was extended to solve problems in turbulent combustion in Ghoniem, Chorin and Oppenheim [6], where the dynamic field produced by the expansion of the reacting flow across the flame is of primary importance. Due to the lack of a proper tool to solve for heat diffusion in such system, the variation of the flame speed with temperature was left out. In this work, an attempt to develop such a technique is presented with emphasis on random walk to solve the diffusion equation.

In general, the process of diffusion affect the transport of momentum or temperature in terms of the Fourier equation. The application of random walk to these processes is developed here first in the case of one dimensional flow or heat conduction utilizing the analogy between them, and then extended to a two dimensional space. The analogy between the two processes is presented in table 1, in which the dependent variable of a diffusion equation is shown to fall into three differential orders,

1. the primary variable, e.g. the velocity, u, or the temperature, T.
2. the secondary variable defined as the gradient of the primary variable, e.g. the vorticity, ξ, or the heat flux, q.
3. the integral variable determined as either the line integral of the primary variable around the boundary of an area element or the area integral of the secondary variable over this element (the two being equivalent according to the Stokes Theorem), e.g. the circulation γ, or the internal energy, e.

In the following, the fundamental idea of random walk is introduced. It is then extended to a hierarchy of diffusional processes, determined by the imposed boundary condition, to cover all possible problems of interest. Finally, the method is expanded to solve the heat conduction equation in two-dimensions.

Fundamental Idea

The fundamental idea of stochastic random walk solution to the one-dimensional diffusion equation is presented in table 2. It is based on the fact that the formal solution of the Fourier equation of diffusion, subject to the initial condition of a Dirac delta function in a free space, $\phi(0,0) = \delta(0)$, is identical to the probability density function of a Gaussian random variable with a zero mean and a variance of $(2\alpha t)$. Since the sum of Gaussian variables is a Gaussian variable with a variance equal the sum of the variances, the solution can be constructed by a sequence of displacements of a set of discrete elements of the diffusing component, $\delta\Phi_i$, each displacement is drawn from a set of Gaussian variables with a variance of

$2\alpha\ \Delta t$. These elements are generated by partitioning the initial distribution of the primary variable into a number of elements of equal strength, $\delta\Phi$ where $\Sigma\ \delta\Phi_i = 1$. The solution, $\phi(y,t)$, is obtained at any moment by local sampling over an area element δy around y, representing the gradient of the diffusing element. By global sampling over the whole field, the value of the diffusing element $\Phi(y,t)$ can be evaluated.

Application of Boundary Conditions

The three fundamental differential orders of the dependent variable of the diffusion equation shown in table 1 lead to a hierarchy of diffusion processes, as presented in table 3. Since the equation of diffusion is invariant under differentiation, the primary variable and the diffusing element are determined according to the initial condition or the distribution given at $t = 0$. The element with an initial distribution of a Dirac delta function represents the secondary variable, while its integral is the primary variable. Consequently, elements of the secondary variable have a local field that can be presented in the solution by a Dirac delta function, while elements of the primary variable, being the integrals of the secondary variable, impose an extended field of a Heavyside step function distribution.

Conditions on the boundaries of the field are applied in terms of the behavior of diffusing elements, $\delta\phi$, if they cross these lines. There are three possibilities:

1. ϕ = const., a reflective boundary where elements are reflected back into the field to satisfy the conservation of ϕ.
2. $\frac{\partial\phi}{\partial y} = 0$, a line of symmetry where elements are reflected back with negative signs, representing the effect of a mirror-image field.
3. $\frac{\partial\phi}{\partial t}$ = const., a source boundary where elements are generated on it by integrating the source strength expression, this case is analysed in column 4 of table 3.

Since the diffusion equation is linear, complex boundary conditions are applied using the principle of superposition.

Extension to Two Dimensions

Table 4 presents the extension of random walk to two dimensional diffusion. The analysis here is restricted to heat conduction problems, since two dimensional diffusion of vorticity involves the solution of the full Navier-Stokes equation, a task that is addressed by the random vortex method. For an initial temperature distribution of a Dirac delta function, the solution is obtained in terms of elements of energy, δe_i, and the temperature distribution is evaluated by local sampling among these elements within elements of area, δA_i.

The case of finite temperature distribution on the boundary requires using a hybrid scheme of temperature elements, δT_i near the wall and energy elements in the interior as shown in table 5. Inside a thickness δ_s, of the order of magnitude of the standard deviation of the Gaussian random numbers, diffusion is assumed to be one-dimensional in the direction normal to the wall and the results of the diffusion problem defined in column 2 of table 3 are used. In the interior, energy elements generated by the displacements of the temperature elements outside the layer δ_s move unbaisedly in two directions, a problem corresponds to the case described in table 4. As shown in the last row of table 5, the motion of an energy element into this layer raises its temperature and decerease δT_i accordingly, thus keeping both the energy and temperature fields continuous.

Results and Discussion

In order to check the validity and accuracy of the method, problems in diffusion that have exact solutions were solved using the above procedure. Figure 1 to figure 4 represent solutions for one dimensional diffusion of either heat or vorticity, while figure 5 and figure 6 show solutions for two dimensional diffusion of heat.

Figure 1 presents the vorticity distribution over a suddenly accelerated plane, Stoke's first problem, using 1000 and 100 vortex elements after 50 computational time steps of 0.1 each. Figure 2 displays the velocity distribution evaluated for the same problem, using 100 and 10 vortex elements. The reduction of the statistical error in the calculations of the velocity profiles is due to the integration, performed by the global sampling, over the vorticity elements and illustrates the advantage of diffusing elements of gradients over elements of components. Figure 3 and Figure 4 describe the solution of non-self-similar developing flow fields. Velocity profiles and kinetic energy profiles of the Couette flow are presented in figure 3, using 100 vortex element and a time step of .001, time is non-dimensionalized with respect to the channel height and the molecular diffusivity. The energy profiles, presented by broken lines, coincide with the exact curves, they represent the second integral of the vorticity elements as derived in column 3 in talbe 3. Both boundaries are taken as reflective lines to satisfy the conversation of total circulation. In figure 4, velocity profiles for the Poiseuille flow are shown, where at $y = 0$ vortex elements are generated by the action of the pressure force, while at $y = 0.5$, representing the center line of the channel, vortex elements are reflected with a negative sign to satisfy the symmetry conditions. Element strength is taken as 0.001 and the steady-state number of elements which are accumulated in the field oscillates between 491 to 504, resulting in a center-line velocity of 0.491 to 0.504, the exact value is 0.5. The corresponding heat conduction problems to these flow problems are described in the last row of table 3.

Figure 5 and figure 6 show temperature distributions for two-dimensional heat conduction in a corner and a square, respectively, where the profiles are calculated along the diagonal in both cases. A thickness of twice the standard deviation is used as δ_s and 25 temperature elements are employed to impose the constant temperature boundary condition. The time step for the self-similar case, the corner, is 0.001, while it is 0.0001 for the non-similar case to achieve almost the same accuracy.

Conclusions

A random element method for the numerical solution of the Fourier diffusion equation of vorticity and heat is presented here for one-dimensional and two dimensional systems. Its applications to problems with exact solutions are used to demonstrate the procedure as well as to prove the validity of the algorithm. This method is obviously not recommeded for such simple cases, but for systems in which a number of transport processes take place simultaneously. Of special interest is when diffusion of an active species has a dynamic effect such as natural convection with strong interaction between temperature and density, or turbulent combustion with significant thermal expansion.

The most attractive feature of random walk methods is their Lagrangian nature; functions are represented in terms of discrete elements. Thus they lend themselves to the treatment of problems dealing with diffusional flow processes where chemical reactions proceed along particle path, as the case of turbulent combustion.

References

1. Chorin, A.J. (1973) "Numerical Studies of Slightly Viscous Flow," J. Fluid Mech., 57, 785-796.

2. Chorin, A.J. (1978) "Vortex Sheet Approximation of Boundary Layers," J. Comp. Phys., 27, 428-442.

3. Chorin, A.J. (1980) "Vortex Models and Boundary Layer Instability," SIAM J. Scientific Stat Comp., 1, 1-24.

4. Einstein, A. (1926) Investigation on the Theory of the Brownian Movement. Translation, Methuen and Co., Ltd., London: (Reprint, Dover Publications, Inc., New York, 1956).

5. Courant, R., Friedrichs, K., and Levy, H (1928) On the Partial Difference Equations of Mathematical Physics (Translation from Mathematische Annalen, 100) AEC Computing Facility, Institute of Mathematical Sciences, New York University, 1956.

6. Ghoniem, A.F., Chorin, A.J. and Oppenheim, A.K., "Numerical Modeling of Turbulent Flow in a Combustion Tunnel" Phil. Trans. Royal Soc. Lond., A 304, 303-325.

This work was supported by NASA on Grant NAG 3-131 and by the Office of Energy Research, Basic Energy Science Division of the U.S. Department of Energy Under Contract No. W-7405-ENG-48.

TABLE 1 Analogy between Diffusional Processes

Conserved Quantity	Momentum	Energy
Primary variable	u	T
Diffusion equation	$\frac{\partial u}{\partial t} = \nu \frac{\partial^2 u}{\partial y^2}$	$\frac{\partial T}{\partial t} = \alpha \frac{\partial^2 T}{\partial y^2}$
Diffusing element	$f = \int \rho u dA$	$e = \int \rho c T dA$
Conserved quantity	Vorticity	Heat
Secondary variable	$\xi = - \frac{\partial u}{\partial y}$	$q = - k\, \partial T/\partial y$
Diffusion equation	$\frac{\partial \xi}{\partial t} = \nu \frac{\partial^2 \xi}{\partial y^2}$	$\frac{\partial q}{\partial t} = \alpha \frac{\partial^2 q}{\partial y^2}$
Diffusing element	$\gamma = \int \xi dA$	$\dot{e} = \int q dA$

Note: In 1-D

$dA = dy$;

$\dot{e} = de/dt$

TABLE 2 Fundamental Idea

Differential Equation	$\frac{\partial \phi}{\partial t} = \alpha \frac{\partial^2 \phi}{\partial y^2}$
Boundary Conditions	$\phi(y,0) = \delta(0)$; $\phi(\pm\infty,t) = 0$
Constraint	$\Phi = \int_{-\infty}^{\infty} \phi\, dy = 1$
Formal Solution	$\phi = \frac{1}{\sqrt{2\pi}\sigma} \exp[-(y/\sigma)^2]$; $\sigma = \sqrt{2\alpha t}$
Stochastic Solution	$\Sigma\, \delta\Phi_i = 1$ $y_i(t + \Delta t) = y_i(t) + \eta_i$ $E(\eta_i) = 0$; $E(\eta_i^2) = 2\alpha\, \Delta t$
Local Sampling	$\phi = \frac{1}{\delta y} \Sigma\, \delta\Phi_i\, \delta(y-y_i)$
Global Sampling	$\Phi = \Sigma\, \delta\Phi_i\, H(y-y_i)$

Nomenclature

H - Heavyside step function

δ - Dirac delta function

E - Expected value

η - Random variable

σ - Standard deviation

TABLE 3 Hierarchy of Diffusional Processes

No	1	2	3	4
Differential Equation	$\frac{\partial \phi}{\partial t} = \alpha \frac{\partial^2 \phi}{\partial y^2}$	$\frac{\partial \phi}{\partial t} = \alpha \frac{\partial^2 \phi}{\partial y^2}$	$\frac{\partial \phi'}{\partial t} = \alpha \frac{\partial^2 \phi'}{\partial y^2}$	$\frac{\partial \phi}{\partial t} = \alpha \frac{\partial^2 \phi}{\partial y^2} - 1$
Initial Condition	$\phi(y,0) = \delta(0)$	$\phi'(y,0) = \delta(0)$	$\phi''(y,0) = \delta(0)$	$\phi(0,t) = 0$
Constraint	$\Phi = 1$	$\phi = 1$ or $f(t)$	$\phi' = 1$ or $f(t)$	$\phi''(0,t) = 1/\alpha$
Diffusing Element	$\delta \Phi_i$	$\delta \phi_i$	$\delta \phi'_i$	$\delta \phi_i$
Elements of Solution	$\phi = \frac{1}{\delta y} \Sigma\, \delta \Phi_i\, \delta(y-y_i)$ $\Phi = \Sigma\, \delta \Phi_i\, H(y-y_i)$	$\phi' = \frac{1}{\delta y} \Sigma\, \delta \phi_i\, \delta(y-y_i)$ $\phi = \Sigma\, \delta \phi_i\, H(y-y_i)$	$\phi' = \Sigma\, \delta \phi'_i\, H(y-y_i)$ $\phi = \Sigma\, \delta \phi'_i (y-y_i)\, H(y-y_i)$	$\delta \phi_i = \alpha \phi''(0,t) \delta t$ $\phi = \Sigma\, \delta \phi_i\, H(y-y_i)$
Examples	Instantaneous deposition of energy	Prescribed temperature on the boundary (Couette flow)	Prescribed heat flux on the boundary	Uniformly distributed source of energy (Poiseuille flow)

Note: $\phi' = \partial \phi / \partial y$ etc.

TABLE 4 Extension to Two-Dimensions

Differential Equation	$\frac{\partial T}{\partial t} = \alpha \nabla^2 T$
Boundary Conditions	$T(x,y,0) = \delta(0,0)$
Constraint	$\int_{\pm\infty}\int T \, dxdy = 1$
Formal Solution	$T = \frac{1}{\sqrt{2\pi\sigma}} \exp[-(x/\sigma)^2] \cdot \frac{1}{\sqrt{2\pi\sigma}} \exp[-(y/\sigma)^2]$
Stochastic Solution	$\Sigma \, \delta e_i = 1$ $\underline{r}_i(t + \Delta t) = \underline{r}_i(t) + \underline{\eta}_i$ $E(\eta_i) = 0 \;\; ; \;\; E(\eta_i{}^2) = 2\alpha \, \Delta t$
Local Sampling	$T = \frac{1}{\delta x \delta y} \; \Sigma \, \delta e_i \; \delta(x-x_i) \; \delta(y-y_i)$

TABLE 5 A Hybrid Scheme for Two-Dimensions

Domain	Boundary $y < \delta_s$	Interior $y > \delta_s$
Differential Equation	$\frac{\partial T}{\partial t} = \alpha \frac{\partial^2 T}{\partial y^2}$	$\frac{\partial T}{\partial t} = \alpha \nabla^2 T$
Boundary Condition	$T = 1$	$T = \delta(y-\delta_s)$
Diffusing Element	δT_i	δe_i
Coupling	$\delta e_i = \delta T_i * (y_i-\delta_s)$	$\delta T_i = \pm \, \delta e_i/\delta_s$
Stochastic Solution	$\underline{r}_i(t+\Delta t) = \underline{r}_i(t) + \eta_i$	$\underline{r}_i(t+\Delta t) = \underline{r}_i(t) + \underline{\eta}_i$
Sampling	$T = \frac{\Sigma \, \delta e_i}{\delta_s} + \Sigma \, \delta T_i \; H(y-y_i)$	$T = \frac{1}{\delta A_i} \Sigma \, \delta e_i \; \delta(\underline{r}-\underline{r}_i)$

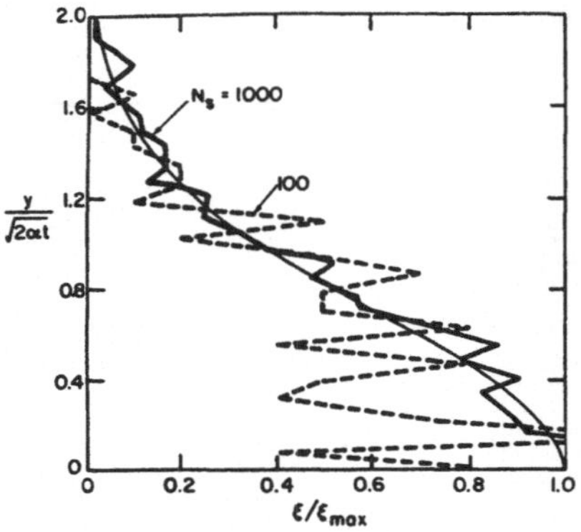

Fig. 1 Vorticity distribution in Stoke's problem

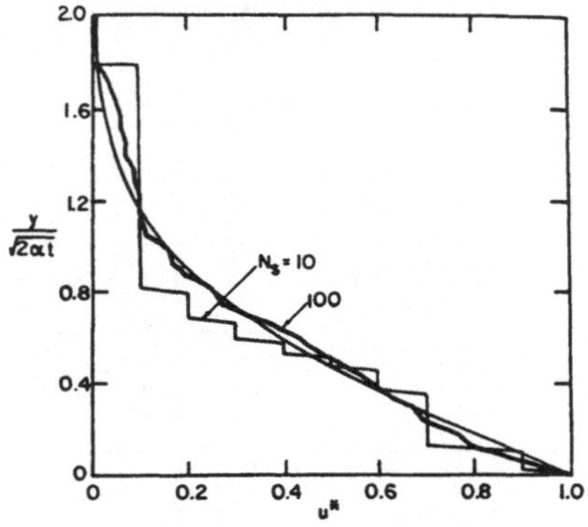

Fig. 2 Velocity profiles in Stoke's problem

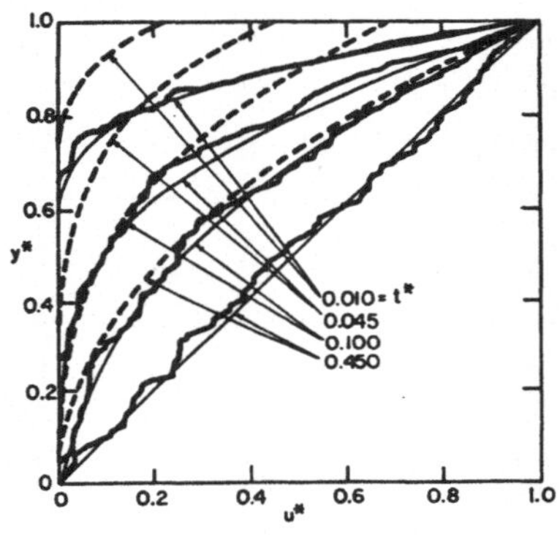

Fig. 3 Velocity and energy profiles in Couette flow

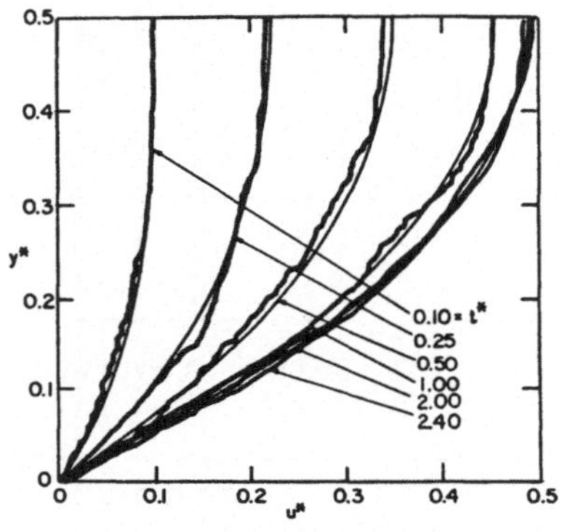

Fig. 4 Velocity profiles in Poiseuille flow

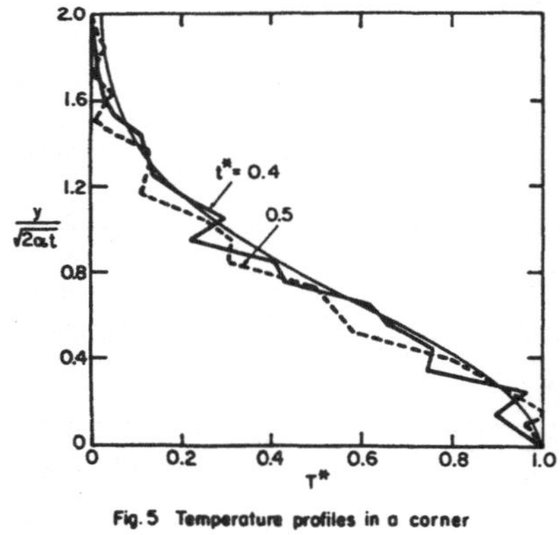

Fig. 5 Temperature profiles in a corner

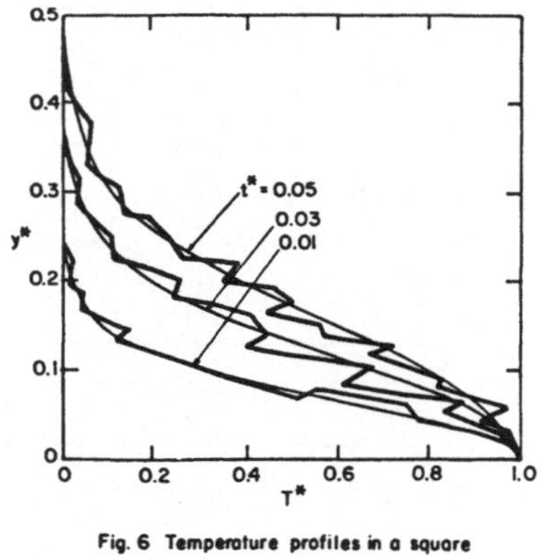

Fig. 6 Temperature profiles in a square

A HIGH REYNOLDS NUMBER FLOW WITH CLOSED STREAMLINES

by E.W. HADDON

University of East Anglia, Norwich, England.

INTRODUCTION

In an earlier paper Lyne (1971) has considered the flow in a pipe of circular cross-section, with radius a, which is coiled in a circle, of radius L (where L >> a) when a pressure gradient along the pipe is varying sinusoidally with frequency ω and zero mean. He shows that the motion depends upon two independent parameters

$$\epsilon^2 = U^2/La\omega^2 \qquad \text{and} \qquad R = U^2a/L\omega\nu, \tag{1}$$

where U is a typical velocity along the pipe and ν is the kinematic viscosity of the fluid. Lyne defines $\beta = (2\nu/\omega a^2)^{\frac{1}{2}} = (2\epsilon^2/R)^{\frac{1}{2}}$ as the ratio of the Stokes layer thickness to the radius of the pipe and then develops a solution in the limit $\beta \to 0$ with $R = O(1)$. There is no mean flow along the coiled pipe but at $O(\beta)$ there is a time independent motion in the pipe cross-section which consists of two recirculating regions symmetrically disposed about that diameter of the cross-section which lies in the plane of the coiled pipe. The flow is directed towards the centre of the coil along that diameter. For $\beta << 1$ the steady streaming is induced by the action of Reynolds stresses within the Stokes shear layer at the pipe wall. Outside the shear layer it can be shown that the Reynolds stresses are zero at this order and that the flow is driven by the steady velocity which persists at its edge. The equations governing the streaming in this region are the steady two-dimensional Navier-Stokes equations with R as Reynolds number. Lyne discusses the steady streaming outside the shear layer by a series expansion for $R << 1$ and by an approximate method for $R >> 1$ which is based upon the premise, Batchelor (1956), that the motion has an inviscid core of closed streamlines surrounded by viscous boundary layers of thickness $O(R^{\frac{1}{2}})$.

In this paper we solve the Navier-Stokes equations numerically for this configuration over a sequence of values of R up to 3000. Our aims are (i) to confirm the effectiveness of Lyne's approximate method for which there is no formal justification, but which gave encouraging results when compared with numerical calculations by Kuwahara and Imai (1969); (ii) to demonstrate the effectiveness of the numerical method of solution of the Navier-Stokes equations introduced by Dennis and Hudson (1978), and (iii) to investigate the method of Dennis and Hudson when applied to a stretched grid which places more points in the boundary layer at the pipe wall where both the vorticity and its radial gradient become unbounded as $R \to \infty$.

FORMULATION AND NUMERICAL METHOD

Lyne (1971), in his theoretical development, expands the flow variables in powers of β. He finds, at $O(\beta)$, that Reynolds stresses acting within the Stokes shear layer of thickness $O(a\beta)$ at the pipe wall are responsible for the generation of a time-independent streaming component in the flow. The tangential steady streaming persists to the edge of the Stokes layer where it is responsible for driving a steady circulation, again at $O(\beta)$, which is governed by the steady two-dimensional Navier-Stokes equations with Reynolds number R. Thus we have, in cylindrical polar coordinates centred in the pipe cross-section, with the coil centre lying along $\theta = \pi$,

$$\nabla^4\psi + \frac{R}{r}\frac{\partial(\psi,\nabla^2\psi)}{\partial(r,\theta)} = 0, \tag{2}$$

where ψ is a stream function related to the velocity components in the (r,θ) directions respectively by

$$u = r^{-1}\partial\psi/\partial\theta, \qquad v = -\partial\psi/\partial r, \tag{3}$$

and the boundary conditions are

$$\psi = 0 \text{ and } v = -\partial\psi/\partial r = -\frac{1}{4}\sin\theta \text{ on } r = 1. \tag{4a,b}$$

In solving (2) numerically we adopt the commonly used split-operator method, introducing ζ to denote vorticity, and thus (2) yields

$$\nabla^2\zeta - R\left(\frac{v}{r}\frac{\partial\zeta}{\partial\theta} + u\frac{\partial\zeta}{\partial r}\right) = 0, \qquad \nabla^2\psi = -\zeta. \tag{5a,b}$$

The boundary condition for (5b) is, from (4a), simply $\psi = 0$ on $r = 1$. For (5a) we use the slip-velocity condition (4b) and a method due to Woods (1954) to derive an expression for ζ at $r = 1$. Since the flow is symmetrical about the pipe diameter shown in Figure 1 we confine the calculation to the semi-circle $0 \le \theta \le \pi$ which introduces the conditions $\psi = \zeta = 0$ on $\theta = 0$ and $\theta = \pi$.

The numerical scheme employed for the solution of (5a) is due to Dennis and Hudson (1978). This method has been further exploited by Dennis (1980) in his study of steady flow through a curved pipe and by Haddon and Riley (1983) in a study of the mean circulation in standing waves. Finite-difference equations are developed from (5a) which are not only accurate to second-order but also yield an iteration matrix which is diagonally dominant. Standard central-difference approximations are used for (5b). We separate (5a) into

$$\frac{\partial^2\zeta}{\partial r^2} + \frac{1}{r}\frac{\partial\zeta}{\partial r} - Ru\frac{\partial\zeta}{\partial r} = A(r,\theta), \qquad \frac{\partial^2\zeta}{\partial\theta^2} - Rrv\frac{\partial\zeta}{\partial\theta} = -r^2A(r,\theta), \tag{6a,b}$$

where $A(r,\theta)$ is unknown. In (6a), along a line $\theta = \theta_0$, we write

$$\zeta = F(r,\theta_0)\exp\{-f(r,\theta_0)\} \qquad \text{where } f(r,\theta_0) = -\tfrac{1}{2}R\int_{r_0}^{r} u(z,\theta_0)dz, \tag{7}$$

which gives, for F,
$$\frac{\partial^2F}{\partial r^2} + \frac{1}{r}\frac{\partial F}{\partial r} + \tfrac{1}{2}R\left(\frac{\partial u}{\partial r} + \frac{u}{r} - \tfrac{1}{2}Ru^2\right)F = A(r,\theta_0)e^f. \tag{7a}$$

Similarly in (6b), along a line $r = r_0$, we write

$$\zeta = G(r_0,\theta)\exp\{-g(r_0,\theta)\} \qquad \text{where } g(r_0,\theta) = -\tfrac{1}{2}Rr\int_{\theta_0}^{\theta} v(r_0,z)\,dz, \tag{8}$$

giving, for G,

$$\frac{1}{r^2}\frac{\partial^2 G}{\partial\theta^2} + \frac{R}{2r}\left(\frac{\partial v}{\partial\theta} - \tfrac{1}{2}Rrv^2\right)G = -A(r_0,\theta)e^g. \tag{8a}$$

Equations (7a) and (8a) are next discretised using central-differences about the point (r_0,θ_0). Since $\exp\{f(r_0,\theta_0)\} = \exp\{g(r_0,\theta_0)\} = 1$, addition eliminates the unknown $A(r_0,\theta_0)$. In the usual notation this discretisation introduces the values F_1,F_3,G_2,G_4 at the grid points $(r_0 + h,\ \theta_0)$, $(r_0 - h,\ \theta_0)$, $(r_0,\ \theta_0 + k)$, $(r_0,\ \theta_0 - k)$ respectively, where h and k are the grid spacings in the r,θ directions. The resulting difference equation will be second-order accurate in h,k. We now use (7) to replace F_1 by $\zeta_1 e^{f_1}$ and F_3 by $\zeta_3 e^{f_3}$, and similarly for G_2,G_4, and further we replace $u(z,\theta_0)$ in (7) by a Taylor series about (z_0,θ_0). Integrating the series and setting $r = r_0 + h$ yields

$$f_1 = -\tfrac{1}{2}R\left\{hu_0 + \frac{h^2}{2}\left(\frac{\partial u}{\partial r}\right)_0 + \frac{h^3}{6}\left(\frac{\partial^2 u}{\partial r^2}\right)_0 + \ldots\right\}. \tag{9}$$

On substituting (9) into the exponential series expansion we obtain

$$e^{f_1} = 1 - \tfrac{1}{2}Ru_0h - \left\{\tfrac{1}{4}R\left(\frac{\partial u}{\partial r}\right)_0 - \tfrac{1}{8}R^2u_0^2\right\}h^2 + O(h^3), \tag{10}$$

and similarly for e^{f_3}, e^{g_2} and e^{g_4}. The resulting difference equation in ζ can be reduced by use of the continuity equation $\partial u/\partial r + u/r + r^{-1}\partial v/\partial\theta = 0$ and the observation that since $h(\zeta_1 - \zeta_3) = O(h^2)$ and $\zeta_1 + \zeta_3 = 2\zeta_0 + O(h^2)$ certain terms can be eliminated or replaced whilst maintaining second-order accuracy. Finally, from (5a) we obtain the equation

$$c_1\zeta_1 + c_2\zeta_2 + c_3\zeta_3 + c_4\zeta_4 = c_0\zeta_0, \tag{11}$$

where

$$c_1 = 1 + \tfrac{1}{2}h/r_0 - \tfrac{1}{2}Ru_0h + R^2u_0^2h^2/8,\quad c_2 = (h/k)^2(1/r_0^2 - \tfrac{1}{2}Rv_0k/r_0 + R^2v_0^2k^2/8),$$

$$c_3 = 1 - \tfrac{1}{2}h/r_0 - \tfrac{1}{2}Ru_0h + R^2u_0^2h^2/8,\quad c_4 = (h/k)^2(1/r_0^2 + \tfrac{1}{2}Rv_0k/r_0 + R^2v_0^2k^2/8),$$

$$c_0 = 2 + 2h^2/r_0^2k^2 + R^2(u_0^2 + v_0^2)h^2/4.$$

Manipulation of these expressions for the coefficients establishes that $c_1 \geq 0$, $c_2 > 0$, $c_3 \geq 0$, $c_4 > 0$ under all circumstances. Since we also have $c_1 + c_2 + c_3 + c_4 = c_0$ the associated matrix for the system of equations represented by (11) is always diagonally dominant, which is desirable for the iterative method of solution. The standard central-difference approximations used in the discretisation of (5b) also yield a diagonally dominant matrix of coefficients.

Numerical solutions have been obtained on a number of different grids for a range of values of R. For each grid, we chose, as initial solution, the distribution of ψ and ζ satisfying (5a,b) when R = 0, namely

$$\psi = \sin\theta\,(r^3-r)/8, \qquad \zeta = -r\sin\theta, \tag{12}$$

and then iterated to convergence at some finite R. This solution then became the initial solution in the iterative calculation for the solution at some larger value of R. From (11), on dividing through by c_0 and writing $d_i = c_i/c_0$ we obtain the point relaxation procedure

$$\hat{\zeta}_0 = (1-\omega_1)\zeta_0 + \omega_1\{d_1\zeta_1 + d_2\zeta_2 + d_3\zeta_3 + d_4\zeta_4\}, \tag{13}$$

where ω_1 is a suitable relaxation parameter and $\hat{\zeta}_0$ denotes the new value of ζ at point 0. Similarly from (5b) we obtain

$$\hat{\psi}_0 = (1-\omega_2)\psi_0 + \omega_2\{e_1\psi_1 + e_2\psi_2 + e_3\psi_3 + e_4\psi_4 + e_0\zeta_0\}. \tag{14}$$

In using (13) and (14) in an iterative procedure the most recently computed values of the stream function and vorticity are used in the right sides.

The iterative procedure sweeps through all internal grid points using (14) to update ψ at each point and next performs an identical sweep using (13) to update ζ at all internal points. An iterative cycle is completed by the computation of new values for ζ on r = 1 using the difference equation derived from the boundary condition. Convergence of the iterative procedure is assessed through two criteria

$$\Sigma|1 - \psi_0/\hat{\psi}_0| < \eta_1 \qquad \text{and} \qquad \Sigma|1 - \zeta_0/\hat{\zeta}_0| < \eta_2, \tag{15}$$

where the summation extends over all grid points and η_1, η_2 are suitable tolerances which may reflect the total number of points over which the summation is taken.

RESULTS

We have solved (5a,b) as described in the previous section for a range of values of R up to 3000 over a number of different grids. A characteristic feature of flows with closed streamlines is that, as the Reynolds number increases without limit, an inviscid core of uniform vorticity develops, surrounded by viscous boundary layers (see Batchelor (1956)). The approximate method of Lyne (1971) for large R is based upon this feature and enables him to predict the limiting value of the vorticity as $R \to \infty$. In Figures 2 and 3 we show the vorticity along the lines $r = 1/2$ and $\theta = \pi/2$ respectively and we see the development of this inviscid core of uniform vorticity as R increases. In Table 1 we present the value of the vorticity at the point $r = 1/2$, $\theta = \pi/2$ for various values of R on various grids, expressed as the number of grid intervals in the r and θ directions. We also include in the table a value for the vorticity, at each value of R, obtained by Richardson extrapolation which is accurate to fourth-order.

Grid size / R	40,40	80,80	160,160	Extrapolated
1000	-0.5010	-0.5345	-0.5449	-0.5487
1500	-0.4662	-0.5202	-0.5396	-0.5472
2000	-0.4350	-0.5069	-0.5362	-0.5481
3000	-0.3851	-0.4815	-0.5295	-0.5500

Table 1: Vorticity at $r = 1/2$, $\theta = \pi/2$.

From the extrapolated values and the knowledge that the correction to the core vorticity is $O(R^{-\frac{1}{2}})$, we obtain the expression $\zeta \approx -0.56 + 0.4\ R^{-\frac{1}{2}}$ for large R, which represents the values in the table to about ½%. This asymptotic result gives $\zeta = -0.56$ for the limiting value of the vorticity as $R \to \infty$ which is precisely the value given by Lyne. In a sense this is remarkable since Lyne's method is an approximate one for which there is no formal justification in the limit as $R \to \infty$.

The iterative method of solution we have described in the previous section converges without difficulty to a solution even at the higher Reynolds numbers. The only limitation to its use is the grid size, which has to be decreased as the Reynolds number increases in order to resolve the boundary layer at the pipe wall and the free shear layer along the pipe diameter. The computational cost of halving the grid size in both r and θ directions is a four-fold increase in the work on each iterative cycle together with an increase in the number of cycles needed to achieve convergence. In the boundary layer at the pipe wall the vorticity is $O(R^{\frac{1}{2}})$ and its radial gradient is O(R). The third objective of this work was to investigate the use of a stretching function $\rho = \rho(r)$ such that in a stretched grid there are relatively more grid points spanning the boundary layer, in which gradients are large, than in the uniform core where gradients are low. We anticipated that this approach would yield better results for the same total number of grid points because of the improved resolution in the boundary layer. Initial results are, however, contrary to our expectations; for the same number of grid points we find no significant difference in the number of cycles of iteration required to meet the same convergence criteria when compared with the unstretched grid and the numerical solutions compare unfavourably with those from the unstretched grid.

REFERENCES

Batchelor, G.K., J.Fluid Mech., 1 (1956) 177.

Dennis, S.C.R. and Hudson, J.D., Proc.1st Int.Conf.on Numerical Methods in Laminar and Turbulent Flow, Pentech Press (1978) 69.

Dennis, S.C.R., J.Fluid Mech., 99 (1980) 449.

Haddon, E.W. and Riley, N. To be published in Wave Motion, (1983).
Kuwahara, K. and Imai, I., Phys.Fluids Supplement II, 12 (1969) 94.
Lyne, W.H., J.Fluid Mech., 45 (1971) 13.

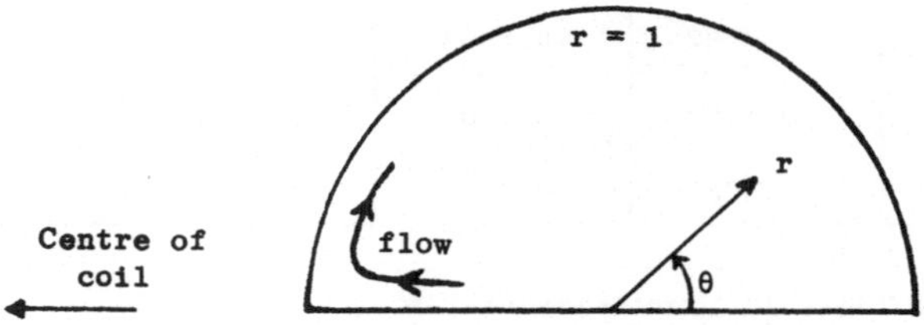

Figure 1. Half cross-section of the pipe.

Grid: 40,40; 80,80; 160,160

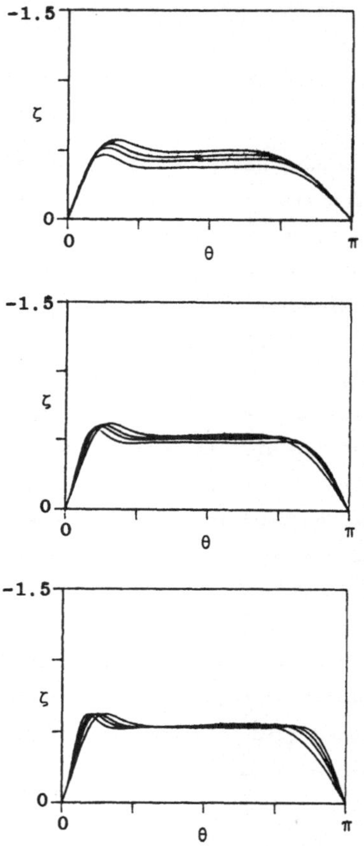

Figure 2. Along $r = 1/2$.

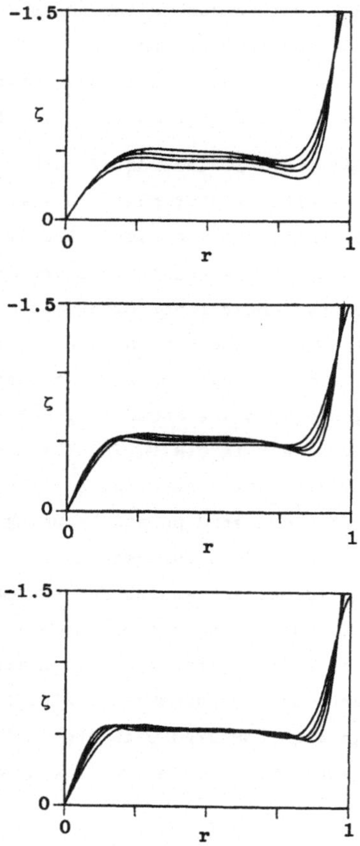

Figure 3. Along $\theta = \pi/2$.

Each diagram shows the vorticity for Reynolds numbers 1000, 1500, 2000 and 3000. The lowest curve in each diagram corresponds to $R = 3000$ (see Table 1).

COMPRESSIBLE SWIRLING FLOW INTO A CONSTANT VOLUME CYLINDER

W. Hautermann, H.-J. Thies, N. Peters
Institut für Allgemeine Mechanik,
RWTH Aachen, West-Germany

Abstract

The adiabaticly compressed swirling flow into a cylinder is calculated numerically in the low Mach number limit. In this limit the spatial derivatives of pressure and density vanish, such that only the time variation of density in the continuity equation is taken into account. The velocity is decomposed into an irrotational compressible and a rotational incompressible part. This leads to two Poisson equations for the potential function and the stream function, respectively, which are solved by a relaxation method. In addition to these the vorticity equation and the equation for the tangential velocity are solved by an ADI-technique. The Poisson equation for the potential function has a constant r.h.s. and is solved before the time-dependent calculation. Therefore only three time-dependent p.d.e.s are to be solved rather than four as in the formulation in primitive variables.

1. Introduction

If a compressible fluid flows with a prescribed velocity into a constant volume vessel, the flow field within the vessel is subject to two major effects: 1. the compression of the existing fluid by the addition of mass, 2. the interaction of the momentum of the inflowing stream with the momentum of the existing fluid and the viscous forces at the boundaries. The first effect leads to an increase in pressure and density within the vessel. If the inflowing velocity is sufficiently small compared to the speed of sound, rapidly propagating pressure waves of small amplitude tend to equilibrate the pressure within the vessel. If adiabatic boundary conditions are introduced, the energy equation degenerates to the law of adiabatic compression which relates the pressure to the density. However, the formulation of the Navier-Stokes equations for compressible flow in primitive variables u, v, w and p tends to hide the nature of the physical process. The flow is better understood, if the two effects are separated by dividing the flow field into an irrotational compressible and a rotational incompressible part. The first part represents a sink flow, which results, if the rotational motion of the fluid and viscous forces at the boundaries are neglected. The second part takes these effects into account but neglects the compression. It is the aim of the present paper to show the numerical implications of such a formulation for the case of a swirling axially symmetric flow.

2. Formulation

We consider an axially symmetric flow field in a vertical cylinder with a height/radius ratio $L/R = 3.33$. The inlet velocity enters through a circular slit of $h/L = 0.1$ where the ratio of the radial and the tangential velocity component is fixed as $(u/v)_{in} = 0.5$ and the axial component w_{in} is zero. For adiabatic compression in the low Mach number limit the density ρ is only a function of the time. The continuity equation is then written

$$\frac{1}{r}\frac{\partial}{\partial r}(ur) + \frac{\partial w}{\partial z} = -\lambda(t)$$
$$\lambda(t) = \frac{1}{\rho}\frac{\partial \rho}{\partial t} \quad . \tag{1}$$

The velocity field u, w, v (radial, axial, tangential) is decomposed into an irrotational compressible part (index Φ) and a rotational incompressible part (index ψ)

$$u = u_\psi + \lambda\, u_\Phi = \frac{1}{r}\frac{\partial \psi}{\partial z} + \lambda \frac{\partial \Phi}{\partial r}$$
$$w = w_\psi + \lambda\, w_\Phi = -\frac{1}{r}\frac{\partial \psi}{\partial r} + \lambda \frac{\partial \Phi}{\partial z} \tag{2}$$
$$v = v_\psi \quad .$$

Introducing these into (1) yields two Poisson equations, the equation for the stream function ψ

$$D(\psi) \equiv \frac{\partial^2 \psi}{\partial r^2} - \frac{1}{r}\frac{\partial \psi}{\partial r} + \frac{\partial^2 \psi}{\partial z^2} = \zeta(r,z,t) \tag{3}$$

and the equation for the potential function Φ

$$\frac{\partial^2 \Phi}{\partial r^2} + \frac{1}{r}\frac{\partial \Phi}{\partial r} + \frac{\partial^2 \Phi}{\partial z^2} = -1 \quad . \tag{4}$$

The vorticity ζ is defined as

$$\zeta = r\left(\frac{\partial u_\psi}{\partial z} - \frac{\partial w_\psi}{\partial r}\right) \tag{5}$$

and calculated from the equation

$$\frac{\partial\zeta}{\partial t} + u\,\frac{\partial\zeta}{\partial r} + w\,\frac{\partial\zeta}{\partial z} - \zeta(\lambda + \frac{2u}{r}) - \frac{\partial v^2}{\partial z} = \nu D(\zeta) \tag{6}$$

The tangential velocity $v = \varphi/r$ is calculated from

$$\frac{\partial\varphi}{\partial t} + u\,\frac{\partial\varphi}{\partial r} + w\,\frac{\partial\varphi}{\partial r} = D(\varphi) \tag{7}$$

The boundary conditions are:

<u>At $z = 0$ and $z = L$:</u>

$$\psi = 0\ , \quad \frac{\partial\Phi}{\partial z} = 0\ , \quad \zeta = \frac{\partial^2\psi}{\partial z^2}\ , \quad \varphi = 0 \tag{8}$$

together with the non-slip condition

$$u = \frac{1}{r}\,\frac{\partial\psi}{\partial r} + \lambda\,\frac{\partial\Phi}{\partial r} = 0 \tag{9}$$

<u>On the axis $r = 0$:</u>

$$\psi = 0\ , \quad \frac{\partial\Phi}{\partial r} = 0\ , \quad \zeta = 0\ , \quad \varphi = 0 \tag{10}$$

<u>At $r = R$:</u>

$$\psi = 0\ , \quad \zeta = \frac{\partial^2\psi}{\partial r^2} - \frac{1}{R}\,\frac{\partial\psi}{\partial r} \tag{11}$$

	$z > L - h$	$z < L - h$	
$\lambda\ \partial\Phi/\partial r$	u_{in}	0	(12)
φ	$R\ v_{in}$	0	(13)

together with the non-slip condition

$$w = -\frac{1}{R}\,\frac{\partial\psi}{\partial r} + \lambda\,\frac{\partial\Phi}{\partial z} = 0 \tag{14}$$

With $u_{in}(t)$ prescribed, the relative mass increase $\lambda(t)$ in the cylinder may be determined using Gauss' integral relation

$$\int_\sigma \text{grad}\ \Phi\ d\sigma = \int_V \nabla\ \Phi\ dV$$

with $\text{grad}\ \Phi = -\,u_{in}/\lambda$ $(L > z > L - h)$, $d\sigma = 2\pi Rh$ and $\nabla\ \Phi = -\,1$, $V = \pi R^2 L$ as

$$\lambda = \frac{2h}{RL} u_{in} \tag{15}$$

3. Discretisation of the boundary conditions

A major difficulty of the stream function-vorticity formulation lies in the discretisation of the boundary conditions of the vorticity equation. The discretisation used in this analysis will be demonstrated for the boundary condition at $z = L$ $(j = M)$. If an external point with index 0 is introduced and the indices M-1 and M-2 represent points at one and two mesh sizes in the interior, the gradients of ψ are discretized at the radius r with index k as

$$\left.\frac{\partial\psi}{\partial z}\right|_M = \frac{1}{6\Delta z}(2\psi_{0,k} + 3\psi_{M,k} - 6\psi_{M-1,k} + \psi_{M-2,k})$$
$$\left.\frac{\partial^2\psi}{\partial z^2}\right|_M = \frac{1}{(\Delta z)^2}(\psi_{0,k} - 2\psi_{M,k} + \psi_{M-2,k}) \tag{16}$$

The external point is eliminated and $\zeta_{M,k}$ is written

$$\zeta_{M,k} = \left.\frac{\partial^2\psi}{\partial z^2}\right|_M = \frac{-7\psi_{M,k} + 8\psi_{M-1,k} - \psi_{M-2,k}}{2\,\Delta z^2} - \frac{3r}{\Delta z}\lambda\left.\frac{\partial\Phi}{\partial r}\right|_M \tag{17}$$

where eq. (9) has been used.

Similarly one obtains for $z = 0$ $(j = 1)$

$$\zeta_{1,k} = \frac{-7\psi_{1,k} + 8\psi_{2,k} - \psi_{3,k}}{2\,\Delta z^2} + \frac{3r}{\Delta z}\lambda\left.\frac{\partial\Phi}{\partial r}\right|_1 \tag{18}$$

and at $r = R$ $(k = N)$ with eq. $(11)_2$ and eq. (14)

$$\zeta_{j,N} = \frac{-7\psi_{j,N} + 8\psi_{j,N-1} - \psi_{j,N-2}}{2\,\Delta r^2} + \left(\frac{3}{\Delta r} - 1\right)\lambda\left.\frac{\partial\Phi}{\partial z}\right|_N \tag{19}$$

4. Results

The equations have been nondimensionalized with the radius R and a reference inflow velocity $v_{in,ref}$, which is the constant inflow velocity attained at time t_c after a linear initial increase from zero to $v_{in,ref}$ during the interval $0 < t < t_c$. The Reynolds number was 500. A grid point system of 100 equally spaced points in z-direction and 30 points in r-direction was used. The two Poisson equations were solved by successive overrelaxation using a nine-point (five points in both directions) discretisation /1/. At one point from the boundary, the external point was eliminated using a five point formula at the boundary. For the vorticity equation and the tangential velocity a conventional ADI-method /2/ was used. The nondimensional time was $t_c = 10$ and the calculation was continued with a constant nondimensional velocity $v_{in} = 1$ until $t = 30$. At that time an asymptotic decay to a time-independent solution was observed. The surprising result that a time-independent solution exists for a steadily compressed flow is consistent with the differential equations for time-independent boundary conditions and constant Reynolds number. The second condition may be viewed as an approximation which requires a particular temperature dependence of the dynamic viscosity. In Fig. 1 lines of constant values of the potential function are shown. The value zero was assigned to the point $j = 90$, $k = 30$ underneath the inlet slot. Fig. 1 illustrates the effect of the sink of the potential function due to the compression. In Fig. 2-4 the combined values for the velocity components at $t = 30$ are shown over r for various heights in the cylinder. From Fig. 2 it is seen that the maximum of the tangential velocity shifts from the right at $z = 3.0$ (lower part of the inlet slot) to the center at the bottom of the cylinder. Fig. 3 shows that the radial velocity is large at the bottom of the cylinder, where a slowly rotating stagnation type flow appears. Fig. 4 shows the downward flow in the center and the upward flow at the side in the lower part of the cylinder. The non-slip boundary condition at $r = 1$ is satisfied for the tangential velocity by definition and for the radial and axial velocities reasonably except for $z = 0.23$ in Fig. 4. A movie film illustrates the motion of particles issued with the incoming flow.

References

/1/ W. Hautermann: Numerische Berechnung einer kompressiblen, rotationssymmetrischen Drallströmung in einen Zylinder mit seitlichem Einlaß, Diplomarbeit, Institut für Allgemeine Mechanik, RWTH Aachen, 1979.

/2/ F. Bartels: Rotationssymmetrische Strömungen im Spalt konzentrischer Kugeln, Dissertation, Aachen 1978.

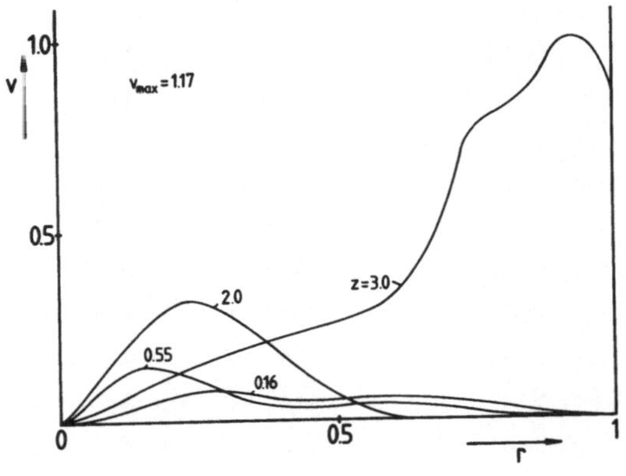

Fig. 2: Tangential velocity

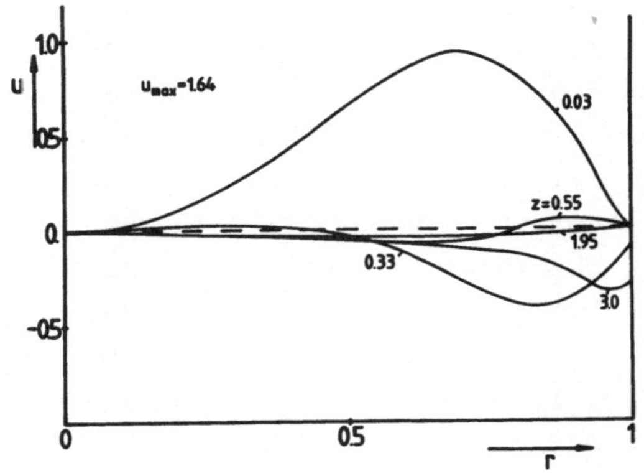

Fig. 3: Radial velocity

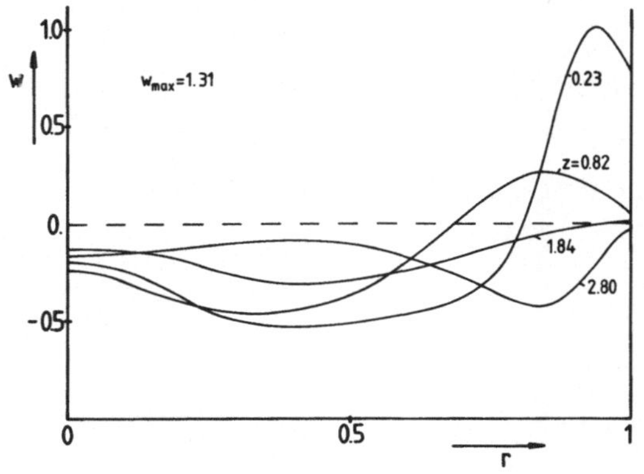

Fig. 4: Axial velocity

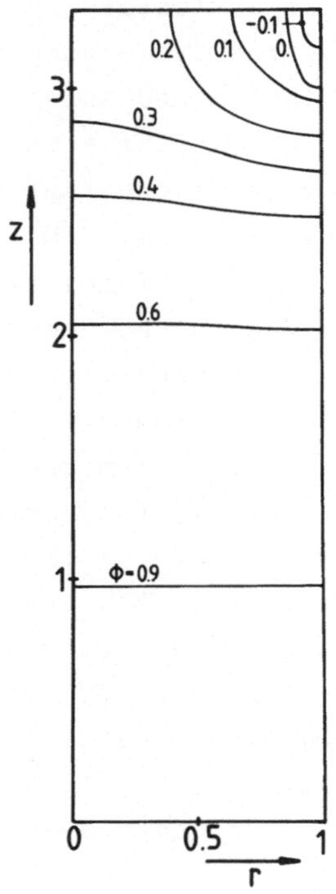

Fig. 1: Potential function

Techniques for Efficient Implementation of Pseudo-Spectral Methods and Comparisons with Finite Difference Solutions of the Navier-Stokes Equations

by

Richard S. Hirsh, Thomas D. Taylor, Margaret M. Nadworny, and J. L. Kerr
The Johns Hopkins University
Applied Physics Laboratory
Laurel, Maryland 20707

The solution of the incompressible Navier-Stokes equations by pseudo-spectral methods will unquestionably yield superior accuracy when compared to standard finite difference methods for equivalent numbers of grid/interpolation points. However, to be an acceptable tool, the pseudo-spectral technique must also be competitive with previously used methods from the standpoint of computational efficiency and reliability. The present investigation examines a number of factors contributing to the overall usefulness of pseudo-spectral methods that have not been reported previously. Among these are: time integration of the equations; the Poisson solver and its application in time dependent problems; and the method of computing spectral representations in pseudo-spectral calculations, especially in multi-dimensional problems. Due to the time integration scheme chosen, the same code can be used with either finite difference or pseudo-spectral methods, providing a direct check of comparative efficiency.

In order to conduct the study, a time dependent, two dimensional, incompressible, stratified, inviscid flow problem was chosen which can be described by the Navier-Stokes equations, subject to the Boussinesq approximation for the density. As a result, the problem solved is unsteady flow in confined region, see Figure 1, with no flow allowed through the boundary. This implies that there will be no energy loss within the region of interest. The stream function-vorticity formulation has been employed in the form

$$\frac{D\zeta}{Dt} = \frac{1}{F^2}\rho_y \qquad (1)$$

$$\frac{D\rho}{Dt} = \psi_y \qquad (2)$$

$$\nabla^2\psi = \zeta \qquad (3)$$

where $$F^2 = \frac{U_{ref}^2}{N^2 L_{ref}^2}$$

with N the natural oscillation frequency of the system. Any initial conditions on vorticity or density perturbation can be specified.

Time Integration

The first phase in the solution of these equations was to determine an explicit, second order accurate time integration scheme that would yield energy conservation for long integration times. This involved looking at Lax-Wendroff, MacCormack, and Adams-Bashforth methods, as well as a relatively neglected scheme presented by Abarbanel and Zwas [1] or Gary [2]. This last scheme (denoted by AZ) was found to be very appropriate for our incompressible problem, although it was originally tested for flows with shocks and subsequently rejected [1]. The method is an explicit, two level, multistep procedure which iterates to a time centered result.

Using a general time dependent problem as an example

$$f_t = g$$

the AZ method solves for the value at the new time step (n+1) from

$$\frac{f^{n+1,k+1} - f^n}{\Delta t} = \frac{1}{2}(g^{n+1,k} + g^n) \qquad (4)$$

where k is the iteration parameter. The stability of the procedure depends on the number of iterations, k, in an odd way, see Table 1. Even more interesting is the dissipativity of the method. A simple one dimensional wave propagation problem was solved using the AZ integration scheme, and the energy in the system was calculated. Table 2 shows the increase or decrease from the initial energy at the end of one thousand time steps as a function of the number of iterations. It can be seen that the amount of dissipation can be easily controlled by the number of iterations used, allowing energy conservation to any specified criterion.

The fact that the AZ method is an iterative procedure makes it very suitable with the usage of the Poisson solvers described below for the unsteady problems. In incompressible time dependent problems an exact solution of the Poisson equation is required at each time step. This is normally achieved either by iterating to machine error, or directly solving the Poisson equation. However, since the AZ method is already iterative, another, separate iteration scheme can be used for the Poisson equation (3). Only a fixed, small number of steps are taken, and then convergence of the overall time integration is checked. If the solution has not converged, then the (outer) time step iteration, which calls the (inner) Poisson iteration is performed again.

Pseudo-Spectral Technique

In order to compute derivatives or fluxes in real space by pseudo-spectral methods, the coefficients of a polynomial expansion of a function must be determined. Chebyshev polynomials have been chosen for the present study, so the expansion of any function f(x) in terms of the polynomials $T_n(x)$ is

$$f_n(x_i) = \sum_n a_n T_n(x_i) \qquad (5)$$

where the a_n are the coefficients needed for the calculations, and the interpolation points x_i are chosen as the nodes of the first omitted polynomial, i.e.,

$$x_i = (i-1) \cos \theta$$

Equation (5) can be thought of as a linear system of equations

$$Ta = f \qquad (6)$$

for the coefficients a_n. The method previously used to determine the a_n's has been the Fast Fourier Transform (FFT). This is simply an extremely efficient means of inverting the matrix T, of Equation (6). It requires only order NlogN operations after some initial overhead penalty, but restricts the number of interpolation points, N, usually to some power of 2.

We have previously shown [3] that for one dimensional problems alternate methods are available which are as efficient as the FFT for the number of interpolation points less than 64, but with no restrictions on N. For two dimensional problems we have used a method which directly solves Equation (6) by using Crout's reduction method. Although there is an operations penalty to be paid for the initial reduction, once this is accomplished the solution to

Equation (6) can be obtained in order N^2 operations, which, depending on the relative expense of the overhead involved, can be competitive with FFT's. Table 3 shows the results of such a comparison. Two FFT's were used, one, FFT1, directly from a textbook, and the other, FFT2, a highly optimized FFT used for signal processing. The result is clearly that up to 128 x 128 points, the direct solver is as efficient as the FFT. It is this Crout reduction technique that was used in all the solutions to be discussed.

Most previous pseudo-spectral approaches which require the solution of a Poisson equation have performed the calculation in spectral space, not in real space where the actual solution is required. That is, the solution is not for the function itself, but rather the coefficients of the solution. The boundary conditions are also imposed in spectral space, leading to a solution procedure which is effectively Lanczos' Tau method. This can lead to difficulties since the highest order term is being used to satisfy the boundary conditions [4].

There is, however, another procedure which yields a pseudo-spectral solution to Poisson's equation, but operates only in real space [5, 6]. The method is an iterative procedure which uses a finite difference predictor, followed by a spectral corrector. Thus, Equation (3) is solved as

$$L_{fd}\psi^{m+1} = L_{fd}\psi^{m} + \alpha(\zeta - L_{sp}\psi^{m}) \qquad (7)$$

where L_{sp} is the spectral operator, and L_{fd} is a finite difference operator which approximates L_{sp}. The optimal iteration parameter, α , has been determined by Orszag [5] to be 4/7. The solution of the finite difference problem given by Equation (7) was accomplished by the use of the Crout reduction algorithm previously employed in finding the Chebyshev coefficients. There are other more efficient means than the Crout scheme, but its use was convenient since it was already available. This Poisson solution procedure can solve Equation (3) to machine accuracy in a specified number of iterations, m=M, if desired, but since the Poisson solution is just the inner iteration in the AZ time integration, only m=6 iterations were performed in the inner loop before returning to the outer loop and checking overall convergence. If it is necessary to do another Poisson solve before advancing to the next time level, then the previous Poisson solution effectively preconditions the solution procedure, Equation (7) since L_{sp} operating on ψ is a good approximation to ζ.

Finite Difference Technique

In order to compare the pseudo-spectral calculations with finite difference calculations, Equations (1) through (3) were also solved using finite difference techniques. Second order finite difference representations of all derivatives were used; however, to algebraically conserve energy in the computational domain, some care was necessary. For the potential energy, which is proportional to ρ^2 , a version of Piacsek-Williams [7] differencing suitable to the non-staggered grid was used on the advective terms in the form

$$(u\rho)_x = \frac{1}{4\Delta x}\left[(u_{i+1} + u_i)\rho_{i+1} - (u_i + u_{i-1})\rho_{i-1}\right] \qquad (8)$$

In order to conserve kinetic energy, which is found from the product $\zeta\psi$, standard conservative differencing was used, i.e.,

$$(u\zeta)_x = \frac{1}{2\Delta x}\left[u_{i+1}\rho_{i+1} - u_{i-1}\rho_{i-1}\right] \qquad (9)$$

It can be shown that both (8) and (9) algebraically conserve energy, and the computational results bear this out. Other combinations of Piacsek-Williams and conservative differencing were tried in both Equations (1) and (2), but the form given above was the only one which conserved total energy over long integration times.

The solution of the Poisson equation was accomplished by a standard ADI procedure. Again, as this represented an iteration loop within the AZ time integration, only a small number (typically 6 to 8) iterations were performed for the solution ψ. More iterations did not change the results of the outer loop here, or in the pseudo-spectral calculations, and only increased the computation time.

Direct Code Comparisons

The iterative time integration offered an excellent way to test the relative efficiencies of pseudo-spectral and finite difference methods, due to the method of implementation the scheme (4). Rewriting Equations (1) and (2) as

$$\frac{\partial \zeta}{\partial t} = \zeta RHS$$

$$\frac{\partial \rho}{\partial t} = \rho RHS$$

where the RHS terms include everything other than the time derivative, i.e., advective and linear terms, the time integration using the AZ method is

$$\frac{\zeta^{n+1,k+1}}{\Delta t} = \frac{1}{2}\,(\zeta RHS^{n+1,k} + \zeta RHS^{n})$$

$$\frac{\rho^{n+1,k+1} - \rho^{n}}{\Delta t} = \frac{1}{2}\,(\rho RHS^{n+1,k} + \rho RHS^{n})$$

$$\nabla^2 \psi^{n+1} = \zeta^{n+1,k+1}$$

To advance in time all that is required is the evaluation of the RHS's. Thus, a basic code was used which had subroutine calls to give the evaluations of the RHS's and Poisson solution by either finite difference methods or pseudo-spectral methods. The remainder of the code structure remained unchanged.

Results

To demonstrate the comparative capabilities of each method a simple case of a nonstratified flow will be given. An initial vorticity distribution was specified, and allowed to move around the closed square domain. The finite difference solution was performed on a 33 x 31 grid. The spectral solution used 15 x 15 interpolation points. Figure 2 shows the initial vorticity distribution. The time step taken in the calculations was the same physical time in each case. After 100 time steps the two solutions are shown in Figure 3a. The pseudo-spectral solution has developed some grid scale oscillations generated by the function interpolation between the nodes. To control these a local smoother, previously used by Myers et al [8], was incorporated into the pseudo-spectral solution. The result after 100 steps is shown in Figure 3b. In either the smoothed or unsmoothed case, the finite difference and spectral solutions are indistinguishable from each other. After 800 steps the solutions are shown in Figure 4. Again, the two solutions are almost equivalent, despite the difference in solution points utilized. The difference could be due to the lower accuracy of the finite difference solution which does have some third order truncation errors. It should be noted though, that the finite difference solution does not exhibit any noticeable damping or phase shifts as some previously depicted second order accurate solutions have shown.

The normalized finite difference time for the 800 steps is 1.0. The pseudo-spectral time is 3.5. However, for equivalent accuracy, the finite difference method would require at least 60 x 60 points, which would make its computation time 4.0, more than that of the pseudo-spectral method. There are still further improvements in the pseudo-spectral method, which can improve its efficiency, such as using banded or ICCG solvers in the Poisson solution of Equation (7), or matrix evaluations of the coefficient sum shown in Equation (5) to get the real space variables.

Conclusions

The results of this study have shown that: (i) direct matrix inversions are competitive with FFT's for calculating pseudo-spectral representations; (ii) all real space Poisson solvers are possible with the use of a predictor-corrector procedure; (iii) the iterative time integration scheme is very attractive for use in incompressible flows; (iv) energy conservation in stratified flow using finite difference techniques can be accomplished using a combination of conservative and Piacsek-Williams differencing; and (v) computation times for pseudo-spectral calculations are faster than finite difference calculations of equivalent accuracy.

REFERENCES

1. Abarbanel, S., and Zwas, G.: An iterative finite difference method for hyperbolic systems, Math. Comp., 23, 549-565, 1969.

2. Gary, J.: On certain finite difference schemes for hyperbolic systems, Math. Comp., 18, 1-18, 1964.

3. Taylor, T. D., Hirsh, R. S., Nadworny, M. M., and Kerr, J. L.: FFT vs conjugate gradient method for solution of flow equations by pseudo-spectral methods, in Proceedings of the Fourth GAMM Conference on Numerical Methods in Fluid Mechanics, H. Viviand, ed., pp 311-325, F. Vieweg, Braunschweig, 1982.

4. Peyret, R. and Taylor, T. D.: Computational Methods for Fluid Flow, Springer, New York, 1982.

5. Orszag, S.: Spectral methods for problems in complex geometries, in Numerical Methods for PDE's, pp 273-305, Academic Press, New York, 1979.

6. Morchoisne, Y.: Pseudo-spectral space-time calculations of incompressible viscous flows, presented at AIAA 19th Aerospace Sciences Meeting, St. Louis, January 1981.

7. Piacsek, S. A. and Williams, G. P., Conservation properties of Convection Difference Schemes, J. Comp. Phys., 6, 392-405, 1970.

8. Myers, R. B., Taylor, T. D., and Murdock, J. W.: Pseudo-spectral simulation of a two-dimensional vortex flow in a stratified incompressible fluid, J. Comp. Phys., 43, 180-188, 1981.

Table 1. Stability properties of AZ method

Iterations	1	2	3	4	5	6	7	8	9	10	11	. . .
Stable			x	x			x	x			x	

Table 2. Energy conservation properties of AZ method

Iterations	2	3	4	5	6	7	8	9
Energy	$+10^{-2}$	-10^{-2}	-10^{-4}	$+10^{-4}$	$+10^{-6}$	-10^{-6}	-10^{-8}	$+10^{-8}$

Table 3. Time required for matrix solution

	16x16	32x32	64x64	128x128
FFT1	4.2	14.3	59.3	252.7
FFT2	3.1	8.3	15.4	104.4
Crout	2.3	6.3	15.5	117.0

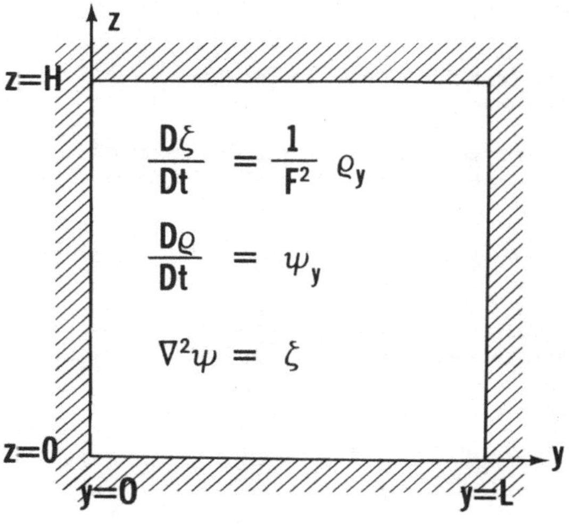

Figure 1
Geometry & Equations

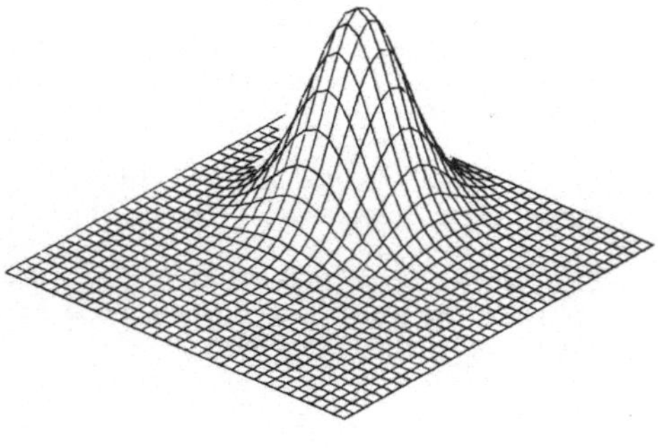

Figure 2
Initial vorticity distribution

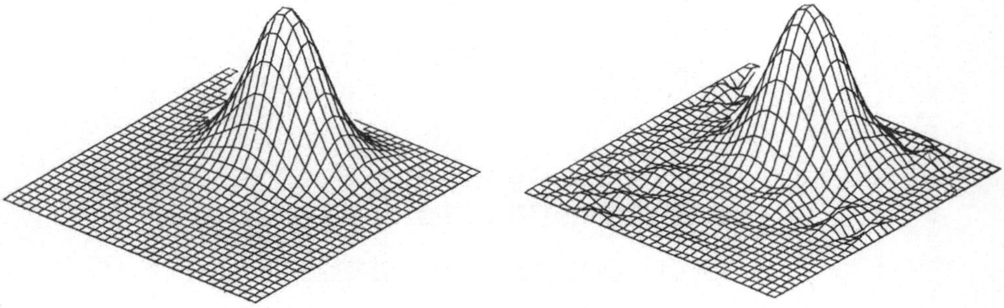

Finite difference solution Unsmoothed pseudo-spectral solution

Figure 3a
Vorticity distribution after 100 time steps

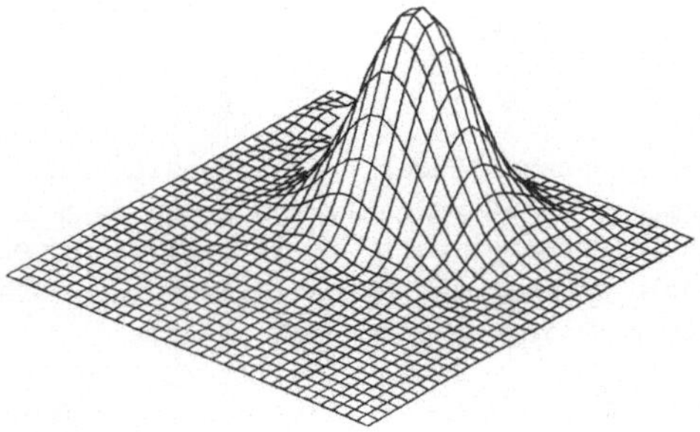

Figure 3b
Smoothed vorticity distribution after 100 time steps

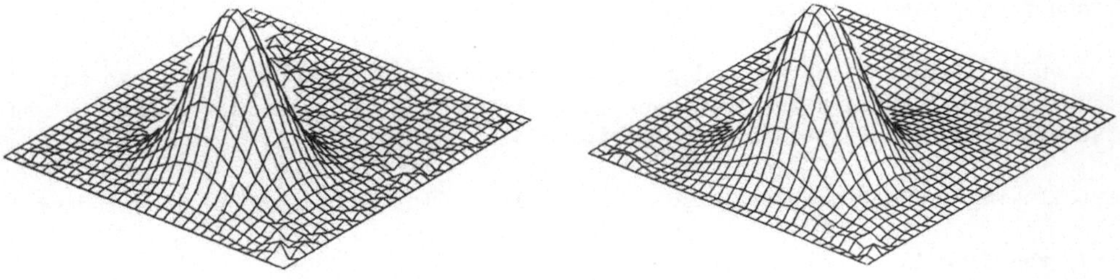

Finite difference solution Smoothed pseudo-spectral solution

Figure 4
Vorticity distribution after 800 time steps

NUMERICAL SIMULATION OF THE FLUID FLOW IN FRONT OF THE SCOOP OF A GAS CENTRIFUGE

Marc HITTINGER (*) - Maurice HOLT (**)

(*) CISI, BP.24 - 91190 GIF-s/-YVETTE. (France)

(**) UNIVERSITY of CALIFORNIA, BERKELEY. (USA)

INTRODUCTION.

In most mechanically driven gas centrifuges, the scoop is either separated from the main chamber by a diaphragm, or is numerically simulated by a disc rotating at a slightly slower speed than the chamber wall. It is believed that the separative power of a gas centrifuge could be improved by optimising the fluid flow. For ultra-centrifugation the fluid is rotating at very high speed.

Near the wall high values of Mach number are reached. The effect of the scoop, which is then a stationary obstacle, is very violent. A shock wave is formed in front of the scoop and vortices are formed behind the scoop. This is a rather complicated problem to solve. We studied a simplified version of the problem as shown in Fig.2.

In the model proposed, we assume that the region between the shock wave and the scoop is narrow so that the region could be assumed to be almost plane. The scoop is then represented by a cylinder as shown in Fig.2.

The flow is divided into two regions by a stagnation line, one passing above the scoop, the other beneath it. The lower flow is like that in a nozzle, where the gas is accelerated by the base of the centrifuge.

The following assumptions were also made concerning the gas in the region studied :

(i) the flow is steady
(ii) the volumic viscosity is neglected
(iii) the viscosity effects are neglected ($\nu = 0$)

This is true only in this small region in front of the scoop. Victosity effects become important above and behind the scoop since they induce the motion of the gas (the gas is driven by the wall by means of frictions).

(iv) heat conduction effects are neglected ($k = 0$)
(v) the gas is treated as ideal.

DESCRIPTION OF THE METHOD.

Our model is two dimensional and, under the assumptions stated above, the gas flow is governed by the following equations in an x - y system of coordinates :

Continuity

$$\rho \left[\frac{\partial u}{\partial x} + \frac{\partial v}{\partial y} \right] + u \frac{\partial \rho}{\partial x} + v \frac{\partial \rho}{\partial y} = 0$$

X - Momentum

$$\rho \left[u \frac{\partial u}{\partial x} + v \frac{\partial u}{\partial y} \right] + \frac{\partial P}{\partial x} = 0$$

Y - Momentum

$$\rho \left[u \frac{\partial v}{\partial x} + v \frac{\partial v}{\partial x} \right] + \frac{\partial P}{\partial y} = 0$$

The thermal equation is replaced by Bernoulli's equation, which is easier to use in our method

$$\frac{\gamma P}{(\gamma - 1)\rho} + \frac{1}{2} \left[u^2 + v^2 \right] = \text{constant}$$

We introduce the stream function by

$$\frac{\partial \psi}{\partial x} = - \rho v$$

$$\frac{\partial \psi}{\partial y} = \rho u$$

and the entropy function by

$$\Phi(\psi) = \frac{1}{\rho} P^{1/\gamma}$$

The flow field is calculated by two different methods in each of the regions divided by the stagnation line :

(i) in the upper region, the flow is similar to the flow behind on shock wave formed in front of a blunt body and a modification of Telenin's method is used there (GILINSKII, TELENIN, TINYAKOV (1964).

(ii) in the lower region, the flow is similar to that observed in a transsonic nozzle. The method of lines is used for this part (HOLT (1977), JONES, SOUTH, KLUNKER (1972), KLOPPER, HOLT (1976))

The flow pattern is computed for the two methods for a particular stagnation line. A comparison of the velocities is made along this line, if they are different, the stagnation line is automatically modified in order to decrease the difference by an optimisation algorithm due to POWELL (1964).

The two methods (i.e.Telenin's method and the Method of lines) are very similar. The first basic idea is to transform the equations so that the derivatives of the unknowns with respect to one variable are expressed as functions of the derivatives in the second variable.

If the x_2 - derivatives are known, we have a simple system of ordinary differential equations to integrate between the shock wave and the scoop. This integration is performed using a 5th order Runge Kutta method.

Telenin's method and the Method of lines give us the values of the x_2 - derivatives of the unknowns at a step n knowing the values of these functions at a step n - 1. This is achieved by fitting a polynomial through the values of the variables at step n - 1, for Telenin's Method, and by using a finite difference scheme for the Method of lines.

The system is always of the type :

$$\left\{\begin{aligned} \frac{\partial v}{\partial x_1} &= A\frac{\partial v}{\partial x_2} + B\frac{\partial u}{\partial x_2} + C \\ \frac{\partial u}{\partial x_1} &= D\frac{\partial v}{\partial x_2} + E\frac{\partial u}{\partial x_2} + F \\ \frac{\partial P}{\partial x_1} &= G\frac{\partial v}{\partial x_2} + H\frac{\partial u}{\partial x_2} + I \end{aligned}\right.$$

To which can be added a similar equation for $\frac{\partial \psi}{\partial x_1}$

The second basic step in the calculation is to modify the boundary conditions on the initial line (shock wave) in order to minimize deviations from boundary conditions on the final line (scoop surface) - HOLT (1977), TELENIN, GILINSKII, TINYAKOV (1964).

PRELIMINARY FORMULATION OF THE MODAL PROBLEM.

This model was already presented by the authors. We will summarize the derivation. It is of interest to transform the governing equations in terms of an x, ψ system instead of using au x, y system, where ψ is the stream function, the physical problem is transformed as shown on Fig.3. A reduced variable ξ is used where :

In the uppon region,

$$\xi = \frac{\psi - \psi_\beta(x)}{\psi_s(x) - \psi_\beta(x)}$$

ψ_β = value on the scoop
ψ_s = value on the shock wave

The integration is performed from $\xi = 1$ to $\xi = 0$ (i.e.from the shock line to the stagnation/scoop line.

In the lower region,

$$\xi = \frac{\psi}{\psi_A} \qquad \text{where} \qquad \psi_A = \text{value on the stagnation/scoop line}$$

The integration is there carried out from $\xi = 0$ to $\xi = 1$ (i.e.from the bottom of the centrifuge to the stagnation/scoop line).

The ξ deratives can thereafter be expressed as functions of the x-derivatives. The formula are rather complicated and were already given in a previous paper (HITTINGER, HOLT and SOUBBARAMAYER (1981)).

The initial conditions for the upper part are the conditions behind a shock wave. Since we used as data the value of the variable upstream of the shock wave (they correspond to different slices) once a new shock wave is determined.

The velocity and pressure distributions behind the shock wave are given by the Rankine Hugoniot equations :

$$\left\{\begin{array}{ll} \rho_1 u_{n_1} = \rho_2 u_{n_2} = m & (1) \\ P_1 + m u_{n_1} = P_2 + m u_{n_2} & (2) \\ u_{n_1}{}^2 + \frac{\gamma}{\gamma - 1}\frac{P_1}{\rho_1} = \frac{u_{n2}}{2} + \frac{\gamma}{\gamma - 1}\frac{P_2}{\rho^2} & (3) \\ u_{T_1} = u_{T_2} & (4) \end{array}\right.$$

Here the index 1 denotes the value in front of the shock wave and the index 2 the values behind.

u_N is the velocity component normal to the shock wave

u_T is the velocity component tangent to the shock wave.

These equations can be solved in u, v, p, ρ as functions of the velocity at infinity $V\infty$, the Mach number $M\infty$, the density at infinity $\rho\infty$ and the shock wave angle σ.

Once the shock wave shape has been determined by Telenin's Method in the upper part (for a given stagnation line) and assuming that the shock wave is normal to the bottom of the centrifuge, we can have a good approximation of shock wave shape in the lower part and therefore the velocity and pressure distribution behind.

We then assume a velocity distribution at the bottom of the centrifuge and use it as initial data.

In the upper part, the stagnation line is a stream line, therefore $d\psi = 0$. This can also be expressed by the following relation :

$$- v + u\frac{dy}{dx} = 0$$

Furthermore, the velocity must be tangent to the scoop surface. Then, in both parts, the condition to be satisfied is the same i.e. :

$$u \frac{dy}{dx} - v = 0$$

This is also true in the lower part.

Unfortunately the results with this first model were not very good. This was due to several factors, some of which only became clear after obtaining the solution from the second model.

With this method, convergence is rapidly reached only if we start from a good approximation of the solution (i.e. with a shock wave close to reality). In the absence of good estimates and with the two values of Mach number (M = 2 and M = 7) chosen convergence was difficult to achieve.

With the shock wave shape assumed, the x, ψ approximation did not appear to be appropriate. The shock wave rises steeply with in front the scoop and therefore some points for the polynominal fitting were very close while others were very distant (See Fig.4). This led to very instable results and optimisation was often impossible since the pressure computed was negative and the results unsignificant. For this reason we decided to use a different model at least in the upper part.

DERIVATION FOR THE SECOND MODEL.

In order to get a more regular type of grid, we decided to use an (r, θ) derivation in the upper part and a new set of equations were derived. The derivation is long but straight foward. We used the reduced variable :

$$\xi = \frac{r(\theta) - r_\beta(\theta)}{r_s(\theta) - r_\beta(\theta)}$$

where
r_s is the radial coordonate on the shock wave
r_β is the radial coordonate on the scoop

Then

$$\left\{\begin{array}{l} \dfrac{\partial v}{\partial \xi} = A \dfrac{\partial u}{\partial \theta} + B \dfrac{\partial v}{\partial \theta} + C \\[2ex] \dfrac{\partial u}{\partial \xi} = D \dfrac{\partial u}{\partial \theta} + E \dfrac{\partial v}{\partial \theta} + F \\[2ex] \dfrac{\partial P}{\partial \xi} = G \dfrac{\partial u}{\partial \theta} + H \dfrac{\partial v}{\partial \theta} + I \\[2ex] \dfrac{\partial \psi}{\partial \xi} = - \rho v \epsilon \end{array}\right.$$

Where

$$\left\{\begin{array}{l} D = \dfrac{\epsilon r}{\Delta} (2\, uv + \dfrac{b}{r} (r'_\beta + \epsilon' \xi) \\[2ex] E = - \dfrac{\epsilon r}{\Delta} b \\[2ex] F = - \dfrac{\epsilon r}{\Delta} [(v - cr)(uv + \dfrac{b}{r} (r'_\beta + \xi \epsilon')] + du \\[2ex] A = - \dfrac{(r'_\beta + \epsilon' \xi)}{r} D + \dfrac{\epsilon}{r} \\[2ex] B = - \dfrac{(r'\beta + \epsilon' \xi)}{r} E \end{array}\right.$$

$$\left\{\begin{aligned} C &= -\frac{(r'\beta + \epsilon'\xi)}{r}\, F - \frac{v\epsilon}{r} + c\,\epsilon \\ G &= \frac{\epsilon P^{1/\gamma}}{\phi(\psi)} \left[-\frac{v}{r} + \left(\frac{v}{r} \frac{(r'\beta + \epsilon'\xi)}{\epsilon} - \frac{u}{\epsilon} \right) D \right] \\ H &= \frac{\epsilon P^{1/\gamma}}{\phi(\psi)} \left(\frac{v}{r} \frac{(r'\beta + \epsilon'\xi)}{\epsilon} - \frac{u}{\epsilon} \right) E \\ I &= \frac{\epsilon P^{1/\gamma}}{\phi(\psi)} \left(\frac{v}{r} \frac{(r'\beta + \epsilon'\xi)}{\epsilon} - \frac{u}{\epsilon}\, F + \frac{v^2}{r} \right) \end{aligned}\right.$$

Where

$$\left\{\begin{aligned} a &= \gamma\, P^{(\gamma-1)/\gamma}\, \phi(\psi) - u^2 \\ b &= \gamma\, P^{(\gamma-1)/\gamma}\, \phi(\psi) - v^2 \\ c &= \frac{\gamma}{\gamma-1} \frac{P}{\phi(\psi)} \frac{d\phi(\psi)}{d\,\psi} \\ d &= \gamma\, P^{(\gamma-1)/\gamma}\, \phi(\psi) \\ \Delta &= ar^2 + 2uvr\,(r'\beta + \epsilon'\xi) + b\,(r'\beta + \epsilon\,\xi)^2 \end{aligned}\right.$$

In the lower region where results has not yet been obtained, the first formulation in (x, ψ) system of coordinates was kept.

The stagnation line is assumed to be diping some angle θ starting normal to the shock wave. The shape will be modified in order to have the same velocity distributions in the lower and the upper regions.

RESULTS FOR THE SECOND MODEL.

In order to validate our program, we decided to first compute symmetric cases (horizontal stagnation line), with air ($\gamma = 1.4$) for which results have been published by BELOTSERKOVSKII (1964) using the method of Integral Relations.

The results shown in Fig.5 are for a Mach number of 4, and the agreement is rather good . We then compute the same problem but with UF_6 ($\gamma = 1.065$) convergence was first very difficult to reach using θ between 0 and 60°. This was mainly due to an important supersonic region. When we reduced the calculations to a range of value between 0 and 40° we got very good results.

If we compare the mechanical properties of UF_6 to those of air, we can see that the coefficient γ is very low $\gamma = 1.065$ ($\gamma = 1.4$ for air) and that the density is about ten times that of air. For this reason the deviation for a given shock wave at a given Mach number is much larger and therefore the shock wave for an given position of normal shock point is stronger (steeper).

In Fig.6, we show the results for different values of the Mach number. As the Mach number increases the shock wave get closer to the scoop and the subsonic region smaller.

Unsymmetric cases were also tried for a Mach number of 4. For those case the shock wave is further from the scoop and steeper.

We still have to compute the flow in the lower region by the Method of Lines but we are confident since Klopfer's results were good for nozzle flows.

We then will have to match the two regions by minimizing the difference between the velocities above and beneath the stagnation line.

CONCLUSION.

We derived an efficient algorithm to compute shock waves in front of blunt bodies even with very low values of γ.

ACKNOWLEDGEMENT.

This work was sponsored by the Département du Génie Isotopique du CEA SACLAY.

REFERENCES.

BELOTSERKOVSKII O.M., Prik.Mat.Mekh 22, 206-219 (1958)

GILINSKII, S.M., TELENIN, G.F., TINYAKOV, G.P.; IZV. Akad. Narck, SSSR Mekh. Mash. 4 9-28 (1964) (Translated as NASA TT F297)

HITTINGER, M., HOLT, M., SOUBBARAMAYER. "Numerical solution of the flow field near a gas centrifuge scoop". Proc 4th Workshop on Gases in Strong Rotation, OXFORD (1981)

HOLT, M. "Numerical methods in Fluid Dynamics", Springer, Verlag BERLIN (1977)

JONES, D.J., SOUTH, J.C., KLUNKER, E.B. J.Comp.Phys. 9, 496-527 (1972)

KLOPFER, G.H., HOLT, M., Proc.Symposium Transsonicum II, GÖTTINGEN 1975, pp.376-383 BERLIN, Springer (1976)

POWELL, M.J.D. Computer Journal, Vol.7, pp.303-307 (1964).

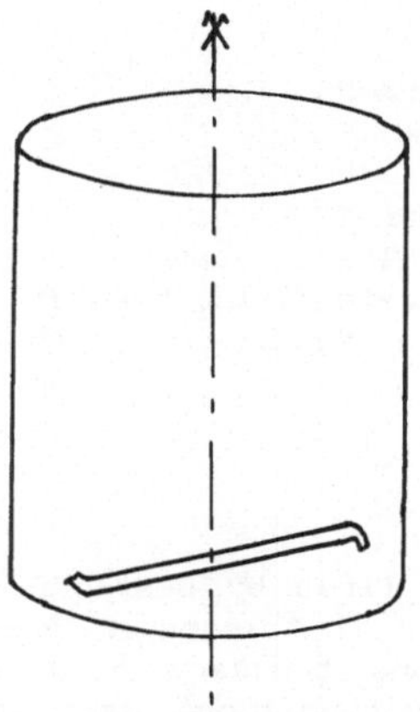

Fig.1 - Type of centrifuge studied

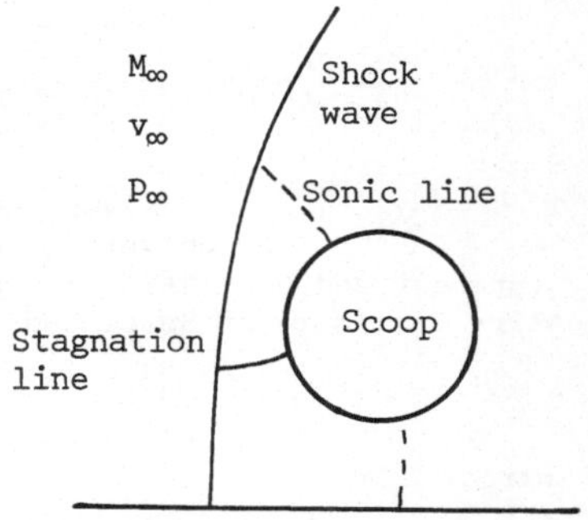

Fig.2 - Description of the flow near the scoop

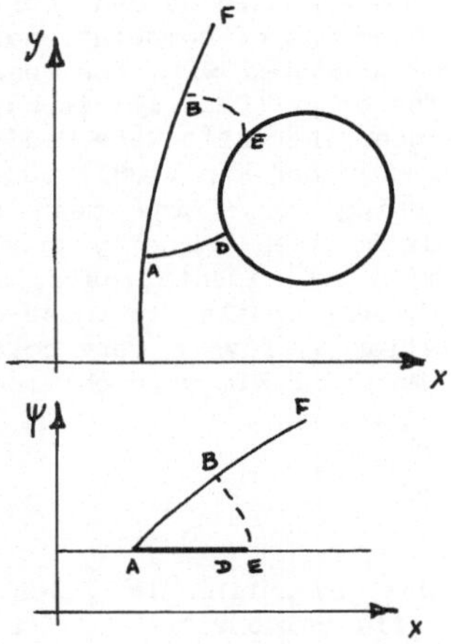

Fig.3 - Transformation of the problem to (x, ψ) plane

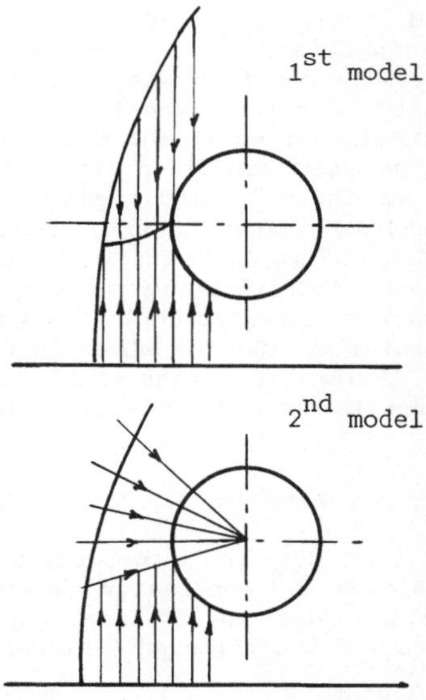

Fig.4 - Two models proposed

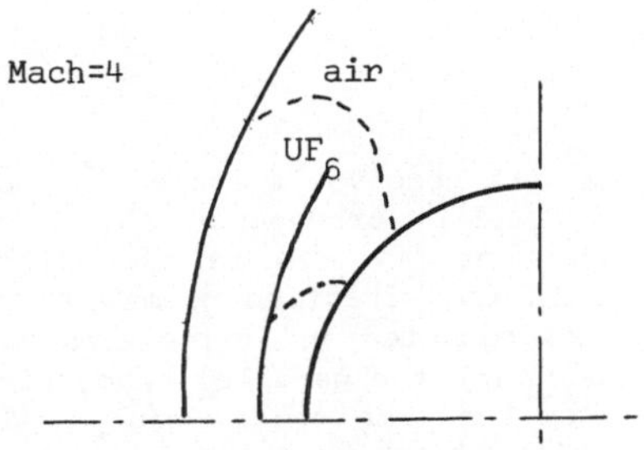

Fig.5 - Symmetric flows for air (γ = 1.4) and for UF_6 (γ = 1.065)

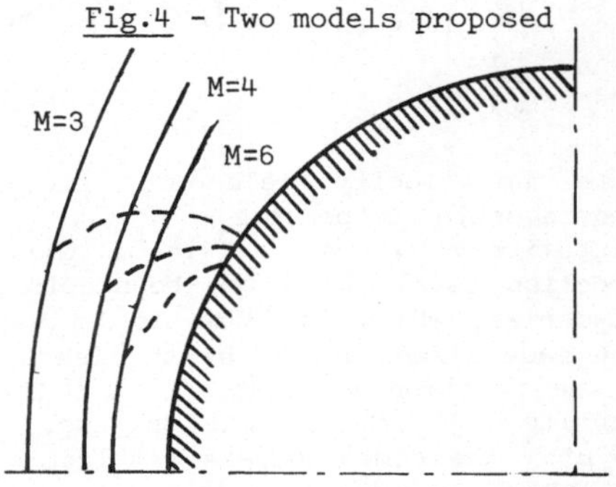

Fig.6 - Symmetric flows for UF_6 for different values of the Mach number.

STUDY OF INCOMPRESSIBLE TURBULENT CHANNEL FLOW BY LARGE EDDY SIMULATION

Kiyosi Horiuti*, Kunio Kuwahara**
*The University of Tokyo, Bunkyo-ku, Tokyo, Japan
(present address: NASA Ames Research Center, Moffett Field, CA94035, USA)
**The Institute of Space and Astronautical Science, Meguro-ku, Tokyo, Japan

§1 Introduction

Large eddy simulation was successfully applied first by Deardorff (1970) to simulate the three dimensional turbulent channel flow. In this paper, the same flow is studied by using improved numerical methods and boundary conditions. In downsteam and spanwise directions, cyclic boundary conditions were imposed for velocity and pressure. Two types of boundary conditions near the wall were used and compared. The one (CASE I) is based on the logarithmic law of the mean velocity near the wall and has a slip boundary condition which was used by Deardorff, where molecular-viscous term was neglected. Large improvements of computational scheme especially in solving the Poisson equation were done compared with the Deardorff's calculation, and burst was observed numerically. The other (CASE II) is based on a no-slip boundary condition, where fine mesh spacing near the walls were used to take account of the molecular viscosity. Very recently, Moin and Kim (1981) studied the same problem under no-slip boundary condition using very fine mesh such as 63x64x128. They got the results that the calculated mean velocity profile and turbulence statistics are in very good agreement with experimental data, and that the detailed time-dependent flow structures are strikingly similar to those observed experimentally. On the other hand, present calculation employs a very coarse grid such as 16x16x21 but essentially same results with Moin and Kim were obtained. The final results show not only burst but sweep also.

§2 Filtered Momentum and Continuity Equations.

We consider an incompressible flow whose time evolution is given by the Navier-Stokes and continuity equations for the velocity components u_i , $i = 1,2,3$ and the pressure p . After applying filtering operator to these equations, the equations for the filtered field are obtained as follows,

$$\frac{\partial \bar{u}_i}{\partial t} + \frac{\partial}{\partial x_j}\overline{u_i u_j} = -\frac{1}{\rho_0 u^{*2}}\frac{\partial \bar{P}}{\partial x_i} + \frac{1}{Re}\nabla^2 \bar{u}_i \; , \tag{1}$$

$$\frac{\partial \bar{u}_i}{\partial x_i} = 0 \; , \tag{2}$$

where the velocity field u_i are decomposed into its resolvable-scale $\bar{u}_i$ and subgrid-scale components u_i' . Here, $i,j = 1,2,3$ corresponds to x,y,z respectively, where x is in the downstream direction, y is in the lateral direction parallel to the boundaries, and z is in the direction normal to the boundaries, which is taken to be vertical. The co-ordinates and velocities have been made dimensionless by the length scale h separating the parallel boundaries, by the friction velocity $u^* = (\tau/\rho_0)^{1/2}$, and by the time scale h/u^* . The quantity τ is the wall stress; ρ_0 , the density which is assumed constant; $p/\rho_0 u^*$, the dimensionless resolvable pressure; Re , the Reynolds number u^*h/ν ; ν , the kinematic viscosity.

In the present study, Reynolds' averaging assumption has been applied, then

$$\overline{u_i u_j} = \overline{u_i}\,\overline{u_j} + \overline{u_i' u_j'} \; . \tag{3}$$

The method of evaluation used here is

$$\overline{u_i'u_j'} - \frac{1}{3}\delta_{ij}\overline{u_k'u_k'} = -K\left(\frac{\partial \overline{u}_i}{\partial x_j} + \frac{\partial \overline{u}_j}{\partial x_i}\right) \tag{4}$$

where K is the SGS eddy coefficient. In this study, the assumption of Smagorinsky et al.(1965) for K is used:

$$K = (c\Delta)^2 \left[\frac{\partial \overline{u}_i}{\partial x_j}\left(\frac{\partial \overline{u}_i}{\partial x_j} + \frac{\partial \overline{u}_j}{\partial x_i}\right)\right]^{1/2}, \tag{5}$$

where c is a dimensionless constant and Δ is a representative grid interval. We denote the horizontal average of a quantity by angular brackets: $\langle\ \rangle$,and deviation from the horizontal average by ". The horizontal average of (1) for i=3 may be subtracted from (1) to give

$$\frac{\partial \overline{u}_i}{\partial t} = \overline{Q}_i - \frac{\partial}{\partial x_i}(\overline{p}'' - 2x_1) + \frac{1}{Re}\nabla^2\overline{u}_i, \tag{6}$$

where

$$\overline{Q}_i = -\frac{\partial}{\partial x_j}\Big(\overline{u}_i\overline{u}_j + \overline{u_i'u_j'} - \frac{1}{3}\delta_{ij}\overline{u_k'u_k'}$$

$$-\delta_{i3}\delta_{j3}\langle\overline{u}_3^{\,2} + \overline{u_3'^2} - \frac{1}{3}\overline{u_k'u_k'}\rangle\Big), \tag{7}$$

$$p'' = \left(\overline{p}/\rho_0 u^{*2} + \frac{1}{3}\overline{u_k'u_k'} + 2x_1\right) - \langle\overline{p}/\rho_0 u^{*2} + \frac{1}{3}\overline{u_k'u_k'} + 2x_1\rangle, \tag{8}$$

The quantity $\frac{\partial}{\partial x_i}2x_1$ above is the dimensionless gross downstream pressure gradient, and $\overline{u_k'u_k'}/3$ has been added to and subtracted from the Reynolds-stress terms.

§3 Numerical Method for CASE I

The region between boundaries to be treated has a downstream length of 3.2h and a lateral width of 0.8h. The downstream length was subdivided into 32 equal grid intervals, the lateral width into 16 and the height into 20 equal grid intervals.

The grid points for the different dependent variables do not coincide but are staggered. The positions of the grid points nearest to the wall are in the logarithmic layer, the viscous sublayer is therefore not treated explicitly. And the molecular viscous terms in (6) are neglected. The pressure p is calculated from the Poisson equation.

$$\nabla^2\overline{p}'' = \frac{\partial \overline{Q}_i}{\partial x_i} - \frac{\partial}{\partial t}\left(\frac{\partial \overline{u}_i}{\partial x_i}\right). \tag{9}$$

Although the last term in (9) vanishes because of (2), it is retained here to cancel out numerical errors.

The Adams-Bashforth method was employed for the time differencing of the marching equation (6) with respect to the advective terms in Q . Time step was set at Δt=0.0025 , a value sufficiently small to avoid the conditional numerical instabiltiy of the advective type. In the Poisson equation for p , the right-hand-side is expressed so as to satisfy the continuity equation at the next step, as it should be with incompressibility, but retaining that of n-step, where n t=t .

In downstream and spanwise directions, cyclic boundary conditions were imposed for velocity and pressure. No-slip boundary conditions cannot be used here at walls, so at $z = 1/2\Delta z$ and $1 - 1/2\Delta z$, which is the distance from the wall to the first grid points for u and v , the mean flow should obey the logarithmic law of the wall. The boundary conditions were given according to the logarithmic law of the wall. The initial conditions were given as the superposition of random fluctuation, whose magnitudes were twice as large as those observed experimentally by Laufer (1950), to the mean velocity profile observed also by Laufer. The finite difference form in (9) employed by Deardorff was the compact 7-point operator, and the method used for solving (9) was successive point over-relaxation. In the present study, (9) is approximated by the use of discrete-Fourier transformation, which made the present calculation much more accurate and effective.

§4 Numerical Method for CASE II

This chapter is devoted to the calculation that takes account of the no-slip boundary condition. In order to include the molecular viscous terms, several mesh points must be placed into the viscous sublayer. For the sake of the efficient calculation, nonuniform mesh system in vertical direction was used. And non-iterative Crank-Nicolson type implicit scheme was adopted.

The filtered Navier-Stokes and continuity equations are rewritten as follows.

$$\overline{u}_t = -D\overline{u} - \overline{p}_x'' + (2\langle K\rangle + \frac{1}{Re})\overline{u}_{xx} + (\langle K\rangle + \frac{1}{Re})\overline{u}_{yy} + (\langle K\rangle + \frac{1}{Re})\overline{u}_{zz} + 2 ,$$

$$\overline{v}_t = -D\overline{v} - \overline{p}_y'' + (\langle K\rangle + \frac{1}{Re})\overline{v}_{xx} + (2\langle K\rangle + \frac{1}{Re})\overline{v}_{yy} + (\langle K\rangle + \frac{1}{Re})\overline{v}_{zz} ,$$

$$\overline{w}_t = -D\overline{w} - \overline{p}_z'' + (\langle K\rangle + \frac{1}{Re})\overline{w}_{xx} + (\langle K\rangle + \frac{1}{Re})\overline{w}_{yy} + (2\langle K\rangle + \frac{1}{Re})\overline{w}_{zz} ,$$

$$\overline{u}_x + \overline{v}_y + \overline{w}_z = 0 , \tag{10}$$

where in order to avoid the convolution sum later, eddy viscosity K is decomposed into to parts $\langle K\rangle$ and K'' . Here, time integrations for the non-linear terms $D\overline{u}$, $D\overline{v}$ and $D\overline{w}$ are approximated by Adams-Bashforth method and the others are approximated by Crank-Nicolson type implicit scheme.

If we try to solve the equations as CASE I, some ambiguity about the boundary conditions for pressure arises because of the property of the implicit scheme. So it is more appropriate to solve the equations totally than to determine the pressure field separately. Firstly, the coupled linear ordinary differencial equations are obtained by the use of discrete-Fourier transformation in downstream and spanwise directions, where the velocity and pressure are represented as before. Here, to use this transformation, regular mesh system is adopted instead of staggered system.

Thus, a system of linear coupled difference equations with complex coefficients are obtained. The velocity fields are set to be zero at the walls and the pressure at one mesh outside of the walls are determined by extrapolation. Near the walls, Smagorinsky model is not appropriate, and eddy viscosity should be equal to zero at walls. Therefore, van Driest-type modification is adopted. Initial conditions are given as same as CASE I.

§5 Numerical Results for CASE I

In Fig.1 the mean velocity profile and the vertical profiles of total stress are shown for several typical steps. For several hundreds steps from the beginning, the mean velocity increases compared with initial profiles as Deardorff. About 700 steps, the mean velocity begins to be retarded. As a steady state, very flat profile, which agrees with the experimental data was obtained.

The isopleths of $\overline{u}''$, $\overline{w}$, $\overline{v}$ and $\overline{p}''$ in x-z plane are shown in Fig.2. Note that the very crowded contour lines of pressure near the walls appear, which means the large pressure gradients at the walls. In Deardorff (1970), the longitudinal eddies of u-component were found to be elongated in the downstream direction. In the present calculation, such strong tendencies are not observed but both u and v eddies have distinct downstream tilts as Deardorff's. Fig.3 shows the equi-shear lines. After about 700 steps, wavy oscillations near the walls and their stretching into the central region is observed. Fig.4 shows the side views of time lines for every 10 steps from 2550 to 2600 steps. The direction of the mean flow is from left to right, and the lines are released from the wire at $y = 0.4$, The retarded part appears near the walls and their ejection into the central region -- burst -- are simulated which agrees well with experimental observations near the wall by Kim et al (1971).

§6 Numerical Results for CASE II

The downstream length was sudivided into 16 equal grid intervals, the lateral into 16 and the vertical into 21 hyper-tangent type non-uniform intervals, and Re and t was set at 1280, 0.002 respectively.

Results corresponding to Fig.1 and Fig.2 are shown in Fig.5 and Fig.6 respectively. Flattened mean velocity profile was obtained. Fig.7 shows the close-up of the side views of time lines for every 5 steps from 3160 to 3185 steps. The rapid movements of particles toward the wall are recognized in the Figure, and the sweeping process is evidently simulated, which seems to be due to the 'no-slip' instead of 'slip' boundary condition. Fig.8 shows the top views for every 10 steps from 3240 to 3280 steps. The lines are released from the line set at $x = 0.2$, $y = 0 - 0.8$ and $z = 0.05$. The lift-up of horseshoe type vortex begins due to overtaking of the part which has a relatively higher velocity profile with a slower part, whose phase is nearly half a spanwise wave-length shifted from the faster part. Similar patterns are observed in CASE I also. Fig.9 gives the energy balance, which shows very good agreement with Moin and Kim (1981).

§7 Discussions and Conclusions

1. The results reported in Deardorff (1970) are yet transient stage especially due to the lack of accuracy in the solution of the Poisson equation.

2. The molecular viscosity does not play a direct role to the generation of bursts, because they are qualitatively simulated with 'slip' boundary condition as shown in CASE I.

3. There are two types of explanation for the formation of streaks prior to bursts, one is due to the stationary streamwise eddy (Kline et al. (1967)), and the other to the wave pattern (Morrison et al. (1971)). Morrison et al. found a wave propagation in the viscous sublayer. In Fig.8, such waves are observed. In the time lines released upper than those in Fig.8, the patterns are somewhat different from those in Fig.8, and the overtaking patterns are not observed clearly. Instead, an alternating array of high and low speed regions is clearly revealed. Therefore, we can conclude that the waves have proper wave-numbers and convection velocities in downstream and spanwise directions and the inclination angles to downstream direction, which depend on the distance from the walls. The angle is larger in sublayer than in buffer or logarithmic layer, which is in good agreement with the results of Morrison et al.

4. Molecular viscosity plays an impotant role in the process of sweep, which is ascertained by the consideration of continuity of momentum. With slip boundary condition as CASE I, the fluids near the wall have a relatively higher momentum than those with no-slip boundary condition. Therefore, in CASE I, compensation for the lack of momentum due to the ejection of low momentum part to the central region can always be accompanied by itself.

5. Present study shows good agreement with Moin and Kim (1981) concerning about the statistical values, even if the mesh of the present study is as coarse as 16x16x21. Therefore, large-eddy simulation is considered to be a very effective and excellent tool for turbulent flow calculation.

References

1. Deardorff, J.W.(1970): J.Fluid Mech. **41** p453
2. Moin, P. and Kim, J.(1981): NASA TN-81309
3. Smagorinsky,J., Manabe,S. and Holloway,J.L.(1965): Mon. Weath. Rev. **93** p727
4. Laufer,J.(1950): NASA TN-1053
5. Kline,S.J., Reynolds,W.C., Schraub,F.A. and Runstadler,P.W. (1967): J. Fluid Mech. **50** p136
6. Morrison,W.R.B., Bullock,K.J. and Kronauer,R.E.(1971): J.Fluid Mech. **47** p639

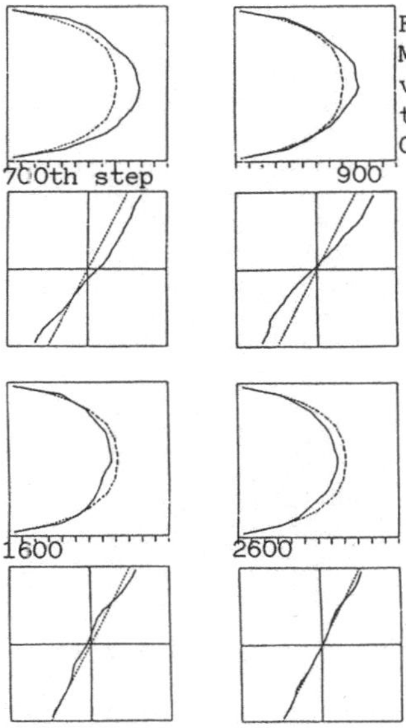

Fig.1
Mean profiles of
velocity and
total stress
CASE I

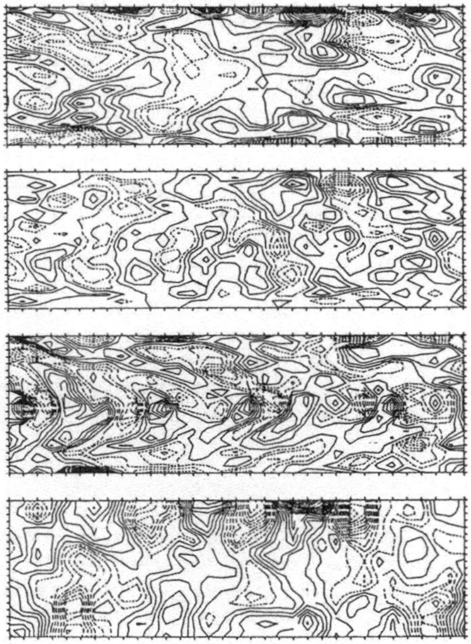

Fig.2
The isopleths for
$\bar{u}''$, $\bar{v}$, $\bar{w}$, $\bar{p}''$
in x-z plane at
700th step
CASE I

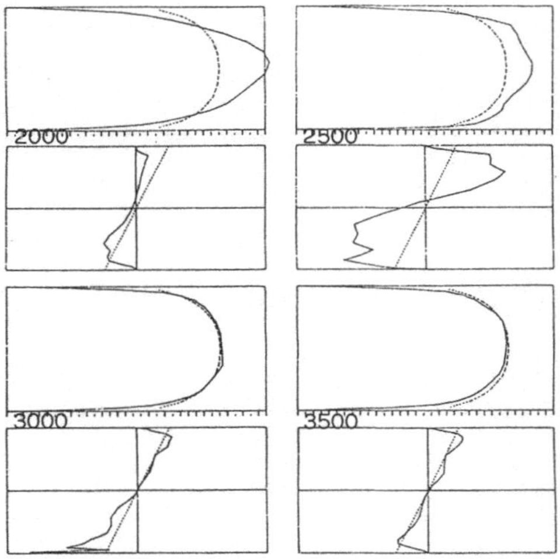

Fig.5 Velocity and total stress , CASE II

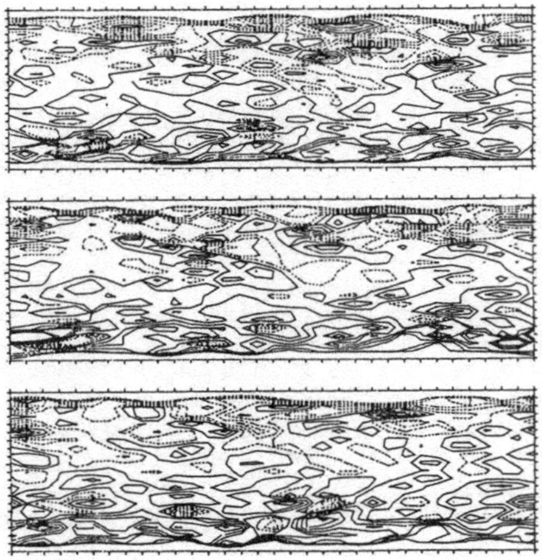

Fig.3 Equi-shear lines, CASE I

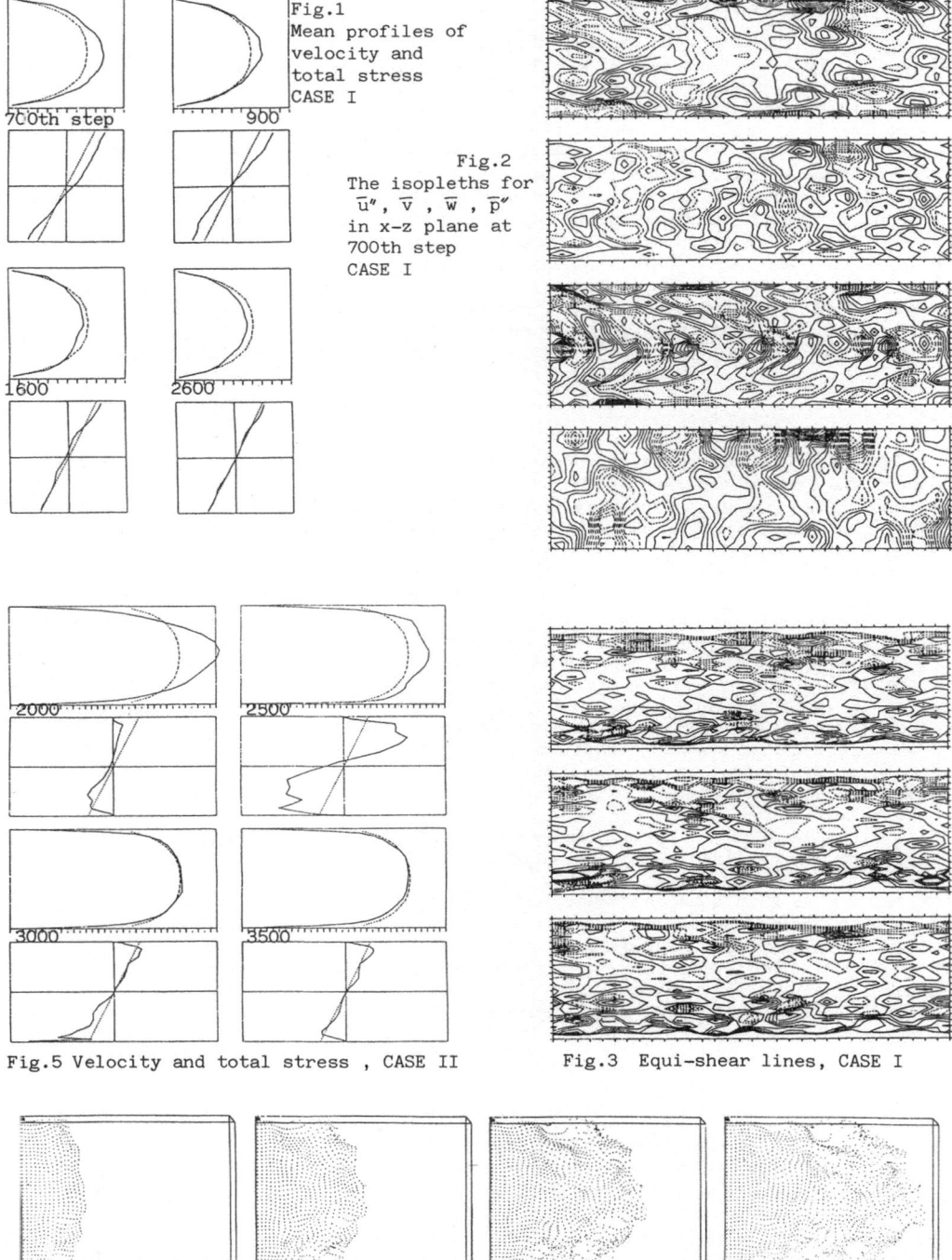

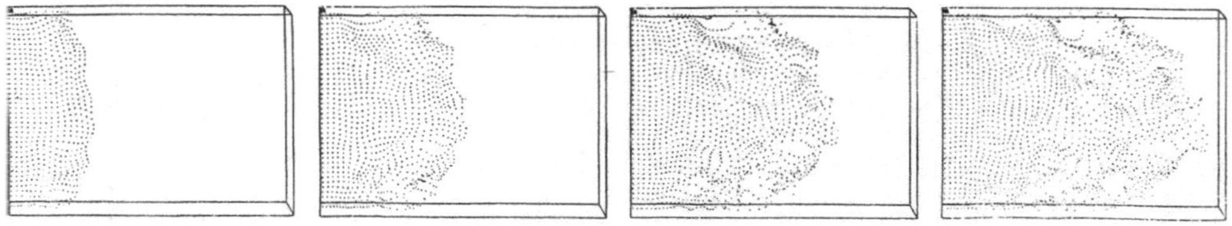

Fig.4 Side view of the time lines, CASE I

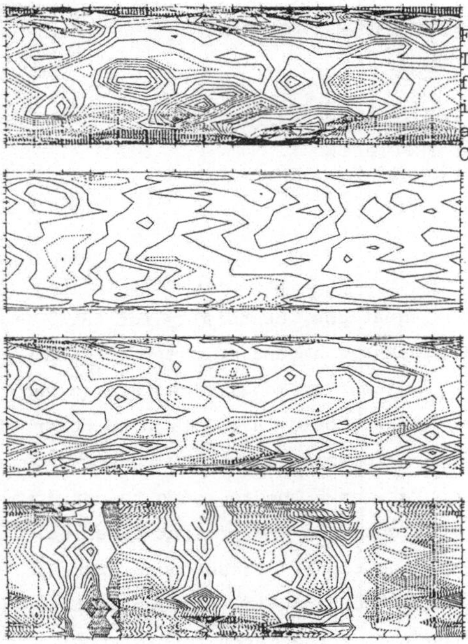

Fig.6
The isopleths
for $\overline{u}'', \overline{v}, \overline{w}, \overline{p}''$
in x-z plane
at 3000th step
CASE II

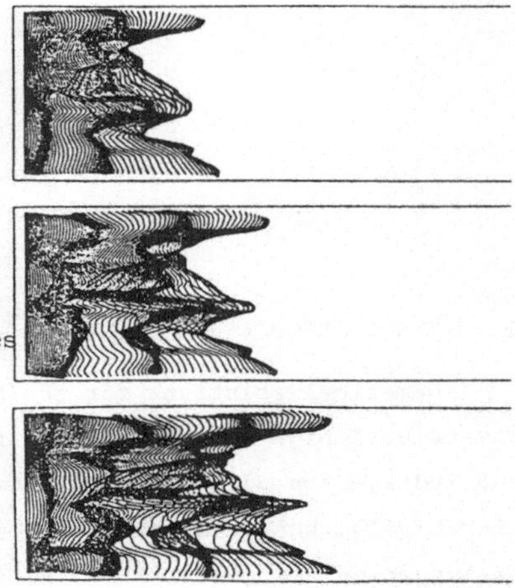

Fig.8
Top view of
the time lines
CASE II

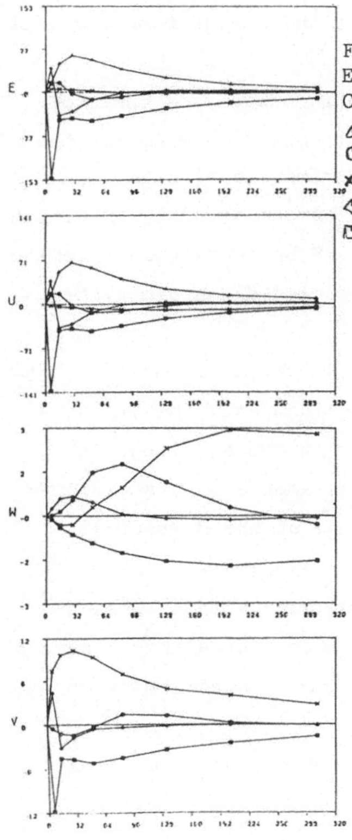

Fig.9
Energy balance
CASE II
△ production
○ velocity transport
× pressure transport
◇ viscous diffusion
□ dissipation

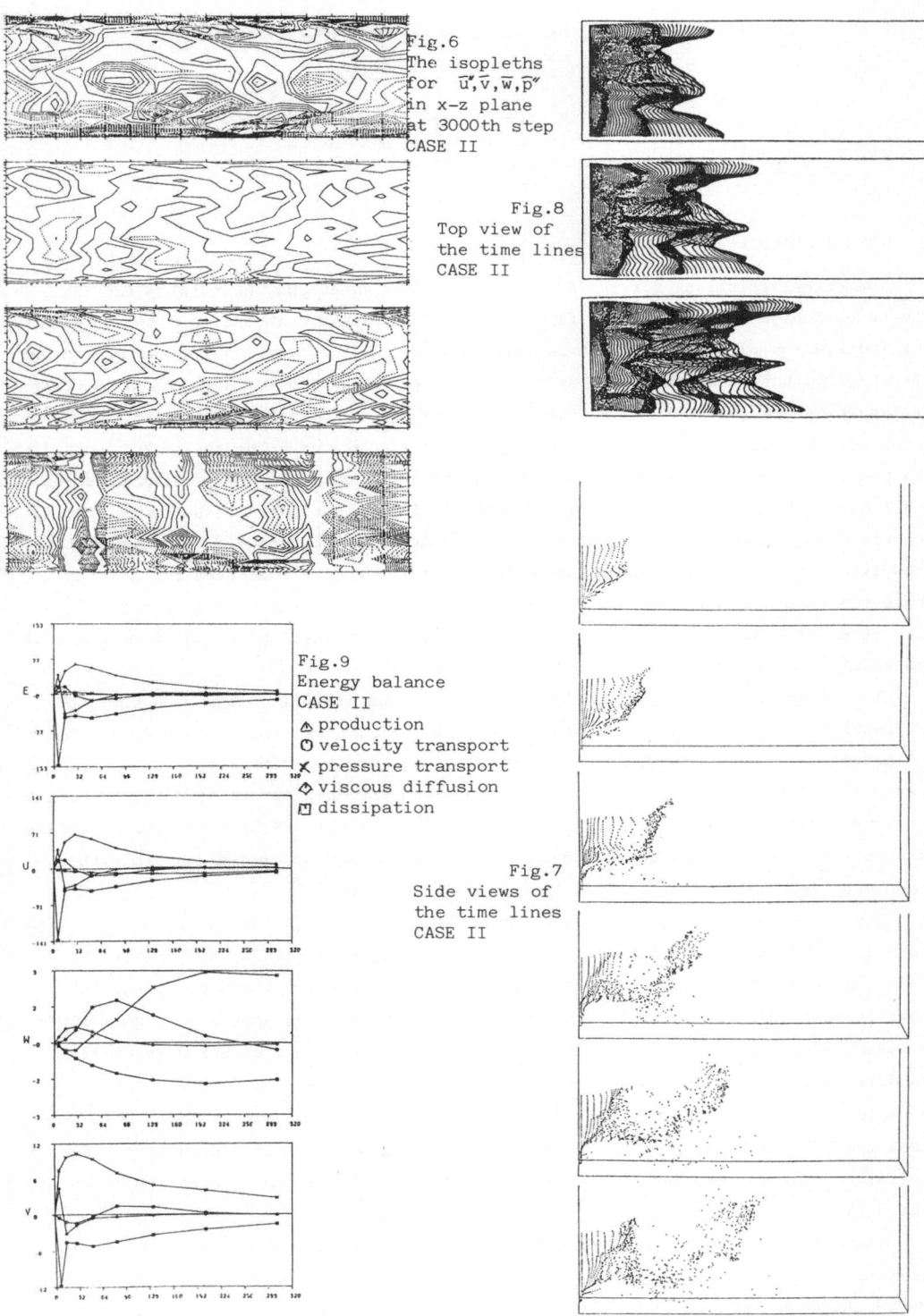

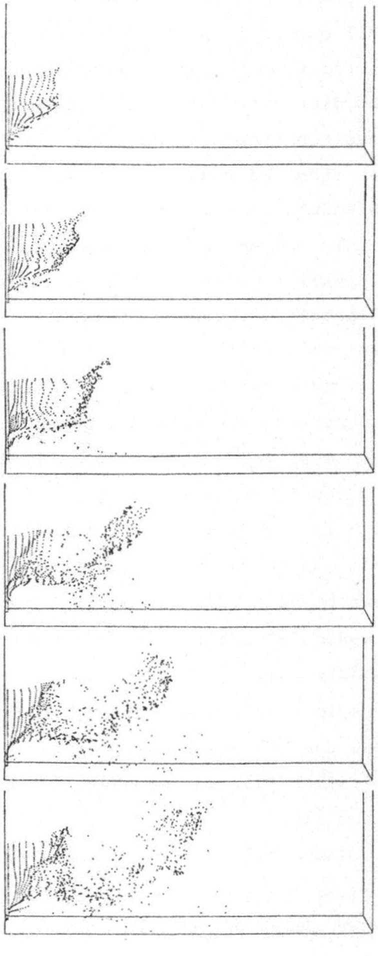

Fig.7
Side views of
the time lines
CASE II

NUMERICAL SOLUTION AND BOUNDARY CONDITIONS FOR BOUNDARY LAYER LIKE FLOWS

M. Israeli and A. Lin
Department of Computer Science
Technion - Israel Institute of Technology
Haifa, Israel

1. INTRODUCTION

Numerical solutions for the Navier-Stokes (NS) equations, especially for three dimensional high Reynolds number flows, require considerable amount of computer time and storage for adequate resolution. Since usually multidimensional flows are solved iteratively, the velocity and pressure fields have to be stored in addition to other fields which are required by the iterative procedure. In some cases slow convergence rates were encountered, resulting in very long computations. One approach to cut down the computer requirements is to use approximate equations. In the present paper we shall consider the "Parabolized Navier-Stokes" (PNS) equations. These equations are derived from the Navier-Stokes equations by neglecting the diffusion effects in the flow direction (in agreement with boundary layer theory). Procedures for solving these equations use marching techniques (also through singular points) along the main direction and result in considerable savings both in storage and computation time, if the solution after one marching sweep is acceptable.

It was shown [1,2] that the PNS equation have an elliptic nature and therefore the initial value problem is not well posed. However, a well posed problem for the PNS equations, can be formulated by specifying, for the two dimensional case, two upstream conditions (for the velocities, say) and one down stream boundary condition (for the pressure, say). Difference approximations to the PNS boundary value problem must reflect its elliptic-parabolic nature; the resulting coupled algebraic system must be solved globally (usually by iterative techniques).

The multiple marching sweep as a global iteration method has the advantage that the velocity fields are generated during the marching process and only the pressure field has to be stored from sweep to sweep. In cases where the PNS equations are known to be a suitable approximation to the NS equations this approach is attractive. An essential ingredient for this to be true is a proper choice of the downstream boundary conditions. In some cases the PNS equations may be inaccurate, like in regions of non-uniformity, while being a fair approximation elsewhere. In such cases the PNS solution can still be used to advantage by incorporation into a numerical scheme for the NS equations on a course grid; this is known as the "Booster" method [3].

Numerical experience shows that the rate of convergence of the usual multiple marching sweep deteriorates quickly as Δx decreases and eventually diverges. In the present paper we develop a new and fast method to solve the PNS (and NS) equations iteratively, and discuss the effect of downstream boundary conditions.

2. THE GOVERNING DIFFERENCE EQUATIONS AND MARCHING PROCEDURE

We consider two dimensional steady incompressible laminar viscous flow along a flat plate. The governing PNS equations are [1]

$$U_x + V_y = 0 \tag{1.1}$$

$$L(U) = -P_x \tag{1.2}$$

$$L(V) = -P_y \tag{1.3}$$

where the conservative form of the operator L is used, i.e.:

$$L(f) = \frac{\partial}{\partial x}(Uf) + \frac{\partial}{\partial y}(Vf) - \frac{1}{R}\frac{\partial^2 f}{\partial y^2} \; . \tag{1.4}$$

R is the Reynolds number based on the plate length. Numerical solutions of eqs. (1) are obtained by spreading a grid over the computational domain. Let us assume that the grid points are distributed evenly along the x and y coordinates with the spacing Δx and Δy respectively. Using the conservative form (1.4) eqs. (1) are differenced at the point (i,j) as follows:

$$U_{ij} + U_{ij-1} - U_{i-1,j} - U_{i-1,j-1} + 2\frac{\Delta x}{\Delta y}(V_{ij} - V_{ij-1}) = 0 \tag{2.1}$$

$$U^2_{ij} - U^2_{i-1,j} + \frac{\Delta x}{2\Delta y}[(UV)_{ij+1} - (UV)_{ij-1}] - \frac{\Delta x}{R\Delta y^2}(U_{ij+1} - 2U_{ij} + U_{ij-1}) = $$
$$= P_{ij} - P_{i+1,j} \tag{2.2}$$

$$(UV)_{ij} + (UV)_{ij+1} - (UV)_{i-1,j} - (UV)_{i-1,j-1} + 2\frac{\Delta x}{\Delta y}[(V_{ij+1})^2 - (V_{ij})^2] +$$
$$+ \frac{2}{R\Delta y}(U_{ij+1} - U_{ij} - U_{i-1,j+1} + U_{i-1,j-1}) = 2\frac{\Delta x}{\Delta y}(P_{ij} - P_{ij+1}) . \tag{2.3}$$

Here a forward difference is used for the P_x term in the U momentum equation (1.2) [1], where $P_{i+1,j}$ is treated as known. This scheme results in an unconditionally stable ("departure free") marching. During the marching the non-linear algebraic system is solved at every station i by the full Newton method [1]. Obviously, the results after one marching sweep are strongly dependent on the initial pressure field.

3. GLOBAL ITERATIVE METHODS

Numerical experiments with the present computer code show that the solution after one marching sweep is not close to the final solution of the PNS equations, when the initial pressure field is constructed using the boundary layer assumption $P_y = 0$. The simplest iterative technique to solve the equations is by multiple marching sweeps with the primitive equations (2), where only the pressure field is kept from iteration to iteration [1]. Numerical experiments also show that for certain nets this procedure diverges. The divergence occurs also for the linearized version of eqs. (2).

Fig. (1) presents the residium of the pressure field as function of the global iteration's sweep number for a 21 ×11 field. A jump is encountered every 10 iterations (probably related to the arrival of a boundary pressure pulse, traveling at the numerical scheme speed) leading to ultimate divergence. However, convergence is reported when combining the above procedure with Multigrid techniques [4]. It was thought that the replacement of one of the momentum equations by the Poisson equation for the pressure will improve the convergence properties. Indeed, the two resulting schemes converged at a satisfactory rate, but the solution did not satisfy the replaced momentum equations. A successful implementation of the marching technique is derived in the next section.

4. THE QUASI-PRIMITIVE ITERATIVE SCHEME

The first step is the derivation of a Laplace equation for the pressure from a constant coefficient version of the system of difference equations (2). The vectors U_m, V_m, P_m, contain the N values of the corresponding variables on the m-th line (x = constant) of the marching sweep (including the specified boundary values). Elimination of V_m between the continuity and V-momentum equations will result in:

$$P_m = D(U_m - U_{m-1}) + E(U_m - 2U_{m-1} + U_{m-2}). \tag{3}$$

On the other hand the U-momentum can be written in the form:

$$P_{m+1} - P_m = R_m = F(DU_m + E(U_m - U_{m-1})) \qquad m = 2,3,.... \tag{4}$$

Here D,E, F are square matrices of order N.

From (3) and (4) we obtain: $FP_m = R_m - R_{m-1}$. (5)

Subtracting successive U-momentum equations and using (4) gives:

$$P_{m+1} - (2I + F)P_m + P_{m-1} = 0, \qquad m = 3,4,.... \tag{6}$$

which is Laplace's equation. The first equation of (4) can be used as a derivative condition at the left boundary namely: $P_3 - P_2 = R_2$. (7)

We now apply the SLOR Scheme to (6) and (7) (ignoring temporarily the downstream boundary condition) to get the downstream marching form:

$$-P_2^* + P_3^{(k-1)} = R_2 \tag{8}$$

$$P_{m-1}^{(k)} - (2+F)P_m^* + P_{m+1}^{(k-1)} = 0, \qquad m = 3,4,.... \tag{9}$$

where $P_m^{(k)} = \omega P_m^* + (1-\omega)P_m^{(k-1)}$, ω is the over relaxation factor, and the superscript denotes the iteration sweep number. In order to recover the primitive variable formulation we relate the velocity field in (5) to the starred pressure field, i.e.

$$FP_m^* = R_m - R_{m-1} . \tag{10}$$

Substitution in (9) gives: $P_{m-1}^{(k)} - 2P_m^* + P_{m+1}^{(k-1)} = R_m - R_{m-1}$, $m = 3,4,....$ (11)

Successive summations of (8) with (9) gives: $P_{m+1}^{(k-1)} - P_m^* = R_m + S_m$, $m = 2,3,...$ (12)

which is the primitive variable marching form of the U-momentum equation. The source term S_m in (12) satisfies, $S_2 = 0$ and:

$$S_m = S_{m-1} + (P^*_m - P_m^{(k-1)}) + (P^*_{m-1} - P_{m-1}^{(k)}), \qquad m = 3,4,... \tag{13}$$

S_m vanishes upon convergence. The computational form of (12) for $m = 3,4,...$ is:

$$-2P^*_m = R_m + \tilde{S}_m; \quad \tilde{S}_m = \tilde{S}_{m-1} - P_{m+1}^{(k-1)} + 2P^*_{m-1} - P_{m-1}^{(k)}; \quad \tilde{S}_2 = P^*_2 - P^*_3 . \tag{14}$$

Thus, the theory of over relaxation can be applied exactly to the constant coefficient case of system (2).) For the non-linear case this theory can serve as a guide to the choice of ω. Alternately, one can choose $\omega = 1$ and apply the multigrid procedure.

5. BOUNDARY CONDITIONS

The problems considered here simulate boundary layer like flows, where the inflow conditions (at $X/L = 0$) are assumed to be known. The edge conditions should be determined by viscous-inviscid interaction with the potential flow. However, we assume here that the edge conditions can be determined from local relations, and concentrate on the end (downstream) conditions. These conditions should allow the flow far downstream to reattach and behave like a boundary layer.

We start with the PNS system where only a downstream boundary condition for the pressure is needed. Since P_y is small in boundary layers, a possible condition would be:

$$P(L,y) = P(L,\infty), \tag{15}$$

which is equivalent to omission of the normal pressure gradient in the normal momentum equation (2.3). But, this is not a trivial assumption since the P_y term is of the same order of magnitude as the other terms in (2.3). A more suitable boundary condition can be derived by writing the P_x terms in eq. (2.2) in the form:

$$P_x(L,y) = P_x(L,\infty) - \int_0^y P_{xy} dy .$$

Since the integral term is small compared to other terms in eq. (2.2), we get for $X = L$ the end condition: $L(U) = - P_x(X = L,\ y = \infty)$. (16)

The boundary conditions for the NS equations that will best approximate the present situation and result in a well posed problem are not obvious. Motivated by boundary layer scaling, we can choose (15) or (16) as one of the downstream conditions. The choice of the second boundary condition is motivated by vorticity transport considerations; the vorticity is advected downstream by the main flow while the back diffusion is small. It follows that specifying an incorrect downstream vorticity will modify the flow only locally, in a thin boundary layer near the exit. In such a boundary layer the scale will be of the order of the inverse of the Reynolds number and the main contribution to the vorticity will come from the x derivatives. We expect that specification of $V_{xx}=0$; $U_{xx}=0$ at the exit will minimize the local generation of vorticity. This amount to using the PNS equations at $X = L$ together with (15) or (16).

6. RESULTS

First we will establish the need for the pressure field iteration procedure. For a 41×41 grid, fig. 2 shows that the values of the maximum V velocity after the first sweep are 3 to 5 times smaller than those of the converged solution. This figure depicts also the effect of the downstream boundary condition. The maximum V velocities for the two boundary conditions (15) and (16) agree initially, however, the difference grows to about 27% at the plate's end. The maximum differences in the wall shear is about 15%.

The edge is assumed to be in the potential region where two conditions out of the following three can be chosen: (i) specified streamwise velocity, (ii) static pressure (iii) stagnation pressure. It turns out that the difference between the various choices for the edge condition is small (since the magnitude of the square of the normal velocity is negligible).

The location of the maximum of the normal velocity, V_{max}, can be considered to be the edge of the viscous region or the contact point between the lower and the middle deck in the triple deck theory. Thus, it is reasonable to compare the V_{max} values of the PNS equations to those of the Blasius solution. It can be seen from fig. 2 that V_{max} of the Blasius solution changes more rapidly with x than that of the PNS equations; the first behaves like $x^{-\frac{1}{2}}$, while the last behaves exponentially for large x.

Fig. 3 compares the pressure overshoot for the boundary conditions (15) and (16). There is a maximum in both pressure profiles due to the viscous interaction, this maximum is closer to the wall than that of V. We note that the considerable difference in the pressure for the two boundary conditions is felt throughout the whole computational domain.

Fig. 4 shows the average rate of convergence as a function of the overrelaxation factor ω, for problem (1) and for its constant coefficient version. Figures (4a) and (4b) correspond to the end conditions (15) and (16).

For the linearized version with Dirichlet end condition the dependence on ω agrees with the theory of SLOR for Poisson equation. For the nonlinear case the curves are shifted to the right while keeping a similar shape. There is a strong dependence on Δx and a weak dependence of Δy (compare the curves for the 21×3 grid to those of 21×11), thus low resolution runs can be used to estimate ω. The end condition (16) gives results similar to those of the previous case but the curves are shifted to the left.

Fig. 5 can serve as a justification for the use of the PNS equations. It can be seen that even for R=100, the difference between the NS and PNS solutions is quite small away from the end stations.

ACKNOWLEDGEMENT

The first author was supported by Air Force contract no. 77 - 3405

REFERENCES

[1] Rubin, S.G. and Lin, A., (1980), Marching with the Parabolized Navier Stokes Equations, Israel J. of Tech., 18.

[2] Israeli, M., Reitman, V., Salomon, S., Wolfshtein, M., (1981), On the Marching Solution of the Elliptic Equations in Viscous Fluid Mechanics, Proc. of 2nd Int. Conf. on Numerical Methods in Laminar and Turbulent Flows, C. Taylor et al. editors, Venice.

[3] Ungarish, M. and Israeli, M., (1980), Improvement of Numerical Schemes by Incorporation of Approximate Solutions Applied to Rotating Compressible Flows, IC7NMFD, Standord.

[4] Rubin, S.G. (1982), Incompressible NS and PNS Solution Procedures and Computational Techniques. Von Karman Institute Lecture Notes.

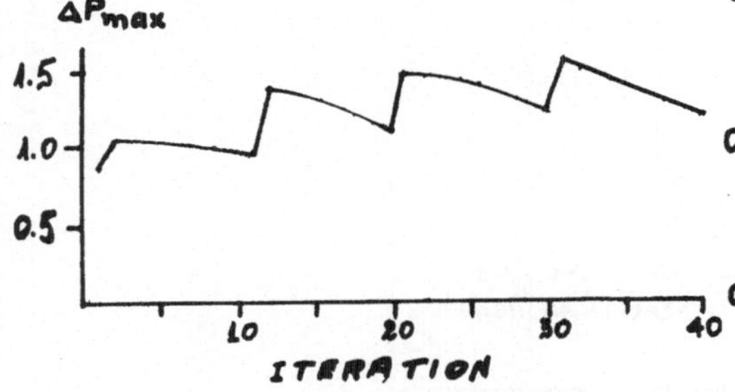

Fig.1 Pressure residue vs the global iteration number.

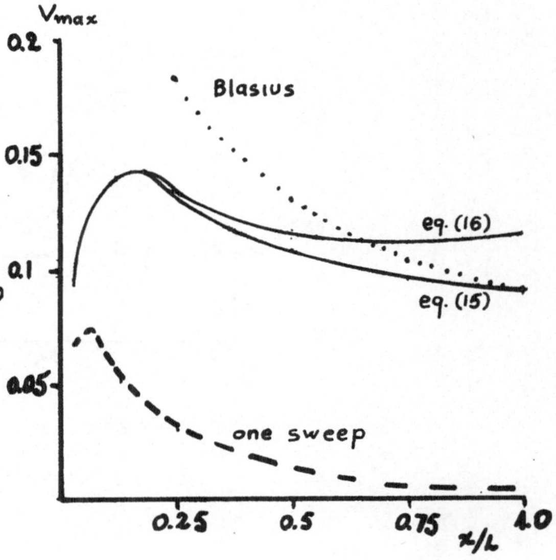

Fig.2 Converge V_{max} profiles.

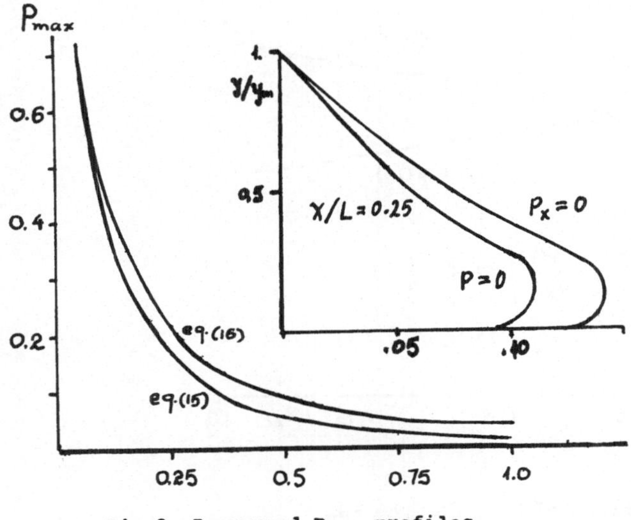

Fig.3 Converged P_{max} profiles.

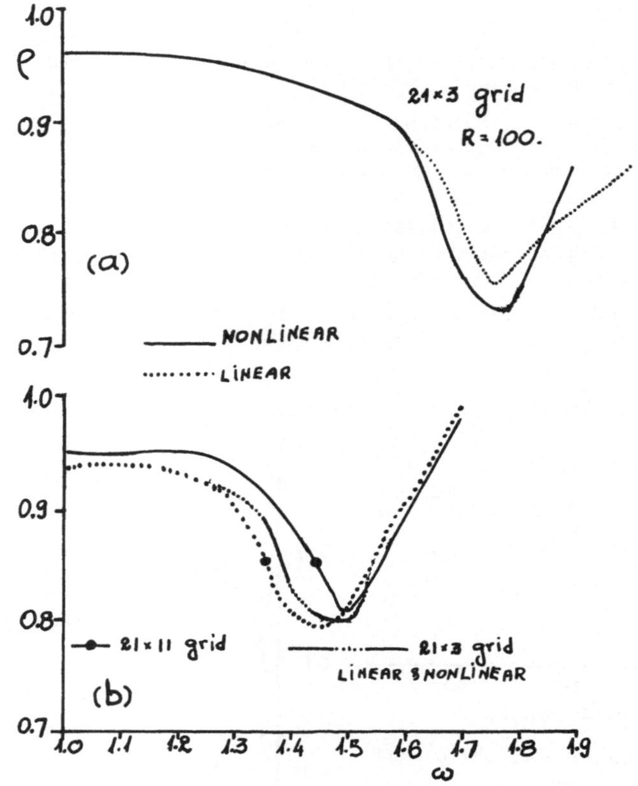

Fig. 4 Rate of Convergence for coarse and fine grids
(a) eq. (15), (b) eq. (16).

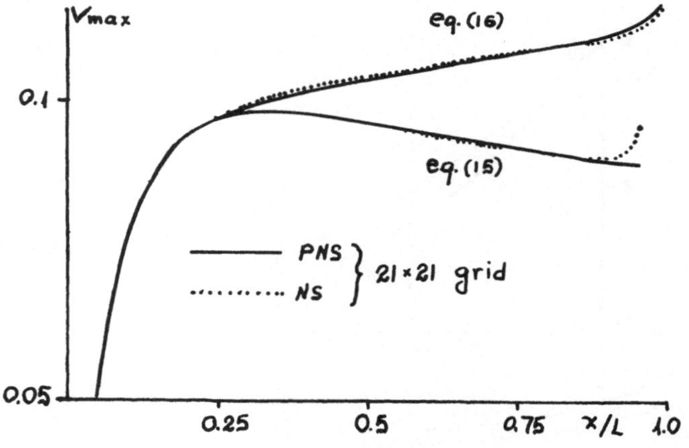

Fig.5 Comparison of V_{max} for the PNS and NS equations.

RELAXATION SOLUTION OF THE FULL EULER EQUATIONS

GARY M. JOHNSON
NASA LEWIS RESEARCH CENTER
CLEVELAND, OHIO 44135, USA

SUMMARY

A numerical procedure for the relaxation solution of the full steady Euler equations is described. By embedding the Euler system in a second-order surrogate system, central differencing may be used in subsonic regions while retaining matrix forms well suited to iterative solution procedures and convergence acceleration techniques. Hence, this method allows the development of stable, fully-conservative differencing schemes for the solution of quite general inviscid flow problems. Results are presented for both subcritical and shocked, supercritical internal flows. Comparisons are made with a standard time-dependent solution algorithm.

INTRODUCTION

Techniques for the time-accurate solution of the unsteady Euler equations are well known and have relatively firm theoretical basis. The methodology for obtaining steady solutions to the Euler equations is in a more formative state and has undergone considerable evolution in the recent past.

Given steady boundary conditions and assuming that a unique steady solution exists, one may solve the unsteady Euler equations in a non-time-accurate fashion by means of an algorithm with a stability bound in excess of the CFL condition or by using local time stepping. These constitute the simplest form of pseudo-unsteady solution procedure. Other methods fix the total enthalpy at its steady-state value or use enthalpy damping, as proposed by Jameson, Schmidt and Turkel (1981), to accelerate convergence. Ni (1981) and Steger (1981) have developed multiple grid schemes for pseudo-unsteady solution of the time-dependent Euler equations. Approaches which modify the unsteady Euler equations by adding either time-dependent terms or time-dependent equations which enhance convergence have been developed by Essers (1980) and Viviand (1981).

Attempts to deal directly with the steady first-order Euler system meet with immediate difficulties. Only centered difference operators will lead to numerical schemes which are simultaneously stable for the entire system in subsonic flow. However, such schemes lead to ill-conditioned matrices which defeat most iterative solution procedures.

Semi-direct methods using Newton iteration to avoid this numerical difficulty have been developed by Rizzi (1979) for the homoenthalpic Euler equations, by Rizzi and Sköllermo (1981) for the full potential equation written as a first-order system, and by Shubin, Stephens and Glaz (1981) for the quasi-one-dimensional Euler equations. The large matrices used by such methods may be an impediment when they are generalized to the full Euler system, or to higher dimensions.

An alternative approach to solving the steady Euler equations involves either embedding them in a higher-order steady system or preconditioning their finite difference representations to enhance their compatibility with iterative procedures. Chattot, Guiu-Roux and Laminie (1981) used a variational approach to transform the first-order system representing the potential equation into an equivalent second-order system which was solved by a conjugate gradient algorithm. Lomax (1981), Desideri and Lomax (1981) and Lomax, Pulliam and Jespersen (1981) described a strategy for solving the Euler equations by preconditioning the finite difference equations, choosing a stable relaxation procedure, and accelerating its convergence by a multiple grid technique. Preconditioned relaxation solutions were obtained for two-dimensional subsonic flow by Desideri and Lomax, while Lomax, Pulliam and Jespersen used a multiple grid procedure to solve a quasi-one-dimensional supersonic flow.

In Johnson (1981) we developed an approach whereby a first-order partial differential system is embedded in a second-order surrogate system which may then be solved by means of the same sort of numerical techniques routinely used on the potential equation. We obtained results with the full Euler system for both supersonic and subcritical, subsonic flow and with the transonic small disturbance equations for both subcritical and supercritical flow. The embedding used with the Euler equations assumed the invertability of flux-vector Jacobian matrices and was thus not suitable for use in transonic flow computations. The present paper presents the details of an improved surrogate equation technique which is capable of treating the full steady transonic Euler equations.

SURROGATE EQUATION TECHNIQUE

Given a first-order partial differential system, we embed this system in a second-order surrogate system, apply additional constraints to restrict the solution set of the surrogate, and solve the resulting partial differential problem by means of a conventional iterative procedure. This method maintains the generality of the Euler equations, while allowing the use of the same sort of relaxation procedures developed for the efficient solution of second-order equations.

Consider a first-order system written in conservation law form, such as

$$\left[\frac{\partial}{\partial x}(A\ ) + \frac{\partial}{\partial y}(B\ ) \right] q = 0$$

where q is an n-component vector and A and B are n x n matrices. We embed this system in a second-order surrogate of the form

$$\left[\frac{\partial}{\partial x}(M\ ) + \frac{\partial}{\partial y}(N\ ) \right] \left[\frac{\partial}{\partial x}(A\ ) + \frac{\partial}{\partial y}(B\ ) \right] q = 0$$

This system preserves the conservation law form of the original first-order system. The nature of the second-order partial differential operator is controlled by the choice of the matrices M and N. For example, the choice $M = A^T$, $N = B^T$ symmetrizes the coefficients of the terms of highest order and causes the surrogate system to be elliptic, while the choice $M = A^T$, $N = -B^T$ results in a non-elliptic system which may be upwind differenced. Alternative choices for M and N are possible but will not be discussed here. The problem specification is completed by requiring that, in addition to satisfying the original boundary conditions of the underlying first-order system, the solution to the surrogate system must also satisfy the first-order equations themselves at the boundaries. This is done to insure uniqueness. Additionally, in the case where we employ an elliptic surrogate system to compute a supercritical flow, this boundary treatment allows the introduction of dissipative terms for proper shock capture.

Observe that, by switching the second-order operator from elliptic to hyperbolic type when the flow changes from subsonic to supersonic, it is possible to create a type-dependent differencing scheme for the surrogate system. Such a scheme could provide an alternative means for insuring the correct shock capture and thus relax some of the constraints on the boundary treatment. While initial computations indicate that this may be a viable approach, the results to be presented subsequently were obtained using the choice $M = A^T$, $N = B^T$ everywhere in the domain. Notice that, in this case, the embedding operator is a close relative of the formal transpose of the Euler operator:

$$- \left[A^T \frac{\partial}{\partial x} + B^T \frac{\partial}{\partial y} \right]$$

A discrete representation of this operator has been independently proposed as a preconditioning operator by Desideri and Lomax. Because they operate on the finite difference equations, their approach is in several additional respects distinct from the one discussed here. For example, preconditioning the finite difference equations results in an effective non-compact differencing of the second-order system. Permutation of the resulting matrices is required to restore compact structure and reduce bandwidth. Furthermore, the surrogate equation technique appears to offer more flexibility in the treatment of boundary conditions than is available with the preconditioned finite difference equation approach.

The second-order partial differential problem, being compatible with iterative techniques, may be solved by a variety of methods. For demonstration purposes, we use fully-conservative differencing together with the well-established successive line relaxation method.

RESULTS

We compute subcritical and shocked, supercritical flows in a straight channel with a 10% half-thick circular arc airfoil mounted on its lower wall. The second-order partial differential problem is illustrated in Fig.1. As physical boundary conditions, we require that flow tangency be satisfied at solid walls, we specify total pressure, total temperature and flow angle at the inlet, and we specify the exit static pressure. We require that the first-order Euler equations be satisfied at each boundary to provide the additional boundary conditions necessary to completely pose the problem and to insure the correct shock capture. As a standard of comparison for the accuracy of the results presented here, we have recomputed all cases using the explicit MacCormack (1969) algorithm.

The subcritical test case had an isentropic inlet Mach number of 0.5. Fig.2 shows the comparative upper and lower surface Mach number distributions. Isomach contours are plotted in Fig.3. The supercritical test case was run at an inlet Mach number of 0.675, producing a shocked but unchoked flow. The comparative surface Mach number distributions and isomachs are shown in Figs.4 and 5, respectively. The sonic line is dashed in the isomach plots. Comparison of the results of the surrogate equation algorithm with those of the MacCormack algorithm is encouraging. Minor discrepancies may be attributed, in part, to differences in the truncation error of the two algorithms or to the lack of complete annihilation of low frequency error modes.

Representative convergence histories are shown in Fig.6. The surrogate equation algorithm, using successive line overrelaxation on the second-order system, converges more rapidly than the MacCormack algorithm, using local time stepping at 0.9 of the local CFL limit. Since the residuals are defined quite differently for the two methods, the error measure used for this comparison is the correction to the vector of conservation variables. Consequently, the relative convergence rates are more significant than the indicated levels of error. Having demonstrated the capability of the surrogate equation technique to solve the full Euler equations by relaxation, it should prove relatively straightforward to further accelerate convergence.

While all results presented here are two-dimensional, the extension of the surrogate equation technique to three dimensions presents no essential difficulties. Furthermore, the use of this technique as an inviscid component of a zonal procedure for the iterative solution of the steady Navier-Stokes equations appears feasible.

CONCLUSIONS

We may obtain a solution to the full steady transonic Euler equations by using a surrogate second-order system together with the original Euler physical boundary conditions and additional constraints obtained from the first-order Euler system.

This surrogate equation technique provides a means for formulating problems involving the full steady Euler equations in such a way as to allow the use of stable, fully-conservative differencing and relaxation solution procedures. Hence, we may solve either irrotational or rotational flow problems across the entire spectrum of subsonic, transonic and supersonic conditions without resort either to derived dependent variables, semi-direct methods, or to an unsteady formulation.

Embedding the Euler equations in a second-order system allows the application of the many convergence acceleration techniques which have been developed for other second-order systems. Thus, the surrogate equation technique provides an opportunity for the construction of fast and efficient numerical procedures for the solution of the full steady Euler equations.

ACKNOWLEDGEMENT

The encouragement and advice of Prof. J.L. Steger are gratefully acknowledged.

REFERENCES

Chattot, J.J., Guiu-Roux, J. & Laminie, J.: Lect. Notes in Phys., Vol 141, pp 107-112, Springer, 1981.

Desideri, J.A. & Lomax, H.: AIAA Paper 81-1006, 1981.

Essers, J.A.: Compu. Fluids, Vol 8, pp 351-368, 1980.

Jameson, A., Schmidt, W. & Turkel, E.: AIAA Paper 81-1259, 1981.

Johnson, G.M.: Lect. Notes in Phys., Vol 141, pp 236-241, Springer, 1981.

Lomax, H.: AIAA Paper 81-0994, 1981.

Lomax, H., Pulliam, T.H. & Jespersen, D.C.: AIAA Paper 81-1027, 1981.

MacCormack, R.W.: AIAA Paper 69-354, 1969.

Ni, R.H.: AIAA Paper 81-1025, 1981.

Rizzi, A.: Lect. Notes in Phys., Vol 90, pp 460-467, Springer, 1979.

Rizzi, A. & Sköllermo, G.: Lect. Notes in Phys., Vol 141, pp 349-353, Springer, 1981.

Shubin, G., Stephens, A. & Glaz, H.: J. Comput. Phys., Vol 39, pp 364-374, 1981.

Steger, J.L.: NASA CR 3415, 1981.

Viviand, H.: Lect. Notes in Phys., Vol 141, pp 44-54, Springer, 1981.

FIGURES

Tangency Condition

$$\left[\frac{\partial}{\partial x}(A\ ) + \frac{\partial}{\partial y}(B\ ) \right] q = 0$$

p_o specified
T_o specified
v/u specified

$$\left[\frac{\partial}{\partial x}(A\ ) + \frac{\partial}{\partial y}(B\ ) \right] q = 0$$

$$\left[\frac{\partial}{\partial x}(A^T\ ) + \frac{\partial}{\partial y}(B^T\ ) \right] \left[\frac{\partial}{\partial x}(A\ ) + \frac{\partial}{\partial y}(B\ ) \right] q = 0$$

p specified

$$\left[\frac{\partial}{\partial x}(A\ ) + \frac{\partial}{\partial y}(B\ ) \right] q = 0$$

Tangency Condition

$$\left[\frac{\partial}{\partial x}(A\ ) + \frac{\partial}{\partial y}(B\ ) \right] q = 0$$

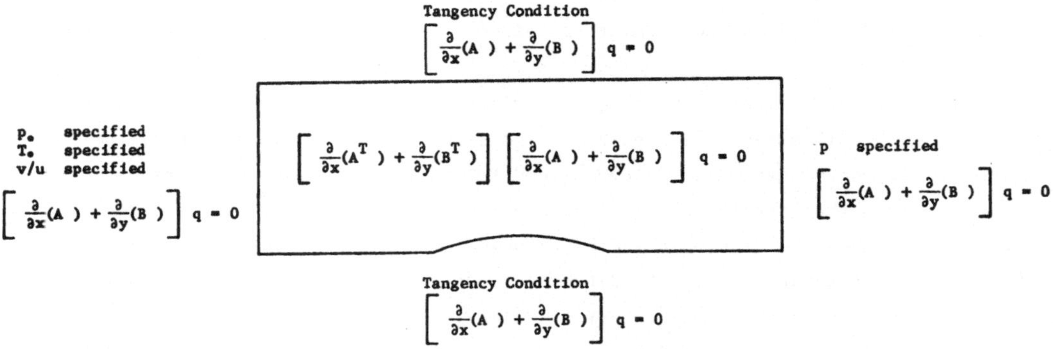

Fig. 1 Partial Differential Problem

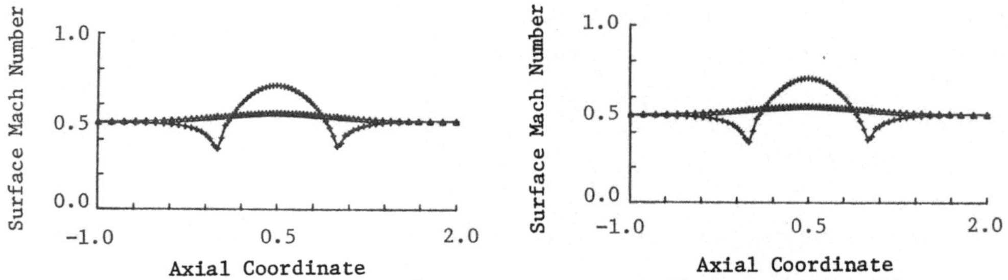

Surrogate Equation Algorithm

MacCormack Algorithm

Fig. 2 Subcritical Mach Number Distributions

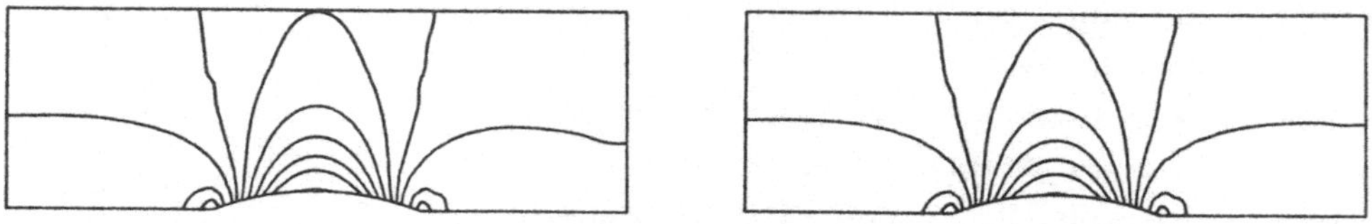

Surrogate Equation Algorithm   MacCormack Algorithm

Fig. 3 Subcritical Isomach Contours

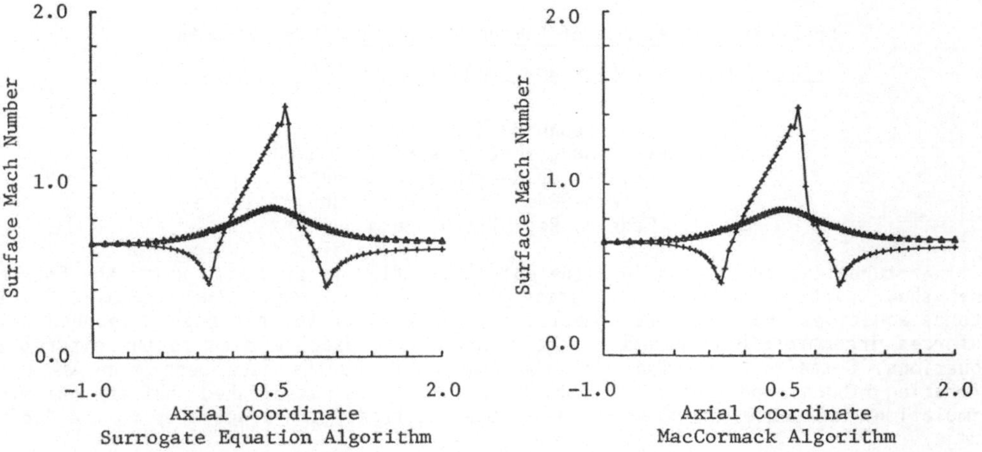

Fig. 4 Supercritical Mach Number Distributions

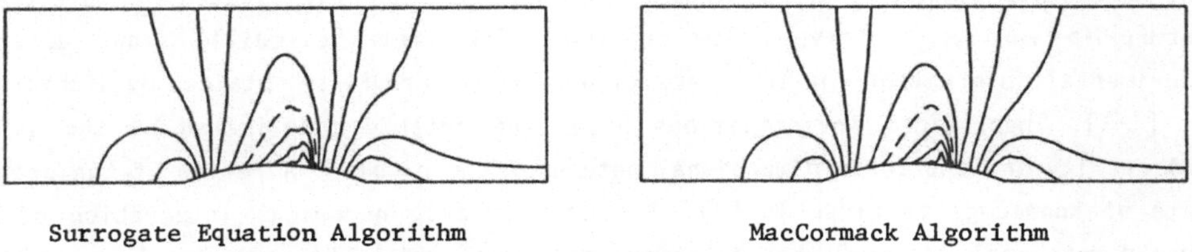

Fig. 5 Supercritical Isomach Contours

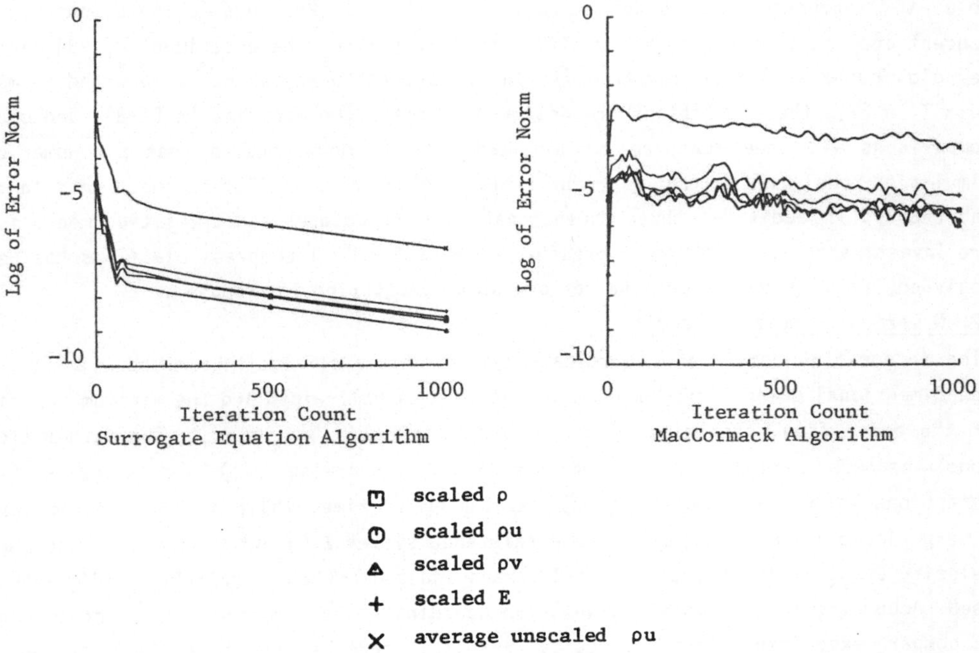

Fig. 6 Convergence Histories

Spectral Simulations of Laminar-Turbulent Transition in Plane Poiseuille Flow and Comparison with Experiments

Leonhard Kleiser
Kernforschungszentrum Karlsruhe GmbH
Institut für Reaktorentwicklung
Postfach 3640, D-7500 Karlsruhe
Federal Republic of Germany

Laminar-turbulent transition in plane Poiseuille flow is simulated using the Fourier-Chebyshev spectral method to integrate the three-dimensional time-dependent Navier-Stokes equations. For pressure computation, a new algorithm has been developed which enforces incompressibility and boundary conditions exactly even in the discretised equations. Detailed comparisons of the numerical results have been made with the vibrating-ribbon experiments of Nishioka et al. It is established that the numerical simulations reproduce the experimentally observed transition process up to the "spike" stage.

1. Introduction

The transition of laminar flows into turbulence has been investigated for more than a century in various prototype flow problems. For plane Poiseuille flow, detailed experimental observations of the transition process were first obtained by Nishioka et al. [1-3]. Theoretical investigations have been greatly complicated by the strong nonlinearity and the three-dimensional nature of the problem. A review of the present state of knowledge is given in [4]. In recent years, numerical integration of the three-dimensional Navier-Stokes equations has been used to investigate the transition problem. However, it has not yet been generally accepted that numerical simulations are able to reproduce the experimental observations. Previous calculations [5,6] concentrated on a parameter range different from that of the experiments [1-3] (lower Reynolds numbers, larger amplitudes). In the present work, which is reported in more detail in [7], the simulations are designed to match the experiments [1-3], and close comparisons with the measurements are made. It is demonstrated that the numerical simulation model used in fact reproduces flow fields quite similar to those seen in the laboratory. In addition, three-dimensional flow structures and transition mechanisms are investigated. The problem of pressure computation in incompressible flows has been analysed, and a general algorithm for pressure computation is proposed.

2. Numerical simulation model

The integration domain and coordinate system of the simulation are shown in fig. 1. Non-dimensional quantities are based on the channel half-width and the maximum velocity of the basic flow. The basic plane Poiseuille flow is $U(x_3)=1-x_3^2$. The mathematical model assumes periodicity in the horizontal streamwise (x_1) and spanwise (x_2) directions with disturbances $\underline{u}(\underline{x},t)$ developing in time, while in the experiments a time-periodic flow develops in x_1. A reference system $\xi_1=x_1-Ct$ moving with the phase velocity $C=c_{TS}$ of the two-dimensional ribbon-induced Tollmien-Schlichting (TS) wave is used. According to this model, downstream coordinate and time have to be interchanged to compare experimental and theoretical results. There has been much discussion on the relevance of this model for the transition occuring in the experiments.

In numerical computations of incompressible flows the treatment of continuity equation and pressure presents special problems. For three-dimensional flows it is common to use the primitive variables and to compute the pressure from a Poisson equation. However, the pressure boundary values are not known a priori. Numerous techniques have been proposed to overcome this difficulty in connection with various special discretisations. In [7] a new algorithm is proposed which is not restricted to a special geometry or numerical discretisation technique. More precisely, the algorithm applies to the Stokes problem arising after time-discretisation of the Navier-Stokes equations with explicit treatment of the nonlinear terms. It is based on the fact that the continuity equation may be replaced equivalently by the Poisson equation for the pressure and the condition div $\underline{u} = 0$ on the boundary of the domain. From this condition, the correct pressure boundary conditions are obtained in the discretised equations via an influence matrix technique. The application of the algorithm to the present channel flow problem is described in [8].

The spatial discretisation is based on a spectral method [9] with Fourier expansions in the horizontal directions

$$\underline{u}(\underline{x},t) = \Sigma\ \Sigma\ \hat{\underline{u}}(k_1,k_2,x_3,t)\ \exp(ik_1\alpha_1x_1+ik_2\alpha_2x_2) \qquad (1)$$

(truncated at $|k_j|<N_j/2$, $j=1,2$), and Chebyshev polynomial expansions of the Fourier coefficients $\hat{\underline{u}}$ in x_3. Nonlinear terms are calculated by the pseudospectral approximation [9]. In time, the Crank-Nicolson method is employed for the viscous terms and the second order Adams-Bashforth method for the nonlinear terms. A variety of other schemes has been used for comparison as well. The numerical method is implemented in the code CHANSON which has been carefully tested by various verification calculations [7].

3. Numerical results

Numerical results are presented for the Reynolds number Re = 5000 and wavenumbers $\alpha_1=2\pi/L_1=1.12$ and $\alpha_2=2\pi/L_2=2.1$ adjusted to the experiments [1-3]. At time t=0 an initial disturbance $\underline{u}° = A_{TS}\underline{u}_{TS}+A_{3D}\underline{u}_{3D}$ is superimposed consisting of a two-dimensional TS wave and a pair of three-dimensional oblique waves with wavenumbers $(\alpha_1,\pm\alpha_2)$ as in the Benney-Lin model [4]. The normalisation of the linear eigensolutions is $\max|u_{1,TS}|=\max|u_{1,3D}|=1$ and zero phase difference of $u_{1,TS}$ and $u_{1,3D}$ at t=0. Both disturbance parts $u_{1,TS}$ and $u_{1,3D}$ are antisymmetric in x_3. The period of one TS-oscillation is $T_{TS}\approx20$. A_{TS} varies from 0.7% to 3%, while A_{3D} is set to 0.1% . The horizontal resolution is $N_1=N_2=8$ for parametric studies which has been shown to be fully sufficient for the early stages of transition at the present low initial amplitudes. In the simulation with $A_{TS}=3\%$ reported below the resolution was $N_1=N_2=16$ for $t\le116$ and $N_1=N_2=32$ for $t>116$. In the normal direction $N_3+1=41$ Chebyshev polynomials are retained in all cases. $\bar{u}_1$ denotes the mean disturbance (where the average is taken over one streamwise wavelength $0\le\xi_1\le L_1$), $u_1'=[\overline{(u_1-\bar{u}_1)^2}]^{\frac{1}{2}}$ the rms value of the fluctuation about this mean and $u_1'{}_{max}$ its maximum over x_3.

Fig. 2 shows the timewise development of the maximum fluctuation $u_1'_{max}$ at the "peak position" $x_2=0$ for different initial amplitudes A_{TS}. We observe the same threshold behaviour as is seen in the experiment [1,fig.15]. The threshold value is $u_1'_{max}\approx 1.5\%$ in accordance with [1]. In the lower amplitude cases all disturbances are damped. The run with $A_{TS}=3\%$ leads to transition with the "spike" stage beginning at $t\approx 122$. Breakdown of the well-ordered wave motion occurs at $t\approx 128$. With the present spatial resolution the simulation cannot reproduce the fine-scaled details appearing in the experiment at this stage (but it does not become unstable). For this run detailed comparisons with experimental measurements have been made.

Fig. 3 shows the spanwise distribution of $u_1'_{max}$ at subsequent stages, illustrating the formation of the "peak-valley-structure". The times t=78 and t=110 are selected such that the calculated and experimental values agree at the peak position $x_2=0$ (in the experiment [3] measurements were made only at one fixed streamwise position at which different stages of development were produced by increasing the ribbon amplitude). At t=78 there is very good agreement, but also at the later stages the agreement is still reasonably good.

The total mean velocity $U+\bar{u}_1$ at different wall distances is shown in fig. 4. Again the agreement is quite good. Results for the instantaneous velocity distribution at the peak position are shown in fig. 5. The total velocity profiles (fig. 5a) develop inflection points and strong shear layers around $x_3\approx \pm 0.6$ as in the experiments. The development of the instantaneous disturbance $u_1(\xi_1)$ at $x_3=-0.59$ (extended periodically to two wavelengths L_1) is shown in fig. 5b. The abscissa increases to the left to resemble the oscilloscope traces taken in the experiment. For t>80 the initially sinusoidal oscillation becomes nonlinearly distorted and strong narrow low-velocity pulses (the spikes) appear finally. The traces at t=122 and t=125 are similar to the experimentally observed ones [3] at the 1-spike and 2-spike stage, respectively, though somewhat less narrow.

Fig. 6 illustrates the three-dimensional structure of the flow in the lower half channel at t=122. We see a periodic array of "horseshoe" vortices similar to those observed in boundary layer flow visualisation experiments. In the downstream loops of these vortices, at the peak positions, low-speed fluid accumulates which leads to the spikes of the instantaneous velocity signals.

A key to the understanding of the transition process is obtained by considering the timewise evolution of the various Fourier amplitudes in (1). Fig. 7 shows (a) the maximum (over x_3) amplitudes $\hat{u}_{1max}(k_1,0) := 2\max|\hat{u}_1(k_1,0,x_3,t)|$ of the two-dimensional wave ($k_1=1$) and its streamwise harmonics ($k_1>1$), and (b) the amplitudes $\hat{u}_{1max}(1,k_2) := 4\max|\hat{u}_1(1,k_2,x_3,t)|$ of the oblique wave ($k_2=1$) and its spanwise harmonics ($k_2>1$). After an initial transient period $t\leq T_{TS}$ the dominating two-dimensional wave and its harmonics attain a nearly stationary state. This is because its amplitude nearly coincides with the threshold of the two-dimensional theory [4]. The two-dimensional modes are not affected by the three-dimensional disturbance until $t\approx 80$. The three-dimensional modes increase with nearly constant growth rate. This indicates a linear three-dimensional instability of the two-dimensional wave-modified flow which in fact has been found by a linear stability analysis [4,6]. When the three-dimensional modes have grown to the magnitude of the two-dimensional modes, strong interactions take place and all components grow rapidly until finally breakdown occurs. This transition mechanism has been veryfied in various simulations using different three-dimensional initial distributions and amplitudes [7].

4. Conclusions

Numerical simulations of the laminar-turbulent transition process have been carried out and compared in detail with experimental measurements of mean velocity, rms-values of fluctuation and instantaneous velocity distribution. Very satisfactory agreement has been found up to the spike stage. In particular it has been shown that the model of x_1-periodic, timewise amplified disturbances is able to reproduce the experimental observations. The three-dimensional flow structure is dominated by an array of horseshoe-vortices in the wall region similar to those seen in boundary layer flow visualisations. In accordance with [4,6], the mechanism responsible for transition has been shown to be a secondary instability of the two-dimensional ribbon-induced finite amplitude wave against infinitesimal three-dimensional disturbances. Thus it has been demonstrated that direct numerical simulation is able to reproduce the laminar-turbulent transition process in plane Poiseuille flow up to the spike stage in good agreement with experiments. It is an important progress that now theoretical means are available to investigate this highly nonlinear, three-dimensional process with the prospect of much future progress.

References

[1] M. Nishioka, S. Iida, Y. Ichikawa: An experimental investigation of the stability of plane Poiseuille flow. J.Fluid Mech. 72 (1975) 731-751

[2] M. Nishioka, S. Iida, S. Kanbayashi: An experimental investigation of the subcritical instability in plane Poiseuille flow (in Japanese). Proc. 10th Turbulence Symposium, Inst. Space Aeron. Sci., Tokyo Univ., 1978, p. 55-62

[3] M. Nishioka, M. Asai, S. Iida : An experimental investigation of the secondary instability. Proc. IUTAM Symposium, Stuttgart, 1979 (ed. R. Eppler, H. Fasel) Springer, Berlin 1980, 37-46

[4] Th. Herbert: Stability of plane Poiseuille flow - theory and experiment. Report VPI-E-81-35, to be published in Fluid Dynamics Transactions

[5] S.A. Orszag, L.C. Kells: Transition to turbulence in plane Poiseuille and plane Couette flow. J.Fluid Mech. 96 (1980) 159-205

[6] S.A. Orszag, A.T. Patera: Subcritical transition to turbulence in plane channel flows. Phys.Rev.Lett. 45 (1980) 989-993

[7] L. Kleiser: Numerische Simulationen zum laminar-turbulenten Umschlagsprozeß der ebenen Poiseuille-Strömung. Dissertation, Karlsruhe 1982 (report KfK3271)

[8] L. Kleiser, U. Schumann: Treatment of incompressibility and boundary conditions in 3-D numerical spectral simulations of plane channel flows. Proc. 3rd GAMM-Conference on Numerical Methods in Fluid Mechanics (ed. E.H. Hirschel), Vieweg Verlag, Braunschweig 1980, 165-173

[9] D. Gottlieb, S.A. Orszag: Numerical analysis of spectral methods: Theory and applications. NSF-CBMS Monograph 26, SIAM, Philadelphia 1977

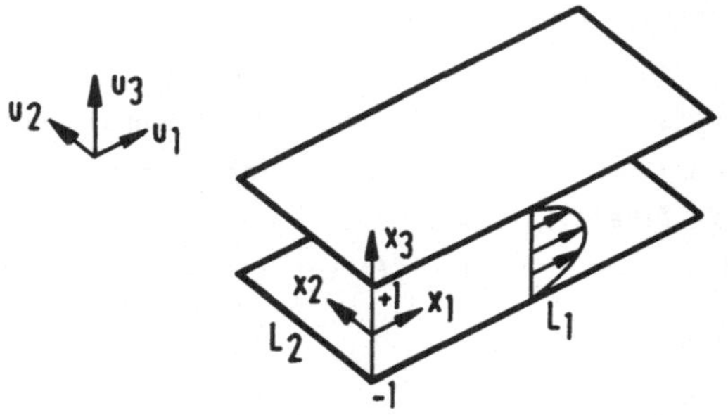

Fig.1
Geometry and coordinates

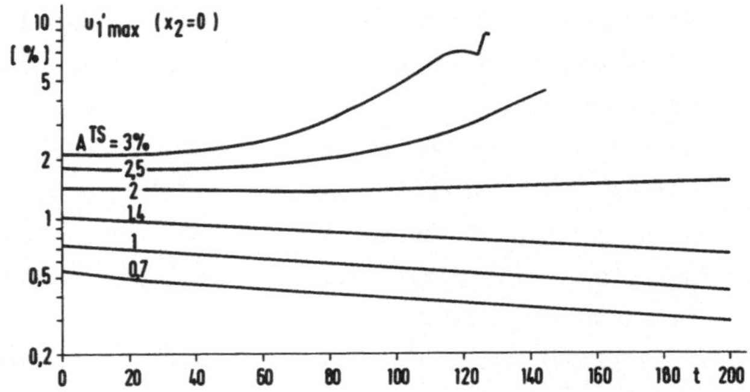

Fig.2 Timewise development of maximum u_1-fluctuation for different initial amplitudes A_{TS}

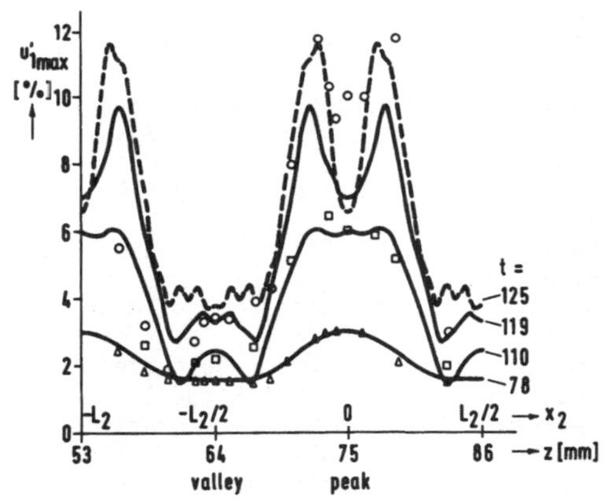

Fig.3 Development of peak-valley-structure
△ □ ○ Experiment [3]
═══ Numerical simulation

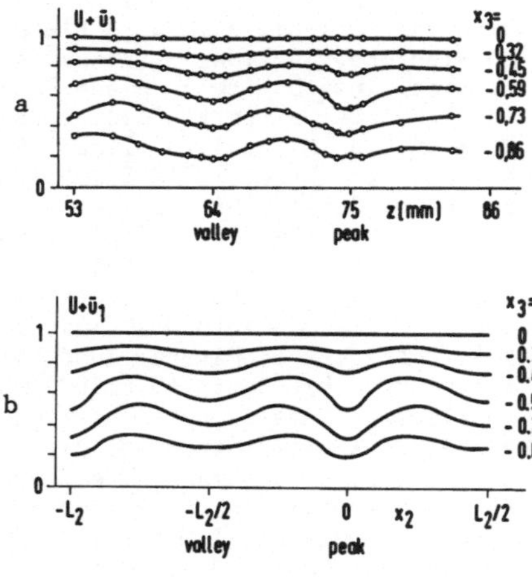

Fig.4 Total mean velocity $U+\bar{u}_1$
a Experiment [2], 1-spike stage
b Numerical simulation, t=122

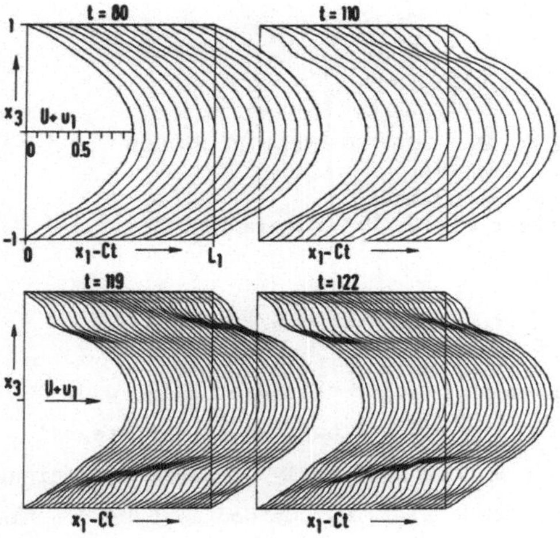

Fig.5a Instantaneous velocity profiles at peak over one period

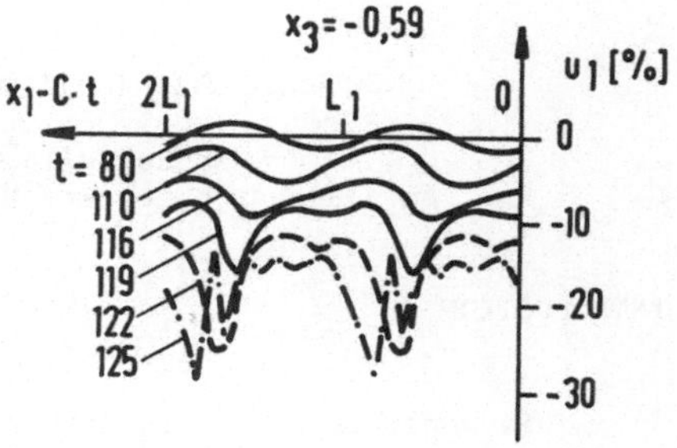

Fig.5b Calculated instantaneous u_1-disturbance at peak

Fig.6 Sketch of the three-dimensional horseshoe vortex structure at t=122

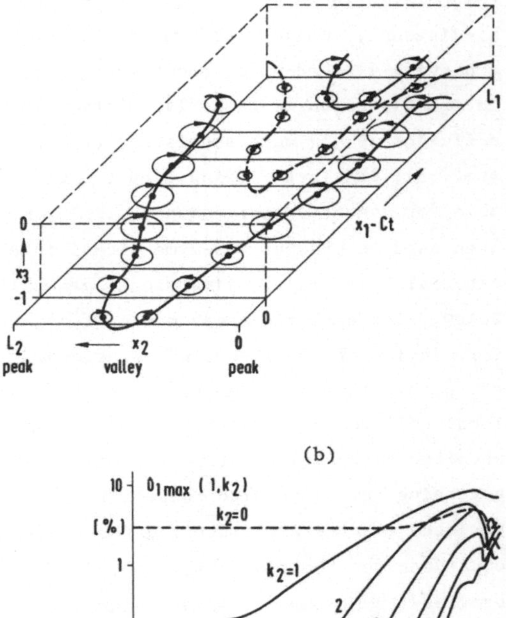

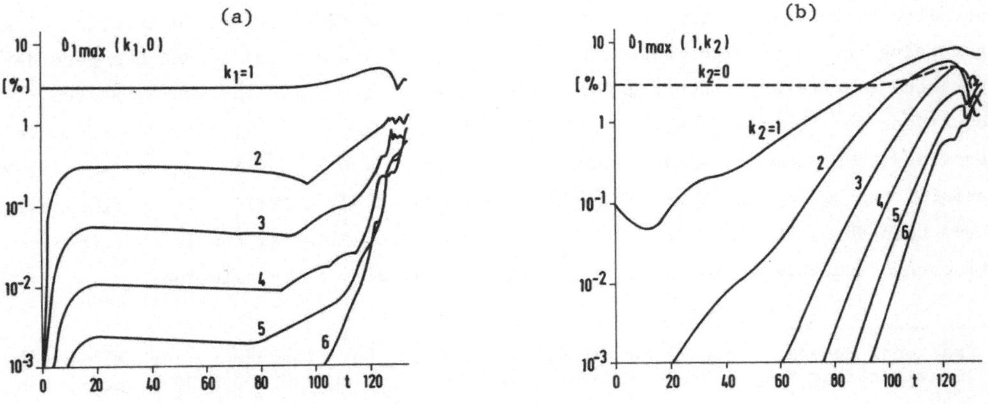

Fig.7 Development of (a) two-dimensional and (b) three-dimensional maximum Fourier amplitudes $û_{1max}$

TRANSONIC-FLOW COMPUTATION USING AN EXPLICIT-IMPLICIT METHOD

W. Kordulla* and R. W. MacCormack**
Ames Research Center, NASA, Moffett Field, California USA

INTRODUCTION

This paper is concerned with the integration of time-dependent Euler and Reynolds-averaged Navier-Stokes equations for transonic flows, using an integral or "finite-volume," formulation of the second author's recently presented efficient explicit-implicit predictor corrector method (ref. 1). That method is based on his well-proven , unsplit, second-order-accurate, explicit predictor corrector method (ref. 2). However, since explicit methods can result in extensive computer times (especially in turbulent flow cases) the latest approach (ref. 1) incorporates a bi-diagonal, implicit procedure into the predictor corrector sequences, thus achieving a speed-up of about two orders of magnitude for the shock-boundary-layer interaction problem on a flat plate. The scheme in reference 1 is at least as efficient as the more sophisticated hybrid explicit-characteristics-implicit scheme (ref. 3), the latter being competitive with fully implicit methods but less suitable for use on vector computers. The method, described in reference 1, has since been used to predict two-dimensional internal supersonic flows (refs. 4, 5). For external flows in two-dimensional generalized coordinates, the method is outlined below. The explicit steps can be recognized as an approximation to the integral formulation of the governing equations. The implicit steps, which yield the change of $\underline{\hat{q}}$ per time-step, where $\underline{\hat{q}}$ is the conservative solution vector including the local cell volume, are best viewed as an approximation to the differential equation, obtained by differentiating the conservative partial differential formulation of the governing equations with respect to time. After the numerical scheme has been described, results are presented for the solution of the Euler and Navier-Stokes equations for the flow about the 12%-thick Boeing VR-7 (TAB) airfoil (coordinates provided by McCroskey, Ames Research Center) at $M_\infty = 0.3$, $\alpha = 5°$, and $Re = 4 \times 10^6$, and for two transonic flows about the 12%-thick RAE2822 airfoil. These flow cases have been chosen because of the absence of separated flow regions so that the influence of turbulence modeling is not overly crucial for the solution.

---

*National Research Council Senior Research Associate. Permanent address: DFVLR-Institut fuer Theoretische Stroemungsmechanik, Bunsenstr. 10, D-3400 Goettingen, FRG.

**Current permanent address: Department of Aeronautics and Astronautics, FS-10, University of Washington, Seattle, Wash., 98195, U.S.A.

GOVERNING EQUATIONS

The governing equations, arbitrarily nondimensionalized with free-stream quantities, for a volume V with the boundary ∂V and with the computational coordinates $\tau = t$, $x^i = x^i(x^{j'},t)$, i, j' = 1(1)3, see Fig. 1 for two dimensions, read:

$$\frac{\partial}{\partial \tau} \int_V \underline{q}\, (g)^{\frac{1}{2}}\, dv + \oint_{\partial V} (\underline{q}\underline{v} + \underset{\sim}{b}) \cdot \underline{n}\, dA = 0, \tag{1}$$

where only pressure and viscous forces are considered, and where

$$\underline{q} = (\rho, \rho u, \rho v, \rho w, e)^T, \quad e = f_{pe} \frac{p}{\gamma - 1} + f_c \frac{\rho}{2} (\underline{u})^2 ,$$

$$f_{pe} \equiv 2(\gamma - 1)/[2 + \gamma(\gamma - 1)M_\infty^2], \quad f_c \equiv \gamma M_\infty^2 f_{pe}, f_{pu} = (\gamma M_\infty^2)^{-1} ,$$

$$\underline{v} = \underline{u} - \underline{u}_{mesh} ,$$

$$b = (\underline{b}_\rho, \underline{\underline{b}}_m, \underline{b}_e)^T ,$$

$$\underline{b}_\rho = 0 , \quad \underline{\underline{b}}_m = f_{pu}\, p\, \underline{\underline{I}} + Re^{-1} \underline{\underline{\tau}} ,$$

$$\underline{\underline{\tau}} = -\lambda \text{ div } \underline{u}\, \underline{\underline{I}} - \mu[(\text{grad } \underline{u}) + (\text{grad } \underline{u})^T] ,$$

$$\underline{b}_e = -\gamma(Re\, Pr)^{-1} \mu \text{ grad } T + f_{pe}\, p\underline{u} + f_c Re^{-1} \underline{\underline{\tau}} \cdot \underline{u} .$$

Here $(g)^{\frac{1}{2}}$ is the volume of a computational cell dv or the inverse of the Jacobian of the transformation relations. In finite-volume formulations transformations are not needed explicitly, and serve solely to conveniently generate meshes and to provide an ordering of the solution sweeps. The usual definition of the flow quantities is used in the notation. Cartesian velocity vectors have been chosen in the solution vector to reduce the computational effort, which otherwise increases considerably for general coordinates (ref. 6). The partial differential equation corresponding to equation (1) is

$$\frac{\partial}{\partial \tau} [\underline{q}(g)^{\frac{1}{2}}] + \frac{\partial}{\partial x^\ell} [(\underline{q}\, \underline{v} + \underset{\sim}{b})(g)^{\frac{1}{2}} \underline{g}^\ell] = 0 \tag{2}$$

or

$$\hat{\underline{q}},_\tau + {}^\ell\hat{\underline{F}},_\ell = 0 ,$$

where $(g)^{\frac{1}{2}} \underline{g}^\ell$ denotes the contravariant surface normals of surfaces $x^\ell = \text{constant}$ (Fig. 1). Differentiated with respect to time, equation (2) provides the basis for the implicit algorithm:

$$\tilde{\hat{\underline{q}}},_\tau + ({}^\ell\hat{\underline{\underline{A}}}\, \tilde{\hat{\underline{q}}}),_\ell = 0 , \quad \tilde{\hat{\underline{q}}} \equiv \hat{\underline{q}},_\tau , \quad {}^\ell\hat{\underline{\underline{A}}} \equiv \frac{\partial\, {}^\ell\hat{\underline{F}}}{\partial \hat{\underline{q}}} . \tag{3}$$

All equations have to be supplemented with appropriate initial and boundary conditions and with the constitutive thermodynamic relations, as well as with a turbulence model. Here the scalar eddy viscosity model of Baldwin and Lomax (ref. 7) is being used, where the pressure gradient correction according to Cebeci has been included in the wall damping term.

NUMERICAL SCHEME

Let $\Delta\tau$ be the time-step, n the time-level, $\underline{U}$ the numerical solution vector, ${}^{\ell}\underline{\hat{F}}$ the flux across the cell surface x^{ℓ} = constant, and L_i the one-dimensional factors of the implicit operator; the predictor corrector sequences are then the following:

Predictor

$$\Delta\underline{U}^n = -\Delta\tau\ (\Delta_+\ {}^{\ell}\underline{\hat{F}}/\Delta x^{\ell})\ , \qquad \ell = 1(2)3,$$

$$(L_1L_3)_+\ CFL_2\ \delta\underline{\bar{U}} = CFL_2\ \Delta\underline{U}^n\ , \tag{4}$$

$$CFL_1 \equiv \min\ [1.0,\ 0.5\ CFL/CFL_{i,k}],\ CFL_2 \equiv 1 - CFL_1\ ,$$

$$\underline{\bar{U}} = \underline{U}^n + CFL_2\ \delta\underline{\bar{U}} + CFL_1\ \Delta\underline{U}^n\ .$$

Corrector

$$\Delta\underline{\bar{U}} = -\Delta\tau\ (\Delta_-\ {}^{\ell}\underline{\bar{\hat{F}}}/\Delta x^{\ell})\ , \qquad \ell = 1(2)3,$$

$$(\bar{L}_1\bar{L}_3)_-\ \overline{CFL}_2\ \delta\underline{U}^{n+1} = \overline{CFL}_2\ \Delta\underline{\bar{U}}\ , \tag{5}$$

$$\underline{U}^{n+1} = \frac{1}{2}\ [\underline{U}^n + \underline{\bar{U}} + \overline{CFL}_2\ \delta\underline{U}^{n+1} + \overline{CFL}_1\ \Delta\underline{\bar{U}}]\ ,$$

where the subscripts indicating the spatial location have been omitted for clarity. The plus and minus subscripts indicate forward and backward two-point differences, for example, $\Delta_+(\ )/\Delta x^{\ell} = [(\ )_{\ell+1} - (\ )_{\ell}]/(x^{\ell+1} - x^{\ell})$. The overbars in the corrector sweeps indicate that the quantities are determined with predictor values.

The above scheme differs from the one presented originally in reference 1 as follows. First, it was found advantageous to retain in the final solution as much of the explicit contribution as possible, more for low CFL numbers than for large ones. The weighting provides a smooth blending of the explicit and implicit operators. The value of CFL in the weighting terms is usually 0.9, and $CFL_{i,k}$ is the local two-dimensional CFL number. Second, additional numerical dissipation was added in both the explicit and implicit portions. The original method (ref. 1), when applied to the shock-boundary-layer interaction on a flat plate, required only one smoothing term proportional to the increment of entropy per time-step. For the more complicated airfoil flow computations it is found advantageous to add additional numerical dissipation, as indicated below, if large time-steps are to be taken. Third, the treatment of the wall boundary condition in the predictor step at the end of the L3 sweep is changed. There the wall flux is cancelled immediately, and not carried over to the corrector step as suggested for the original method. These modifications enhance robustness and convergence; however, they also increase the computational effort, compared with that of the original method (ref. 1).

Compared with conventional fully implicit methods, the computational effort of the method used is reduced in three ways, as indicated already in reference 1. First, because of two-point differencing, the implicit operators yield bi-diagonal coefficient matrices with comparatively small computational effort. Second, it is observed in reference 1 that it is sufficient to consider the Eulerian Jacobians instead of the viscous ones, correcting the former by adding some representative viscous coefficients to the Eulerian eigenvalues. Third, the implicit procedure is skipped, whenever locally the explicit stability conditions are satisfied, for example, far away from the body where the mesh spacing is large, a feature enabled by the bi-diagonal structure of the matrices. Note that for stability reasons, absolute values of the eigenvalues are used, so that the physics of the implicit portion has to be considered an approximation to that of the governing equations, even for the Euler equations. This is of no importance for steady-state computations but has to be checked for time-accurate predictions.

The boundary conditions are handled as described in reference 1, and at the downstream plane extrapolation is used in spite of the subsonic flow conditions. Note that along the coordinate line $k = 2$ (fig. 2), which divides the upper and lower wake, an average of the fluxes corresponding to the flow at both sides of that line is used for the sweeps away from the line.

RESULTS

The computations of the flow about airfoils are based on C-type meshes wrapped around the airfoil (fig. 2). Following Deiwert's approach (Ames Research Center, private communication) the mesh is constructed using (piecewise) straight coordinate lines, one family of which is normal to the surface within the viscous layer or, in the case of inviscid flow, in the first cell layer away from the body. The results given have to be considered preliminary in the sense that the addition of numerical dissipation has not yet been fine-tuned to cope with oscillations observed in some solutions. The choice of appropriate smoothing, which has to be added in virtually any finite-difference scheme, is well known to be a delicate matter if viscous details are to be retained. The additional dissipation introduced into the predictor-corrector sequences is the same for all results given, and is of nominally third-order small with the derivatives of velocity and pressure in the sweeping directions as coefficients. Near the nose of the airfoil an additional term of the same type is introduced for the body surface normal sweeps with the increment of the surface normal velocity across the viscous layer as coefficient, where the local surface normals are considered. This term is used only within the viscous region and its influence is exponentially reduced with increasing distance in downstream direction, such that it does not affect the solution.

Distributions of the pressure coefficient Cp for viscous and inviscid flow predictions, based on different methods and meshes, are compared in figures 3 and 4 for the Boeing airfoil featuring a flat-plate-like trailing edge (the help of P. Buning, Ames Research Center, in adopting his version of DISSPLA's graphic package to the present need is gratefully acknowledged). The transition point was arbitrarily fixed at $x_T/6 = 0.144$. The present 108×32 cells C-mesh solution agrees well with the 123×60 points O-mesh fully implicit Navier-Stokes solution by Mehta (Ames Research Center, unpublished results) (fig. 3), except for the discrepancies near the trailing edge which are due to the poor resolution of the coarser mesh. Both solutions use the complete viscous terms on the explicit right-hand side. Mehta also uses all terms on the implicit left-hand side. Figure 4 shows a comparison of the Cp distributions of a full potential flow (ref. 8) (148×30 O-mesh, open trailing edge) with the present Euler solution (126×32, C-mesh). Apart from the unresolved trailing-edge region and the leading-edge expansion the overall agreement is good, in particular for the lower surface. As could be expected for this low-speed, unseparated flow, inviscid and viscous solutions agree fairly well with each other except in the upper-surface trailing-edge region where the boundary layer effectively reduces the curvature of the concave surface, and near the nose.

The cases 8621, test 1 and 9 (see also ref. 9), of the AFOSR-HTTM-Stanford Conference on Complex Turbulent Flows have been chosen to test the present solution in the transonic flow regime. Figures 5 to 7 present results for shock-free supercritical flow with the nominal free-stream conditions $M_\infty = 0.676$, $\alpha = 2.4°$, and $Re = 5.7 \times 10^6$; figures 8 to 12 display results for shocked flow at $M_\infty = 0.73$, $\alpha = 3.19°$, and $Re = 6.5 \times 10^6$. Transition is introduced at $x_T/C = 0.11$ and 0.03, respectively. Note that nominal free-stream conditions are used since matching the lift to obtain corrected free-stream conditions is still quite costly for Navier-Stokes solutions. In the case of the shock-free transonic flow, the pressure distributions, as predicted by Euler and full-potential flow (ref. 8) solutions, agree very well with each other except for the front region on the upper surface where the fine O-mesh full-potential-flow prediction yields stronger expansion, see fig. 5, where the experimental data from ref. 9 are given as well. Since neither of the two solutions agree with these experimental data, viscous or wind tunnel effects are important. The Navier-Stokes solution in figure 6 is in good agreement with experiment, with the discrepancies on the lower surface pointing to a small angle-of-attack correction. A run with $\alpha = 1.89°$, however, according to the suggested wind-tunnel correction, indicates too low a lift and drag and moment coefficients that are too high; the run with the nominal $\alpha = 2.4°$ yielded lift, drag, and moment coefficients that were too high, as compared with experiment. In figure 7, wall shear-stress distributions Cf_e and Cf_∞ are plotted in comparison with experimental data. The experimental Cf_∞ values have been obtained from Cf_e given in ref. 9, using isentropy assumptions. The use of Cf_∞ values as results of Navier-Stokes solutions, in particular for generalized coordinates, is believed to be more reliable, since they do not depend on the definition of the edge of the viscous layer. There is good qualitative agreement.

The shocked supercritical flow about the RAE airfoil poses a more difficult challenge because of the strong shock. Note that there was no separation observed in experiment (ref. 9). In figure 8 the Cp distribution obtained with the Euler solution is shown in comparison with experiment. The typical overexpansion on the upper surface can be noticed, and the shock is predicted farther downstream than observed experimentally. If necessary the oscillations caused by the shock could be largely removed by second-order small smoothing terms proportional to the normalized second derivative of the pressure profile. A comparison, with the conservative full-potential solution by TAIR is not meaningful, since a shock would be predicted near the trailing edge that would have too large a shock Mach number for the potential-flow approximation to be valid. Chordwise Cp distributions of Navier-Stokes solutions for two different meshes, one almost twice as fine as the other (216×60 versus 108×32), are plotted in figure 9. There are two major discrepancies. One is the difference in the location of the shock, which is, or course, smeared considerably for the coarse mesh solution; the other discrepancy occurs at the upper-surface trailing edge. The fine-mesh solution developed oscillations in the supersonic flow region, which can be noticed more clearly in figure 10. The comparison with experimental Cp values in figure 10 yields excellent agreement with respect to the shock location and the supersonic expansion peaks. The shock strength, however, is somewhat overpredicted with agreement recovering toward the trailing edge. The discrepancy on the lower surface calls for a reduction of the angle of attack, which would however influence the shock location. The Cf distributions, predicted with the fine- and coarse-mesh Navier-Stokes solutions, are compared with the experimental values in figure 11. The higher expansion near the nose on the upper surface for the fine-mesh solution is reflected in the larger shear stress after transition has been initiated. The shear on the front portion of the airfoil is predicted well with both grids, with the higher shear in front of the shock for the fine mesh resulting in higher shear on the aft portion of the airfoil as well, where there is less good agreement. Both solutions indicate separation which is not observed in experiment. However, the separation bubble for the fine mesh is thinner than the distance at which the first velocity values are measured in experiment.

CONCLUDING REMARKS

First applications of the explicit-implicit method of reference 1 to flows past airfoils are reported. These applications required the modification of the original scheme as described above. By comparison with results obtained with different methods and meshes, it is shown that inviscid and viscous flows about an airfoil at low speed can be predicted reasonably well. This holds for the transonic flow regime as well, if the free-stream conditions are correct and if a suitable mesh is used.

REFERENCES

1. MacCormack, R. W., 1981. A Numerical Method for Solving the Equations of Compressible Viscous Flow. AIAA Paper 81-110 (see also: Numerical Solution of Compressible Viscous Flows at High Reynolds Numbers, Advances in Fluid Mechanics, ed. E. Krause, Springer Lecture Notes in Physics, Vol. 148, 1981, pp. 254-267).

2. MacCormack, R. W., 1969. The Effect of Viscosity in Hypervelocity Impact Cratering. AIAA Paper 69-354.

3. MacCormack, R. W., 1976. An Efficient Numerical Method for Solving the Time Dependent Compressible Navier-Stokes Equations at High Reynolds Number. Computing in Applied Mechanics, AMD Vol. 18, The American Society of Mechanical Engineers.

4. von Lavante, E., and Thompkins, W. T., 1982. An Implicit Bi-diagonal Numerical Method for Solving the Navier-Stokes Equations. AIAA Paper 82-63.

5. Kumar, A., 1981. Some Observations on a New Numerical Method for Solving the Navier-Stokes Equations. NASA TP-1934.

6. Hirschel, E. H., and Kordulla, W., 1981. Shear Flow in Surface-Oriented Coordinates. Notes on Numerical Fluid Mechanics, Vol. 4, Vieweg Verlag, Braunschweig.

7. Baldwin, B. S., and Lomax, H., 1978. Thin Layer Approximation and Algebraic Model for Separated Turbulent Flows. AIAA paper 78-257.

8. Dougherty, F. C., Holst, T. L., Gundy, K. L., and Thomas, S. D., 1981. TAIR - A Transonic Airfoil Analysis Computer Code, NASA TM-81296.

9. Cook, P. H., McDonald, M. A., and Firmin, M.C.P., 1979. Aerofoil RAE 2822 — Pressure Distributions, and Boundary Layer and Wake Measurements. Experimental Data Base for Computer Program Assessment, AGARD-AR-138, A6-1 to A6-77.

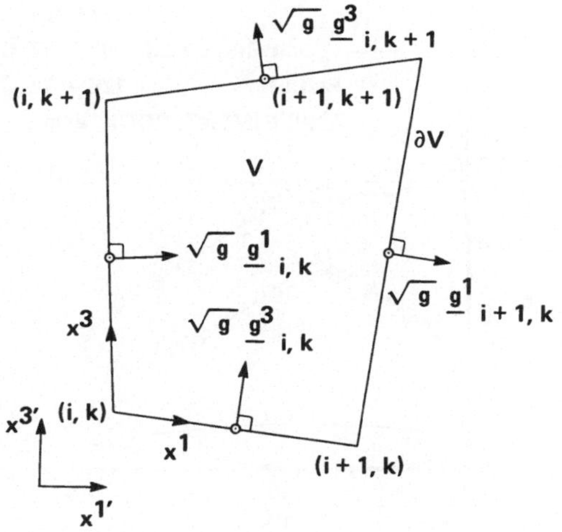

Figure 1. Sketch of a two dimensional mesh cell.

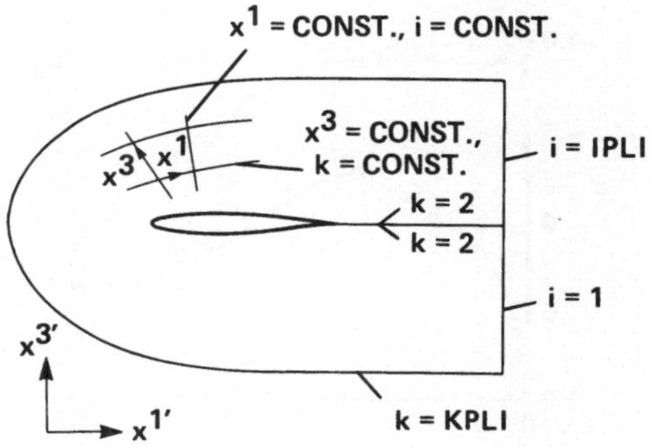

Figure 2. Sketch of C-type mesh around an airfoil.

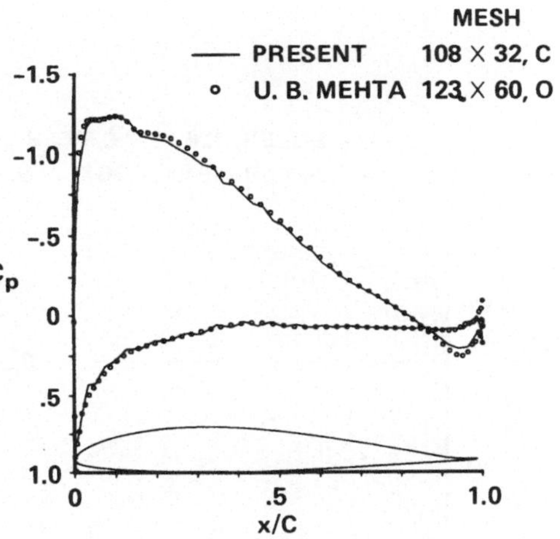

Figure 3. Comparison of Cp distributions of different Navier-Stokes solutions for the Boeing VR-7 (TAB) airfoil: $M_\infty = 0.3$, $\alpha = 5°$, $Re = 4 \times 10^6$, $x_T/C = 0.144$, $t/C = 0.12$.

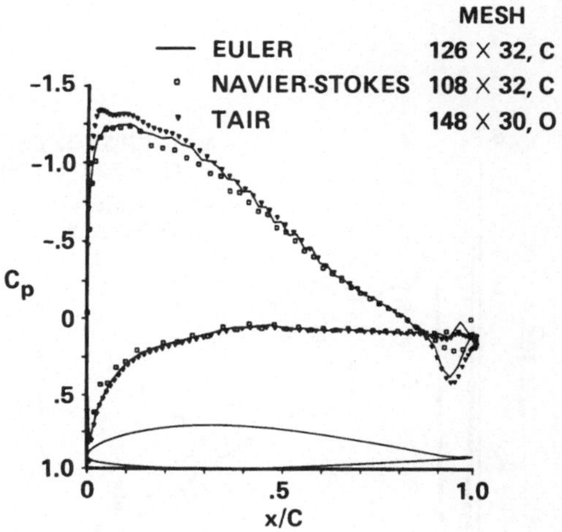

Figure 4. Pressure coefficient of inviscid and viscous flow solutions for the Boeing VR-7 (TAB) airfoil: $M_\infty = 0.3$, $\alpha = 5°$, $Re = 4 \times 10^6$, $x_T/C = 0.144$, $t/C = 0.12$.

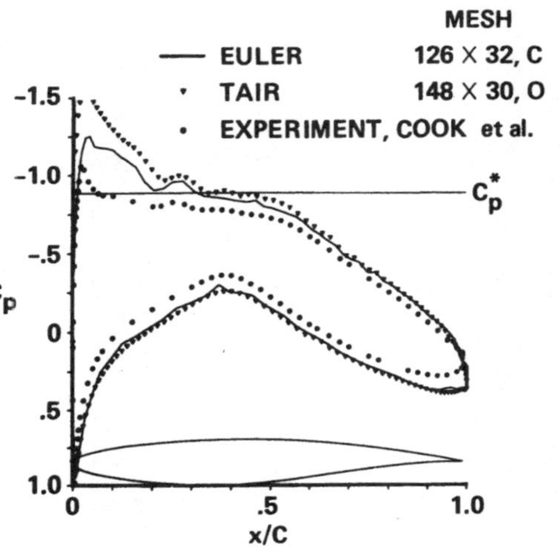

Figure 5. Comparison of Cp distributions of inviscid flow solutions and experiment for the RAE2822 airfoil: $M_\infty = 0.676$, $\alpha = 2.4°$, $Re = 5.7 \times 10^6$, $x_T/C = 0.11$, $t/C = 0.12$.

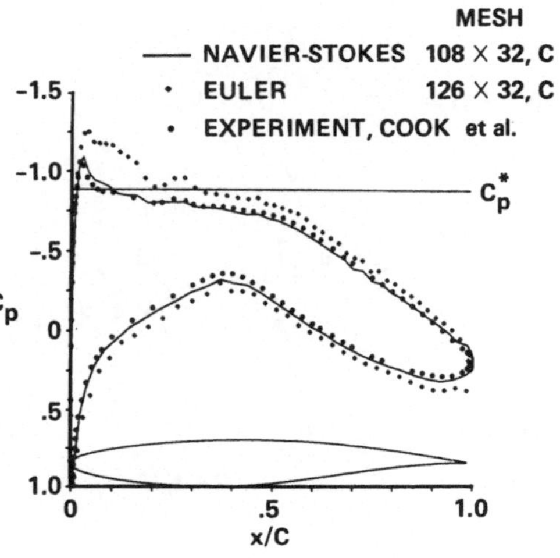

Figure 6. Cp distribution of inviscid and viscous solutions and experiment for the RAE2822 airfoil: $M_\infty = 0.676$, $\alpha = 2.4°$, $Re = 5.7 \times 10^6$, $x_T/C = 0.11$, $t/C = 0.12$.

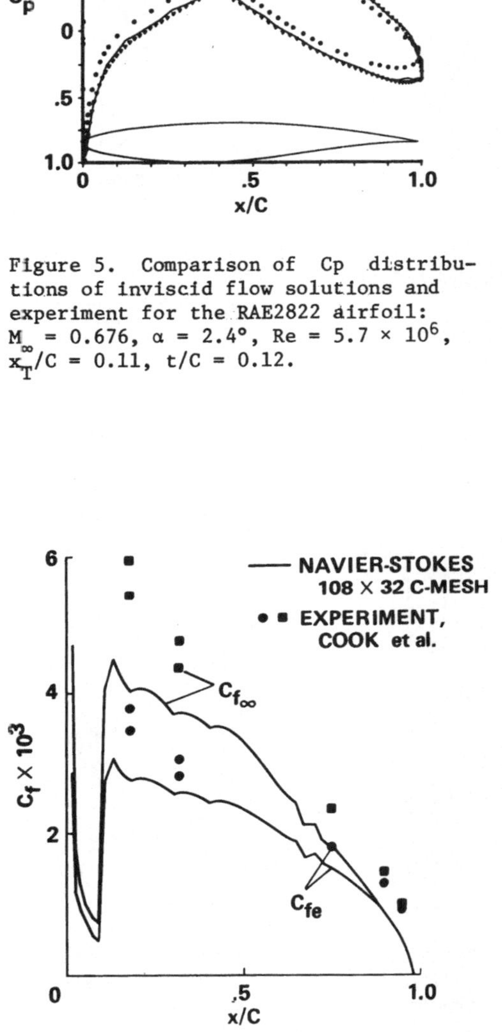

Figure 7. Experimental and predicted wall shear-stress distributions Cf on the upper surface of the RAE2822 airfoil: $M_\infty = 0.676$, $\alpha = 2.4°$, $Re = 5.7 \times 10^6$, $x_T/C = 0.11$, $t/C = 0.12$.

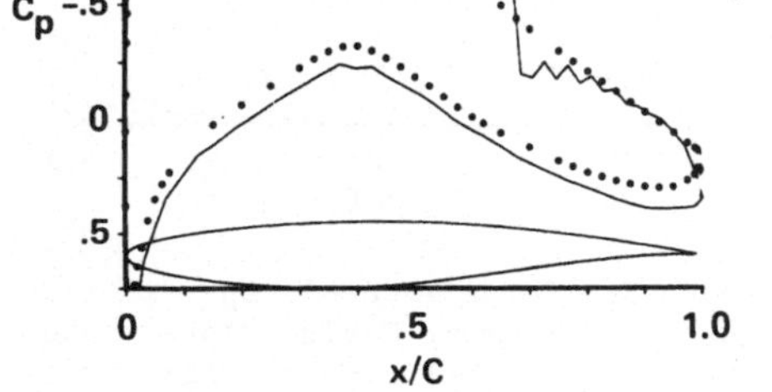

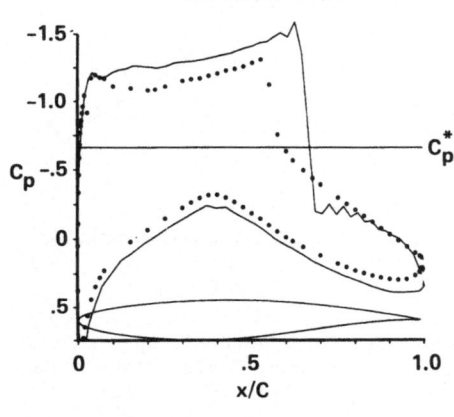

Figure 8. Inviscid flow and measured pressure coefficient for the RAE2822 airfoil: $M_\infty = 0.73$, $\alpha = 3.19°$, $t/C = 0.12$, $Re = 6.5 \times 10^6$, $x_T/C = 0.03$.

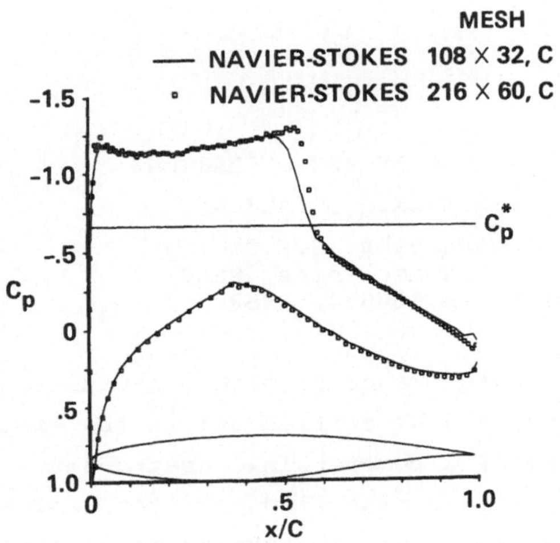

Figure 9. Pressure coefficient of Navier-Stokes solutions for fine and coarse meshes for the RAE2822 airfoil: $M_\infty = 0.73$, $\alpha = 3.19°$, $Re = 6.5 \times 10^6$, $x_T/C = 0.03$, $t/C = 0.12$.

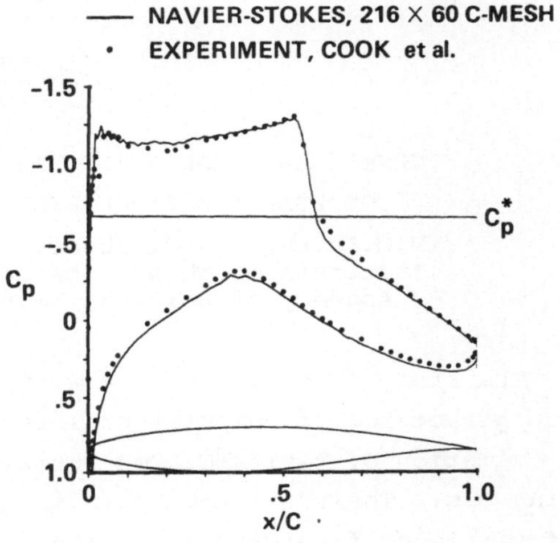

Figure 10. Comparison of Cp of the Navier-Stokes solution on a fine mesh with experiment for the RAE2822 airfoil: $M_\infty = 0.73$, $\alpha = 3.19°$, $Re = 6.5 \times 10^6$, $x_T/C = 0.03$, $t/C = 0.12$.

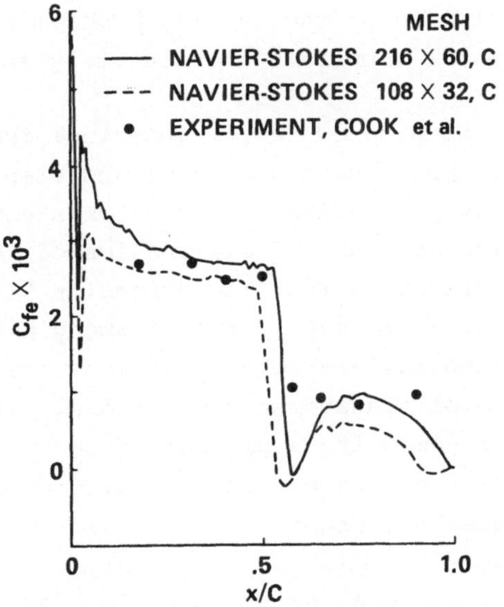

Figure 11. Comparison of predicted and experimental Cf distributions on the upper surface of the RAE2822 airfoil: $M_\infty = 0.73$, $\alpha = 3.19°$, $Re = 6.5 \times 10^6$, $x_T/C = 0.03$, $t/C = 0.12$.

NUMERICAL MODELING OF VISCOUS FLOWS IN APPROXIMATION OF FULL AND SIMPLIFIED NAVIER-STOKES EQUATIONS

V.M.Kovenya, A.S.Lebedev, G.A.Tarnavsky, S.G.Cherny
Institute of Theoretical & Applied Mechanics, USSR
Academy of Sciences, Novosibirsk 630090, USSR

The study of viscous flows within a wide range of flow parameters and geometries of computational regions can be carried out in the approximation of Navier-Stokes equations of a compressible heat-conducting gas. Their solution even in the case of the flow past the bodies of simple configurations contains zones of sharp inhomogeneities parallel with the regions of a smooth flow. The detailed study of such zones requires a great number of points and as a consequence large expenditures of computer time and memory. The division of a computational region into a number of subregions with a sufficient accuracy for practical applications can be found within the framework of simplified equations obtained from the complete Navier-Stokes equations under various assumptions about the flow character. Such a complex approach combining several different models enables one to simplify significantly the solution of the initial problem and to find its solutions using late model computers at relatively small expenditure of computer time.

The present paper is devoted to the numerical modeling of the steady flow problems in the approximation of full and simplified Navier-Stokes equations. The steady parabolized Navier-Stokes equations obtained from the full equation by means of elimination of the second derivatives (repeated and mixed) in the flow direction [1,2] are chosen as a simplified model. Being the most complete among a chain of simplified models, the parabolized Navier-Stokes equations are valid to describe flows in the presence of transverse separations. At the same time, their application for computing supersonic viscous flows enables one to reduce the dimension of the problem, i.e., considerably reduce the expenditure of computer resources.

Numerical Algorithms

Numerical methods employed for the solution of a separate problem or a class of problems should meet certain requirements such as accuracy, conservation, simplicity in operation and economy, these requirements depending on the objectives of the study. A wide application of results of numerical computations in practice emphasizes a main requirement of the algorithm: economy and accuracy. The existing numerical methods are guided, as a rule, by the solution of equations of a certain type and their generalization to other classes of equations can lead to a loss of economy or accuracy. For example, the explicit schemes are sufficiently economical for solving the equations of a hyperbolic type but their application to solve parabolic equations is ineffective because of the rigid limitation imposed on the stability. An effective solution of the classes of problems within the framework of different models requires the development of universal algorithm which should possess sufficient accuracy and be economical and easily adaptive to the different types of equations. This leads to the need to emply a single approach by constructing numerical algorithms. Below as the main method of solution is chosen the method of splitting in terms of the physical processes and the spatial directions [1].

The splitting method enables one
to reduce the solution of multidimensional problems to a set of their one-dimensional analogues that renders it economical;
to construct the difference and iteration schemes for problems of different dimension (the splitting in terms of the spatial directions) and for the physical-mathematical models (the splitting in the terms of the physical processes) within the framework of a single approach.

The accuracy of the method required can be achieved either by employing schemes of a higher order of accuracy or by employing the moving network adapted to the flow grids. The steady solution of Navier-Stokes equations is found on the basis of the relaxation principle. Marching methods employed for the solution of parabolized equations in a supersonic flow region, and in the local subsonic zones the regularization method [2, 4] is employed. Their combination enables one to find the solution of the initial problem throughout the whole computational region using a single algorithm.

The description of the splitting method for solving the different problems, the realization of boundary conditions and the analysis of stability is presented in papers [1-3]. Below we shall dwell on some peculiarities of the algorithm of solution of the simplified steady Navier-Stokes equations.

The main assumption made in the simplified equations is the one about the smallness of longitudinal components of viscous stresses tensor as compared with normal and azimuthal ones. Therefore, a system of coordinates in which the initial equations are considered should be orthogonal in a region where the influence of viscosity is important, i.e., the boundary layer and shock wave region. In this paper the initial equations are written down in an arbitrary curvilinear system of coordinates, and as components of velocity vector are chosen its expansion in a cylindrical system of coordinates. This fact makes it possible to widen the classes of body geometries under study and to avoid the loss of accuracy in the difference approximation of initial equations. Note that the operating of the difference and iteration schemes of splitting [1-3] is reduced to scalar sweeps that renders the algorithm economical.

II. Results

Within the framework of complete and simplified Navier-Stokes equations the supersonic flow of a compressible gas past a sphere-cone and a sphere-cylinder is studied. The computational region (Fig. 1) was subdivided into a number of subregions: the flow over a lateral body surface (segment 1) , the flow in a near wake (II) and a far wake (III). In region (I) the solution of the problem has been found in the approximation of complete and simplified Navier-Stokes equations, in region (II) in the approximation of complete Navier-Stokes equations and in region III in the approximation of parabolized Navier-Stokes equations. To decrease the influence of approximate setting of boundary conditions, the computational regions were partially overlapped. The flow parameters obtained from the solution of complete Navier-Stokes equations for the flow past a sphere-cone body were set as boundary conditions on the left border of the region I. On the right borders of regions I and II "soft" conditions were set, i.e., ti was assumed that the second derivatives in the flow direction were zero. The right border of region II was chosen at a sufficient distance from a body end in such a way that the flow over it should be supersonic. The bow shock was singled out in the process of solution. Calculations were carried out at M=10, Re=3.5 10^4 (Reynolds number was calculated using the body length), Pr=0.72, γ =1.4, ω =0.75, T_w =6T_D, α_1=10°, α_2=0°.

Fig. 2 represents the pressure distribution along the body surface and the location of the bow shock. Fig.3 represents the distributions of the heat flux q_w and the coefficient of local friction in the region I. Solid lines (*) are results obtained with the help of Navier-Stokes (parabolized) equations. A maximum difference in the flow parameters is observed in the vicinity of the salient point and accounts for up to 20% in pressure and up to 25% in heat flux. Outside the vicinity of the salient point these differences are significantly less.

The difference in the calculation results is explained by the difference in the models. In particular, the upstream influence of viscous terms is excluded in a model of parabolized equations.

A zone of reversed flow arises in a region behind the body. Its length is $\approx 0.9D$ where D is the diameter of the cone end. Fig. 4 shows the velocity vectors in a near wake region. At a distance $\approx 3.6D$ the flow behind the body becomes supersonic throughout the whole computational region, however, the influence of repeated and mixed derivatives in the flow direction is of importance up to the cross section $x \approx 6D$. At $x > 6D$ the approximation of the parabolized Navier-Stokes equations holds. Calculations in region III carried out with the help of complete and simplified equations yield practically coincident results.

Of course, the solution of Navier-Stokes equations gives a more realistic picture of the flow field. However, the difference in the coefficients of wave resistance C_d and a full coefficient of friction resistance C_{f_n} for various models does not exceed 1.8% and 2%, respectively (the calculation is carried out without regard to the base pressure as in the near wave region a model of the simplified equations is not valid). At the same time the ratio of computer time expenditures when solving with the use of complete and simplified equations for region I increased by roughly a factor of 10, for region III 100 times.

In the next series of calculations supersonic flow past a sphere-cylinder body for the same values of parameters as above, at angle of attack $\alpha = 20^{\circ}$, is considered in the approximation of parabolized Navier-Stokes equations using the marching method. On the leeward side there arises a region of reversed flow (transverse vortices). Fig. 5 illustrates this fact with the distribution of longitudinal C_{f_1} and transversal C_{f_2} coefficients of friction and heat flow q_w in the cross section equal to R (R is the cylinder radius). In the region of separation the coefficient C_{f_2} is negative. Fig. 6 represents the azimuthal velocity component in the same cross section. A zone of separated flow is $\approx$ I R. In all the calculations the difference grids were made nonuniform approximating the body according to the exponential law. The accuracy of calculations was checked by comparing with calculations of other authors and by change of the grid steps in all directions.

References:

1. Kovenya V.M., Yanenko N.N. Metod rasshchepleniya v zadachakh gazovoi dimamiki, Novosibirsk, Nauka, 1981.
2. Kovenya V.M., Cherny S.G. Metod resheniya statsionarnykh uproshchennykh uravnenii vyazkogo gaza. Preprint ITPM SO AN SSSR, Novosibirsk, 1981, No.42, p.51.
3. Kovenya V.M., Tarnavsky G.A., Yanenko N.N. Neyavnaya raznostnaya skhema dlya chislennogo resheniya prostranstvennykh uravnenii gazovoi dinamiki, Zhurn.Vychisl.Matem.i Matem.Physiki, 1980, vol. 120, No.6, pp.1466-1482.
4. Yanenko N.N., Kovenya V.M., Tarnavsky G.A., Cherny S.G. Economical Method for Solving the Problems of Gas Dynamics. Lecture Notes in Physics, No.141, pp.448-453.

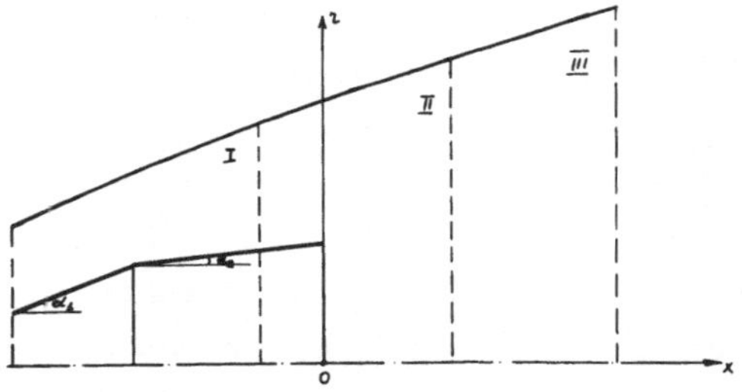

Fig. 1

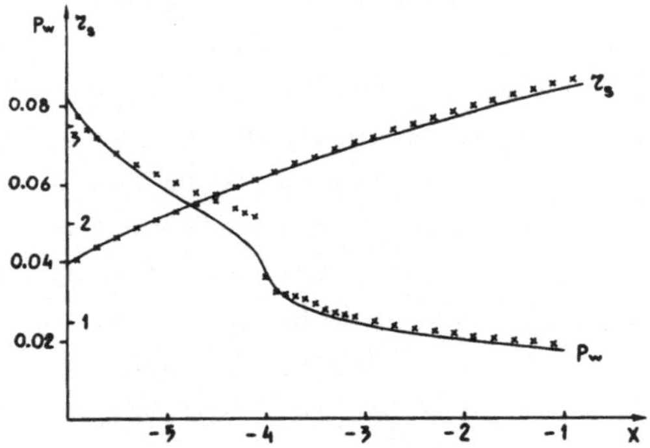

Fig. 2

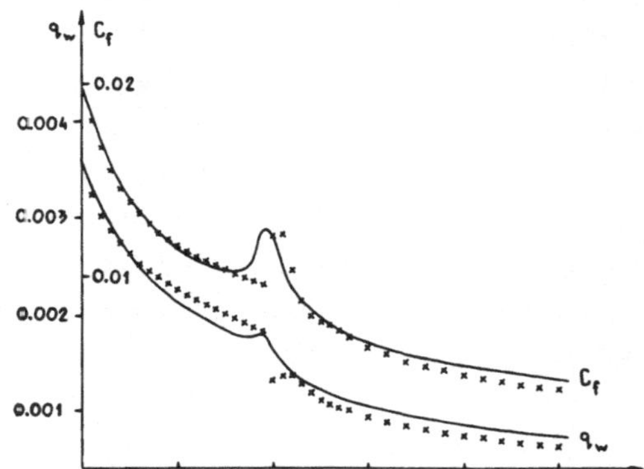

Fig.3

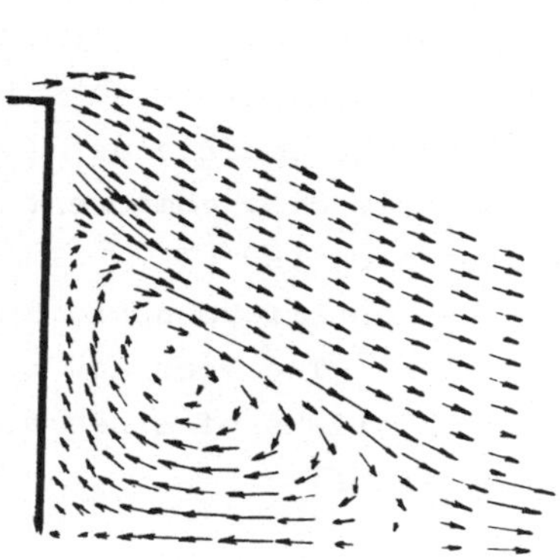

Fig.4

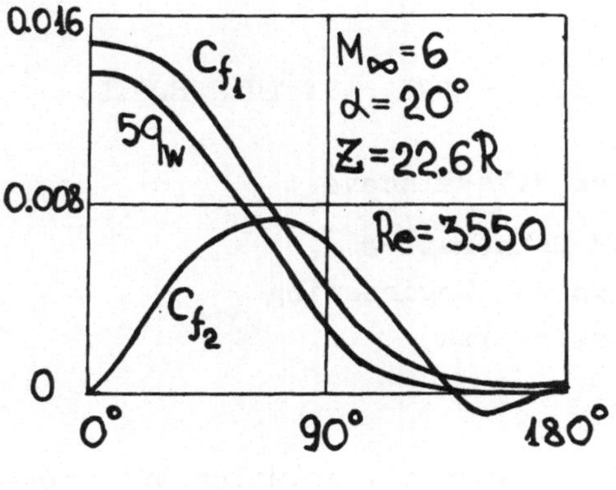
0.016
0.008
0
0°
90°
180°
C_{f_1}
$5q_w$
C_{f_2}
$M_\infty=6$
$\alpha=20°$
$Z=22.6R$
$Re=3550$

Fig.5

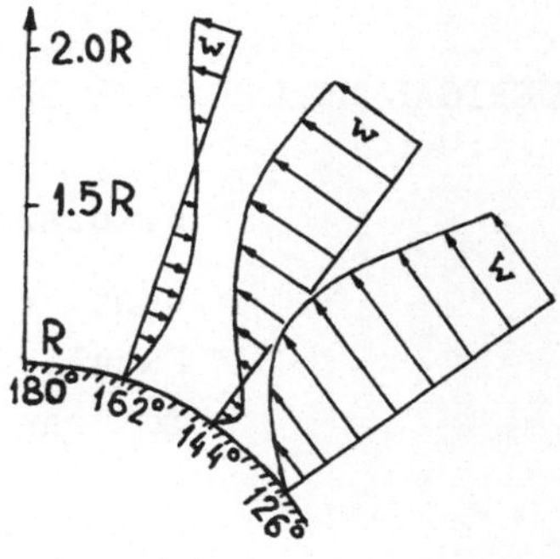
2.0R
1.5R
R
w
180°
162°
144°
126°

Fig.6

NUMERICAL SOLUTION OF TRANSONIC SHEAR FLOWS PAST THIN BODIES

K.Kozel,J.Polášek,M.Vavřincová

Dept. of Applied Mathematics
Faculty of Mechanical Engineering
TU Prague,Czechoslovakia

The work deals with a method of numerical solution of three-dimensional inviscid steady shear flows past thin bodies in a tunnel or through a cascade with upstream Mach number $M_\infty = M_\infty(z), z \in \langle 0, LZ \rangle$. A governing equation is derived by perturbation theory and is an extension of the classical perturbation potential equation.

The problem is solved by a method of finite differences,the system of difference equations is solved by a modification of succesive line relaxation method.

Several numerical results of transonic shear flows past thin body,a system of thin bodies in a tunnel or through a cascade with M_∞ changing from $M_\infty(0)<1$ to $M_\infty(LZ)>1$ are presented and compared with other numerical results.

I.Mathematical model

Consider steady three-dimensional transonic shear flows past thin body with upstream velocity vector $\vec{U}_\infty = (U_\infty(z), 0, 0)$ and a_0=const. (a-speed of sound).The flow is described by continuity equation

$$\frac{\partial(\rho U_i)}{\partial x_i} = 0 , \quad (i = 1,2,3) , \tag{1}$$

Euler equation

$$U_j \frac{\partial U_i}{\partial x_j} = - \frac{1}{\rho} \frac{\partial p}{\partial x_i} \quad (i,j = 1,2,3) \tag{2}$$

and adiabatic relation for pressure p and density ρ .Multiplying (2) by U_i and summing we obtain

$$U_i U_j \frac{\partial U_i}{\partial x_j} = - \frac{U_i}{\rho} \frac{\partial p}{\partial x_i} = - a^2 \frac{U_i}{\rho} \frac{\partial \rho}{\partial x_i} . \tag{3}$$

Using (1) we can change the right hand side (3) to obtain

$$U_i U_j \frac{\partial U_i}{\partial x_j} = a^2 \frac{\partial U_i}{\partial x_i} \tag{4}$$

where a^2 is known from the Bernoulli equation.

Consider perturbation velocity vector $\vec{u} = (u_1, u_2, u_3)$ defined by the following relations($U_\infty = |\vec{U}_\infty|$)

$$U_1 = U_\infty (1 + \delta^{2/3} M_\infty^m u_1) , \quad U_2 = U_\infty \delta M_\infty^m u_2 , \quad U_3 = U_\infty \delta M_\infty^m u_3 , \tag{5}$$

where $u_i = u_i(x_1, \tilde{x}_2, x_3) = u_i(x, \tilde{y}, \tilde{z}) , \tilde{x}_j = \delta^{1/3} x_j \, (j = 2,3)$, δ is ratio thickness, m=const. .Using (5) in (4) and condition $\mathrm{rot}\,\vec{u} = 0$ we can derive an approximation (4) for perturbation velocity potential φ ($\vec{u} = \text{grad}\,\varphi$) in the form

$$V\varphi_{xx} + \varphi_{\tilde{y}\tilde{y}} + \varphi_{\tilde{z}\tilde{z}} + E\varphi_{\tilde{z}} = 0, \tag{6}$$

$$V = K - (\varkappa+1)M_\infty^{m+2}\cdot\varphi_x\,,\ K = \frac{1-M_\infty^2}{\delta^{2/3}}\,,\ E = \frac{1}{M_\infty}\Big[m + \frac{1-M_\infty^2}{1+\frac{\varkappa-1}{2}M_\infty^2}\Big]\frac{dM_\infty}{d\tilde{z}},$$

$\varkappa$ is Poisson´s constant, $m \in \langle 0,-2)$.
Conservative form of this equation (6) has the form

$$P_x + Q_{\tilde{y}} + R_{\tilde{z}} = 0, \tag{7}$$

$$P = H(M_\infty(\tilde{z}))[K\varphi_x - \tfrac{1}{2}(\varkappa+1)M_\infty^{m+2}\varphi_x^2]\,,\ Q = H\varphi_{\tilde{y}}\,,\ R = H\varphi_{\tilde{z}}.$$

Function $H(M_\infty(\tilde{z}))$ is a positive solution of the following equation

$$\frac{dH}{d\tilde{z}} = \frac{dH}{dM_\infty}\cdot\frac{dM_\infty}{d\tilde{z}} = HE \Rightarrow H = C\cdot M_\infty^{1+m}[m + \tfrac{\varkappa-1}{2}M_\infty^2]^{-\frac{1}{2}\frac{\varkappa+1}{\varkappa-1}}\,,\ C>0. \tag{8}$$

Boundary conditions along a body surface are considered in linearized form similar to three-dimensional case with M_∞= const. .

<u>Remark 1</u>:For the case of a blade with a torsion we can consider an extension of the described model.Let $\vec{U}_\infty = (U_{\infty 1}, U_{\infty 2}, 0)$ in (x,y,z) coordinate system.Consider transformation (x,y,z)→(s,n,z) where $\vec{U}_\infty = (S_\infty(\tilde{z}),0,0)$ or $\vec{U}_\infty = (S_\infty(\tilde{z}), N_\infty(\tilde{z}), 0)$, $|N_\infty/S_\infty| \ll 1$, $S_\infty > 0$ .The governing equation (6) transforms to the "same" equation but x→s,y→n. Boundary conditions along a body surface are changed by the similar way.

II.<u>Differential and difference problem</u>

By a weak solution of our problem we mean function φ fulfilling an integral relation

$$\oint\!\!\oint_{\partial D} P\,d\tilde{y}\,d\tilde{z} + Q\,dx\,d\tilde{z} + R\,dx\,d\tilde{y} = 0 \tag{9}$$

along boundary ∂D of every closed subdomain D of domain of solution Ω and an appropriate boundary conditions,periodicity conditions in the case of the flows in cascades ,initial conditions for x⟶∞ and $M_\infty < 1$ and a modification of Kutta condition(wake condition).

Assuming the flows between two parallel walls(z=0,z=LZ) we consider $\varphi_x = K_1, \varphi_{\tilde{y}} = K_2, \varphi_{\tilde{z}} = 0$ for x⟶∞ and downstream M < 1.In the case of the flow in a tunnel $K_1 = K_2 = 0$,in the case of the flow in a cascade $K_i = K_i(\gamma(z))(i=1,2)$ similar to two-dimensional problems or $K_i = K_i(G)$, $G = \int_0^{LZ}\gamma(\tilde{z})d\tilde{z}$ etc.

<u>Remark 2</u>: a)Define a value $\gamma(x,\tilde{z}) = -\oint_c \varphi_x dx + \varphi_{\tilde{y}} d\tilde{y}$, z=const. as a modification of two-dimensional circulation of the perturbation velocity vector along a "profile".Let $\gamma(x,\tilde{z}) = \gamma(\tilde{z})$ and $\gamma(\tilde{z}) \neq 0$,then function φ

has a discontinuity

$$\gamma(\tilde{z}) = \varphi(x, 0+, \tilde{z}) - \varphi(x, 0-, \tilde{z}),\ x \geq x_{TE} \tag{10}$$

b)Our definition of the weak solution implies that the following is valid

$$\langle P\rangle(d\tilde{y}d\tilde{z})_{s_i} + \langle Q\rangle(dxd\tilde{z})_{s_i} + \langle R\rangle(dxd\tilde{y})_{s_i} = 0 \tag{11}$$

for the solution along a shock wave s_i .Here $\langle P\rangle = P_1 - P_2$ stands for the difference of limit values from either sides of the surface s_i ,index s_i denotes the projection of surface element in the direction $1 \to 2$.

The problem was solved by the method of finite differences mostly with a regular orthogonal grid,in the case offlow in cascades in two "periods".The difference scheme is an approximation (9) along the boundary of each computational cell $\partial D_{i,j,k}$ written in the form

$$(P_{i+1/2,j,k} - P_{i-1/2,j,k})\Delta\tilde{y}\Delta\tilde{z} + (Q_{i,j+1/2,k} - Q_{i,j-1/2,k})\Delta x\Delta\tilde{z} + $$
$$+ (R_{i,j,k+1/2} - R_{i,j,k-1/2})\Delta x\Delta\tilde{y} = 0. \tag{12}$$

Values φ_x , φ_y , φ_z in Q , R are approximated by central differences of the second order,values $\varphi_x|_{i+1/2,j,k}$ in P are approximated by central differences of the second order for $V_{i,j,k} = K-(\varkappa+1)M_\infty^m(\varphi_{i+1,j,k} - \varphi_{i-1,j,k})/(2\Delta x)$ and for $V_{i,j,k}<0$ by backward differences of the first order.The same is true for $\varphi_x|_{i-1/2,j,k}$ .The difference scheme is an extension of the difference scheme used in [1] for two-dimensional problems and in [2] for three-dimensional problems with M_∞=const. Described difference scheme was modified to the form used in this work

$$(1-\mu_{i,j,k})\tau_x^2 V_{i,j,k}(\varphi_{i+1,j,k} - 2\varphi_{i,j,k} + \varphi_{i-1,j,k}) + \mu_{i-1,j,k} V_{i-1,j,k}\tau_x^2 . (\varphi_{i,j,k}$$
$$- 2\varphi_{i-1,j,k} + \varphi_{i-2,j,k}) + \tau_y^2(\varphi_{i,j+1,k} - 2\varphi_{i,j,k} + \varphi_{i,j-1,k}) +$$
$$\frac{1}{2}\left(1 + \frac{H_{k+1}}{H_k}\right)\varphi_{i,j,k+1} - \left(1 + \frac{H_{k+1}+H_{k-1}}{2H_k}\right)\varphi_{i,j,k} + \frac{1}{2}\left(1+\frac{H_{k-1}}{H_k}\right)\varphi_{i,j,k-1} = 0, \tag{13}$$

$$\tau_x = \frac{\Delta\tilde{z}}{\Delta\tilde{x}},\ \tau_y = \frac{\Delta\tilde{z}}{\Delta\tilde{y}},\ \mu_{i,j,k} = \begin{cases} 0 & \text{for } V_{i,j,k} > 0 \\ 1 & \text{for } V_{i,j,k} < 0 \end{cases},\quad H_k = H(\tilde{z}_k).$$

Boundary conditions along the body surface and the walls are fulfilled in the similar way as in [1] , [2] , [6] .

The system of described difference equations is solved by sucesive line relaxation method in the planes $z=z_k(k=1,\dots,NZ)$ similar to two-dimensional problems in [1] with fixed values $\varphi_{i,j,k+1}$, $\varphi_{i,j,k-1}$ from previous iteration cycle.Iteration process begins near hub($z=z_1$) and finishes near tip(z=NZ) and begins with values $\varphi^0_{i,j,k} = \varphi^0_x|_{i,j,k} = 0$.

III. Numerical results

A) Fig.1 shows results of numerical solution of transonic shear flows past a NACA 0012 profile in the tunnel with $M_\infty = 0{,}73 + 0{,}6z/LZ$, $z \in \langle 0, LZ \rangle$ in the form of distribution of Mach number along lines $y=0, z=z_k (k=1,\dots,NZ)$. We can say that maximal velocity on the body surface increases with increasing M_∞ but shock wave appeared practically in the region of the chocked flow near the trailing edge. This means that the shock wave appears for higher M_∞ and is thinner as compared with two-dimensional or three-dimensional case and $M_\infty =$ const. The shock wave is in this case influenced by changing M_∞ or $H(M_\infty)$. Distance of the point with $M=1$ situated on detached shock wave for $M_\infty > 1$ from the leading edge is greater in comparison with two-dimensional case for the same M_∞.

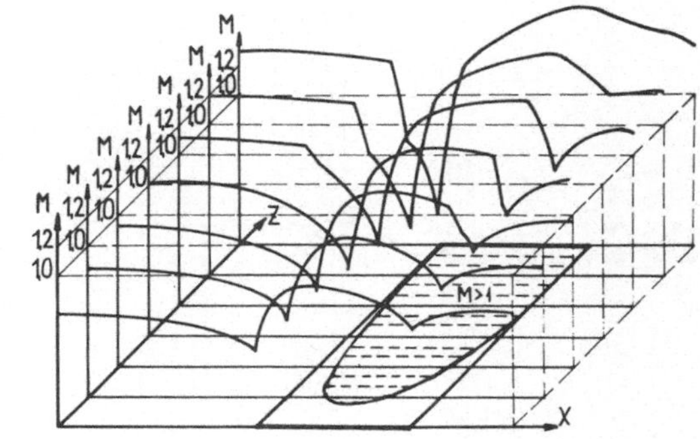

Fig.1

B) Fig.2a, 2b show results of transonic shear flows through cascade with stagger angle $\beta = \frac{\pi}{2}$, $p/c = 2$ with symmetrical 8% parabolic blades and $M_\infty = 0{,}65 + 0{,}55z/LZ$ (see [3]) using distribution of Mach number along lines $y=0$ (fig.2a). Fig.2b shows distribution of z-component of velocity vector U_3 using distribution $\varphi_{\tilde{z}}$ along the body surface and near the body surface for $y=0$.

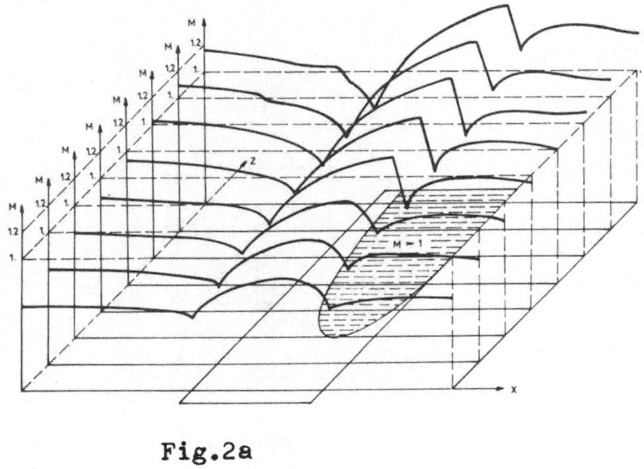

Fig.2a

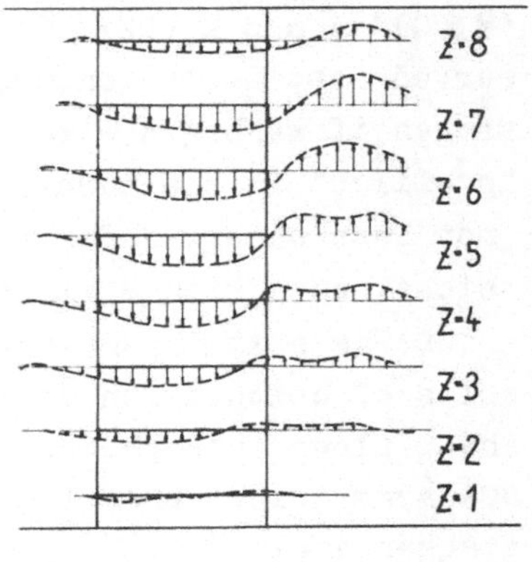

Fig.2b

Good agreement of our results with results of Oliver and Sparis [3] (fig.3a,3b) can be observed. This has been achieved by numerical solution of full system of equations of inviscid flows.

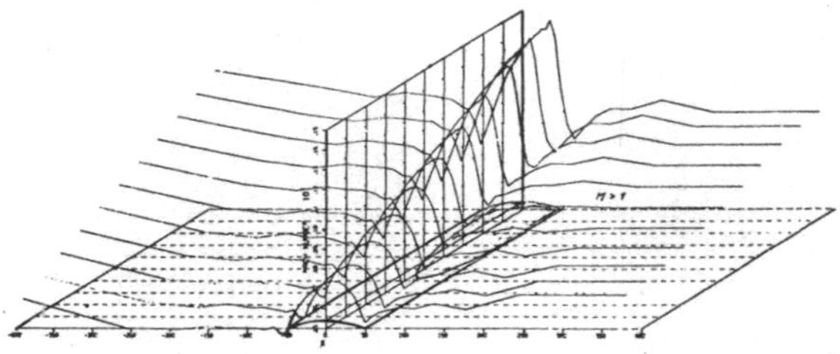

Fig. 3a

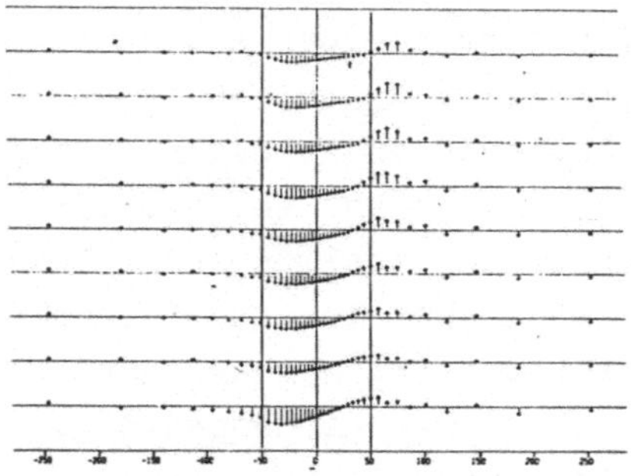

Fig. 3b

C) In the third case we computed transonic shear flow past the system of bodies in the tunnel with cascade geometry (see fig.4). The bodies are symmetrical parabolic arcs, at the hub $\delta=8\%$ and then it linearly changes to $\delta=4\%$ near the tip. Using this body we changed walls for $y=\pm\ell$, $\ell>0$, as it is seen in fig.4. We considered $M_\infty = 0,7+0,6z/LZ$, $z \in \langle 0,LZ \rangle$. Fig.4 shows interesting results of computation using lines with M=const. in the planes $z=z_k=\Delta z/2+(k-1)\Delta z$ (k=6, 7,8) for increasing upstream Mach numbers $M_\infty(z_k)>1$. In this case ($M_\infty(0)<1$ and $M_\infty(LZ)>1$) we observed that iteration process converges if we begin with all three "profiles" in the tunnel. This has not been observed for two-dimensional case with $M_\infty>1$ ([4]).

D) The next fig.5a shows the results of computation of transonic shear flows through cascade from DCA symmetrical profiles with $\delta=8\%$, stagger angle $\beta=\pi/2$, $\alpha=2°$ and $M_\infty=0,7+0,6z/LZ$ using lines M=const.

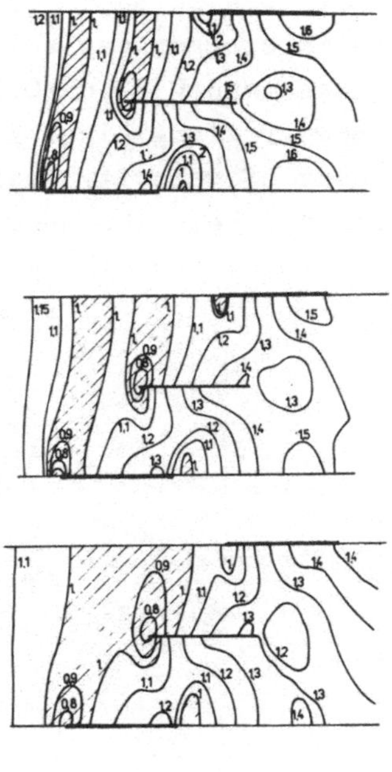

Fig. 4

in the planes $z=z_k$ (k=6,7) for M_∞>1.Our numerical results are compared (qualitatively) with two-dimensional interferometric experimental results(fig.5b)published in [5](dotted line ~ sonic line). Upstream Mach numbers in two-dimensional experiment are lower (1,035;1,073) in comparison with computation.

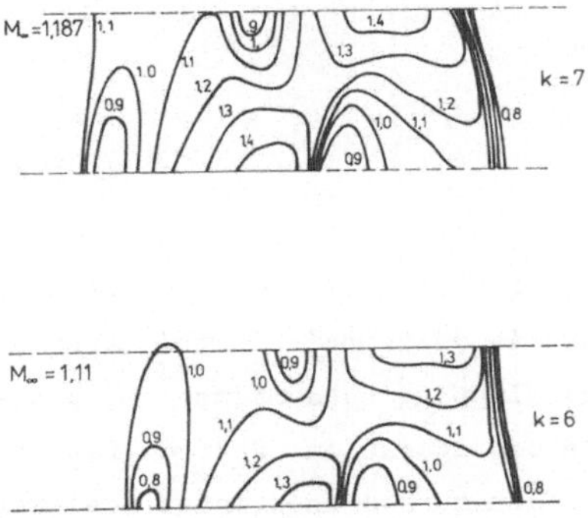

Fig.5a

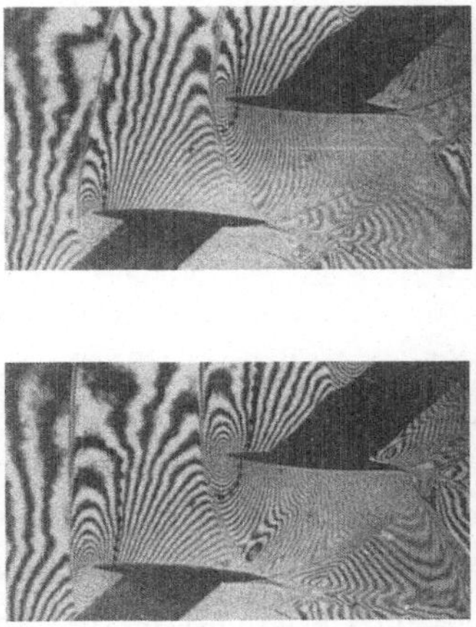

Fig.5b

References

[1] Kozel,K.;Polášek,J.;Vavřincová,M.:Numerical Solution of Transonic Flow Through a Cascade with Slender Profiles,Proceedings of VI.International Conference on Numerical Methods in Fluid Dynamics,Tbilisi ,1978

[2] Kozel,K.;Polášek,J.;Vavřincová,M.:Über eine Relaxationsmethode zur Berechnung von ebenen transonischen Strömungfeldern,ZAMM, T 204-206,1980

[3] Oliver,D.A.;Sparis,P.:Computational Studies of Three-Dimensional Transonic Shear Flow,NACA CR-1816,1971

[4] Kozel,K.;Rozsypal,P.:Computation of Transonic Flow Past a System of Thin Profiles in a Channel,Strojnický časopis No.6, 1981 (in Czech)

[5] Dvořák,R.:On the Development and Structure of Transonic Flow in Cascades,Symposium Transsonicum II,Göttingen,1975

[6] Kozel,K.;Rozsypal,P.;Vavřincová,M.:Numerical Solution of Three-Dimensional Transonic Flow Past Thin Body,Strojnický časopis, 1982 (in Czech)

[7] Sator,F.G.:Computation of Transonic Flow with Detached Bow-Shocks Through Two-Dimensional Turbomachinery Cascades,ICAS Paper,No 76-40,1976

TRANSONIC POTENTIAL FLOW CALCULATION ABOUT COMPLEX BODIES BY A TECHNIQUE OF OVERLAPPING SUBDOMAINS

by T.H. Lê

Office National d'Etudes et de Recherches Aérospatiales (ONERA)
92320 Châtillon (France)

ABSTRACT

A method for constructing a grid system for calculating the transonic flow field about complex configurations with multiple components is described. In this approach the computational domain is divided into multiple overlapping subdomains, which are defined according to the different components of the configuration. Surface fitted grids are generated separately for the different components and then interfaced with each other in such a way that accuracy of the solution algorithm is maintained. In each subdomain the transonic full potential equation in non conservative form is solved by using the S.L.O.R. iteration scheme. As a test case, the present method has been applied to subcritical, supercritical, lifting and non lifting flows over a three-dimensional configuration with two components, a wing-fuselage combination.

1. INTRODUCTION

The tremendous improvement of algorithms for the solution of the transonic full potential equation, e.g. refs. [1] to [7], allows to consider their applications about aerodynamic configurations of increasing geometrical complexity up to a realistic aircraft configuration.

The main problem posed by complex geometries is the generation of a suitable grid which allows to satisfy the exact surface boundary conditions and which does not lead to convergence difficulties. Moreover, mesh refinements at the appropriate places should not increase too much the total number of mesh points. The generation of a single conforming or quasi-conforming grid for a complex configuration gives rise to great difficulties because the natural grid systems for the different aircraft components (wing, fuselage, nacelle or tail) are not in general compatible between themselves.

To alleviate these difficulties the present method adopts an approach based upon a subdomain technique similar to component-adaptive grid interfacing used by Atta in the two-dimensional case [8]. The computational space is divided into several interconnected boxes or overlapping subdomains, in such a way that the problem of mesh generation for each subdomain will be simpler than the problem of generating a single mesh for a complete flow field. The matching of the different subdomains can be achieved by different ways. A very general one is to make two adjacent domains overlap and to use tridimensional interpolation to relate the two solutions in the overlapping region. In this way the meshes in the two subdomains are completely uncoupled. In the present method we require the two meshes to have two surfaces in common, so that only two-dimensional interpolations are necessary to relate the two solutions.

The subdomain approach has also been applied for the solution of Navier-Stokes equations [9] ; it is also attractive in that the equations solved need not be the same for each subdomain, thus more accurate treatment for localized region of the flow field can be achieved [10].

2. SUBDOMAIN APPROACH FOR A WING-FUSELAGE COMBINATION

The problem considered is that of the transonic flow past a wing-fuselage combination in free air. The flow field domain is divided into two overlapping subdomains (fuselage and wing). The overlapping region is limited by two vertical planes R_W and R_f which are respectively part of the boundaries of the wing subdomain and of the fuselage subdomain. R_w and R_f are mesh surfaces for both subdomains but with different distributions of mesh points for each subdomain. Figure 1 shows a sketch of the decomposition of the flow field of the wing-fuselage combination. Then in each subdomain grids are generated separately.

2.1 Mesh generation in wing subdomain

The grid system in wing subdomain is a "C" type mesh, the generation of which is based upon a parabolic transformation [11].

The main features of grid generation are as follows :

- a first family of coordinate surfaces consists in spanwise plane sections
- in each section, C type mesh is generated by a parabolic mapping
- in the three (ξ , η, φ) mesh directions stretching functions are introduced to produce adequate point repartition
- the root plane section is the R_w boundary.

2.2 Mesh generation in fuselage subdomain

The discretization of the fuselage subdomain is based on a family of mesh planes all orthogonal to a given axial direction which is the fuselage axis. Fig. 2 shows typical cross section geometries, respectively upstream of the wing, in the leading edge region, near midchord and in the trailing edge region. In each cross section plane, full lines [A, B, (B'), C] and [D, E, (D')] are body contours and dashed lines [B, D] or [B, E] are cut lines. In these cross sections, the R_f boundary and the R_w boundary, are vertical adjacent mesh lines.

For generating the computational grid we use an analytical transformation, $x = \xi$ $y = f(\xi, \eta, \varphi)$; $z = g(\xi, \eta, \varphi)$, based upon a sequence of simple conformal and shearing transformations which are suitable for all typical cross sections of fuselage geometry. Then in surfaces ζ= constant the computational domain looks like the one shown in fig. 3 where the corresponding R_w and R_f boundaries are η = const. lines separated by one $\Delta\eta$ mesh.

A uniform rectangular mesh in this domain is then mapped back to the physical domain to give the cartesian coordinates of the grid points. Figures 4 b), c) and d) show the meshes in the three sections. Figure 4 a) shows a partial view of the wing-fuselage combination mesh system.

Fig. 5 shows a partial view of the meshes in the R_f plane for the wing subdomain (full lines) and for the fuselage subdomain (dashed lines). The meshes in the R_w plane are of course similar.

2.3 Solution algorithm

The full potential equation in non conservative form is solved in both subdomains by a finite difference method elaborated at ONERA by J.J. Chattot and al..Details on the numerical scheme and on the treatment of boundary conditions can be found in refs. [6], [7].

The coupling of the numerical solutions in the two subdomains is achieved by means of Schwarz's alternating method [12].

This method has been chosen among other decomposition techniques [13], [14], owing to its simplicity, and also to the fact that it seems well-suited to vectorized calculation.

The solution process is performed in cycles as follows : it starts by calculating a solution for the flow field in wing subdomain. After a number of SLOR iterations the distribution of velocity potential is interpolated in the boundary R_f of fuselage subdomain. This distribution is used as a Dirichlet condition for fuselage subdomain. Calculation of a solution in fuselage subdomain is then started and a number of SLOR iterations are performed. The distribution of the velocity potential is then interpolated in R_f and used as a Dirichlet condition for wing subdomain. This constitutes one cycle and the whole process is then repeated.

3. COMPUTED RESULTS

The present method is evaluated by presenting numerical results for subcritical, supercritical, lifting and non lifting flows. The model selected consists of the W_A B_2 configuration of RAE [15].

Calculations are performed using a (71 x 20 x 36) mesh for the fuselage and a (88 x 20 x 17) mesh for the wing. The number of iterations performed in each subdomain can vary typically between 10 to 30, for one cycle and 40 to 60 cycles are necessary to decrease the maximal residual error value below 10^{-6} in both subdomains. Calculations in wing subdomain were initialized with a calculation in a (44 x 20 x 10) coarse grid.

The W_AB_2 configuration was calculated at a freestream Mach number (M_∞) of 0.8 and an angle of attack (α) of 0 and 2 degrees, and at $M_\infty = 0.9$ and $\alpha = 0°$.

Results of calculations are shown at two spanwise stations in fig. 6 and are compared to experiments [15]. The agreement is rather good, the discrepancy being more noticeable near the velocity peak on the upper surface, probably because of the mesh being not, fine enough. Further calculations will be performed in finer mesh size (78x20x54) (176x20x32) for example. Fig. 6e shows pressure distribution calculated along the body at angular location $\varphi = 90°$ in the case $M_\infty = 0.8$ and $\alpha = 0°$ which is in general agreement with the experimental evolution.

REFERENCES

[1] CAUGHEY, D.A. and JAMESON, A. : AIAA Paper 79-1513 July 1979.

[2] YU, N.J., : AIAA Paper 80-1391 July 1980.

[3] LEE, K.D. and RUBBERT, P.E. : Lecture Notes in Physics, Vol 141, Springer Verlag (1981)(pp. 266-271).

[4] BAKER, T.J. : AIAA Paper 81-1015 (June 1981).

[5] HECKMANN, G., : Subsonic/Transonic Configuration Aerodynamics, AGARD CP-285, Münich 1980, (pp. 3-1, 3-11).

[6] CHATTOT, J.J., COULOMBEIX, C., MANIE, F., and SCHMITT, V. : D.G.L.R. Symposium Transonic Configuration Bad Harzburg, June 1978 ONERA T.P. n° 1978-67.

[7] CHATTOT, J.J., COULOMBEIX, C., TOME, C.S. : La Recherche Aérospatiale 1978-4 pp. 143-159.

[8] ATTA, E. : AIAA Paper 81-0382.

[9] METIVET, B. and MORCHOISNE, Y. : Proceed. of 4th GAMM Conference on Numerical Methods in Fluid Mechanics, Paris 7-9 Octobre 1981, Notes on Numerical Fluid Mechanics, Vol. 5 (H. Viviand, Ed.), Vieweg (1982) (pp. 207-219).

[10] CAMBIER, L., GHAZZI, W., VEUILLOT, J.P. and VIVIAND, H. : 5ème Colloque International sur les méthodes de calcul scientifique et technique de l'INRIA, ONERA T.P. n° 1981-143.

[11] JAMESON, A. : Comm. Pure Appl. Math., Vol. 27, 1974 (pp. 283-309).

[12] SCHWARZ, H.A. : Gesammelte Mathematische Abhandlungen Vol. 2.

[13] HELLER, D. : Siam Reivew Vol. 20, n° 4, October 1978.

[14] MORICE, P. : IRIA Rep. 7214/72017 1972.

[15] TREADGOLD, D.A., JONES, A.F., and WILSON, K.H. : Experimental Data Base for Computer Program Assessment AGARD AR-138 (pp. B4-1, B4-25).

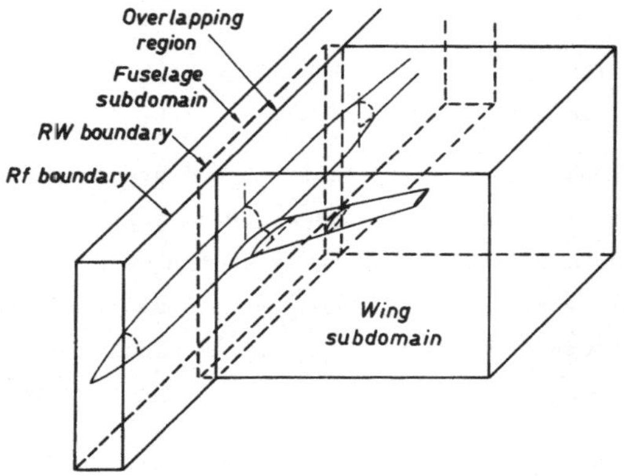

Fig. 1 : Subdomain decomposition technique : wing/fuselage configuration

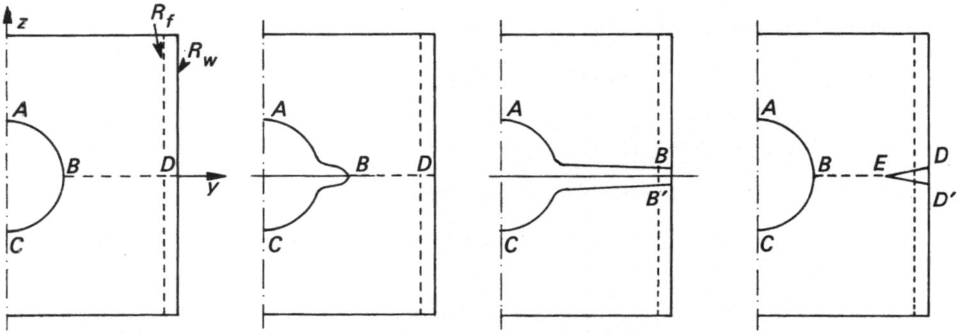

Fig. 2 : Schematic of cross-sections in fuselage sub-domain (x = const. sections)

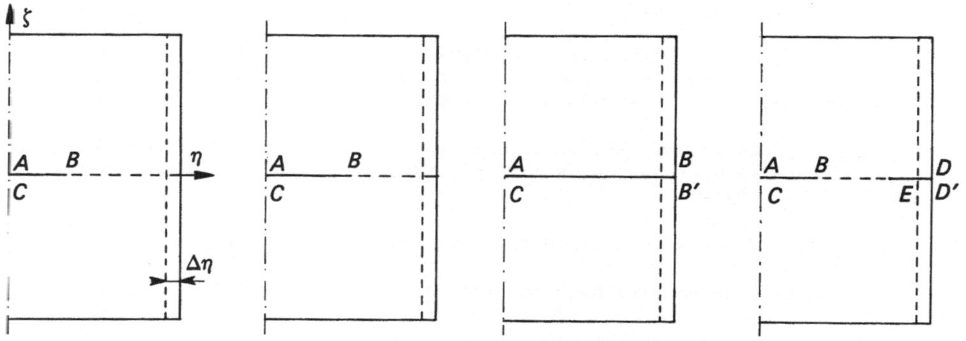

Fig. 3 : Mapped plane

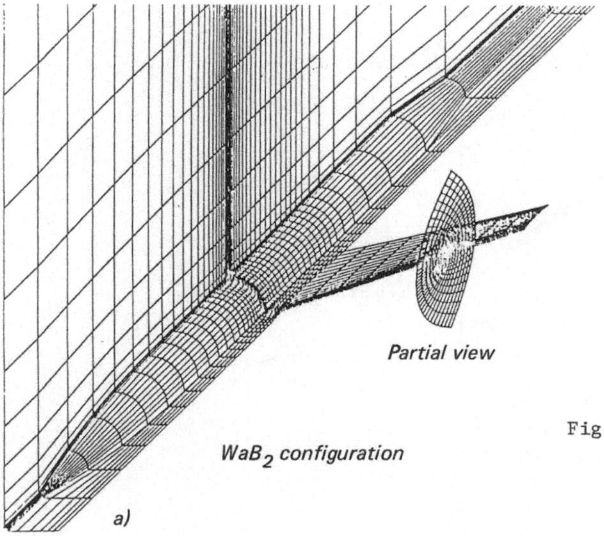

Fig. 4 : Wing-body combination mesh

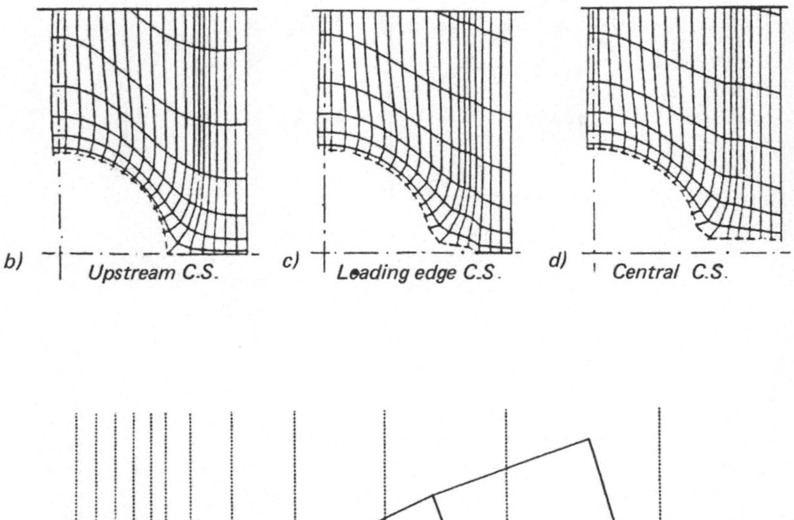

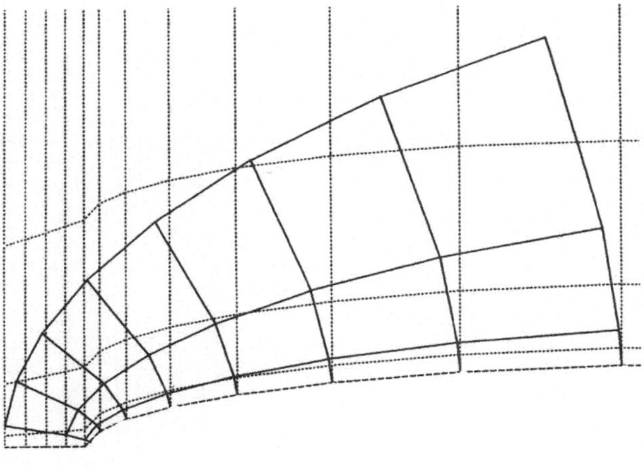

Fig. 5 : Mesh systems in R_f boundary plane

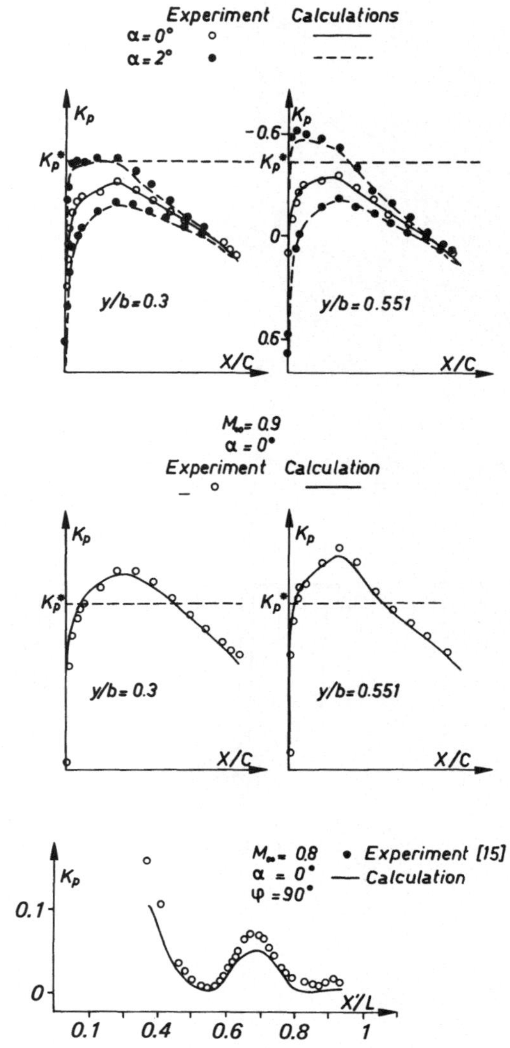

Fig. 6 : Pressure distribution comparisons calculations - experiments

CALCULATION METHOD FOR

TRANSONIC SEPARATED FLOWS OVER AIRFOILS INCLUDING SPOILER EFFECTS*

by J.C. LE BALLEUR**

Office National d'Etudes et de Recherches Aérospatiales (ONERA)
92320 Châtillon (France)

INTRODUCTION

Based on viscid-inviscid interaction concepts, composite numerical solvers are now developed for viscous flows. The objective is to retrofit a viscous capability into the efficient inviscid solvers, with the accuracy of the boundary layer-like techniques, without the limitations of the boundary layer theory. A composite solver for the full Navier-Stokes equations has been for example suggested by Khosla, Rubin [1], with a numerical technique where the viscid and inviscid-like equations are solved almost coupled.

The present composite solver [4] assumes a "Defect Formulation" [2, 3, 5] of the viscous equations for the viscid-inviscid splitting, introduces "Thin Layer" approximations, and solves uncoupled the inviscid and viscous-like components. A coupling algorithm, based on explicit-like coupling relaxation techniques, "Direct" and "Semi-Inverse", is used to converge the viscid-inviscid interaction with consistent schemes at each coupling node (strong coupling).

The capability of the inverse boundary layer methods in separated flows, Catherall, Mangler [6], Carter [7, 8], is here joined to progress in viscid-inviscid splitting and numerical coupling [2 -5]. With fully overlapping domains for the viscous and inviscid computations, the Defect Formulation implements very simply an allowance on the normal pressure gradient inside the viscous layer . The first approximation identifies with the overlapping inviscid pressure gradient and generates Defect Integral methods similar to the classical boundary layer methods [4, 5]. The Defect Formulation also provides the recovery of the supersonic upstream influence[2, 3, 5], and insures a full capability in shock-waves boundary -layer interaction. This paper summarizes equations detailed in ref. [4]. The solver couples viscous and inviscid techniques of minimal complexity, potential codes and interacted Defect integral methods for boundary layers or dissymmetrical wakes. The composite solution is shown to be yet representative of complex flows at high lift, with rear-loading or trailing-edge separations. In addition, we report here the progress with a new simple modelling which, combined with the basic solver, provides a flow simulation over transonic airfoils with a trailing-edge base or a spoiler flap.

*Study conducted with DRET support

**Senior Scientist - ONERA

DEFECT FORMULATION OF THE VISCOUS EQUATIONS

Denoting $\bar{U}_{(x,y)}$ the unknown vector in the viscous field, $\bar{F}(\bar{U})$ and $\bar{G}(\bar{U})$ the flux-terms in the conservation equations, $\bar{H}(\bar{U})$ the source-terms, (x,y) the curvilinear body-coordinates, we assume an overlapping inviscid flow solution $U_{(x,y)}$, which identify with the real viscous flow far from the boundary layer and wake. We denote $F(U)$, $G(U)$, $H(U)$ the flux and source terms of the inviscid equations. The viscid-inviscid splitting is assumed as :

Inviscid equivalent flow
$$\frac{\partial F}{\partial x} + \frac{\partial G}{\partial y} = H \tag{1}$$

Viscous Defect equations
$$\frac{\partial (F-\bar{F})}{\partial x} + \frac{\partial (G-\bar{G})}{\partial y} = H - \bar{H} \tag{2}$$

Matching Coupling conditions
$$0 = \lim_{y \to \infty} \left[U - \bar{U} \right]_{(x,y)} \tag{3}$$

The coupling conditions (3) insure a continuous matching of the viscous and inviscid solutions at large y, without requirement of any patching boundary. The lack of asymptotic expansion in (3) allows to recover a Strong Coupling, which removes the boundary layer theory limitations in flow separation or shock-waves interaction, even when Thin Layer approximations are involved in (2). The Defect equations (2) and coupling relations (3) are the most appropriate to introduce an integral method of solution, based on a y-integration of the Defect flux-balance (2), between y = 0 and $y \to \infty$. No assumption of small inviscid gradients inside the viscous layer is required. From the Defect continuity equation, we get exactly, denoting ρ and $q = (u,v)$ the inviscid density and velocity, $\bar{\rho}$ and $\bar{q} = (\bar{u}, \bar{v})$ the viscous-ones :

$$\rho v_{(x,0)} = \frac{\partial}{\partial x}\left[\rho q \delta^* \right]_{(x,0)} + \bar{\rho}\bar{v}_{(x,0)} \tag{4}$$

$$\delta^* . \rho q_{(x,0)} = \int_0^\infty \left[\rho u_{(x,y)} - \bar{\rho}\bar{u}_{(x,y)} \right] . dy \tag{5}$$

These relations are no-more subject to boundary layer approximations, but recover the usual wall-transpiration formula (4) because of the unusual definition of the Defect displacement thickness δ^* in (5).

APPROXIMATE THIN LAYER DEFECT EQUATIONS

The Defect equations (2) are also very appropriate to assume Thin Layer approximations in the momentum equations. However, due to the Defect Formulation where the viscous pressure field $\bar{p}$ is matched to the inviscid-one p, the viscous approximation never degenerates into the Prandtl assumptions $\bar{p}(x,y) = \bar{p}(x) = p(x,0)$. At first order, the y-momentum equation reduces only to $\bar{p}(x,y) = p(x,y)$. Then, the Defect x-momentum integral equation may be written in the usual Von Karman form :

$$\left[\frac{d\theta}{dx} + \frac{\delta^* + 2\theta}{q}\frac{\partial q}{\partial x} + \frac{\theta}{\rho}\frac{\partial \rho}{\partial x} = \frac{C_f}{2} + \frac{\bar{\rho}\bar{v}}{\rho q}\left(1 - \frac{\bar{u}}{q}\right) \right]_{(x,0)} \tag{6}$$

$$\left[\delta^* + \theta \right] . \rho q^2_{(x,0)} = \int_0^\infty \left[\rho u^2_{(x,y)} - \bar{\rho}\bar{u}^2_{(x,y)} \right] . dy \tag{7}$$

as soon as (7) is selected to define a generalized Defect momentum-thickness $\theta(x)$.

A second order modelling of the y-momentum equation is only used to deduce a corrected viscous pressure :

$$\frac{\partial(p-\bar{p})}{\partial y} = -K^*(x)\cdot\left[\rho u^2-\bar{\rho}\bar{u}^2\right](x,y) \tag{8}$$

$$\left[p-\bar{p}\right](x,0) = -K^*(x)\cdot\left[\delta^*+\theta\right]\cdot\rho q^2(x,0) \tag{9}$$

The curvature K^* (x) is different from the wall or the wake-center-line curvatures usually involved in boundary layer theory, and is based on a y-averadged "induced" curvature of the interacted inviscid streamlines (strong viscous coupling). The equations (4) and (9) provide the inviscid boundary conditions across the inviscid wake-cut y = 0, if they are applied both to the upper and lower half-wakes. The displacement condition (4) is exact, but the curvature effect (9) is subject to the interacting second order modelling. The wake dissymmetry may be fully taken into account through the geometry positioning of the minimal velocity locus y = 0, through the distinct upper-lower induced curvatures K^* (x, 0±), through the full computation of distinct upper-lower Defect thicknesses $\delta^*(x\pm)$, $\theta(x\pm)$ for the two half-wakes.

TURBULENT INTEGRAL METHOD MODELLING-COMPOSITE SOLUTION

An analytical velocity profiles modelling $\tilde{u}$ is used to close the Defect integral equations (4) (6). The Coles' formula is used to combine the "Law of the Wall" with an original "Law of the Wake". The Wake component is an analytical similar solution deduced from a mixing length closure, with some additional modelling when the shape parameter of the velocity profile is very high (extensive separations) see [4]. Finally, the viscous field $\bar{q}$ (x,y) is assumed to be deduced from the overlapping inviscid-one q (x,y) with the following composite solution, involving the local Mach and Reynolds numbers, and the two viscous free-parameters computed with the integral equations, thickness $\delta(x)$ and shape A (x) :

$$\bar{q}(x,y) = q(x,y)\cdot\tilde{u}\left[x,\frac{y}{\delta(x)},A(x),M(x),R_{\delta}(x)\right] \tag{10}$$

The required integral equation added to (4) (6) is the entrainment equation, a collocation with the local x-momentum equation at (x, δ) :

$$\frac{d\delta}{dx}-\left[\frac{v}{u}\right](x,\delta) = \left[\frac{1}{\rho u}\,\frac{\partial\tau/\partial y}{\partial(u-\bar{u})/\partial y}\right](x,\delta) = E(x) = \frac{\tilde{\tau}}{\tilde{\tau}_{eq}}\cdot E_{eq} \tag{11}$$

In case of turbulent equilibrium flows, the entrainment term E_{eq} (x) has been connected to the velocity profiles through algebraic modelling considerations for the Reynolds stress $\tau(x,y) = \tau_{eq}(x,y)$, see [4]. For general non-equilibrium flows, the velocity profiles are assumed unchanged, and the departure from equilibrium of the Reynolds stress or entrainment is assumed to be deduced from an approximate system of turbulent transport equations. This system computes y-averadged estimates for the Reynolds stress $\tilde{\tau}(x)$, the turbulent kinetic energy $\tilde{k}_{(x)}$ and the local dissipation $\tilde{\varepsilon}(x)$, as a departure from their equilibrium counterparts $\tilde{\tau}_{eq}(x)$, $\tilde{k}_{eq}(x)$, $\tilde{\varepsilon}_{eq}(x)$, which have been connected with the velocity profiles [4] :

$$\begin{aligned} \tau(x,y) &\simeq \frac{\tilde{\tau}(x)}{\tilde{\tau}_{eq}(x)} \cdot \tau_{eq}(x,y) \\ \frac{D\tilde{k}}{Dt} &= \frac{\tilde{\varepsilon}\,\tilde{\delta}}{\tilde{k}} \left[\frac{\tilde{\tau}}{\tilde{\tau}_{eq}} \frac{\tilde{\varepsilon}_{eq}}{\tilde{\varepsilon}} \tilde{k}_{eq} - \tilde{k} \right] \\ \frac{D\tilde{\tau}}{Dt} &= 1.5 \frac{\tilde{\varepsilon}\,\tilde{\delta}}{\tilde{k}} \left[\frac{\tilde{k}^2}{\tilde{k}_{eq}^2} \frac{\tilde{\varepsilon}_{eq}}{\tilde{\varepsilon}} \tilde{\tau}_{eq} - \tilde{\tau} \right] \\ \frac{\tilde{\varepsilon}}{\tilde{\varepsilon}_{eq}} &= \left[\frac{\tilde{k}}{\tilde{k}_{eq}} \right]^{\frac{3}{2}} \end{aligned} \qquad (12)$$

The Defect integral system (4) (6) (11) (12) is straightforwardly used for dissymmetrical half-wakes, after vanishing the law of the wall in the velocity profiles, and prescribing the continuity of the viscous field at y = 0.

VISCOUS AND INVISCID NUMERICAL TECHNIQUES

The potential relaxation solver of Chattot, Coulombeix [9] is used for the outer flow with an adaptative C-mesh for the wake cut. Each 5 to 10 cycles, the different Coupling relaxation techniques control the viscous boundary conditions (4) (9) of the inviscid solution. The viscous differential equations are marched with an implicit technique, see [4]. The inviscid pressure is prescribed for attached viscous layers. An inverse method of solution is switched on before separation according to the shape parameter of the viscous layer. Then, for boundary layers, the inviscid transpiration velocity $v_{(x,0)}$ is prescribed. For dissymmetrical wakes, the inverse method prescribes the instantaneous coupling conditions at y = 0, displacement and curvature, through the angular discontinuity of the velocity vector $\left[\frac{v}{q}(x,0+) + \frac{v}{q}(x,0-) \right]$ and the velocity jump $\left[q(x,0+) - q(x,0-) \right]$. The viscous upstream influence is fully recovered because the viscid-inviscid coupling is converged with centered or downwind discretizations of the inviscid gradient $\frac{\partial q}{\partial x}(x,0)$ at supersonic nodes. Each 250 cycles, a mesh update with interpolation adjusts the wake-cut over the viscous center-line geometry.

DIRECT RELAXATION COUPLING TECHNIQUES (ATTACHED)

At coupling cycle, the q^n prescribed viscous solutions provide a predictor $\tilde{v}_{(x,0)}$ to update the inviscid boundary conditions. For stability, this predictor has to be relaxed with a local x-dependent coefficient. This coefficient is deduced from an approximate stability analysis [2-5], assuming an inviscid Prandtl-Glauert perturbation field coupled with the overall influence function of the viscous integral method :

$$\left[\frac{\delta}{q} \cdot \frac{\partial q}{\partial x} \right]^{\pm}_{(x,0)} = a_1^{\pm}(x) \cdot \left(\frac{v}{q} \right)^{+}_{(x,0)} + a_2^{\pm}(x) \cdot \left(\frac{v}{q} \right)^{-}_{(x,0)} + a_3^{\pm}(x) \qquad (13)$$

$$\left[\frac{\partial q}{\partial x}\right]^{+}_{(x,0)} - \left[\frac{\partial q}{\partial x}\right]^{-}_{(x,0)} = a_4(x) \tag{14}$$

For a boundary layer the simplified relation (13) (without the lower quantities) provides an estimate of the coupling amplification coefficient $\mu_D(x,\alpha)$ at wave-number α , and consequently the local relaxation $\omega_{opt}(\mu_D)$ required for an over-relaxation-like procedure ($0 < \omega < 2$), see [2-5], (R = 0) :

$$\left[v^{n+1} - v^n\right]_{(x,0\pm)} = \omega \cdot \omega_{opt}(\mu_D^{\pm}) \cdot \left[\tilde{v} - v^n \pm R\right]_{(x,0\pm)} \tag{15}$$

The stability may be controlled with the maximum wave number α_{max} on the coupling grid, [2-5]. For a dissymmetrical wake, the stability of the displacement effect coupling is controlled for perturbations where a frozen curvature effect is assumed. Then, equation (14) may be joined with the relations (13) to extract a purely upper (or lower) half-wake influence function. The corresponding amplification coefficients $\mu_D^{\pm}$ are used in the relaxation formulae (15), where the non-zero residual $R\left[\tilde{v}(x,0\pm), v^n(x,0\pm)\right]$ is now the angular error for the wake geometry positioning at cycle n, [4].

SEMI-INVERSE RELAXATION COUPLING TECHNIQUES (SEPARATED)

The v^n- prescribed viscous solution provides a predictor of the velocity $\tilde{q}$, which is locally compared to q^n at each coupling node, to correct iteratively v^n. The technique is assessed on the coupling stability analysis [4,5], computes at each node the relaxation coefficient, and is overrelaxation-like ($0 < \omega < 2$) :

subsonic node : $\left[v^{n+1} - v^n\right]_{(x,0\pm)} = \omega \cdot \frac{\sqrt{1-M^2}}{\alpha} \cdot \omega_{opt}(\mu_I^{\pm}) \cdot \left[\frac{\partial \tilde{q}}{\partial x} - \frac{\partial q^n}{\partial x}\right]_{(x,0\pm)}$

supersonic node : $\left[v^{n+1} - v^n\right]_{(x,0\pm)} = \omega \cdot \frac{\sqrt{M^2-1}}{\alpha^2} \cdot \omega_{opt}(\mu_I^{\pm}) \cdot \left[\frac{\partial^2 \tilde{q}}{\partial x^2} - \frac{\partial^2 q^n}{\partial x^2}\right]_{(x,0\pm)}$ (16)

The amplification coefficient $\mu_I^{\pm}$ corresponds to the inverse coupling iteration. The stability may still be controlled with the maximum wave-number $\alpha = \alpha_{max}$. For coupling dissymmetrical wakes, only the local estimate of the amplification and relaxation coefficients is different from the boundary layer semi-inverse coupling, see [4].

MODELLING FOR SPOILER FLAP OR TRAILING-EDGE BASE

The spoiler flap is assumed to be hinged on the wall and the ignored dead-air region below the spoiler is looked as a part of the equivalent airfoil section, fig. 1. The exact geometry of this equivalent airfoil (G) includes two backward-facing step stations, at spoiler-tip and trailing-edge, poorly compatible with the airfoil solver. A smooth approximate geometry (G') is then substituted to (G), and the interacting inviscid boundary conditions are transferred from (G) to (G') as in small perturbations techniques. The simplified geometry (G') has been assumed to be the airfoil geometry without spoiler, with a trailing-edge closure involving a few percent chord and preserving the chord length.

As a first order expansion of the interacted inviscid solution from (G) to (G'), the inviscid velocity vectors $\vec{q}$ (x) along (G), and $\vec{q}'$ (x') along (G'), at corresponding stations x and x', are identified. Denoting $\vec{x}$ and $\vec{x}'$ the unit vector tangent to (G) and (G'), the transpiration velocity along (G) corresponds to a distribution of injection angle $(\vec{x}, \vec{q}) \simeq (\vec{x}, \vec{q}')$, deduced from the viscous equations coupled with the outer velocity $|\vec{q}(x)| \simeq |\vec{q}'(x')|$. The injection angle $(\vec{x}', \vec{q}')$ along (G') is however different from the viscid-inviscid interaction contribution $(\vec{x}, \vec{q}')$:

$$(\vec{x}', \vec{q}') = (\vec{x}', \vec{x}) + (\vec{x}, \vec{q}') \tag{17}$$

The complementary term $(\vec{x}', \vec{x})$ represents the geometrical slope error of (G') with respect to (G) at station x. The modified interacted transpiration velocity of (17) only takes into account the slopes discontinuities of (G) at the singular backward-facing step stations. The height H of the step has been assumed to change suddenly the displacement thickness and the viscous layer. This viscous jump is calculated simply in assuming a local control volume analysis with equal upstream and downstream pressures. From mass and momentum conservation :

$$\delta^*(x+) = \delta^*(x-) + H \qquad \theta(x+) = \theta(x-) \tag{18}$$

The control volume relations (18) and the interacted equivalent injection (17) may be used jointly with the basic airfoil solver, both with and without separated flow. Separation bubbles may be present at the spoiler hinge or at the trailing-edge. Separation may be induced by relations (18) at the spoiler tip or at the trailing-edge.

EXAMPLES OF RESULTS

a) Subsonic flow - High lift : The flow is the so-called "Stalled airfoil case" of the 1981- Stanford Conference, and corresponds approximately to the maximum lift of the NACA 4412 section. The computed trailing-edge bubble, fig. 2, shows approximately a separation at 75% chord, with a rear stagnation point in the wake. A fully dissymmetrical wake calculation with converged positioning is used. The non-equilibrium turbulent modelling and the wake-curvature correction are switched off.

b) Transonic flow - Trailing edge Base : The fig. 3 compares to experiment the pressure distributions over the CAST 7 airfoil at supercritical conditions. The potential solver is the non-conservative one. The full two-equations turbulent modelling and the wake curvature correction are used. The computation without base modelling assumes an approximate geometry to close the trailing edge, but does not use the modified transpiration velocity of (17). The full modelling provides a noticeably improved agreement with experiment. The base thickness is 0.5% chord.

c) Supercritical airfoil with spoiler flap : The RA 16 SC1 airfoil of Aerospatiale has been studied with a 10° spoiler flap on the upper surface, fig. 4. The agreement with experiment of the pressure distributions is very promising, although the optimal incidence correction for wind-tunnel interference was not carefully adjusted. The control volume relations (18) downstream of the spoiler induce the separation, and the upper boundary layer does not reattach before the trailing edge. An acceptable plateau pressure is predicted. The two-equations turbulent modelling is used, as well as the dissymmetrical wake calculation and positioning, but the curvature correction is switched off. The potential solver is the conservative-one. A trailing edge base is present. We notice that a negative lift due only to the spoiler is recovered in the computation, with a supersonic area on the lower side, and a large separated flow region downstream the spoiler. The restitution of the composite viscous solution is shown with the Mach-lines and streamlines contours on figs. 5 and 6.

Acknowledgements

The author wishes to thanks H. Consigny and J.J. Philippe for providing the experimental data of the spoiler flow, and D. Blaise for his help.

REFERENCES

[1] KHOSLA P.K., RUBIN, S.G. - AIAA Paper n° 82-0099 (1982).

[2] LE BALLEUR, J.C., PEYRET, R., VIVIAND, H. - Computers and Fluids, Vol. 8, n° 1, p. 1-30 (1980).

[3] LE BALLEUR, J.C. - AGARD - CP-168, Introduction Paper, n° 1 (1981).

[4] LE BALLEUR, J.C. - La Recherche Aérospatiale n° 1981-3, English edition, p. 21-45 (1981).

[5] LE BALLEUR, J.C. - V.K.I. Lecture Series 1982-04, Computational Fluid Dynamics (1982).

[6] CATHERALL, D., MANGLER, K.W. - J.F.M., Vol. 26, Part 1 (1966).

[7] CARTER, J.E. - NASA TR-R-447 (1975).

[8] CARTER, J.E. - AIAA Paper n° 81-1241 (1981).

[9] CHATTOT, J.J., COULOMBEIX, C., TOME, C. - La Recherche Aérospatiale n° 1978-4, p. 143-159 (1978).

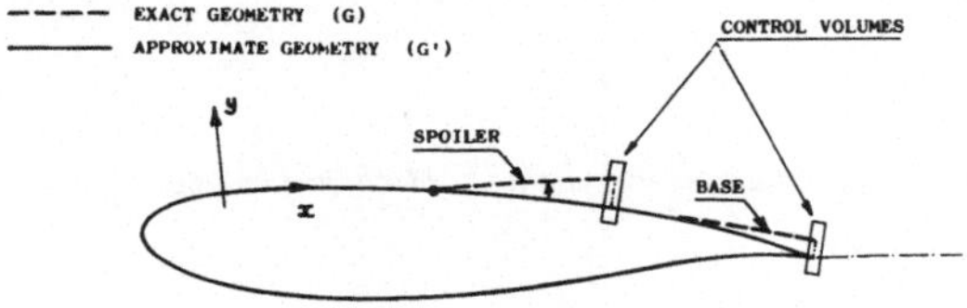

Fig. 1 – Spoiler modelling.

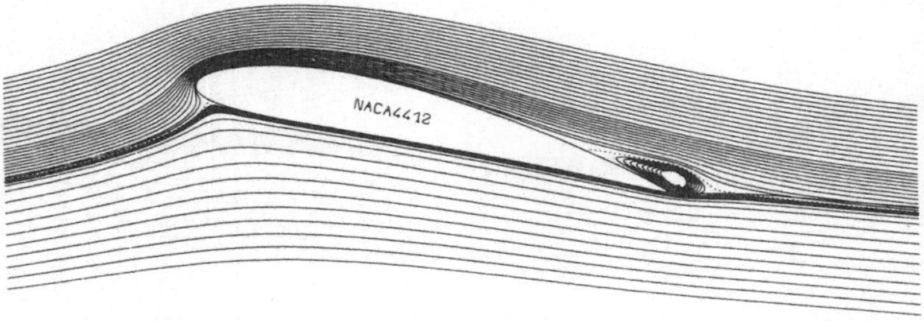

Fig. 2 – Low speed – High lift – Composite viscous solution with trailing edge bubble – NACA 4412 ($\alpha = 13.6°$, $R = 1.5 \times 10^6$, $V = 20$ m/s).

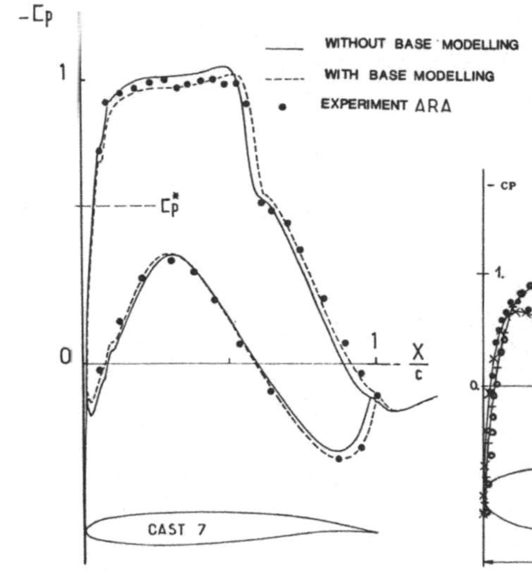

Fig. 3 – Influence of base modelling at trailing edge – Pressure on CAST 7 airfoil ($M = 0.76$, $R = 6 \times 10^6$, $C_L = 0.61$).

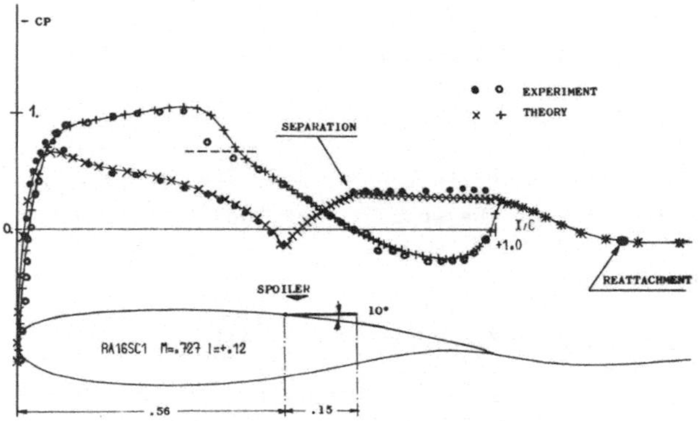

Fig. 4 – Pressure on the supercritical RA16SC1 airfoil with spoiler flap ($M = 0.727$, $R = 4.2 \times 10^6$, $\alpha = 0°$, $\alpha_{spoiler} = 10°$).

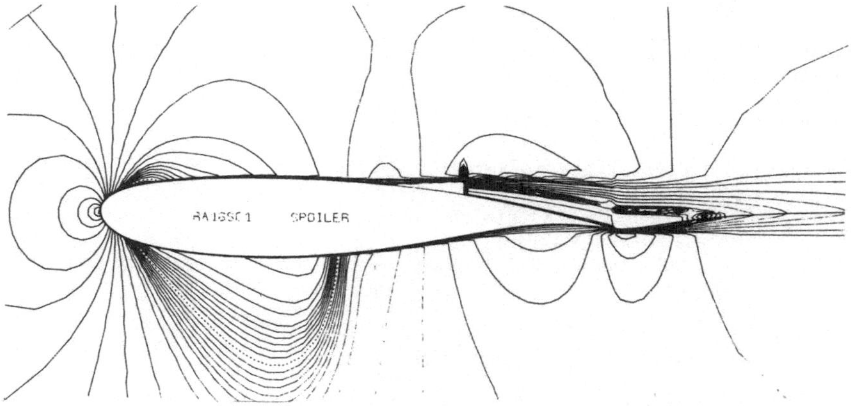

Fig. 5 – Composite viscous solution on RA16SC1 airfoil with spoiler – Mach-lines contours.

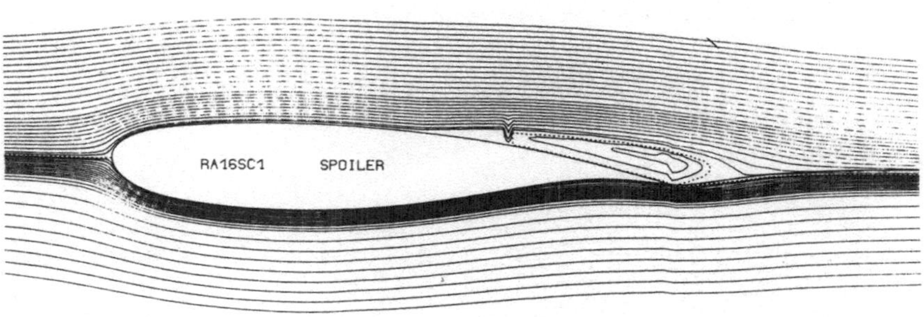

Fig. 6 – Composite viscous solution on RA16SC1 airfoil with spoiler – Streamlines contours.

ON THE USE OF SEVERAL COMPACT METHODS FOR THE STUDY OF UNSTEADY INCOMPRESSIBLE VISCOUS FLOW FOR OUTER PROBLEMS (II)

Y. Lecointe and J. Piquet

Ecole Nationale Supérieure de Mécanique
Nantes, France.

INTRODUCTION

The efficiency of several Navier-Stokes schemes has been discussed for inner flow problems in [1] and [2] . In what follows, a similar study is performed for outer flow problems, completing [3]. Work is concerned with the Navier-Stokes equations in their vorticity stream function formulation discretized spatially by compact algorithms.

CHARACTERISTICS OF THE ALGORITHMS

The main features of the methods are as follows. The Poisson equation is solved by an optimized A.D.I. method. The "Mehrstellen" discretization is defer-corrected so that sixth order accuracy is attainable on ψ. The vorticity equation can be time dif--ferenced either with the help of a Beam & Warming two-step, quasi one leg method which is Δt^2-accurate and A-stable or with a more common A.D.I. method (Peaceman-Rachford). Both methods allow the splitting of the space operator so that unidimensional problems are solved sequentially.

Four different space discretizations for the vorticity are compared : a standard upwind scheme with second order corrections [4] has been implemented in connection with a A.D.I. method in time, either in conservative form [UCh^2-scheme] or in convec--tive form [UTh^2-scheme]. Two formally fourth order accurate schemes of "mehrstellen" type have also been implemented with an approximate factorization scheme of Beam & Warming type in time. One is the "standard" centered "mehrstellen" scheme [CMeh-h^4 scheme] used by Krause *et Al* and Ciment *et Al* which is known to be mesh Reynolds num--ber restricted. The other is the unrestricted OCI scheme (Berger *et Al*) with optimal parameters [UOCI-h^4 scheme] . This scheme builds in a monotonicity property as a result of the fact that it automatically switches its form from a h^4- accurate Numerov discretization when advection is not present to a second order accurate upwind scheme when the mesh Reynolds number becomes high, the upwinding direction depending on the sign of the velocity components.

RESULTS

Comparisons of the aforementionned schemes have been mainly performed for the impulsively started circular cylinder (symetric configuration). The following points arise: (i) The importance of the wiggles determine the quality of the mesh resolution. (ii) The upwinding must be switched off at the rear stagnation point in order to pro--tect the symetry of the flow. (iii) The flow is usually considered to be irrotational on the outer boundary, but for " short time" integrations, a two-step Lax-Wendroff scheme can be used both on the vorticity and on the stream function, so that eddies are convected outwards. (iv) A second order accurate wall vorticity condition is used with UCh^2 and UTh^2 , a fourth order accurate condition is used with CMeh-h^4 and UOCI-h^4. (v) Except with CMeh-h^4 , the global lift coefficient always appears to be less than 10^{-4}, which implies no deterioration of the symetry. (vi) Integral vorticity cons--traints are correctly satisfied in the steady limit.

Comparisons with experiments from Coutanceau & Bouard [5] lead to the following results:
(i) Re=200. Lines of constant vorticity at [t=4, r_∞ =20 a, 51x61 exponentially stret--ched mesh, Δt= 1/100] are shown in Fig.1. CMeh-h^4 appears non symetric due to wig--gling perturbations which develop between t=2 and t=2.5 leading to a lack of conver--gence on Poisson equation and to a failure of the calculation for t > 4. Among the three other schemes, UOCI-h^4 gives the least shift in time (Fig. 2) and the best characteristics of the primary eddy (see e.g. b/2a).
(ii) Re=550. Fig. 3 shows streamlines and lines of constant vorticity in the following conditions: [t=6, r_∞ = 20 a, 61x121 exponentially stretched mesh, Δt = 1/100]

The best characteristics of the primary eddy are given by UTh^2 but secondary vortices appear on streamlines only with UCh^2 and $UOCI-h^4$ so that the best compromise results from $UOCI-h^4$ (Fig. 4).
(iii) Several preliminary results have been obtained for Re=3000,5000,8500 with the following data: [r_∞ = 8a , 61x121 exponentially stretched mesh, Δt = 5/1000]. Only UCh^2 was able to converge but wiggles on the vorticity field indicate a strong lack of radial resolution, advocating for the consideration of the half cylinder problem.

Several numerical simulations of the Karman vortex street have been performed at Re = 200. Because of the high value of Δt = 5/100, centered second order schemes are found to diverge. Results given in Fig. 5 show the time evolution of the Karman vortex street with UCh^2 [r_∞ = 83 a , 46x91 exponentially stretched mesh]. Fig. 6 shows the time evolution of drag C_x, lift coefficient C_y and of the velocity components u_r and u_θ on the axis of the flow, two diameters downstream of the cylinder. While the radial component u_r is indifferent to the direction of the vortices and thus oscillates at $2S_o$, the circumferential component oscillates at S_o. A frequency analysis [6] has been performed on $C_x(t)$ and $C_y(t)$ in order to measure the free Strouhal number So and to investigate the coherence of the flow. The same type of analysis has been performed with $UOCI-h^4$ and, here again, best results are obtained with $UOCI-h^4$ (see Table 1).

Numerical experiments with UCh^2 have been conducted for periodic surging and heaving motions of a circular cylinder in order to exhibit locking-on at Re = 200. The periodic motion has been introduced at t = 64.25 when the "free" Karman vortex street appears to be roughly established. Locking-on in a surging motion is obtained for reduced frequencies F in the vicinity of $2S_o$. Fig. 7 shows that intermittent locking-on - identified by a lack of longitudinal coherence - arises for a/D=.1 , F=.32 while locking-on is obtained for a/D=.3 , F=.4. Frequency characteristics of the obtained regimes are gathered in Table 1. The case of a heaving motion is also considered (Fig. 8). Here again, locking-on is obtained for a/D=.3 , F=.2 .

The case of a heaving elliptic airfoil at Re=1000 has been investigated with UCh^2. Results shown in Fig. 9 give streamline plots and Fig. 10 details drag and lift coef--ficients which are compared with results [7] obtained with the same scheme but with a mesh resolution different from the one used here: [r_∞ = 20a, 51x81 exponentially stret--ched mesh, a is the radius of the circle after mapping], emphasizing the problem of computation of forces. For the sake of clarity, no filtering or integral vorticity control has been used.

Some preliminary results have been obtained for the flow round an impulsively started from rest NACA0012 airfoil in the following conditions:[Re=5000, 61x121 expo--nentially stretched mesh, R_∞=5.5c; c=chord length] . When the steady state is obtained (at $t=t_o$), a pitching motion is introduced where $\alpha = 10° [1 - \cos f(t-t_o)]$;$f=2U_\infty k/c$; k=.5. Fig.11 gives streamline plots for α =0° and α = 6°80.

Finite difference methods of TOMCAT type can hardly lead to an orthogonal mesh because Cauchy-Riemann equations cannot be simultaneously fulfilled on the wall and on the outer boundary. Fig.12 shows that the orthogonalization procedure introduced (by an iterative modification of the location of the mesh points on the outer bounda--ry) gives very bad results in the vicinity of the trailing edge, advocating for a new method resting on the combination of singularity methods and conformal mapping (Fig.13) The obtained mesh leads to a separable Poisson equation and can include attraction along the wall and in the wake of the airfoil.

BIBLIOGRAPHY

[1] Lecointe,Y. & Piquet,J. Numerical Methods in Laminar and Turbulent Flow,Ed. Pine--ridge Press, 1981, pp.53-64.
[2] Lecointe,Y. & Piquet,J. Numerical Properties and Methodologies in Heat Transfer-Proc.2nd.Nat.Symposium (1981)pt.2.1 Ed.T.M.Shih-Hemisphere
[3] Lecointe,Y. & Piquet,J. International Symp.on refined modelling of flows,(1982) AIRH-IAHR-EdF, to be held in Paris (7-10 sep.1982)
[4] Ta Phuoc Loc (1980) Journ. Fluid Mech. 100,pt.1,pp111-128.
[5] Coutanceau,M. & Bouard,R. (1977) Journ. Fluid Mech. 79,pt.1,pp.257-272
(1980) Journ. Fluid Mech.101,pt.3,pp.583-607.
[6] Sulmont,P. & Rajaona,D.R. (1982) session ATMA + personal communication of routines.
[7] Ta Phuoc Loc & Daube,O. (1981) W.A.M. of ASME-Chicago-Vortex flows; pp.155-171.
Authors gratefully acknowledge financial support of DRET through contract 81/075.

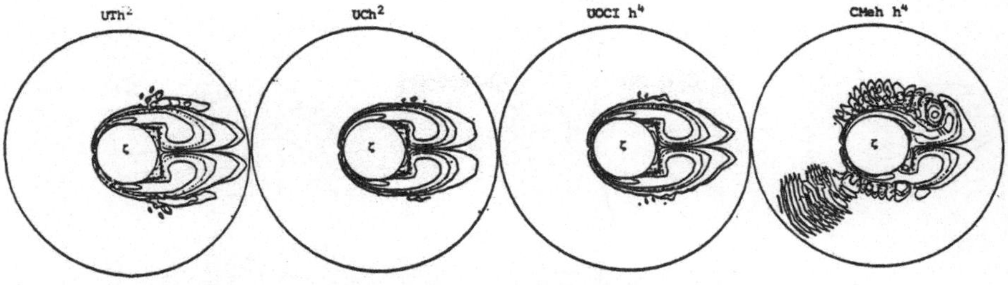

Fig. 1 Comparison of lines of constant vorticity for the studied schemes at t=4, for Re=200

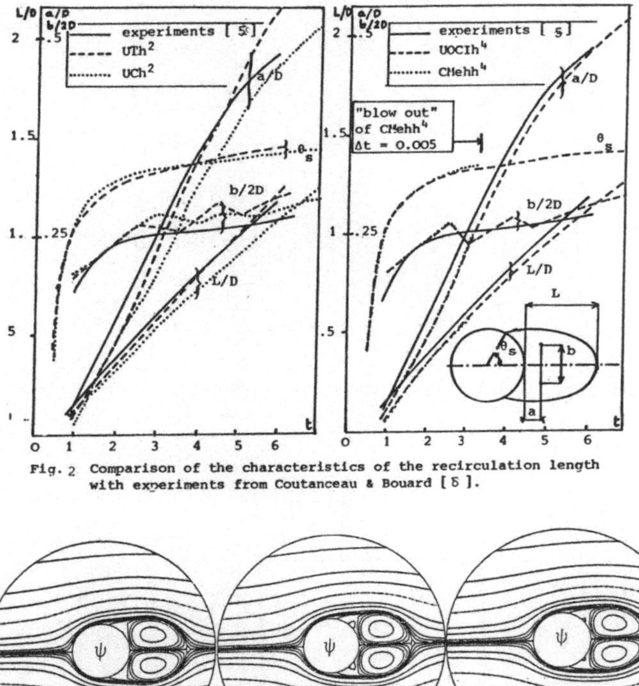

Fig. 2 Comparison of the characteristics of the recirculation length with experiments from Coutanceau & Bouard [5].

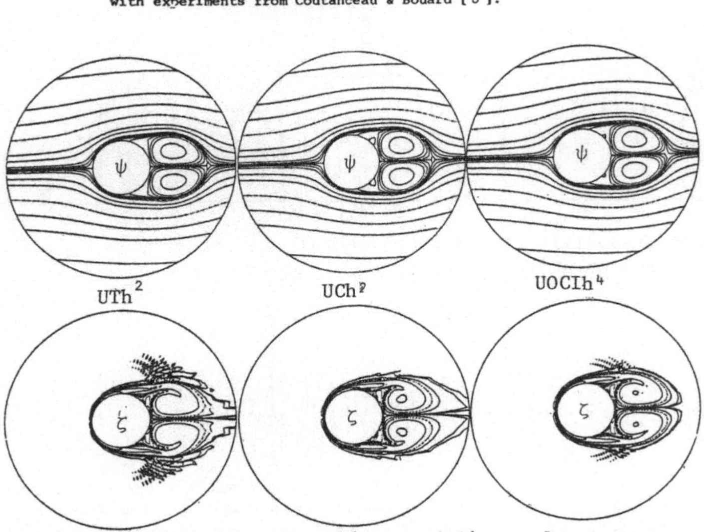

Fig. 3. Comparisons of streamlines and lines of constant vorticity for the studied schemes at t=6, for Re=550

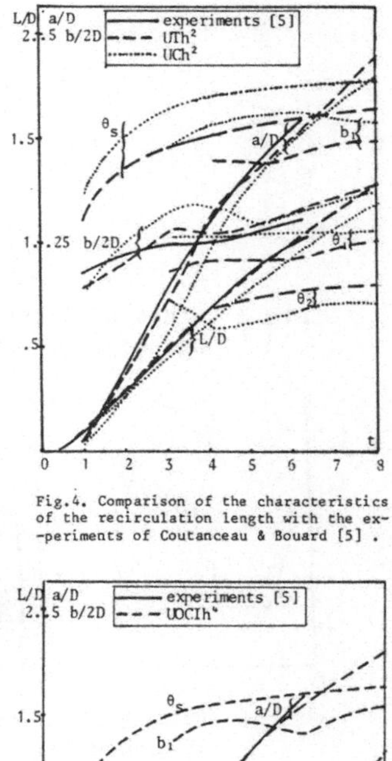

Fig. 4. Comparison of the characteristics of the recirculation length with the experiments of Coutanceau & Bouard [5] .

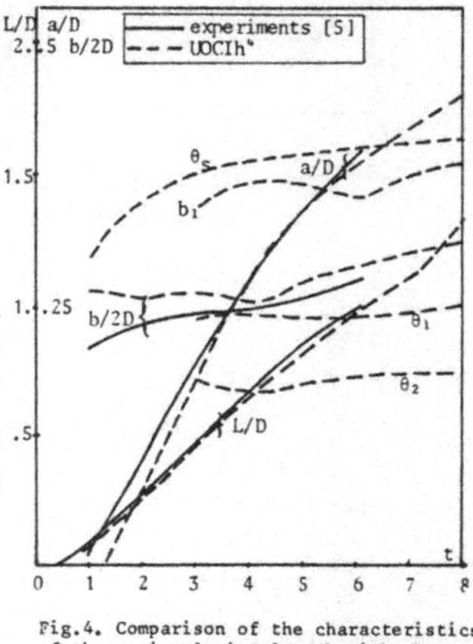

Fig. 4. Comparison of the characteristics of the recirculation length with the experiments of Coutanceau & Bouard [5] .

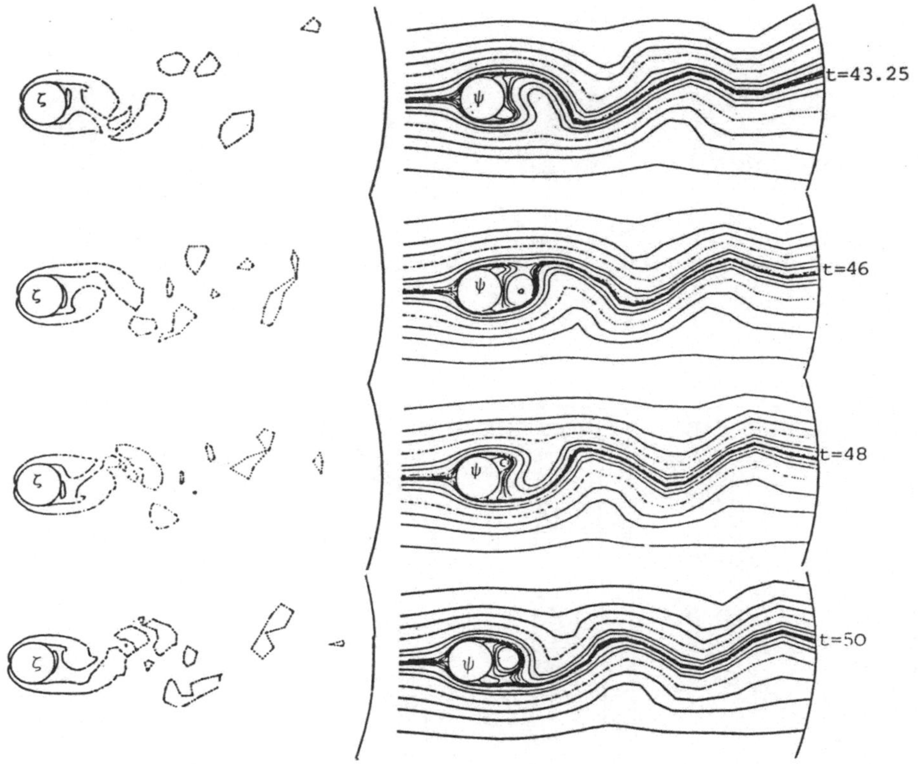

Fiq.5. Lines of constant vorticity and streamlines at different times. Unsteady results.

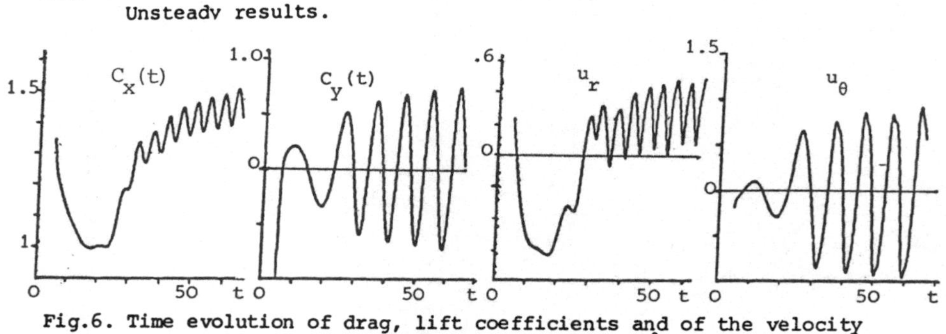

Fig.6. Time evolution of drag, lift coefficients and of the velocity components at a fixed point in the wake. UCh^2-scheme.

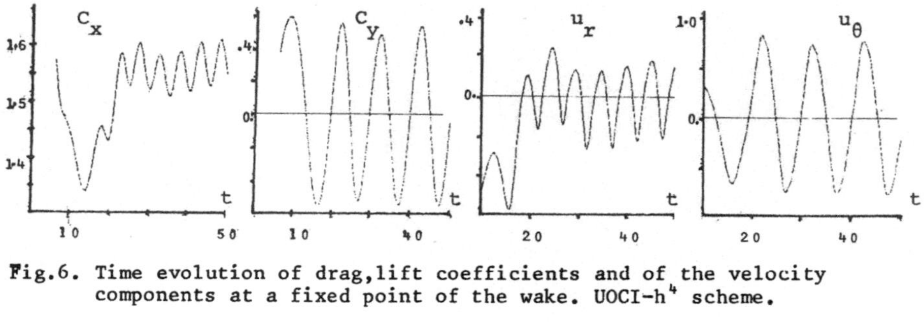

Fig.6. Time evolution of drag,lift coefficients and of the velocity components at a fixed point of the wake. UOCI-h^4 scheme.

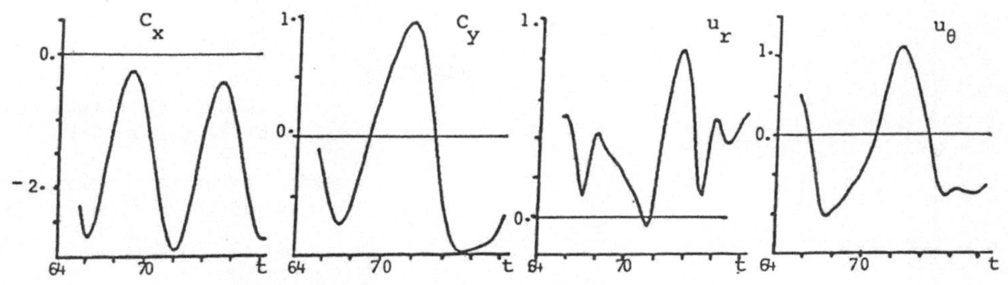

surging motion : a/D = .1   F = .32

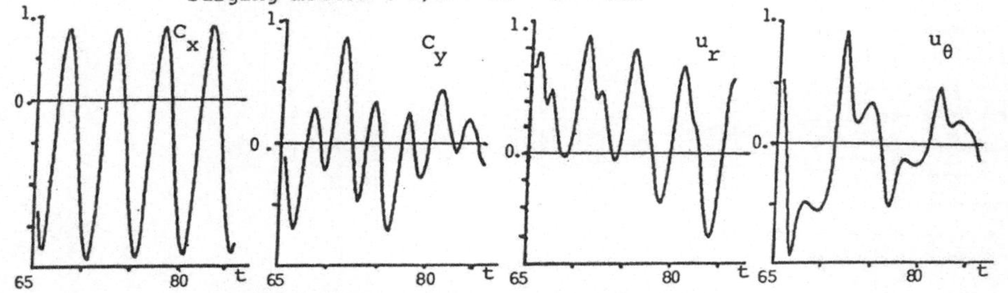

surging motion : a/D = .3   F = .4

Fig.7. Study of locking on in a surging motion.

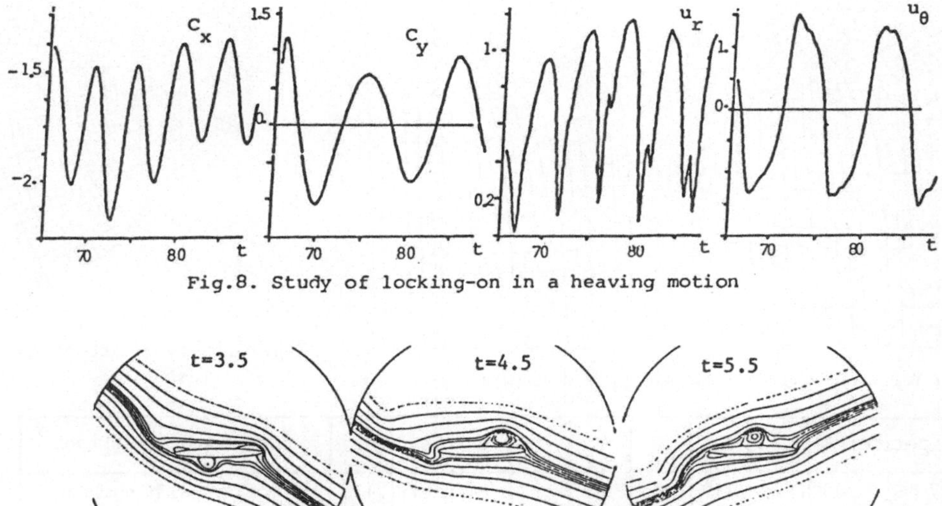

Fig.8. Study of locking-on in a heaving motion

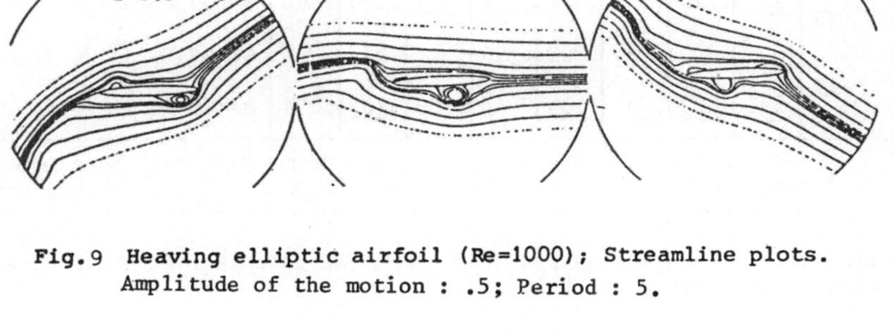

Fig.9 Heaving elliptic airfoil (Re=1000); Streamline plots.
Amplitude of the motion : .5; Period : 5.

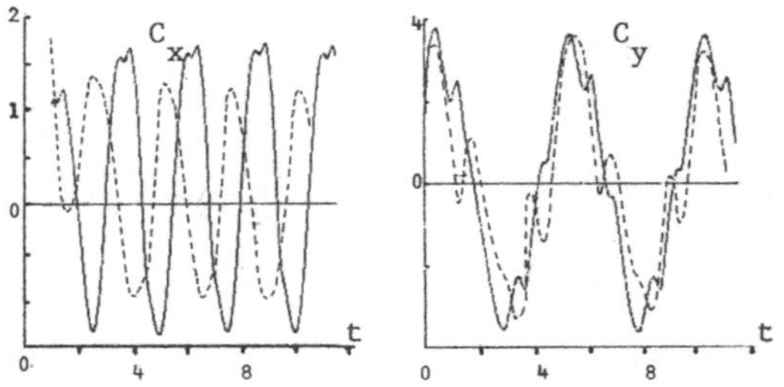

Fig.10. Drag and Lift coefficients on a heaving elliptic airfoil.

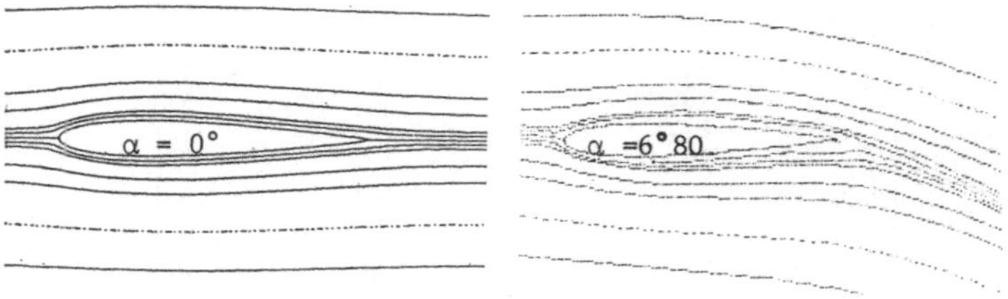

Fig.11. Streamline plots round a NACA0012 airfoil. Re=5000

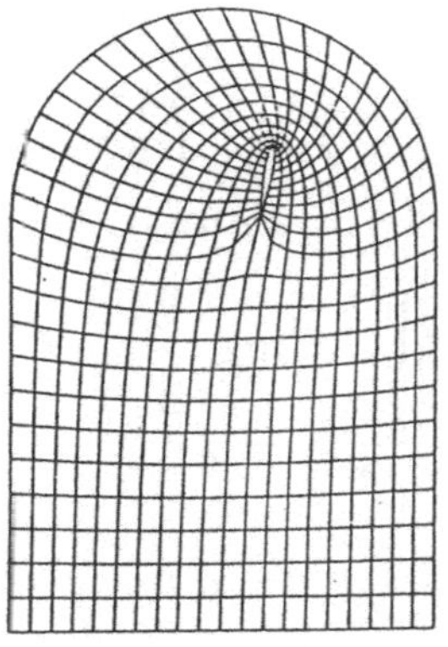

Fig. 12 Mesh obtained with TOMCAT.

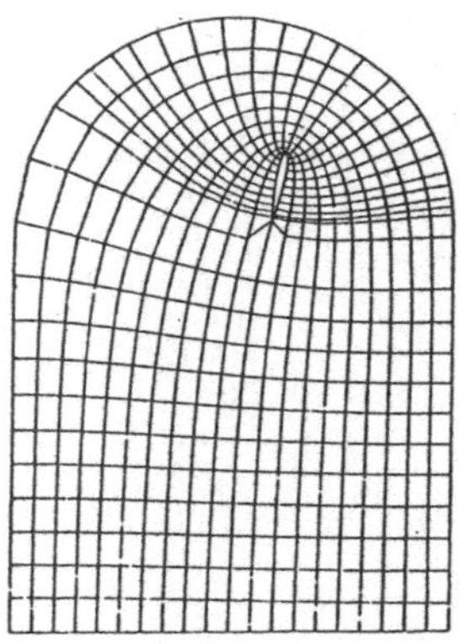

Fig. 12 Mesh obtained after iterated orthogonalization procedure.

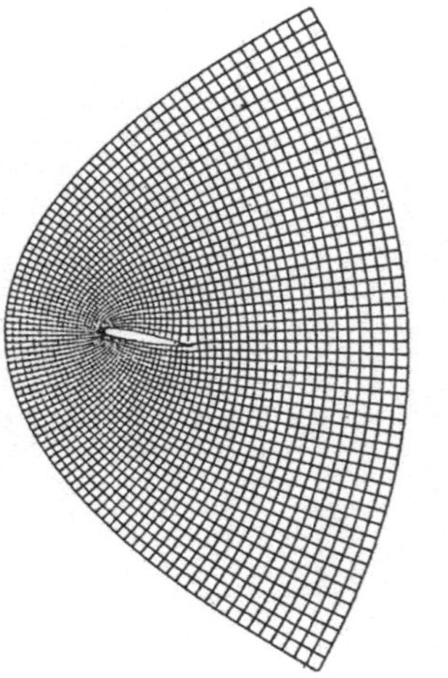

Fig.13. C-type orthogonal mesh

experiments	C_x	C_y	u_1	v_1	ζ_1	u_2	v_2	ζ_2	numerical conditions †
$.17 < S_o < .19$	.224	.119	.235	.119	.123	.099	.105	.115	no motion UCh^2-scheme
	.194	.097	.194	.097	.097	.039	.089	.084	no motion $UOCIh^4$-scheme
shedding† frequency	.15	.115	.247	.128	.157	.159	.120	.108	surge F=.32 $\frac{a}{D}$=.1 UCh^2
	.2	.3	.196	.302	.295	.2	.297	.75	surge F=.4 $\frac{a}{D}$=.3 UCh^2
	.195	.083	.201	.104		.059	.103	.074	heave F=.2 $\frac{a}{D}$=.3 UCh^2
†	Strouhal values given are obtained from frequencies adimensionalized with U_∞/D.								$F=n_m D/a$; n_m: frequency of the motion.

Table. Results of frequency analysis of numerical calculations of the unsteady flow round a circular cylinder.

MESH GENERATION STRATEGIES FOR CFD ON COMPLEX CONFIGURATIONS

S. Leicher, W. Fritz, J. Grashof, J. Longo
Dornier GmbH Friedrichshafen, West Germany

Abstract

An overview of different mesh generation methods in use and under further development at Dornier will be given. Some presented examples (e.g. air intakes, fuselages, wings, wing-fuselage combinations and other related technical combinations) demonstrate the capability of mesh generation techniques to make possible complex three-dimensional flowfield computations. All methods described have in common contour-conformal meshes, i.e. the body surface is a coordinate surface. From the topological point of view two grid types are discerned: single block grids and multi-block structured grids. For both grid types different mesh types like H-H, C-H, O-O or C-O meshes can be used. Each mesh can be generated by one of the described methods.

Introduction

Successful computational algorithms for the transonic full potential equation and the Euler equations have evolved to solve three-dimensional problems. A very effective algorithm is the finite volume method for both, full potential equation and Euler equations (refs. 1 - 4). The finite volume approach allows the treatment of complex configurations like air intakes, fuselages and wing-fuselage combinations. The accuracy of finite volume results depends mainly on the representation of the configuration. Therefore, contour-conformal meshes (the body surface is a coordinate surface) are commonly used for the discretization. To improve the efficiency, e.g. minimize the total number of mesh points, different grid and mesh types can be used.

Mesh generation by interactive graphics or interpolation

The mesh points are constructed more or less by hand taking advantage of a computer-graphics systems software. An example is given by fig. 1. (See also ref. 5). The advantage consists in the users control, especially for complicated situations; on the other hand this grid-generation may becomes laborous and tedious, and it is not applicable to time-dependent grids.

The grid lines can also be computed using interpolating functions with special properties, e.g. stretching, periodicity etc. This method has the advantage that it is programmable and thus the grid can be adapted to configurations by a simple change of some characteristic parameters. As an example the grid for a calculation of a supersonic inlet flow is shown in fig. 2. The grid is time dependent and aligned to the bow shock in front of the inlet (see refs. 3, 6).

Mesh generation by complex functions

Methods using complex functions are restricted on two-dimensional situations or on three-dimensional configurations which can be composed by a series of two-dimensional grids. In this category conformal mapping is the common procedure.

An alternative way is the transformation by an appropriate basic complex function resembling the mesh of the stated problem and discretization in the transformed plane. The discretized body contour and the outer part of the mesh boundary are transformed point by point. Then a set of geometric partial differential equations of the Poisson type are solved in the transformed plane yielding the transformed mesh, which finally is transformed back into the physical plane. An example for this procedure is a Maxwell-type mesh for inlet calculations. (See ref. 7). The complex transformation

$$z = \frac{a}{\pi} \cdot (w+1+e^{w}) \quad , \quad z = x+i\cdot y, \; w = u+i\cdot v$$

reveals a picture according to fig. 3, where the parameter curves u = const. and v = const. are shown. This transformation would yield a nice computational grid for an inlet with zero thickness (see strip $v = \pi$). Now, the Maxwell-grid is modified especially near $v = \pi$ in order to imbed the shape of an inlet section. Finally, the transformed u(i,j) and v(i,j) are improved by solving a set of geometric partial differential equations. This corrected values u(i,j) and v(i,j) are transformed back to the physical grid coordinates x(i,j) and y(i,j). A sample of the final grid is the three-dimensional grid in fig. 4, which is composed by a set of two-dimensional Maxwell-grids.

Quasi-Optimization of gridpoint distribution

This technique is mainly applied to bodies alone and wing-body combinations. For simple configurations as wing alone or wings with cylindrical bodies analytical coordinate transformations can be used (refs. 1, 2). For more complex geometries, a single-block structured mesh with mixed analytical and numerical transformation is used (ref. 8). Figures 5 and 6 illustrate the relationship between physical and computational domain. First, suitable surface grids have to be generated. This is done by a square root coordinate transformation (ref. 1) for the wing, the solution of a 2-D partial differential equation system for the symmetry plane and farfield tip section and by user-specified discretization of the other farfield planes. The fuselage surface grid is constructed by the intersections of the fuselage-surface with those planes normal to the symmetry plane which connect the points on the wing and the corresponding farfield points. These radial grid lines are devided into a given number of intervals with a smooth variation in arc length. Figure 7 shows the basic mesh layout on the wing and the body surface. The 3-D grid then is obtained by solving a set of Poisson equations of the form

$$\vec{U}_{xx} + \vec{U}_{yy} + U_{zz} = \vec{P}(u,v,w,)$$

with the computational coordinates $\vec{U} = (u,v,w)$ and the source terms $\vec{P} = (P,Q,R)$ to control the interior grid spacings (ref. 9). For a fighter-type aircraft the surface grid computed by the grid generation code is shown in figure 8.

Multi-block structured grid generation

For complex configuratons with multi-components single-block structured grids can become highly skewed and generally undesirable. Another approach is the grid imbedding technique (refs. 10, 11), which is flexible in treating complex geometries, especially with multi-components. However, current implementations forgo the advantages of surface fitting. An alternative approach is the use of multi-block structured grids (refs. 12, 13, 14, 17). Figure 9 shows a typical multi-block structured grid about a wing-fuselage-pylon-nacelle configuration. The computational domain is devided into multiple rectangular blocks which can be defined arbitrarily to produce surface fitted grids whose following the natural lines of the configuration. The next step is a one-dimensional perimeter discretization along the block perimeters. This provides boundary conditions for a subsequent two-dimensional grid generation producing grids covering the block surfaces. These in turn serve as boundary conditions to produce three-dimensional volume grids filling each block. The main advantage of multi-block structured grids is their adaptability to complex configurations. Figure 10 shows a multi-block structured grid around a wing-body combination. However, multi-block structured grids introduce special problems in the flow solution algorithms like so-called fictitious corners, collapsed edges and non-analytic block boundaries (ref. 13).

Final Remarks

For each of the presented examples flow calculations have been carried out. The related results are published in the references 3, 4, 5, 15, and 16. At present, the main effort is concerned to the numerical solution of the flow equations using block-structured grids.

References

1. Jameson, A., Caughey, D.A., "A Finite Volume Method for Transonic Potential Flow Calculations." Proceedings, AIAA 3rd Computational Fluid Dynamics Conference, 1977

2. Caughey, D.A., Jameson, A., "Recent Progress in Finite Volume Calculations for Wing-Fuselage Combinations." AIAA paper 79-1513, 1979

3. Rizzi, A.W., Schmidt, W., "Study of Pitot-Type Supersonic Inlet-Flowfields Using the Finite Volume Approach". AIAA paper 78-1115, 1978

4. Grashof, J., "Investigation of the Three-Dimensional Transonic Flow around an Air Intake by a Finite Volume Method for the Euler Equations." In: Transport Phenomana in Fluid Mechanics, Springer Verlag, Berlin/Heidelberg/New York, to be issued autumn 1982

5. Seibert, W., "Anwendung des Euler-Verfahrens zur Bestimmung der Umströmung von Fahrzeugen am Beispiel einer Testgeometrie." Dornier note BF 30-2198/82, 1982

6. Grashof, J., Schmidt, W., Rizzi, A.W., "Verfahren zur Berechnung der reibungsfreien Strömung an Oberschalleinläufen." Dornier Report 80/44 B, 1980

7. Grashof, J., "Inlet Mesh Generation by Using a Maxwell Transformation." Dornier note, 1982

8. Yu, N.J., "Grid Generation and Transonic Flow Calculations for Wing-Body Configurations." Boeing Report D6-45277/1980, 1980

9. Middlecoff, J.F., Thomas P.D., "Direct Control of the Grid Point Distribution in Meshes Generated Elliptic Equations." Proceedings, AIAA 4th Computational Fluid Dynamics Conference, 1979

10. Boppe, C.W., "Computational Transonic Flow about Realistic Aircraft Configurations." AIAA 16th Aerosapce Sciences Meeting, Huntsville, Ala., paper 78-104, 1980

11. Boppe, C.W., Stern, M.A., "Simulated Transonic Flows for Aircraft with Nacelles, Pylons and Winglets." AIAA 18th Aerospace Sciences Meeting, Pasadena, Cal., paper 80-0130, 1980

12. Lee, K.D., Rubbert, P.E., "Transonic Flow Computations using Grid Systems with Block Structure." Paper in "The Seventh International Conference on Numerical Methods in Fluid Dynamics", Stanford University, California, June 1980

13. Lee, K.D., "3-D Transonic Flow Computations Using Grid Systems With Block Structure." AIAA paper 81-0998, 1981

14. Fritz, W., "Außenlastinterferenz im Transschall". Dornier Report 82 BF/9B, 1982

15. Jameson, A., Schmidt, W., "Euler solutions as Limit of Infinite Reynolds-Number for Separated Flows and Flows with Vortices." Paper in: Proceedings of the 8th ICONMFD, Aachen, West-Germany, 1982

16. Leicher, S., "Vergleich der verschiedenen ZKPF4 Nachrechnungen." Dornier note BF 30-2211/82, 1982

17. Longo, J.M., "Body-Grid-Generation", Dornier note BF 30/FAA-2092/82, 1982.

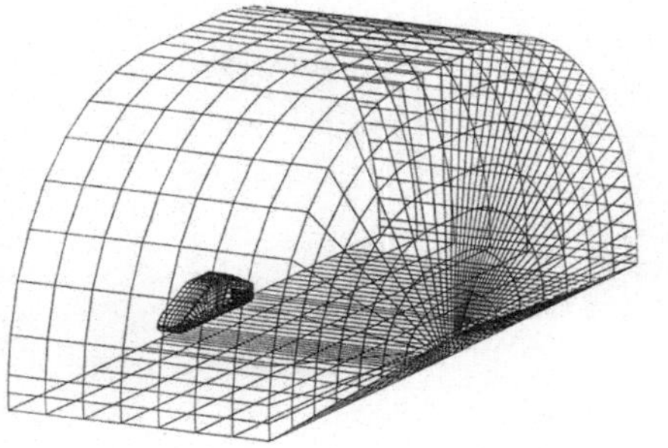

Figure 1. Grid Generated by Interactive Graphics

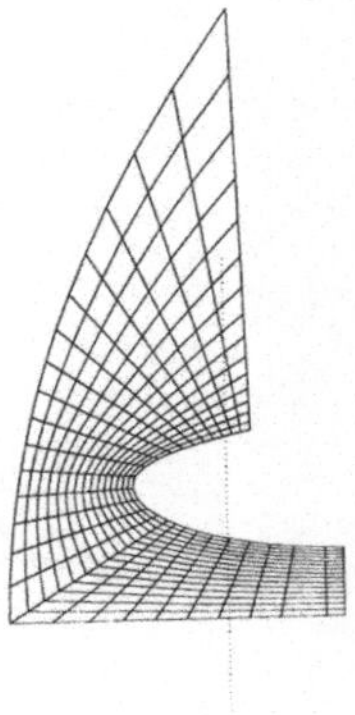

Figure 2. Grid Generated by Interpolation

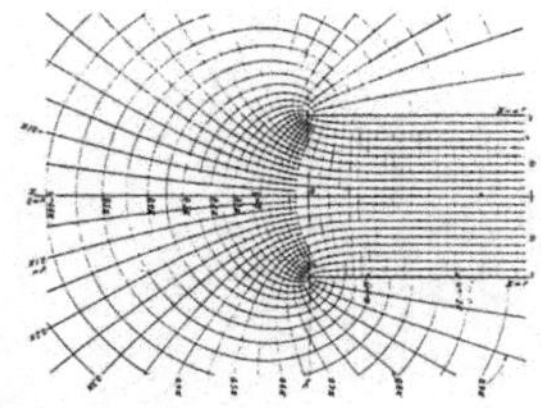

Figure 3. Maxwell Transformation

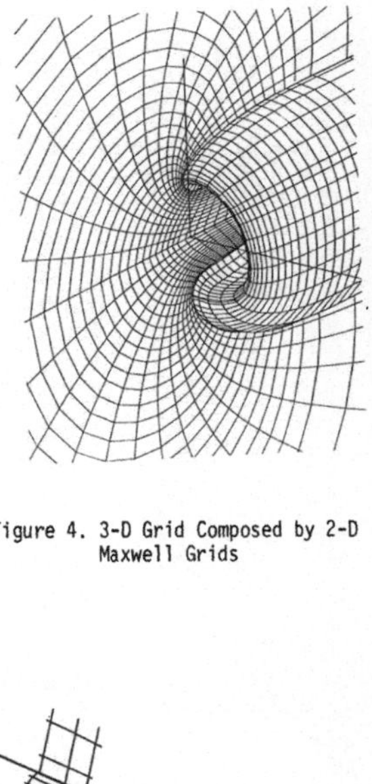

Figure 4. 3-D Grid Composed by 2-D Maxwell Grids

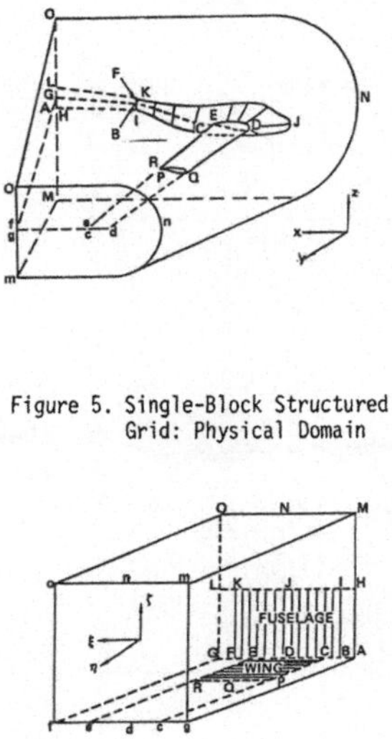

Figure 5. Single-Block Structured Grid: Physical Domain

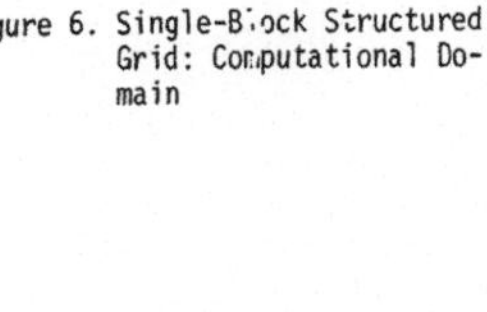

Figure 6. Single-Block Structured Grid: Computational Domain

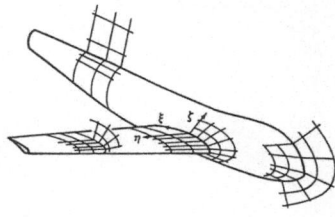

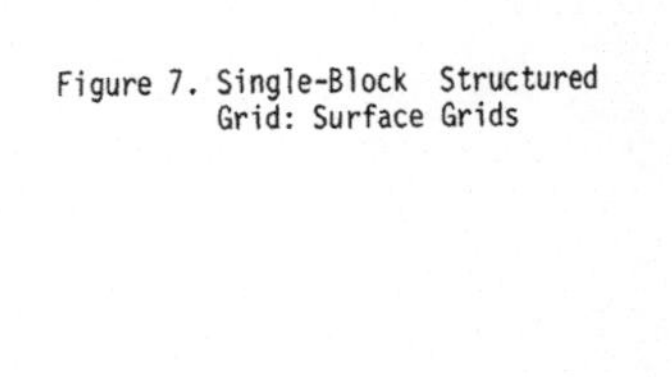

Figure 7. Single-Block Structured Grid: Surface Grids

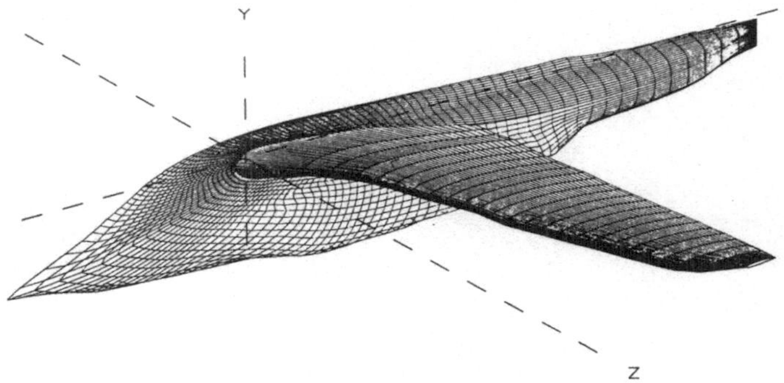

Figure 8. Surface Grid for a Fighter Aircraft

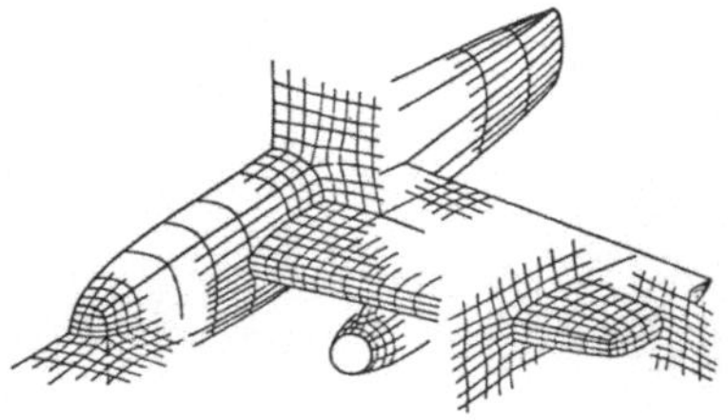

Physical Domain

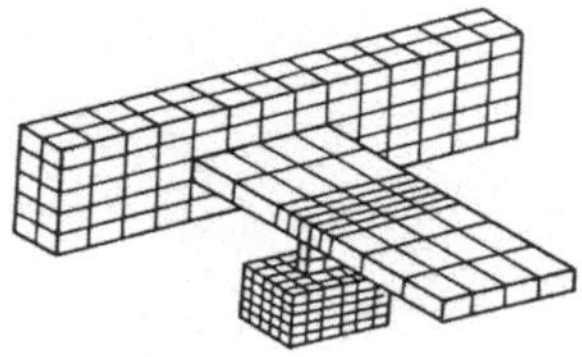

Computational Domain

Figure 9. Block Structuring of a Complex Configuration

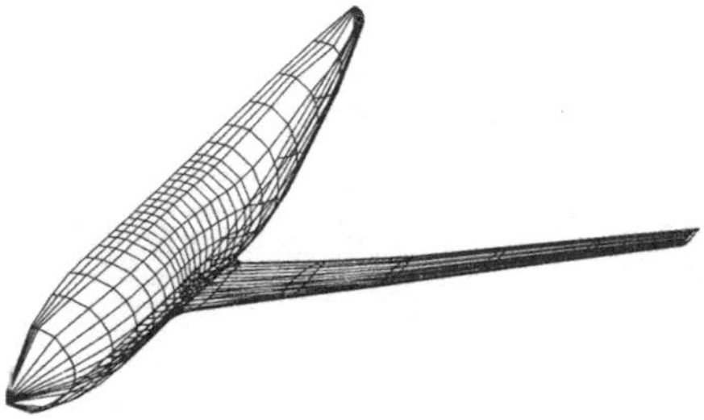

Surface Grids

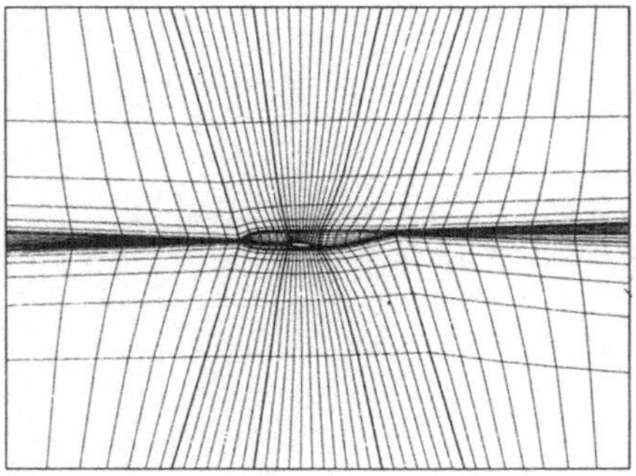

Figure 10. 3-D Grid Generation for a Wing/Body Combination

A NEW NUMERICAL METHOD FOR THE SIMULATION OF THREE-DIMENSIONAL FLOW IN A PIPE

A. Leonard and A. Wray
NASA Ames Research Center
Moffett Field, CA 94035 U.S.A.

INTRODUCTION

In the last few years, major advances have been made in the numerical simulation of wall-bounded transitional and turbulent shear flows. So far, most of the emphasis has been on flows with planar boundaries. Transitional flows in a flat-plate boundary layer (ref. 1) and in a channel (ref. 2), as well as turbulent channel flow (ref. 3), have been investigated without modeling the near-wall region; Moin (ref. 4) gives a critique of these investigations. Except for the study of Patera and Orszag (ref. 5) of axisymmetric pipe flow and the simulation of turbulent flow in annuli by Schumann (ref. 6), using modeled boundary conditions, very little work has been reported on nonplanar flows.

In this paper we present a new numerical technique for simulating three-dimensional, unsteady, incompressible pipe flows and demonstrate its utility and accuracy. Each vector function in the expansion of the velocity field is divergence-free and satisfies the boundary conditions for viscous flow. Some of the benefits of the expansion technique are as follows: (1) pressure is eliminated from the dynamics, (2) only two unknowns per "mesh point" are required, (3) it provides implicit treatment of the viscous terms at no extra computational cost, and (4) no fractional time-steps are required.

In addition, the method uses spectral expansions: Fourier series in the azimuthal (θ) and streamwise (x) directions and global polynomials in the radial (r) direction. Thus, for smooth velocity fields, we expect rapid convergence of our expansions, independent of boundary constraints, as long as the radial polynomials are eigenfunctions of a singular Sturm-Liouville problem (ref. 7). In general, Chebychev or Legendre polynomials are good candidates, but we show that for cylindrical geometry a certain choice of the Jacobi polynomials is particularly advantageous in minimizing the coupling of the resulting equations for the expansion coefficients while satisfying the analytical behavior of the flow variables near $r = 0$. As discussed below, the method has been tested on the linear stability problem for Poiseuille flow.

MATHEMATICAL FOUNDATIONS

The governing equations are the incompressible Navier-Stokes equations for the velocity $\underline{u}$ and the disturbance pressure p,

$$\frac{\partial \underline{u}}{\partial t} + \underline{\omega} \times \underline{u} = -\nabla(p + \frac{1}{2} u^2) - \frac{dP}{dx} \hat{e}_x + \frac{1}{Re} \nabla^2 \underline{u} \tag{1}$$

$$\nabla \cdot \underline{u} = 0 \tag{2}$$

Here $\underline{\omega} = \nabla x \underline{u}$ is the vorticity, Re is the Reynolds number, $\frac{dP}{dx}$ is the constant mean pressure gradient, and the density $\rho = 1$ everywhere. The boundary condition at the pipe wall ($r = 1$) is $\underline{u}|_{r=1} = 0$. We assume periodic boundary conditions in the x direction with period L.

We proceed formally by defining the projection operator $\mathscr{P}$ which projects an arbitrary vector field into the space of divergence-free fields satisfying tangency at the boundary; that is, if $\underline{\Omega}$ is an arbitrary vector field, then we have the unique decomposition,

$$\underline{\Omega} = \underline{\psi} + \nabla\varphi \tag{3}$$

where $\underline{\psi}$ satisfies

$$\nabla \cdot \underline{\psi} = 0 \quad , \quad \underline{\psi} \cdot \underline{n}|_{\text{boundary}} = 0 \tag{4}$$

and $\mathscr{P}$ is the operator that accomplishes this decomposition,

$$\mathscr{P}\underline{\Omega} = \underline{\psi} \tag{5}$$

By applying this operator to the momentum equation, we obtain the time-evolution of the divergence-free velocity field as (ref. 8)

$$\frac{\partial \underline{u}}{\partial t} + \frac{1}{Re} \mathscr{P}(\nabla \times \nabla \times \underline{u}) = -\mathscr{P}(\underline{\omega} \times \underline{u}) - \frac{dP}{dx} \hat{e}_x \tag{6}$$

Our overall strategy is to expand $\underline{u}$ in terms of divergence-free vector functions satisfying the viscous boundary conditions and periodicity in x and θ. We then substitute this expansion into equation (1) and apply a weighted residual method which mimics the application of the projection operator to obtain evolution equations for the expansion coefficients.

EXPANSION METHOD

We write the velocity field $\underline{u}$ as the expansion

$$\underline{u}(r,\theta,x,t) = \sum_{n,k,\ell} a_{n,k,\ell}^{(t)} \underline{\chi}_n(r) \exp(ikx + i\ell\theta) \tag{7}$$

where each expansion vector satisfies

$$\nabla \cdot \left[\underline{\chi}_n(r) \exp(ikx + i\ell\theta)\right] = 0 \tag{8}$$

and

$$\underline{\chi}_n(1) = 0 \tag{9}$$

We derive a system of ordinary differential equations (ODE's) for the coefficients $a_{n,k,\ell}$ by substituting the above expansion into the momentum equation and taking the inner product of the result with a set of weight vectors which are divergence-free,

$$\nabla \cdot \left[\underline{\xi}_m(r) \exp(-ikx - i\ell\theta)\right] = 0 \tag{10}$$

and satisfy the inviscid boundary condition,

$$\underline{\xi}_m(1) \cdot \hat{e}_r = 0 \tag{11}$$

[The (k,ℓ) dependence of $\underline{\chi}_n$ and $\underline{\xi}_m$ is suppressed.] This weighted residual method mimics the application of the projection operator. For example if $\underline{\zeta} = \underline{\xi}_m(r) \exp(-ikx - i\ell\theta)$ then

$$\int_V \underline{\zeta} \cdot \nabla\varphi \, dV = -\int_V (\nabla \cdot \underline{\zeta})\, \varphi \, dV + \int_S \varphi (\underline{\zeta} \cdot \underline{n}) \, dS \tag{12}$$

$$= 0$$

for an arbitrary scalar field φ with periodic boundary conditions in x.

The result for each wave vector (k,ℓ) is a system of ODE's

$$A\,\dot{\underline{a}} + \frac{1}{Re} B\,\underline{a} = \underline{f} \tag{13}$$

where

$$A_{mn} = \int_0^1 \underline{\xi}_m \cdot \underline{\chi}_n \, r dr \tag{14a}$$

$$B_{mn} = \int_0^1 \underline{\xi}_m \cdot \widetilde{\nabla \times \nabla \times \underline{\chi}_n}\, r dr \tag{14b}$$

$$f_m = -\int_0^1 \underline{\xi}_m \cdot \left[\widetilde{\underline{\omega} \times \underline{u}} + 2\pi L \frac{dP}{dx} \hat{e}_x \right] r dr \tag{14c}$$

and where $\sim$ denotes double-Fourier transformation in x and θ. Thus, except for the nonlinear term, the coupling of the equations occurs only through the radial modes.

The choice of radial functions in $\underline{\chi}_n$ and $\underline{\xi}_m$ must be made carefully to (1) minimize the coupling between radial modes (that is, obtain a banded structure for A and B if possible); (2) allow construction of A and B with relative ease; (3) obtain efficient computation of $\underline{f}$; and (4) obtain rapid convergence while satisfying the constraints imposed on $\underline{\chi}_n$ and $\underline{\xi}_n$, including the correct behavior as $r \to 0$.

We find that the sequence of expansion vectors $\left\{\underline{\chi}_{-1}, \underline{\chi}_o^+, \underline{\chi}_o^-, \ldots, \chi_n^+, \chi_n^-, \ldots\right\}$ defined in the following satisfies the above requirements. For $n \geq 0$ the (r,θ,x) components are

$$\underline{\chi}_n^\pm = \begin{pmatrix} \chi_{n,r}^\pm \\ \chi_{n,\theta}^\pm \\ \chi_{n,x}^\pm \end{pmatrix} = \begin{pmatrix} \pm\, ikq_n^{\ell\pm1} \\ kq_n^{\ell\pm1} \\ \mp\frac{1}{r}\frac{d}{dr}(rq_n^{\ell\pm1}) - \frac{\ell}{r} q_n^{\ell\pm1} \end{pmatrix} \quad (k \neq 0) \tag{15}$$

where

$$q_n^\ell = r^{|\ell|}(1 - r^2)^2 g_n^{(|\ell|)}(r^2) \tag{16}$$

and $g_n^{(\ell)}(y)$ is the shifted Jacobi polynomial (ref. 9),

$$g_n^{(\ell)}(y) = P_n^{(o,\ell)}(2y - 1) \tag{17}$$

satisfying the orthogonality condition

$$\int_0^1 y^\ell g_m^{(\ell)}(y)\, g_n^{(\ell)}(y)\, dy = C_m^\ell \delta_{m,n} \tag{18}$$

For the case $k = 0$, the above expansion vectors clearly are not complete and must be replaced by an alternative set. A convenient choice is given by ($n \geqslant 0$)

$$\underline{\chi}_n^- = \begin{pmatrix} -\dfrac{i\ell q_n^\ell}{r} \\ \dfrac{dq_n^\ell}{dr} \\ 0 \end{pmatrix} \quad \underline{\chi}_n^+ = \begin{pmatrix} 0 \\ 0 \\ q_n^\ell \end{pmatrix} \tag{19}$$

It is a simple matter to verify that the expansion vectors defined above yield the correct behavior of $\underline{\tilde{u}}(r,k,\ell)$ as $r \to 0$; for example, if $\ell > 0$,

$$\tilde{u}_r \to \gamma r^{\ell-1}$$

$$\tilde{u}_\theta \to i\gamma r^{\ell-1}$$

$$\tilde{u}_x \to \beta r^\ell$$

where γ and β are complex constants. The additional vector $\underline{\chi}_{-1}$ is required because $\tilde{u}_\theta(k \neq 0)$ and $\tilde{u}_x(k = 0)$ would otherwise have a double zero at the wall.

The corresponding weight vectors are essentially the curl of the $\underline{\chi}$'s. More specifically, if $k \neq 0$ the weight vectors may be expressed as

$$\underline{\xi}_m^\pm = \widetilde{\nabla \times \nabla \times} \begin{pmatrix} \mp i q_n^{\ell\pm1} \\ q_n^{\ell\pm1} \\ 0 \end{pmatrix} \tag{20}$$

while the $\underline{\chi}_n^\pm$ have the form

$$\underline{\chi}_n^\pm = \widetilde{\nabla \times} \begin{pmatrix} -iq_n^{\ell\pm1} \\ \mp q_n^{\ell\pm1} \\ 0 \end{pmatrix} \tag{21}$$

As a result, the (+) vectors are uncoupled from the (-) vectors. The resulting matrices A and B are nonadiagonal, except for an additional nonzero row and a column owing to the vector $\underline{\chi}_{-1}$. The limited bandwidth of the viscous matrix B results from the particular choice of the polynomials $g_n^{(\ell)}$ given by equation (17). In particular, the Laplacian operator in the (r,θ) plane is equivalent to a tridiagonal matrix in the following sense:

$$\nabla^2\left[q_n^\ell(r)\exp(i\ell\theta)\right] = \left(\frac{1}{r}\frac{d}{dr}\,r\,\frac{d}{dr} - \frac{\ell^2}{r^2}\right)q_n^\ell(r)\exp(i\ell\theta)$$

$$= r^\ell\left[b_n^\ell\, g_{n-1}^{(\ell)}(r^2) + c_n^\ell\, g_n^{(\ell)}(r^2) + d_n^\ell\, g_{n+1}^{(\ell)}(r^2)\right]\exp(i\ell\theta)$$

CONVERGENCE TESTS

As a test, the method described above was applied to the problem of determining the time eigenvalues for linearized flow in a pipe. The calculations were performed on a CDC 7600 computer. We assume $u \sim \exp(\lambda t)$ and order the eigenvalues such that Real $(\lambda_1) \geqslant$ Real $(\lambda_2) \geqslant$ .... The results for λ_1 with Re = 9600, $\ell = 1$, $k = 1$, are given in the table below, where $N + 2$ is the number of radial modes in the expansion.

N	λ	
20	-0.02312	- i 0.95050
25	-0.02317074	- i 0.95048142
30	-0.023170795769	- i 0.950481396659
35	-0.023170795764	- i 0.950481396668

Note that the convergence is exponential in N or some power of N, typical of spectral methods (ref. 7), and that there is no indication of significant round-off errors. The results agree with that of Salwen et al. (ref. 10), who obtained $\lambda_1 = 0.02317 - i(0.95048)$, using an expansion in Stokes' eigenfunctions to solve the linear stability problem.

In figure 1 we show convergence of some of the higher eigenvalues for the case $k = 1$, $\ell = 1$, Re = 3000, a Reynolds number in the range where a number of interesting transitional phenomena have been observed experimentally. Note that a large number of eigenvalues are predicted accurately for 30 to 35 radial expansion functions. In figure 2, the amplitudes of the coefficients of $\underline{\chi}_n^-$ for the eigenvalue λ_1 are shown as a function of n for four Reynolds numbers with $k = 1$ and $\ell = 1$. Each of the $\underline{\chi}_n^-$ are normalized so that

$$\int_0^1 |\underline{\chi}_n^-|^2\, r\,dr = 1$$

and the radial expansion $N = 35$ (37 expansion functions) was used for all cases. Again, the coefficients approach zero exponentially in some power of n.

SUMMARY

A new numerical method has been developed to investigate three-dimensional, unsteady pipe flows using a new velocity-vector expansion method. Each vector function in the expansion set is divergence-free and satisfies the boundary conditions for viscous flow. Other features of the general technique are as follows: (1) pressure is eliminated from the dynamics; (2) only two unknowns per "mesh point" are required; (3) there is rapid convergence of spectral methods; (4) there is implicit treatment of the viscous term at no extra computational cost; and (5) no fractional time-steps are required. In the present application of the method to flow in a pipe, the behavior of each flow variable near the computational singular point is treated rigorously and expansions in Jacobi polynomials have been shown to be particularly advantageous. The method has been tested on the linear stability problem for Poiseuille flow and has demonstrated rapid convergence of the eigenvalues and eigenfunctions as the number of radial modes is increased.

ACKNOWLEDGMENT

The authors wish to thank Dr. Parviz Moin and Mr. Robert Moser for many helpful discussions.

REFERENCES

1. Wray, A., and Hussaini, Y.: Numerical Experiments in Boundary Layer Stability, AIAA Paper 80-0275, Pasadena, Calif., 1980.

2. Orszag, S. A., and Kells, L. C.: Transition to Turbulence in Plane Poiseuille and Plane Couette Flow, J. Fluid Mech., Vol. 96, 1980, p. 159.

3. Moin, P., and Kim, J.: Numerical Investigation of Turbulent Channel Flow, J. Fluid Mech., Vol. 118, 1982, p. 341.

4. Moin, P.: Numerical Simulation of Wall-Bounded Turbulent Shear Flows, Proceedings 8th International Conference on Numerical Methods in Fluid Dynamics (this issue).

5. Patera, A. T., and Orszag, S. A.: Finite-Amplitude Stability of Axisymmetric Pipe Flow, J. Fluid Mech., Vol. 112, 1981, p. 467.

6. Schumann, U.: Subgrid Scale Model for Finite Difference Simulations of Turbulent Flows in Plane Channels and Annuli, J. Comp. Phys., Vol. 18, 1975, p. 376.

7. Orszag, S. A.: Spectral Methods for Problems in Complex Geometries, J. Comp. Phys., Vol. 37, 1980, p. 70.

8. Chorin, A. J., and Marsden, J. E.: A Mathematical Introduction to Fluid Mechanics. Springer-Verlag, New York, 1979.

9. Handbook of Mathematical Functions. M. Abramowitz and I. A. Stegun, eds., NBS AMS 55, Sec. 22, 1968.

10. Salwen, H., Cotton, F. W., and Grosch, C. E.: Linear Stability of Poiseuille Flow in a Circular Pipe, J. Fluid Mech., Vol. 98, 1980, p. 273.

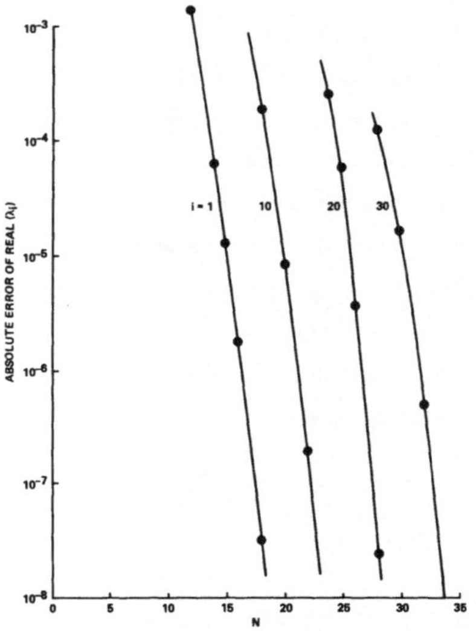

Fig. 1. Convergence of Real (λ_i); $k = \ell = 1$, Re = 3000.

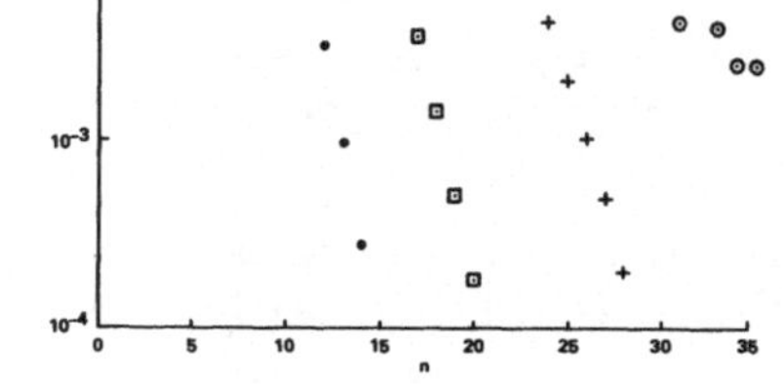

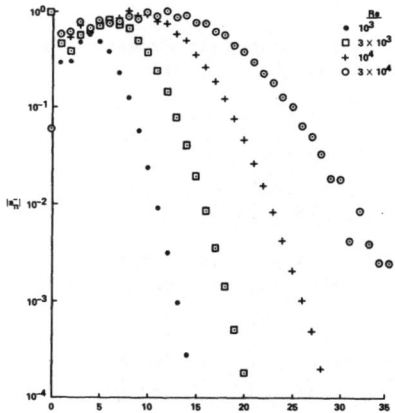

Fig. 2. Amplitudes of the expansion coefficients for the eigenvector corresponding to λ_1 ; $k = \ell = 1$.

AN IMPLICIT FINITE-VOLUME METHOD FOR SOLVING THE EULER EQUATIONS

by A. Lerat*, J. Sidès and V. Daru

Office National d'Etudes et de Recherches Aérospatiales (ONERA)
92320 Châtillon (France)

INTRODUCTION

Various explicit methods have been developed to solve the Euler equations for steady or unsteady transonic flows. These methods allow the prediction of supercritical flows around airfoils without restriction on the strength and motion of shock waves, but they are expensive in computing time due to the severe limitation on the time step required for numerical stability. In order to remove this limitation, we present an implicit method for the solution of the Euler equations. The method is second-order accurate and noniterative like the implicit methods proposed by Beam and Warming [1] and MacCormack [2]. An important feature of the present method is that the implicit terms are of the order of the truncation error, which permits some simplifications and easy treatments of boundary conditions.

The method has been expressed in the finite-volume formulation and applied to steady and unsteady transonic flows. Accurate results have been obtained with an important reduction in computing time over explicit methods.

BASIC METHOD FOR A HYPERBOLIC SYSTEM IN ONE-SPACE VARIABLE

The basic method has been constructed in [3] (see [4], ch. 6 for more details) for a hyperbolic system of conservation laws in one-space variable :

$$w_t + f\ (w)_x = 0 \tag{1}$$

where w and f (w) are m-component vectors. It is an implicit method which possesses the following properties :

P1 : conservation form (suitably defined for implicit schemes)

P2 : second-order accuracy in time and space

P3 : spatial symmetry

P4 : computation of the unknowns by solving a block-tridiagonal linear system $\mathcal{L}$

P5 : linear stability in L^2, for any time step

P6 : dissipation in the sense of Kreiss [5], for any time step

P7 : strict diagonal dominance of $\mathcal{L}$ (case where m = 1), for any time step.

*Ecole Nationale Supérieure d'Arts et Métiers, 75640 Paris Cédex 13 and Laboratoire de Mécanique Théorique, Université Paris VI. Consultant at ONERA.

The construction has been made in a systematic way within the framework of difference schemes using three points in space and two-time levels. The resulting time differencing can be written in the form :

$$\Delta w + \alpha \Delta t\,[A(w)\,\Delta w]_x + \beta\,\frac{\Delta t^2}{2}\,[A^2(w)\,(\Delta w)_x]_x + \gamma\,\frac{\Delta x^2}{2}\,(\Delta w)_{xx} = -\Delta t\, f(w)_x + (1-2\alpha)\,\frac{\Delta t^2}{2}\,[A(w)\, f(w)_x]_x \qquad (2)$$

where $w = w^n$ is the numerical solution at time $n\Delta t$, $\Delta w = w^{n+1} - w^n$, $A(w)$ is the $m\times m$ matrix $df(w)/dw$ and α, β, γ are three parameters (real numbers). Note that the choice $\alpha = 1/2$, $\beta = \gamma = 0$ corresponds to the Beam and Warming scheme. When the spatial derivatives are approximated by three-point centered formulae, the following results have been proved :

a) the class of schemes (2) satisfies the property P1 to P4.

b) The subclass of schemes which meets also the additional requirements P5 to P7 is defined by the inequalities :

$$\alpha < \frac{1}{2}, \quad \gamma < \frac{1}{2}, \quad \beta \leqslant \alpha - \frac{1}{2} \quad \text{and } \beta < \frac{\alpha^2}{4(\gamma-1)} \qquad (3)$$

In a more precise way, property P6 holds everywhere except at a point where the matrix $A(w)$ has a zero eigenvalue, such as a stagnation or sonic point for the Euler equations.

c) A reduction of the truncation error of the above subclass of schemes can be achieved for large time step, by setting :

$$\beta = \alpha - \frac{1}{2} \qquad (4)$$

For the sake of simplicity, the parameters have been chosen such as :

$$\alpha = 0, \quad \beta = -\frac{1}{2} \quad \text{and} \quad \gamma = 0 \qquad (5)$$

Since for $\alpha = 0$, the right-hand side of (2) is equal to a Δw given by an explicit scheme of second-order accuracy, the basic method (2), (5) can be implemented as follows :

- Explicit or physical stage : compute a provisional vector $\hat{w}^{n+1}$ by an explicit scheme of second-order accuracy and set $\Delta\hat{w} = \hat{w}^{n+1} - w^n$.

- Implicit or mathematical stage : ensure unconditional stability by solving

$$\Delta w - \frac{\Delta t^2}{4}\,[A^2(w)\,(\Delta w)_x]_x = \Delta\hat{w} \qquad (6)$$

that is a linear algebraic system $\mathcal{L}$ for Δw, after spatial differencing.
The physical stage may be the one-step Lax-Wendroff scheme as in [3], or a predictor-corrector scheme $\mathcal{S}^{\alpha}_{\beta}$ [6] as suggested in the study [7].
Concerning the mathematical stage, we observe that the implicit term is of the order

of the truncation error. This stage can be greatly simplified if we replace (6) by :

$$\Delta w - \frac{\Delta t^2}{4} [\rho_A^2 (w) (\Delta w)_x]_x = \Delta \hat{w} \quad (7)$$

where the scalar ρ_A denotes the spectral radius of the matrix A. It can be shown that the properties P1 to P7 remain true with the mathematical stage (7), when the space differencing uses the formula :

$$\{[\rho_A^2 (w) (\Delta w)_x]_x\}_i \approx [(\rho_A^2)^n_{i+\frac{1}{2}} (\Delta w_{i+1} - \Delta w_i) - (\rho_A^2)^n_{i-\frac{1}{2}} (\Delta w_i - \Delta w_{i-1})]/\Delta x^2 \quad (8)$$

where the subscript i refers to the point $x = i\Delta x$ and

$$(\rho_A^2)^n_{i+\frac{1}{2}} = \rho_A^2 (\frac{1}{2} w^n_i + \frac{1}{2} w^n_{i+1})$$

In the linear system $\mathcal{L}$, the blocks are now diagonal matrices, so that the property P4 and P7 are improved as follows :

P'4 : computation of the unknowns by solving a tridiagonal linear system $\mathcal{L}$

P'7 : strict diagonal dominance of $\mathcal{L}$, for any time step.

An efficient solution of the linear system $\mathcal{L}$ can be achieved by using the LU factorisation method, also called the double sweep method. However, it should be noted that the replacement of (6) by (7) yields a slight increase of the truncation error.

EXTENSION TO THE CASE OF SEVERAL-SPACE VARIABLES

We have extended the method to the system of m-conservation laws :

$$w_t + f (w)_x + g (w)_y = 0 \quad (9)$$

by splitting the method again in two stages : a physical one based on an explicit scheme approximating (9) with a second-order accuracy and a mathematical one constructed with the A.D.I. technique (a review of which can be found in [8]). The latter can be written as :

$$\begin{aligned} \Delta w^* + \beta \frac{\Delta t^2}{2} [\rho_A^2 (w) (\Delta w^*)_x]_x &= \Delta \hat{w} \\ \Delta w + \beta \frac{\Delta t^2}{2} [\rho_B^2 (w) (\Delta w)_y]_y &= \Delta w^* \end{aligned} \quad (10)$$

where $\Delta\hat{w}$ is given by the physical stage and ρ_A and ρ_B denote the spectral radii of $A = df/dw$ and $B = dg/dw$, respectively.

The advantage of the A.D.I. technique is that we need only the solution of two tridiagonal linear systems at each time step. However, to achieve this purpose we have dropped the mixed-second derivatives of Δw in the mathematical stage, which might introduce a stability condition. A linear stability analysis has been performed in the scalar case (m = 1) when the physical stage is the Lax-Wendroff scheme [9] and when centered formulae similar to (8) are used in the mathematical stage. It has been found that the expressions (10) ensure unconditional stability if

$$\beta \leqslant -1 \tag{11}$$

In the case where $\Delta x B = \Delta y A$, then the inequality (11) is necessary for unconditional stability.

For the solution of the system (9) on a curvilinear mesh, the method has been expressed in the finite-volume formulation. The space domain is now divided into a number of quadrilateral cells $\Omega_{i,j}$ with boundaries $\Gamma_{i,j}$ composed of sides $\Gamma_{i\pm\frac{1}{2},j}$ and $\Gamma_{i,j\pm\frac{1}{2}}$, and we denote by $\nu_{i,j}$ the unit-outward normal to $\Gamma_{i,j}$. In the case of a fixed mesh, the mathematical stage becomes an approximation to the integral system :

$$\int_{\Omega_{i,j}} \Delta w^{*} \, d\Omega + \beta \frac{\Delta t^2}{2} \int_{\Gamma_{i+\frac{1}{2},j} \cup \Gamma_{i-\frac{1}{2},j}} \rho^2 A^{\nu}(w) \frac{\partial(\Delta w^{*})}{\partial \nu_{i,j}} \, d\Gamma = \int_{\Omega_{i,j}} \Delta w \, d\Omega \tag{12.a}$$

$$\int_{\Omega_{i,j}} \Delta w \, d\Omega + \beta \frac{\Delta t^2}{2} \int_{\Gamma_{i,j+\frac{1}{2}} \cup \Gamma_{i,j-\frac{1}{2}}} \rho^2 A^{\nu}(w) \frac{\partial(\Delta w)}{\partial \nu_{i,j}} \, d\Gamma = \int_{\Omega_{i,j}} \Delta w^{*} \, d\Omega \tag{12.b}$$

where the right-hand side of (12.a) comes from the physical stage and $A^{\nu}=\nu^{x}A+\nu^{y}B$, ν^{x} and ν^{y} being the cartesian components of $\nu_{i,j}$.

In the present calculations, the physical stage is the explicit finite-volume method described in [10], that is an extension of the optimal $\mathcal{S}^{\alpha}_{\beta}$ scheme [6] to the case of two-space dimensions, and we have taken $\beta = -1/2$ in (12).

APPLICATION TO TRANSONIC FLOW CALCULATIONS

a) Steady transonic flow. The implicit method has been applied to solve the Euler equations for the flow over the NACA 0012 airfoil at Mach 0.8 and an angle of attack of 1°25. Taking advantage of the steady nature of the flow, we have replaced the unsteady energy equation by the condition of constant total enthalpy, we have used a local time step and we have performed a pseudo-unsteady calculation starting from a uniform flow defined by the upstream infinity conditions.

The boundary conditions are taken into account in the physical stage of the method as reported in [10]. In the mathematical stage, the implicit terms are of the order of the truncation error and thus it is easy to treat them on the boundary. One must only avoid that this treatment ruins the stability of the method or the good structure of the algebraic systems to be solved. In the present computation, we have introduced a row of fictuous cells outside the domain, in which we have simply prescribed $\Delta w^{*} = 0$ or $\Delta w = 0$.

On a fine C mesh made of 188 x 24 cells, the steady state was reached after 1000-time iterations for a mean value of $|\Delta w/\Delta t|$ (over the components of Δw and all the mesh cells) lower than 5×10^{-4}. The local time step was 5.7 times greater than the maximal value allowed for the stability of the explicit method used in [10]. The computing time was 24 mn CPU on a CYBER 750 computer. For the same pseudo-unsteady evolution calculated in the same mesh, the explicit method of [10] had taken 5000-time iterations and 1h15 mn (the convergence test is executed only every 50-time iterations in both methods). The isobar and iso-Mach lines obtained by the present method are shown on figure 1.

b Unsteady transonic flow. We have also calculated a real unsteady flow over the NACA 0012 airfoil oscillating in translation with a great amplitude at a low reduced frequency in transonic regime. This problem simulates the flow over a section of a helicopter rotor blade. It involves important displacements of shock waves. The freestream Mach number is $M_\infty = 0.536$ and there is no incidence, so that the flow is symmetric. The Mach number relative to the airfoil is defined by

$$M_{r,\infty} = M_\infty + M_0 \sin (kt)$$

where $M_0 = 0.327$ and the reduced frequency $k = 0.185$.

A comparison has been made between the numerical solutions computed by the explicit method of [10] with a Courant-Friedrichs-Lewy number CFL = 1 and by the present method with CFL = 5 and 10, on the same moving mesh (94 x 24 cells around the half airfoil). Figure 2 shows the isobar contours plotted for kt = 60° and 120° corresponding both to $M_{r,\infty} = 0.819$. The comparative timings are given below for an evolution until the same physical time (CYBER 750-CPU time).

These numerical results illustrate the efficiency of the present method for practical applications.

METHOD	CFL NUMBER	NUMBER OF TIME INCREMENTS	COMPUTING TIME
explicit [10]	1	8954	1 h 22 mn 22 s
implicit	5	1791	24 mn 58 s
implicit	10	904	12 mn 39 s

REFERENCES

[1] BEAM, R.M. and WARMING, R.F. - J. Comput. Phys. 22, p. 87-110 (1976).

[2] MacCORMACK, R.W. - AIAA Paper n° 81-0110 (1981).

[3] LERAT, A. - C.R. Acad. Sci. Paris 288 A, p. 1033-1036 (1979).

[4] LERAT, A. - ONERA publication n° 1981-1 (1981).

[5] KREISS, H.O. - Comm. Pure Appl. Math. 17, p. 335-353 (1964).

[6] LERAT, A. and PEYRET, R. - La Rech. Aérosp. n° 1975-2, p. 61-79 (1975).

[7] HOLLANDERS, H. and PEYRET, R. - La Rech. Aérosp. n° 1981-4, p. 287-294 (1981).

[8] BRILEY, W.R. and McDONALD, H. - J. Comput. Phys. 34, p. 54-73 (1980).

[9] LAX, P.D. and WENDROFF, B. - Comm. Pure Appl. Math. 17, p. 381-398 (1964).

[10] LERAT, A., and SIDÈS, J. - IMA Conf. Reading Univ., March 81, to appear in Academic Press.

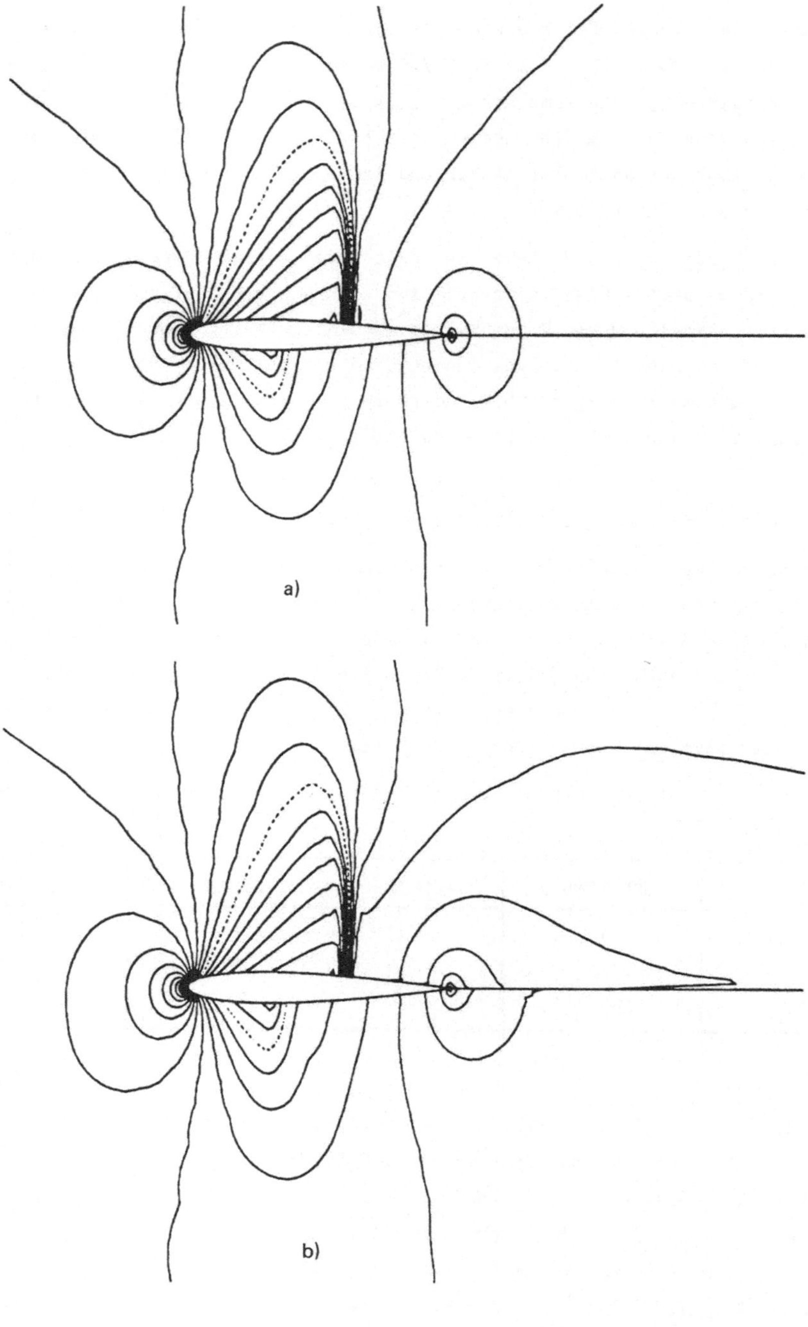

Fig. 1 : (a) Isobar lines and (b) Iso-Mach lines over the NACA 0012 airfoil computed by the present method ($M_\infty = 0.8$, $i = 1°25$).

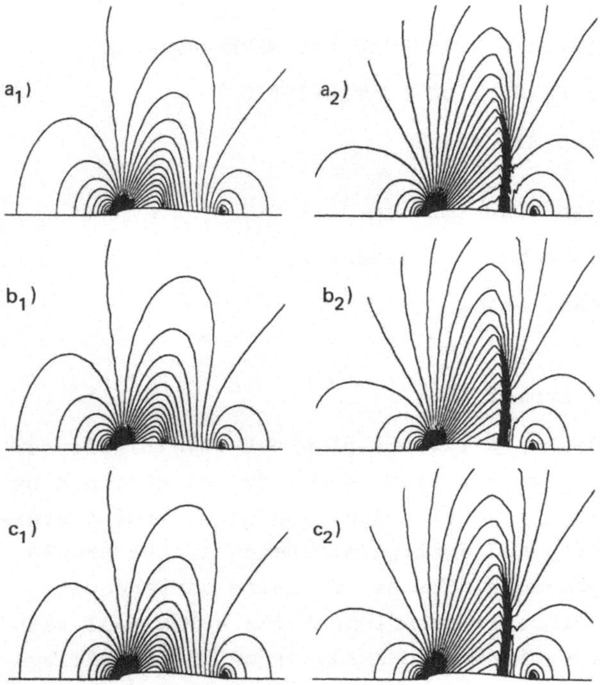

Fig. 2 : Isobar lines over the NACA 0012 airfoil oscillating in translation :

(a_j) explicit method [10] with CFL = 1

(b_j) present method with CFL = 5

(c_j) present method with CFL = 10

(j = 1 for kt = 60°, j = 2 for kt = 120°)

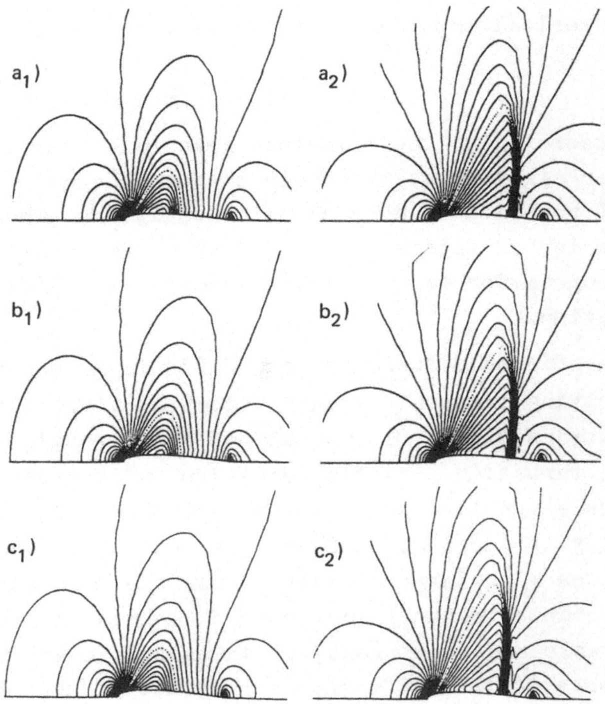

Fig. 3 : Iso-Mach lines over the NACA 0012 airfoil oscillating in translation (corresponding to fig. 2)

FINITE DIFFERENCE COMPUTATION OF PRESSURE AND WAVE-DRAG OF SLENDER BODIES OF REVOLUTION AT TRANSONIC SPEEDS WITH ZERO-LIFT

Li Shou-ying and Luo Shi-jun
North-western Polytechnical University,
Xi'an, China

Abstract

The pressure, the wave-drag and the positions of shock-wave of slender bodies of revolution at transonic speeds with zero-lift are obtained by solving the transonic axisymmetric potential equation with large disturbance in the free stream direction and small disturbance in the transverse direction, using the Murman-Cole schemes of finite differences. The computed results for three different configurations agree well with known wind tunnel test results. A linearized analysis of the stability and the convergence of line overrelaxation of the difference equations for steady axisymmetric small perturbation potential flow is made. The numerical experiences do agree with the theoretical conclusions.

1. Introduction

According to the transonic area rule, the zero-lift wave-drag of a slender aircraft at transonic speeds is equal to that of its equivalent body of revolution under certain conditions. Therefore, it is important for design of aircraft to build up a practical and convenient method and program for calculating transonic zero-lift wave-drag of a body of revolution.

The finite difference method [1] is an effective numerical method to solving transonic steady potential flow. References [2,3,4] have calculated well the flow about bodies of revolution by solving classical small perturbation potential equation and boundary conditions. However, nearby the stagnation points, the disturbance in the free stream direction usually is large. Therefore, the potential equation and the boundary conditions with large disturbance in the free stream direction and small disturbance in the transverse [5] are used in this paper. To make computation stable and convergent, it is useful to analyse the stability and convergence of the difference equations. Of course, the analysis are only made under the linearized condition for the complexity of nonlinear problem.

2. Basic Equations and Formulas

A body-fixed cylindrical coordinate system with origin at the nose of the body is used as illustrated in Fig.1.

It is assumed that disturbance is large in the x-direction and small in the r-direction. The perturbation velocity potential equation is

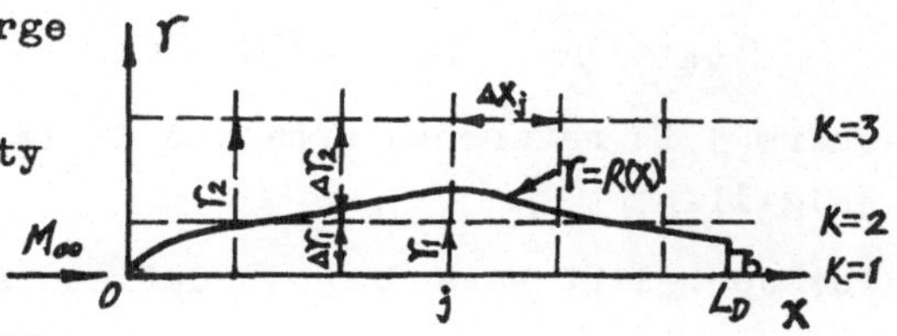

Fig.1 Illustration of model and mesh

$$(1-M^2)\varphi_{xx}+(r\varphi_r)_r/r=0 \tag{1}$$

$$1-M^2=\frac{1-M_\infty^2-\frac{\gamma+1}{q_\infty}M_\infty^2\varphi_x-\frac{\gamma+1}{2q_\infty^2}M_\infty^2\varphi_x^2}{1-\frac{\gamma-1}{q_\infty}M_\infty^2\varphi_x-\frac{\gamma-1}{2q_\infty^2}M_\infty^2\varphi_x^2} \tag{2}$$

where M is the local Mach number; q_∞ and M_∞ represent the uniform oncoming velocity and Mach number respectively, γ is the ratio of specific heats.

Assume that the body is slender. The boundary condition on the body including sting are transfered to the body axis, i.e.,

$$[r\varphi_r(x,r)]_{r=0}\approx[r\varphi_r(x,r)]_{r=R}=R(x)R'(x)[q_\infty+\varphi_x(x,R)] \tag{3}$$

where R(x) is the radius of the body. The boundary condition at far-field is approximated by that at infinite:

$$\varphi=0\ . \tag{4}$$

Set $\Delta r_2=2\Delta r_1$. The boundary condition on the body (Fig.1) is embedded into the potential equation as follows:

$$\left.\frac{\partial(r\varphi_r)}{\partial r}\right|_{K=2}=\frac{1}{\Delta r_2}\left\{\varphi_{j,3}-\varphi_{j,2}-RR'[q_\infty+\varphi_x(x,r_2)]\right\}+O(\Delta r^2)$$

where j and k are the x-grid and r-grid ordinal numbers respectively. The difference equations for the potential are solved by the line-overrelaxation along the r-axis with Seidel iteration. Overrelaxation operation is defined by

$$\varphi_{j,K}^{(n)}=\varphi_{j,K}^{(n-1)}+\omega(\bar{\varphi}_{j,K}^{(n)}-\varphi_{j,K}^{(n-1)}) \tag{5}$$

where ω is relaxation factor and $\bar{\varphi}_{j,k}^{(n)}$ is the solution in the nth iteration before the overrelaxation.

The pressure coefficient on the body is given by the exact Bernoulli's equation:

$$C_p=\frac{2}{\gamma M_\infty^2}\left\{\left[1-(\gamma-1)M_\infty^2\left(\frac{\varphi_x}{q_\infty}+\frac{\varphi_x^2}{2q_\infty^2}+\frac{R'^2}{2}\left(1+\frac{2\varphi_x}{q_\infty}+\frac{\varphi_x^2}{q_\infty^2}\right)\right)\right]^{\frac{\gamma}{\gamma-1}}-1\right\} \tag{6}$$

where φ_x is obtained according to the slender-body theory:

$$\varphi_x \approx (\varphi_x)_{i,2} + (R'^2 + RR'')\, q_\infty \ln(R/r_2)$$

The pressure drag coefficient excluding the base drag is given by the following formula

$$C_{X0} = \frac{2\pi}{S} \int_0^{L_D} C_p R(x) R'(x) dx \qquad (7)$$

where S is reference area and L_D is the length of the truncated body (Fig.1).

The zero-lift wave-drag is defined as

$$C_{XB0} = C_{X0} - C_{X0}\Big|_{M_\infty = 0.85}$$

3. Computational Procedure and Results

The x and r mesh used for the calculations is 62x16 with Δx equal to $L_D/40$ along the body and $\Delta r_1 = L_D/80$. The grid spacings along r-axis as well as x-axis ahead and behind the body after two equal spaced steps increase monotonicly with a constant ratio 2. The x-mesh extends about $10L_D$ ahead and behind the body. The r-mesh extends about $200L_D$ laterally. The relaxation factors are as follows:

When $M_\infty < 1$

$$0.9 \leqslant \omega_b \leqslant 1.7\ , \qquad 0.9 \leqslant \omega_p \leqslant 1.0$$

and when $M_\infty \geqslant 1$

$$0.8 \leqslant \omega_b \leqslant 0.9\ , \qquad 0.8 \leqslant \omega_p \leqslant 0.9$$

where ω_b and ω_p represent the relaxation factor at subsonic and supersonic points respectively.

The computed results are presented for three bodies of revolution having different configurations and compared with experimental data [6,7]. The agreement is good. The pressure distributions on the body surface and the zero-lift wave-drag for a parabolic-arc body of revolution having fineness ratio of 12 are given in Fig.2(a),(b) respectively. The results are shown in Fig.3 for an equivalent body of revolution having fineness ratio of 8.3 to a swept-wing and curved-body combination and for the isolated body of the combination having fineness ratio of 9.8. The pressure distributions on and some distances to the body for the combination at $M_\infty = 0.98$ are presented in Fig.3(a). The positions of the shock-wave which are obtained by Fig.3(a) are shown in Fig.3(c). The zero-lift wave-drag for the combination and for the isolated body are given in Fig.3(b).

Set $q_\infty = 1$ and $L_D \approx 1$. Iterative runs required with $|\Delta\varphi|_{max} \leqslant 10^{-4}$ or $|\Delta C_{XB0}| \leqslant 0.0001$ are about 150 for $M_\infty < 1$, 40 for $M_\infty > 1$ and 300 for M_∞

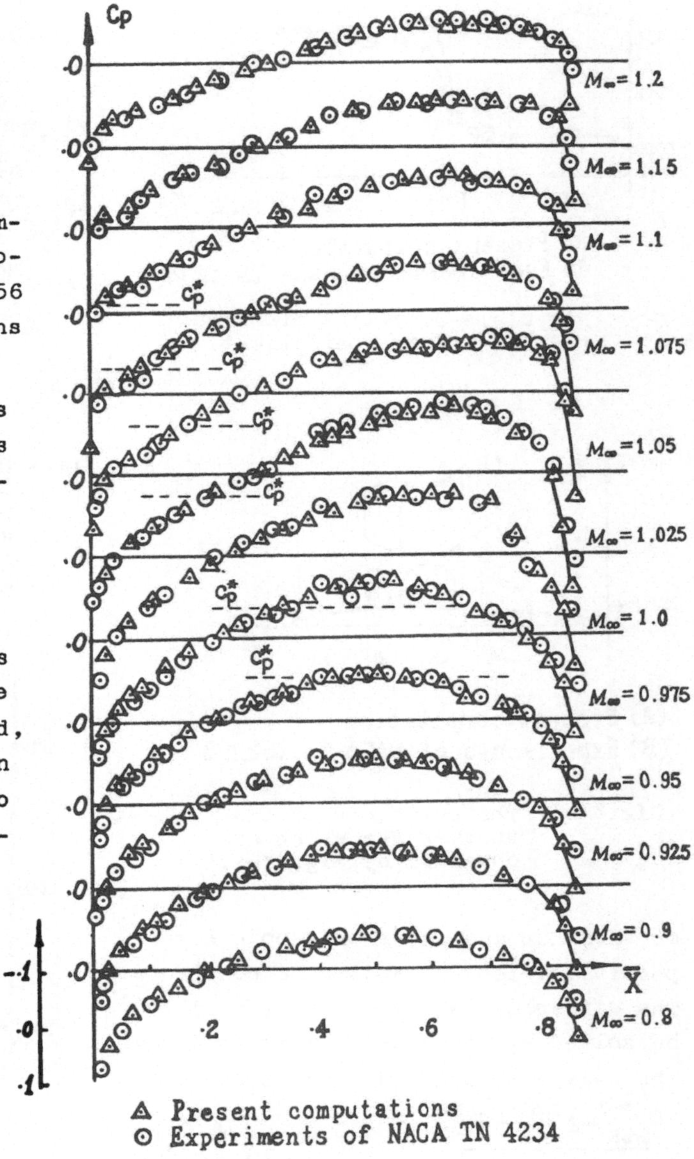

Fig.2(a)- The surface pressure distribution of a parabolic-arc body of revolution having fineness ratio of 12.

near by unit. All iterations are initiated from zero perturbation potential. The computer time for one body at $M_\infty = 0.8$ through 1.2 with Mach numbers totalled 11-12 is about 2 hours on Felix C-256 or half an hour on Siemens 7760.

The numerical experiences are that the stability is quite good when zero initial fields are used and the computation is unstable when Jacob iterations are used in supercritical cases. As M_∞ is very close to unit or the mesh spacing is shortened, the iterative computation of φ does not converge to the usual degree of accuracy (while the computations of the surface pressure do converge to the usual degree of accuracy). This might be explained by the vanishing of the artificial viscosity of the potential difference equations at locally supersonic points, that is $(1-M^2)\varphi_{xxx}\Delta x \to 0$ when $M \to 1$ or $\Delta x \to 0$.

4. Linearized Analysis for Stability and Convergence

Aussume that the coefficient of φ_{xx}, $1-M^2$, is a constant. Thus, Eq.(1) is linearized and can be transformed into

$$\pm\varphi_{xx} + (r\varphi_r)_r/r = 0 \qquad \begin{matrix}(8a)\\(8b)\end{matrix}$$

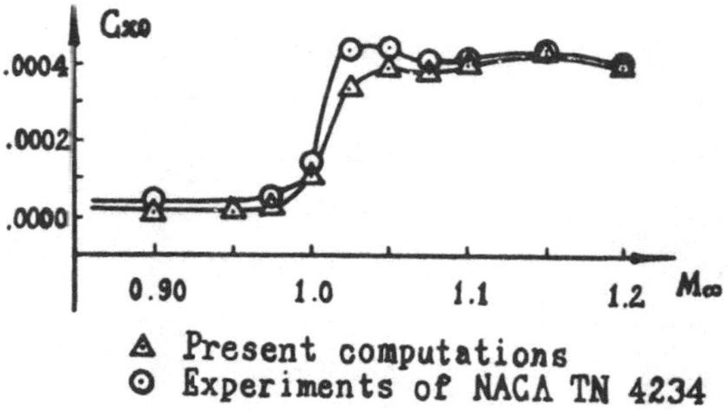

Fig.2(b)- Zero-lift pressure drag coefficient.

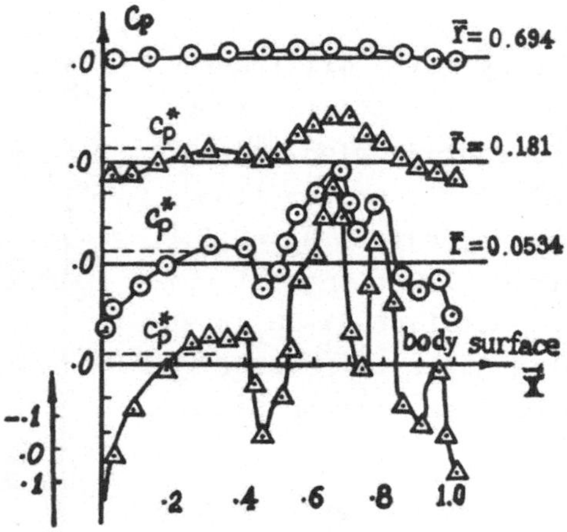

Fig.3(a)- Computed pressure distributions of a body [7] of revolution equivalent to a swept-wing and curved-body combination at $M_\infty = 0.98$.

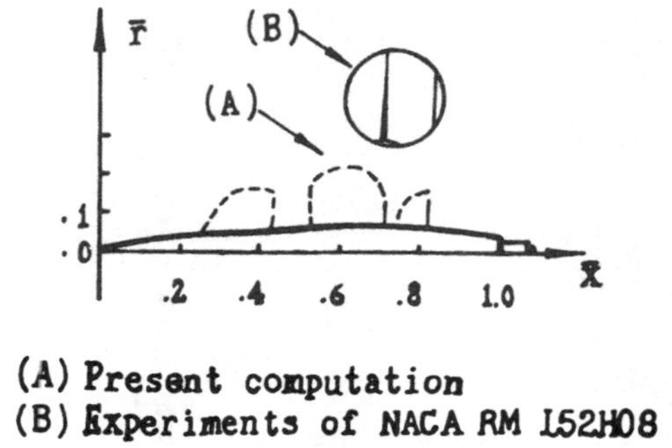

Fig.3(c)- The positions of the shock-waves which are obtained by Fig.3(a).

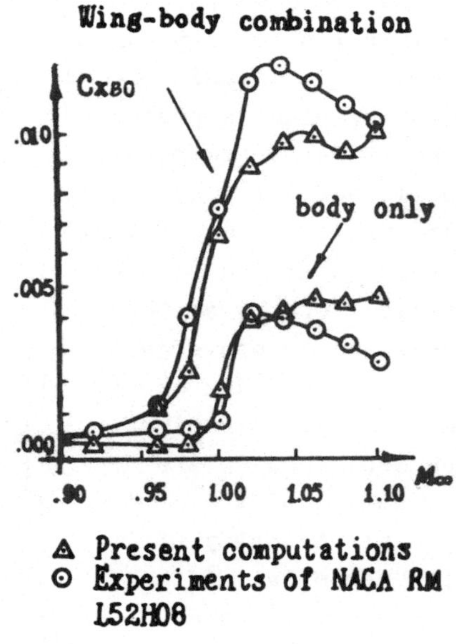

Fig.3(b)- Zero-lift wave drag coefficients for a body of revolution equivalent to a swept-wing and curved-body combination and for a isolated body.

at subsonic and supersonic points respectively. Let Δx and Δr be constants. The difference equations of (8a,b) to be solved by line-overrelaxation along the r-axis with Seidel iterations are

$$\left.\begin{array}{l} \dfrac{\varphi^{(n-1)}_{j+1,k}-2\bar{\varphi}^{(n)}_{j,k}+\varphi^{(n)}_{j-1,k}}{\Delta x^2} \\ -\dfrac{\bar{\varphi}^{(n)}_{j,k}-2\varphi^{(n)}_{j-1,k}+\varphi^{(n)}_{j-2,k}}{\Delta x^2} \end{array}\right\} + \frac{\bar{\varphi}^{(n)}_{j,k+1}-2\bar{\varphi}^{(n)}_{j,k}+\bar{\varphi}^{(n)}_{j,k-1}}{\Delta r^2}$$

$$+ \frac{\bar{\varphi}^{(n)}_{j,k+1}-\bar{\varphi}^{(n)}_{j,k-1}}{r_k\cdot 2\Delta r} = 0 \qquad \begin{array}{r}(9a)\\(9b)\end{array}$$

Eqs.(9a,b) are equivalent to the following differential equations respectively:

$$\varphi_{xx}+\varphi_{rr}-\frac{\Delta t}{\Delta x}\varphi_{xt}+\frac{1}{r}\varphi_r-\frac{\Delta t}{\Delta x^2}\left(\frac{2}{\omega}-1\right)\varphi_t=0 \qquad (9a')$$

$$-\varphi_{xx}+\varphi_{rr}+\frac{1}{r}\varphi_r-\frac{\Delta t}{\Delta x^2}\left(\frac{1}{\omega}-1\right)\varphi_t=0 \qquad (9b')$$

Both Eqs.(9a') and (9b') are hyperbolic and hence there are stability problems. The solution of Eqs.(9a) and (9b) may be expressed as $\varphi(x,r,t)+\delta(x,r,t)$, where $\varphi(x,r,t)$ is the exact solution and $\delta(x,r,t)$ is the rounding error. $\delta(x,r,t)$ also satisfies Eqs.(9a,b) and can be represented by

$$\delta(x,r,t)=e^{\alpha t+i(\beta_1 x+\beta_2 r)} \qquad (10)$$

where $i=\sqrt{-1}$, β_1 and β_2 are arbitrary real numbers, α is function of β_1 and β_2. Stability requires that $|e^{\alpha\Delta t}|\leq 1$. Using Eqs.(5), (9a) and (10), we obtain $0<\omega\leq\omega_1<2$ for $M<1$, where ω_1 decreases as the mesh spacing decreases.

By convergence, we mean that the solutions of Eqs.(9a',b') converge to those of Eqs.(8a,b) respectively.

Set $\tau=t+\frac{1}{2}\frac{\Delta t}{\Delta x}x$. Eq.(9a') can be changed into

$$\varphi_{xx}+\varphi_{rr}-\frac{1}{4}\left(\frac{\Delta t}{\Delta x}\right)^2\varphi_{\tau\tau}+\frac{1}{r}\varphi_r-\frac{\Delta t}{\Delta x^2}\left(\frac{2}{\omega}-1\right)\varphi_\tau=0 \qquad (11)$$

The solution of Eq.(11) is

$$\varphi(x,r,\tau)=G_0(x,r)+\sum_{m=1}^{\infty}\left(A_m e^{-p_m\tau}+B_m e^{-q_m\tau}\right)G_m(x,r) \qquad (12)$$

where $G_m(x,r)$ are the eigen functions of the boundary problems of following equations

$$G_{xx}+G_{rr}+G_r/r+k_m^2G=0$$

k_m^2 is the corresponding eigen values, $k_1^2<k_2^2<\cdots<k_{m-1}^2<k_m^2<\cdots$; $G_0(x,r)$ is the solution of Eq.(8a); A_m and B_m are constants;

$$p_m,\,q_m=\frac{2}{\Delta t}\left(\frac{2}{\omega}-1\right)\mp\sqrt{\frac{4}{\Delta t^2}\left(\frac{2}{\omega}-1\right)^2-4\left(\frac{\Delta x}{\Delta t}\right)^2k_m^2}$$

Obviously, when $0<\omega<2$, $\lim_{t\to\infty}\varphi(x,r,t)=G_0(x,r)$ for $M<1$.

All similarly derived results are summarized in the following table:

Table 1. Stability and Convergence Conditions for r-line Overrelaxation Iterations

	Seidel		Jacob	
Conditions	$M<1$	$M>1$	$M<1$	$M>1$
Stability	$0<\omega\leq\omega_1<2$	$0<\omega_2<\omega\leq 1$	$0<\omega\leq\omega_1<2$	Always Unstable
Convergence	$0<\omega<2$	$0<\omega<1$	$0<\omega$	

The numerical experiments show that these conditions are sufficient for stability and convergence.

5. Concluding Remarks

As compared with classical small perturbation equation, the potential equation with large disturbance in the free stream direction and small disturbance in the transverse direction is well applicable to the flow with stagnation, and the amount of the labor is not increased to solve it. The zero-lift wave drag which agrees well with experiment can be obtained by integrating the pressure on the body.

It is being investigated that, as M_∞ is very close to unit or the mesh spacing is shortened, the iterative computation of φ does not converge to the usual degree of accuracy.

References

[1] Murman,E.M., et al. AIAA J. 9(1971) 114-121
[2] Krupp,J.A., et al. AIAA J. 10(1972) 880-886
[3] Bailey,F.R. NASA TN D-6582(1971)
[4] Stahara,S.S., et al. AIAA J. 18(1980) 63-71
[5] Luo,S.J., et al. Computer Methods in Applied Mechanics and Engineering 27(1981) 129-138
[6] Taylor,R.A., et al. NACA TN 4234(1958)
[7] Whitcomb,R.T. NACA RM L52H08(1952)

NUMERICAL SOLUTION OF VISCOUS FLOW IN UNBOUNDED FLUID

C. H. Liu
NASA Langley Research Center, Hampton, VA

L. Ting
New York University, New York, NY

Abstract

Incompressible viscous flow fields induced by initial vorticity distributions with bounded support or exponential decay in the far field are investigated. A numerical scheme for the solution of the vorticity distribution and the velocity field is presented with special emphasis on the treatment of the boundary data. The efficiency of the scheme is demonstrated. The present method has been applied to the study of the merging and collision of vortex rings.

1.0 Introduction

Incompressible, viscous flow fields induced by initial vorticity distributions are investigated with special emphasis on the treatment of the boundary data for the numerical solution. The numerical methods developed in this investigation can be used to study the vortex core structure, vortex self-merging, and the intersection of multiple vortex filaments. These types of problems appear frequently in nature and in flow fields behind aircraft [1 to 4].

Let $\vec{V}$ and $\vec{\omega}$ represent the velocity and vorticity vectors, respectively. The governing equations are

$$\nabla \cdot \vec{V} = 0 \tag{1}$$

$$\vec{\omega}_t + (V_i \vec{\omega})_{x_i} + (\omega_i \vec{V})_{x_i} = \nu \Delta \vec{\omega} \tag{2}$$

$$\vec{\omega} = \nabla \times \vec{V} \tag{3}$$

subject to the initial condition

$$\vec{\omega}(\vec{x},0) = \vec{\omega}_0(\vec{x}) \tag{4}$$

and boundary condition

$$\vec{V}(\vec{x},t) \to 0 \quad \text{as} \quad |\vec{x}| \to \infty \tag{5}$$

The initial data $\vec{\omega}_0(\vec{x})$ are considered to be of either bounded support or assumed to decay exponentially with r, where $r = |\vec{x}|$. One can then impose the far-field condition

$$\vec{\omega} \text{ decays exponentially with } r \tag{6}$$

Because condition (6) on $\vec{\omega}$ is much stronger than (5) on $\vec{V}$, the vorticity evolution equation (2) is used with equation (6) as the boundary condition on a finite domain. The vector velocity potential $\vec{A}$ is then introduced [4], where $\vec{V} = \nabla \times \vec{A}$ and $\nabla \cdot \vec{A} = 0$. $\vec{A}$ fulfills the Poisson equation for all t

$$\Delta\vec{A} = -\vec{\omega} \tag{7}$$

The exact solution of equation (7) subject to (5) is

$$\vec{A}(\vec{x},t) = \frac{1}{4\pi} \iiint_{-\infty}^{\infty} \frac{\vec{\omega}(\vec{x},t)}{|\vec{x} - \vec{x}'|} dx_1' \, dx_2' \, dx_3' \tag{8}$$

This approach was employed for two-dimensional problems by Wu and Thompson [5] and Lo and Ting [6]. However, the evaluation of the Poisson integral (8) can be extremely time consuming even when "standard" analytical methods for solving singular integral equations are used. For a three-dimensional problem with N grid points in each direction, the number of operations is $O(N^6)$. Even if a Poisson solver [7] is used for the interior domain and the Poisson integral is used to evaluate the boundary data, the number of operations is still $O(N^5)$. Of course, if the primitive condition (5) is imposed on the finite boundary, a very large domain with a correspondingly large N would be needed.

Equation (8) is used herein to develop the far-field behavior of $\vec{A}$ which in turn will be used as the boundary data. The number of operations for the establishment of the boundary data is then reduced to $O(N^3)$.

When the vorticity distribution is concentrated almost entirely in a sphere of radius a (around the origin), the far-field behavior of $\vec{A}$ can be represented by a power series in (a/r) where the coefficients are moments of the vorticity distribution. The first moments and linear combinations of higher moments have been shown to be time invariant [7, 8]. It was suggested by Ting [10, 11] that these results could be used to determine the boundary data for the numerical calculations and also to check the accuracy of the numerical solutions. These general concepts have been applied herein and appropriate modifications have been introduced to take into account the special geometry of each specific problem.

When the vorticity distribution is concentrated (almost entirely) in a slender "tube-like" region with reference cross-sectional radius a much less than the radius of curvature R, the use of the same far-field representation for the boundary data requires that the linear dimension of the domain for the numerical solution be much larger than R. To take advantage of the fact that $a \ll R$, a different "far-field" representation has been developed so that the linear dimension of the domain is much larger than a, but remains, at most, of the order of R.

Since the numerical integration of the vorticity evolution equation is much faster than that of the Poisson equation, one can further increase the efficiency of the scheme by solving the vorticity evolution equation in a large domain D and the Poisson equation in a small sub-domain D_1. The velocities on the boundaries and throughout the complementary domain $(D - D_1)$ are defined by the far-field representations. In Section 3, it will be demonstrated that the "two-domain method" has an accuracy comparable to that for the single, larger domain D, but consumes little additional computational time over that for the smaller domain D_1.

2.0 Appropriate Domain and the Boundary Data

To illustrate how to choose the appropriate finite domain for the numerical solution of the Navier-Stokes equation and how to generate the boundary data for different types of vorticity distributions, only axisymmetric problems will be considered here. There are two distinct types:

(1) The vorticity distribution is concentrated almost entirely in a sphere with an effective radius a (centered at the origin), e.g., the self-merging of a vortex ring. The far field is defined by the condition $|\vec{x}| \gg a$. The vorticity decays exponentially with distance. The three-dimensional invariant conditions [8,9] are applicable and are employed to develop the far-field behavior of the velocity field [11]. Due to axisymmetry, the domain can be a square of side 2mR centered at the origin in an azimuth plane, the r-z plane. The number m must be chosen so that the error in the boundary data generated from the far-field behavior is within the required accuracy of the numerical solution and hence is of the order of the error of the finite difference scheme. The generation of the boundary data is described in subsection 2.1.

(2) The vorticity distribution is concentrated almost entirely inside a torus whose cross-sectional radius a is much smaller than its centerline radius R, e.g., in the merging or collision of two vortex rings with small core radii. In the cross-sectional plane, the vorticity, which has a length scale a, decays exponentially with ρ, where ρ is the radial distance to the centerline of the torus. Due to axial symmetry, the domain can be a circle with $\rho \ll ma$ or a square with side 2ma. Since the vorticity does not decay within the length scale of a, the three-dimensional invariant conditions are not applicable for the development of the far-field behavior, i.e., for $\rho \gg a$. The far-field behavior must be developed directly from the Poisson integral for the axisymmetric problem. Again, the number m will be determined from the constraint on the error of the boundary data and the required accuracy of the computation. Depending upon the order of magnitude of R/a, two different expansions for the Poisson integral are developed. In one case, II(a),

$a \ll \rho = O(R)$ and in the other case, II(b), $a \ll \rho \ll R$. The generation of the boundary data for these two cases is described in subsections 2.2.1 and 2.2.2, respectively.

2.1 Vorticity Concentrated Near the Origin

The vector velocity potential, which depends only on the scalar stream function, can be expressed by a power series of (a/r) for this type of problem. The stream function $\psi(r,z,t)$ can be expressed in terms of vorticity distribution ζ [4,10]

$$\psi = \frac{r}{4\pi} \int_{-\infty}^{\infty}\int_{0}^{\infty}\int_{0}^{2\pi} \frac{\zeta(r',z',t)}{\rho} r'\cos\theta \, d\theta \, dr' \, dz' \tag{9}$$

where $\rho^2 = (z-z')^2 + r^2 + r'^2 - 2\,rr'\cos\theta$. To obtain the final result, the behavior of ψ at large distances $d = (z^2+r^2)^{1/2}$ is needed. Note that

$$\rho^{-1} = d^{-1}\left\{1 + c_1 d^{-1} + 1/2\,(3c_1{}^2 - c_2)d^{-2} + 1/2\,(5c_1{}^3 - 3c_1c_2)\,d^{-3} + O(d^{-4})\right\} \tag{10}$$

where $c_1 = (z/d)\,z' + (r/d)\,r'\cos\theta$ and $c_2 = z'^2 + r'^2$. From (9), the far-field behavior of ψ is

$$\psi(r,z,t) = \frac{1}{4d}\left(\frac{r}{d}\right)^2 \left\{\langle\zeta r'^2\rangle + \frac{3}{d}\left(\frac{z}{d}\right)\langle\zeta r'^2 z'\rangle + \frac{3}{2d^2}\left[\left(5\left(\frac{z}{d}\right)^2 - 1\right)\langle\zeta r'^2 z'^2\rangle + \left(\frac{5}{4}\left(\frac{r}{d}\right)^2 - 1\right)\langle\zeta r'^4\rangle\right]\right\} + O(d^{-4}) \tag{11}$$

where

$$\langle\zeta r'^m z'^n\rangle \equiv \int_{-\infty}^{\infty}\int_{0}^{\infty} \zeta(r',z',t)\, r'^m z'^n \, dr' \, dz' \tag{12}$$

Since the second moment of vorticity distribution $\langle\zeta r'^2\rangle$ is conserved, there are three integrals to be evaluated at each time step in (11).

2.2 Vorticity Concentrated in the Small Neighborhood of a Ring

For this case, the behavior of ψ (fig. 1) is needed for a vortex ring at $\tau_1/\delta \gg 1$. The point $G(R,Z)$ is the reference center of the vorticity distribution with the image point G'. $P(r',z')$ is the vortex ring of strength $\zeta(r',z')dr'dz'$ and $Q(r,z)$ is the far-field point in space. The scaling lengths of this problem are $\sigma/R = O(\delta/R) \ll 1$, $\delta/\tau_1 \ll 1$, and $\tau_1 < R$.

2.2.1 Size of the Domain ~ Radius of the Ring

In this instance, the requirement remains that $\tau_1 < R$ and $\tau_2 > R$, but τ_1 can now be of the order of R. Note that no assumption is made requiring $\rho_1/\rho_2 \ll 1$. The expansion parameter of this analysis is σ/τ. It can be shown that

$$2\pi\psi(X,Y) = a_{0,0}\,\langle\zeta\rangle + a_{1,0}\,\langle\zeta x\rangle + a_{0,1}\,\langle\zeta y\rangle + O(\delta^2/L^2) \tag{13}$$

with $\delta \ll L < R$ and L is the size of the domain. The symbol $\langle\zeta\rangle \equiv \Gamma$ is the total circulation of the vorticity field. The coefficients are

$$a_{0,0} = (\tau_1 + \tau_2)\,G(\lambda_o)$$

$$a_{1,0} = (X/\tau_1 + X/\tau_2)\,[-G(\lambda_o) + 2\lambda_o\,G'(\lambda_o)]$$

$$a_{0,1} = -\,[Y/\tau_1 - (Y+2R)/\tau_2]\,G(\lambda_o) + \lambda_o(\tau_1+\tau_2)\left\{1/R + 2[Y/\tau_1 - (Y+2R)/\tau_2]/(\tau_1+\tau_2)\right\}G'(\lambda_o)$$

where $G(\lambda_o) = F(\lambda_o) - E(\lambda_o)$, $\lambda_o = 4R(Y+R)/(\tau_1+\tau_2)^2$ and F and E are the complete elliptic integrals of the first and second kind, respectively. Although only the terms up to $O(\delta^2/L^2)$ are given in (13), the actual analysis was carried out to $O(\delta^4/L^4)$, which was then used in the numerical calculation.

2.2.2 Size of the Domain << Radius of the Ring

This is the special case where $\delta \ll \tau_1 \ll R$. The expansion parameters of this analysis are δ/τ_1 and τ_1/τ_2. The far-field behavior of ψ can be obtained as

$$\frac{2\pi}{\sqrt{(Y+R)R}}\psi(X,Y) = a_{0,0}\langle\zeta\rangle + a_{1,0}\langle\zeta x\rangle + a_{0,1}\langle\zeta y\rangle + o(\ell n\Lambda\ \Lambda^m\ (\delta/L)^n),\ m+n=2 \qquad (14)$$

where

$$a_{0,0} = 1/2\ \ell n\ (16/\Lambda^2) - 2$$

$$a_{1,0} = \frac{X}{\tau_1^2}\left(1 - \frac{1}{2R}\right) - \frac{X}{\tau_2^2}$$

$$a_{0,1} = \frac{Y}{\tau_1^2}\left(1 - \frac{1}{2R}\right) - \frac{(Y+2R)}{\tau_2^2} + \left(\frac{1}{4}\ \ell n\ \frac{16}{\Lambda^2} - 1\right)\frac{1}{R}$$

where $\Lambda = \tau_1/\tau_2 = O(L/R)$. Note that the leading term contains $\ell n(16/\Lambda^2)$ which is the familiar local two-dimensional result. As in Section 2.2.1, the actual analysis of (14) was carried out to terms of $o(\ell n\Lambda\ \Lambda^m(\delta/L)^n)$ with $m + n = 4$.

It should be mentioned that the corresponding integral invariant for the type II problem is

$$\langle\zeta\rangle + \frac{2}{R}\langle\zeta y\rangle + \frac{1}{R^2}\langle\zeta y^2\rangle = \text{constant} \qquad (15)$$

3.0 Two-Domain Method

The appropriate far-field representations for types I and II derived in the previous section are used for numerical examples. The accuracy and the efficiency of the scheme are compared in Table I with schemes using the Poisson integral to evaluate the boundary data. The corresponding two-dimensional comparisons were made by Weston and Liu [12].

Table I. Test of Approximate Boundary Data

Region: $r = [5,15]$, $z = [-5,5]$

Initial Condition $\zeta(r,z,t_o) = \sum_{i=1}^{2} \frac{\Gamma_i}{4\pi\nu t_o} e^{-[(z-Z_i)^2 + (r-R_i)^2]}$

$\Gamma_i/\nu = 100$, $R_i = 10$, $Z_1 = 1.4$, $Z_2 = -1.4$

Region	Boundary Condition	Relative Time Per Step	Relative ζ_{max} at $t = 2t_o$	Relative Integral Invariant	
				$t = t_o$	$t = 2t_o$
$D_1 = D$	Equation (8)	1.0	1.0	1.0	0.9994
$D_1 = D$	Equation (13)	0.0095	1.0016	1.0	0.9994

These results demonstrate that the accuracy of the scheme is comparable to that of using the Poisson integral to evaluate the boundary data, but the computational time is reduced by two orders of magnitude.

Since the numerical integration of the vorticity evolution equation is much faster than that of the Poisson equation, the efficiency of the scheme can be further increased by solving the vorticity evolution equation in a large domain D and the Poisson equation only in a small sub-domain D_1. Note that the integrals of moments of the vorticity distribution $\langle\zeta r^m z^n\rangle$ that appeared in (11), (13), and (14) are functions of time in general, but some of them are conserved. However, the coefficients of these integrals are functions of space variables only. The velocities on the boundaries and throughout the complementary domain $(D - D_1)$ are defined by the far-field representations. The accuracy and the savings in computational time are shown in Table II.

Table II. Test of "Two-Domain Method"

Far-Field Representation - Type I

Initial Condition $\zeta(r,z) = 16\exp[-(r-2)^2-z^2]$

Region	Grid Size		Relative Time Per Step	$t/t_o = 2$	
	D	D_1		Relative ζ_{max}	Relative Integral Invariant
$D_1 = D$	41x41	41x41	1	1	0.98
$D_1 \subset D$	41x41	24x31	0.60	1.002	1.02

Similar results are obtained for type II computations.

The time-dependent deviation of the integral invariant decreases as the size of the domain increases. Therefore, the integral invariant can be used to gauge the error due to the finite size of the domain. Numerical results also demonstrate that the "two-domain method" has the same degree of accuracy as that for the single, larger domain D, but consumes little additional computational time over that for the smaller domain D_1. The two-domain method, coupled with the development of appropriate far-field representations for the velocity fields, is an efficient numerical scheme for the analysis of an unsteady, unbounded, viscous flow field.

4.0 Numerical Examples of Collision of Vortex Rings

The present scheme is applied to the motion and decay of a pair of coaxial vortex rings; in particular, one example is presented of the interaction of two identical vortex rings of opposite sense in a "head-on collision" (fig. 2). The initial vorticity distribution at $t_o = 1$ is given by

$$\zeta(r,z) = \Gamma_1(4\pi\nu)^{-1}\exp\{-[(z+Z_o)^2 + (r-R_o)^2]\} + \Gamma_2(4\pi\nu)^{-1}\exp\{-[(z-Z_o)^2 + (r-R_o)^2]\}$$

The length and the time scales are so chosen that $4\nu = 1$. The two vortex rings are of equal and opposite strength, i.e., $\Gamma_1 = -\Gamma_2 = 16\pi$ with vortex core centers at $(20,\pm 2)$. The initial radius of each effective viscous core is equal to unity, i.e., $\delta_o = 1$.

The computed trajectories of the colliding vortex rings are shown for $z < 0$ in figure 3 along with those results obtained by inviscid theory and matched asymptotic analysis. Figure 4 shows that the maximum vorticity begins to deviate from the asymptotic solution when $t/t_o \simeq 2$. The errors of the asymptotic solutions are 12.6, 74, 168, and 233 percent at $\bar{t} = t/t_o = 4$, 8, 12, and 16, respectively. The inviscid theory actually predicts that $\zeta_{max}(t)$ increases with time. Figure 5 displays the decay of circulation of the different theories, for the left half plane $z < 0$, versus $Z(t)/R_o$. The Navier-Stokes solution indicates that the vortex merging occurs at about $\bar{t} = 3.5$. The axial position $|Z(t)|$, corresponding to the location of $\zeta_{max}(t)$, increases for $\bar{t} > 3.5$ due to the cancellation of vorticity in the overlap region. Figure 6 shows the vorticity contours of the Navier-Stokes solution for $\bar{t} = 15.56$. The core radii obtained by inviscid and asymptotic theories, which overlap the $z = 0$ line, are also shown in this figure. In the inviscid theory, the core size becomes smaller due to vortex stretching, but the distance between the two vortex core centers becomes less than the core diameter when the two vortices come close to each other. Because the Navier-Stokes solution more accurately simulates the viscous effect, the vortex rings diffuse much faster than predicted by the extrapolation of the asymptotic analysis. Navier-Stokes results for a head-on collision of two vortex rings also show the eventual annihilation of circulation of the vorticity in the rings.

Additional numerical studies of the interaction of two vortex rings are being carried out and will be reported elsewhere.

5.0 Concluding Remarks

The two-domain method, coupled with the development of appropriate far-field representations for the velocity fields, is an efficient numerical scheme for the analysis of an unsteady, unbounded, viscous flow field.

Using the examples of axisymmetric merging and interaction of vortex rings, the following have been demonstrated: (1) the use of the matched asymptotic solutions [10] to study the decay in the vortical cores and to create the initial data prior to the merging; (2) the selection of the appropriate finite domain for the numerical solution of the Navier-Stokes equations; and (3) the generation of the appropriate boundary data for two distinct types of initial vorticity distribution. For Types I and II, the size of the vortical core is respectively of the order of and much smaller than the ring radius.

The boundary data for Type I were obtained as a specialization from a three-dimensional problem [11] to an axisymmetric one. The derivation of the boundary data for Type II makes use of the axisymmetry of the problem and hence cannot be extended to the three-dimensional problem. For a corresponding three-dimensional problem, for example, in the study of the intersection of two vortex filaments with small effective vortical core radii, the filaments overlap only along small segments of their entire length. Since the extent of the overlap region is small comparable to the reference radius of curvature of the filaments, the matched asymptotic solution [10] remains valid to describe the motion and the decay of the filament away from the overlap region but is not applicable in the overlap region. For this relatively small region, numerical solution of the Navier-Stokes equations must be obtained with the initial data and the boundary data provided by the matched asymptotic solution. Research on this type of problem is in progress.

Acknowledgements

The research of the second author was supported by NASA Langley Research Center and by the Office of Naval Research.

References

1. Brown, C. W., "Aerodynamics of Wakes Vortices," AIAA J., vol. 11, April 1973, pp. 531-536.
2. Greene, G. C., "Wake Vortex Alleviation," AIAA Paper 81-0798, May 1981.
3. Leonard, A., "Vortex Methods for Flow Simulation," J. of Comp. Physics, vol. 37, no. 3, October 1980.
4. Lamb, H., Hydrodynamics. Dover, 6th ed., New York, 1945.
5. Wu, J. C., and Thompson, J. F., "Numerical Solutions of Time-Dependent Incompressible Navier-Stokes Equations Using an Integro-Differential Formulation," Computer and Fluids, vol. 1, 1973, pp. 197-215,
6. Lo, R.K.C., and Ting, L., "Studies of the Merging of Vortices," The Physics of Fluids, vol. 19, no. 6, June 1976, pp. 912-913.
7. Swarztrauber, P. N., and Sweet, R. A., "ALGORITHM 541, Efficient FORTRAN Subprograms for the Solution of Separable Elliptic Partial Differential Equations [D3]," ACM Transactions on Mathematical Software, vol. 5, no. 3, September 1979, pp. 352-364.
8. Truesdell, C., The Kinematics of Vorticity. Indiana University Press, Bloomington, 1954.
9. Howard, L. N., "Divergence Formulas Involving Vorticity. Archive for Rational Mechanics and Analysis, vol. 1, no. 1, 1957, pp. 113-123.
10. Ting, L., "Studies on the Motion and Decay of Vortex Filaments", Advances in Fluid Mechanics, vol. 148, Springer Verlag, 1981, pp. 67-105.
11. Ting, L., "Integral Invariants and Decay Laws for Vorticity Distributions," APS Paper No. DH5, Fluid Dynamics Conf., Monterey, California, November 1981.
12. Weston, R. P., and Liu, C. H., "Approximate Boundary Condition Procedure for the Two-Dimensional Numerical Solution of Vortex Wakes," AIAA Paper 82-0951, 1982.

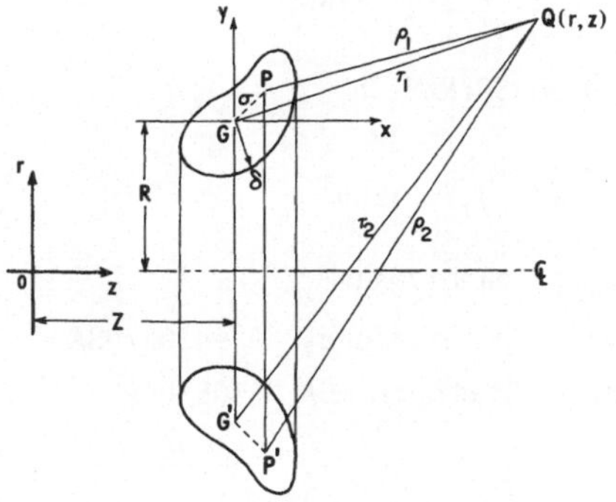

Figure 1. Coordinate system for an axisymmetric vortex ring.

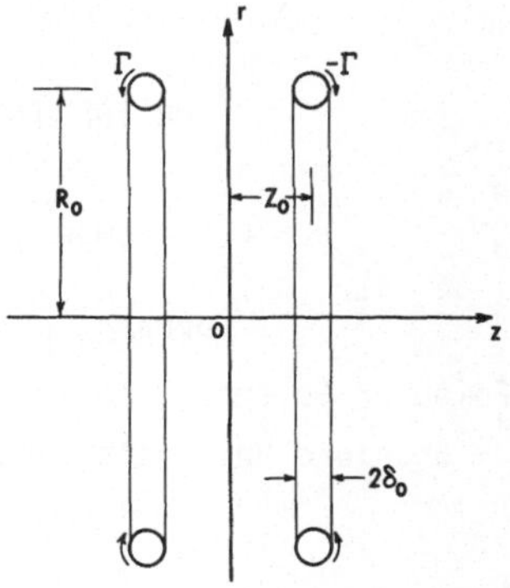

Figure 2. Initial geometry for two coaxial vortex rings.

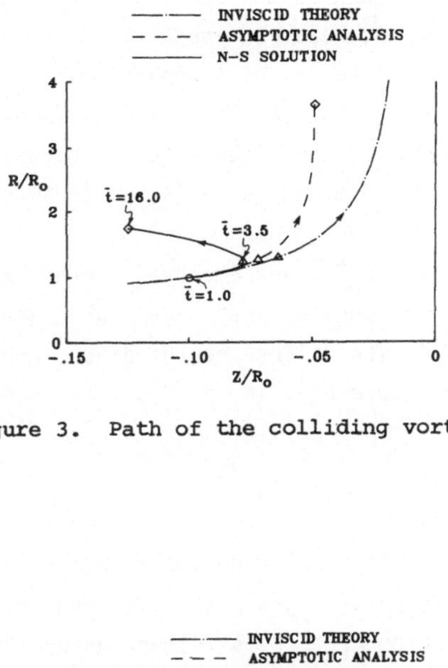

Figure 3. Path of the colliding vortex.

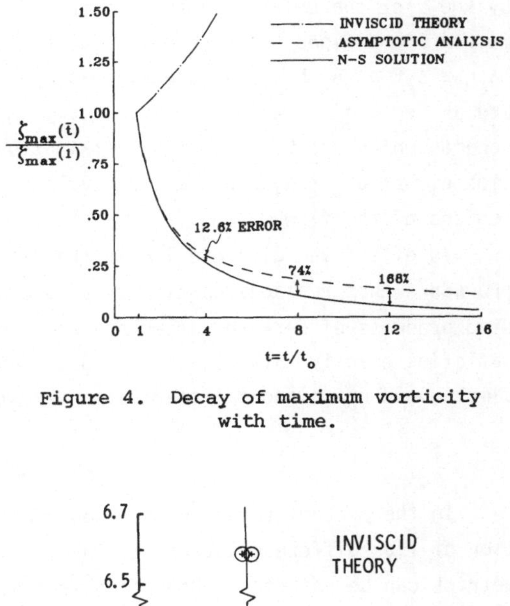

Figure 4. Decay of maximum vorticity with time.

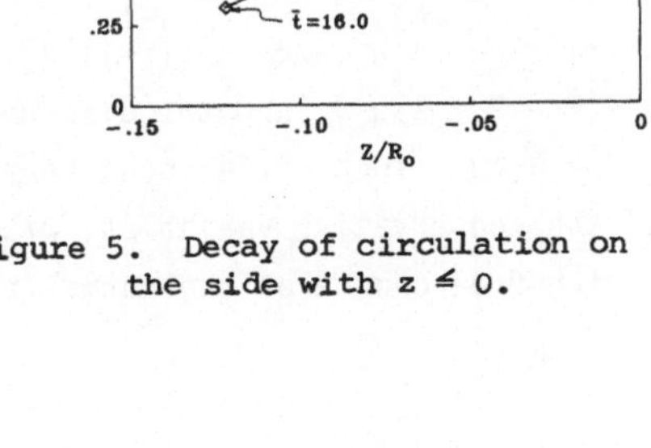

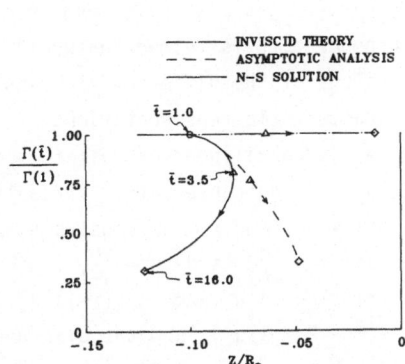

Figure 5. Decay of circulation on the side with $z \leq 0$.

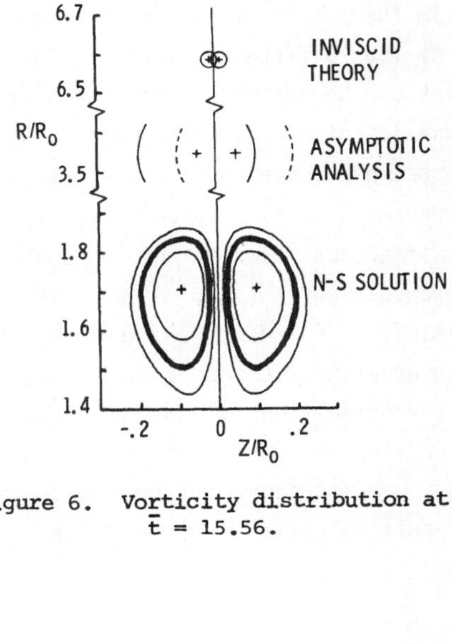

Figure 6. Vorticity distribution at $\bar{t} = 15.56$.

A NATURAL CONSERVATIVE FLUX DIFFERENCE SPLITTING

FOR THE HYPERBOLIC SYSTEMS OF GASDYNAMICS

C.K. Lombard*, Joseph Oliger** and J.Y. Yang†

*PEDA Corporation, Palo Alto, CA 94301/USA

**Computer Science Department, Stanford University, Stanford, CA 94305/USA

†Stanford University and Informatics, Inc., Stanford, CA 94305/USA

Introduction

An attractive framework for constructing globally conservative upwind difference schemes with accurate, sharp discontinuity capturing capability has been presented by Roe[1] for the Euler equations. Central to the approach is the construction of a matrix $\tilde{A}$ defined on the simple spatial difference interval Δ_ξ such that $\tilde{A}\Delta_\xi q \equiv \Delta_\xi F$; $\tilde{A}$ has real eigenvalues $\tilde{\lambda}_i$ and a complete set of eigenvectors $\tilde{e}_i$, and as $q_L \to q_R$, $\tilde{\lambda}_i \to \lambda_i$, $\tilde{e}_i \to e_i$, $\tilde{A} \to A = \frac{\partial F}{\partial q}$. Thus stable upwind difference schemes which are based on the distribution of pieces of the eigenvector decomposition of $\tilde{A}\Delta_\xi q = \sum_i \tilde{\lambda}_i \tilde{a}_i \tilde{e}_i$ are equivalent to the method of characteristics in smooth regions of the flow.

To effect the discrete linear transformation of $\Delta_\xi F$, Roe employed a mean value procedure that replaces certain triad products in F such as ρu^2 and $\rho u H$ with diad products of derived variables $\rho^{\frac{1}{2}}u$ and $\rho^{\frac{1}{2}}H$. The differences of diad products satisfies exactly $\Delta(a\, b) = \bar{a}\,\Delta b + \bar{b}\,\Delta a$ and in the case of triad products no such simple, and unambiguous linearization exists.

CSCM Splitting

In the present paper we show how exact linearization and eigenvector decomposition of flux differences that may involve spatially varying gas properties and grid metrics can be effected. Departing from Roe, the procedure[2,3] is termed "natural" because it follows discrete mean value difference analogs of the differential transforms between conservative, nonconservative, and characteristic representations.

Where $\Delta_\xi f$ is a symbolic representation for the ξ curvilinear coordinate spatial difference terms of the "nonconservative equations" in nonconservative variable differences $\Delta q' = (\Delta\rho\ ,\ \bar{\rho}\,\Delta u\ ,\ \Delta P)$ in quasi 1-D flow (Appendix), we construct $\Delta_\xi f$ to satisfy $\Delta_\xi F = \bar{M}\,\Delta_\xi f$, where $\bar{M}$ is the average of $M = \frac{\partial q}{\partial q'}$ and $P = \frac{p}{\gamma-1}$. The nonconservative difference vector $\Delta_\xi f$ involves terms like $\Delta_\xi u$ and $\Delta_\xi[(\gamma-1)\,P]$. From the definitions of the conservative variables $(E = P + \frac{1}{2}\rho u^2)$ and the diad mean value difference relations $\Delta\rho u = \bar{\rho}\,\Delta u + \bar{u}\,\Delta\rho$, $\Delta\frac{u^2}{2} = \bar{u}\,\Delta u$, such differenced terms of $\Delta_\xi f$ can be expressed exactly in differences of the conservative variables, with the resulting relation $\Delta_\xi f = \tilde{M}^{-1}(\Delta_\xi)q$. Finally, with free choice of the normalizing transform $\overline{T}^{-1}$ defining characteristic variable differences, (we choose $\frac{1}{\overline{\rho}}$ for $\Delta\rho$, $\frac{1}{\overline{c}}$ for Δu , and $\frac{1}{\overline{\gamma P}}$ for ΔP) we are able to express the flux differences

$$\Delta_\xi F = (\overline{M}\ \overline{T})(\overline{T}^{-1}\ \tilde{M}^{-1}\ (\Delta_\xi)q) \tag{1}$$

The right hand side matrix relation which is in the form $\tilde{A}\ (\Delta_\xi)q$ has much in common with Roe's. Indeed, except for a scaling multiple of $\overline{\rho}$, the columns of the first bracket matrix $(\overline{M}\ \overline{T})$ appear formally the same as his eigenvectors e_i but with simpler averaged definitions of u , c , H (respectively the velocity, sound speed, and total enthalpy). The second bracketed matrix expression in (1) contains the eigenvalues and characteristic variable differences corresponding to Roe's λ_i and a_i .

Because of variable gas properties and variable metrics which may be introduced inside the difference operators the eigenvalues λ_i and functions a_i are difficult to compute. We don't need explicit values of the a_i for our formulation and approximate the λ_i using $\Lambda = (\overline{u}\ ,\ \overline{u} + \overline{c}\ ,\ \overline{u} - \overline{c})$. These approximate values are only used to construct a partitioning based upon their sign. Errors in this partition will only occur near zero when directional dependence is unimportant and the contributions small. The matrix representation is split by introducing a diagonal truth matrix D composed of 1's and 0's between the bracketed expressions such that $D^+ = (|\Lambda| + \Lambda)/2|\Lambda|$ and $D^- = (|\Lambda| - \Lambda)/2|\Lambda|$, $D^+ + D^- = I$ and call the resulting split matrix pieces $\tilde{A}(D^+) = \tilde{A}^+$, $\tilde{A}(D^-) = \tilde{A}^-$. Then following the general prescription of Roe, stable difference schemes are constructed by "sending" the associated split flux difference pieces $\tilde{A}^+(\Delta_\xi)q$, appropriately weighted, forward to mesh nodes and $\tilde{A}^-(\Delta_\xi)q$ backward. We have named explicit and implicit methods involving this splitting Conservative Supra-Characteristics Methods (CSCM) partly in recognition of the related nonconservative Split Coefficient Matrix Method (SCM) of Chakravarthy, Anderson and Salas[4].

The splitting is the basis for both explicit and implicit upwind interior point and characteristic boundary point approximations. A fully coupled first order implicit interior point, boundary point scheme that satisfies Oliger's sufficient condition[5] for unconditional stability of the discrete initial boundary value problem is

$$\begin{aligned}(I + \tilde{A}^+\nabla_\xi + \tilde{A}^-\Delta_\xi)\delta q_j &= -(\tilde{A}^+\Delta_\xi q)_{j-1} - (\tilde{A}^-\Delta_\xi q)_j \\ (\tilde{\tilde{A}}{}_1^{\prime -} + \tilde{A}_1^-\Delta_\xi)\delta q_1 &= -(\tilde{A}^-\Delta_\xi q)_1 \\ (\tilde{\tilde{A}}{}_N^{\prime +} + \tilde{A}_{N-1}^+\nabla_\xi)\delta q_N &= -(\tilde{A}^+\Delta_\xi q)_{N-1}\ .\end{aligned} \tag{2}$$

Here at the left inflow boundary $j=1$ and right outflow boundary $j=N$ the well posed characteristic boundary point matrices $\tilde{\tilde{A}}^- = (\overline{M}\ \overline{T}\ D^-)_1(T^{-1}\ M^{-1})_1$ and $\tilde{\tilde{A}}^+ = (\overline{M}\ \overline{T}\ D^+)_{N-1}(T^{-1}\ M^{-1})_N$ are augmented (as needed to complete the set of eigenvectors) by linearized time differenced boundary condition relations after the approach of Kentzer[6]. As described in reference 7, for boundary conditions that are functionally nonlinear in the dependent variables, the linearization causes the boundary conditions to drift unless corrected. In op cit a Newton-Raphson iteration procedure is described that operates effectively on the boundary point matrices $\tilde{\tilde{A}}'$ to reduce the boundary condition error to within predetermined bounds while not disturbing the associated characteristic variable differences predicted from the interior.

For many flow situations involving steep gradients, such as weak oblique shock systems where no eigenvalue changes sign along the chosen coordinate lines or the boundary layer in higher Reynolds number viscous compressible flow, it is necessary to use higher order schemes to avoid excessive numerical viscosity. At the same time, in the immediate vicinity of discontinuities we want to avoid introducing errors in domain of dependence that arise in differencing across discontinuities. In reference 3 we have shown how second order upwind and third order biased upwind schemes can be constructed of flux difference split pieces such that the method automatically drops down to first order in the appropriate vicinity of singularities.

The process which we illustrate for a shock is effected through the use of simple algebraic switches that derive from the elements d_i of the truth functions D^+ and D^- of the simple interval flux difference splittings. The presence of a shock between the (j-1)th and jth mesh intervals accompanies a change in the third eigenvalue $\overline{u} - \overline{c}$. This is detected by the truth function $(d_S)_j = (d_3^+)_{j-1}(d_3^-)_j$. Contributions to methods of higher than first order can then be cancelled automatically by the use of the truth function complement $(1 - d_s)$. Accordingly, the following operators provide: (1) respectively, second and third order accuracy in smooth regions of the flow, (2) effective switching at shock transitions and (3) result in unconditionally stable methods when added to the right hand side of the implicit interior point equation (2)

$$-\tfrac{1}{2}(1-d_s)_{j-1}((\tilde{A}^+\Delta_\xi q)_{j-1} - (\tilde{A}^+\Delta_\xi q)_{j-2}) - \tfrac{1}{2}(1-d_s)_{j+1}((\tilde{A}^-\Delta_\xi q)_j - (\tilde{A}^-\Delta_\xi q)_{j+1}) \quad (3)$$

$$\begin{aligned}-(1-d_s)_{j-1}(1-d_s)_j(1-d_s)_{j+1}(\tfrac{1}{3}((\tilde{A}^+\Delta_\xi q)_j + (\tilde{A}^-\Delta_\xi q)_{j-1}) - \tfrac{1}{6}((\tilde{A}^+\Delta_\xi q)_{j-1} \\ + (\tilde{A}^+\Delta_\xi q)_{j-2} + (\tilde{A}^-\Delta_\xi q)_j + (\tilde{A}^-\Delta_\xi q)_{j+1}))\end{aligned} \quad (4)$$

Extensions

The CSCM method is a precursor and provides insight for two conservative flux difference splitting methods whose pieces have the simple form $\tilde{\tilde{A}}^+\Delta_\xi F$ and $\tilde{\tilde{A}}^-\Delta_\xi F$, $\tilde{\tilde{A}}^+ + \tilde{\tilde{A}}^- = I$. In the first case we take functionally $\tilde{\tilde{A}}^\pm = \overline{M}\,\overline{T}\,D^\pm\,\overline{T}^{-1}\,\overline{M}^{-1}$ where $\overline{M}$ and $\overline{T}^{-1}$ and their inverses are defined the same as for CSCM. Then

$$\tilde{\tilde{A}}^+\Delta_\xi F = \tilde{\tilde{A}}^+(\tilde{A}^+(\Delta_\xi)q + \tilde{A}^-(\Delta_\xi)q) = \tilde{A}^+(\Delta_\xi)q \quad \text{since } D^+D^+ = D^+ ,\ D^+D^- = 0 \quad (5)$$

and conversely for $\tilde{\tilde{A}}^- \Delta_\xi F$. The first form which we label CSCMF is closely related to what could be termed the modified Lax-Wendroff upwind difference scheme presented by Huang[8] for the isenthalpic equations and recently surveyed by Harten, Lax and van Leer[9]. In the second case, called Characteristics Flux Difference Splitting Averaged (CFDSA) as formulated by Yang[10], see also Yang, Bershader and Lombard[11], $\tilde{\tilde{A}}^\pm = (\overline{M\ T\ D^\pm\ T^{-1}\ M^{-1}})$.

Results

For results with CSCM we show solutions obtained with backward Euler implicit versions for two quasi 1-D nozzle flow problems: Shubin's supersonic-subsonic and Blottner's subsonic-subsonic nozzle flows. Both of these problems have been studied previously and are described in more detail by Yee, Beam and Warming[12]. The effectively exact solutions shown on our graphs in 51 points connected by solid line were developed by Warming.

In Figure 1 we show a density plot for the Shubin's nozzle problem run implicitly with the first order scheme equations (2) and a maximum mesh CFL number of 100. For engineering purposes with the fairly linear imposed subsonic outflow pressure boundary condition, boundary condition correction is not required and was not used to obtain the result shown.

In Figure 2 we show convergence histories of the root mean square residual of density for the Shubin's nozzle problem run (solid line) explicitly at CFL 0.8 and with overprescribed outflow boundary conditions, and implicitly at CFL numbers of 10 (circles) and 100 (triangles) with well posed characteristic outflow boundary conditions. The latter case corresponds to Figure 1. For practical purposes the approach to convergence is monotone with increasing CFL but the residuals do not fall as low at the highest CFL number

In Figure 3 we show a density plot for Blottner's nozzle problem run at CFL 50 and also without (subsonic) boundary condition correction at either inflow or outflow. Again the outflow boundary is effectively computed but quite evidently the nonlinear subsonic inflow boundary conditions of entropy and total enthalpy have drifted and overall solution accuracy is markedly deteriorated. In Figure 4 we show a similar density plot but from a run with boundary condition correction to within a relative error of 10^{-5}, which takes about two iterations.

In Figure 5 we show the time history of absolute error in an important measure of overall solution accuracy the mass flux at outflow: * for uncorrected boundary conditions and points connected by solid line for corrected boundary conditions.

In Figure 6 we show a corrected result for the third order interior point scheme for equations (2) and (4). The result shown in the synoptic of the paper was in error. In Figure 7 we show a second order result obtained with equations (2) and (3) and on a 31 point mesh, rather than 51 point as shown in the other figures.

Finally, for results with CFDSA we show a Mach reflection calculation for a shock incident at 42° with the plane and Mach number 4.23. The numerical method was explicit with an adaptation of the second order zero phase error method of Fromm[13] (also van Leer[14]). The second order method was found to reduce the smearing of the oblique shocks markedly over the simple first order upwind method.

References

(1) Roe, P.L. (1981). Lecture Notes in Physics 141, 354-359.
(2) Lombard, C.K. (1982). NASA CR166307.
(3) Lombard, C.K.; Oliger, J. and Yang, J.Y. (1982). AIAA 82-0976.
(4) Chakravarthy, S.R.; Anderson, D.A. and Salas, M.D. (1980). AIAA 80-0286.
(5) Oliger, Joseph (1980). Numerical Methods for Partial Differential Equations, Seymour Parter, ed., Academic Press.
(6) Kentzer, C.P. (1971). Lecture Notes in Physics 8, 108-113.
(7) Lombard, C.K.; Oliger, J.; Yang, J.Y. and Davy, W.C. (1982). AIAA 82-0837.
(8) Huang, Lan Chieh (1981). JCP 42, 195-211.
(9) Harten, Amiram; Lax, Peter D. and van Leer, Bram (1982). ICASE Report 82-5.
(10) Yang, Yaw Yen (1982). Ph.D. Dissertation, Aeronautics-Astronautics, Stanford University, Stanford, CA/USA.
(11) Yang, J.Y.; Bershader, D. and Lombard, C.K. (1982). SIAM Meeting, Stanford U. July, 1982.
(12) Yee, H.C.; Beam, R.M. and Warming, R.F. (1981). AIAA 81-1009-CP
(13) Fromm, Jacob E. (1968). JCP 3, 176-189.
(14) van Leer, Bram (1974). JCP 14, 361-370.

Appendix

Nonconservative $\Delta_\xi f$	Conservative $\Delta_\xi F$	
$\overline{\rho}\Delta_\xi \hat{W}_\xi + \overline{\hat{W}}_\xi \Delta_\xi \rho$	$\Delta_\xi(\rho \hat{W}_\xi)$	$\hat{W}_\xi = u\hat{\xi}_x$
$\overline{\rho\hat{W}_\xi}\Delta_\xi u + \tilde{\hat{\xi}}_x \Delta_\xi((\gamma-1)P)$	$\Delta_\xi(\rho u \hat{W}_\xi) + \tilde{\hat{\xi}}_x \Delta_\xi((\gamma-1)P)$	$\tilde{\hat{\xi}}_x = \dfrac{\overline{\hat{W}}_\xi}{\overline{u}} \cong \overline{\hat{\xi}}_x$
$\overline{\hat{W}}_\xi \Delta_\xi P + \overline{\gamma P}\Delta_\xi \hat{W}_\xi$	$\Delta_\xi(\rho H \hat{W}_\xi)$	$H = \dfrac{\gamma P}{\rho} + \dfrac{1}{2}u^2$

$$\overline{M} = \begin{array}{ccc} 1 & 0 & 0 \\ \overline{u} & 1 & 0 \\ \overline{\dfrac{u^2}{2}} & \overline{u} & 1 \end{array} \qquad \overline{T}^{-1} = \begin{array}{ccc} -\dfrac{1}{\overline{\rho}} & 0 & \dfrac{1}{\overline{\gamma P}} \\ 0 & \dfrac{1}{\overline{\rho}\overline{c}} & \dfrac{1}{\overline{\gamma P}} \\ 0 & -\dfrac{1}{\overline{\rho}\overline{c}} & \dfrac{1}{\overline{\gamma P}} \end{array}$$

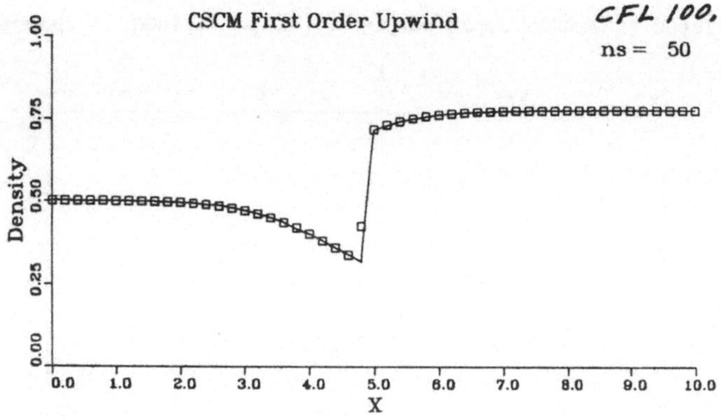

Figure 1. Density plot, Shubin's nozzle, CSCM method.

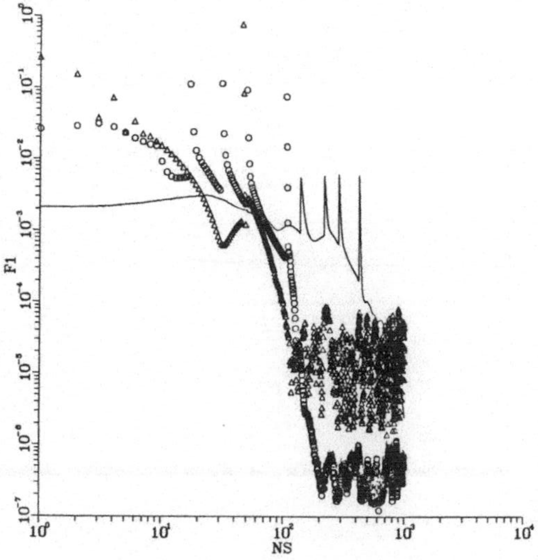

Figure 2. Density convergence history, Shubin's nozzle — CFL 0.8, o CFL 10., Δ CFL 100.

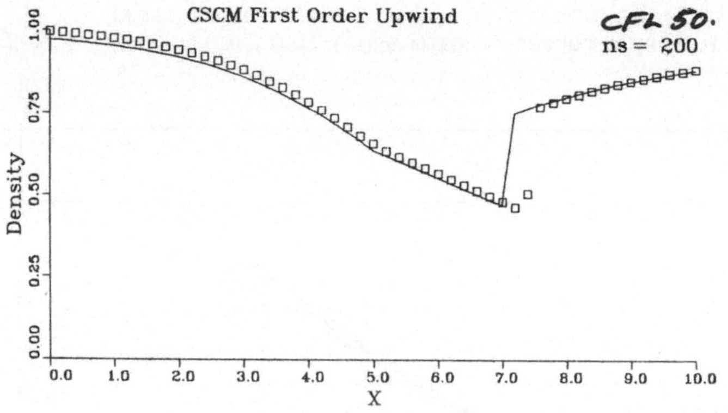

Figure 3. Density plot without boundary condition correction, Blottner's nozzle.

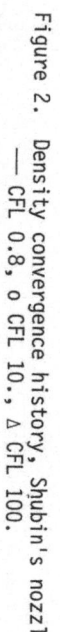

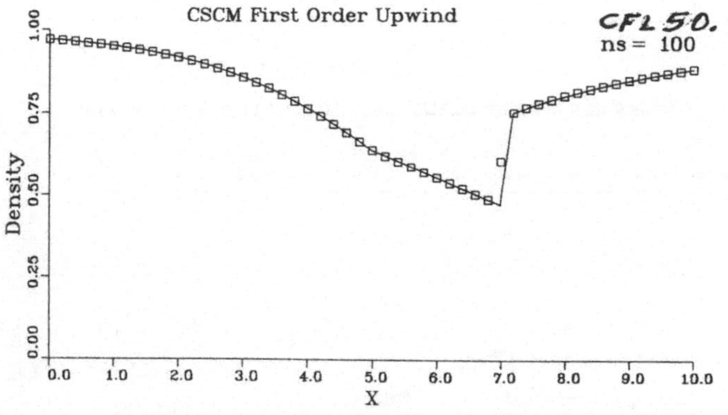

Figure 4. Density plot with boundary condition correction, Blottner's nozzle.

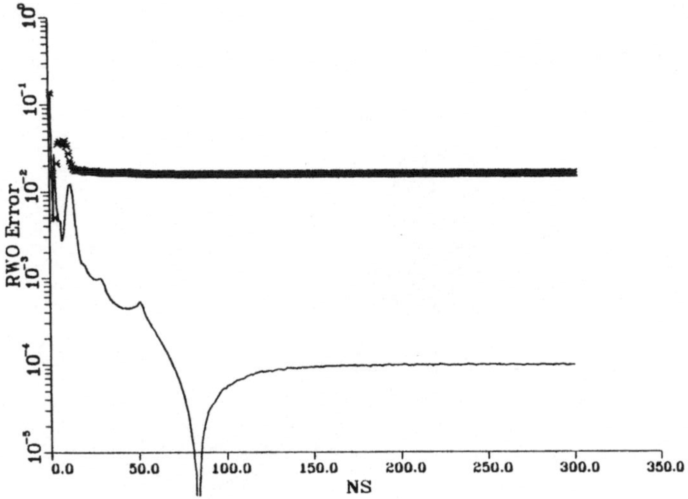

Figure 5. Time history of error in outflow mass flux, — with b.c. correction, * without.

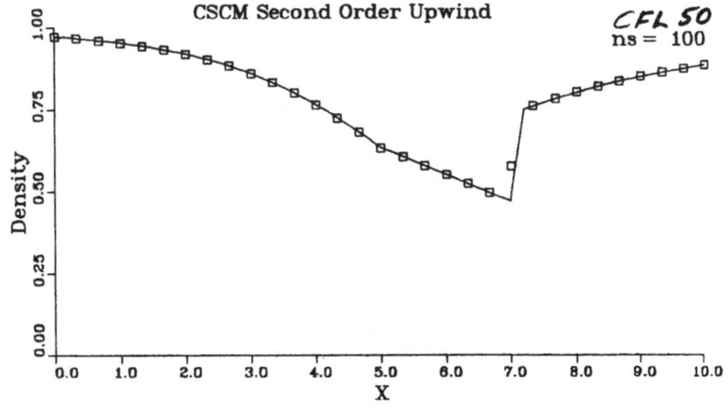

Figure 7. Density plot for second order scheme 31 points rather than 51.

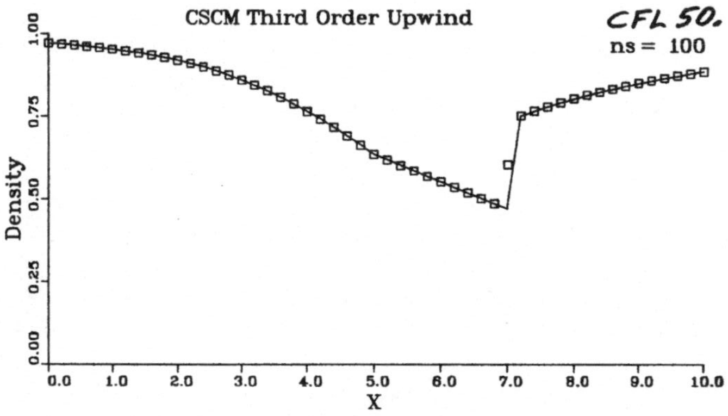

Figure 6. Density plot for third order scheme.

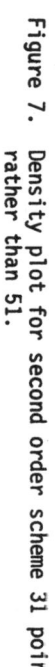

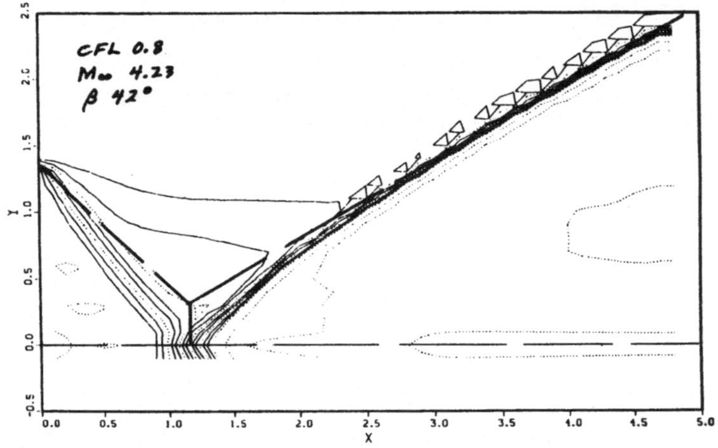

Figure 8. Mach reflection, CFDSA method, second order zero phase error.

Philip S. Marcus
Steven A. Orszag
Anthony T. Patera
Department of Mathematics
Massachusetts Institute of Technology
Cambridge, MA 02139 USA

SIMULATION OF CYLINDRICAL COUETTE FLOW

We have developed, tested, and implemented a numerical code for calculating the viscous, three dimensional flow between two differentially rotating cylinders (cylindrical Couette flow). In this paper we describe the methods that we have used to apply efficiently pseudo-spectral techniques to a complicated (i.e. non-Cartesian) geometry with real (i.e. viscous, no-slip) boundaries and the tests that we have used to determine the accuracy of our code.

I. OBSERVED STATES OF COUETTE FLOW

When the inner cylinder is rotating and the outer cylinder is held stationary, four non-chaotic flows have been observed experimentally. We define the Reynolds number as $RE = \Omega R_1 \ (R_2 - R_1)/\nu$ where ν is the viscosity, Ω is the angular velocity of the inner cylinder, and R_1 and R_2 are the radii of the inner and outer cylinders. We also define the radius ratio ξ to be R_1/R_2. For a given geometry, as the Reynolds number is increased to RE_{crit}, the stable flow changes from its laminar profile to N axisymmetric Taylor vortices stacked on top of each other. Near the endplates the vortices disappear. The upper surface boundary condition (rigid or free) strongly influences whether N is even or odd. A rigid upper surface tends to make N even. The Couette system selects the number N so that the axial wavelength of the cells, λ, is of order twice the radial gap separation (which is nearly the critical axial wavelength for the onset of the Taylor vortices). The number N is not unique for a fixed Reynolds number and geometry and is determined by the past history of the system. As the Reynolds number is increased further, azimuthal traveling waves [proportional to $\exp(im_1\theta)$] form on the vortices. Each state in this regime can be identified by N and m_1. The waves travel with a speed c_1. If this state is observed in the proper rotating frame, it appears as a steady-state. Although the speed c_1 depends on the radii of the inner and outer cylinders, it appears to be almost independent of m_1. For Reynolds numbers near critical, the wave speed is a strong function of RE, but for RE greater than ~6 times the critical value, the wave speed approaches a constant value. The wave speed is also a function of the axial wavelength. This non-axisymmetric, steady (in the proper rotating frame) flow becomes unstable as the Reynolds number is increased further and two traveling waves appear. The second traveling wave has azimuthal number m_2 and phase speed c_2. The second wave speed depends upon the radius ratio and weakly depends on N, m_1, m_2, and the Reynolds number. When the flow has one or two traveling waves, the states are meta-stable and the Reynolds number alone does not determine uniquely which particular state the flow will be. The numbers (N,m_1,m_2) that characterize the flow are not unique functions of the Reynolds number but depend upon the past history of the flow. The flow with two traveling waves does not appear as a steady state to any observer. When the Reynolds number is greater than a second critical value, the time spectrum of the flow develops a broad component centered around a frequency f_b. In this case, the motions are aperiodic (weakly chaotic). There is an increase in the amount of small scale structure visible in the flow as the Reynolds number is increased further. The azimuthal traveling waves eventually disappear but the Taylor vortex cells persist to the highest values of Reynolds number examined.

II. EQUATIONS AND BOUNDARY CONDITIONS

Our primary motivation for writing an initial-value code rather than a steady-state solver is that we are interested in computing the transitions among the four non-chaotic flows. One of the flows is quasi-periodic and not a steady-state. The initial-value equation that we solve is the Navier-Stokes equation in cylindrical coordinates with the boundary conditions that the radial and axial components of the velocity at the radial boundaries vanish and that the azimuthal components of the velocity match the cylinder rotation speed, Ω, at the walls. We assume periodicity in the axial direction and require that the velocity be divergence-free.

In a reference frame rotating with angular velocity C with respect to the inertial frame, the Navier-Stokes equation is

$$\frac{\partial v_c}{\partial t} = (v_c + v_{LAM} - C\, r\hat{e}_\theta) \times (\omega_c + \omega_{LAM}) + RE^{-1}\nabla^2 v_c - \nabla P_c \tag{1}$$

where the velocity seen by an observer in the inertial frame, $\hat{v}$, is

$$\hat{v}(r,\theta,z,t) \equiv v_c(r,\theta+ct,z,t) + v_{LAM}(r) \tag{2}$$

where v_{LAM} is the laminar velocity of Couette flow (as seen in the inertial frame)

$$v_{LAM}(r) = \frac{\xi}{1-\xi^2}[\frac{1}{r(1-\xi)} - r(1-\xi)]\hat{e}_\theta \tag{3}$$

where ω_{LAM} is the vorticity of the laminar flow

$$\omega_{LAM} = -2\xi/(1+\xi)\ \hat{e}_z \tag{4}$$

and where ω_c is the relative vorticity

$$\omega_c(r,\theta,z,t) \equiv \nabla \times v_c(r,\theta,z,t). \tag{5}$$

We use the relative velocity $v_c(r,\theta,z,t)$ as our dependent variable rather than the full velocity because $v_c(r,\theta,z,t)$ obeys the homogeneous boundary condition:

$$v_c(r,\theta,z,t) = 0 \quad \text{at the radial bounaries.} \tag{6}$$

In equation (1) the velocity is in units of ΩR_1 and the length is in units of (R_2-R_1). Our code is designed so that the speed of the rotating frame of the observer, C, can be easily changed at any timestep (in fact we generally automatically update C so that one of the two traveling waves appears steady). Equation (1) is solved spectrally and we adopt the notation that each variable $Q(r,\theta,z,t)$ is written as the spectral sum:

$$Q(r,\theta,z,t) = \sum_m \sum_k \tilde{Q}(r,m,k,t)\exp[i(m\theta+2\pi kz/\lambda)] \tag{7}$$

Note that $\hat{e}_r \cdot v_c(r,m=0,k=0,t)$ must always be identically equal to zero.

The radial dependence of each quantity is evaluated at the Chebyshev collocation points. Radial derivatives are found by fast-Fourier transforming into Chebyshev space, taking the spectral derivative, and inverse transforming back into physical radial space. Azimuthal and axial derivatives are evaluated spectrally. The nonlinear terms are computed by transforming into axial and azimuthal physical space, multiplying the values at the collocation points, and then inverse transforming. We do not remove aliasing errors.

III. THE TIME-SPLITTING STEPS

The nonlinear terms are computed in rotation form using a second-order Adams-Bashforth method. The velocity at the end of the nonlinear fractional step that goes from timestep N to timestep N+1 is:

$$v^{N+1/3}(r,\theta,z,t) = \frac{\Delta t}{2}[3(v_c^N + v_{LAM} - C\, r\hat{e}_\theta) \times (\omega_c^N + \omega_{LAM}) + v_c^N$$
$$- (v_c^{N-1} + v_{LAM} - C\, r\hat{e}_\theta) \times (\omega_c^{N-1} + \omega_{LAM})] \tag{8}$$

(For the remainder of this section, we assume that the angular speed of the observer, C, is zero to simplify the notation and the presentation of the equations. We omit the subscript, C, from the velocity.) The stability of the explicit nonlinear step in equation (8) is governed by the Courant condition. We can modify equation (8) to allow bigger timesteps by observing that even in a rotating frame, the largest velocity component is the z-independent, axisymmetric (m=0,k=0) component of the azimuthal velocity. In cylindrical Couette flow the mean azimuthal velocity is analogous to the mean temperature gradient in convection; it drives the advective instability and has a profile that is modified by order unity due to the nonlinear interactions. Like the mean temperature gradient, the mean velocity becomes modified so that in the interior of the flow the mean profile is nearly neutrally stable. Typically, $|\tilde{v}'| < 50.0\ |\tilde{v}_\theta(r,0,0,t)|$ where $\tilde{v}' \equiv \tilde{v} - \tilde{v}_\theta(r,0,0,t)\ \hat{e}_\theta$. We therefore treat the nonlinear contribution of $\tilde{v}_\theta(r,0,0,t)$ and $\tilde{\omega}_\theta(r,0,0,t)$ implicitly.

Adding the gradient of the pressure to $v^{N+1/3}$ makes the velocity divergence-free at the end of the next fractional step:

$$v^{N+2/3} = v^{N+1/3} - \nabla\Pi^{N+1} \tag{9}$$

The pressure head, Π^{N+1}, is computed by requiring that $v^{N+2/3}$ be divergence-free at the interior points. At the radial boundaries we require that $\partial v_r^{N+1}/\partial r = 0$. This last condition is equivalent to requiring that $\nabla\cdot v^{N+1} = 0$ at the radial boundaries. (We do not impose the boundary condition that $\partial v_r^{N+2/3}/\partial r = 0$ because we have found that it leads to a slow instability.) Since v^{N+1} is unknown during this fractional step, we write

$$\Pi^{N+1} = \Pi_{invis}^{N+1} + \Phi^{N+1} \tag{10}$$

where

$$\nabla^2\Pi^{N+1} = \nabla\cdot v^{N+1/3} \quad \text{in the fluid interior} \tag{11}$$

$$\nabla^2\Phi^{N+1} = 0 \quad \text{in the fluid interior} \tag{12}$$

and where Π_{invis}^{N+1} satisfies the inviscid boundary condition for the pressure

$$\frac{\partial\Pi_{invis}^{N+1}}{\partial r} = \hat{e}_r \cdot v^{N+1/3} \quad \text{at the boundaries} \tag{13}$$

The viscous step is the final fractional step and gives v^{N+1} :

$$v^{N+1} = v^{N+2/3} + \frac{\Delta t}{2RE}\nabla^2(v^N + v^{N+1}) \quad \text{in the fluid interior} \tag{14}$$

The operator $(1 - \frac{\Delta t}{2RE}\nabla^2)$ is inverted by requiring $v_r = v_\theta = v_z = 0$ at the radial boundaries. The Greens function, ϕ^{N+1}, is evaluated at the end of the viscous step. It is computed by evaluating a (diagonal) capacitance matrix once in a pre-processing stage and inverting the capacitance matrix after each viscous step.

Using the Greens function with the fractional step method produces a global temporal error of $\mathcal{O}(\Delta t^2)$. Instead of using Greens functions we can reduce the time-splitting error by solving the pressure equation with the inviscid boundary condition and using a Richardson extrapolation. For this particular geometry we have found that the use of Greens functions require no more storage than a Richardson extrapolation, but uses significantly less computing time. Furthermore, the Richardson extrapolation error is $\mathcal{O}(\Delta t^{3/2})$ in the velocity and larger than $\mathcal{O}(\Delta t)^{3/2}$ for the radial derivatives of v_r at the boundary. In particular, the error in $\frac{\partial^2 v_r}{\partial_r^2}$ using a Richardson extrapolation remains order unity.

IV. RESULTS

In addition to an initial-value code, we have written an eigenvalue code that solves the linearized sixth-order Navier-Stokes equation without the use of fractional steps. (The divergence of the velocity is everywhere zero including the boundaries). We have compared the linear eigenvalues (both the real and imaginary parts) computed from our linear eigenvalue-solver to the growth rates and velocities computed from the initial-value solver. For large Reynolds numbers, the errors in the growth rates and velocities from the initial-value solver decrease as Δt^2 (until the timestep is so small that truncation errors begin to dominate). The decrease confirms that the code is second-order accurate in time and shows that in the linear equations the splitting error is small. For the smallest timestep the errors in the growth rate and velocity are one part in 10^6 . We have also compared these eigenvalues to those calculated by Chandrasekhar (1961) for axisymmetric modes in a viscous fluid with $\xi = 0.5$ and to those of Kreuger, Gross, and DiPrima (1966) for non-axisymmetric, viscous flows in a thin gap with $\xi = 0.95$. We agree with these eigenvalues within one part in 10^4 .

We have several checks of consistency within our code. When a solution settles down to a quasi-periodic, periodic, or steady state, we calculate the divergence of the velocity. The divergence is a good measure of the time splitting error. We have found that for all modes (k,m) $(\nabla\cdot\tilde{v})/|\tilde{v}|\ \min(k,m/r)$ is less than 10^{-6} not only at the fluid interior but also at the radial boundaries. For flows with one traveling wave, we also compute the curl of $(v\times\omega + RE^{-1}\nabla^2 v)$ which should be zero (when observed in the rotating frame of the traveling wave). We find that $\nabla\times(v\times\omega + RE^{-1}\nabla^2 v)$ (in units of the average enstrophy) is of order 10^{-6}. In the steady state the rate of angular momentum flux in the radial direction

$$F = \frac{2\pi}{RE}\, r^3\, \frac{\partial r\tilde{v}_\theta(r,0,0,t)}{\partial r} - \int r^2 v_r v_z\, d\theta dz / \int dz \tag{15}$$

should be independent of radius and equal to the torque at the boundary. We find that the fractional variation in flux over the entire radius is about 0.003%. We have also solved the same initial-value problem (i.e. the same radius ratio and Reynolds number) several times while forcing the code to compute the solutions in different rotating reference frames. In one rotating frame the flow (with one traveling wave) appears as a steady-state but in the other frames it is periodic in time. Comparison of these solutions allows us to measure the temporal accuracy of the time-splitting when the nonlinear terms are large. We find that even for large nonlinear terms, the time stepping error remains small.

We have also measured the initial angular momentum of the flow. The initial angular momentum plus the temporally integrated torque should be equal to the angular momentum of the fluid in its final state. This type of consistency measure very accurately tests the torque and the structure of the solution near the boundary. The fractional difference between the actual angular momentum and that predicted by integrating the torques for ~ 4 rotation periods of the inner cylinder is of order 10^4 .

We have compared our numerically calculated torques with those measured experimentally by Donnelly and Simon (1960) for axisymmetric Taylor vortex flow with a radius ratio of $\xi = 0.5$. and for Reynolds numbers up to 4 times the critical value. Our agreement is within experimental error. One difficulty in comparing the torques with experiments is that there is a slight dependence of torque on the axial wavelength. The torque measurements of Donnelly and Simon (and almost all other experimentalists) are done with opaque cylinders so that the axial

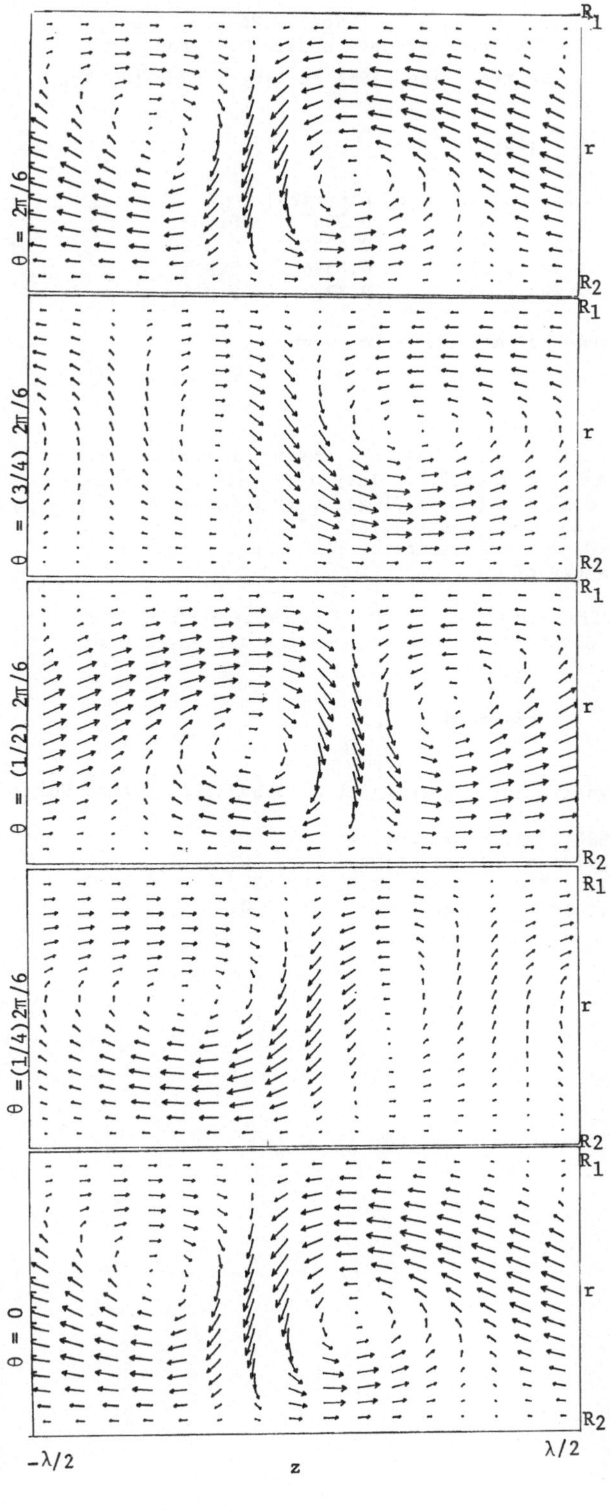

wavenumber (and often the azimuthal wavenumber) is not known and must be assumed. We have also found good agreement between our torques computed from an initial-value solver and those torques computed by Meyer-Spasche and Keller (1980)using a steady-state solver.

The most severe test that we have performed is the calculation of the wave speeds. We have found that at large Reynolds number (10 times RE_{crit}) where the wave speed is insensitive to the exact value of the Reynolds number, one can be somewhat cavalier with the treatment of the boundary conditions and obtain the correct result for the wave speed. At small values of the Reynolds number (2 times RE_{crit}) , the boundaries must be treated very carefully. Table 1 shows the experimentally measured values of the wave speed c_1, for $RE=459.8 \approx 3.96$ times RE_{crit} with radius ratio $\xi = 0.875$ and $m = 6$ for the two extreme values of the axial wavelength (i.e. flows whose axial wavelengths are not between these two numbers are unstable or nearly unstable and difficult to maintain experimentally). Table 1 also includes the experimental value of c_1 for $RE = 230.2 \approx 2.00$ times RE_{crit} with $\xi = 0.868$ and $m = 6$ for two extreme values of the axial wavelength. Included in the table are our numerical values computed with 33 radial collocation points, 32 axial collocation points per Taylor cell pair, and 16 aximuthal points per $2\pi/6$ radians. Our calculated wave speeds agree with the laboratory measurements within the experimental uncertainty. We have agreement to 3 significant digits in the wave speed. The experimentally measured values of the wave speed are only accurate to 1% due to the uncertainties in the viscosity (or Reynolds number) while the experiments are being performed. The experimental measurements of c_1 presented here were done by King and Swinney (1982). We have found similar agreement between our numerically computed wave speeds c_1 and c_2 and the experiments for flows with two traveling waves at larger ($RE \approx 10\ RE_{crit}$) Reynolds numbers.

TABLE 1

RE_{crit}	RE/RE_{crit}	λ	c_1 lab	c_2 numerical	ξ
115.1	2.05	3.00	0.365	0.3647	0.868
115.1	2.00	2.14	0.403	0.4028	0.868
116.1	3.96	2.32	0.359	0.3596	0.875
116.1	3.96	3.90	0.339	0.3397	0.875

wave speeds for m=6 traveling wave

All four of the flows presented in Table 1 have a symmetry in addition to being 6-fold symmetric about the axis of rotation. They also have the symmetry $v_r(r,\theta,z) = v_r(r,\theta+\frac{2\pi}{12} - z)$, $v_\theta(r,\theta,z) = v_\theta(r,\theta + \frac{2\pi}{12},z)$, $v_z(r,\theta,z) = -v_z(r,\theta + \frac{2}{12} - z)$. We show this symmetry by plotting the azimuthal modulation of a pair of Taylor cells. Figure 1 shows the v_z and v_r components of the velocity in the (z-r) plane for five different values of θ.

This work was supported by the Office of Naval Research Contract #N00014-82-C-0451, a National Science Foundation Fellowship #SPI-8009181, and the Air Force Office of Scientific Research under Grant # 77-3405.

REFERENCES

Chandrasekhar, S., (1960), Hydrodynamic and Hydromagnetic Stability, Oxford University Press.
Donnelly, R. J., & Simon, N.J., 1960, J.F.M., 7, 401.
King, G., & Swinney, H.L., 1982, private communication.
Kreuger, E. R., Gross, A., & DiPrima, R.C., 1966, J.F.M., 24,521.
Meyer-Spasche, R., & Keller, H. B., 1980 J. Comp. Phys., 35, 100.

SOLUTION-ADAPTIVE GRID FOR THE CALCULATION OF

THREE-DIMENSIONAL LAMINAR AND TURBULENT BOUNDARY LAYERS

W.L. Melnik
University of Maryland at College Park
and David Taylor Naval Ship R&D Center

Maryland 20742, U.S.A.

INTRODUCTION

The main objective in the development of optimal coordinates for the calculation of three-dimensional boundary layers is to resolve (within a prescribed accuracy) the resulting flow field with a minimal number of grid points. Since the largest variation of flow properties usually occurs normal to the body surface, the transformation of only this coordinate will be considered in the realization of this objective. Incidentally, no attempt will be made to determine node locations for the best possible approximation with a given number. It will be sufficient to construct a reasonably good (and cheap) approximation with a nonuniform mesh which is defined at the outset of the calculation. It is, of course, recognized that even a problem of such limited scope must depend to some degree on the other coordinates as well.

In streamline coordinates, a Levy-Lees type of transformation may scale the normal coordinate to accommodate the thickness of three-dimensional boundary layers as suggested by the results of Ref. 1 for two-dimensional flows. However, in body oriented coordinates this transformation can easily become singular at various locations, and the boundary layer growth can be accounted for along only one of the two surface coordinates (Ref. 2).

This paper introduces a direct transformation of the normal coordinate which is completely solution adaptive to capture just the domain of interest and enhance the resolution of the large changes of flow properties in the wall layer. Moreover, the governing equations automatically reduce to similarity type where appropriate. Initial velocity profiles obtained from their solution are fully consistent with the continuation of the solution (of the general equations). This computational method is illustrated by results for the boundary-layer development over prolate spheroids at various angles of attack.

GOVERNING EQUATIONS

With no essential loss of generality, the development of this calculation is expressed in prolate spheroidal coordinates (Fig. 1). The governing equations are those for the thin shear layer approximation of the Reynolds-averaged Navier-Stokes equations, with the turbulent stresses modelled by an isotropic eddy viscosity.

The normal coordinate is transformed by two length scales to facilitate the numerical solution. It will be shown that Δ_2 rectangularizes computational space and that Δ_1 enlarges the wall layer. With

$$\eta = \frac{y}{\Delta_1} \qquad \text{for} \quad 0 \le \eta \le 1 \tag{1a}$$

$$= 1 + \frac{y - \Delta_1}{\Delta_2 - \Delta_1} \qquad \text{for} \quad \eta > 1 \tag{1b}$$

the governing equations transform to

continuity

$$\frac{1}{\Delta_\ell}\frac{\partial g}{\partial \eta} = \frac{V}{\Delta_\ell}\frac{\partial \Delta_\ell}{\partial \xi} + \pi_1 u + \pi_2 w + \frac{u_e}{h_1}\frac{\partial u}{\partial \theta} + \frac{w_e}{r}\frac{\partial w}{\partial \omega} \tag{2}$$

θ-comp of momentum:

$$\frac{1}{\Delta_\ell}\frac{\partial}{\partial \eta}[(1+\varepsilon)s] + gs + \pi_3(1-u^2) + \pi_4(1-uw) + \pi_5(1-w^2) = V\frac{\partial u}{\partial \xi} \tag{3}$$

ω-comp of momentum:

$$\frac{1}{\Delta_\ell}\frac{\partial}{\partial \eta}[(1+\varepsilon)t] + gt + \pi_6(1-uw) + \pi_7(1-w^2) = V\frac{\partial w}{\partial \xi} \tag{4}$$

shear variables:

$$s = \frac{1}{\Delta_\ell}\frac{\partial u}{\partial \eta} \qquad \text{and} \qquad t = \frac{1}{\Delta_\ell}\frac{\partial w}{\partial \eta} \tag{5a,b}$$

where

$$-g = v + \Delta_\ell \, V\frac{\partial \eta}{\partial \xi} \tag{6}$$

$$V\frac{\partial}{\partial \xi} = \frac{u_e u}{h_1}\frac{\partial}{\partial \theta} + \frac{w_e w}{r}\frac{\partial}{\partial \omega} \tag{7a}$$

$$h_1 = \sqrt{1 - e^2\cos^2\theta} \tag{7b}$$

and $\Delta_\ell = \Delta_1$ for $0 \le \eta \le 1$; $\Delta_\ell = \Delta_2 - \Delta_1$ for $\eta > 1$

The eccentricity e of the spheroid is expressed in terms of its thickness ratio τ by

$$e = \sqrt{1 - \tau^2} \tag{7c}$$

The physical components of velocity have been made dimensionless by their local values at the outer edge of the boundary layer, except for the normal component, $v = \sqrt{Re_L}\,\tilde{v}/u_{ref}$ and $y = \tilde{y}\sqrt{Re_L}/L$, where the Reynolds number is based on free stream properties and the characteristic length of the body, L = 2a. The pressure-gradient terms π's are evaluated from the potential-flow solution given in Lamb (Ref. 3) thus neglecting effects of boundary-layer separation. Eqs. (2)-(5) are to be solved subject to the boundary conditions of no-slip $u = w = 0$ on an impermeable surface at $\eta = 0$ where $g = 0$ and the inviscid flow velocities attained as $\eta \to \infty$.

The length scales Δ_1 and Δ_2 must be determined simultaneously with the solution of the equations from auxiliary conditions still to be prescribed. The essential computational domain is captured by the requirement that the dimensionless velocity V at the outer boundary, say $\eta = 2$, be held fixed to a preassigned value $V = V^*$. With $V^* = 0.995$, Δ_2 is essentially a measure of the boundary layer thickness. Although only one length scale may suffice to resolve the velocity profile of a laminar boundary layer, turbulent flow is typically characterized by still another, quite different length scale near the wall, associated with Δ_1. The normal coordinate through this layer is expanded by requiring the dimensionless velocity at $\eta = 1$ be held at some specified value $V = V_{w1}$ (say $V_{w1} = 0.2$).

The Maise-McDonald (Ref. 4) mixing length model is especially well suited for evaluation of the turbulent eddy viscosity ε,

$$\varepsilon = \ell^2 \sqrt{Re_L [(u_e s)^2 + (w_e t)^2]} \tag{8}$$

since its value ℓ_o in the wake-like layer is directly related to the boundary layer thickness Δ_2, i.e.,

$$\ell = \min \{\ell_i, \ell_o\} \tag{9a}$$

where

$$\ell_i = 0.41y \, [1 - \exp(\frac{-y}{A})] \tag{9b}$$

$$\ell_o = 0.089 \, \Delta_2 \tag{9c}$$

The outer boundary conditions $u = w = 1$ were applied at $2.4 < \eta_e < 2.8$. Since s and t at η_e were usually $O(10^{-4})$ or smaller, the value of η_e was considered sufficiently large. However, the outer boundary conditions can be replaced by equivalent conditions at $\eta = 2$ taken from Shalman's asymptotic solution (Ref. 5), which yields

$$u = w = V^* \tag{10a,b}$$

$$s = t = V^*(1-V^*) \, [\frac{7}{3\lambda^*} + \lambda^*] \tag{11a,b}$$

$$\text{where } \lambda^* = \Delta_1 \int_0^1 \frac{V}{V_e} d\eta + (\Delta_2 - \Delta_1) \int_1^2 \frac{V}{V_e} d\eta \tag{12}$$

Since the auxiliary condition duplicates in part the boundary conditions (10a,b), it was replaced by the requirement that either (11a) or (11b) be satisfied by the solution. Not only did results using Eqs. (10)-(12) confirm those obtained with the usual boundary conditions, they appeared to be more accurate in the neighborhood of separation.

It should be noted that the governing equations reduce to similarity type if either or both velocity components at the outer edge of the boundary layer vanish. Their solutions yield initial data along intersecting planes for the continuation of the solution by the general box method. For example, at a three-dimensional stagnation point, $u_e = w_e = 0$, the equations naturally reduce to a set of ordinary differential equations in the only remaining independent variable η.

The locus defined by $u_e = 0$, which separates the downstream from the upstream inviscid flow, is herein identified as the dividing line. In general it does not coincide with any plane of symmetry, although it may for special geometries and flow conditions (eg the sphere in global coordinates). As $u_e \rightarrow 0$ Eq. (3) becomes:

$$u_{e_\omega}(1-uw) + rK_2 w_e(1-w^2) = 0 \tag{13}$$

which requires that $u = w$, a result entirely consistent with the boundary conditions. Consequently, the flow becomes coplanar in θ = const planes at the dividing line. The resulting equations exhibit a dependence only on η and ω and serve to advance the solution from one meridian to the next. Wang (Ref. 6) marched his solution in a similar manner based on considerations of the domain of dependence of the governing equations but did not comment on the peculiar qualities of the flow near the dividing line.

METHOD OF SOLUTION

The governing equations are solved using Keller's box method (Ref. 7). The rationale of this method is to replace the equations by an equivalent set which contains only first-order derivatives. Consequently, difference approximations are naturally centered and of second-order accuracy, even with a variable mesh, since values at only two successive points are required for the difference formulas. With the application of appropriate difference approximations, Eqs. (2)-(5) are replaced by equivalent nonlinear algebraic equations, centered about the centroid of the difference molecule. These equations are solved by iteration about a trial solution using Newton's method. This fully coupled system of equations is solved in terms of Δ_1 and Δ_2 by an extension of the block elimination method described in Ref. 8 to a system of five unknowns. Finally, the length scales Δ_1 and Δ_2 are obtained simultaneously from the auxiliary equations. When evaluation of the eddy viscosity lagged during the iterations, the rate of convergence was only linear, even though the equations were fully coupled. However with dependence of the eddy viscosity properly included in the linearization about a trial solution, a quadratic rate of convergence was obtained for turbulent flows as well.

Starting from the forward stagnation point the calculations were marched along meridians, first forward toward a location very near the nose of the body, then in the downstream direction from the stagnation point along the same meridian, and terminated when negative cross flow or separation was encountered. The calculation was continued in a similar manner along adjacent meridians, starting from the dividing line, until its end at the nose was reached. From this point the calculation was marched circumferentially to the adjacent meridian and then downstream. When negative crossflow was encountered, the calculation could be continued only in the circumferential direction using a difference molecule equivalent to scheme (4) of Wang.

RESULTS AND DISCUSSION

The computer program was checked out on the Burroughs B7700 with the test case of a sphere ($\tau = 1$) at various angles of incidence between the coordinate axis and free stream. The number of grid points N1 and N2 across each layer of the computational domain and values of V_{w1} are indicated for each example. The calculations were considered converged when skin friction agreed to 4 or 5 significant figures between successive iterations.

The skin friction along various meridians is plotted as a function of the polar angle ϕ measured from the stagnation point of the sphere in Fig. 2 and 3. The results are clearly axisymmetric with respect to ϕ even though the flow was fully three-dimensional in the prolate-spheroidal coordinates. Laminar boundary separation was located at $\phi = 105.1^o$ as compared to 105.9^o predicted by the results of Ref. 9, while turbulent separation was located at $\phi = 131^o$ (see Fig. 3) with $Re_L = 826{,}000$.

In the final example, the laminar skin friction along selected meridians is shown in Fig. 4 for a prolate spheroid with $\tau = 1/4$ at 6° incidence to the free stream. Accurate resolution of the large negative overshoot which developed in the circumferential velocity profile required more grid points in the wall layer than in the outer layer of the computational domain. Agreement with Wang's results (shown only along the plane of symmetry for sake of clarity) is excellent over the nose region. However, based on the results for the test case (Fig. 2), it would appear that the velocity profiles for $\theta > 0.5$ are more accurately resolved by the present calculation.

It is concluded from these examples that velocity profiles may be accurately resolved using as little as 15 to 30 nodes in the calculation of solutions to three-dimensional boundary layer flows.

REFERENCES

1. Carter, J.E., Edwards, D.E. and Werle, M.J., NASA-CP-2166, 197-212, 1980.
2. Cebeci, T., Chang, K.C. and Kaups, K., Ocean Engineering, 7, 229-289, 1980.
3. Lamb, Horace, Hydrodynamics, 6th Ed. Cambridge Univ. Press, 139-156, 1953.
4. Maise, G. and McDonald, H., AIAA J., 6, 73-80, 1968.
5. Shalman, E. Yu., Izv. Adad, Nauk USSR, MZG, No. 5, 155-157, 1979.
6. Wang, K.C., J. Fluid Mech., 72, 49-65, 1975.
7. Keller, H.B., Numerical Solution of Partial-Differential Equations II, J. Bramble (ed.), Academic Press, N.Y., 327-350, 1970.

8. Cebeci, T. and Bradshaw, P.: Momentum Transfer in Boundary Layers. McGraw-Hill/ Hemisphere Publishing Corp., Washington, 1977.

9. Smith, A.M.O. and Clutter, D.W., AIAA J. 1, 2062-2071, 1963.

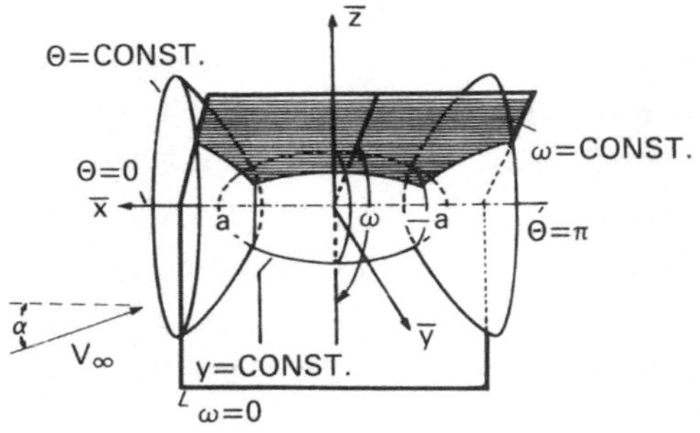

Figure 1 - Prolate Spheroidal Coordinates (Θ, ω, y)

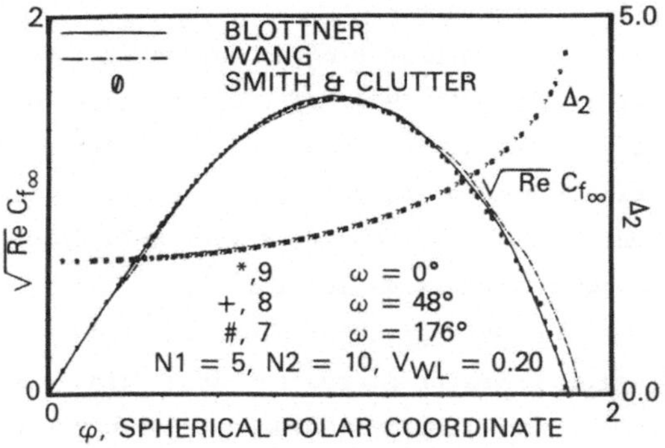

Figure 2 - Skin Friction Distribution and Thickness of Laminar Boundary Layer Over a Sphere

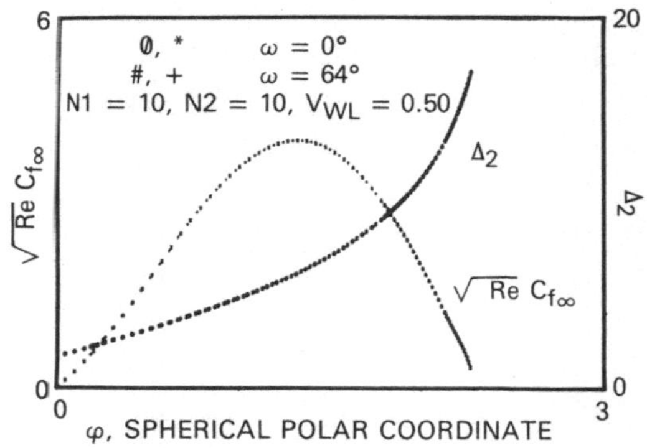

Figure 3 - Skin Friction Distribution and Thickness of Turbulent Boundary Layer Over a Sphere

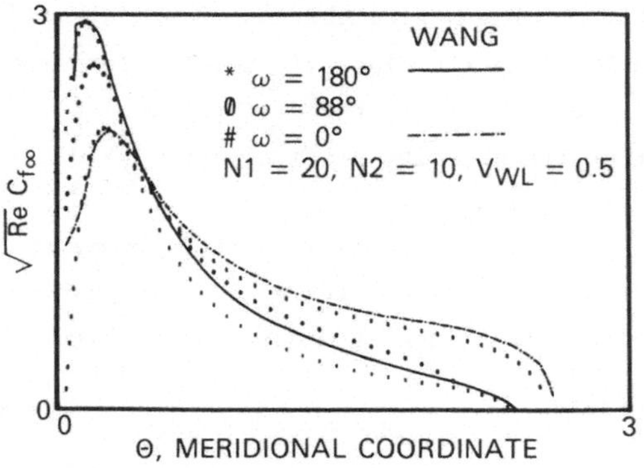

Figure 4 - Distribution of Skin Friction Magnitude Over a Prolate Spheroid, $\tau = 1/4$ with 6° Incidence

Calculation of Pressure Statistics in Turbulent Free Shear Flows by Direct Numerical Simulation

Ralph W. Metcalfe
James J. Riley
Flow Research Company
Kent, WA 98031/USA

Introduction

The objectives of the work described in this paper are to perform accurate numerical simulations of turbulent free shear flows, and to use the simulations to study the behavior of certain critical turbulence quantities which have been difficult to measure in the laboratory. Since some of these terms, such as the pressure-velocity correlations, can have a significant influence on turbulence models, an accurate description of their behavior is essential. In our simulations, we directly solve the three dimensional, time dependent, nonlinear Navier-Stokes equations using pseudo-spectral numerical methods. No turbulence modeling or ad hoc closure assumptions are made. The simulations are performed at sufficiently high Reynolds numbers so that the flows are strongly nonlinear and turbulent, but low enough so that all scales of motion containing significant energy are adequately resolved. The evolution of the complex velocity field is computed in time; relevant statistical quantities are computed either during the course of the calculation or later from data fields saved periodically. As long as the simulations are sufficiently accurate, they are somewhat analogous to laboratory experiments with the advantage that a very detailed description of the flow fields can be obtained. The principal disadvantage is the limited spatial resolution which effectively determines the maximum possible Reynolds number. However, to the extent that free turbulent shear flows behave in a manner consistent with Reynolds number similarity, much valuable information can be obtained from a careful study of moderate Reynolds number flows.

Background

The calculations described here were performed by the Direct Numerical Simulation technique. This involves solving the Navier-Stokes equations:

$$\frac{\partial u}{\partial t} = \underset{\sim}{u} \times \underset{\sim}{\omega} - \nabla\pi + \nu\nabla^2 \underset{\sim}{u}$$

$$\nabla \cdot u = 0$$

where $\underset{\sim}{u} = (u,v,w)$ is the velocity field, $\underset{\sim}{\omega} = \nabla x \underset{\sim}{u}$ is the vorticity, $\pi = p+1/2u^2$ is the pressure head, with p being the pressure, and ν the kinematic viscosity. The velocity field is expanded in a three dimensional Fourier series as

$$u(\underset{\sim}{x},t) = \sum_{|k_j|<N} u_j(\underset{\sim}{k},t) \exp\left[2\pi i\left(\frac{k_1 x_1}{L_1} + \frac{k_2 x_2}{L_2} + \frac{k_3 x_3}{L_3}\right)\right]$$

where L_i is the computational domain size in the ith direction. The key idea of the pseudo-spectral method (Orzsag, 1971; Gottlieb and Orszag, 1977) employed here is that the nonlinear product terms, such as $\underset{\sim}{u} \times \underset{\sim}{\omega}$, are computed pointwise in physical space, which can be done very efficiently since this is a local operation. Likewise, the spatial derivative terms are most naturally computed locally in Fourier space.

The Fourier to physical space transformations necessary to compute these terms are most efficiently computed using the fast Fourier transform. Although aliasing errors are introduced by the computation of the nonlinear terms, they tend to be largest at the highest wave numbers, and these modes have relatively little energy in accurate turbulence simulations. We have used periodic boundary conditions since these are consistent with the flow fields we are simulating and since they permit use of a faster, higher resolution code. Periodic boundary conditions also allow easy computation of the pressure. Adams-Bashforth or leap-frog time differencing has been used to compute the nonlinear terms, and Crank-Nicolson time differencing has been used on the viscous terms. Our calculations have been performed on 32x32x32 point computational grids and take about 4 sec/time step to run on the CDC 7600 at NASA Ames. Thus, a typical 400 time step 32^3 calculation requires about 25 minutes of machine time.

We have made use of laboratory data to initialize our calculations. The mean velocity field is based on the self-similar profiles found in the experimental data. The turbulence field is initialized by the method suggested in Orszag and Pao (1974). In this technique, the desired energy spectrum is specified and the initial field has the proper spatial distribution of turbulence intensity in a statistical sense. However, since the phases of the Fourier modes are random, none of the important velocity correlations such as the Reynolds stresses and skewnesses have the proper values. These are allowed to build up over the first several dozen time steps of the simulation.

Results

One of the objectives of this work is to use the technique of direct numerical simulations to gain insight into the behavior of some of the critical terms in turbulence modeling and to understand why these models often fail to accurately predict the behavior of the turbulent flows. Many turbulence models used in the numerical simulation of complex turbulent flows are based on the transport equations for the Reynolds stresses. A critical aspect of turbulence model development is defining reasonable approximations to terms such as the triple moments and pressure-velocity correlations. This is made difficult by the lack of good experimental data or theoretical analysis regarding the behavior of these terms.

The Reynolds stress transport equation can be written as

$$\frac{\partial}{\partial t}\overline{u_i'u_j'} + \overline{u_i'u_k'}\,\overline{\frac{\partial u_j}{\partial x_k}} + \overline{u_j'u_k'}\,\overline{\frac{\partial u_i}{\partial x_k}} + \frac{\partial}{\partial x_k}\overline{u_i'u_j'u_k'} = -\frac{1}{\rho}\left[\overline{u_i'\frac{\partial p'}{\partial x_j}} + \overline{u_j'\frac{\partial p'}{\partial x_i}}\right] + \nu\left[\frac{\partial^2\overline{u_i'u_j'}}{\partial x_k^2} - 2\,\overline{\frac{\partial u_i'}{\partial x_k}\frac{\partial u_j'}{\partial x_k}}\right] \tag{A1}$$

We have examined the behavior of some of these terms in several complex turbulent shear flows. For example, the pressure interaction terms are often factored into diffusion and return-to-isotropy components either as

$$\overline{u_k'\frac{\partial p'}{\partial x_i}} + \overline{u_i'\frac{\partial p'}{\partial x_k}} = \underbrace{\frac{\partial}{\partial x_i}\overline{u_k'p'} + \frac{\partial}{\partial x_k}\overline{u_i'p'}}_{\text{Diffusion}}\ \underbrace{-\ \overline{p'\left(\frac{\partial u_i'}{\partial x_k} + \frac{\partial u_k'}{\partial x_i}\right)}}_{\text{Return to Isotropy}} \tag{A2}$$

(Launder, Reece and Rodi, 1975) or

$$\overline{u_k'\frac{\partial p'}{\partial x_i}} + \overline{u_i'\frac{\partial p'}{\partial x_k}} = \underbrace{\frac{2}{3}\,\delta_{ik}\frac{\partial}{\partial x_i}\overline{p'u_j'}}_{\text{Diffusion}} + \underbrace{\overline{u_k'\frac{\partial p'}{\partial x_i}} + \overline{u_i'\frac{\partial p'}{\partial x_k}} - \frac{2}{3}\,\delta_{ik}\frac{\partial}{\partial x_j}\overline{p'u_j'}}_{\text{Return to Isotropy}} \tag{A3}$$

(Lumley and Khajeh-Nouri, 1974).

We have performed simulations of the evolution of a turbulent wake with and without swirl present. The presence of swirl is of interest since it adds significant physical complexity to the flow field and since many turbulence models have had trouble predicting flows with swirl. Figure 1 shows the initial mean axial velocity $\overline{U}_z$, mean angular velocity $\overline{U}_\theta$, and rms turbulence intensity, normalized by the peak mean axial velocity, as functions of distance from the axis of the wake. Note that the peak mean swirl velocity is about half the peak mean axial velocity and that the peak turbulence intensity is about 1/3 of the peak mean axial velocity. In Figure 2 we plot the rms pressure and the turbulence energy for the case without swirl after the flow has developed for some time ($t = 17$, where t is time non-dimensionalized by the half radius of the initial mean axial velocity profile and the peak mean axial velocity). The pressure is normalized by $\rho \overline{u_z'^2}\max$ and the energy by $\overline{u_z'^2}\max$. Note that the rms pressure fluctuations are of the same order as the turbulence intensity and that the pressure decays more slowly away from the wake axis. In Figure 3, the same quantities are plotted, but in the presence of swirl. This shows that the swirl tends to enhance the amplitude of the rms pressure levels. The second order correlation $\overline{u_\theta' u_z'}/\overline{u_z'^2}$ max is plotted in Figure 4 for the swirling wake simulation. This quantity was measured experimentally by Morse (1979) for a swirling jet flow. Although some turbulence models predict that this term should be uniformly negative, our results are consistent with his experimental data.

In the case of the temporally evolving turbulent wake, a comparison of the relevant diffusive terms in the turbulent energy equation

$$\frac{\partial}{\partial t} \frac{\overline{u_i'^2}}{2} + \frac{1}{r}\frac{\partial}{\partial r} r \left(\overline{u_r' \frac{u_i'^2}{2}} + \overline{u_r' p'} \right) + \ldots\ldots = 0$$

is shown in Figure 5 for the non-swirling wake. Note that $\overline{u_r'p'}$ is countergradient and of the same order as $u_r u_i^2$, which is consistent with several current turbulence models. For the off-diagonal Reynolds stress components, the diffusive components are different for the models in equations A2 and A3 above. For model A2, the relevant terms in the equation are

$$\frac{\partial}{\partial t} \overline{u_r'u_z'} + \frac{1}{r}\frac{\partial}{\partial r} r \left(\overline{u_r'^2 u_z'} + \overline{u_z'p'} \right) + \ldots$$

while for model A3, they are

$$\frac{\partial}{\partial t} \overline{u_r'u_z'} + \frac{1}{r}\frac{\partial}{\partial r} \left(\overline{r\, u_r'^2 u_z'} \right) + \ldots$$

As can be seen in Figure 6, the $\overline{u_z'p'}$ term actually dominates the velocity diffusion term $u_z' u_r'^2$. Thus, model A2 would produce a countergradient diffusive effect because of the $\overline{u_z'p'}$ term, and this is inconsistent with the physics of the problem. Model A3, in which the pressure/velocity term does not appear in the diffusive component would seem to be the preferable approach here. A similar analysis for the $\overline{u'^2_r}$ equation shows that again the model A3 is more consistent with the physics of the simulation than is model A2.

We are currently performing some higher resolution (64^3 point) simulations of a temporally evolving turbulent mixing layer. Preliminary results indicate that the pressure velocity correlation terms, as well as some of the other higher moments, are very sensitive to the presence and behavior of large scale structures in the flow field. There is some indication that the large scale structures can actually cause the correlation terms to change sign. If this turns out to be more universally the case, it would suggest that conventional turbulence modeling of such flows may be very difficult and that some sort of direct numerical calculation of the large eddies in such flows may be necessary.

Acknowledgements

The authors would like to thank Steven Orszag, Morris Rubesin, and Robert Rogallo for useful discussions, NASA Ames Research center for support under Contract No. NAS2-9855 and for computer time on their CDC 7600, and the Office of Naval Research for support under Contract No. N00014-78-C-0346.

References

Gottlieb, D. and Orszag, S. A., "Numerical Analysis of Spectral Methods: Theory and Applications," NSF-CBMS Monograph No. 26 Soc. Ind, App. Math., Philadelphia, 1977.

Launder, B. E., Reece, G. J., and Rodi, W., "Progress in the Development of a Reynolds Stress Turbulence Closure," J. Fluid Mech. 68, 537 (1975).

Lumley, J. L., and Khajeh-Nouri, B., "Computational Modeling of Turbulent Transport," Adv. Geophys. 18A, 169-192 (1974).

Morse, A., "Axisymmetric Turbulent Shear Flows With and Without Swirl," Ph.D. Thesis (University of London), 1979.

Orszag, S. A. 1971, "Numerical Simulation of Incompressible Flows Within Simple Boundaries: Accuracy," J. Fluid Mech., Vol. 49, pp. 75-112.

Orszag, S. A., and Pao, Y. H. 1974, "Numerical Computation of Turbulent Shear Flows," Adv. Geophys., Vol. 18A, pp. 225-236.

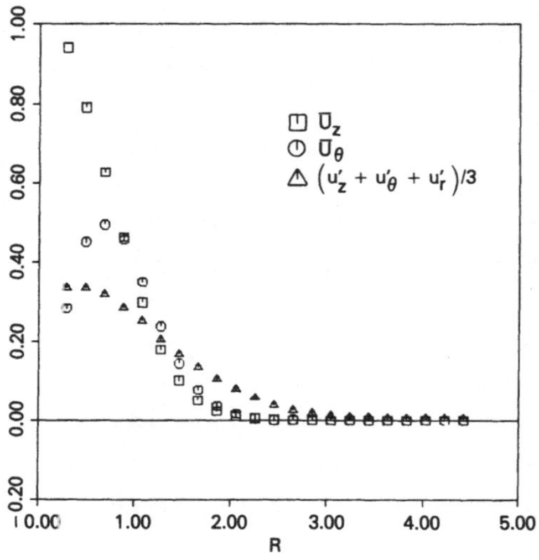

Figure 1. Mean Velocity Profiles and Turbulence Intensity for Swirling Wake at t = 0.

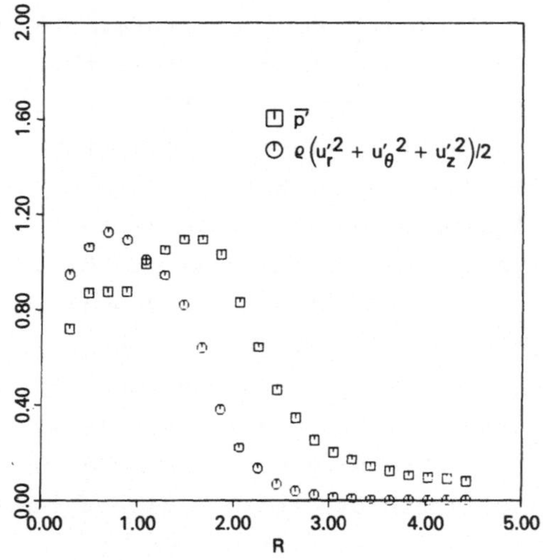

Figure 2. RMS Pressure and Turbulence Intensity for Swirling Wake at t = 17.

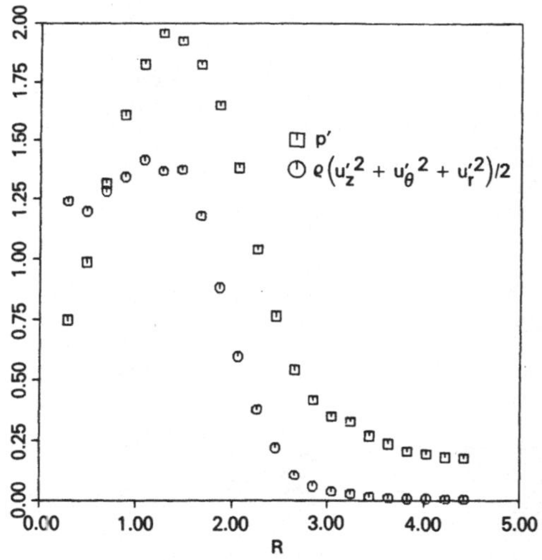

Figure 3. RMS Pressure and Turbulence Intensity for Non-swirling Wake at t = 17.

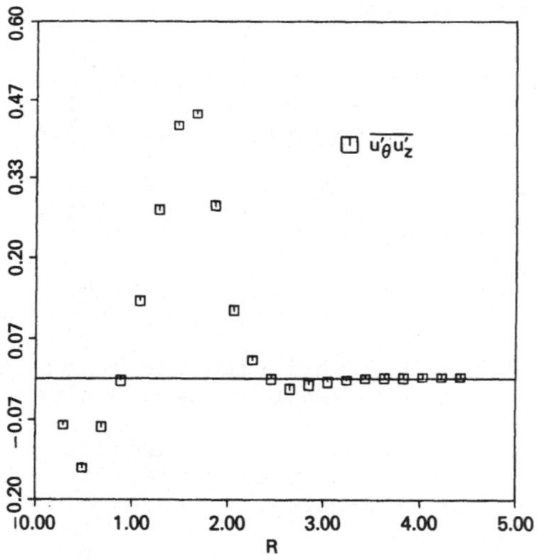

Figure 4. $\overline{u'_r u'_\theta}$ for Swirling Wake at t = 17.

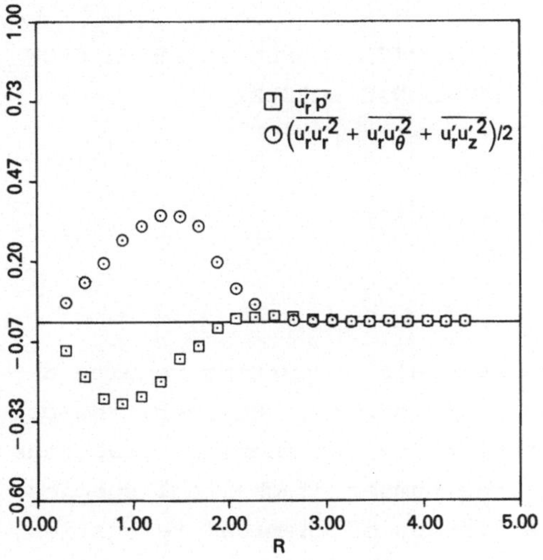

Figure 5. Comparison of Pressure and Velocity Diffusion Terms for Turbulent Wake Without Swirl at t = 17.

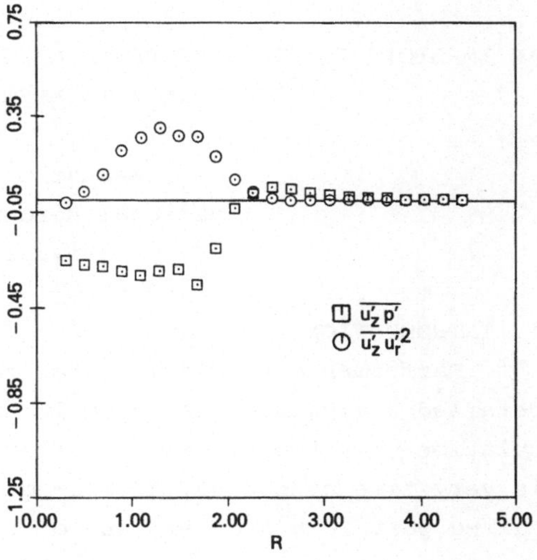

Figure 6. Comparison of Pressure and Velocity Diffusion Terms for Turbulent Wake Without Swirl at t = 17.

AN IMPLICIT FINITE DIFFERENCE METHOD FOR CHEMICAL NONEQUILIBRIUM FLOW THROUGH AN AXISYMMETRIC SUPERSONIC NOZZLE

Kazuhiro Nakahashi
National Aerospace Laboratory
Miyagi, Japan

1. INTRODUCTION

Performance predictions for rocket propulsion systems require a detailed analysis of the flow in the exhaust nozzle. However, the calculation of a two-dimensional flow with finite-rate chemical reactions is very time consuming, because of a large number of chemical species taking part in a problem, and of the stiffness of chemical relaxation equations in a near equilibrium flow region. The stiffness of equations makes it impractical to use a standard explicit integration method.

A commonly used method for analyzing the two-dimensional nonequilibrium nozzle flow is an 'explicit-implicit method'[1-4] in which the fluid dynamic equations (equations of global continuity, axial momentum, radial momentum and energy) are integrated using explicit methods, while the species conservation equations are integrated using implicit methods in order to remove the unstable property of the calculation of chemical relaxation equations. However, we can not regard the explicit-implicit method as a sufficiently stable one, because the iteration, which is needed to couple the solutions of the fluid dynamic equations to those of the species conservation equations, often suffers from the ill convergency in the near equilibrium region. This trouble is mainly caused by the large differences in the properties between the fluid dynamic equations and the species conservation equations.

In this report, a noniterative finite-difference algorithm has been developed for the stiff equations of a chemically reacting nonequilibrium flow. It consists of a linearization of the system of equations and an implicit treatment of the axial derivative terms. In order to make the algorithm simple and to save computer storage, the radial derivative terms are treated explicitly. After linearizing and differencing, the equations are reduced to a linear algebraic equation system of order Ns+4, where Ns is the number of chemical species.

2. CONSERVATION EQUATIONS

Following assumptions are made in the present study. (a) The computational region is the supersonic flow field downstream of the nozzle throat and, throughout the region, there are no shock waves.

(b) Each component of the gas is a perfect gas. (c) The flow is assumed to be everywhere in instantaneous translational, rotational, and vibrational equilibrium.

Under these assumptions, a hyperbolic system of equations for inviscid flow of reacting gas mixture is written using the axisymmetric coordinate system(x,r). (Fig.1)

$$u\frac{\partial\rho}{\partial x} + v\frac{\partial\rho}{\partial r} + \rho\left(\frac{\partial u}{\partial x} + \frac{\partial v}{\partial r}\right) + \frac{1}{r}\rho v = 0 \qquad (1)$$

$$u\frac{\partial u}{\partial x} + v\frac{\partial u}{\partial r} + \frac{a^2}{\gamma}\frac{\partial P}{\partial x} = 0 \qquad (2)$$

$$u\frac{\partial v}{\partial x} + v\frac{\partial v}{\partial r} + \frac{a^2}{\gamma}\frac{\partial P}{\partial r} = 0 \qquad (3)$$

$$u\frac{\partial P}{\partial x} + v\frac{\partial P}{\partial r} - \frac{\gamma}{\rho}\left(u\frac{\partial \rho}{\partial x} + v\frac{\partial \rho}{\partial r}\right) - \frac{\gamma K}{a^2} = 0 \qquad (4)$$

$$u\frac{\partial Ci}{\partial x} + v\frac{\partial Ci}{\partial r} - \frac{1}{\rho}\omega_i = 0, \quad i=1,\cdots,N_s \qquad (5)$$

and

$$P = \ln P, \quad p = \rho RT, \quad a^2 = \gamma p/\rho$$
$$K = \gamma\sum_{i=1}^{Ns}\left(\frac{\bar{R}}{W_i}\right)\left(\frac{\omega_i}{\rho}\right)T - (\gamma-1)\sum_{i=1}^{Ns} h_i\left(\frac{\omega_i}{\rho}\right) \qquad (6)$$

where C_i is the species mass fraction and ω_i is the species production rate.

In order to transform the physical plane(x,r) to a rectangular plane(ξ,η), the following coordinate transformation is employed(Fig.1).

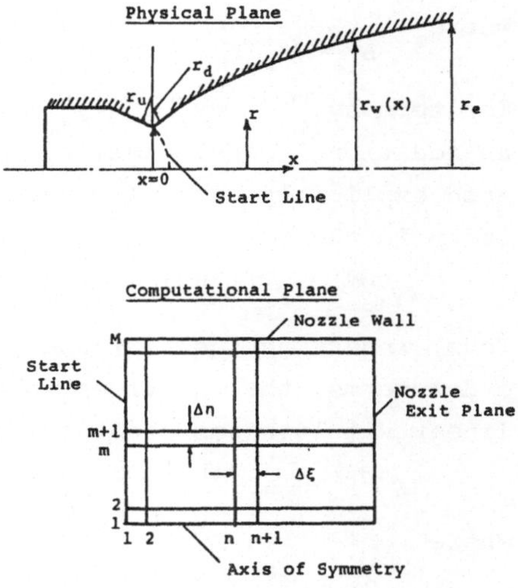

Fig.1 Coordinate system

$$(\xi,\eta) = (x, r/r_w) \qquad (7)$$

Applying this transformation to Eqs.(1) to (5) yields the final form of governing equations as

$$\frac{\partial y_i}{\partial \xi} = f_i, \quad i=1,2,\cdots,N \qquad (8)$$

where y_i represent the velocity components (u,v), natural log of pressure P, temperature T, and species mass fraction (C_j, $j=1,\cdots,N_s$), respectively. f_i is the function of these unknowns as well as (ξ,η). N_s is the number of chemical species and $N(=N_s+4)$ the total number of unknowns.

3. FINITE-DIFFERENCE PROCEDURE

When an implicit difference approximation is applied to the ξ-derivative in Eq.(8), we obtain

$$\Delta y_i^{(n+1)} = \frac{1}{2}\Delta\xi^{(n+1)}\left[\, f_i^{(n)} + f_i^{(n+1)} \right] + O(\Delta\xi^3) \qquad (9)$$

where

$$\Delta y_i^{(n+1)} = y_i^{(n+1)} - y_i^{(n)}, \qquad \Delta\xi^{(n+1)} = \xi^{(n+1)} - \xi^{(n)}$$

and the superscript n refers to a finite-difference step for ξ.

As the nonlinear function $f_i^{(n+1)}$ includes the unknowns $y_j^{(n+1)}$ (j=1,$\cdots$,N) implicitly, we need to consider the linearization of $f_i^{(n+1)}$ in $y_j^{(n+1)}$ so as to make the scheme noniterative. This linearization can be obtained using a local Taylor expansion, regarding f_i as a function of ξ, y_j and $y_{j\eta}$ (j=1,$\cdots$,N), where subscript η denotes differentiation.

$$f_i^{(n+1)} = f_i^{(n)} + A_i^{(n)}\Delta\xi^{(n+1)} + \sum_{j=1}^{N} [\, B_{ij}^{(n)}\Delta y_j^{(n+1)} + C_{ij}^{(n)}\Delta y_{j\eta}^{(n+1)} \,] + O(\Delta\xi^2) \qquad (10)$$

where

$$A_i = \partial f_i/\partial\xi \,, \quad B_{ij} = \partial f_i/\partial y_j, \quad C_{ij} = \partial f_i/\partial y_{j\eta} \qquad (11)$$

The term $\Delta y_{j\eta}^{(n+1)}$ $(=y_{j\eta}^{(n+1)} - y_{j\eta}^{(n)})$ appearing in the above equation requires an additional consideration. In the proposed scheme this term is evaluated explicitly, assuming that this is independent of the unstable property in the calculation for the near equilibrium flow region. Namely,

$$\Delta y_{j\eta}^{(n+1)} = y_{j\eta\xi}^{(n)}\Delta\xi^{(n+1)} + O(\Delta\xi^2) = f_{j\eta}^{(n)}\Delta\xi^{(n+1)} + O(\Delta\xi^2) \qquad (12)$$

Thus, inserting Eqs.(10) and (12) into Eq.(9) and neglecting the third-order terms, the equations to be solved are reduced to the following linear simultaneous equations of the order N at each grid point.

$$\Delta y_i^{(n+1)} - \frac{1}{2}\Delta\xi^{(n+1)} \sum_{j=1}^{N} B_{ij}^{(n)}\Delta y_j^{(n+1)} = \frac{1}{2}\Delta\xi^{(n+1)} \, [\, f_i^{(n)} + \alpha_i^{(n)} \,] \qquad (13)$$

where

$$\alpha_i^{(n)} = f_i^{(n)} + A_i^{(n)}\Delta\xi^{(n+1)} + \sum_{j=1}^{N} C_{ij}^{(n)} f_{j\eta}^{(n)}\Delta\xi^{(n+1)} \qquad (14)$$

Derivatives with respect to η included in $f_i^{(n)}$ and $f_i^{(n+1)}$ are approximated by a noncentered difference scheme in which the forward differences are used in $f_i^{(n)}$ and the backward differences in $B_{ij}^{(n)}$ and $\alpha_i^{(n)}$ of Eq.(13). This noncentered scheme was found to be superior to the centered one in stabilizing the solution.

By evaluating $\Delta y_{j\eta}^{(n+1)}$ explicitly in the proposed scheme, the value of ξ-step ($\Delta\xi$) is restricted by the CFL condition. A full implicit method whose step size is not restricted by the CFL condition is possible if $\Delta y_{j\eta}^{(n+1)}$ is evaluated implicitly as

$$\Delta y_{j\eta,m}^{(n+1)} = (\Delta y_{j,m+1}^{(n+1)} - \Delta y_{j,m-1}^{(n+1)})/2\Delta\eta \qquad (15)$$

where m is the index of a radial grid. Replacing Eq.(12) with this difference form, the final equation corresponding to Eq.(13) is a block tridiagonal system for each axial plane. However this full implicit method was not employed in the present study, because it requires a large memory size and is less efficiency as discussed later.

Boundary conditions along the nozzle wall can be obtained using characteristic equations along stream lines and left running characteristics. After deriving the difference forms of the characteristics equations, a linearization scheme similar to Eq.(10) is applied to the nonlinear functions appearing in the difference equations. This linearization reduces the equations to a final form of simultaneous algebraic equations whose unknowns are $V^2(=u^2+v^2)$, P, T, C_i, $i=1,\cdots,Ns$.

Construction of an initial data line suitable for the calculation of the supersonic flow field is made using the analytical solution of Kliegel and Levine[5] for the transonic flow field of the nozzle throat. The detailes for starting the finite-difference calculation from the initial data line is found in reference 6.

4. OPERATIONAL COUNT

In calculation using an implicit scheme, most of the computational time is consumed in solving large order simultaneous equations. Therefore, the computational efficiency of each scheme may be made clear by investigating the arithmetic operational counts required in the calculation of a set of simultaneous equations constructed in each scheme. The comparison is made for the following three schemes, the proposed scheme, the explicit-implicit scheme and the full implicit scheme. In the following discussion, we consider the operational counts only for the calculation of a set of simultaneous equations, and other arithmetic operations are neglected.

(A) The proposed scheme: A set of simultaneous equations of the order Ns+4 is constructed at each grid point and can be solved using the Gauss elimination method. Thus the operational count for each grid point is[7]

$$(Ns+4)^3/3 + (Ns+4)^2 - (Ns+4)/3 \quad \text{ops.} \tag{16}$$

(B) The explicit-implicit scheme: The species conservation equations are solved implicitly constructing at each grid point a set of simultaneous equations of the order Ns. In addition, this scheme needs iterations to couple the solutions of fluid dynamic equations to those of species conservation equations. Thus, the operational count must be multiplied by the number of iterations, I, and we get the real count per grid point as[7]

$$I(Ns^3/3 + Ns^2 - Ns/3) \quad \text{ops.} \tag{17}$$

(C) The full implicit scheme: Handling the term $\Delta y_{j\eta}^{(n+1)}$ implicitly as Eq.(15), this scheme constructs a block-tridiagonal system of equations with each block having dimensions $(Ns+4)\times(Ns+4)$ at each axial plane. Denoting the total number of the grid points for radial direction by M, the count for a grid point in this scheme is[7]

$$(3 - 2/M)[(Ns+4)^3 + (Ns+4)^2] \quad \text{ops.} \tag{18}$$

Fig.2 shows these counts as a function of the number of chemical species, Ns, considered in a problem. Line A indicates the count of the proposed method, and line C, which is about ten times as large a count as line A, the count of the full implicit method. Although the unconditionally stable property of the full implicit method make it possible to adopt a larger ξ-step size, this size cannot be more than ten times the step size of the proposed method due to the accuracy requirement in practical problems. Line B indicates the count for the implicit part of the explicit-implicit method. The number of iterations, I, required in the practical computation is about two to three for the nearly frozen region and five to ten or more for the near equilibrium region. Thus the operational count of line B (I=3 or 10) becomes greater than the proposed scheme's count.

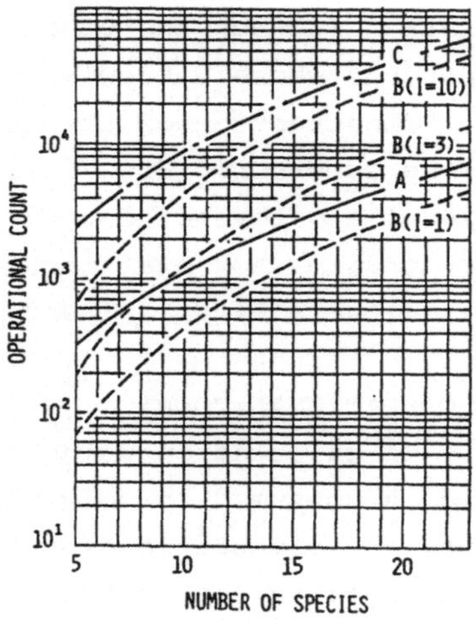

Fig.2 Arithmetic operational count required in each scheme

5. COMPUTATIONAL RESULTS AND DISCUSSIONS

In Figures 3 and 4, computational results are shown for a bell-shaped rocket nozzle. The accuracy of the proposed scheme was checked by comparing the pressure distributions along the nozzle wall with experimental results[8](Figure 4), as well as with those of other numerical methods for nonreactive and reactive gas flows. Numerical stability was checked by computational experiments, and it was confirmed that the CFL condition is the stability limit.

The computational times required in the proposed scheme and the explicit-implicit scheme[3] for the calculation of the flow field of a practical rocket nozzle are compared in Figure 5. Hydrogen-oxygen propellant was assumed and the six chemical species were taken into account for a H-O system. Additionally, in order to examine the effect of the number of species on computational time, assumed argon-like nonreactive gases, whose molecular weights are equal to argon species, were added to the propellant. By this figure, along with the discussion of arithmetic operational counts, it is well demonstrated that the proposed scheme has high computational efficiency, especially for problems with a large number of chemical species.

Although the present study is confined to the supersonic flow fields, it may be applied to subsonic-transonic flow problems with finite-rate chemical reactions, by combining it with a time-dependent method. The time-dependent method requires much computational time and a very large amount of computer storage. Thus, much greater advantages of the proposed scheme can be expected in those reacting gas flow problems.

REFERENCES

1) Nickerson, G. R., Coats, D. E. and Bartz, J. L., NASA CR-152999, 1977.
2) Ratliff, A. W., NASA CR-3121, 1979.
3) Nakahashi, K., Kisara, K., Moro, A. and Miyajima, H., in Proc. of 24th Space Science and Technology, Japan, 1980.
4) Stiles, R. J. and Hoffman, J. D., AIAA Paper No.81-1432, 1981.
5) Kliegel, J. R. and Levine, J. N., AIAA J., Vol.7, No.7, pp.1375-1378, 1969.
6) Nakahashi, K., Moro, A. and Miyajima, H., Technical Report of National Aerospace Laboratory, Japan, NAL TR-634, 1980.
7) Isaacson, E. and Keller, H. B., "Analysis of Numerical Methods", John Weley & Sons, 1966
8) Miyajima, H. et al., Technical Report of National Aerospace Laboratory, Japan, NAL TR-662, 1981.

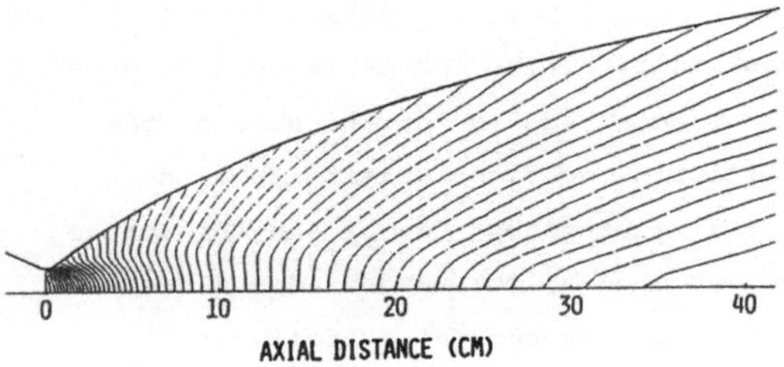

Fig.3 Constant Mach number lines in a bell nozzle, AR=140, H/O propellant

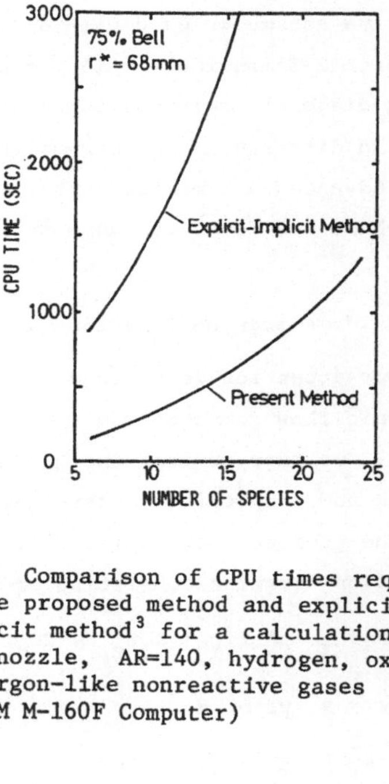

Fig.5 Comparison of CPU times required in the proposed method and explicit-implicit method[3] for a calculation of a bell nozzle, AR=140, hydrogen, oxygen and argon-like nonreactive gases (FACOM M-160F Computer)

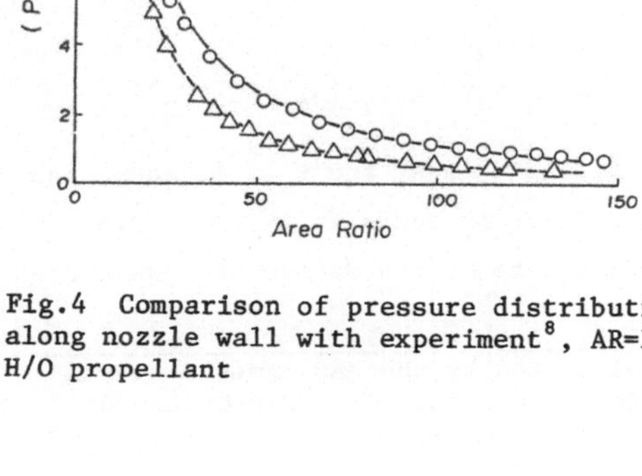

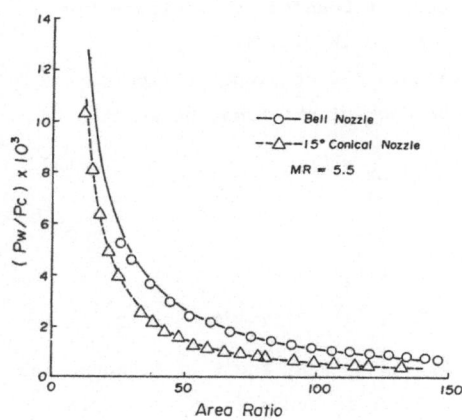

Fig.4 Comparison of pressure distribution along nozzle wall with experiment[8], AR=140, H/O propellant

NONITERATIVE GRID GENERATION USING PARABOLIC DIFFERENCE EQUATIONS
FOR FUSELAGE-WING FLOW CALCULATIONS

S. Nakamura
The Ohio State University
Mechanical Engineering Department
206 West 18th Avenue, Columbus, Ohio 43210
U.S.A.

1. Introduction

This paper describes a fast method of generating 3-dimensional grids for fuselage-wing transonic flow calculations using parabolic difference equations, as an extension of the author's previous work [1]. The word "noniterative" in the title indicates that no iterative scheme is used in the 3-dimensional sense. With the proposed method, grids are generated from one grid surface to the next, starting from the fuselage surface and completing at the outer boundary without global iterations. The computational procedure is similar to the solution of 2-dimensional heat conduction equation with the backward differencing on the time domain, which uses an iterative scheme at each time step. Although an iterative scheme is used to generate each 2-dimensional grid plane, its convergence rate is fast because of the strong diagonal dominance inherent in the elliptic equations arising from the backward differencing of parabolic differential equations. Use of the ADI scheme which advances a time step after each sweep was studied but was abandoned because the scheme was found to cause unsmoothness of grids around the arifoil.

2. Grid Generation Equations and Numerical Scheme

Our attention is focused on grid generation for 3-dimensional fuselage-wing transonic flow computations. Denoting the coordinates of the computational domain by (X,Y,Z), we suppose that the grids to be generated are bounded by fuselage-wing surface and a specified outer flow boundary as plotted in Fig. 1.

The proposed scheme is derived from the standard 3-dimensional elliptic grid generation equation consisting of three equations each of which may be written as

$$A_1 r_{XX} + A_2 r_{YY} + A_3 r_{ZZ} + 2B_1 r_{XY} + 2B_2 r_{YZ} + 2B_3 r_{ZX} = 0 \qquad (1)$$

where r = x, y, or z,

$$A_1 = (x_Y{}^2 + y_Y{}^2 + z_Y{}^2)(x_Z{}^2 + y_Z{}^2 + z_Z{}^2) - (x_Y x_Z + y_Y y_Z + z_Y z_Z)^2$$
$$B_1 = (x_Y x_Z + y_Y y_Z + z_Y z_Z)(x_Z x_X + y_Z y_X + z_Z z_X) - (x_X x_Y + y_X y_Y + z_X z_Y)(x_Z{}^2 + y_Z{}^2 + z_Z{}^2)$$

A_2, A_3, B_2, and B_3 are obtained by rotating X, Y, and Z in A_1 and B_1. Although the spacing control terms such as in refs. 2 and 3 are not included in Eq.(1), grid spacing may be controlled by modifying the coefficients of the difference equations as described next.

Funds for the support of this study have been allocated by the NASA-Ames Research Research Center, Moffett Field, California, under Interchange No. NCA2-OR565-101.

In deriving the difference equations for Eq.(1), usually the grid spacing on the computational domain is set to unity for all the directions. However, this assumption is not necessary at least until the grids are used for actual flow calculations. Therefore, it is assumed here that the grid spacings on the computational domain, denoted by f, g, and h as shown in Fig.2, are locally nonuniform. Using the nonuniform grid spacings, difference equations for Eq.(1) are first derived. With a modification to the difference term for r_{ZZ}, the proposed grid generation equation is written in the form,

$$\begin{aligned}
&A_1[(r_{i-1,j,k}-r_{i,j,k})/f_{i-1}+(r_{i+1,j,k}-r_{i,j,k})/f_i] \\
&+A_2[(r_{i,j-1,k}-r_{i,j,k})/g_{i-1}+(r_{i,j+1,k}-r_{i,j,k})/g_i] \\
&+A_3[(r_{i,j,k-1}-r_{i,j,k})/h_{k-1}+(r_{i,j,K}-r_{i,j,k})/H_k] \\
&+2B_1(r_{i+1,j+1,k-1}-r_{i+1,j-1,k-1}-r_{i-1,j+1,k-1}+r_{i-1,j-1,k-1})/(f_{i-1}+f_i)/(g_{j-1}+g_j) \\
&+2B_2(r_{i,j+1,K}-r_{i,j-1,K}-r_{i,j+1,k-1}+r_{i,j-1,k-1})/(g_{j-1}+g_j)/(h_{k-1}+H_k) \\
&+2B_3(r_{i+1,j,K}-r_{i-1,j,K}-r_{i-1,j,k-1}+r_{i+1,j,k-1})/(f_{i-1}+f_i)/(h_{k-1}+H_k) = 0 \qquad (2)
\end{aligned}$$

where K is the value of k on the outer boundary, H_k is the distance between (i,j,k) and (i,j,K) on the computational domain. The specific equations for x, y, and z are obtained by replacing r in the above equation by x, y and z, respectively. Notice that Eq.(2) does not involve the values on the (k+1)th grid plane, but rather uses the values of r on the outer boundary surface, k=K. Thus, Eq.(2) is a parabolic difference equation. The outer boundary condition $r_{i,j,K}$ in Eq.(2) may be gradually altered as k is increased. The direction of grid lines in the direction of k may be controlled by altering $r_{i,j,K}$ in Eq.(2).

In order to derive an ADI solution algorithm, Eq.(2) is rewritten as

$$\begin{aligned}
Lr_{i,j,k} \equiv\ & a^L r_{i-1,j,k} + a^R r_{i+1,j,k} - (a^L+a^R)r_{i,j,k} \\
& + a^B r_{i,j-1,k} + a^T r_{i,j+1,k} - (a^B+a^T)r_{i,j,k} - a^C r_{i,j,k} = S_{i,j,k} \qquad (3)
\end{aligned}$$

where

$$a^L = A_1/f_{i-1},\quad a^R = A_1/f_i,\quad a^B = A_2/g_{j-1},\quad a^T = A_2/g_j,\quad a^C = A_3(1/h_{k-1} + 1/H_k)$$

and S includes all the terms in Eq.(2) that do not explicitly appear in Eq.(3). Equation (3) may be written in the matrix form as

$$[H+V+D]_k \vec{r}_k = \vec{S}_{r:k} \qquad (4)$$

where $\vec{r}_k = \vec{x}_k$, $\vec{y}_k$ or $\vec{z}_k$. Since Eq.(4) does not involve $r_{i,j,k+1}$, it may be solved for $r_{i,j,k}$ if $r_{i,j,k-1}$s are all given. Thus, when the initial conditions $r_{i,j,1}$ are all given, the solution may be obtained in the increasing order of k. Equation (4) is a 2-dimensional discrete elliptic equation for each level of k. The ADI scheme for Eq.(4) is written as

$$[D^{-1}(H+D)(V+D)]_k \delta\vec{r}_k^{\,(t)} = \vec{S}_{r:k} - (H+V+D)\vec{r}_k^{\,(t-1)} \qquad (5)$$

where t is the iteration number and $\vec{r}_k^{(o)}$ is the initial guess which is set to $\vec{r}_{k-1}$. The above iterative scheme converges very rapidly (5 to 8 iterations for $\max|\delta r_{i,j,k}/r_{i,j,k}| < 10E-4$) because of the diagonal dominance provided by D.

3. Boundary Conditions

The present algorithm needs the grids on the outer boundary and those on the fuselage surface to be prescribed. In the sample grids shown in this paper, the former are analytically generated. The latter are generated as follows. The fuselage surface is first defined in the θ-r-y cylindrical coordinate system, where θ is the angle around the fuselage axis. An H-grid system is generated on the θ-y retcangular coordinates by using the 2-dimensional version of the present marching grid generation method [1], and then translated to the x-y-z coordinates(see Fig.3).

Symmetry boundary conditions are applied to the vertical center plane (symmetry of the fuselage: j=1 and j=jmax), so the grids there are obtained as a part of the solution.

The grid lines to be located on the wing are preselected. The coordinates of $y_{i,j,k}$ and $z_{i,j,k}$ on the wing are specified as functions of $x_{i,j,k}$ for each grid line, i. During one ADI iteration cycle, the values of $x_{i,j,k}$ on the wing are determined as solution of the grid generation equation, but $y_{i,j,k}$ and $z_{i,j,k}$ on the wing are fixed. They are revised, however, after every iteration by using the prescribed functions and the updated values of $x_{i,j,k}$.

4. Grid Spacing and Angle Control

The grid spacing is controlled by changing f_i, g_j, h_k and H_k as necessary. For example, as f_i is decreased, the distance between $r_{i,j,k}$ and $r_{i+1,i,j}$ becomes smaller. The direction of the line between $r_{i,j,k}$ and $r_{i,j,k+1}$ may be controlled by changing $r_{i,j,K}$ for the outer boundary as k increases, as described in the previous paper [1]. In 3-dimensional grid generation, however, changing the outer boundary conditions, $r_{i,j,K}$, every time k increases is cumbersome, particularly when the wing has both tapered and non-tapered sections.

The difficulty described above is overcome by using the anchor technique introduced next. Suppose the grid, $r_{i,j,k}$, is desired to be bound loosely to a prescribed "anchor point", $\tilde{r}_{i,j,k}$. The anchor constraints are incorporated into Eq.(4) as

$$Lr_{i,j,k} = S_{r:i,j,k} - w_{i,j,k}(\tilde{r}_{i,j,k} - r_{i,j,k}) \tag{6}$$

where w is a user-specified weight for the anchor effect at the grid (i,j,k). Notice that as w is increased, the grid (i,j,k) is more tightly bound by the anchor point. If $w \to \infty$ as an extreme, Eq.(6) becomes a fixed boundary condition. Use of anchors for a small number of selected grids is easy and useful.

5. Illustration of Generated Grids

The top and front views of the fuselage-wing geometry used in the present calculations are shown in Fig. 4. The wing has the NACA-0012 cross section with thickness ratio of 10%. The leading edge is swept 30 degrees, and its 1/3 span length is tapered. The outer boundary is set at approximately two times the span length of the wing from the fuselage surface. The number of grids in each direction is imax=60, jmax=30 and kmax=K=20, respectively.

Figure 5 shows the grids generated on selected grid planes in the flow field together with the grids on the fuselage, outer boundary and the wing upper surface. Figure 6 shows the grids around the wing at k=5 with and without anchors. The grids shown here are preliminary results and for illustration only, because they need to be further refined with additional grids around the wing before they can be used for a flow calculation.

For each grid surface k, the number of ADI iterations was approximately 5, or equivalently the total computational time was equal to that for 5 SLOR iterations in the 3-dimensional domain. The number of iterations to generate the 3-dimensional grids using the 3-dimensional elliptic grid generation equation would be 50 at least. Thus the present method is estimated to be 10 times faster than the elliptic grid generation method. This is of course a very conservative estimate, because the number of iteration with the 3-dimensional elliptic grid generation would be two to four times larger if grid spacing and orthogonality control terms are used.

The present method does not require 3-dimensional arrays of variables because, essentially as soon as a grid surface is generated for each k, it is stored in a slow memory device such as tape or disk space and do not have to be reused before all the remaining grids are generated. On the other hand, relaxation schemes for the elliptic grid generation equations sweep the entire grids in every iteration cycle, so all the grid coordinates should be stored in the fast core memory. Otherwise, the computational speed would be intolerably slow. The core space required by the elliptic grid generation method for the present configuration is at least 3x60x30x20=108,000 even without counting the arrays necerssary for the working space. The 3-dimensional ADI scheme requires three times of this number.

6. Conclusions

The proposed method is shown to generate 3-dimensional grids for fuselage-wing flow calculations at least 10 times faster than the elliptic grid generation method and with much smaller memory requirement. This efficiency is important when the grids are revised often, before or during the flow solution procedure, for various reasons including the solution adaptive grid generation.

Although only H-grids are demonstrated, no difficulty is anticipated in applying the proposed method to C-grids and O-grids in three dimensions.

The present scheme uses essentially an iterative solution scheme for the 2-

dimensional elliptic equations. It seems quite possible, however, to eliminate the ADI scheme by applying the 2-dimensional version of the present non-iterative grid generation scheme [1] on each grid surface k except with the coupling to the grid surface k-1 and to the outer boundary grid surface K. The computational time will then become still faster by a factor of 5.

REFERENCES

[1] Nakamura, S. "Marching Grid Generation Using Parabolic Partial Differential Equations," in Numerical Generation of Curvilinear Coordinate Systems, Elsevier/North Holland (forthcoming)

[2] Mastin, C. W. and Thompson, J. F., "Transformation of Three Dimensional Regions onto Rectangular Regions by Elliptic Systems," Numer. Math. 29, 397-407 (1978)

[3] Thomas, P. D., "Construction of Composite Three Dimensional Grids from Subregion Grids Generated by Elliptic Systems," Proceedings of AIAA Computational Fluid Dynamics Conference, Palo Alto Calif., 22-23 June (1981)

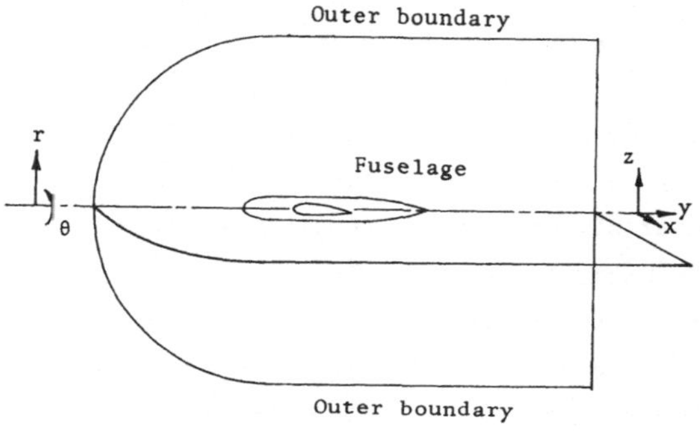

Figure 1 Flow field bounded by fuselage-wing and the outer boundary

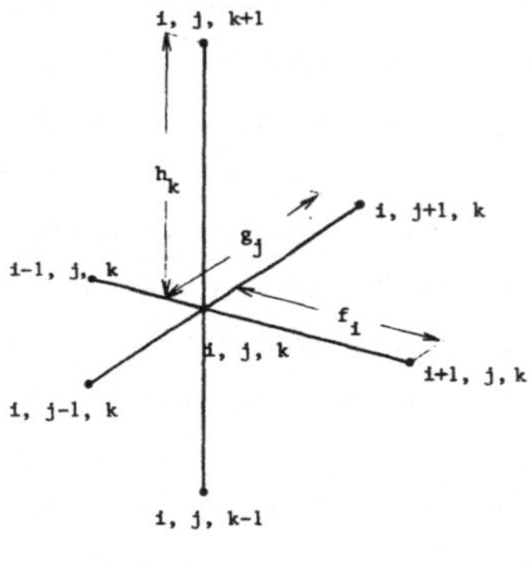

Figure 2 Notations for the grids in the computational domain

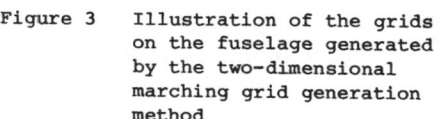

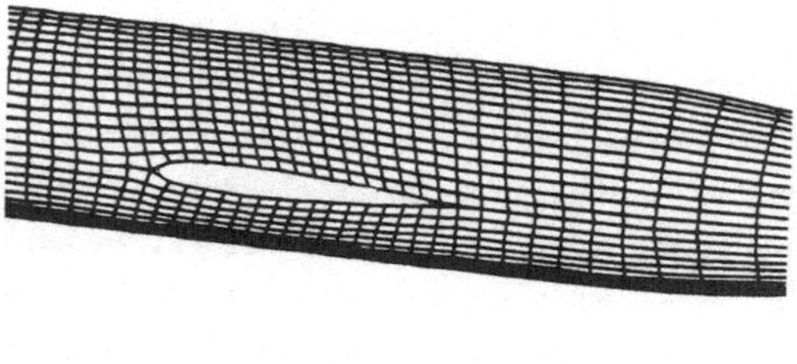

Figure 3 Illustration of the grids on the fuselage generated by the two-dimensional marching grid generation method

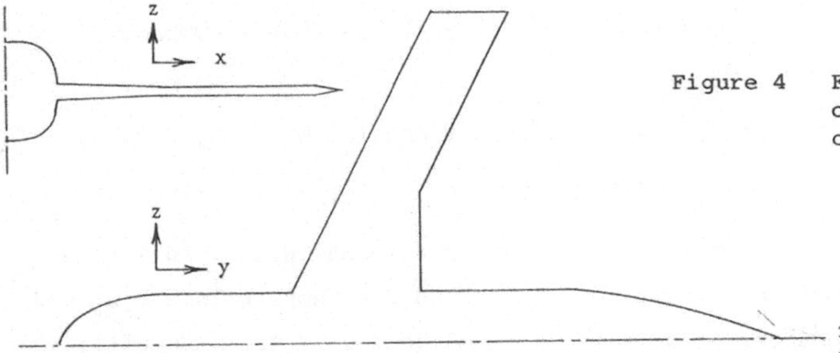

Figure 4 Front and side views of the fuselage-wing configration

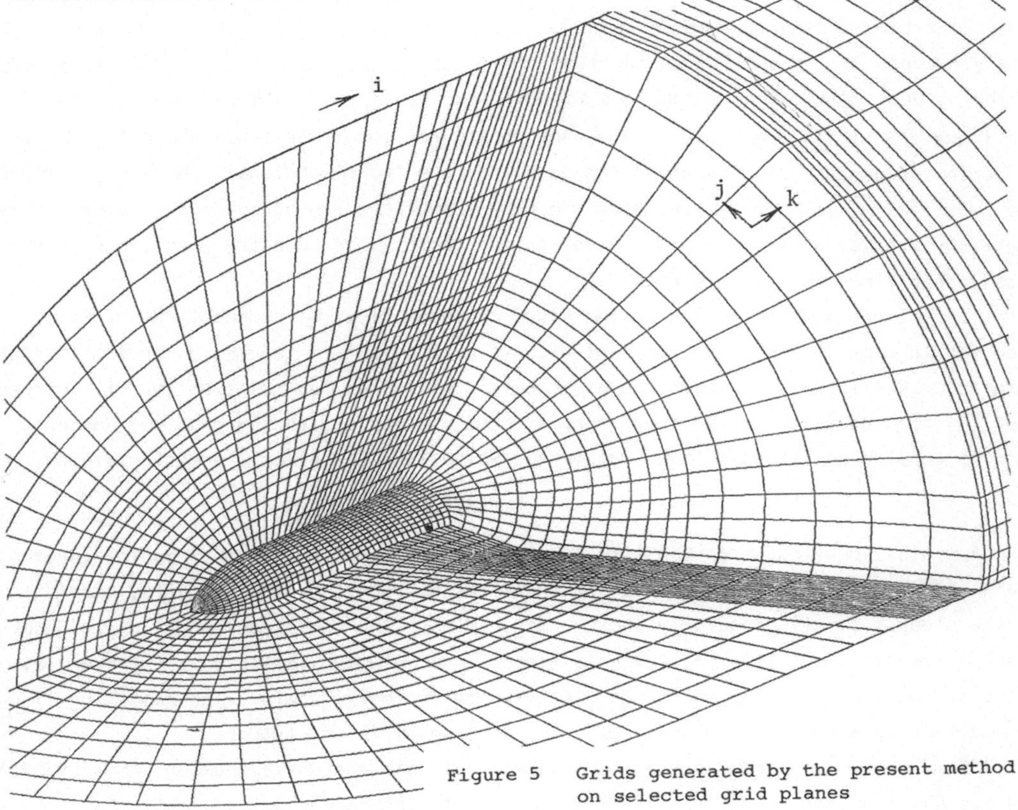

Figure 5 Grids generated by the present method on selected grid planes

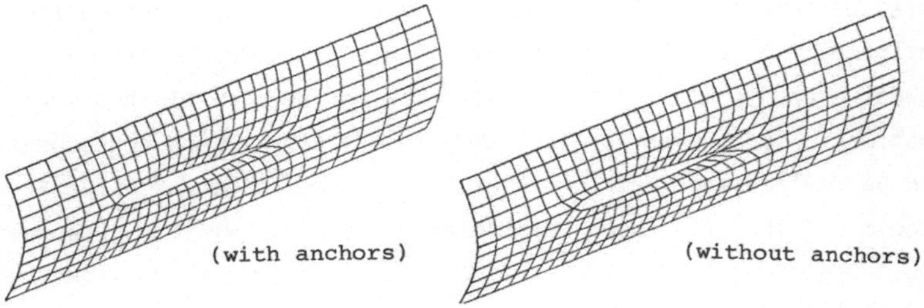

Figure 6 Effect of anchors on the grids around the wing (5th grid plane counted from the fuselage)

THE NUMERICAL CALCULATION OF ROTATING FLUID FLOWS AT LOW ROSSBY NUMBERS

M. A. Page

School of Mathematics and Physics, University of East Anglia, Norwich, England

1. INTRODUCTION

In this paper some numerical methods which can be used to calculate flows in a rapidly rotating fluid are described. The flows considered are those within a closed cylindrical container, rotating about an axis which is aligned with the sidewalls, when there is a slow motion superposed upon the rigid rotation of the fluid, so that the Rossby number Ro is small. In addition, the Ekman number E is considered to be small so that viscous diffusion is confined to thin layers in the fluid (Greenspan, 1968). The flows calculated are those in the geostrophic region of the fluid where the motion is depth-independent and the governing equations are two-dimensional. However, between the geostrophic flow and the walls of the container there are ageostrophic layers of two types; by analysing these layers theoretically, appropriate boundary conditions on the geostrophic flow can be calculated.

The primary motivation behind the study of these flows is to model the effects observed in experiments performed on fluids in cylindrical containers under almost rigid rotation. For this reason both the geometry and parameter regimes examined will correspond to those usually used in laboratory studies; in particular the range of Rossby numbers considered includes those values for which inertial effects are significant.

2. FORMULATION

A typical configuration consists of a cylindrical container with a radial length scale ℓ^* and depth $d\ell^*$, where d is order unity, rotating about an axis, parallel to the sidewalls, with angular velocity $\Omega^*\hat{\underset{\sim}{k}}$. Within the container is a fluid of constant kinematic viscosity ν^*, which is forced into relative motion by a differential velocity $U^*\underset{\sim}{U}$ on the lid, where U^* is a typical velocity scale. The resulting flow is modified by a small topography $h\ell^*$ on the base of the container.

The dimensional parameters are chosen so that both the Rossby number $Ro = U^*/\Omega^*\ell^*$ and the Ekman number $E = \nu^*/\Omega^*\ell^{*2}$ are small, with the former not so small that inertial effects are insignificant. Under these conditions the scaled relative velocity $\underset{\sim}{u}$ satisfies the geostrophic equation $2(\hat{\underset{\sim}{k}} \times \underset{\sim}{u}) = -\underset{\sim}{\nabla}P$, where P is the reduced pressure, to lowest order in most of the fluid. Taking the curl of this equation shows that the motion is independent of the axial coordinate z and consequently a stream function ψ can be defined, so that in radial polar coordinates $ur = -\partial\psi/\partial\theta$ and $v = \partial\psi/\partial r$. The region of the flow where this solution is valid will be referred to as the geostrophic region and it consists of both the interior, where viscous diffusion is negligible, and the $E^{\frac{1}{4}}$-layers, which are parallel to the rotation axis. The solution is, however, not valid in two types of thin ageostrophic regions, namely the Ekman layers on both the top and the bottom of the container and the $E^{1/3}$-layers which, like

ψ_I is not known beforehand and therefore an iterative scheme must be used where $\nabla^2\psi = \zeta_I$ and (4) are solved alternately for ψ_I and ζ_I. This iterative procedure enables a simple approximate method of solving (4) to be used in which $\partial\zeta/\partial r$ is evaluated from the previous iteration and $\partial\zeta/\partial\theta$ is evaluated implicitly; since the flow is generally around the annulus this scheme is likely to be stable. However, there are two difficulties with this approach. Firstly, the difference equations (4) are not diagonally dominant unless the time derivative $\partial\zeta/\partial t = [\zeta(t+\Delta t) - \zeta(t)]/\Delta t$ is reintroduced and secondly, for α large, the streamlines in some parts of the flow tend to follow radial lines. Therefore, an r-implicit integration is required also and this is achieved by using an alternating directions implicit, or ADI, iteration (Peaceman & Rachford, 1955) which is more commonly used with diffusion equations. Since the steady solution is sought a difference scheme accurate to only $O(\Delta t)$ was used and the solutions converged in 20 to 30 iterations with $\Delta t = 1$. An example of a converged solution with $\alpha = 4$, $\lambda = 3$ and $b = 2$ is shown in Figure 2.

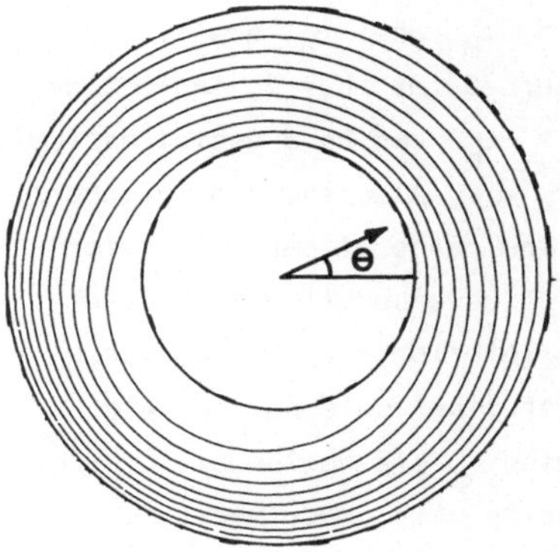

Figure 2.

4. THE $E^{\frac{1}{4}}$-LAYER FLOW

The interior flow described above does not satisfy the no-slip condition on the sidewalls of the container and therefore the complete geostrophic flow has not been found until the flow in the boundary layers on $r = 1, b$ has been calculated. In these layers, which are of thickness $O(\delta)$ where $\delta = (\frac{1}{2}dE^{\frac{1}{2}})^{\frac{1}{2}}$, a scaled radial coordinate is introduced so that $\bar{r} = (r-1)/\delta$ for the layer near the inner wall, along with scaled velocities $(u, v) = (\delta\bar{u}, \bar{v})$. The tangential velocity $\bar{v}$ then adjusts from the slip velocity $\bar{v}_e$ as $\bar{r} \to \infty$, given by $v_I(1, \theta)$, to $\bar{v} = 0$ on the wall, with $\bar{u}$ also vanishing on $\bar{r} = 0$. To lowest order in δ the flow then satisfies the equations (Walker & Stewartson, 1972)

$$\lambda(\bar{u}\frac{\partial\bar{v}}{\partial\bar{r}} + \bar{v}\frac{\partial\bar{v}}{\partial\theta}) = \lambda\bar{v}_e\frac{d\bar{v}_e}{d\theta} + \bar{v}_e - \bar{v} + \frac{\partial^2\bar{v}}{\partial\bar{r}^2}, \quad \frac{\partial\bar{u}}{\partial\theta} + \frac{\partial\bar{v}}{\partial\bar{r}} = 0, \tag{5}$$

which, for $\lambda \neq 0$, resemble the classical boundary layer equations for a non-rotating fluid. This similarity, which is particularly marked when $\lambda >> 1$, indicates that the boundary layer could possibly separate from the wall when the external flow is decelerating. To examine this possibility equations (5) are integrated numerically using the box method (Cebeci & Keller, 1971) with the discretisation in θ chosen so that the values of $\bar{v}_e$ calculated in §3 can be used directly. In the $\bar{r}$ direction an exponential stretching is used with at least 50 gridpoints between $\bar{r} = 0.05$ and $\bar{r} = 25$. Starting from the minimum of $\bar{v}_e$, at which a suitable initial profile can be found providing $\min_\theta(\bar{v}_e) > 0$ (Page, 1981), the numerical integrations can proceed around the boundary-layer and these calculations show that the minimum of the skin friction $\partial\bar{v}/\partial\bar{r}(0,\theta)$ decreases as the topographic height α increases until, at a critical value $\alpha_c(\lambda)$, it vanishes at a point $\frac{1}{2}\pi < \theta_s < \pi$. At this point a singularity, analogous to that examined by Goldstein (1948), appears to be present and the calculations cannot be continued for $\theta > \theta_s$. Using the momentum equation (5) a useful lower bound for α_c can be calculated from the interior flow (Buckmaster, 1969), namely the solution $\alpha_m(\lambda)$ of

$$\min_\theta(d\bar{v}_e/d\theta) = -1/\lambda. \qquad (6)$$

Both α_m and α_c are plotted on Figure 3 for $b = 2$, indicating that α_m can be a useful estimate for α_c. Also shown on Figure 3 is the range of parameters for which reversed flow is present in the interior and the boundary-layer flow could not be calculated. This region of interior reversed flow is also important when comparing these results with experiments, such as those by Maxworthy (1977), since it demonstrates that any reversed flow present need not necessarily be caused by boundary-layer separation. However, for Maxworthy's experiments qualitative arguments (Page, 1982a) do suggest that at least some of the reversed flow is due to separation of the $E^{\frac{1}{4}}$-layer from the inner wall.

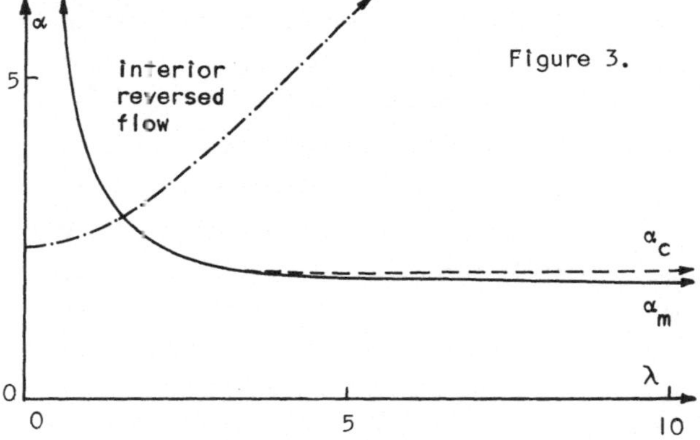

Figure 3.

5. THE COMPLETE GEOSTROPHIC FLOW

In §4 it was shown that inertial effects in the $E^{\frac{1}{4}}$-layers can be sufficiently important to affect the O(1) dynamics of the interior flow through boundary-layer separation. Therefore, for a numerical model to reproduce an experimentally observed flow, it is important that these effects be included and this can best be done by integrating the complete geostrophic flow, governed by (1). This has been done by Beardsley (1973) for the sliced-cylinder configuration and in the following both this geometry and the extension by Hide & Hocking (1979), which includes a free shear layer, will be considered. The geometry, illustrated in Figure 4, consists of a cylindrical container $r \leqslant b$ with a bottom sloping at a small angle β, so that $h = \tan\beta\ r \sin\theta$, with forcing from the portion of the lid where $r \leqslant 1$, so that $\underset{\sim}{U} = (0, rH(1-r), 0)$ and $\zeta_T = 2H(1-r)$. In Beardsley's study, for which $b = 1$, the radial coordinate is stretched according to an algebraic transformation based on a boundary-layer scale of $(E/\tan\beta)^{1/3}$, suitable when $E^{\frac{1}{4}} \ll \tan\beta \ll 1$. However, in the following an exponential transformation $\eta(r)$ is used which resolves the boundary layers of thickness $\delta = (\frac{1}{2}dE^{\frac{1}{2}})^{\frac{1}{2}}$ near $r = 1$ and $r = b$. Apart from this the numerical method when $b = 1$ is essentially the same as Beardsley's with centred finite differences, such that $\Delta\eta = b/40$ and $\Delta\theta = 2\pi/64$, and an integral treatment at the origin $r = 0$. The method involves time-stepping with equation (1), using an ADI method and a boundary value iteration due to Israeli (1970), coupled with a fast Fourier transform solution of the Poisson equation $\nabla^2\psi = \zeta$. Due to the nonlinearity in the problem, an iterative procedure is required at each time step but this converges to 0.1% in 2 or 3 iterations. Since the steady state solution is of most interest the time step Δt is chosen to be equal to the smaller of the two time scales, which are $\frac{1}{2}d/\tan\alpha$ and $\frac{1}{2}d/E^{\frac{1}{2}}$, and the numerical scheme is stable with this choice of Δt. The extension of this method to the geometry considered by Hide & Hocking (1979) is not immediate because in that case there is an ageostrophic $E^{1/3}$-layer separating two geostrophic regions at $r = 1$. To relate the flows on either side of this layer it is necessary to specify matching conditions for ψ, $\partial\psi/\partial r$, ζ and $\partial\zeta/\partial r$ across $r = 1$ and to obtain these the flow in the $E^{1/3}$-layer must be calculated. This can be done for $Ro = 0$ (Stewartson, 1957) giving

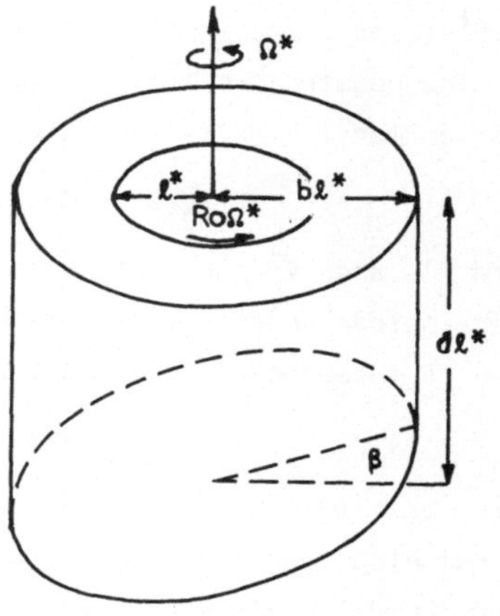

Figure 4.

$$\Delta\psi = \frac{\partial\psi}{\partial r} = \Delta\zeta = 0, \quad \frac{\partial\zeta}{\partial r} = \tfrac{1}{2}\delta^{-2} \tag{7}$$

and this solution is a good approximation when $Ro \ll E^{1/6}\tan\beta$. Providing there is a gridpoint at $r = 1$ the final condition in (7) can be included as a δ-function term in ζ_T at that point (Page, 1981) so that $\zeta_T(1,\theta) = -(\Delta r)^{-1}$, where Δr is the grid spacing at $r = 1$. This, and the continuity of ζ, enables Beardsley's method to be used across the ageostrophic $E^{1/3}$-layer. Typical numerical results are shown in Figure 5 illustrating a remarkable agreement with the experiments considering that $Ro \doteq 5E^{1/6}\tan\beta$. Further results are given in Page (1982b). Finally, as in Beardsley's work there is no evidence that the calculation breaks down near $r = b$ due to separation of the $E^{1/4}$-layer, although an unsteadiness, which may be due to boundary-layer effects, does arise for larger values of Ro still less than unity.

6. CONCLUSIONS

The formulation presented in §2, where both the interior flow and the $E^{\frac{1}{4}}$-layers are treated together as the geostrophic flow, enables the motion in a rotating container to be calculated for larger Rossby numbers than those for which the interior flow alone, as calculated in §3, is accurate. This is demonstrated in §4 where the $E^{\frac{1}{4}}$-layer is seen to separate from the sidewall when the ratio $Ro/E^{\frac{1}{2}}$ is sufficiently large. However in §5, where the influence of the $E^{\frac{1}{4}}$-layer is taken into account directly, there is no evidence of any equivalent breakdown in the solution. In fact, the main restrictions on the parameters in that case are those for which the theoretical analyses in the ageostrophic regions, which surround the geostrophic flow, remain valid.

the $E^{\frac{1}{4}}$-layers, are parallel to the sidewalls (Greenspan, 1968).

The geostrophic flow can be calculated by evaluating higher order terms in the equation for the axial vorticity component, $\zeta = \nabla^2\psi$, which gives (Hide & Hocking, 1979)

$$\partial\zeta/\partial t + Ro\ J(\psi, \zeta) = (2 + Ro\zeta)\partial w/\partial z + E\nabla^2\zeta, \qquad (1)$$

where J and ∇^2 are the two-dimensional Jacobian and Laplacian. The term $\partial w/\partial z$, which is independent of z, can be estimated from the fluxes into the Ekman layers at $z = h, d$ (Greenspan, 1968) as

$$\partial w/\partial z = [E^{\frac{1}{2}}\{\tfrac{1}{2}\zeta_T - \zeta\} - J(\psi, h)]/d, \qquad (2)$$

where $\zeta_T = (\underset{\sim}{\nabla} \times \underset{\sim}{U}).\hat{\underset{\sim}{k}}$. The boundary conditions at the sidewalls are $u = v = 0$, while the matching conditions for ψ, ζ and their normal derivatives across a detached shear layer, where an $E^{1/3}$-layer is present, can, in principle, be calculated theoretically. The formulation above, which is quite general, will now be applied to two specific geometries.

3. INTERIOR FLOW CALCULATIONS

As an example of a calculation of the interior flow for which viscous diffusion is neglected, the motion in an annular container $1 \leqslant r \leqslant b$, shown in Figure 1, will be examined when both Ro and h are $O(E^{\frac{1}{2}})$. In particular the steady flow is sought when $\underset{\sim}{U} = (0, r, 0)$, so that $\zeta_T = 2$, and $h = \alpha E^{\frac{1}{2}}\sin\theta$. Under these conditions equation (1) simplifies to

$$J(\psi_I\ ,\ \lambda\zeta_I + \alpha\sin\theta) = 1 - \zeta_I, \qquad (3)$$

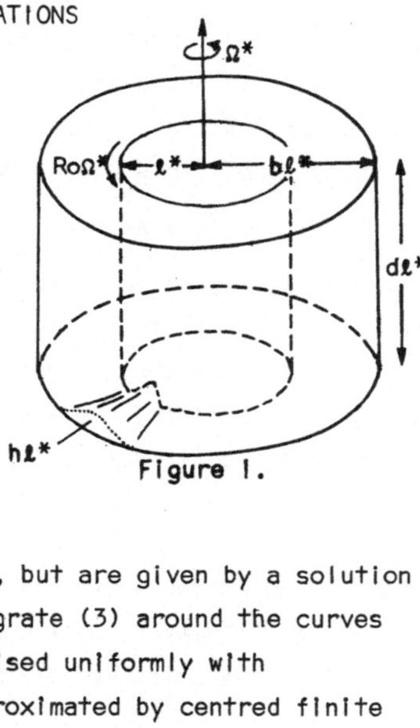

Figure 1.

where $\lambda = Ro\ d/2E^{\frac{1}{2}}$, which is an ordinary differential equation for the vorticity along streamlines. However, since the streamlines are not known *a priori*, but are given by a solution of the equation $\nabla^2\psi_I = \zeta_I$, it is not convenient to integrate (3) around the curves ψ_I = constant. Instead, both r and θ are discretised uniformly with $\Delta r = (b - 1)/40$, $\Delta\theta = 2\pi/64$ and derivatives are approximated by centred finite differences. Assuming ζ_I is known, the Poisson equation for ψ_I can then be solved using a fast Fourier transform technique, with the boundary conditions $\psi_I(1, \theta) = 0$ and $\psi_I(b, \theta) = Q$, where Q is an undetermined constant. This constant is, however, fixed by a circulation condition (Davey, 1978) which on $r = 1$ gives that $\oint v_I\, d\theta = \pi$. Assuming now that ψ_I is known, the vorticity equation (3) can be solved subject to a periodicity condition on ζ_I. The discrete form of this equation is

$$\lambda[u_{i,j}\ (\frac{\zeta_{i+1,j} - \zeta_{i-1,j}}{\Delta r}) + v_{i,j}\ (\frac{\zeta_{i,j+1} - \zeta_{i,j-1}}{r_i\Delta\theta})] = 1 - \zeta_{i,j} - \frac{1}{r_i}\ (v\ \frac{dh}{d\theta})_{i,j} \qquad (4)$$

where the subscripts i,j refer to the point $(r_i, \theta_j) = (1 + i\Delta r, j\Delta\theta)$. In practice

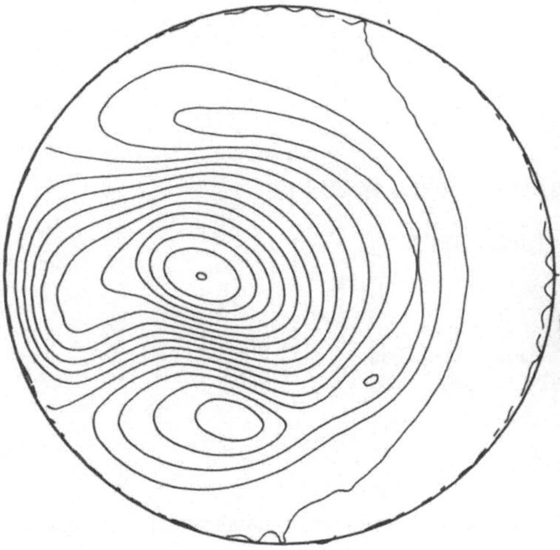

Figure 5. Ro = 0.207, E = 1.83×10^{-4}, b = 1.91, d = 0.984, $\beta = 10^{\circ}$.

7. REFERENCES

Beardsley, R.C. (1973). Proc. NAS Symp. on Num. Models of Ocean Circulation.

Buckmaster, J. (1969). J. Fluid Mech., 42, 481-498.

Cebeci, T. & Keller, H.B. (1971). Lecture Notes in Physics, 8, 92-100, Springer-Verlag.

Goldstein, S. (1948). Quart. J. Mech. Appl. Math., 1, 43-69.

Greenspan, H.P. (1968). Theory of rotating fluids, Cambridge University Press.

Hide, R. & Hocking, L.M. (1979), Geophys, Astrophys. Fluid Dyn., 14, 19-43.

Israeli, M. (1970). Studies in Appl. Math., 49, 327-349.

Maxworthy, T. (1977). Z. angew. Math. Phys., 28, 853-864.

Page, M.A. (1981). Ph.D. Thesis, University of London.

Page, M.A. (1982a). to appear in J. Fluid Mech.

Page, M.A. (1982b). to appear in Geophys. Astrophys. Fluid Dyn.

Peaceman, D. & Rachford, H. (1955). J. Soc. Ind. Appl. Math., 3, 28-41.

Stewartson, K. (1957). J. Fluid Mech., 3, 17-26.

Walker, J.D.A. & Stewartson, K. (1972). Z. angew. Math. Phys., 23, 745-752.

NUMERICAL TECHNIQUES FOR MULTIDIMENSIONAL MODELING OF SHOCK-GENERATED TURBULENCE

J. M. Picone and J. P. Boris
Laboratory for Computational Physics
Naval Research Laboratory
Washington, D.C. 20375

ABSTRACT

We discuss shock-generated turbulence as related to the dynamics of hot gaseous channels produced by laser pulses and electric discharges in gaseous atmospheres. Accurate models of channel development must account for compressible flows, which are associated with shocks that are present at early times, as well as the incompressible residual motion which is responsible for channel cooling. We present an efficient numerical technique which uses the flux-corrected transport algorithm and permits automatic adjustment of the time step size to account for both types of behavior. Comparisons of two-dimensional calculations to theory and to experimental data on laser channels show good agreement. Our numerical results reveal a mechanism by which the distribution of turbulent scale lengths is generated during energy deposition.

THE PHYSICS OF HOT GASEOUS CHANNELS

Recent experimental studies of electric discharges in air and laser pulses in air and pure nitrogen [1-2] have revealed the detailed dynamics of the hot gaseous channels which remain. During and immediately after energy deposition, the resulting hot gas quickly expands to achieve pressure equilibrium with the surrounding atmosphere, producing a shock wave which propagates away in a short time. The channel then cools on time scales which are orders of magnitude faster than those characterizing classical (nonturbulent) thermal conduction. As a channel cools, the radius increases according to the equation

$$R^2(t) = R^2(\tau) + 4\alpha\,(t-\tau), \qquad (1)$$

where R is the radius of the channel, t is time measured from the beginning of a discharge or pulse, $t=\tau$ is the time at which pressure equilibrium is reached, and α is the thermal diffusivity. For electric discharges which deposit $\approx$ 300-600 J/m, measurements at early times give $\alpha \approx 500$ cm^2/s. The experimental CO_2 laser pulses in nitrogen deposit $\approx$ 10 J/m, giving $\alpha \approx 250$ cm^2/s. For nonturbulent thermal conduction, we have $\alpha \approx 1.0$ cm^2/s for air at 800K and 1 atm [3].

As a channel cools, turbulent structure appears first at the boundary and then in the interior. This has led us naturally to search for mechanisms in which long-lived rotational motion is produced. The only successful analysis to date (1,4) relies on the (rigorous) equation for the time development of vorticity, i.e.,

$$\frac{d\xi}{dt} + \xi\nabla \cdot v = \xi \cdot \nabla v + (\nabla\rho \times \nabla P)/\rho^2 \quad , \tag{2}$$

where $\xi = \nabla \times v$ is the vorticity, v is the fluid velocity, ρ is the density, and P is the pressure. According to eq. (2), any misalignment between the pressure and density gradients during expansion of the gas to pressure equilibrium will generate vorticity. This will occur when asymmetries exist in the structure of a pulse and when consecutive pulses are noncollinear. After pressure equilibrium is reached, we may represent the flow field in terms of one or more pairs of vortex filaments of strength $\pm\kappa$, where

$$\kappa = U_m (R_1 - R_0) \ \ell n \ (\rho_\infty/\rho_0) \ f. \tag{3}$$

In eq. (3), U_m is a characteristic expansion velocity; R_0 is the initial channel radius; R_1 is the radius of the channel just after expansion to pressure equilibrium; ρ_∞ is the ambient density; ρ_0 is the density at the center of the channel; and $f(\lesssim 1)$ is a form factor. The number of vortex filament pairs depends on the situation.

NUMERICAL MODEL

An accurate numerical model of channel dynamics is necessary for the solution of problems which are too difficult to treat analytically, as well as for the verification and calibration of our theoretical results [1,4]. We favor the use of a finite difference scheme to integrate the equations for conservation of mass, momentum, and energy. A major complication is the fact that the flows which generate vorticity are supersonic and require short time steps, as dictated by the Courant condition, while the (incompressible) residual flows, which are responsible for channel cooling, could be computed with much longer time steps ($\lesssim$ 100 x). The model which we describe below includes a simple, effective method of numerically separating the rotational flows from the (radial) shock flows at times when the latter are no longer important near the channel. This permits the adjustment of the time step to account for the slower flows, saving considerable computer time.

Our calculations are multidimensional, using time-step splitting in conjunction with the latest version of flux-corrected transport (FCT) [5]. The grid is Cartesian and is uniform and finely spaced in the region near the channel. Outside of this fine grid, the cells increase geometrically so that the boundaries are far from the channel. This permits the shock to decouple completely from the channel before interacting with the boundary. After the shock has moved well away (> 10 channel radii) from the channel, the pressure field near and within the channel is approximately ambient. At this point, we calculate and subtract the average radial velocity field from the total velocity field, leaving only the rotational flows of interest. To compute this average radial velocity field, we use the following self-consistent method, based on digital multichannel analysis:

(1) Relative to the center of the channel, define a set of contiguous radial intervals which encompass the entire differencing grid.

(2) For each interval, determine the grid points whose radial displacements from the channel center fall within the interval. Average the radial velocity components of these grid points.

(3) From this average radial velocity field, denoted $\langle v_r \rangle$, determine the average radial potential function by integrating the Poisson equation,

$$\nabla^2 \phi = \nabla_r \langle v_r \rangle.$$

(4) Interpolate the field $\langle v_r \rangle$ onto the grid by straightforward differencing of the equation $\nabla_r \phi = \langle v_r \rangle$ to maintain consistency with the differencing solution for the total velocity field.

(5) Subtract the field $\langle v_r \rangle$ from the total velocity field at each grid point.

(6) Set the pressure field to the ambient value and set the velocities near and at the shock wave to zero to eliminate all high speed flows and to prevent shock interaction with the boundary.

Immediately after the above sequence, we must increase the size of the time step, since the flows of interest are much slower than the sound speed. Our ability to use FCT, however, requires that the Courant condition be satisfied. We accomplish this by scaling the pressure down by a factor β, so that $P \to P/\beta$ and $c_s \to c_s/\beta^{1/2}$, where the local speed of sound is $c_s = (\gamma P/\rho)^{1/2}$ and γ is the ratio of principal specific heats. The time step will scale roughly as $\Delta t \to \beta^{1/2}\,\Delta t$. An appropriate value for β will insure that the errors introduced by our techniques are small. Setting the pressure field to the ambient value in step (6) above will initially eliminate the minute pressure gradients which maintain the rotational motion of the fluid. This will permit the velocity field in the vicinity of a vortex filament to expand momentarily. The expansion, however, reduces the local density and pressure approximately adiabatically until the pressure gradients again counteract the centrifugal force experienced by a given fluid element. We have found that the effect on the vorticity ξ is of order $M'^2 \equiv v^2/c_s'^2$, where $c_s' \equiv c_s/\beta^{1/2}$ is the speed of sound after scaling and M' is the resulting Mach number. In practice we choose β so that $M' \lesssim 1/5$ to insure that such errors are small. Another important concern is the effect of scaling on the incompressibility of the residual flow field. Since errors related to compressible effects for a subsonic flow field are also of order M'^2 [6], our constraint on β again insures that such errors are less than a few percent. Should β be too large, we would expect some energy to be diverted to the formation of compressional waves, resulting in a reduction in intensity of the rotational motion. Our calculations have verified that the flow pattern remains rotationally stable and accurate for several simulated ms after our procedure is performed. The residual vorticity has proven to be correct to first order, as expected, and no compressional waves have arisen at late times, again supporting our analysis.

NUMERICAL SIMULATIONS

All numerical simulations to date have used a Cartesian grid of 100 x 100 cells. The central region, where energy is deposited, consists of 50 x 50 square cells chosen so that an individual pulse would be contained in an area no smaller than 8 x 8 cells. Outside the uniform central region, the cell dimensions increase geometrically by a factor $(1+\delta)$ from cell to cell, where $\delta<0.3$, so that the grid boundary is far from the channel. All pulses initially have a Bennett pressure profile

$$P(\underset{\sim}{r}) = P_\infty + (P_0-P_\infty)/\left(1+\frac{|\underset{\sim}{r}-\underset{\sim}{r}_0|^2}{a^2}\right)^2 \quad , \qquad (4)$$

where P_0 is the pressure at the center of the pulse, P_∞ is the ambient pressure, $\underset{\sim}{r}_0$ is the position of the center of the pulse, and a is a constant (the "Bennett radius"). The model deposits all energy instantaneously in the form of an increase in internal energy. We use a real air equation of state routine based on the data of Gilmore [7-8].

As our first example, we choose the simple case of noncollinear pulses produced at different times. The first pulse deposits energy at time t=0.0 in the center of the grid, and the second pulse occurs 1 ms later and is displaced to the (reader's) right of the first by a distance equal to the Bennett radius (a=0.5cm). The initial peak overpressures (P_0-P_∞) of the pulses are 4.7 atm and 2.4 atm, respectively. Figure 1 shows density contours and velocity vectors in the finely gridded region at t=1.26 ms and density contours at t=2ms. The contours range from 3.0×10^{-4} g/cm^3 at the center of the grid to 1.1×10^{-3}g/cm^3 at the edge. From Fig. 1a we see that the noncollinearity of the pulses produces a noncircular channel cross section. The velocity plot clearly demonstrates the presence of a vortex filament pair whose flow pattern pulls fluid from the position of the first pulse toward that of the second. Figure 1c shows that the gas at the center of the system has cooled under the influence of the residual flows. These features are in complete agreement with theoretical predictions [4] based on Eq. (2) and (3).

A more important and difficult case is that of a laser pulse with an approximately circular envelope. Although turbulent cooling might appear to be less important in this situation, experimental data for a laser pulse in nitrogen at 1.2 atm give $\alpha \approx 250 cm^2/s$. Laboratory burn patterns have also revealed that nonuniformities ("hot spots") exist in the interior of the laser pulse. We have used this fact in a test calculation in which the actual laser pulse is assumed to consist of seven smaller, identical, simultaneous pulses, as in the pressure contour diagram of Fig. 2a. Again we show only the uniform central region of the grid, which is 1.2 cm x 1.2 cm. The energy deposited per unit length is ≈ 9 J/m, in agreement with experiment. We use the real air equation of state routine (rather than one for nitrogen) and 1.0 atm ambient pressure. The pressure contours range from 1.1×10^6 dyne/cm^2 to 3.0×10^6 dyne/cm^2 while the density contours in the remaining diagrams range from 5.7×10^{-4} g/cm^3 to 1.1×10^{-3} g/cm^3. The rather mild hot spots in Fig. 2a would be marginally detectable by a burn pattern. From Fig. 2a, we see that the shock wave produced by each hot spot will sweep through the density minima occurring at the positions of the other hot spots, producing vorticity according to Eq. (2). By t=56 μs, the channel has achieved pressure equilibrium and a temperature of $\lesssim$ 650K. The envelope is nearly circular, even though nonuniformities persist in the interior of the channel. Because the grid is rather coarse and does not have sixfold symmetry and because we use time-step splitting, the channel properties do not retain the exact sixfold symmetry of Fig. 2a. The use of a finer grid, and perhaps multidimensional FCT [9], would improve the situation, although multidimensional FCT is slower (≈2x) and requires more storage space, as compared to the algorithm used here. At later times, the boundaries distort under the influence of local vortex filament pairs, and the channel dimensions increase. When viewed from a direction perpendicular to the channel axis, the channel should appear to expand, and the boundary distortions should appear as striations parallel to the channel axis. These features are in fact present in Schlieren photographs. Our calculation yields $\alpha \approx 125$ cm^2/s, which is within a factor of 2 of the experimental value.

CONCLUDING REMARKS

This method for calculating the generation of turbulence by shocks and the subsequent cooling of hot gaseous channels has proven to be accurate and efficient. We have reproduced theoretical predictions for consecutive, noncollinear pulses and have verified features observed experimentally for laser channels in air.
Viewing the diagrams in Fig. 2, we find that the scale lengths of nonuniformities within the laser pulse will determine the scale lengths appearing in the turbulent structure. The resolution provided by the numerical grid, however, places a lower limit on the distribution and could affect the interactions between turbulent structures of various sizes.

ACKNOWLEDGEMENTS

We gratefully acknowledge the support of the Defense Advanced Research Projects Agency and the Office of Naval Research. Conversations with M. Lampe, R. Greig, M. Raleigh, R. Fernsler, M. Fry, and R. Guirguis have proven most helpful.

REFERENCES

1. J.M. Picone, J.P. Boris, J.R. Greig, M. Raleigh, and R.F. Fernsler, J. Atmos. Sci., 38 (9), 2056-62 (1981).
2. J.R. Greig, R.E. Pechacek, M. Raleigh, and K.A. Gerber, NRL Memo Rep. 4826 (1982)
3. G.K. Batchelor, An Introduction to Fluid Dynamics (Cambridge University Press, New York, 1967), Appendix 1, p. 594.
4. J.M. Picone and J.P. Boris, "Vorticity Generation by Asymmetric Energy Deposition in a Gaseous Medium," submitted to Phys. Fluids.
5. J.P. Boris and D.L. Book, Methods in Computational Physics, Vol. 16 (Academic Press, New York, 1976), pp. 85-129.
6. A.H. Shapiro, The Dynamics and Thermodynamics of Compressible Fluid Flow, Vol.1 (Wiley, New York, 1953), Chapter 10.
7. F.R. Gilmore, RAND Corp. Rep. RM-1543 (1955).
8. F.R. Gilmore, Lockheed Missile and Space Co. Rep. DASA 1917-1 (1967).
9. S.T. Zalesak, J. Comput. Phys., 31 (3), 335-362 (1979).

Fig. 1 (a) 1.26 ms (b) 1.26 ms (c) 2.0 ms

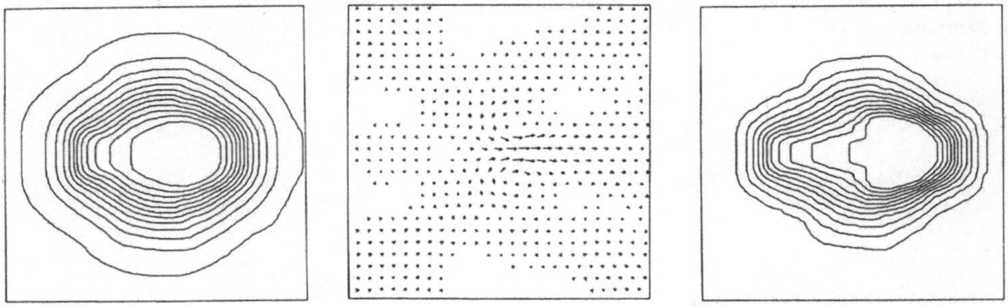

Fig. 2 (a) 0.0 μs (b) 56 μs (c) 1.24 ms

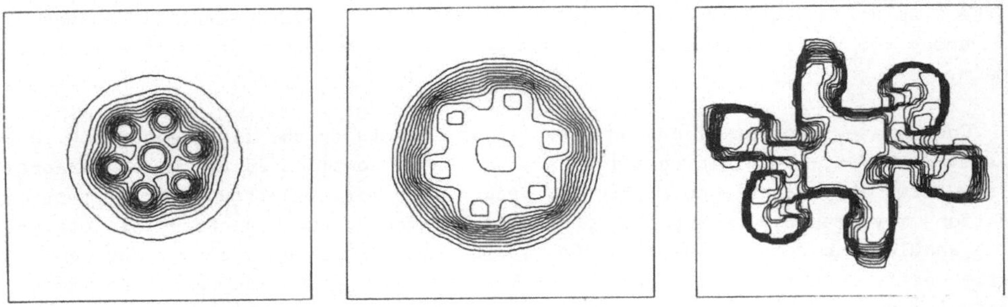

ADAPTIVE CURVILINEAR GRIDS FOR LARGE REYNOLDS NUMBER VISCOUS FLOWS

by

R. Piva, A. Di Carlo, B. Favini, G. Guj

University of Rome, Rome, Italy

1. INTRODUCTION

The adoption of more efficient spatial discretizations plays a central role for the improvement of computational schemes for viscous flow problems, particularly at increasing Reynolds numbers. Perhaps too little attention to this subject has been paid in recent years, if compared to the large effort dedicated to other computational issues of like importance, as time integration or iterative procedures. Curvilinear grids have been introduced primarily to provide FD methods with the geometric versatility inherent in other numerical methods. In fact boundary fitted curvilinear coordinates allow the field equations defined on a complicated domain to be solved on a uniform FD mesh in the transformed space. In addition, variable mesh spacing may be easily handled by suitable mappings, when local refinements are needed for an accurate solution. The elliptic grid generation, first proposed by Thompson, Thames et al. [1,2] has been probably the most successful tool for the accomplishment of the above mentioned aims. In two dimensions the mapping of the computational domain in the transformed plane (ξ,η) onto the field domain in the physical plane (x,y) is derived by solving the following set of quasi-linear equations

$$(1.1)\qquad \begin{aligned} \alpha x_{\xi\xi} - 2\beta x_{\xi\eta} + \gamma x_{\eta\eta} &= p \\ \alpha y_{\xi\xi} - 2\beta y_{\xi\eta} + \gamma y_{\eta\eta} &= q \end{aligned}$$

where

$$(1.2)\qquad \alpha = x_\eta^2 + y_\eta^2 , \quad \beta = x_\xi x_\eta + y_\xi y_\eta , \quad \gamma = x_\xi^2 + y_\xi^2$$

while p and q are ad hoc source terms introduced, for each problem, on the base of the resolution required in different regions of the field domain. The values of x and y (or their derivatives) are assigned on the boundary of the transformed domain.

Coordinate transformations have two major effects on the field equations to be solved. First, the field equations are obviously complicated by the appearance of new terms and variable coefficients. Second, and more important, the *difference* equations may possibly be exposed to drastic changes in their accuracy properties, which should be carefully analyzed. We consider in particular second order convection-diffusion problems, where the well known loss of accuracy experienced at increasing Re is caused by the predominance of first order with respect to second order terms [3]. Our purpose is to define a rational criterion for the selection of appropriate coordinate transformations, ensuring that the resulting set of difference equations is not dominated by first derivative terms. For the details concerning the discretization schemes in curvilinear grids we refer to [4,5]. In the following sections we examine first the steady state transport equation of a scalar quantity, then the Navier-Stokes equation (for steady and unsteady problems), which both lead to the same kind of velocity-driven coordinate transformation.

2. SCALAR TRANSPORT EQUATION

Let us first consider the steady-state transport equation of a scalar field θ

$$K\Delta\theta - \vec{v} \cdot \nabla\theta + b = 0 \tag{2.1}$$

where $\vec{v}$ is an assigned velocity field, K a diffusivity and b a source. In a Cartesian coordinate system, eq. (2.1) is represented by

$$K\theta,_{ii} - v_k\theta,_k + b = 0 \tag{2.2}$$

where contributions to first partial derivatives come only from the convective term $\vec{v} \cdot \nabla\theta$. In general curvilinear coordinates†, partial derivatives have to be replaced by covariant ones, and the proper generalization of eq.(2.2) reads

$$K\, \theta|^{i}_{i} - v^{k}\, \theta|_{k} + b = 0 \tag{2.3}$$

where

$$\theta|_k = \theta,_k$$

$$\theta|^{i}_{i} = g^{i}\, \theta|_{ij} = g^{ij}\, (\theta,_{ij} - \Gamma^{k}_{ij}\, \theta,_k) \tag{2.4}$$

g^{ij} are the contravariant components of the metric tensor and Γ^{k}_{ij} the Christoffel symbols of the second kind. After substitution of eqs. (2.4) into eq.(2.3) we obtain the partial differential equation

$$Kg^{ij}\, \theta,_{ij} - (v^{k} + Kg^{ij}\Gamma^{k}_{ij})\, \theta,_k + b = 0 \tag{2.5}$$

which explicitly exhibits, when compared with eq. (2.2), the effects of the coordinate transformation on the representation of the field equation (2.1). In particular, the diffusive term $K\Delta\theta$ now contributes also to the first partial derivatives, in a way depending on the geometry of the curvilinear coordinate system.The numerical performance of finite difference analogs of eq.(2.5) may, as a consequence, be strongly affected by the distortion of the underlying grid.

3. VELOCITY-DRIVEN COORDINATE TRANSFORMATIONS

Given a convective velocity $\vec{v}$, if a coordinate transformation

$$x = \tau(\xi^{h}) \tag{3.1}$$

---

† For concepts and rules of general tensor calculus we refer to [6].

is devised, such that

$$v^k + Kg^{ij}\Gamma^k_{ij} = 0 \tag{3.2}$$

the first derivatives $\theta,_k$ disappear from the field equation (2.5), which simplifies to a form extremely suitable for numerical approximation:

$$Kg^{ij}\,\theta,_{ij} + b = 0 \tag{3.3}$$

A more transparent form of eq. (3.2) may be obtained through contracted multiplication by the covariant base vectors

$$\vec{g}_k \equiv \tau,_k \tag{3.4}$$

Noting that

$$\Gamma^k_{ij}\,\vec{g}_k = \tau,_{ij} \tag{3.5}$$

we obtain

$$\vec{v} + Kg^{ij}\,\tau,_{ij} = \vec{0} \tag{3.6}$$

By introducing the Cartesian components

$$\begin{aligned} \vec{v}(\xi^h) &= u^k(\xi^h)\,\vec{e}_k \\ \tau(\xi^h) &= o + \hat{x}(\xi^h)\,\vec{e}_k \end{aligned} \tag{3.7}$$

the scalar representation follows

$$u^k + Kg^{ij}\,\hat{x}^k,_{ij} = 0 \tag{3.8}$$

In two dimensions, eq. (3.8) expands into

$$\begin{aligned} u + (K/g)\,(\alpha x_{\xi\xi} - 2\beta x_{\xi\eta} + \gamma x_{\eta\eta}) &= 0 \\ v + (K/g)\,(\alpha y_{\xi\xi} - 2\beta y_{\xi\eta} + \gamma y_{\eta\eta}) &= 0 \end{aligned} \tag{3.9}$$

where $u \equiv u^1$, $x_{\xi\xi} \equiv \hat{x}^1,_{11}$, ..., and

$$g \equiv \alpha\gamma - \beta^2 \tag{3.10}$$

is the determinant of the metric tensor, while α, β, γ are defined according to eqs. (1.2). Eqs. (3.9) thus coincide with eqs. (1.1), provided that the source term is equated to $-g/K$ times the convecting velocity. Different source terms, empirically introduced for the control of mesh spacing, may adversely affect the accuracy, producing relatively large first order terms. In particular, the purely diffusive case ($\vec{v} \equiv \vec{0}$) should be treated with homogeneous transformation equations. A simple example of a curvilinear grid produced by eqs. (3.9) is given in Fig. 1a. This 10x10 cells grid covering a rectangle with aspect ratio 1:2 is generated by

$$(4.2) \qquad \nu v^i|^j_j \vec{g}_i - v^k \vec{v},_k - \rho^{-1} \nabla p + \vec{b} = \vec{0}$$

The viscous terms may be reorganized as follows

$$(4.3) \qquad \begin{aligned} v^i|^j_j \vec{g}_i &= [\,(g^{j\ell} v^i|_\ell),_j + g^{jm} v^\ell|_m \Gamma^i_{\ell j} + g^{\ell m} v^i|_m \Gamma^j_{\ell j}]\, \vec{g}_i \\ &= (g^{ij} \vec{v},_i),_j + g^{ik} \Gamma^j_{ij} \vec{v},_k \end{aligned}$$

if use is made of the relationship

$$(4.4) \qquad \vec{g}_i,_j = \Gamma^k_{ij} \vec{g}_k$$

Furthermore, Ricci theorem implies that

$$(4.5) \qquad g^{ik},_i + g^{ik}\Gamma^j_{ij} + g^{ij}\Gamma^k_{ij} = 0$$

Hence, eq.(4.3) transforms into

$$(4.6) \qquad v^i|^j_j \vec{g}_i = g^{ij}(\vec{v},_{ij} - \Gamma^k_{ij} \vec{v},_k)$$

and eq.(4.2) may be rewritten in a form completely analogous to eq.(2.5):

$$(4.7) \qquad \nu g^{ij} \vec{v},_{ij} - (v^k + \nu g^{ij}\Gamma^k_{ij}) \vec{v},_k - \rho^{-1} \nabla p + \vec{b} = \vec{0}$$

Following the approach described in Sect. 3, the solution of eq.(4.1) may therefore be split into two phases: the search of a coordinate transformation τ such that

$$(4.8) \qquad \vec{v} + \nu g^{ij} \tau,_{ij} = \vec{0}$$

and the solution of the field equation

$$(4.9) \qquad \nu g^{ij} \vec{v},_{ij} - \rho^{-1} \nabla p + \vec{b} = \vec{0}$$

However, the transformation and field equations are now coupled through the unknown velocity field. Note also that scalar representations of the vector equation (4.9) should be obtained through a constant base, e.g. the local tangent affine base [7], otherwise first partial derivatives of velocity components would reappear within field equations.

The present approach is not restricted to steady state problems. In an unsteady analysis, the coordinate transformation itself depends also on time

$$(4.10) \qquad x = \tau(\xi^h; t)$$

so that the momentum equation has to be written as

$$(4.11) \qquad \dot{\vec{v}} = \nu g^{ij} \vec{v},_{ij} - (v^k - w^k + \nu g^{ij}\Gamma^k_{ij}) \vec{v},_k - \rho^{-1} \nabla p + \vec{b} = \vec{0}$$

where a superposed dot denotes differentiation with respect to time, keeping ξ^h coordinates fixed, and w^k are the contravariant components of the grid velocity

$$(4.12) \qquad \dot{\tau} \equiv w^k \vec{g}_k$$

In conclusion, both the transformation and field equations take the parabolic form

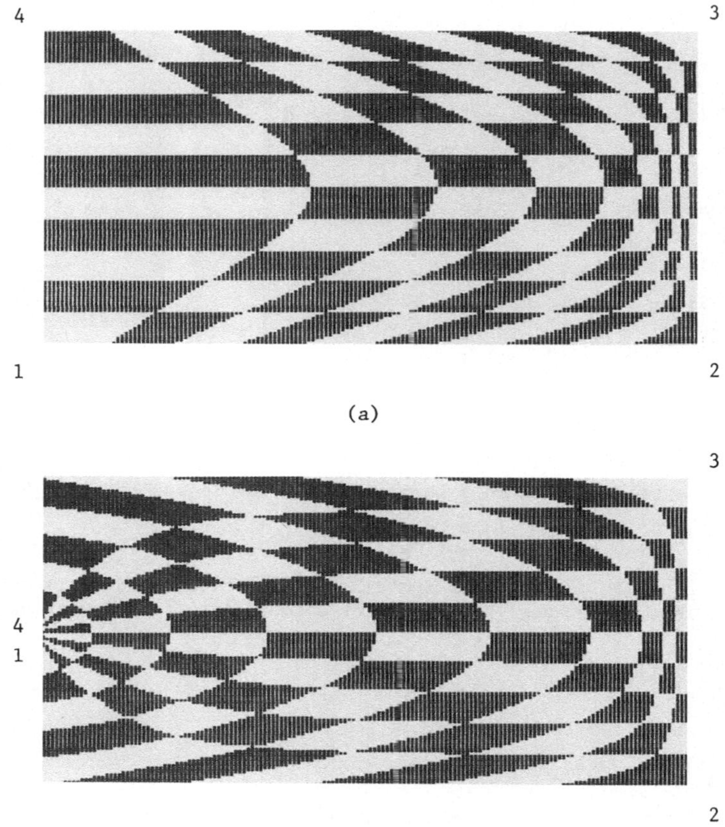

Fig. 1 - Curvilinear grids generated by a uniform velocity field (Pe = 10) with different boundary conditions.

a lengthwise uniform velocity field, for a Péclet number equal to 10 (based on the height as reference length). Each side of the rectangle is mapped onto itself and a uniform spacing is imposed along the boundary. It should be emphasized that different choices for the boundary conditions are still available for a local control of the mesh spacing. Fig. 1b, when compared with Fig. 1a, shows that substantially different boundary conditions (the corners of the computational rectangle are mapped into points 1,2,3,4) can largely affect the local resolution.

4. NAVIER-STOKES EQUATION

The results obtained in the previous sections may be shown to be valid also for a vector transport equation. Let us consider first the steady state momentum conservation equation for a viscous incompressible fluid written in the form

$$\nu v^i|^j_j - v^k v^i|_k - \rho^{-1} p|^i + b^i = 0 \tag{4.1}$$

or, after contracted multiplication by $\vec{g}_i$

(4.13) $\dot{\tau} = \vec{v} + \nu g^{ij}\tau_{,ij}$

(4.14) $\dot{\vec{v}} = \nu g^{ij}\vec{v}_{,ij} - \rho^{-1}\nabla p + \vec{b}$

As a final remark, it can be said that the essential idea underlying the present approach consists in the splitting of the solution of convection-diffusion problems into two phases. The coupling between these two phases plays a central role in adapting, along the transient, the grid to the evolving velocity field Therefore, the numerical integration of the system (4.13, 4.14) deserves a careful consideration in the implementation of the method. Secondly, the method clearly shifts the numerical problems associated with large Reynolds numbers from the field equation to the coordinate transformation equation. This leads to severe difficulties in the numerical generation of a *single* velocity-driven coordinate system over the entire field domain. Hence, to be valid as a general tool, the present method should be used, at large Reynolds numbers, for the generation of *local* adaptive grids.

REFERENCES

1. F.C. Thames, J.F. Thompson, C.W. Mastin, R.L. Walker: "Automatic Numerical Generation of Body Fitted Curvilinear Coordinate System for Field Containing Any Number of Arbitrary Two-Dimensional Bodies", J. Comput. Phys. 15 (1974), 299-319.
2. J.F. Thompson, F.C. Thames, C.W. Mastin: "Boundary Fitted Curvilinear Coordinate Systems for Solution of Partial Differential Equations of Fields Containing Any Number of Arbitrary Two-Dimensional Bodies", NASA-CR-2729 (1977).
3. A. Di Carlo, R. Piva, G. Guj: "On Coordinate Transformation and Curvilinear Cell Models for Large Reynolds Number Flows". In: *Boundary and Interior Layers - Computational and Asymptotic Methods*, J.J.H. Miller Ed. (Dublin: Boole Press, 1980).
4. R. Piva, A. Di Carlo, G. Guj: "Finite Element MAC Scheme in General Curvilinear Coordinates", Comput. & Fluids 8 (1980), 225-241.
5. A. Di Carlo, R. Piva, G. Guj: "Computational Schemes in General Curvilinear Coordinates for Navier-Stokes Flows". In: *Proceedings of the Third GAMM Conference on Numerical Methods in Fluid Mechanics*, E.H. Hirschel Ed. (Wiesbaden: Vieweg, 1980).
6. A. Lichnerowicz: *Eléments de Calcul Tensoriel* (Paris: Colin, 1950). English translation: *Elements of Tensor Calculus* (London: Methuen, 1962).
7. M. Vinokur: "Conservation Equations of Gasdynamics in Curvilinear Systems", J. Comput. Phys. 14 (1974), 105-125.

SIMULATION OF LARGE TURBULENT VORTEX STRUCTURES WITH THE

PARABOLIC NAVIER-STOKES EQUATIONS

J. V. Rakich,[1] R. T. Davis,[2] and M. Barnett[3]

ABSTRACT

The theoretical basis for well-posed marching of a Parabolic Navier-Stokes (PNS) computational technique for supersonic flow is discussed and examples given to verify the analysis. It is demonstrated that stable computations can be made even with very small steps in the marching direction. The method is applied to cones at large angles of attack in high Reynolds-number, supersonic flow. Streamline trajectories generated from the numerical solutions demonstrate the development of vortex structures on the lee side of the cone.

INTRODUCTION

The computation of steady supersonic viscous flows with one set of equations, valid in both the boundary layer and the inviscid supersonic region, has long been attempted to avoid the difficulties with matching conditions for separate inviscid and boundary-layer equations. Use of the unsteady form of the NS equations has been successful but inefficient, in terms of computer time and memory requirements. Therefore, many authors [1-8] have used steady marching or iterative methods to solve numerically the time-invariant NS equations. Single-pass methods have been used for supersonic flows, where the inviscid region is a well-posed initial-value problem, and where the viscous region is a boundary layer, which is itself amenable to a marching solution. The difficulty with solving the complete inviscid/viscous domain at once lies in the thin subsonic layer near the wall, where the pressure is determined by the solution. The absence of a downstream boundary condition makes the problem ill-posed. This combined problem is basically an interaction of a boundary layer with an inviscid, supersonic stream. Lighthill [9] analyzed this problem and found solutions for the boundary-layer displacement thickness of the form $\delta = a \exp(kx)$. One can interpret this exponential growth as the onset of streamwise expansion or compression, which may lead to separation; these solutions have become known as "departure" solutions corresponding to some undetermined downstream boundary condition. This paper describes an approach for avoiding the exponentially growing solution, and presents results for turbulent flow over a cone at large angles of attack.

THE PNS EQUATIONS

The governing equations are written in general curvilinear conservation-law form as follows [8]:

$$\frac{\partial}{\partial\xi}\{J^{-1}[\xi_x E^* + \xi_y F^* + \xi_z G^*]\} + \frac{\partial}{\partial\eta}\{J^{-1}[\eta_x(E - E_v) + \eta_y(F - F_v) + \eta_z(G - G_z)]\}$$
$$+ \frac{\partial}{\partial\zeta}\{J^{-1}[\zeta_x(E - E_v) + \zeta_y(F - F_v) + \zeta_z(G - G_v)]\} = \frac{\partial}{\partial\xi}\{-J^{-1}[\xi_x P_1 + \xi_y P_2 + \xi_z P_3]\} \quad . \tag{1}$$

---

[1]Research Scientist, NASA Ames Research Center, Moffett Field, California, U.S.A.

[2]Professor of Aeronautical Engineering and Applied Mechanics, University of Cincinnati, Ohio, U.S.A.

[3]Graduate Research Assistant, University of Cincinnati, Ohio, U.S.A.

Here J is the Jacobian of the coordinate transormation; E, F, and G are the inviscid flux vectors; and F_v and G_v are the viscous flux vectors--all written in terms of Cartesian velocity components. Note that the viscous terms are omitted from the ξ derivative term; this is the fundamental parabolizing approximation. The $(\ )^*$ terms are central to the discussion of parabolic marching, and are given by the column vectors

$$\left.\begin{aligned} E^* &= \{\rho u,\ \rho u^2 + \omega p,\ \rho uv,\ \rho uw,\ (\rho e_t + p)u\}^T \ , \\ F^* &= \{\rho v,\ \rho uv,\ \rho v^2 + \omega p,\ \rho vw,\ (\rho e_t + p)v\}^T \ , \\ G^* &= \{\rho w,\ \rho uw,\ \rho vw,\ \rho w^2 + \omega p\ ,\ (\rho e_t + p)w\}^T \ , \end{aligned}\right\} \qquad (2)$$

where

$$\rho e_t = p/(\gamma - 1) + 0.5\rho(u^2 + v^2 + w^2) \ .$$

Here u, v, and w are Cartesian velocity components, p is the pressure, ρ the density, and γ the ratio of specific heats. The P terms on the right side are given by

$$\left.\begin{aligned} P_1 &= \{0,\ (1 - \omega)p,\ 0,\ 0,\ 0\}^T \ , \\ P_2 &= \{0,\ 0,\ (1 - \omega)p,\ 0,\ 0\}^T \ , \\ P_3 &= \{0,\ 0,\ 0,\ (1 - \omega)p,\ 0\}^T \ . \end{aligned}\right\} \qquad (3)$$

Following Vigneron et al. [4], the parameter ω is included with the pressure term in the momentum equations. It was shown that the equations can be made formally parabolic by utilizing an appropriate functional form for ω. Considering the two-dimensional viscous subset of equations (1), the mathematical character of the equations is governed by the eigenvalues of the differential system, leading to the condition

$$\omega \le \gamma M_x^{\ 2}/[1 + (\gamma - 1)M_x^{\ 2}] \ . \qquad (4)$$

If this condition is satisfied, then the equations remain parabolic even in the viscous region near the wall where the Mach number is less than unity; a similar form also governs the inviscid region, but is not pertinent to the present work. The purpose of this paper is to further demonstrate the effectiveness of this approach for interacting supersonic flows where the problem can be solved in a single sweep of the PNS equations.

In previous work with the PNS method, a "sublayer" approach was used, in which the pressure gradient term was completely removed in the subsonic region of flow. This is equivalent to a step function for ω, and this places greater significance on the P terms on the right side (RHS) of the equation. Usually, backward-difference approximations have been used for the pressure gradient term on the RHS. However, it can be shown that a backward difference leads to numerical instability for sufficiently small marching steps (which may be a manifestation of the departure behavior). When the ω function is used, the RHS can be set to zero, to a good approximation, making the numerical computation stable, even for small steps. Thus, the ω function and a zero RHS together permit the investigation of departure behavior, which requires very small streamwise steps.

NUMERICAL METHODS

The present computational results are obtained with the Beam-Warming factored algorithm in delta form. Details can be found in reference [8]. Full second-order accuracy can be maintained by use of central differences in η and ζ, and with a three-point backward difference in ξ. However, for the present results, the Euler implicit method is used for the marching direction, making the method first-order accurate in the ξ direction.

TRIPLE-DECK ANALYSIS

For subsonic flows, that is, elliptic equations, a downstream boundary condition must be specified. When attempting to march such a system without specification of the downstream condition, there are an infinity of possible solutions that correspond to various attached or separated flows. When the viscous layer is thin, the departure solution is governed by the interaction of the viscous and inviscid regions near the wall [9]. Stewartson [10] further studied such flows using a triple-deck analysis consisting of a Lighthill sublayer, an inviscid shear region, and the inviscid external flow. This analysis establishes the order of magnitude of the interaction region.

The pressure is constant across both the lower and middle decks, and the velocity in the middle deck corresponds with the attached velocity profile upstream of the interaction, but displaced by δ. Utilizing linear supersonic theory for the pressure at the top edge of the middle deck, one obtains the following momentum equation for the lower deck:

$$u \frac{\partial u}{\partial x} + v \frac{\partial u}{\partial y} = \nu \frac{\partial^2 u}{\partial y^2} - \rho_e u_e{}^2/[\rho_w(M_e{}^2 - 1)^{\frac{1}{2}}] \frac{d^2\delta}{dx^2} \quad , \tag{5}$$

where the subscript w refers to wall, and e refers to the top edge of the middle deck. Introducing the stream function in (5), and utilizing the Lighthill form for the displacement thickness δ, yields an Airy equation. It is found from the solution that the triple-deck scale is given by

$$\Delta\tilde{x} = 0.8272(M^2 - 1)^{3/4}(T_\infty/T_w)^{1/2} C_f{}^{5/4} Re^{3/8} \; \Delta x = k \; \Delta x \quad , \tag{6}$$

where x is the nondimensional physical distance, Re is the Reynolds number, C_f is the skin friction coefficient, T is the temperature, and M is the Mach number. The pressure gradient that results is

$$dp/dx = C \exp(kx) \quad . \tag{7}$$

The importance of this analysis to computations with the PNS equations is that it provides a test of the effectiveness of methods for damping the unwanted departure solutions. To illustrate, we consider a flat-plate boundary layer. The solution shown in Figure 1 at $x/L = 0.5$ is obtained with the ω parameter turned on, and then ω is set to 1.0, so that the full PNS equations are marched for a short distance. The solution immediately starts to "depart" until at $x = 0.5055$ the ω factor is again turned on, and the solution decays toward the well-known boundary-layer solution. If the triple-deck scale had not been resolved by the step size

Δx, the departure would have appeared as an oscillatory divergence. We note here that other investigators report their methods are unable to suppress the departure behavior for small marching steps, while the present approach is quite effective.

Figure 2 demonstrates that the computed departure solutions ($\omega = 1$) agree with Lighthill's linear analysis. This figure shows the magnitude of the computed pressure gradient as a function of the triple-deck distance scale $\widetilde{x} = kx$. The initial slope of the computed departure agrees with the linear analysis until nonlinear effects take over.

COMPUTATION OF TURBULENT FLOWS

Turbulent computations are performed with the Reynolds averaged form of the PNS equations and an algebraic, or zero-equation, eddy viscosity model. Details are given in reference [5] and will not be repeated here. We note, however, that no special modifications were needed for the present calculations, even with a large crossflow separation on the leeward side.

Figures 3 and 4 compare the circumferential variation of pressure and surface stream angle from the present computations with the experimental results of Rainbird [11]. Generally, good agreement is obtained with experiment. The main differences are caused by an incorrect prediction of the location of separation, believed to be due to the inadequacy of the simple turbulence model used. These results were obtained on a CDC 7600 computer with 50 grid points between the body and the shock, and with 47 unequally spaced meridian planes.

FLOW-FIELD SIMULATION

The simulation of turbulent vortex structures is achieved by tracing particle paths using the velocity field obtained from the PNS solution. Initial positions for the tracer particles are specified at field locations near the cone surface, and $x/L = 0.1$. The particle paths are then determined from a simple Euler predictor-corrector, finite-difference scheme. The velocity field is taken at $x = L$ and is assumed invariant with x for the purposes of this simulation.

The flow simulation is shown in Figures 5a and 5b. Three distinct vortices are observed--one emanating from the primary separation, the second from the secondary separation of the flow coming down the leeward plane of symmetry, and a third below the primary vortex and having the same rotation as the primary vortex. This flow structure is suggestive of a global type of mixing in large turbulent structures, which is significant to the development of turbulence models.

REFERENCES

[1] Davis, R.T., *AIAA J.*, 8 (5), 843-851 (1970).

[2] Rubin, S.G. and Lin, T.C., PIBAL Rept. 72-8, Polytechnic Inst. of Brooklyn, Farmingdale, N.Y. (1971).

[3] Lubard, S.C. and Heliwell, W.S., *AIAA J.*, 8 (7) (1974).

[4] Vigneron, Y.C., Rakich, J.V., and Tannehill, J.C., AIAA Paper 78-1137 (1978).

[5] Rakich, J.V., Vigneron, Y.C., and Agarwal, R., AIAA Paper 79-131 (1979).

[6] Schiff, L.B. and Steger, J.L., AIAA Paper 79-130 (1979).

[7] Kovenya, V.M. and Chernyi, S.G., *Chislenne Method Mechaniki Sploshnoi Sred*, 10 (1) (1979).

[8] Tannehill, J.C., Venkatapathy, E., and Rakich, J.V., *AIAA J.*, 20 (2) (1982).

[9] Lighthill, M.J., *Proc. Roy. Soc. London* (1953).

[10] Stewartson, K., *Adv. in Appl. Mech.*, Academic Press, 14, 145-159 (1974).

[11] Rainbird, W.J., *AIAA J.*, 6 (12) (1968).

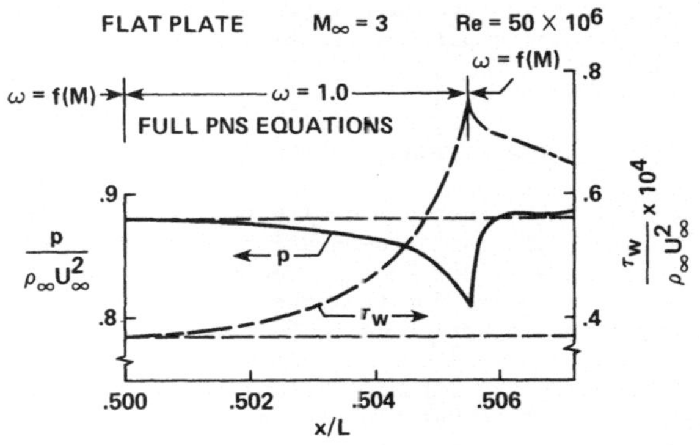

Fig. 1. Influence of "Omega" parameter on departure and recovery.

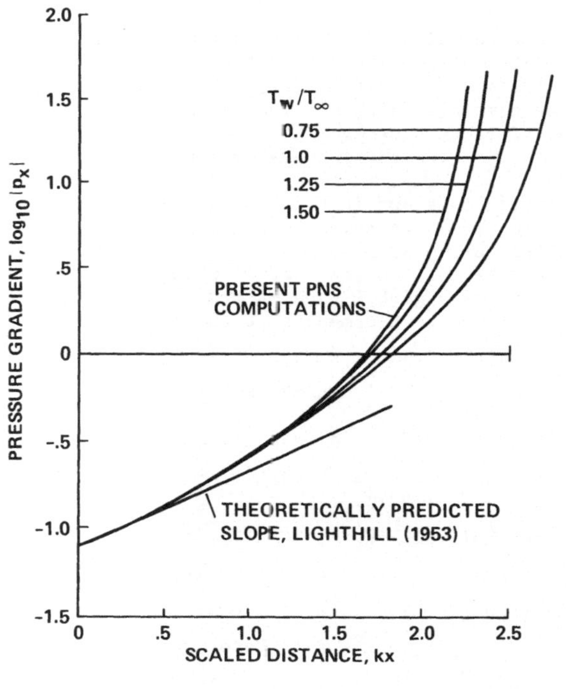

Fig. 2. Departure solution for a flat plate; $M_\infty = 3$, $Re = 50\times10^6$, $T_W = 100$ K.

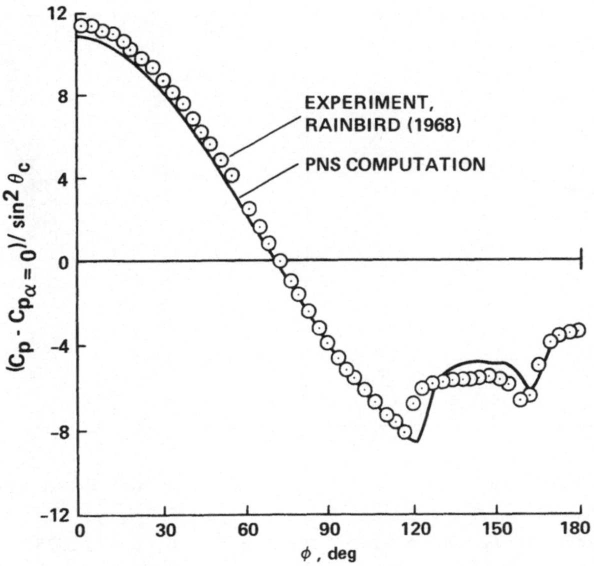

Fig. 3. Circumferential pressure distribution for a 12.5° cone; $M_\infty = 1.8$, $\alpha = 22.7°$, $Re = 25\times10^6$.

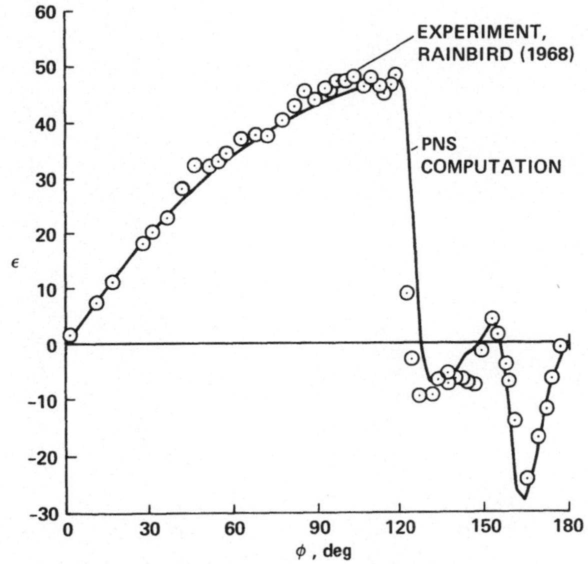

Fig. 4. Surface stream angle relative to the generators; $\theta_c = 12.5°$, $M = 1.8$, $\alpha = 22.75°$, $Re = 25\times10^6$.

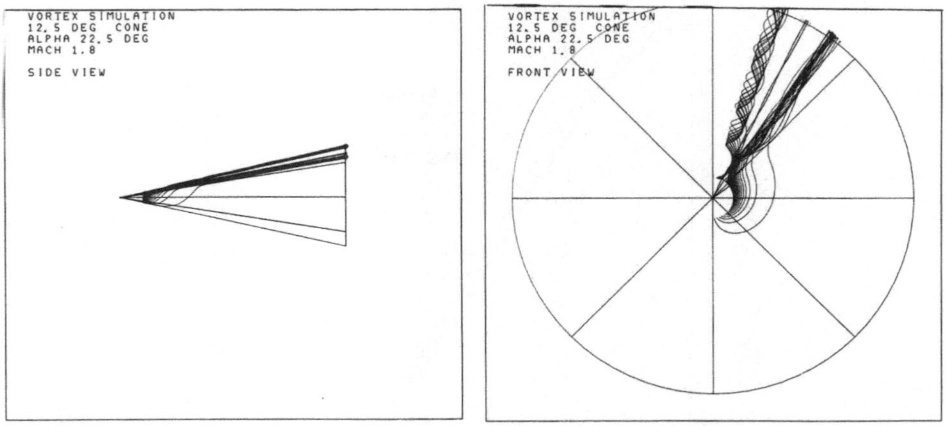

(a) Side view. (b) Front view.

Fig. 5. Streamlines around a 12.5° cone; $M = 1.8$, $\alpha = 22.75°$, $Re = 25\times10^6$.

UNIGRID PROJECTION METHOD TO SOLVE THE EULER EQUATIONS FOR STEADY TRANSONIC FLOW

Arthur Rizzi & Lars-Erik Eriksson
FFA The Aeronautical Research Institute of Sweden
S-161 11 BROMMA, Sweden

INTRODUCTION

We are searching for a method that solves the steady Euler equations more efficiently than, but with the same generality as, the time marching methods currently in use. Primarily due to the non-linearity of the problem, methods to solve the Euler equations are necessarily iterative and can be likened to a time-dependent process with or without true time accuracy. With this viewpoint the task is to reduce the transient error content in order to converge to the steady state. An eigensystem analysis of the convergence problem indicates that it is poorly conditioned in the sense that although the majority of the eigenvalues have large imaginary parts, some of them do have small modulus so that the ratio of largest to smallest modulus is large. Furthermore as the grid is refined for more accurate resolution, those eigenvalues draw nearer to zero and this ratio increases. But the important clue on how to treat the problem is that the long-wavelength and short-wavelength transients are carried separately by the eigenvectors associated respectively with the eigenvalues of small and large modulus. Our previous attack[1] on this problem by successive grid refinement within a conventional time marching approach was an attempt to reduce the large-scale long-wave errors on a coarse mesh where the condition number is not too severe, and then interpolate the coarse solution to a finer mesh in order to improve the fine-scale accuracy, the hope being that the interpolation process would not introduce additional long-wave errors. Although this approach met with some success, in the finest mesh there remained a small-amplitude long-wave transient residue that takes many iterations to eliminate completely because of the knee in the convergence rate which is characteristic of this approach (Fig. 1).

The method we present here is designed purposely to circumvent this problem of slow convergence and consists of two different operations, one suited to treat short-wave transients and the other the long-wave transients. In the first part a time stepping hyperbolic operator acts repeatedly on the initial field and removes the short-wave error so that what remains is a smooth long-wave perturbation upon the steady solution. This long-wave error is then treated in the second part where it is projected to a space which spans the larger one of the smooth error but which is of lower effective order. This leaves us with a matrix equation of greatly reduced order which then can be solved rapidly for the steady solution. Notice that only one grid is implied in this method, the one needed to give the final desired resolution.

NEWTON LINEARIZATION

The time dependent compressible Euler equations in spatially discretized conservation form are

$$q_t = - F(q) \equiv - (\delta_x f + \delta_y g) + \text{boundary conditions} \tag{1}$$

where f and g are fluxes in the finite-volume sense and δ is the usual central difference which when combined with the boundary conditions[1] give us the complete discrete operator F. We seek the steady state q^* that exactly satisfies the difference equations $F(q^*) = 0$ and approach it with the sequence $\{q^n\}$ of time dependent solutions such that $q^n \to q^*$. In order that the sequence number n remain relatively low we must work directly on the transient error, and therefore we linearize Eq. (1) about the current approximation q^n, and where $A \equiv \partial F/\partial q$ we obtain

$$\Delta q_t = - A \Delta q - F(q^n) + \ldots \tag{2}$$

a linear hyperbolic equation for the transient perturbation Δq superimposed upon the known and presumed locally steady solution q^n. Equation (2) can be solved for Δq at a new time level, then a new estimate $q^{n+1} = q^n + \Delta q$ is found, and the entire process is repeated, diminishing Δq until only a low amplitude long-wave transient remains. The time stepping is carried out by

$$\begin{aligned}
\Delta q_o &:= \Delta q && \text{(current level)} \\
\Delta q &:= \Delta q_o - hA\Delta q - hF && \\
\Delta q &:= \Delta q_o - hA\Delta q - hF && \\
\Delta q &:= \Delta q_o - hA\Delta q - hF && \text{(new level)}
\end{aligned} \tag{3}$$

with $h = \Delta t$ which is an explicit three-stage iterative scheme we constructed that approximates the fully implicit backward Euler scheme. It is time accurate to first order and is well suited to the task because it annihilates the modes stemming from eigenvalues with large imaginary parts (Fig. 2) and therefore quickly smooths the error. Just how effectively it does this is indicated in Fig. 3 by the Fourier analysis of the error that remains in the flowfield around an airfoil after a number of Newton steps have been marched by Eqs. (3). At the top we see that the Fourier spectrum of the transient error (i.e. $F(q) = \varepsilon \neq 0$) is full, and so is the spectrum of our current time iterative solution Δq. We ask then how near is this to the true Newton update and the answer is given at the bottom by the Fourier analysis of the residual of our Δq. Here we see that our iterative Δq is inaccurate only at the lower Fourier frequencies, so the error in Δq presumably is smooth.

PROJECTION METHOD

It would be preferable of course to solve Eq. (2) directly for Δq with the time-dependent term set identically to zero, the more common Newton method. That leads unfortunately to a large sparse system and for three-dimensional transonic flow there are presently no very efficient means to solve it. The adoption of this time-dependent perturbation approach is attractive also for two other reasons: first it provides a rather natural way to relax the Newton linearization which, when q^n is far from q^*, may not converge, and second it introduces the smoothing operator (3) which removes the short-wave residual of Δq. The long-wave content however remains and must be handled differently, namely by solving the steady problem

$$A\Delta q = - F \tag{4}$$

itself. We still cannot do this directly, but now, because of the smooth spectrum we reason that Eq. (4) can be solved approximately by projecting it to a space of lower order where the troublesome eigenvalues have larger modulus and where a greater variety of methods can be used effectively because of the reduced size of the problem. The solution of Eq. (4) is postulated as the superposition $\Delta q = \Delta q_o + u$ where Δq_o is the current approximate solution to (4) and u is a low spatial frequency correction. Next we write u, whose length is N (the number of unknowns in the problem), as $u = Pv$ a linear combination of an unknown vector v of length M that defines the reduced solution space. The columns of the rectangular matrix P are a particular choice of base vectors $p_1 \ldots p_M$ that span the original space and characterize the type of the projection. In order to solve for v multiply the last equation by Q^T the transpose of some $N \times M$ matrix and invert. The operator $\pi = P(Q^TP)^{-1} Q^T$ defining the projection is then obtained (Fig. 4). The last step of the procedure is to project either Eq. (2) or (4) with π and solve the resulting reduced set of equations by some appropriate technique and then update the current approximate solution Δq_o. The overall algorithm is summarized in the flowchart in Fig. 5.

Whether a particular projection technique is successful or not lies in its ability to model those modes of the transient error that are not damped by the smoothing operator. Listed in Table 1 are a number of various choices for the base vectors, matrix Q, and solution procedure that we have investigated. The difficulty in assessing the error in the correction to Δq_Q given by the Galerkin type of projection is due to the fact that the matrix A is far from being hermitian and positive definite for the hyperbolic problems considered. Indeed, it is very nearly skew-hermitian since central differencing is used. This means that the projection does not have the "best approximation" property of hermitian positive definite matrices. A direct consequence of this fact is that the choice of base vectors for the reduced problem is much more critical. For example, it is very important that the base vectors span the boundaries of the problem, since it is the boundary conditions that break the skew-symmetry of the matrix A and make the problem well posed. A choice of unsuitable base vectors may in fact hinder rather than help the overall convergence of the calculation. The Galerkin projection with piecewise bilinear base vectors works about as well as, but with less computational effort than, the Petrov-Galerkin projection using Fourier functions, which does have the best approximation property.

TRANSONIC EXAMPLE

We demonstrate the method with a computation that solves the Euler equations for transonic flow $M_\infty = 0.8$, $\alpha = 0^\Theta$ past the NACA 0012 airfoil. The mesh is an O-type having 64 x 14 cells around the full airfoil. The computation begins with the freestream flow as the first approximate solution q^o, and it converges after $n = 13$ cycles. To give an indication of how this solution develops we plot in Fig. 6a with a solid line the pressure coefficient c_p on the airfoil given by the converged solution and with dashed lines three distributions for intermediate value of n prior to convergence. The overall flowfield develops quickly, and it is the region at the foot of the shock that is the slowest to converge. The details of the evolution however are brought out best in Fig. 6b by the reduction in the residuals of the various iteration levels as the computation proceeds. A complete cycle of the entire algorithm is traced by the dashed lines. It begins with the Newton linearization which yields a certain right-side residual, smooths the solution Δq to Eq. (2) by an order of magnitude, projects Δq by the Galerkin technique down to the reduced space (16 x 5) where Eq. (4) is solved iteratively by 10 cycles of scheme (3). Then it projects this advanced reduced solution back up to the full space, smooths it again and updates the Newton linearization leading to a diminished Newton residual. The whole cycle consists of about 75 work units (1 step Eq. (3) = 1 work unit) and consumed about 1½ cpsec on the CYBER 205 vector processor. While not as efficient as a true multigrid method, the projection algorithm is very simple to program and can be vectorized to a high degree. This technique does work and we have verified that it does improve the rate of convergence compared with a corresponding calculation without projection. But the astonishing fact revealed here, and exactly contrary to what one would expect, is that the Newton linearization converges most rapidly when it is far from the solution and hardly at all when it is near the steady state - it suffers the same characteristic knee as the other methods. We have no explanation for this but remark that it is the local details in the stagnation regions and around the shock that fluctuate from one Newton cycle to the next. Evidently this is a feature that cannot be attacked using the linear slope. But perhaps fundamental to our understanding of this problem is to learn what causes the knee in the convergence curves which we see so often?

References

1. Rizzi, A., and Eriksson, L.E., "Transfinite Mesh Generation and Damped Euler Algorithm for Transonic Flow Around Wing-Body Configurations", AIAA Paper No. 81-0999, 1981.

2. Jespersen, D., "Multilevel Techniques for Nonelliptic Problems", Proc. Symposium Multigrid Methods, NASA SP 1981.

BASE VECTORS p_k	PROJECTION TYPE Q	SOLUTION TECHNIQUE
piecewise bilinear	P (Galerkin)	time integration
"	P "	direct solution
Fourier	P "	" "
"	P "	time integration
"	AP (Petrov)	GS relaxation

Table 1. Various combinations of base vectors, projection types, and solution techniques that we have tried.

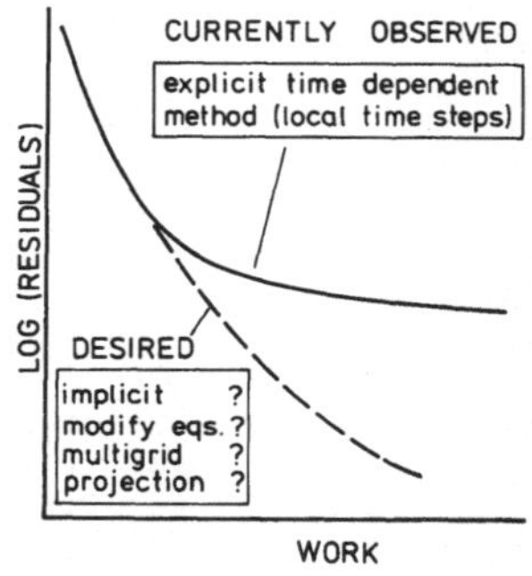

Fig. 1 Slow convergence to the steady state due to the knee in the residual curve observed with explicit methods now in use. Some techniques that mau address this problem are listed.

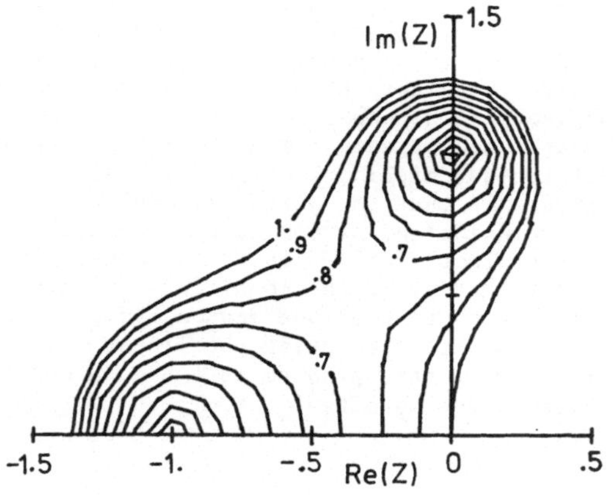

Fig. 2 Contours of constant $|g|$ where $g = 1+z+z^2+z^3$ is the amplication factor of the 3-stage iterative backward Euler scheme (3). A zero is centered on the imaginary axis (cf. Fig. 2 of ref. 2).

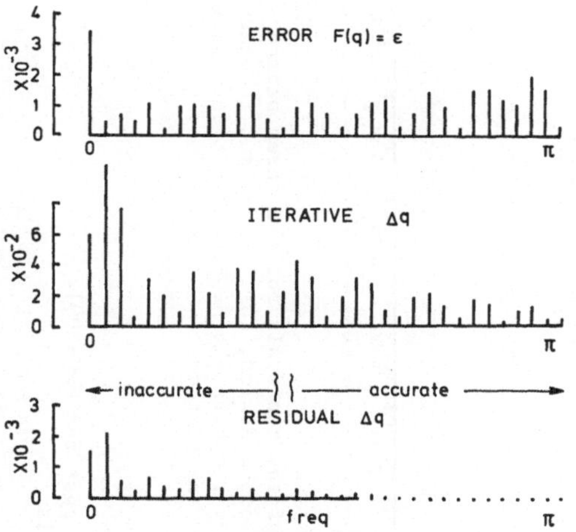

Fig. 3 Fourier analysis of the transient error in F, the iterative solution Δq, and the residual of Δq spatially around an airfoil during the time integration of eq.(2) by scheme (3). The residual is smooth in space.

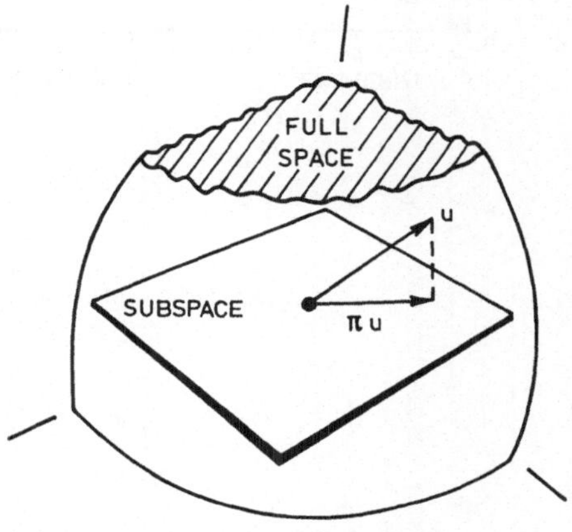

Fig. 4 Representation of the operator $\Pi = P(Q^TP)^{-1}Q^T$ that projects the correction u in the full space to the reduced level Πu in a subspace.

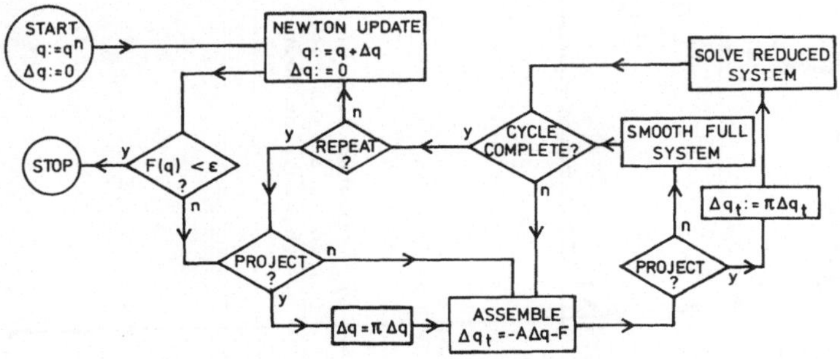

Fig. 5 Flow chart of the projection method whose essential ingredients are: Newton linearization that defines the operator A, projection operator Π , smoothing operator (3), and solution technique for the reduced system.

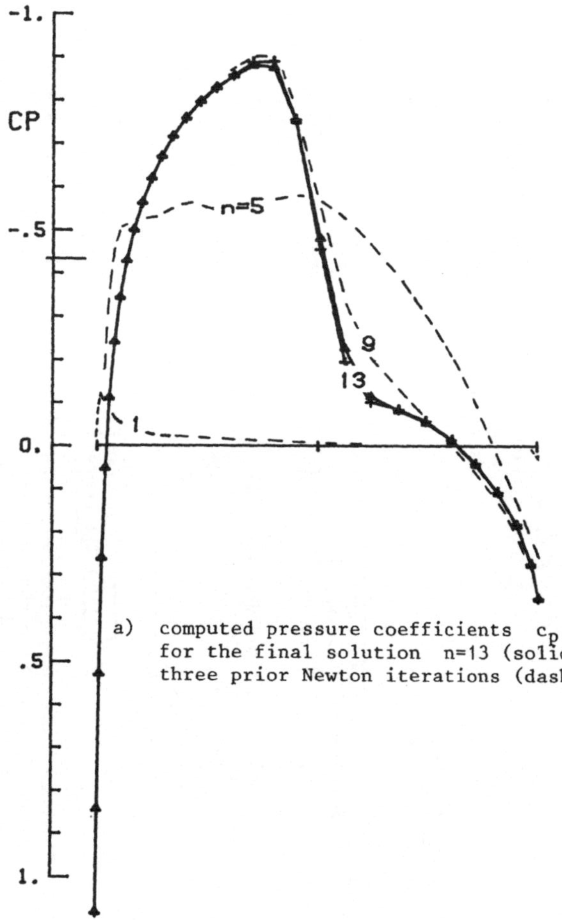

a) computed pressure coefficients c_p on the airfoil for the final solution n=13 (solid line) and three prior Newton iterations (dashed lines).

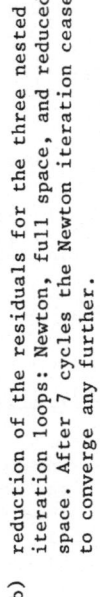

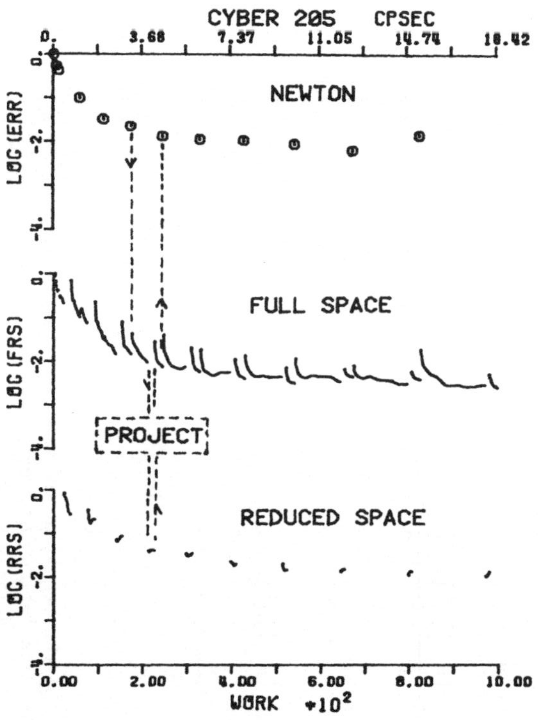

b) reduction of the residuals for the three nested iteration loops: Newton, full space, and reduced space. After 7 cycles the Newton iteration ceases to converge any further.

Fig. 6 Computation by projection method of transonic flow past the NACA 0012 airfoil. $M_\infty = 0.8$ $\alpha = 0.$ O-mesh with 64x14 cells.

NUMERICAL SOLUTION OF VISCOUS FLOW AROUND ARBITRARY AIRFOILS IN A STRAIGHT CASCADE

M. Rosenfeld and M. Wolfshtein
Department of Aeronautical Engineering
Technion - Israel Institute of Technology
Haifa, Israel

1. INTRODUCTION

1.1. General background. A straight cascade of profiles is defined as an array of identical airfoils placed at a constant distance from one another. Although important three dimensional and other effects are neglected, the flow field around a cascade represents a useful approximation to the flow through turbomachines and serves as an important tool for preliminary analysis.

The basic problem is that of a duct flow with curved boundaries and separated regions. Yet, two dimensionality may be assumed in many cases. Non-viscous solutions, which are available for such flows[1] can not predict separation. Viscous solutions are sometimes based on parabolization. Briley[2] uses the parabolic Navier-Stokes equations in ducts. The unsteady compressible thin layer approximation was used by Steger et al.[3], who were able to calculate separated flow regions. However, very few details are given. Shamroth et al.[4] solve the full two dimensional compressible Navier-Stokes equations for a straight cascade. Results are given for a non-staggered uncambered profile at very low angles of attack. It seems that the computation of separated flow fields around staggered cascades with realistic airfoils at medium or high incidence is still a difficult task which needs improvements.

Discretization by finite differences poses serious problems near curved boundaries which are so typical of general cascade problems. Usually, a transformation of coordinates is utilized to map the physical computational region into a rectangular domain where the boundaries coincide with coordinate lines. Conformal mapping is the most obvious choice, but it lacks direct control of the finite-difference mesh distribution. In differential transformations, as proposed by Thompson et al.[5], it is possible to obtain good control over the mesh distribution and to cluster points in desired regions. But the coordinate system is usually non-orthogonal and its generation is expensive. Direct construction of coordinate systems[6] produces a non-orthogonal system as well. However, the mesh generation is more efficient and better control of the mesh distribution may be obtained. Therefore, this approach was chosen in the present work.

1.2. Definition of the problem. In this paper the two-dimensional separated flow field around arbitrary airfoils arranged in a straight cascade is computed. Incompressible, isothermal and laminar flow is assumed. Only steady flow is considered, thus excluding the possibility of resolving unsteady separation which usually persists in high Reynolds numbers and is mostly turbulent. Finite difference solutions are obtained to the elliptic vorticity and stream function equations in non-orthogonal curvilinear coordinates. The emphasis is placed on simplicity and ease of use.

2. PROBLEM FORMULATION

2.1. The coordinate system. The coordinate system is algebraically generated using a method suggested by Eiseman[7]. This method, which does not require a solution of differential equations, produces an "O" type coordinate system. One family of the coordinate lines (ξ) is composed of closed curves encircling the profile, starting with the profile itself and blending into the outer boundary (ABCDA in Fig. 1). The outer boundary consists of two parallel lines (BC and DA) which are spaced one period apart and of two arbitrary curves upstream and downstream of the profile (AB and CD) which close the control volume. The second family of coordinate lines (η) is composed of straight lines normal to the profile. At the periodic boundary the lines leading up and down emerge from identical points (Fig. 1). Coordinate stretching in the η - direction allows mesh refinement near the profile. The main deficiencies of the mesh are that a sharp trailing edge can not be accommodated and the resolution of the trailing wake is not very good. These limitations do not cause serious errors in the present computations.

2.2. The governing equations. The dependent variables were chosen as the vorticity (ω), the stream function (ψ), and the pressure (p). The solution of these quantities is simple and relatively easy. Moreover, this choice requires transformation of scalar variables and not transformation of the velocity vector as in the primitive variables formulation.

The non-dimensional vorticity equation has the form[5]:

$$(\psi_\eta \omega)_\xi - (\psi_\xi \omega)_\eta = (\alpha\omega_{\xi\xi} - 2\beta\omega_{\xi\eta} + \gamma\omega_{\eta\eta} + \sigma\omega_\eta + \tau\omega_\xi)/J\cdot Re \tag{1}$$

where $J = x_\xi y_\eta - x_\eta y_\xi$ is the Jacobian of the transformation and $\alpha = x_\eta^2 + y_\eta^2$, $\beta = x_\xi x_\eta + y_\xi y_\eta$, $\gamma = x_\xi^2 + y_\xi^2$, $\sigma = (y_\xi Dx - x_\xi Dy)/J$, $\tau = (x_\eta Dy - y_\eta Dx)/J$, $Dx = \alpha x_{\xi\xi} - 2\beta x_{\xi\eta} + \gamma x_{\eta\eta}$, $Dy = \alpha y_{\xi\xi} - 2\beta y_{\xi\eta} + \gamma y_{\eta\eta}$. The subscripts ξ and η imply differentiation with respect to these coordinates. The Reynolds number is defined by $Re = u_{in}\, c/\nu$ where c is the chord of the profile, ν is the kinematic viscosity and u_{in} is the x - component of the inlet velocity. The convection terms are expressed in a conservative form. The non-dimensional stream function equation is:

$$\alpha\psi_{\xi\xi} - 2\beta\psi_{\xi\eta} + \gamma\psi_{\eta\eta} + \sigma\psi_\eta + \tau\psi_\xi = -J^2\omega \tag{2}$$

For each velocity field a Poisson equation specifies the pressure field. However, this equation is complex in curvilinear coordinates. A simpler pressure equation is derived by differentiating the first pressure derivatives with respect to each curvilinear coordinate. The two pressure derivatives are obtained from the two momentum equations: $p_\xi = f(\xi,\eta)$ and $p_\eta = g(\xi,\eta)$. Adding up the second pressure derivatives yields a cartesian-like Poisson equation in the curvilinear system, too:

$$p_{\xi\xi} + p_{\eta\eta} = f_\xi + g_\eta = F(\xi,\eta) \tag{3}$$

where F is obtained from the known stream function and vorticity fields. This formulation, already proposed by Ghia et al.[8], allows the use of standard efficient Poisson solvers.

2.3. Boundary conditions. At the profile the no slip and no injection conditions yield: $\psi = \psi_\eta = 0$. At the inflow boundary (AB) the vorticity is assumed to be zero and the stream function is computed from the given uniform inflow velocity. At the periodic boundaries the periodicity of the flow field is utilized.

The boundary conditions at the outflow (CD) requires some attention. The first boundary condition is obtained by neglecting the diffusion terms in the vorticity equations, thus enabling convection of vorticity through this boundary. The outflow condition for the stream function is specified by: $\psi_x = -v_{out}$ where v_{out} is the yet unknown y component of the outlet velocity. It is uniquely determined from the integral momentum balance in the y direction: $v_{out} = v_{in} - F/\Delta\psi$. $\Delta\psi$ is the rate of mass flow between two airfoils and F is the y component of the net force acting on a single profile (as a consequence of the pressure and viscous forces). A similar approach is described by Bosman et al.[9], but there two global iteration loops (for each condition) are required and therefore the solution is more expensive.

3. NUMERICAL SOLUTION

3.1. Discretization. The differential equations are discretized by finite differences over a constant mesh in the computational domain. A nine-point numerical molecule is generated due to the non-orthogonality of the coordinate system. The diffusion terms are approximated by second order central differences. The same approximation of the convective terms leads to severe stability limitations at high Reynolds numbers. Stability may be enforced by using upwind first-order finite-difference approximation for the convection terms.

In order to maintain second order accuracy without loosing the stability, the convective terms are approximated by first order upwind differences and are corrected to second order by deferred corrections[10]. A two step non-iterative method was designed. During the predictor step the upwind first order difference equations are solved till full convergence is reached. In the corrector step the same difference equations are solved again, but this time the estimated first order truncation error (based on the predictor solution) is subtracted from the vorticity equation. Thus, it is seen that the stable upwind difference equations are solved during both steps. Nevertheless, second order accuracy is obtained, although the absolute truncation error is greater than in the usual central difference second order solution.

3.2. Implementation of some of the boundary conditions. The vorticity on the profile is derived from the no slip condition and equation (2). A second order approximation, utilizing the orthogonality of the η - lines to the profile, yields:

$$\omega_w = \left(\frac{\gamma}{J^2}\right)_w \left[-3(\psi_{w+1} - \psi_w)/\Delta\eta^2 - \left(\frac{J^2}{\gamma}\right)_{w+1} \omega_{w+1}/2\right] \qquad (4)$$

where $\Delta\eta$ is the interval in the η direction, w refers to a point on the profile and w+1 to the adjacent point along a constant ξ line.

On the periodic boundaries the vorticity and stream function equations are solved together with the periodicity conditions. Moreover, the derivatives with respect to η on this boundary can not be computed by central difference approximations because the η coordinate is not continuous across this boundary. Therefore, a modified difference molecule is required. This molecule is generated by extending the straight η - lines in the physical domain till they cross one of the coordinate lines of the adjacent profile's grid (see Fig. 2). These points now form a part of the difference molecule for grid points at the periodic boundaries. The mesh thus obtained is not uniform in the η - direction. In order to apply periodicity, the values of the variables at these points are found by interpolation. The finite difference approximations to the second derivatives at the boundaries are first order only, due to the nonuniformity of the mesh.

3.3. Method of solution. The nine-point difference equations for the stream function and vorticity are solved by point Successive Over Relaxation. The five-point difference equations for the pressure are solved by the direct cyclic reduction method.

4. RESULTS

4.1. Test cases. The proposed algorithm was tested by computing the flow field across a cascade of circular cylinders with Re=40. Two cases were considered with the cylinders placed 8 and 16 diameters apart (solidity σ = 1/8 and σ = 1/16). Solutions over refined meshes were used to test the convergence. The convergence of the length of the separation bubble (L_s) behind the cylinder with σ = 1/16 is shown in Fig. 3, where h is proportional to the mesh size. The second order convergence of the scheme is clearly observed. Richardson's extrapolation may be used to improve the accuracy of L_s. Figures 4 and 5 compare L_s and the pressure coefficient distribution (C_p) for the present cases with some experimental and numerical results (most of them for a single cylinder, σ = 0). The influence of the solidity,

σ, was investigated as well. Similar tests were satisfactorily performed for the flow around elliptical cylinders.

4.2. Flow around an airfoil. The full capabilities of the suggested method were tested by computing the flow around a cascade of symmetrical NACA 0012 airfoils with solidity $\sigma = 1$, stagger angle $\lambda = 30^\circ$ and Re = 1000. Four angles of attack were considered between -10° and 15°. The results included the vorticity, the stream function and pressure distributions as well as the turning angle of the flow and the aerodynamic coefficients. Some of the cases show separated regions. No comparison with experiments or computations could be made because relevant data was not available.

Figure 6 shows the streamlines pattern for zero angle of attack. The asymmetry of the flow field due to the non zero stagger angle is easily observed. The pressure coefficient distribution on the profile is shown at Fig. 7. A net force which is created in the negative y direction causes a negative turning angle (-0.7°) of the flow. The nearly singular coordinate system at the trailing edge is responsible for the abrupt changes in the pressure there. Fortunately, these effects are limited to the very near vicinity of the trailing edge and do not affect the solution away from the trailing edge or the net force acting on the airfoil.

Figures 8 and 9 show the streamlines and the pressure coefficient distribution on the profile for an angle of attack of -10°. A steady separation is formed at the suction side, near the trailing edge. A second small separation bubble is created at approximately one-half chord distance from the leading edge. An angle of attack of $+10^\circ$ produces an unseparated flow. It illustrates the influence of the stagger angle on the maximal angle of attack without separation (α_S). When the stagger angle is positive (as in the present case) α_S is greater (in absolute value) for positive angles of attack. Pressure oscillations are found at the separated part of the airfoil (Fig. 9). Similar oscillations were reported by Shamroth et al.[4], even for a zero angle of attack. This may, perhaps, be attributed to second order approximation instabilities of the present numerical scheme. Anyhow, these effects are limited to a small region near the airfoil (a height of 1-2% of the chord length). Above it the pressure remains almost constant, as observed experimentally in separated flow zones. Additional results are not included here for brevity.

5. CONCLUSIONS

The presented method is capable of solving fully viscous flow fields around a staggered cascade of arbitrary airfoils, including separated flows. The method is second order and numerically stable. Good agreement with experimental data is found and reasonable qualitative behaviour is obtained wherever experimental data is not available.

Most of the physical simplifications made (like steady or laminar flow) can be easily removed. The numerical efficiency can be improved by a better selection of the coordinate system and the application of efficient finite difference algorithms to the solution of the equations.

REFERENCES

1. Habashi, G.W., Recent Adv. in Num. Methods in Fluids, 1, 1980, pp. 245-286, Pineridge Press.
2. Briley, W.R. and H. McDonald, AIAA paper 79-1453.
3. Steger, J.L., T.H. Pulliam and R.V. Chima, AIAA paper 80-1427.
4. Shamroth, S., H.J. Gibeling and H. McDonald, AIAA paper 80-1426.
5. Thompson, J.F., F.C. Thames and C.M. Mastin, J. Com. Phys., 15, 1974, pp. 299-319.
6. Eiseman, P.R., J. Comp. Phys., 33, 1979, pp. 118-150.
7. Eiseman, P.R., J. Comp. Phys., 26, 1978, pp. 307-338.
8. Ghia, U., K.N. Ghia, S.G. Rubin and P.K. Khosla, Computers and Fluids, 9, 1981, pp. 123-142.
9. Bosman, C., K.C. Chan and A.P. Hatton, J. of Engineering for Power, 101, 1979, pp. 450-458.
10. Fox, L., Proc. Royal Soc. of London, A190, 1947, pp. 31-58.
11. Dennis, S.C.R. and G.Z. Chang, J. Fluid Mech., 42, 1970, pp. 471-489.
12. Gibeling, H.J. S.J. Shamroth and P.R. Eiseman, NASA CR-2969, 1978.
13. Grove, A.S., F.H. Shair, E.E. Peterson and A. Acrivos, J. Fluid Mech., 19, 1964, pp. 60-80.

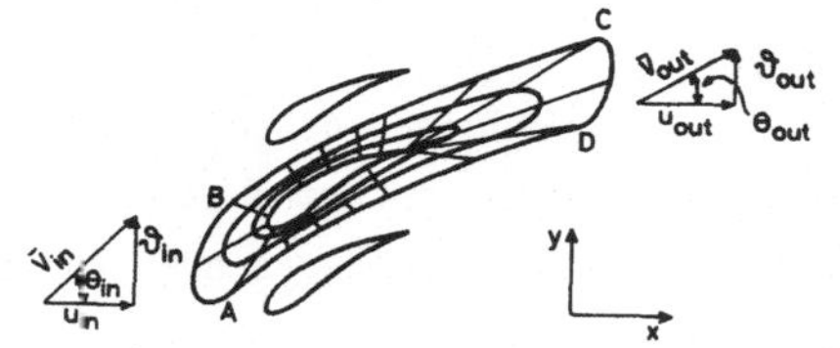

Fig. 1: The cascade and the coordinate system.

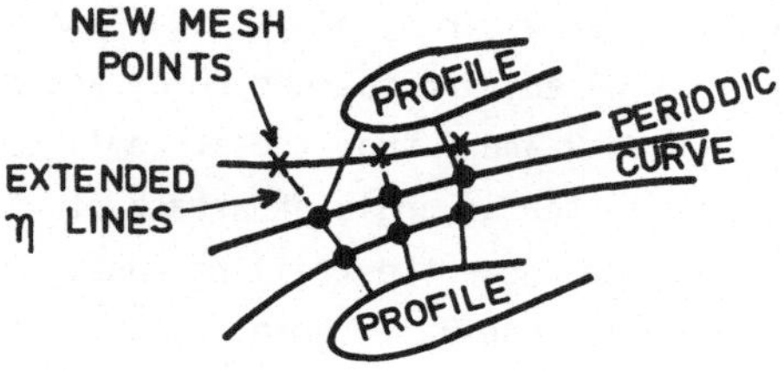

Fig. 2: Extrapolation of the mesh beyond the periodic boundary.

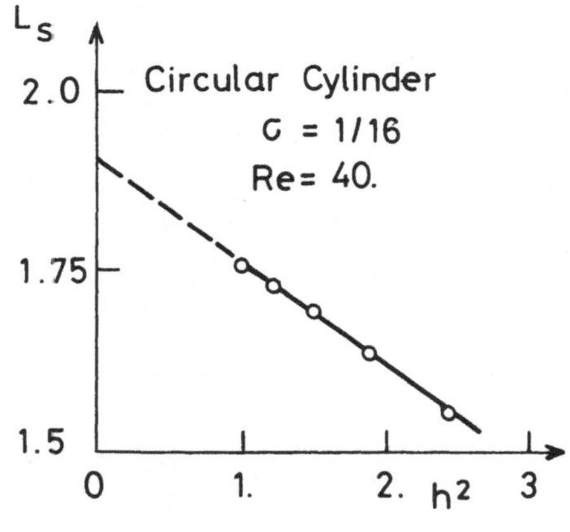

Fig. 3: Convergence of the length of the separation bubble (L_S).

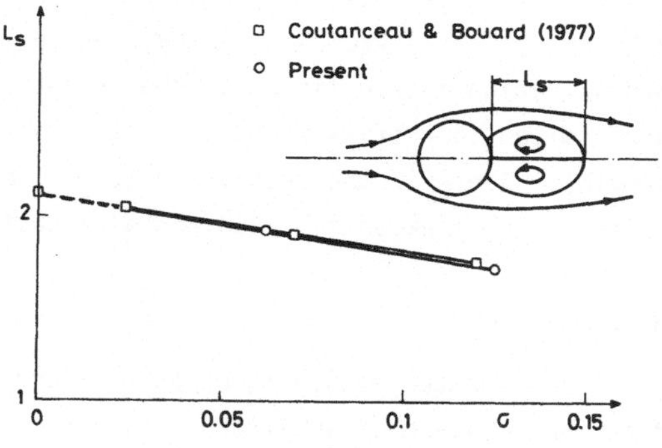

Fig. 4: The influence of the solidity on L_s (Re = 40).

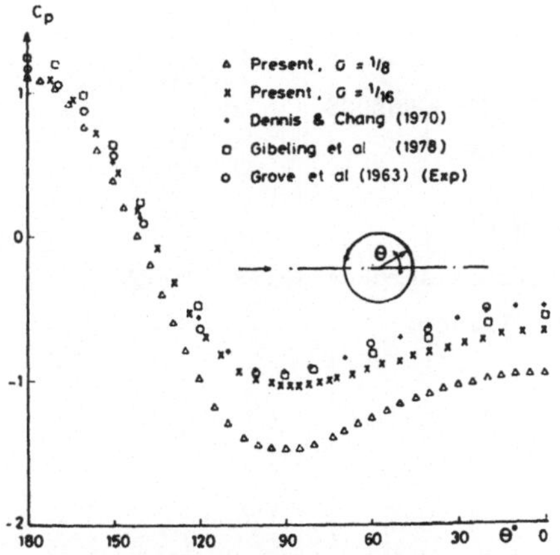

Fig. 5: The pressure coefficient distribution on the cylinder (Re = 40).

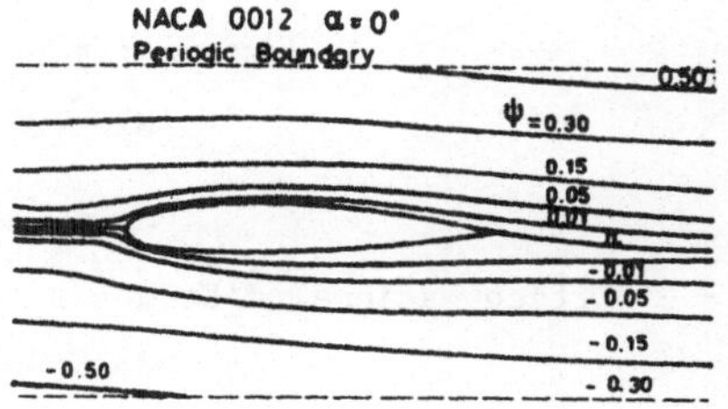

Fig. 6: Non separated streamlines around a symmetric airfoil in a staggered cascade (Re = 1000).

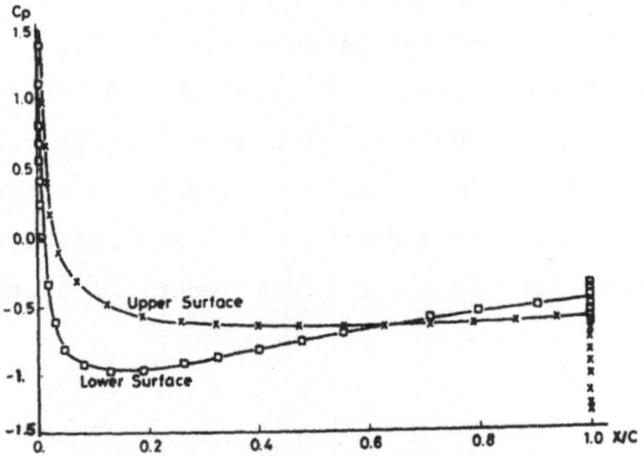

Fig. 7: The pressure coefficient distribution on the airfoil for the non-separated case (Re = 1000).

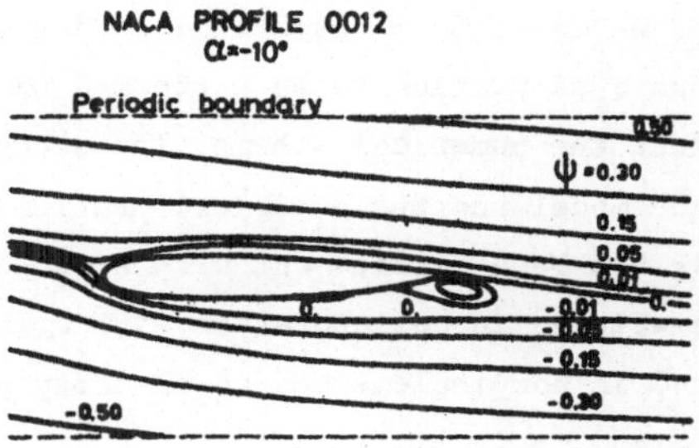

Fig. 8: Separated streamlines around a symmetric airfoil in a staggered cascade (Re = 1000).

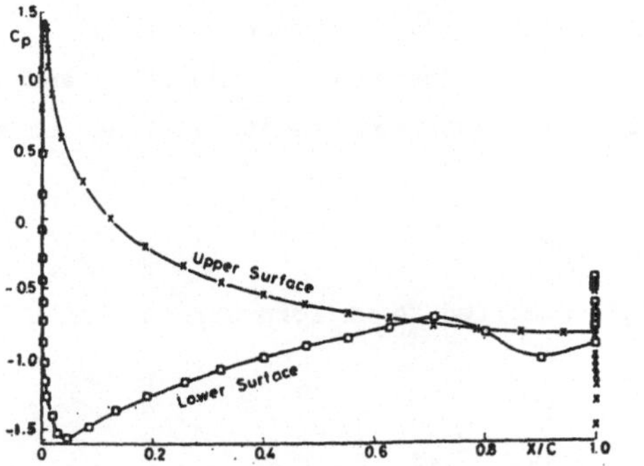

Fig. 9: The pressure coefficient distribution on the airfoil for the separated case (Re = 1000).

NUMERICAL SIMULATION OF HOMOGENEOUS ANISOTROPIC TURBULENCE*

by Philippe ROY
Office National d'Etudes et de Recherches Aérospatiales (ONERA)
92320 Châtillon (France)

ABSTRACT

This research is concerned with results obtained for the simulation of homogeneous turbulence with constant mean velocity gradient. In the first part of the investigation, we describe a formulation deduced from the one proposed by R.S. ROGALLO which allows a simulation to be performed with periodic spectral methods and we also present the numerical scheme. The second and third parts are focused on the study of the model and the numerical problems occuring in these simulations ; the previous study has been carried out with a two-dimensional code. Finally we present a computer code written in collaboration with P. LECA for the three-dimensional simulation. The method is now implemented on an array processor AP 120-b and first numerical simulations will soon be available.

INTRODUCTION

Homogeneous turbulence with constant mean velocity gradient is a step between isotropic homogeneous turbulence and real turbulent flows. It shows the influence of the mean flow deformation upon the turbulent velocity. It has been studied theorically (A. CRAYA [4]) and numerically (R.S. ROGALLO [8] and [9]). Furthermore several experiments have been done to study such flows : for instance S. TAVOULARIS and S. CORRSIN [11] have studied shear flows and J.N. GENCE [5] has studied the succession of two plane strains.

1. MATHEMATICAL FORMULATION AND NUMERICAL SCHEME

We study the flow of an incompressible fluid submitted to uniform stress. Let $\overline{U}$ be the mean flow :

$$U = \overline{U} + V \qquad \overline{U} = A\ (t)\ X$$

$$A\ (t) = (a_{ij}) = (3 \times 3) \text{ matrix} \qquad X = (x,\ y,\ z)$$

$$a_{ii} = 0 \text{ (incompressible fluid)}$$

---

*Work performed with the financial support of Direction des Recherches et Techniques of French Ministry of Defence.

The turbulent velocity V satisfies the following equations :

$$\begin{cases} \frac{\partial V}{\partial t} + (V, \nabla_X) V + (\frac{dA}{dt} + A^2) X + (AX, \nabla_X) V + \nabla_X P = \nu \Delta V \\ \nabla_X V = 0 \end{cases} \quad (1)$$

If $< . >$ denotes the ensemble average, the hypothesis of homogeneity implies :

$$(\frac{dA}{dt} + A^2) X + \nabla < P > = 0$$

So $(\frac{dA}{dt} + A^2)$ must be a symmetrical matrix, i.e. if there is a rotation and if A is constant, the deformation must be in a plane normal to the rotation.

The use of pseudospectral methods with periodic boundary conditions imposes that X must not appear explicitly in the equations. To eliminate the term $(AX, \nabla_X) V$, we perform a Lagrangian change of variables which follows the mean flow : $X = Bx$, where B is solution of the following differential equation :

$$\begin{cases} \frac{dB(t)}{dt} = A(t) B(t) \\ B(0) = B_o \qquad B_o \text{ given} \end{cases}$$

If we set $V = Bv$, the Navier-Stokes equations for v have the following form :

$$\begin{cases} \frac{\partial v}{\partial t} + (v.\nabla_x)v + 2B^{-1}AB\, v + B^{-1}\ {}^tB^{-1}\nabla_x p = \nu({}^tB^{-1}\nabla_x)^2 v \\ \nabla_x . v = 0 \end{cases} \quad (2)$$

where

$$\begin{cases} p = P + f \\ \nabla_x f = (\frac{dA}{dt} + A^2) X \end{cases}$$

($\frac{dA}{dt} + A^2$ is a symmetrical matrix). The equations (2) are solved in space coordinates x using the pseudospectral method introduced by S.A. ORSZAG and G.S. PATTERSON [7]. We study the periodic solutions of (2). For this v is approximated by a truncated discrete Fourier series expansion :

$$u(x, t) = \sum_{k \in \mathbb{K}} \hat{u}(k,t)\, e^{j\, k.x}$$

The derivatives are computed in Fourier space by multiplication, the product of two functions in physical space. The time scheme is a quadrature for the diffusion term and a leap-frog scheme for convection. The evolution of the Fourier coefficients is the following :

$$\left\{\begin{array}{l} \hat{u}_i^{n+1} = -2\,\Delta t\,(j k_1\,\widehat{u_i u_1} + q_{i1}\hat{u}_1^n)\,e^{-\nu\Delta t\, r_{1m} k_1 k_m} \\ \quad - j\, r_{i1} k_1 \hat{p} + \hat{u}_i^{n-1}\quad e^{-2\nu\Delta t\, r_{1m} k_1 k_m} \\ k_i \hat{u}_i^{n+1} = 0 \end{array}\right.$$

with $(q_{i1}) = 2\,(B^{-1}\,AB)^n$ and $(r_{i1}) = (B^{-1t}B^{-1})^n$

The scheme is second order accurate in time. For turbulence simulations, higher order scheme are not used here because they require too many computer resources (see Y. MORCHOISNE [6] and Ph. ROY [10]).

2. MODEL

In three dimensions, a direct simulation of turbulence with 64 points in each direction allows the treatment of flows with $R_\lambda < 100$ (R_λ = Reynolds number based on the Taylor microscale λ). Hence the inertial range is reduced to less than one decade. When flows with higher Reynolds number are simulated, a subgrid scale model must be added.

In 1979, C. BASDEVANT and R. SADOURNY [1] proposed for the simulation of two dimensional turbulence with infinite Reynolds number, a model which consists in replacing the usual dissipation $\nu\Delta u$ by a "super" dissipation $\nu_T\ \Delta^8 u$. The advantage is that this dissipation is much more local in physical and spectral spaces than the usual one. This allows the treatment of a higher number of energetic scales. The power of the Laplacian operator depends on the dimension : it is lower for three dimensional flows.

In strain or shear flow, the ratios between production, velocity-pressure strain and dissipation are important for the study of the Reynolds-stress budget ; therefore it is necessary to keep the usual dissipation term but also to add a "super" dissipation to treat flows with higher Reynolds number.

3. NUMERICAL PROBLEM

The use of a Lagrangian method leads to numerical difficulties. The equations are solved in a cubic reference domain (x coordinates) but the physical domain (X coordinates) is stretched in some directions, shrunk in others and the basis vectors of the domain form angles that do not stay right. The consequence is that the numerical scheme does not achieve the same accuracy in all directions (i.e. the computed turbulent scales are not identical). As for some Lagrangian method, the velocity v must be projected upon a less distorted mesh when the ratio of the scales in two directions is too large.

For instance if we consider a two dimensional shear flow simulation with an initial condition defined on an orthogonal grid, the mesh becomes skewed and the collocation points of the oblique mesh (fig. 1) coincide with the point of the initial mesh (using periodicity) after a time t_o such that $st_o = 1$ (where s is the shear rate) ; in fact this happens each time the value of st is integer. Figure 2 shows that the spectral modes of the two grids are identical in the non hatched area. Hence no aliasing problem appears if the modes of the hatched zones are nullified during the remeshing, but at least one quarter of the modes are lost, corresponding to little scales of turbulence (dissipation range).

During a simulation, the mesh becomes more and more skewed. The simulation is stopped when the mesh is too distorted and the scheme is not enough accurate in the stretched direction. A remeshing is then performed to recover some accuracy and the simulation goes on. In order to estimate the error due to grid distorsion, two kinds of numerical experiments have been carried out : in the first one there is no mean flow and the mesh stays fixed during the whole simulation ; the second one is the shear case and the grid is distorted. These two experiments are performed with a two-dimensional code using the model described in part 2. The number of points is 128 x 128 (64 x 128 complex modes).

a) isotropic turbulence (fig. 3)

Using the same initial condition corresponding to the spectrum S1, two simulations have been made, one using the orthogonal grid M1, the other the skewed grid M2. After 4000 time steps, the two experiments are decorrelated but the stastical properties (Reynolds stress, spectra...) are similar. This is due to the fact that the small scales of the two meshes are different. The more M2 is oblique, the faster is the decorrelation between (M1 ; S1) and (M2 ; S1), and soon the scales present in M2 are too large to treat the energetic scales of the simulation (M1 ; S1) : the simulation is no longer reliable.

If we consider now two simulations carried out on the same grid M1 but starting from initial conditions differing only in the small scales (the spectra are S1 and S2), these simulations will also be decorrelated after 4000 time steps, and the decorrelation velocity between (M1 ; S1) and (M1 ; S2) is similar to the preceding one.

b) shear flow (fig. 4)

Three simulations starting from the same initial condition have been done with a total shear st = 8.

In the simulation R1 (resp. R2, R3) the mesh is distorted during t_o such that $s.t_o=1$ (resp. $st_o = 2$, $st_o = 4$) before remeshing upon the initial orthogonal mesh and all this processus is repeated 8 times (resp. 4, 2).

In the simulation R3, the stretched structures of R1 and R2 do not exist any longer. They correspond to scales not present in the very distorted mesh. In fact the case R2 is the upper distorsion allowed for a statistically correct simulation.
From the two experiments a) and b), we can deduce an empirical rule : no significative change of the turbulence statistical properties is produced as long as the

ratio between the smallest and largest dimensions of the mesh is less than 2.

4. THREE DIMENSIONAL SIMULATIONS

In order to perform high resolution simulations (64 x 64 x 64 points corresponding to 32 x 64 x 64 complex modes) with a low cost , we have implemented the previous method on a computer system (fig. 5) including two array processors AP 120 b and a host (SEL 32/77). This computer system is faster than a CDC 7600.

With only one array processor running, the first version of this code is now operating. The data structure has two levels. It corresponds to the "plane by plane" technique described by C. BASDEVANT [2]. A three-dimensional scalar array is divided in blocks following the directions 2 and 3. (fig. 6). Here, to compute the FFT in the direction 2, 4 blocks must be gathered following this direction (hatched blocks). There is a numeration for the blocks (first level) and one inside each block (second level). The computer time is 2 mn 30 s for each time step.

The new version using the whole system will soon be available. The major improvements are :

- the number of Input and Output is reduced by a better implementation
- the basic time of each Input and Output is reduced by the use of a cache memory (1.5M words)
- the two array processors are used in a multi processor configuration.

The computer time will be around 70 s by time step. Furthermore the data structure has now three levels : a block is divided in subblocks which are dispatched to the processors.

At present time only experiments in decaying homogeneous isotropic turbulence have been made. We present in fig. 7 a three-dimensional specturm corresponding to $R_\lambda = 60$. No model has been used. Here all the production scales have not been treated in order to obtain a long inertial range. (slope -5/3). Near the cut-off, the energy increases because all the dissipation scales are not treated. This phenomena will disappear with the use of the model.

5. CONCLUSION

The two dimensional studies have shown the difficulties raised by the use of the Lagrangian method, and that the subgrid scale model is efficient to treat shear or strain flows.

For three dimensional experiments, we have a tool which allows us to do up to thirty simulations a year. The first planed simulations are the experiments performed by J.N. GENCE [5] in the case of two successive plane strain and the superimposition of a plane strain and a rotation with variable ratio.

ACKNOWLEDGEMENTS

We gratefully acknowledge the very efficient help of P. LECA [3] in developing the code on the parallel structure system. We thank C. BASDEVANT, K. DANG, Y. MORCHOISNE, R. PEYRET and R. SADOURNY for useful discussions.

REFERENCES

[1] BASDEVANT, C., LEGRAS, B. and SADOURNY, R., BELAND, M.: J. ATmos. Sci. Vol. 38.11 (1981).

[2] BASDEVANT, C.: Submitted to J. of Comp. Phys. (1982).

[3] BOISSEAU, J.P., ENSELME, A., GUIRAUD, D., and LECA, P.: La Recherche Aérospatiale n° 1982-1.

[4] CRAYA, A.:P.S.T. n° 345 (1958).

[5] GENCE, J.N.:Thèse d'Etat, Faculté Claude Bernard Lyon (1979).

[6] MORCHOISNE, Y. : AIAA 19th Aerospace Sciences Meeting, St Louis, AIAA Paper n° 81-0109 (1981).

[7] ORSZAG, S.A. and PATTERSON, G.S.:Phys. Rev. Letters, Vol. 28. 2 (1972).

[8] ROGALLO, R.S.: NASA TM-23, 203 (1977).

[9] ROGALLO, R.S.: NASA TM 813 15 (1981).

[10] ROY, Ph.: La Recherche Aérospatiale n° 1980-6.

[11] TAVOULARIS, S., and CORRSIN, S.: J. Fluid Mech. Vol. 104 (1981).

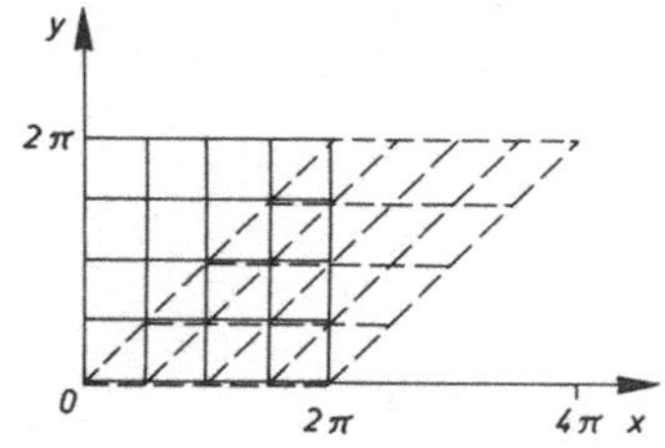

Fig. 1 : Remeshing in physical space

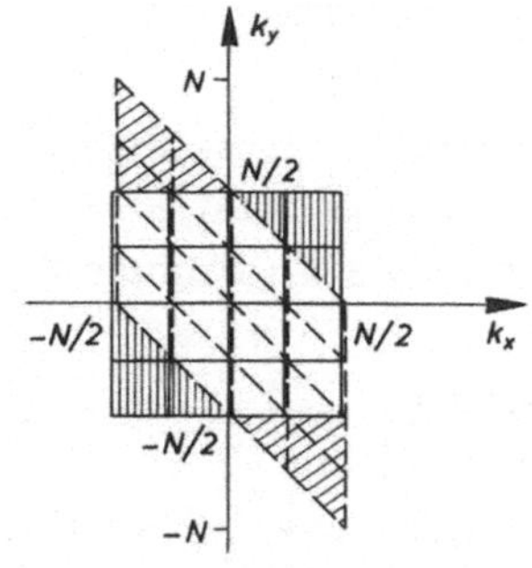

Fig. 2 : Remeshing in spectral space

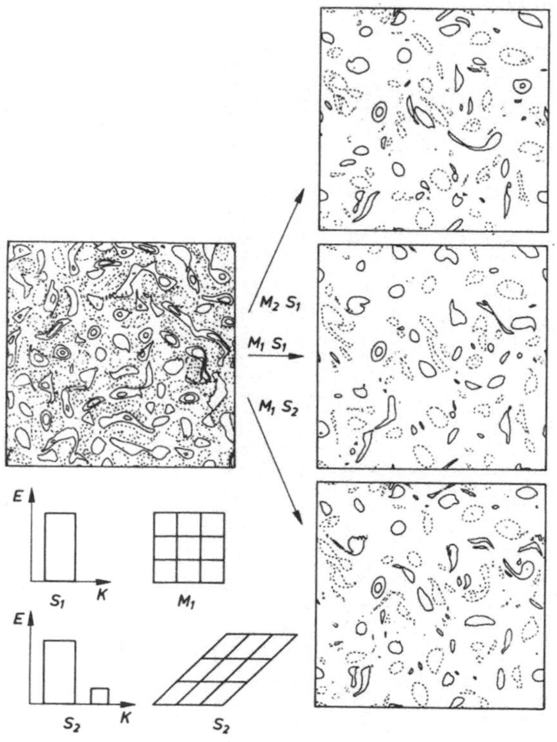

Fig. 3 : Decorrelation for isotropic turbulence. Isovorticity lines after 4000 time steps

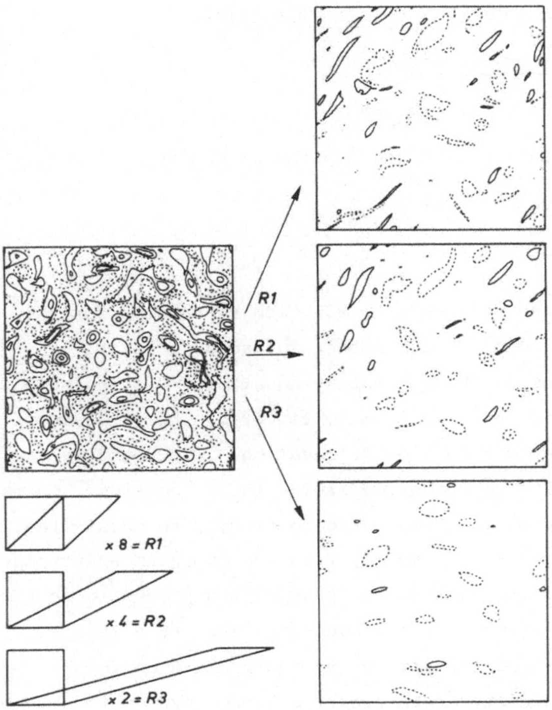

Fig. 4 : Decorrelation for shear turbulence. Isovorticity lines after a total shear st = 8

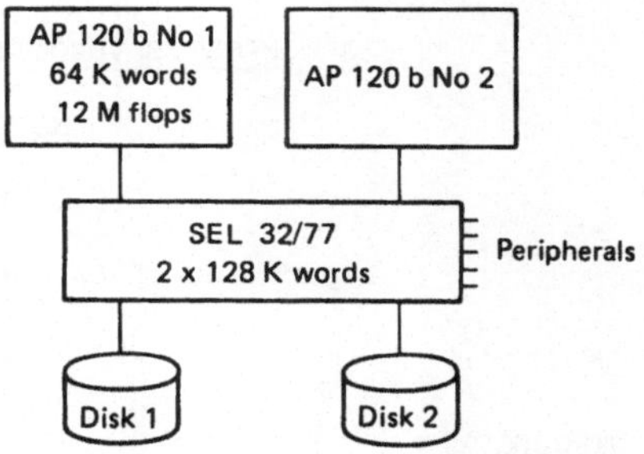

Fig. 5 : Parallel structure system

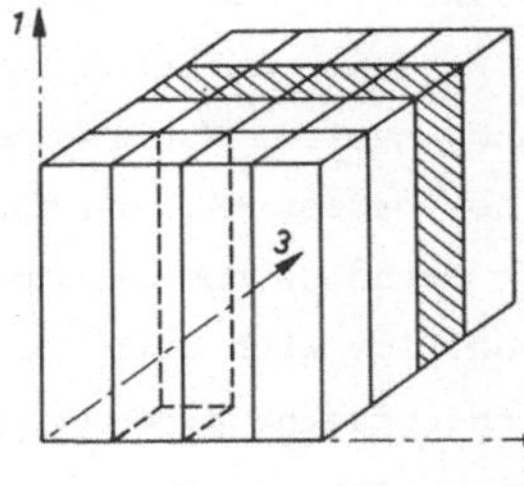

Fig. 6 : Data structure

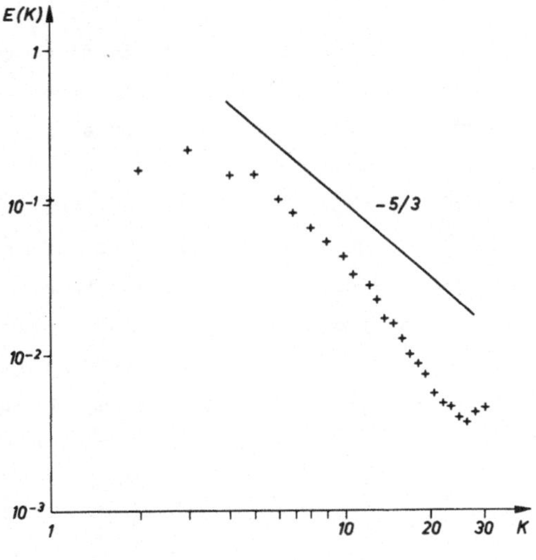

Fig. 7 : Three-dimensional energy spectrum

A COMPOSITE VELOCITY PROCEDURE FOR THE INCOMPRESSIBLE NAVIER-STOKES EQUATIONS

S.G. Rubin and P.K. Khosla

Department of Aerospace Engineering and Applied Mechanics
University of Cincinnati
Cincinnati, Ohio 45221 U.S.A.

INTRODUCTION

One of the major differences between steady-state solution techniques for the Navier-Stokes equations and solution procedures for either inviscid potential or boundary layer problems is the treatment of the continuity equation. For inviscid flow, a potential function is determined entirely from continuity. The pressure is then obtained from the integrated momentum or Bernoulli equation. For boundary layers, the axial momentum and continuity equations determine the velocities. On the other hand, typical Navier-Stokes solvers, in effect, use continuity to obtain the density (or pressure), and the velocities result solely from the momentum equations. For large Reynolds number steady flows, it would appear that such a procedure is in marked conflict with both the asymptotic inviscid and boundary layer theories.

In the present paper, a boundary layer-relaxation procedure based on a new composite-velocity formulation for the incompressible Navier-Stokes system is described. The equations are interpreted and numerically approximated to reflect the composite nature of the flow. The procedure has also been developed independently for subsonic flow [1]. Unlike typical Navier-Stokes procedures that differ significantly from their incompressible flow counterparts, the present developments are essentially identical in both cases. Moreover, the extension to transonic flows should be direct.

The equations are written in a body-fitted orthogonal coordinate system so that arbitrary geometries can be treated. Application to internal and external flows are discussed. Specific geometries include a boattail simulator, the trailing edge of a plate, Joukowski airfoils and a curved channel.

In its final form, the present formulation has some features similar to the velocity-split technique due to Dodge [2]; however, this resemblance is only superficial. In the present analysis, a composite representation of inviscid and viscous region velocities is prescribed in the spirit of matched asymptotic expansions. The complete Navier-Stokes equations are solved. No simplifying approximations are required. The finite-difference form of the resulting equations are solved by a coupled strongly implicit procedure described previously by the authors [3].

COMPOSITE FORMULATION

This formulation is designed for the calculation of large Reynolds number problems with a dominant flow direction, e.g. the ξ-direction. The gradients are largest in surface normal (η) direction. The flow outside the thin viscous region is essentially inviscid and is represented by a potential function ϕ; therefore, the following composite representation of the velocity field to reflect the matched asymptotic boundary layer-inviscid behavior is prescribed.

$$u = \frac{U}{h_1}(1 + \phi_\xi) = u_e U \quad , \qquad v = \frac{1}{h_2}\ \phi_\xi \tag{1}$$

Substitution of these expressions into the Navier-Stokes equations results, after some reorganization, in the following orthogonal system for ϕ, U, G:

$$[h_3\ U(1 + \phi_\xi)]_\xi + (h_3\phi_\eta)_\eta = 0 \qquad \text{(continuity)} \tag{2}$$

$$\frac{\partial u}{\partial t} + \frac{1}{h_1 h_2 h_3}\{[h_2 h_3 u_e^2\ (U^2-U)]_\xi + [h_3 h_1 u_e v(U-1)]_\eta\} + \frac{h_{1_\eta}}{h_1 h_2} u_e v(U-1) + \frac{u_e}{h_1}(U-1)\ u_{e_\xi}$$

$$= -\frac{1}{h_1} G_\xi - \frac{1}{h_1}\bar{G}'(\xi) + \text{viscous terms} \qquad (\xi\text{-momentum}) \tag{3}$$

$$G_\eta = -(U-1)[(\frac{u_e^2}{2})_\eta - \frac{h_{1_\eta}}{h_1} u_e^2\ U] + \text{viscous terms} \qquad (\eta\text{-momentum}) \tag{4}$$

For internal flows $\bar{G}'(\xi)$ is determined from the global conservation of mass. In these equations

$$G = \frac{p}{\rho} + \frac{u_e^2 + v^2}{2} - \bar{G}(\xi) \tag{5}$$

and is similar to the Bernoulli or total pressure for the inviscid region. However, G is not assumed constant, but is determined by the calculation procedure. In an inviscid irrotational region, $U \to 1$ and continuity (2) reduces to the well known potential flow equation; the momentum equations (3, 4) are identically satisfied with G = constant and $\bar{G} = 0$. For internal flows, G = 0 on one wall and $\bar{G}$ is determined by mass conservation. In the viscous region, the ξ-momentum equation determines U, while ϕ and, therefore, the pressure is obtained from the η-momentum equation. The two basic regions pertinent to large Reynolds number flows are appropriately described by this composite set of equations. This method of defining v with a 'potential' was first tested for the flat plate boundary-layer (U,ϕ) equations. The solution of the resulting 2-point boundary value problem reproduced the results obtained with standard methods based on the velocities u and v. It should be pointed out that the present system of equations can be used for the solution of inviscid flows provided that Re = ∞ and U = 1 is enforced at the solid surface. G = 0 decouples the normal momentum equation from the axial momentum and continuity equations and represents the interacting boundary-layer approximation if axial diffusion is neglected. If, in addition, ϕ_ξ and $\phi_{\xi\xi}$ in equations (2) and (3) are replaced with their respective potential flow values, the usual boundary-layer approximation is recovered.

For the calculations of separated flows it is essential that U and ϕ_ξ are coupled in order to eliminate the pressure singularity. G which represents the Bernoulli constant in the inviscid flow can be treated explicitly. Finally, the η-momentum determines G. The necessary coupling between U and ϕ is attained with a coupled strongly implicit procedure for the solution of the algebraic system.

BOUNDARY CONDITIONS

In the present formulation, the physical boundary conditions are represented directly by the following mathematical boundary conditions.

At a solid surface: $U = 0$; $\phi_\eta = 0$, and as $\eta \to \infty$, $\phi \to 0$, $U \to 1$.

For a finite body: ϕ or $\phi_\xi \to 0$ as $\xi \to \pm\infty$, $U(\xi,\eta) \to 1$ as $\xi \to -\infty$,

and $U_{\xi\xi}(\xi,\eta) \to 0$ as $\xi \to \infty$.

For bodies which are infinite in both directions, the inflow and outflow conditions are somewhat ambiguous. However, these conditions must be appropriately specified in order to obtain a meaningful solution. For the boattail geometry ϕ and U are prescribed at the inflow, while $U_{\xi\xi} \to 0$ and $\phi_\xi \to 0$ have been applied as outflow conditions. The ϕ_ξ condition, in the context of the composite formulation, is such as to eliminate the viscous-inviscid interaction. It should be noted that these boundary conditions are consistent with the mathematical character of the equations governing ϕ and U. For example no slip has been satisfied through the boundary-layer like variable U and not through ϕ_ξ.

SOLUTION PROCEDURE

The governing equations have been discretized using second-order accurate central-differencing for all ϕ derivatives. Central or boundary layer-like differencing has been used for U derivatives, except for the $(h_3U)_\xi$ term in the continuity equation; this is backward differenced throughout. The resulting implicit algebraic system of equations has been solved iteratively using a coupled strongly implicit procedure (CSIP). The continuity and ξ-momentum equations for ϕ and U are solved in a coupled fashion, while the η-momentum equation for G is evaluated iteratively. In the ξ-momentum equation, G is treated as known during the iterations. Although G, the "inviscid" total pressure is evaluated explicitly from the η-momentum equation, the static pressure is unknown and depends upon the values of ϕ_ξ and ϕ_η. Since ϕ is evaluated implicitly in the coupled algorithm, this implies that the pressure is also treated implicitly. This circumvents the separation singularity. Explicit artificial viscosity is not required for convergence and the CSIP allows for arbitrarily large values of Δt once the effects of the inviscid initial conditions have been sufficiently smoothed.

COUPLED 2 x 2 SOLUTION ALGORITHM

In an earlier paper, reference [3], the present authors have developed a coupled strongly implicit procedure for the stream function-vorticity form of the Navier-Stokes equations. This algorithm has the distinct advantage of being implicit in both the ξ and η directions, as well as allowing for the coupling of all the boundary conditions. It is this coupling, that eliminates the pressure singularity in the (U, ϕ) formulation. Furthermore, the method is unconditionally stable, allows for arbitrarily large Δt, converges faster than SOR, LSOR, ADI, etc., and is relatively insensitive to grid aspect ratio. The discretized version of the equations for (U, ϕ) can be written as:

$$(A + P)\, V^{n+1} = G + PV^{n}, \quad \text{where V is the } (U, \phi) \text{ solution vector.}$$

P is chosen such that (A+P) can be decomposed into a lower and upper triangular form having a sparsity pattern similar to the original matrix A. This leads to a solution algorithm of the following form:

$$\begin{bmatrix} U_{ij} \\ \phi_{ij} \end{bmatrix}^{n+1} = \begin{bmatrix} GM_{1_{ij}} \\ GM_{2_{ij}} \end{bmatrix}^{n} + \begin{bmatrix} T_{1_{ij}} & T_{5_{ij}} \\ T_{3_{ij}} & T_{7_{ij}} \end{bmatrix}^{n} \begin{bmatrix} U_{i,j-1} \\ \phi_{i,j-1} \end{bmatrix}^{n+1} + \begin{bmatrix} T_{2_{ij}} & T_{4_{ij}} \\ T_{6_{ij}} & T_{8_{ij}} \end{bmatrix}^{n} \begin{bmatrix} U_{i-1,j} \\ \phi_{i-1,j} \end{bmatrix}^{n+1}$$

where n is the iteration index. Although the coupling accelerates the rate of convergence, it also increases the storage requirement by a factor of two. Considerable savings in storage can be realized by re-evaluating some of the coefficients $(T_{i,j})$ during the evaluation of U and ϕ. However, this is achieved at moderate additional computational cost. In its present form the CSIP is slightly different from the one given in reference [3]. In the present case the forward and backward sweeps have been reversed in order to impart a certain degree of marching consistent with boundary layer procedures. As detailed in reference [3], the appropriate recurrence relationships can easily be obtained.

RESULTS

Laminar flow solutions have been obtained for boattail simulator, Joukowski airfoil, finite plate and channel geometries. Reynolds numbers based on typical length scales ranged from 10^3 to 10^5. All the computations were started with arbitrary initial conditions. For the first 40 iterations $\Delta t = 1$, afterwhich Δt was increased to 10^6. Explicit artificial viscosity was not required for what were effectively steady-state calculations.

(a) Boattail Simulator: A typical streamline plot for an axisymmetric boattail geometry is shown in figure 1. The corner angle is approximately 30° and the body radius varies between 1 and 0.5. With 1800 grid points (60 x 30, 12 on the boattail), and Re = 7500, convergence to 10^{-3} is achieved in about 125 iterations. This takes about 10 minutes on the Amdahl 470/V6 or less than 1 minute on the Cray-1 computer.

(b) Joukowski Airfoil: A variety of airfoil thicknesses and Reynolds numbers have been considered. For low Reynolds numbers and small thickness ratio, the flow is unseparated. As either is increased, separation regions appear. Typical streamline pattern and separation location for $t/c = 0.12$, 0.17 and $Re = 10^3$ to 10^4 are shown in fig. 2. As expected, the separation point moves upstream with increasing Reynolds number. This correlates well with solutions obtained with 2nd order boundary-layer theory [4]. Small changes in thickness ratio can lead to large variations on the recirculation region. The solutions are oscillatory for large t/c, Re.

(c) Finite Flat Plate: The flow past the trailing edge of a finite flat plate has been extensively investigated by triple-deck interacting boundary-layer global relaxation procedures [5]. In figure 3 the results of the present formulation are compared with some earlier computations. A 105 x 75 grid has been adequate to provide reasonable agreement.

(d) Internal Flow: Two dimensional straight and curved channels have been investigated by the present technique. The curved channel was generated by using two streamlines of the boattail geometry. A variable ξ and uniform-η (60 x 30) grid was specified for this calculation. Uniform inflow conditions are prescribed. The entrance mass flow rate was 0.4066 and the Reynolds number based on the entrance channel width was Re = 2500. Global mass conservation was insured with the parameter $\bar{G}'(\xi)$. Typical wall pressure distributions and velocity profiles are shown in figure 4.

ACKNOWLEDGEMENT

This research was supported by the Air Force Office of Scientific Research under Grant No. AFOSR 80-0047.

REFERENCES

1. Khosla, P.K. and Rubin, S.G. (1982), AIAA Paper No. 82-0099.
2. Dodge, P.R. and Lieber, L.S. (1977), AIAA Paper No. 77-170.
3. Rubin, S.G. and Khosla, P.K. (1979), Comp. Fluids, Vol. 9, 2, p. 163.
4. Grossman, B. and Rubin, S.G., (1971), ZAMP, Vol. 22, 1, pp. 109-130.
5. Rubin, S.G., (1982), Von Karman Institute Lecture Notes, April 1982.

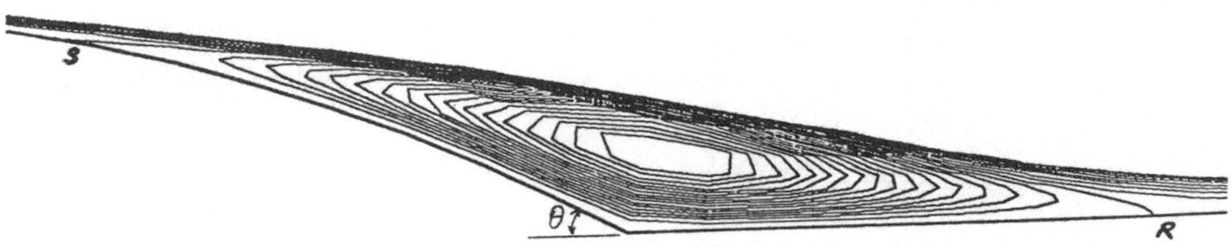

Fig. 1. Streamline Contours for Boattail, $\theta = 30°$, Re = 7500, 60 x 30 Grid.

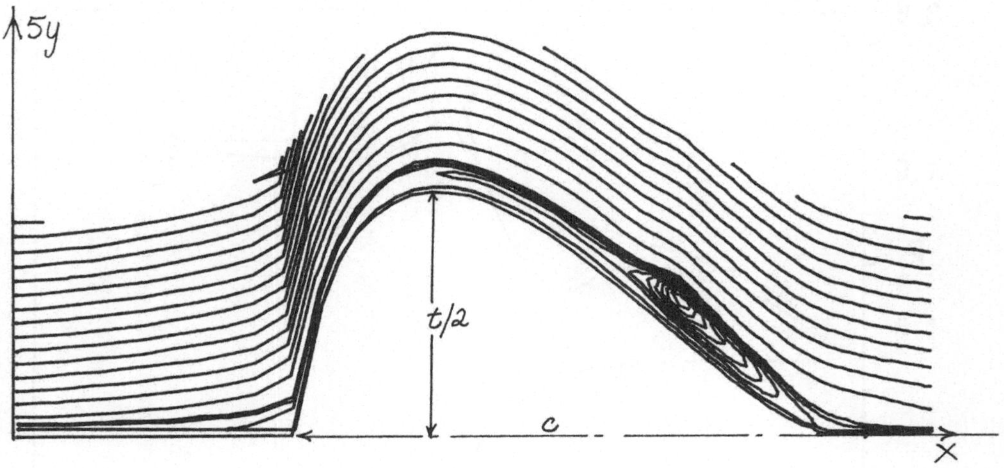

Fig. 2a. Streamline Contours for Joukowski Airfoil: Re = 10,000, t/c = 0.12.

Re / t/c	1,000	5,000	7,500	10,000	10^5
0.12	No Separation	No Separation	0.85	0.80	Unsteady
0.17	No Separation	0.47	Unsteady	Unsteady	--

Fig. 2b. Separation Location for Several Values of t/c, Re.

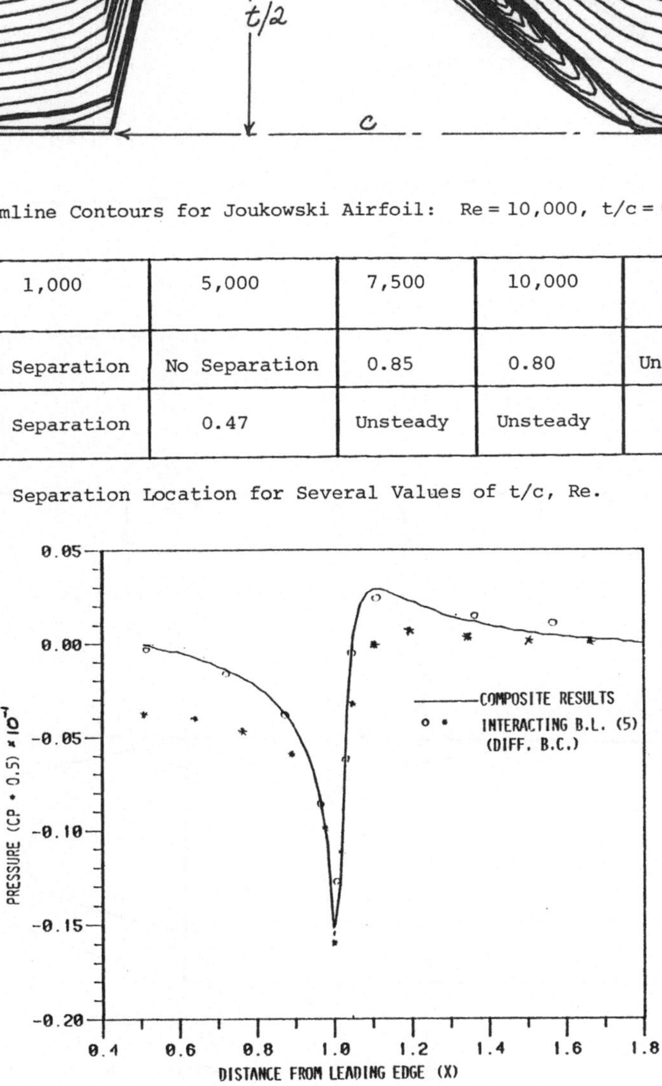

Fig. 3. Pressure Distribution Near Trailing Edge of Flat Plate: Re = 10^5.

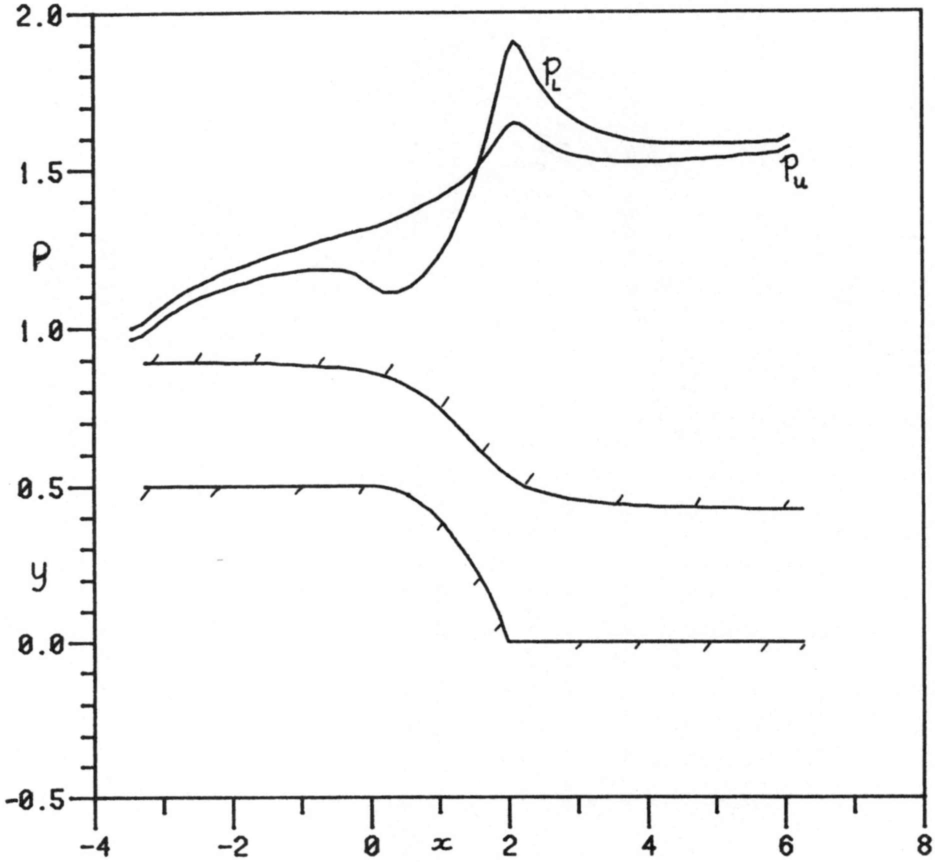

Fig. 4a. Wall Pressures and Channel Geometry.

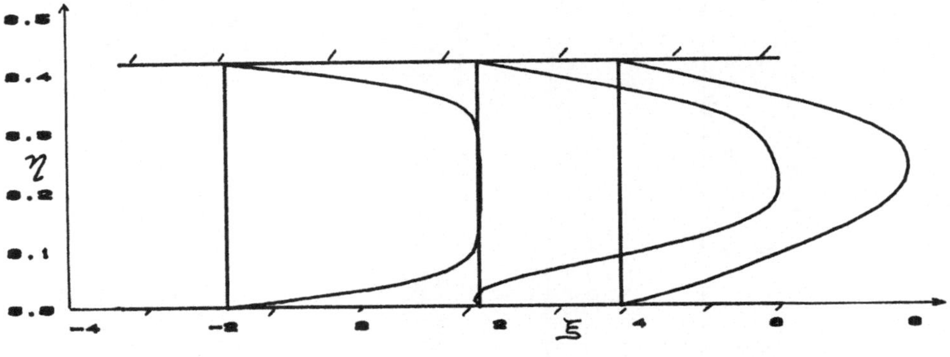

Fig. 4b. Velocity Profiles in Channel-Transformed Plane.

ON CALCULATION ACCURACY IN GAS FLOW PROBLEMS WITH SHOCK TYPE DISCONTINUITIES

V.V. Rusanov, I.V. Besmenov, E.I. Nazhestkina
Keldysh Institute of Applied Mathematics
USSR Academy of Sciences
Moscow I25047, USSR

1. The quality of shock computations by shock-fitting schemes is usually estimated by computing a shock moving with a constant speed against uniform flow. In such a case the plots of all gasdynamics functions are steps moving with the shock velocity. If the scheme is homogeneous and conservative then instead of a step it generates a shock profile which moves as a whole with the same velocity [1, 2].

One of the main factors which influence the form of the shock profile is the order of approximation achieved by the difference scheme on the smooth solutions. Although the real order of approximation of any difference scheme is zero at the discontinuity the difference between the schemes of various orders is distinctly seen in the structure of the "smeared" profile near the discontinuity. On the other hand, for stepwise discontinuity all the difference schemes have an infinite order of approximation in the region where the functions are constant. The computational error arising in this region is due only to the influence of the discontinuity. The distribution of this error was not studied thoroughly probably because the error quickly decreases and becomes unobservable in the usual graphic representation of the profile. Actually the error may vary depending on the scheme by several orders of magnitude. Because of that it is expedient to plot not the functions themselves but logarithms of absolute difference between computed and exact values.

Let $\delta_{f,m} = \lg|\Delta f_m|$, $\Delta f_m = \hat{f}_m - f(x_m)$, where $f(x_m)$ and $\hat{f}_m$ are, respectively, an exact and computed values of the function $f(x)$. If $f(x)$ is discontinuous at the point then $f(\bar{x}) = \{f(\bar{x}+0) + f(\bar{x}-0)\}/2$.

Example 1. The distributions of u_m and $\delta_{u,m}$ in the stationary shock profile are shown in Fig. 1 for the equation

$$(1) \quad u_t + \left[F(u)\right]_x = 0; \quad F(u) = (u + 0.038925)^2/2$$

and shock with parameters

$u_- = 2.593467$ (left), $u_+ = 1.393784$ (right), $D = 2.032550$

The computation was made by the first order difference scheme S_1:

(2) $$u_m^{n+1} = u_m^n - \varkappa\{F(u_{m+1}^n) - F(u_{m-1}^n)\}/2 + \omega\{u_{m+1}^n - 2u_m^n + u_{m-1}^n\}/2$$

where $\varkappa = \Delta t/\Delta x = 0.163997$, $\omega = 0.4$, $x_m = m\Delta x$.

The point $x=0$ coincides with the exact position of the shock $x=\bar{x}(t)$ at the given time. Linear behavior of $\delta_{u,m}$ in the vicinity of $x=0$ shows that the error near the discontinuity has the value

$$C_\pm \exp\{\alpha_\pm |x-\bar{x}|\} = C_\pm \exp\{\hat{\alpha}_\pm |m|\}; \quad \hat{\alpha}_- \approx -1.15, \ \hat{\alpha}_+ \approx -0.580$$

In Fig. 2 u_m and $\delta_{u,m}$ are plotted for the same shock calculated by the third order scheme S_3 given in [3]. The error dependence of $|x-\bar{x}|$ is also exponential with some perturbations due to non-monotonicity. With an increasing distance from the discontinuity the δ_u for both schemes tends to its limiting values depending only on the accuracy of u_-, u_+ and on the round-off error in computations.

2. The error distribution for a system of equations differs from that for a single equation. It is due to the fact that in the case of the system there are characteristics, coming from the shock into the region behind the shock. Along these characteristics the perturbations arising at the shock may propagate. The possibility of this phenomenon was mentioned in [4].

Example 2. Consider the system

(3) $$\rho_t + (\rho u)_x = 0, \quad (\rho u)_t + (\rho^3/3 + \rho u^2)_x = 0$$

which describes the isoenthropic gas flow at k=3. In the continuity region by introducing Riemann invariants $r = u+\rho$, $s = u-\rho$ it splits into two independent equations

(4) $$r_t + (r^2/2)_x = 0; \qquad s_t + (s^2/2)_x = 0$$

Let us consider for system (3) the shock with the parameters $r_-=2.593467$, $s_-=1.080082$, $r_+=1.393784$, $s_+=-1.095049$, $D=2.032550$. The functions δ_r, δ_s are plotted for the stationary shock profiles generated by the schemes S_1 (Fig. 3) and S_3 (Fig. 4). In both cases $\varkappa=0.163997$. It is easy to see that far away from the discontinuity both schemes yield the errors $|\Delta r|$ which on the average are by 1 to 1.5 order of magnitude less than $|\Delta s|$. The errors $|\Delta r|$ are of the same order for the both schemes and $|\Delta s|$ is a little less for the scheme S_3 than for S_1. Thus in the region behind the shock the error arising at the discontinuity has practically the same effect in the schemes S_1 and S_3. If the quality of the scheme is estimated only from the calculation of "a step", the scheme S_1 proves to be even more preferable due to its monotonicity.

3. To provide the monotonicity of the discontinuity profile as well as the high accuracy of the computation in the smooth region hybrid schemes were proposed by some authors [4, 5]. The idea is to use the scheme that has the first order near the discontinuity and the higher order in the smooth region.

Example 3. The computations of the same shock as in Exp. 2 were made by two hybrid schemes of the form $\alpha S_3 + (1-\alpha) S_1$, where α =1 in the smooth region and α =0 at the discontinuity point. Near the discontinuity $0 < \alpha < 1$.

The results of computations are given in Fig. 5 for the scheme proposed in [4], $\alpha = \alpha_H$ and in Fig. 6 for the scheme proposed in [5], $\alpha = \alpha_{HZ}$. It is seen that at $\alpha = \alpha_{HZ}$ the δ_s^{ν} -distribution is more regular. The profiles of z in the hybrid schemes prove to be monotonic and more compressed than in the scheme S_1. Far away from the discontinuity the errors $|\Delta z|$ and $|\Delta s|$ are about of the same order as for the schemes S_1 and S_3.

Thus, in the stationary stepwise shock profile generated by the schemes considered above the errors differ essentially only near the discontinuity. At the distance I0-I5 points from the discontinuity the errors are about of the same order for all the schemes. At the longer distances the errors $|\Delta z|$, $|\Delta s|$ tend to their limiting values which depend not only on the round-off errors but on the jump of invariant s, i.e. $|s_+ - s_-|$. In comparison to $|s_+ - s_-|$ the limiting value of $|\Delta s|$ is about by five orders less.

4. In practical computations the functions at the both sides of the discontinuity usually are not constant. To the error due to the discontinuity influence the truncation error depending on the approximation order is added. To investigate this numerically an exact solution more complex than the stepwise one is needed, namely with the nonconstant values at the both sides of the discontinuity. Such a test based on system (3) was developed in |6|. The r-type shock forms at the time t=1 and travels to the right. The δ_z^{ν}, δ_s^{ν} in these test computations by various schemes are given below for t=I.25 in Figs.7-9.

Example 4. Figure 7 shows the values δ_z^{ν} and δ_s^{ν} for the numerical solution computed by the scheme S_1. The errors for the schemes S_3 and $\alpha_{HZ} S_3 + (1-\alpha_{HZ}) S_3$ are given, respectively, in Figs. 8 and 9. The horizontal lines at the discontinuity point mark the values of $\lg|r_+ - r_-|$, $\lg|s_+ - s_-|$ and the exact position of the shock is at m =0. Figs.7-8 show that $|\Delta r|$ is much less for S_3 than for S_1. In the both cases the error $|\Delta s|$ is larger than $|\Delta r|$ and for the scheme S_1 its decrease is so slow that even far away from the discontinuity it has the same order as the "s" jump at the discontinuity. From Fig. 9 one can see that the distribution of $|\Delta r|$ in the hybrid scheme is similar to that in S_3 when moving away from the discontinuity, and the r-profile is monotonic near the discontinuity. As for $|\Delta s|$, it grows at first with an increasing distance from the discontinuity and becomes by an order of magnitude larger than in the scheme S_3, after which it diminishes.

The most undesirable fact in all the schemes is a slow decrease of $|\Delta s|$ when moving away from the discontinuity. It seems that for all schemes the $|\Delta s|$ value arising near the discontinuity proves to be close to $|s_+ - s_-|$ (it is less than $|s_+ - s_-|$ only by an order of magnitude in the scheme S_3). Since it is impossible to prevent the propagation of $|\Delta s|$ along a characteristic the problem is to maximally reduce the $|\Delta s|$ value with respect to $|s_+ - s_-|$ near the discontinuity. It is difficult to say whether the solution of such a problem is realistic.

References

1. V.V. Rusanov. Non-Linear Analysis of Shock Profile in Difference Schemes. Proc. Second Int. Conf. on Num. Meth. in Fluid Dynamics, In: Lecture Notes in Physics, I970, vol. 8, p.270-278.

2. V.V. Rusanov, I.V. Besmenov. Asymptotics of the discrete discontinuity profile in the shock calculations by difference schemes. DAN SSSR, I981 , v.26I, N 4, p.8I7-820 (in Russian).

3. V.V. Rusanov. Third-order difference schemes for the shock-fitting computations. DAN SSSR, I968, v.I80, N 6, p.1303-1305 (in Russian).

4. A.Harten. The Artificial Compression Method for Computation of Shocks and Contact Discontinuities: III. Self-Adjusting Hybrid Schemes.Math. of Comp. I978, vol. 32, N I42, pp.363-389.

5. A.Harten and G.Zwas. Switched Numerical Shuman Filters for Shock Calculations. Journ. of Eng. Math. I972, vol. 6, N 2, p.207-216.

6. V.V. Rusanov. A Test Case for Checking Computational Methods for Gas Flows with Discontinuities. In: Notes on Numerical Fluid Mechanics. Vol. I. Boundary Algorithms for Multidimensional Inviscid Hyperbolic Flows, Vieweg Vlg., I978, p.100-125.

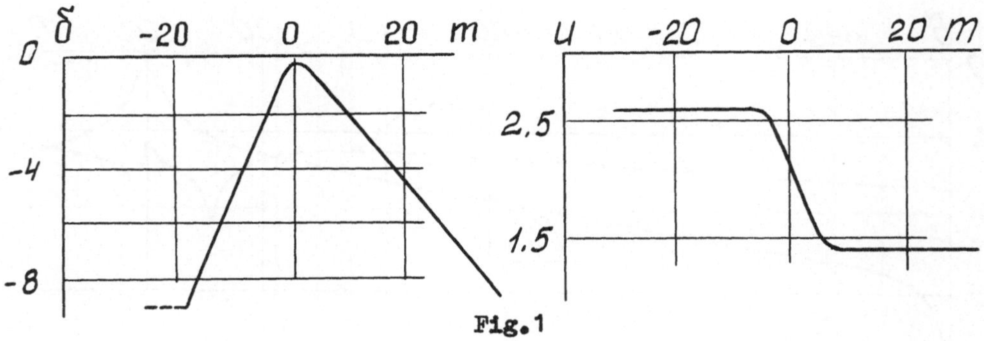

Fig.1

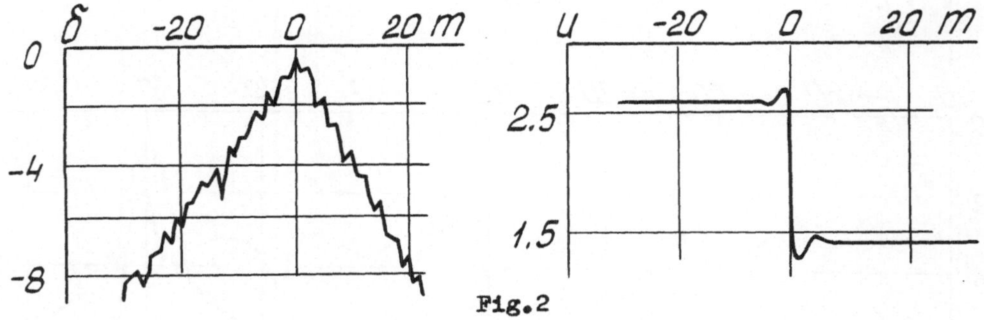

Fig.2

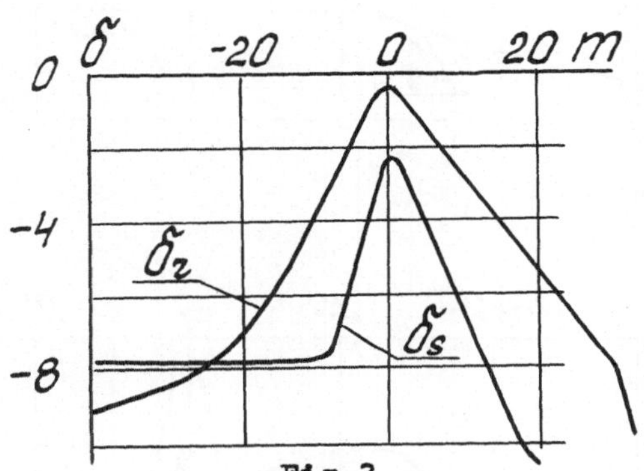

Fig.3

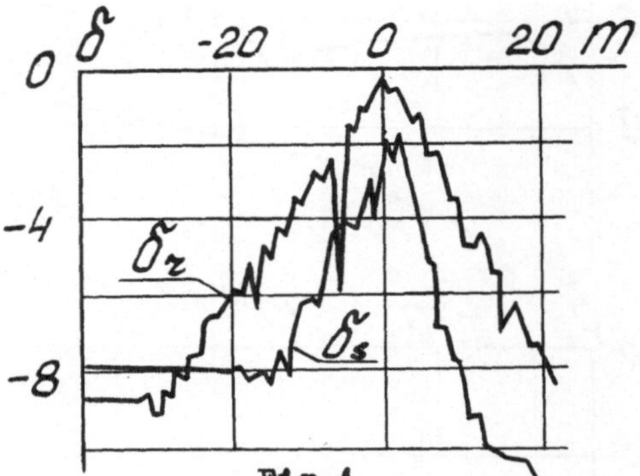

Fig.4

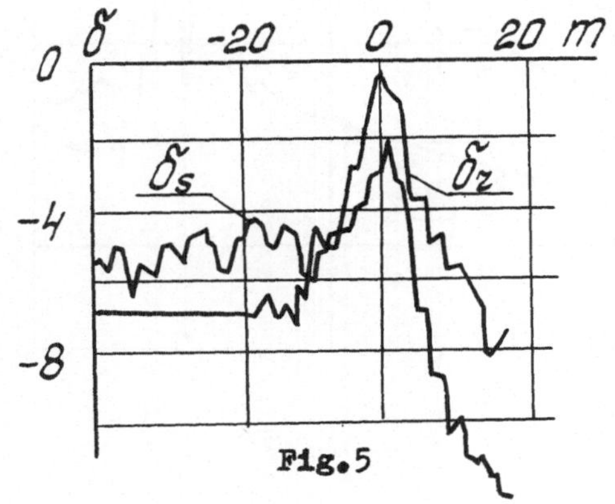

Fig.5

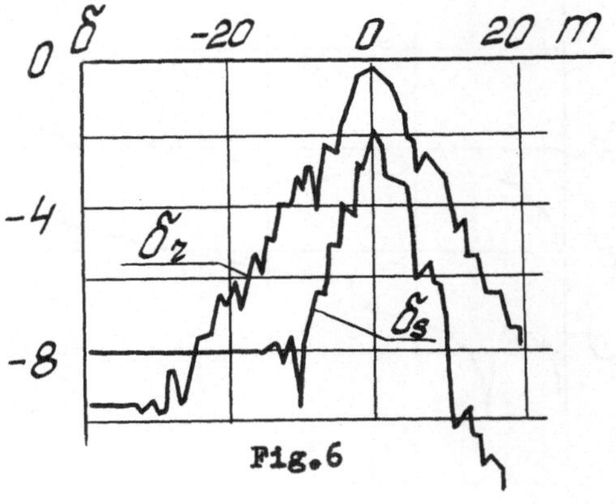

Fig.6

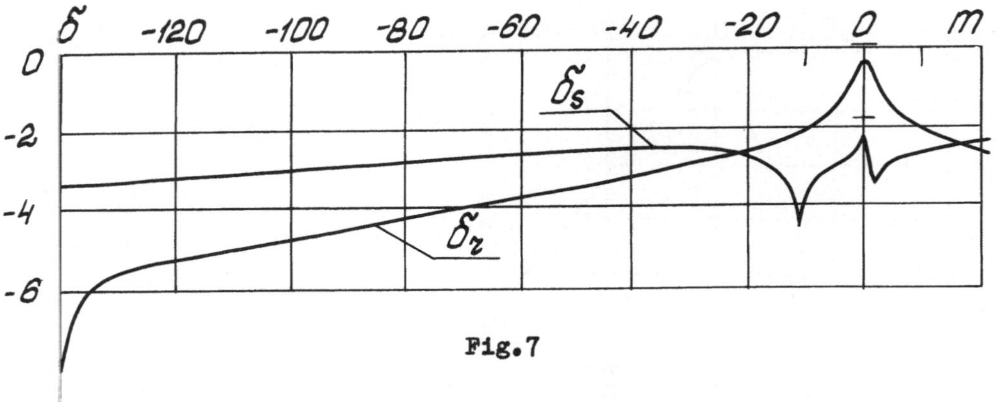

Fig.7

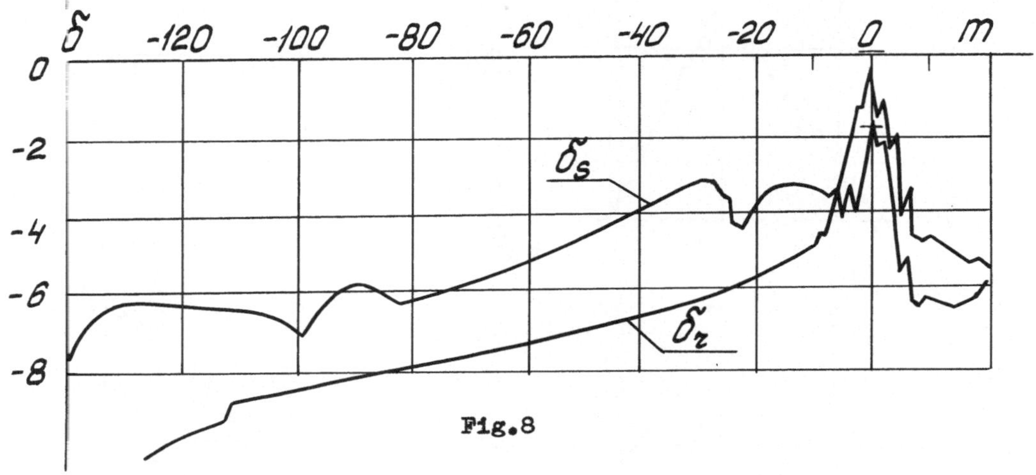

Fig.8

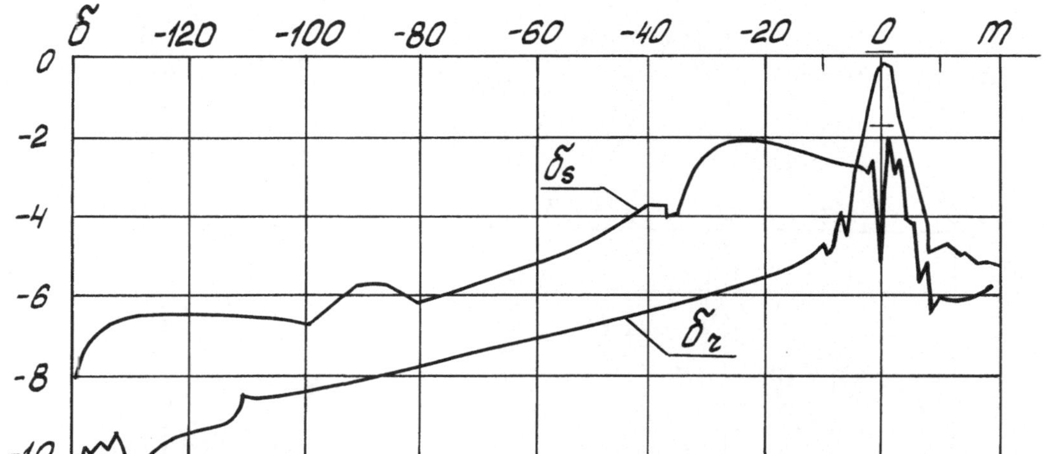

Fig.9

SHOCK-FITTED EULER SOLUTIONS TO SHOCK-VORTEX INTERACTIONS

Manuel D. Salas
NASA, Langley Research Center, Hampton, VA USA

Thomas A. Zang
College of William and Mary, Williamsburg, VA USA

M. Yousuff Hussaini
Institute for Computer Applications in Science and Engineering, Hampton, VA USA

In their recent paper, Pao and Salas [1] presented a finite difference solution to the two-dimensional Euler equations governing the phenomena of shock wave interaction with an isolated vortex. Their study emphasized the acoustic aspects of the problem. Zang, Hussaini, and Bushnell [2] extended this numerical approach to the problem of turbulence amplification in shock wave interactions. In this work it was necessary to resolve rather complex fine-scale structure in order to draw meaningful conclusions about the transient processes that dominate turbulence amplification. The present study is a continuation of these efforts in two directions. First, to develop a highly accurate pseudo-spectral method capable of resolving the crucial small-scale structure on a relatively coarse grid. Second, to gain insight into the nonlinear dynamics of the transient processes involved in the passage of a shock wave over a single vortex, a vortex street and a hot spot.

Spectral methods have been demonstrated [3], [4] to be powerful alternatives to finite difference methods for the numerical solution of smooth flows. Recently, the work of Gottlieb, Lustman, and Orszag [5] and of Zang and Hussaini [6] have shown their applicability to simple one-dimensional compressible flows with shocks. The present paper discusses a Chebyshev pseudo-spectral method that has produced reliable, accurate and efficient solutions to complex, two-dimensional flows with a strong shock. Although the main emphasis is on the spectral technique, solutions to the governing equations obtained by the well known, second-order finite difference method originated by MacCormack are also given. This is the method that was used in [1] and [2]. The finite difference results for the present set of problems were calculated on a very fine grid and are used here for comparison with the solutions obtained with the spectral method.

Statement of the Problem

The physical problem that we model corresponds to an infinite, initially planar normal shock wave moving from left to right into a downstream region containing a flow field representative of one or more vortices, or a hot spot. In order to model the interaction of the shock wave with some given flow field ahead of it, it is only necessary to compute the flow field upstream of the shock. The physical domain, therefore, need consist only of the region between some left boundary $h(t)$, judiciously chosen such that it will be far from the interaction region, and the shock wave front itself $x_s(y,t)$. It is mapped onto the computational domain by the transformation,

$$X = \frac{x - h(t)}{x_s(y,t) - h(t)}, \qquad Y = \frac{\tanh(\alpha y) + 1}{2}, \qquad T = t.$$

The computational domain is thus $(X,Y) \in [0,1] \times [0,1]$. Note the stretching that has been used to handle the infinite extent of the lateral coordinate y. If the relative shock Mach number M_s is sufficiently high ($M_s > 2.08$), the flow upstream of the shock remains supersonic. In this case, the left boundary corresponds to a supersonic inflow, and all dependent variables can be prescribed on it. However, if the relative shock Mach number is low, then radiation-type boundary conditions are used at the left boundary. On the right, the computational region is bounded by the shock wave. Downstream of the shock the flow field is given analytically. The flow field immediately upstream of the shock, as well as the shape and velocity of the shock, are evaluated such that the Rankine-Hugoniot jump conditions and the compatibility condition reaching the shock wave from the upstream side are simultaneously satisfied.

The unsteady, two-dimensional, compressible, Euler equations in the computational plane are written in the form,

$$Q_T + \underline{A}Q_X + \underline{B}Q_Y = 0$$

where $Q = [P,u,v,S]^T$ and

$$\underline{A} = \begin{bmatrix} U & \gamma X_x & \gamma X_y & 0 \\ a^2 X_x/\gamma & U & 0 & 0 \\ a^2 X_y/\gamma & 0 & U & 0 \\ 0 & 0 & 0 & U \end{bmatrix} \qquad \underline{B} = \begin{bmatrix} V & \gamma Y_x & \gamma Y_y & 0 \\ a^2 Y_x/\gamma & V & 0 & 0 \\ a^2 Y_y/\gamma & 0 & V & 0 \\ 0 & 0 & 0 & V \end{bmatrix}.$$

The natural logarithm of the pressure, the speed of sound, and the entropy are represented by P, a, and S, respectively, and γ is the ratio of specific heats. The velocity in the x and y directions are u and v, respectively. All variables are normalized with respect to reference conditions at downstream infinity, as in [1]. The contravariant velocity components are defined by

$$U = X_t + uX_x + vX_y \quad \text{and} \quad V = Y_t + uY_x + vY_y.$$

Subscripts denote partial derivatives with respect to the independent variables.

Solution Techniques

Let k denote the time level and let Δt be the time step increment. The time discretization of eq. (12) is then as follows:

$$\tilde{Q} = [1 - \Delta t\, L^k] Q^k, \qquad Q^{k+1} = \tfrac{1}{2}\,[Q^k + (1 - \Delta t \tilde{L})\tilde{Q}],$$

where the spatial operator L represents an approximation to $\underline{A}\partial_X + \underline{B}\partial_Y$. In the finite difference MacCormack method, the operators L^k and $\tilde{L}$ are evaluated as two points forward and two points backward differences in the predictor (left) and corrector (right) levels, respectively. In the pseudo-spectral method studied here, the solution Q is first expanded as a double Chebyshev series,

$$Q(X,Y,T) = \sum_{p=0}^{M} \sum_{q=0}^{N} Q_{pq}(T)\tau_p(\xi)\tau_q(\eta), \quad \text{where} \quad \xi = 2X - 1 \quad , \eta = 2Y - 1,$$

and τ_p and τ_q are the Chebyshev polynomials of degrees p and q. The derivatives appearing in the spatial operators are then evaluated as

$$Q_X = 2 \sum_{p=0}^{M} \sum_{q=0}^{N} Q_{pq}^{(1,0)} \tau_p \tau_q,$$

where

$$Q_{pq}^{(1,0)} = \frac{2}{c_p} \sum_{\substack{m=p+1 \\ m+p \text{ odd}}}^{M} m Q_{mq}, \quad \text{and} \quad c_0 = 2 \quad , c_p = 1, \ p > 0.$$

The Q_Y derivative is evaluated in a similar fashion.

The evaluation of the shock wave shape and velocity followed the same procedure described in [1], except that in the spectral formulation, the derivatives that must be evaluated on the upstream side of the shock are expressed as Chebyshev expansions. At the left boundary, all variables were specified for supersonic inflow. For the case of subsonic inflow, the two velocity components and the entropy were specified, while the pressure was computed from a quasi-one-dimensional characteristic as described in [8].

The pseudo-spectral method has a tendency to develop slowly growing oscillations. Because of the global nature of this method they are spread over the entire flow field rather than being confined to the vicinity of sharp gradients. The underlying smooth solution can be recovered by a variety of filtering techniques. The results presented here were obtained by applying a von Hann window filter (see [6] for details) every 160 time steps. Another practical consideration is the explicit time-step restriction. The Chebyshev collocation points are clustered near the boundaries. Thus, smaller time-steps must be used in the pseudo-spectral calculations.

Results

Perhaps the simplest interaction to consider is that of a planar shock wave with a hot spot, as shown in Fig. 1. The flow field downstream of the shock wave situated at $x=0$ at $t=0$ is taken as a quiescent field whose temperature distribution σ, is given by

$$\sigma = \frac{\kappa}{2\pi} \exp\left(-((x-x_0)^2 + (y-y_0)^2)/2r^2\right),$$

where for this case $\kappa = 0.25$, $r = .125$, $x_0 = 0.5$, $y_0 = 0.0$, and $M_s = 3$ at $t = 0$. The vorticity distributions obtained by the finite difference method and by the spectral method after the shock wave has passed over the hot spot are shown in Fig. 3. The finite difference solution presented here, and in all cases that follow, was obtained with 75 mesh points in the X direction and 50 in Y. The spectral solutions were all obtained with 33 collocation points in the X direction and 17 in Y. There is very little difference between the two solutions. Both show the two counter-rotating vortices upstream of the shock, which is typical of this interaction. See [2] for more details on the physics.

Figure 2 shows the velocity field for a single vortex about to interact with a shock wave of the same initial strength as in the previous case. The downstream conditions here are obtained by assuming a constant density field, calculating the velocity from the stream function,

$$\phi = \frac{\kappa}{2\pi} \log \sqrt{r^2 + (x-x_0)^2 + (y-y_0)^2}\ ,$$

the pressure from Bernoulli's relation, and the temperature from the equation of state. For the case shown in Fig. 2, the circulation $\kappa=2$ and the softening scale $r=0.1$. This model approaches an idealized incompressible point vortex at large distances but is much smoother near the center. Figure 4 shows the resulting pressure field after the shock wave has passed over the vortex. Overall, the results are qualitatively very similar, although the spectral method seems to be resolving the pressure field more accurately. See [1] for more details on the physics.

Finally, Figs. 5 and 6 show the results for the interaction with the Karman vortex street that simulates the conditions of the experiment reported in [7]. For this case, the stream function representing the vortex is given by the difference of ϕ_+ and ϕ_- where

$$\phi_\pm = \frac{\kappa}{2\pi} \log\left[\cosh\left(\frac{2\pi}{c}\sqrt{r^2 + (y \pm 1/2\, b)^2}\right) - \cos\left(\frac{2\pi}{c}(x \pm 1/2\, c)\right)\right].$$

To match the experiment, the circulation, core radius, shock Mach number and vortex separation parameters were determined as $\kappa = 0.186$, $r = 0.1$, $M_s = 1.3$, $c = .33$, and $b = .048$. For this calculation, the inflow Mach number was subsonic and radiation boundary conditions were applied at the left boundary. The results shown in Fig. 6 are in agreement with the experimentally observed [7] longitudinal compression and lateral elongation of the vortex field after passage through the shock. The finite difference results are noticeably smoother than the spectral ones. However, it is well known that the idealized Karman vortex street is unstable for all but one special ratio of the horizontal and vertical separations. The downstream flow has the stable ratio. Thus, it is likely that the upstream flow is physically unstable. It may be that the spectral results have captured this phenomena, but more computations are needed to settle the issue.

References

[1] Pao, S. P. and Salas, M. D.: A Numerical Study of Two-Dimensional Shock Interaction. AIAA Paper 81-1205. Presented at the AIAA 14th Fluids and Plasma Dynamics Conference, June 23-25, 1981, Palo Alto, CA.

[2] Zang, T. A.; Hussaini, M. Y.; and Bushnell, D. M.: Numerical Computations of Turbulence Amplification in Shock Wave Interactions. AIAA Paper No. 82-0293. Presented at the AIAA 20th Aerospace Sciences Meeting, January 11-13, 1982, Orlando, FL.

[3] Orszag, S. A. and Kells, L. C.: Transition to Turbulence in Plain Poiseuille Flow and Plain Couette Flow, J. Fluid Mech., Vol. 96, 1980, pp. 159-205.

[4] Wray, A. and Hussaini, M. Y.: Numerical Experiments in Boundary-Layer Stability. AIAA Paper No. 80-0275. Presented at the AIAA 18th Aerospace Sciences Meeting, January 14-16, 1980, Pasadena, CA.

[5] Gottlieb, D.; Lustman, L.; and Orszag, S.: Spectral Calculations of One-Dimensional Inviscid Compressible Flows, SIAM J. Sci. Statis. Comput., Vol. 2, 1981, pp. 296-310.

[6] Zang, T. A. and Hussaini, M. Y.: Mixed Spectral-Finite Difference Approximations for Slightly Viscous Flows, Proc. of the 7th Intl. Conf. on Numerical Methods in Fluid Dynamics, Springer-Verlag, 1981, pp. 461-466.

[7] Dosanjh, D. S. and Weeks, T. M.: Interaction of a Starting Vortex as well as a Vortex Street with a Traveling Shock Wave, AIAA J., Vol. 13, 1965, pp. 216-223.

[8] Oliger, J. and Sundstrom, A.: Theoretical and Practical Aspects of Some Initial Boundary Value Problems in Fluid Dynamics, SIAM J. Appl. Math., Vol. 35, 1978, pp. 419-446.

The authors are happy to acknowledge fruitful discussions with D. M. Bushnell. Research of T. A. Zang was supported by NASA Grant No. NAG1-109. Research of M. Y. Hussaini was partially supported by NASA Contracts No. NAS1-16394 and NAS1-15810 while in residence at ICASE, NASA Langley Research Center, Hampton, VA 23665.

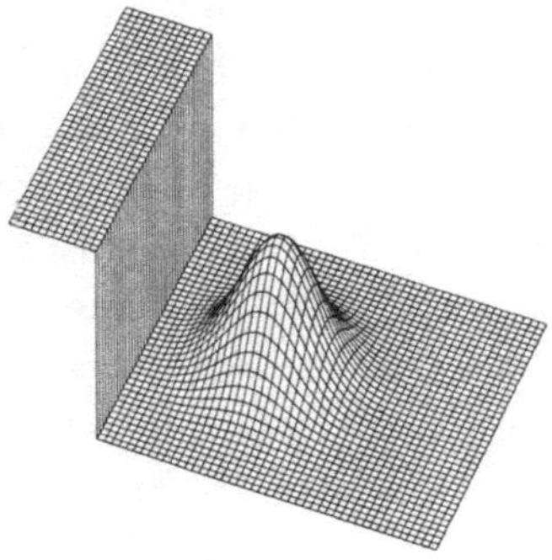

Fig. 1: Surface plot of entropy distribution for a hot spot about to interact with a Mach 3 shock wave.

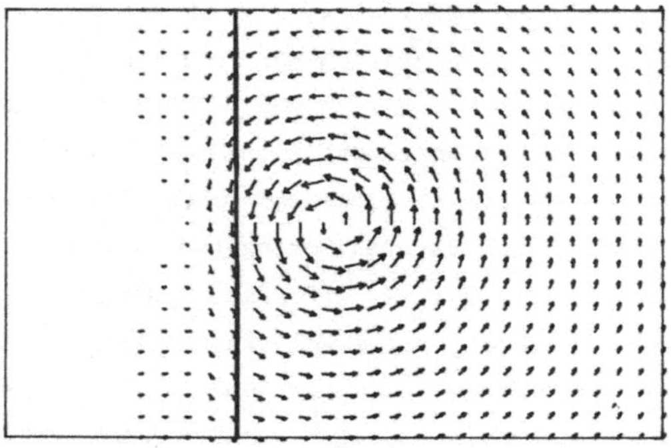

Fig. 2: Velocity field of a single vortex about to interact with a Mach 3 shock wave (solid curve). The velocity vectors represent perturbation from the mean values.

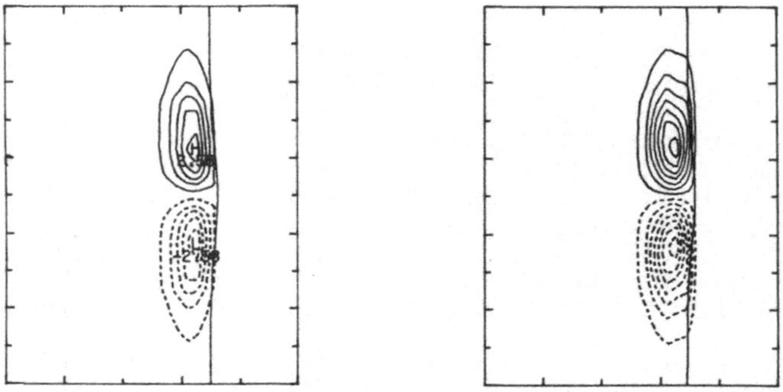

Fig. 3: Vorticity fields at t = 0.20 computed by pseudo-spectral (left) and finite difference (right) methods for a hot spot after interaction with a Mach 3 shock wave (solid curves).

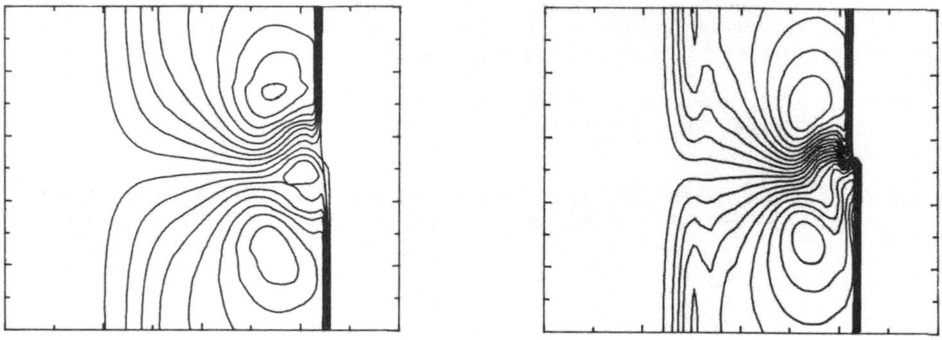

Fig. 4: Isobars at t = 0.20 computed by pseudo-spectral (left) and finite difference (right) methods for a single vortex after interaction with a Mach 3 shock wave.

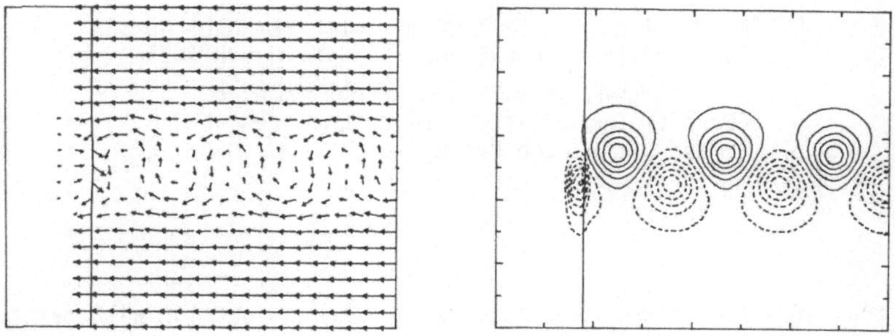

Fig. 5: Velocity field (left) and vorticity contours (right) for a Karman vortex street about to interact with a Mach 1.3 shock wave (solid curves). The velocity vectors represent perturbations from mean values. Negative contour levels are are indicated by dashed lines.

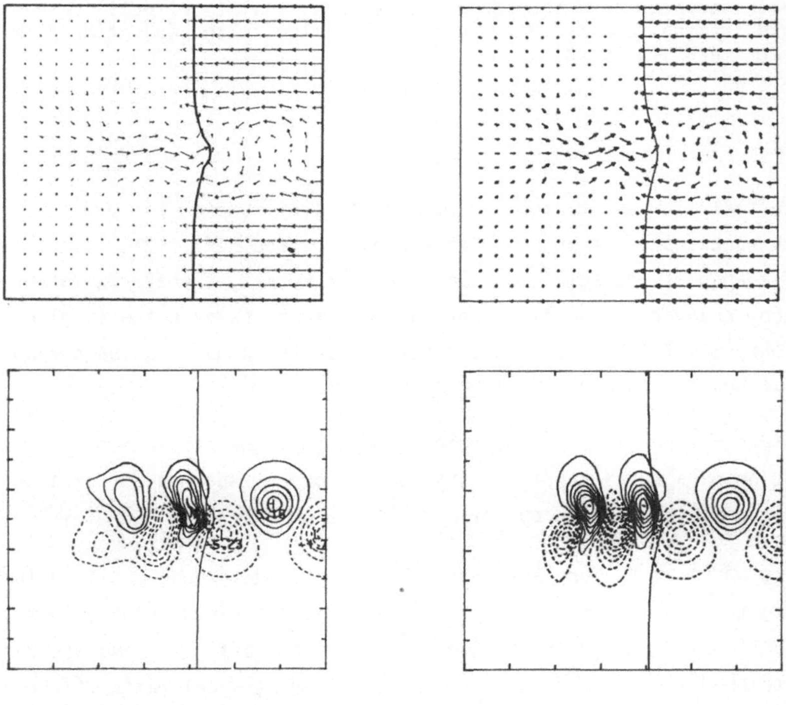

Fig. 6: Velocity fields (top) and vorticity contours (bottom) at t = 0.36 computed by pseudo-spectral (left) and finite difference (right) methods for a Karman vortex street after interaction with a Mach 1.3 shock wave (solid curves). The velocity vectors represent perturbations from mean values. Negative contour levels are indicated by dashed lines.

EULER SOLUTIONS AS LIMIT OF INFINITE REYNOLDS NUMBER FOR SEPARATION FLOWS AND FLOWS WITH VORTICES

Wolfgang Schmidt and Antony Jameson
Dornier GmbH, D-7990 Friedrichshafen, FRG
and Princeton University, N. J., USA

Abstract

A combination of a finite volume discretisation in conjunction with carefully designed dissipative terms of third order, and a fourth order Runge Kutta time stepping scheme, is shown to yield an efficient and accurate method for solving the time-dependent Euler equations in arbitrary geometric domains. Convergence to the steady state has been accelerated by the use of different techniques described briefly. The main attempt of the present paper however is the demonstration of inviscid compressible flow computations as solutions to the full time dependent Euler equations over two- and three-dimensional configurations with separation. It is clearly shown that in inviscid flow separation can occur on sharp corners as well as on smooth surfaces as a consequence of compressibility effects. Results for non-lifting and lifting two- and three-dimensional flows with separation from round and sharp corners are presented.

Introduction

While potential flow solutions have proved extremely useful for predicting transonic flows with shock waves of moderate strength (e.g. see Ref. 1-3), typical of cruising flight of transport and some class of fighter aircraft, the approximation of ignoring entropy changes and vorticity production cannot be expected to give acceptable accuracy when the flight speed is increased into the upper transonic range or the angle of attack is reaching the manoeuvre limit.

However, for lifting flows in potential flow theory the assumption inherent to the specification of a Kutta condition are more important than the error in pressure rise across a shock (pointed out by Lock[4]).

In the first part of the present paper the recently developed finite volume method[5] for solving the time dependent Euler equations is described briefly. Detailed analysis in Ref. 5 - 7 has proven the method to be very efficient and accurate for two- and three-dimensional transonic flows. Different acceleration techniques have been analysed to improve the efficiency further.

Previously reported results on the cylinder in compressible inviscid flow [6,8] indicated that inviscid compressible flow can have solutions with a separated flow region on smooth surfaces if a shock produces vorticity and total pressure losses.

The studies in Ref. 6 - 7 presented results for two- and three-dimensional lifting flows which proved that no explicit Kutta condition is needed to get unique lifting Euler solutions.

The main part of the present paper shows results with no Kutta condition needed at round airfoil trailing edges, inviscid separation at sharp corners, and inviscid transonic flows for slender transonic wing-body configurations with leading edge vortices. These results confirm first results obtained by the authors in Oct. 1981 and by Rizzi[10].

Euler Equation Method

The numerical method used to solve the time-dependent Euler equations is described in detail in Ref. 5. The version used for all cases discussed in the present paper is the unsplit four stage two level scheme with enthalpy forcing and local time stepping. A blend of second and fourth order differences is used to construct dissipative terms of a filter type. Additional acceleration techniques have been studied as reported on Ref. 7, but will not be discussed here.

The far field boundary conditions have been modified with respect to Ref. 5, but are constructed to be non-reflecting. All solid surfaces have no-flux boundary conditions, the wall pressure being extrapolated from the field. No special treatment is given to any wakes or vortices emanating from leading or trailing edges.

Mesh Generation

Two- and three-dimensional contour-conformal grids are constructed using standard O- or C-type procedures as reported in a comparison paper[11]. In three-dimensional flow a mesh refinement technique has been adapted using submeshes of the actual fine mesh to accelerate convergence.

Results

The efficiency and the accuracy of the Euler solver have been confirmed by many numerical experiments. Results have been obtained on IBM and CRAY-1 computers. Some typical results are presented here. Since lifting flows for airfoils with sharp trailing edge have been reported in Ref. 6,7 we will show only one example for lifting two-dimensional flow, with a round trailing edge. Fig. 1 presents streamlines and isobars for such a transonic airfoil computed for an 128 x 32 O-type mesh. Without any explicit condition the trailing edge stagnation point is resolved.

Separation on a smooth surface for the circular cylinder has been presented in Ref. 6,8. Similar results on the upper surface of a supercritical compressor cascade blade have been obtained by Haase[9]. Fig. 2 presents results for a rearward facing step at M = 0.5. The mesh is especially constructed to resolve the region behind the step accurately. The velocity vector plot as well as the streamlines nicely show the recirculating results for this inviscid flow computation. The mechanism for this final result without any supersonic point in the converged solution is similar to the one for the cylinder or the trailing edge flow. Due to compressibility any flow around the corner would produce a shock such that the only possible solution in the converged steady state is the one with the flow leaving the corner. Similar results can be obtained for a cavity in compressible inviscid time-dependent computations solving the Euler equations.

Three-dimensional lifting results have been obtained for a variety of wing-body combinations including ONERA M6 and DFVLR-F4. Fig. 3 presents some of the results obtained for a slender wing-body combination with leading edge vortex flow at subsonic and transonic speed with round and sharp leading edges. Compared with the wind tunnel results [12] both Mach number and leading edge type effects are nicely predicted by the computational method. Velocity vector plots nicely show the vortex position and roll-up behind the wing. A first analysis proved that the interaction of the trailing edge wake and leading edge vortex is nicely predicted within the capturing capabilities of the 81 x 31 x 17 mesh being used.

Conclusions

The paper has presented an efficient solver for the full inviscid time dependent compressible Euler equations giving solutions in two- and three-dimensional flow with separation. It has been shown that this type of separated flow can occur on both round surfaces and sharp corners. In all cases compressibility is needed to allow for these solutions. All these results raise the question of the comparison between the exact inviscid solution and the limit of Navier Stokes solutions if the Reynolds number is increased to infinity.

References

1. Jameson, A.; Caughey, D. A.: AIAA Paper 77-635, 1977
2. Boppe, C. W.: NASA-CR 3030, July 1980
3. Schmidt, W.: AGARD-CP-285, Paper 9, 1980
4. Lock, R. L.: AGARD-CPP-291, 1980
5. Jameson, A.; Schmidt, W.; Turkel, E.: AIAA Paper 81-1259, 1981
6. Schmidt, W.; Jameson, A.; Whitfiled, D.: AIAA Paper 81-1265, 1981
7. Schmidt, W.; Jameson, A.: VKI Short Course on CFD Brussels, 1982
8. Salas, M. D.: AIAA 5th CFD Open Forum Paper 10, 1981

9. Haase, W.: Dornier FB 82/BF 8 B, Dec 1981

10. Rizzi, A. W.; Ericson, L. E.: GAMM Conf. on Num. Methods in Fluid Mechanics, Paris 1981

11. Leicher, S.; Fritz, W.; Grashof, J.; Longo, M.: 8th ICNMFM, Aachen, 1982

12. Manro, M. E., et al: NASA-CR-2610, 1976

Figures

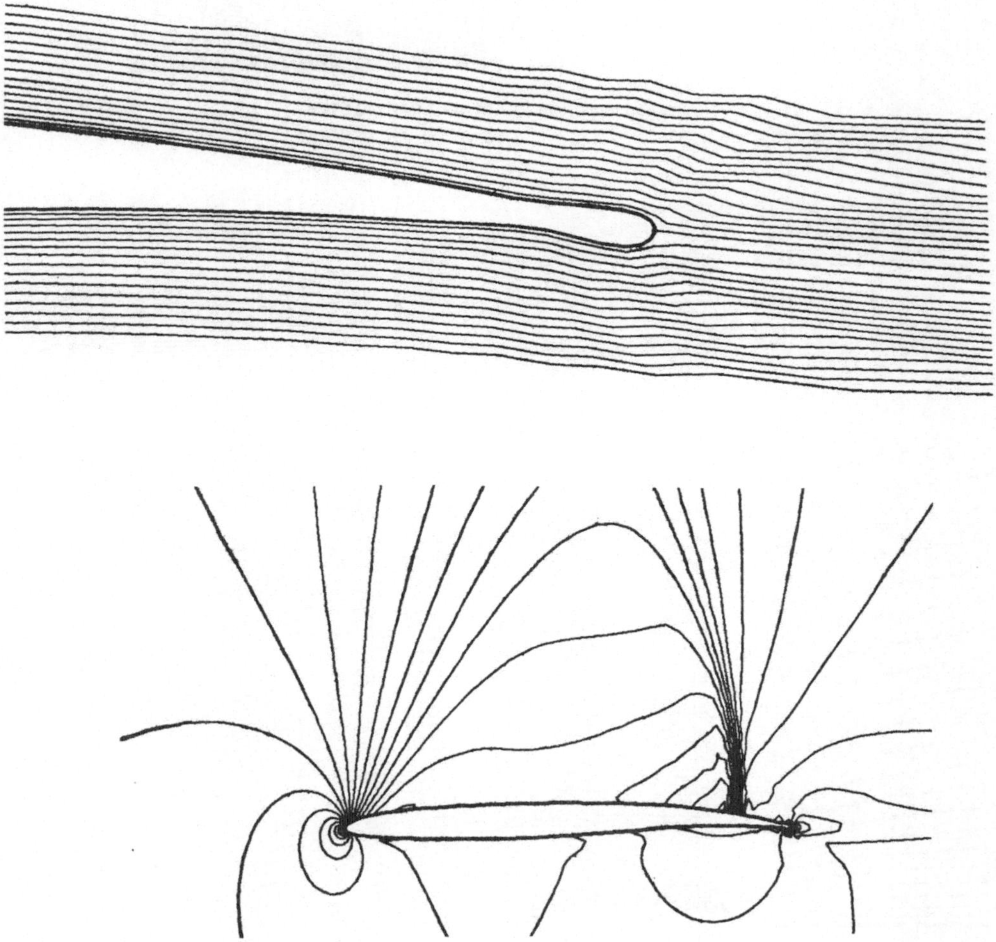

Fig. 1: Computed streamlines and isobars for an airfoil with round trailing edge

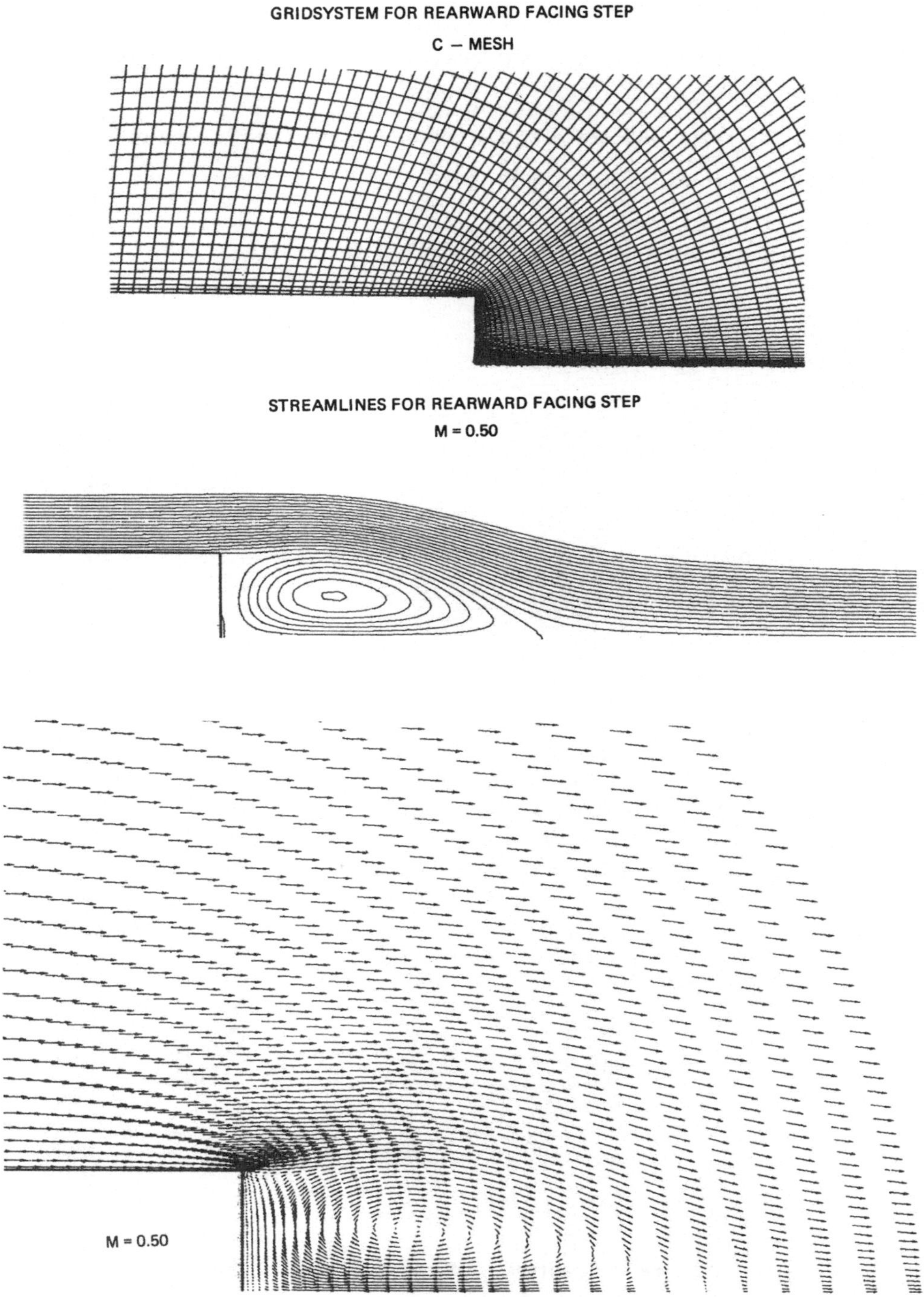

Fig. 2: Euler results for inviscid flow over a rearward facing step

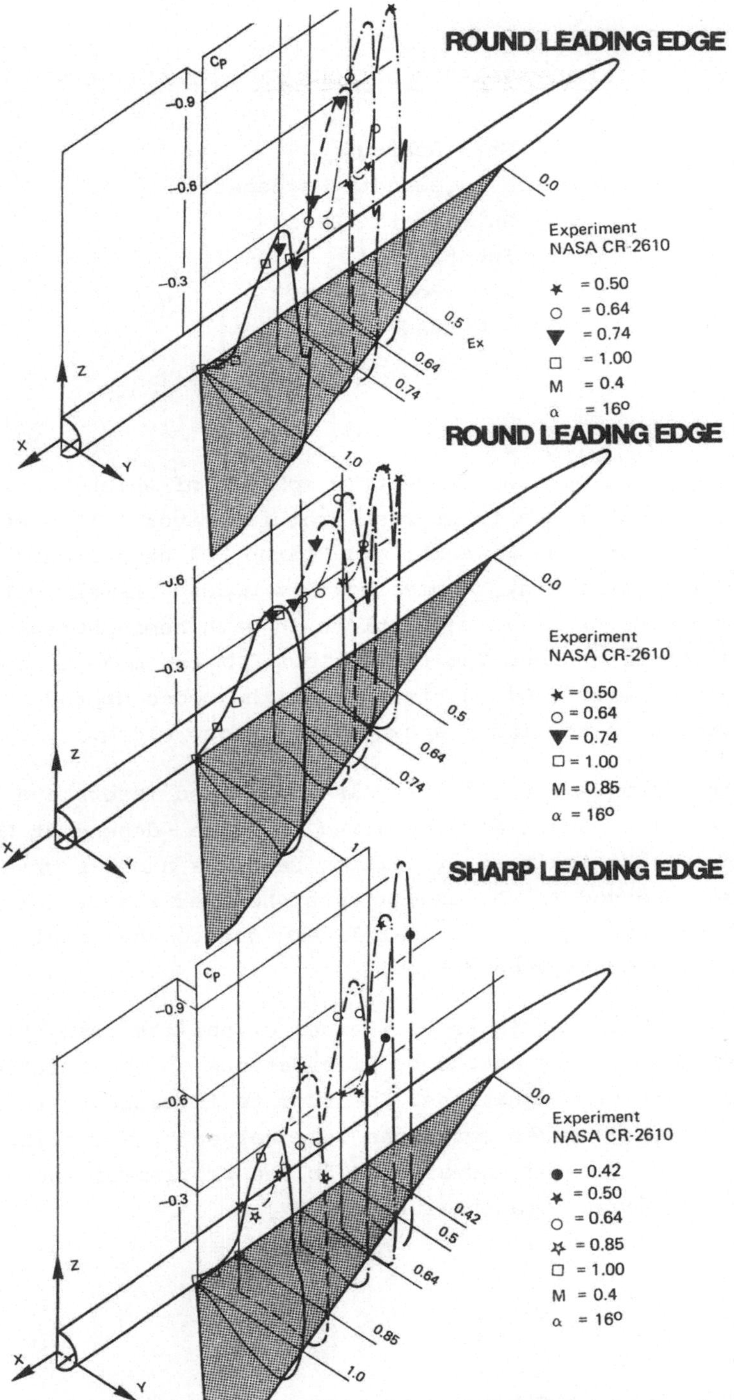

Fig. 3: Comparison of Euler results with experiments for leading edge vortex flow at subsonic and transonic speed

BRANCHING OF NAVIER - STOKES EQUATIONS IN A SPHERICAL GAP *)

Géza Schrauf
Institut für Angewandte Mathematik
der Universität Bonn
Beringstraße 4 - 6
5300 Bonn 1
GERMANY

Introduction

The flow in the gap between two concentric spheres of which the inner one rotates and the outer one is at rest exhibits several flow modes. For the radius ratio of 0.85 Sawatzki and Zierep [9] have observed in their experiments three steady, symmetric flow modes - namely without any, with one, and with two Taylor - vortices in each hemisphere - and two unsteady, asymmetric ones. The transitions from one mode to another were determined by Wimmer [10]. He found that the modes depend on the angular acceleration of the inner sphere and the gap width.

Bonnet & Alziary de Roquefort [3], Bartels [1,2], and Yavorskaya [11] investigated this flow numerically by using the time - dependent Navier - Stokes equations. The calculated transition Reynolds numbers were comparable to those observed in the experiments. However the transitions themselves were different. This is most likely due to the equatorial symmetry imposed as boundary condition.

Another question of interest is as to whether or not the transitions from one mode to another are caused by bifurcations of the solutions of the time - independent flow equations. In order to investigate this problem the steady Navier - Stokes equations were solved by a continuation method proposed by Keller [6] and applied to the problem of the flow in cylindrical gaps by Meyer - Spasche & Keller [7].

---

*) This research has been carried out in cooperation with Prof. E. Krause, Aerodynamisches Institut der RWTH Aachen.
It has been supported by the Deutsche Forschungsgemeinschaft in the Sonderforschungsbereich 72 at the University of Bonn.

Formulation of the problem

If it is assumed that the flow is incompressible with constant viscosity and symmetric with respect to the axis of rotation and the equatorial plane, the governing equations and the boundary conditions read :

$$v^r = \frac{1}{r^2 \sin\theta} \frac{\partial}{\partial\theta} \psi \quad , \quad v^\theta = \frac{-1}{r^2 \sin\theta} \frac{\partial}{\partial r} \psi \quad , \quad v^\phi = \frac{1}{r \sin\theta} \Phi \quad , \tag{1.1}$$

$$(\nabla \times \vec{v})^\phi = \frac{1}{r \sin\theta} \zeta \quad , \tag{1.2}$$

$$\frac{1}{r^2 \sin\theta} \left\{ \psi_\theta \, \Phi_r - \psi_r \, \Phi_\theta \right\} = \frac{1}{Re} D^2 \Phi \quad , \tag{2.1}$$

$$\frac{1}{r^2 \sin\theta} \left\{ \psi_\theta \, \zeta_r - \psi_r \, \zeta_\theta - \zeta \left[\frac{2}{r} \psi_\theta - 2 \cot\theta \, \psi_r \right] + \Phi \left[\frac{2}{r} \Phi_\theta - 2 \cot\theta \, \Phi_r \right] \right\} = \frac{1}{Re} D^2 \zeta \quad , \tag{2.2}$$

$$D^2 \psi = - \zeta \quad , \tag{2.3}$$

$$D^2 = \frac{\partial^2}{\partial r^2} + \frac{1}{r^2} \frac{\partial^2}{\partial \theta^2} - \frac{\cot\theta}{r^2} \frac{\partial}{\partial\theta} \quad , \tag{3}$$

$$(\ )_r = \frac{\partial}{\partial r} \quad , \quad (\ )_\theta = \frac{\partial}{\partial\theta} \quad .$$

$$\Phi(1,\theta) = \sin^2\theta \quad , \quad \zeta(1,\theta) = - \frac{\partial^2}{\partial\theta^2} \psi(1,\theta) \quad , \quad \psi(1,\theta) = 0 \quad ,$$

$$\Phi(\alpha,\theta) = 0 \quad , \quad \zeta(\alpha,\theta) = - \frac{\partial^2}{\partial\theta^2} \psi(\alpha,\theta) \quad , \quad \psi(\alpha,\theta) = 0 \quad ,$$

$$\Phi(r,0) = 0 \quad , \quad \zeta(r,0) = 0 \quad , \quad \psi(r,0) = 0 \quad ,$$

$$\frac{\partial}{\partial\theta} \Phi(r,\tfrac{\pi}{2}) = 0 \quad , \quad \zeta(r,\tfrac{\pi}{2}) = 0 \quad , \quad \psi(r,\tfrac{\pi}{2}) = 0 \quad .$$

$$1 \le r \le \alpha \quad , \quad 0 \le \theta \le \frac{\pi}{2} \quad ,$$

where $\frac{1}{\alpha} = \eta = \frac{R_1}{R_2}$ is the radius ratio, and the Reynolds number is defined as

$$Re = \frac{\omega \, R_1^2}{\nu} \quad . \tag{4}$$

Method of solution

The equations (2) are discretized by central differences of second order. The resulting system of nonlinear difference equations is

$$\begin{bmatrix} C_h(\vec{\psi},\vec{\Phi}) \\ C_h(\vec{\psi},\vec{\zeta}) + B_h(\vec{\psi},\vec{\zeta}) - \Delta r^2 B_h(\vec{\Phi},\vec{\Phi}) \\ 0 \end{bmatrix} + \frac{1}{Re} \begin{bmatrix} D_h^2 & & \\ & D_h^2 & \\ & 1 & D_h^2 \end{bmatrix} \begin{bmatrix} \vec{\Phi} \\ \vec{\zeta} \\ \vec{\psi} \end{bmatrix} = 0 \ . \qquad (5)$$

The solution vector contains the grid functions of the transformed circumferential velocity component $\vec{\Phi}$, the vorticity component $\vec{\zeta}$, and the stream function $\vec{\psi}$.

C_h , B_h , D_h^2 represent the finite difference approximations of the convective - , curvature - , and frictional terms respectively.

The system (5) is solved by using Keller's pseudo - arclength - continuation method [6]. In this method the solution branches are parametized by an approximation s of the arclength. If a solution is obtained for $s = s_o$, the branch can be continued in its tangential direction to provide an initial guess for a Newton iteration at $s = s_o + \Delta s$.

All calculations were performed on a mesh with 11 gridpoints in radial - and 61 in meridional direction.

Results

First the branch of the basic flow, i.e. the flow without Taylor - vortices, was calculated with the low Reynolds number approximation given in [8] as an initial guess. No bifurcation points could be detected along this branch. For $Re \approx 850$ the basic flow becomes hydrodynamically unstable [10], and can only be calculated by a continuation method. The solution exhibits turning points A at $Re \approx 1600$ and B at $Re \approx 1290$ (see Fig. 1.).

In a second series of calculations the solution branch for the flow with two Taylor-vortices was determined. The initial guess was provided with Bartels' solution [1] for $Re = 900$. At $Re \approx 850$ a turning point C was found, beyond which the flow becomes hydrodynamically unstable, corresponding to a Leray - Schauder index of -1 [4]. If one now follows the branch for increasing Reynolds numbers, the vortex just above the equator disappears and only one vortex remains. At $Re \approx 1020$ a second turning point D was observed at which the flow becomes stable again.

The vortex becomes smaller with decreasing Reynolds numbers. A third turning point E appears at $Re \approx 770$ beyond which the flow remains in an unstable one - vortex mode.

If the numerical integration is started with a one vortex solution as obtained by Bartels, the branch described before cannot be reproduced exactly but differs from it by a few percent. This discrepancy can perhaps be explained through the fact that the solution branch of the differential problem is split up into several of the finite - difference problem.

For $1020 \leq Re \leq 2100$, no one - vortex solution for steady flow conditions could be found. Recently this was experimentally confirmed by Bühler [5] for a radius ratio of 0.87 and $1100 \leq Re \leq 2200$.

It can be concluded from these results that for the ratio of radii of 0.85 the branches of the flow modes with Taylor - vortices do not bifurcate from the basic branch as is true for flows in cylindrical gaps [7], and that transition from the basic - to other modes can only be achieved by either unsteadiness or asymmetry.

References

[1] BARTELS, F., Rotationssymmetrische Strömungen im Spalt konzentrischer Kugeln, Dissertation RWTH Aachen, 1978.

[2] BARTELS, F., KRAUSE, E., Taylor vortices in spherical gaps. In: Eppler, R., Fasel, H. (Ed.), Laminar - turbulent transition. Proceedings, IUTAM Symposium, 16.- 22.9.1979. Stuttgart, Springer, Berlin/Heidelberg/New York 1980.

[3] BONNET, J.-P., ALZIARY de ROQUEFORT, T., Ecoulement entre deux sphères concentrique en rotation, Journal de Mécanique 15,3 (1976), 373 - 397.

[4] BENJAMIN, T.B., Applications of Leray - Schauder degree theory to problems of hydrodynamic stability, Math. Proc. Camb. Soc. 79, (1976), 373 - 392.

[5] BÜHLER, K., private communication, June 1982.

[6] KELLER, H.B., Numerical solutions of bifurcation and nonlinear eigenvalue problems. In: Rabinowitz, P. (Ed.), Applications of bifurcation theory, Academic Press, New York, 1977, 359 - 384.

[7] MEYER - SPASCHE, R., KELLER, H.B., Computations of the axisymmetric flow between rotating cylinders, J. Comp. Physics 35 (1980), 100 - 109.

[8] PEARSON, C.E., A numerical study of the time - dependent viscous flow between two rotating spheres, J. Fluid Mech. 28 (1967), 323 - 336.

[9] SAWATZKI, O., ZIEREP, J., Das Stromfeld im Spalt zwischen zwei konzentrischen Kugelflächen, von denen die innere rotiert, Acta Mechanica, 9 (1970), 13 - 35.

[10] WIMMER, M., Experiments on a viscous fluid flow between concentric rotating spheres, J. Fluid Mech. 78 (1976), 317 - 335.

[11] YAVORSKAYA, I.M., ASTA'EVA, N.M., Numerical analysis of the stability and non-uniqueness of spherical Couette flow. In: Hirschel, E.M. (Ed.), Proceedings of the third GAMM-conference on numerical methods in fluid mechanics, DFVLR, Cologne 10. - 12.10.1979, Vieweg, Braunschweig / Wiesbaden 1980.

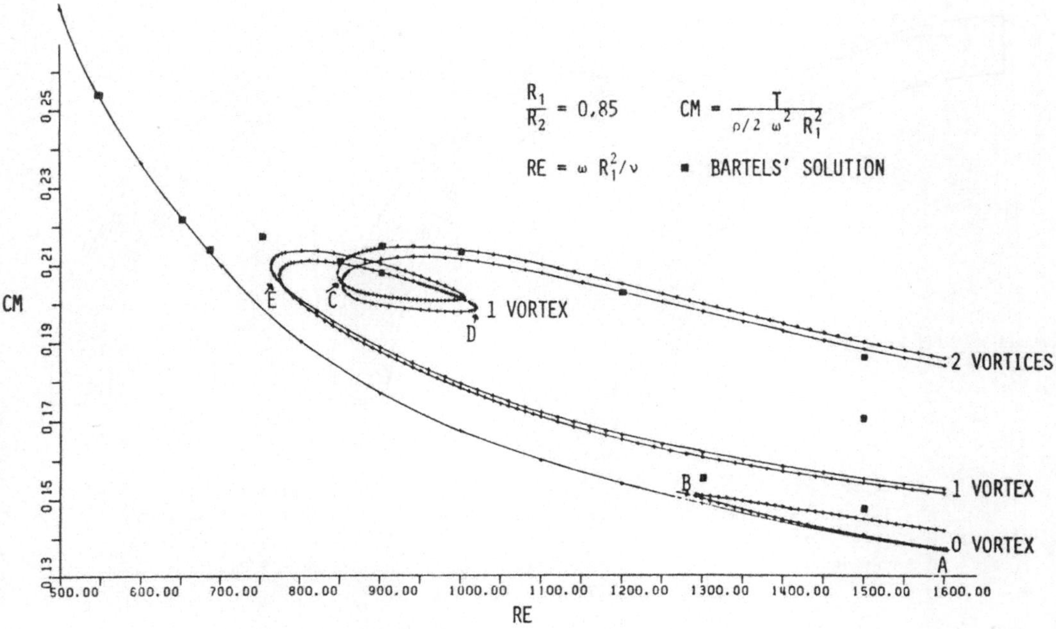

FIGURE 1. TORQUE COEFFICIENT AS A FUNCTION OF REYNOLDS NUMBER.

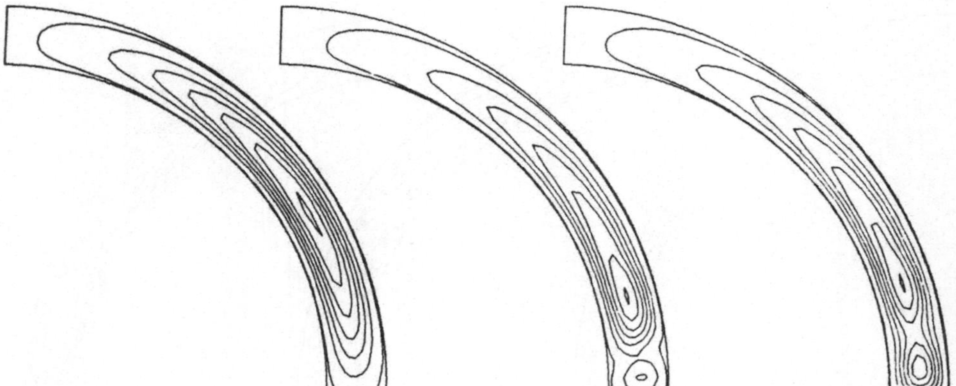

Figure 2.
Basic flow. Re = 500.

Figure 3.
Basic flow. Re = 1604.
(Turning point A).

Figure 4.
Basic flow. Re = 1291.
(Turning point B).

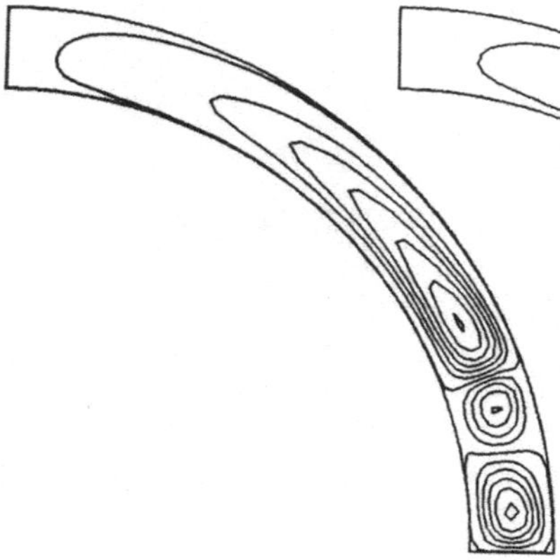

Figure 5.
Flow mode with two
Taylor vortices.
Re = 1600.

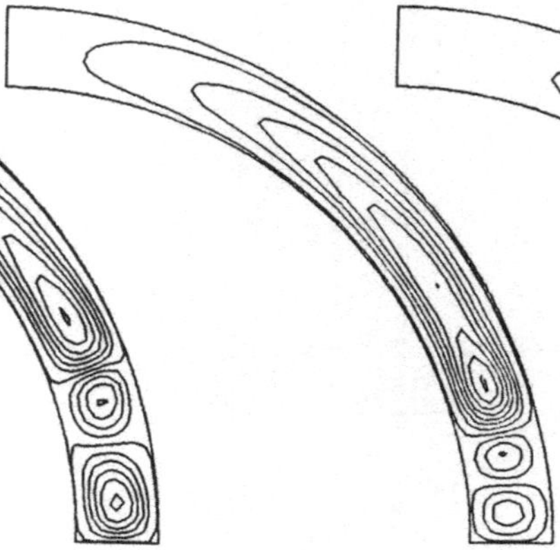

Figure 6.
Flow mode with two
Taylor vortices.
Re = 851.
(Turning point C).

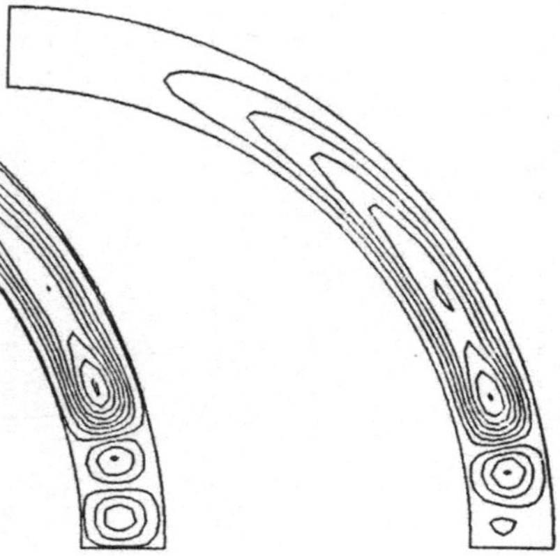

Figure 7.
Instable flow with
two Taylor vortices.
Re = 900.

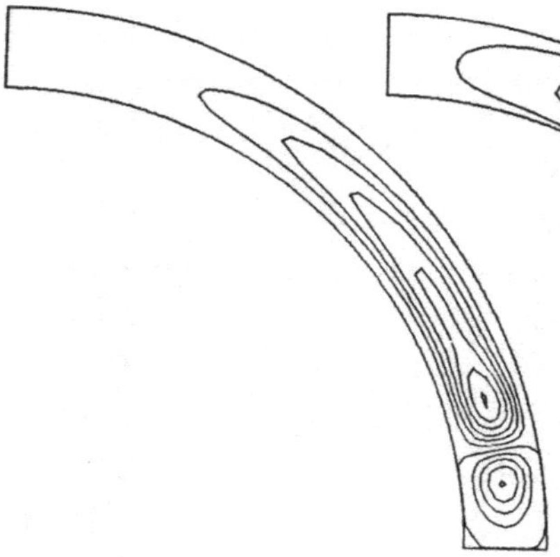

Figure 8.
Flow mode with one
Taylor vortex.
Re = 1019.
(Turning Point D).

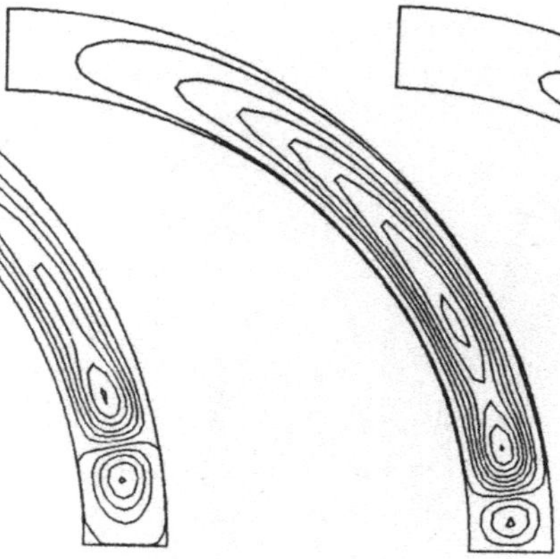

Figure 9.
Flow mode with one
Taylor vortex.
Re 772.
(Turning point E).

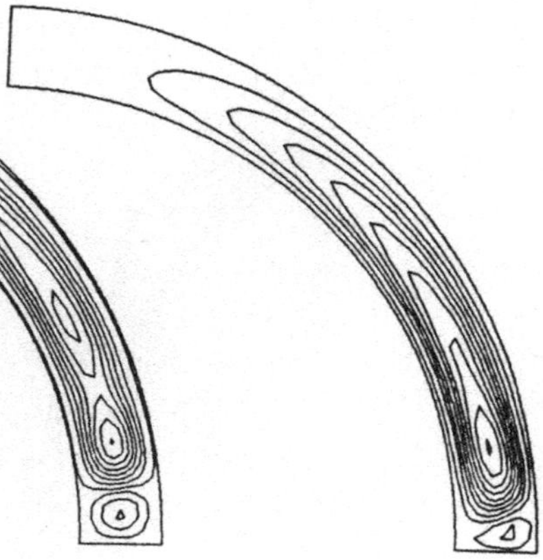

Figure 10.
Unstable flow with one
Taylor vortex.
Re = 1600.

ON THE CONSTRUCTION OF ADAPTIVE ALGORITHMS FOR UNSTEADY PROBLEMS OF GAS DYNAMICS IN ARBITRARY COORDINATE SYSTEMS

Yu.I. Shokin, A.I. Urusov
Institute of Theoretical and Applied Mechanics
USSR Academy of Sciences,
Novosibirsk 630090, U.S.S.R.

At present nonuniform grids including moving grids in which the nodes disposition depends on time are used increasingly for the solution of fluid mechanics problems with the aid of difference schemes (DS). This is explained by the fact that the nonuniform grids enable one to improve the approximation of boundary conditions in the domains of complex configuration and make it possible to take into account the behaviour of flow singularities. An important requirement imposed on the moving grid is that of the nodal concentration in the domain of large gradients. Numerical computations carried out for model problems on the basis of moving grids shrinking in the domain of large gradients show that in this case it is possible to improve the accuracy of the solution obtained [1].

The use of explicit DS on the nonuniform grid for solution of evolution problems is hindered by the stability conditions that make the mathematician generally employ an extremely small time step (as compared to the "mean" step in spatial variables) in the case of the nodal concentration in the domains with large gradients.

Let us find out the requirements which the moving nonuniform grid should meet in order that the time step for explicit DS on these grids be as large as possible. For the sake of simplicity let us consider the case of one space variable.

Let there be a nondegenerate smooth transformation of the coordinate system (t, x) in the coordinate system (t, ξ), which satisfies next conditions:

$$t=t, \quad x = x(t,\xi), \quad \frac{\partial x}{\partial \xi} \geq \alpha > 0, \quad \alpha = \text{const}. \qquad (1)$$

Let in the coordinate system (t, ξ) be given a uniform grid $\{(t_n, \xi_j)\}$ $\xi_j = jh$. We shall call an image of this grid by mapping $(t,\xi) \to (t,x)$ a moving grid and its nodes will be denoted by (t_n, x_j^n) where $x_j^n = x(t_n, \xi_j)$.

Let us consider the DS

$$\frac{w_j^{n+1} - w_j^n}{\tau} + \frac{g_{j+1}^n - g_{j-1}^n}{2h} = \frac{w_{j+1}^n - 2w_j^n + w_{j-1}^n}{2\tau} \tag{2}$$

approximating the equation

$$u_t + f_x = 0 \ , \ f = u^2/2 . \tag{3}$$

Here $w = x_\xi u$, $g = f - u x_t$, $w_j^n = w(t_n, x_j^n)$, $g_j^n = g(t_n, x_j^n, w_j^n)$.

Π - form of the first differential approximation [2] of the DS (2) has the form

$$\begin{aligned} w_t + g_\xi = \frac{\tau}{2} \frac{\partial}{\partial \xi} \Big\{ \Big[\frac{1}{\varkappa^2} - \Big(\frac{dg}{dw}\Big)^2 \Big] w_\xi - \\ - \Big[\Big(\frac{x_t}{x_\xi}\Big)_t - \Big(\frac{x_t}{x_\xi}\Big)_\xi \Big] w \Big\} . \end{aligned} \tag{4}$$

From this we obtain the necessary stability condition of the DS (2) as follows

$$u - \frac{1}{\varkappa} x_\xi \le x_t \le u + \frac{1}{\varkappa} x_\xi \qquad (\varkappa = \tau / h) \tag{5}$$

Thus the velocity of the nodal motion of the grid (x_t) cannot be arbitrary and should satisfy the restriction (5). So if the method of grid construction on the layer t_{n+1} is used which does not employ the information on the nodal distribution on the layer t_n , then it is advantageous to apply the implicit DS since in the case the satisfaction of the inequality (5) for explicit DS can lead (and leads in practice) to extremely small time steps. More detailed analysis of the differential approximation (4) leads to the following statement: in the region of continuous solution of the Cauchy problem for the equation (3) it is always possible to construct moving grid on which one can employ the DS (2) with any time step; in the case of appearance in the solution of strong discontinuities the time step cannot exceed the value

$$\tau^* = \min_j \frac{x_{j+1}^n - x_{j-1}^n}{u_{j-1}^n - u_{j+1}^n} \tag{6}$$

which is equal to time of gradient catastrophe.

In the case of gas dynamics equations the greatest time step which is possible to obtain for the DS on moving grid analogous to (2) is determined by the value

$$\tau^* = \min_j \frac{x_{j+1}^n - x_{j-1}^n}{2C_j^n + |u_{j+1}^n - u_{j-1}^n|} , \qquad (7)$$

where C_j^n is sound speed, u_j^n is medium speed

Note an important feature of the quantities τ^* in (6) and (7), namely that the value τ^* is invariant under the Galilean transformation.

All conclusions which were made on the basis of the investigation of the DS (2) indeed are valid for a wider class of explicit DS, for example, the DS

$$\frac{w_j^{n+1} - w_j^n}{\tau} + \frac{f_{j+1}^n - f_{j-1}^n}{2h} - \frac{(x_t)_{j+1/2}^n u_{j+1/2}^n - (x_t)_{j-1/2}^n u_{j-1/2}^n}{h} = \qquad (8)$$
$$= \frac{\lambda_{j+1/2}^n (u_{j+1}^n - u_j^n) - \lambda_{j-1/2} (u_j^n - u_{j-1}^n)}{2\tau} ,$$

where $\tau = t_{n+1} - t_n$, $(x_t)_j^n = (x_{j+1}^n - x_j^n)/\tau$, λ is parameter, $(x_\xi)_j^n = (x_{j+1}^n - x_{j-1}^n)/2h$, $\varphi_{j+1/2} = (\varphi_{j+1} + \varphi_j)/2$.

Numerical results presented below are obtained on the basis of the DS (8), parameter λ was chosen analogously to the case of first order Rusanov's DS [3] .

Let us now pass to the problem of construction of the moving grid for explicit DS on which the stability condition $\tau \le \tau^*$ can be realized, where τ^* is determined by the formula (6) (for the equation (3)) or (7) (for the gas dynamics equations). One of possible approaches to the solution of this problem based on the minimization of the value $\max_j |(u_j^n - (x_t)_j^n)/(x_\xi)_j^n|$ is described in [4] . Here we present another method which is based on the one of widely used methods of obtaining the equations of the nonuniform grid, namely, the problem of the minimization of some functional is formulated after that the Euler equations are derived which describe the nonuniform grid required. Usually this is made in the following way. Let the value $\tilde{\alpha}$ be small in the cases when the velocity of the grid points motion is close to the velocity of the particles motion of the medium; the value $\tilde{\beta}$ is small in the domains where gradients are small and is large in the domains where the gradients of flow parameters are large; the value $\tilde{\gamma}$ is small when the grid is close to the orthogonal grid and is large in opposite case. Let $\varphi(\tilde{\alpha}, \tilde{\beta}, \tilde{\gamma})$ be a monotonic increasing function with respect to each argument. Let us then require that nonuniform grid (that is corresponding map $(t,\xi,\eta) \to (t,x,y)$)

minimizes the functional $\Phi(x^{n+1}, y^{n+1}) = \iint \varphi(\tilde{\alpha}, \tilde{\beta}, \tilde{\gamma})\, d\xi\, d\eta$.
Depending on the choice of the functions $\tilde{\alpha}, \tilde{\beta}, \tilde{\gamma}, \varphi$ it is possible to get different methods of construction of the nonuniform grid.

Let us choose $\tilde{\alpha}, \tilde{\beta}, \tilde{\gamma}, \varphi$ as follows:

$$\tilde{\alpha} = a\left[\left(\frac{x^{n+1}-x^{n}}{\tau} - u\right)^2 + \left(\frac{y^{n+1}-y^{n}}{\tau} - v\right)^2\right],$$

$$\tilde{\beta} = b\left[(x_\xi^{n+1})^2 + (y_\xi^{n+1})^2 + (x_\eta^{n+1})^2 + (y_\eta^{n+1})^2\right],$$

$$b = \alpha + \beta\left(\left|\frac{\partial f}{\partial x}\right| + \left|\frac{\partial f}{\partial y}\right|\right),$$

$$\tilde{\gamma} = c\,(x_\xi^{n+1} y_\xi^{n+1} + x_\eta^{n+1} y_\eta^{n+1})^2,$$

$$a > 0, \quad c > 0, \quad \alpha > 0, \quad \beta > 0.$$

If $\varphi(\tilde{\alpha}, \tilde{\beta}, \tilde{\gamma}) = \tilde{\alpha} + \tilde{\beta} + \tilde{\gamma}$, then we get the following Euler system equations:

$$\frac{\partial}{\partial \xi}\left[b x_\xi^{n+1} + cp\, y_\xi^{n+1}\right] + \frac{\partial}{\partial \eta}\left[b x_\eta^{n+1} + cp\, y_\eta^{n+1}\right] = \frac{a}{\tau}\left(\frac{x^{n+1}-x^{n}}{\tau} - u\right),$$

$$\frac{\partial}{\partial \xi}\left[b y_\xi^{n+1} + cp\, x_\xi^{n+1}\right] + \frac{\partial}{\partial \eta}\left[b y_\eta^{n+1} + cp\, x_\eta^{n+1}\right] = \frac{a}{\tau}\left(\frac{y^{n+1}-y^{n}}{\tau} - v\right),$$

which is elliptic if $b > cp$, $p = x_\xi^{n+1} y_\xi^{n+1} + x_\eta^{n+1} y_\eta^{n+1}$.
Since α, β, a, c are arbitrary parameters, it is possible to choose them in such a way that the ellipticity condition is satisfied. It is naturally supposed that the grid (x_{ij}^{n}, y_{ij}^{n}) is already constructed at the moment of time t_n.

In the Figures 1,2 the results of numerical computations of the Cauchy problem for equation (3) with the initial data

$$u(0,x) = \begin{cases} 0, & \text{if } x \le 3, \\ -5(x-3), & \text{if } 3 \le x \le 4, \\ -5, & \text{if } 4 \le x \le 5, \\ 5(x-6), & \text{if } 5 \le x \le 6, \\ 0, & \text{if } 6 \le x. \end{cases}$$

are presented. In Fig.1 the position of grid points at different times is shown, in Fig.2 the graphs of approximate (solid line) and exact (broken line) solutions for t=0.755 are presented. Note that the Courant number changed in computations from 10 to 2, that is $10 \geq |\tau u_j^n / (x_j^n - x_{j-1}^n)| \geq 2$.

References

1. Yanenko N.N., Danaev N.T., Liseikin V.D. - O variaczionnom metode postroenija setok.- Chislenii metody mechaniki sploshoy sredy. Novosibirsk, 1977, 8, N°4, p. 157-163
2. Shokin Yu.I. The method of differential approximation. Novosibirsk, "Nauka", 1979.
3. Rusanov V.V. Rachet vzaimodejstvia nestaszionarnyh udarnyh voln s prepjatstvijamy.- J. Vychisl. Matem. Matem. Physiki, 1961, 1, N°2, p. 267-279.
4. Shokin Yu.I., Urusov A.I. Ob odnom metody postroenja podviznyh setok dlja reshenia uravnenii hyperbolicheskogo typa.- Novosibirsk, 1981.

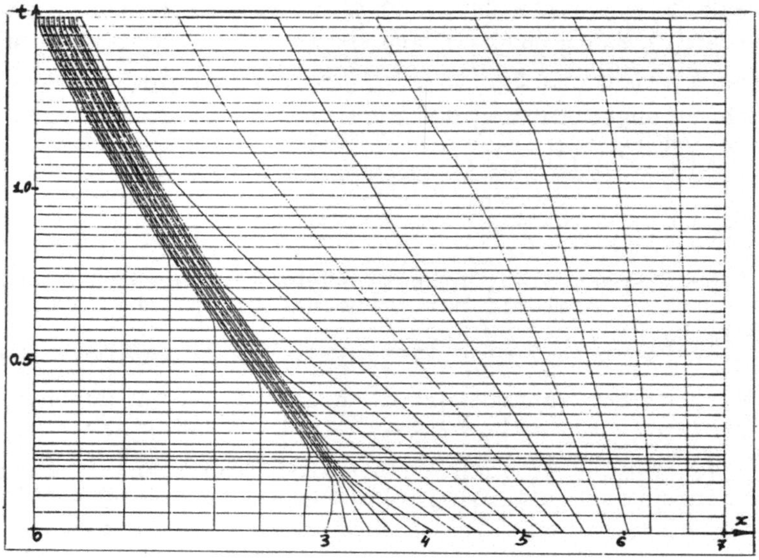

Fig. 1

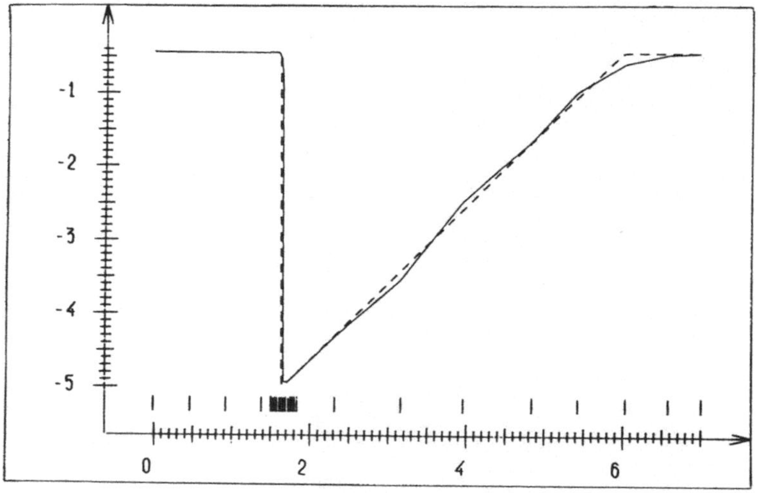

Fig.2

ANALYTICAL TESTS FOR NUMERICAL METHODS IN GAS DYNAMICS

A. F. Sidorov

Institute of Mathematics and Mechanics
Ural Scientific Center, Sverdlovsk, USSR

Verification of the accuracy of various numerical methods in gas dynamics can be effectively realized by exact solutions of gas dynamics equations. In this case it is desirable that these solutions contain the features of shock-wave type, contact boundaries, large value gradient regions, relaxation zones, vacuum regions etc. If for one-dimensional problems such solutions are rather numerous (automodel solutions, Riemann's waves etc.), their number for multi-dimensional flows is quite scarce.

The report deals with several types of steady and unsteady plane and spatial flows of inviscid gas for which analytical representations have been achieved and which contain characteristics of different types.

1. Triple and Double-wave Class Potential Flows.

A. Class of exact unsteady solutions of three-wave type may be represented in the form [1]

$$c = a_0 + (\bar{a}\cdot\bar{u}) \ , \ |\bar{a}|^2 = 0.75(\gamma-1)^2(2-\gamma)^{-1}, \ 1<\gamma<2 \qquad (1.1)$$

$$\Pi = \sum_{i=1}^{3}\left(-0.5(\gamma-1)u_i^2 + T_i(\xi_i)\right) \ , \ \xi_i = (\bar{\alpha}_i\cdot\bar{u}) \qquad (1.2)$$

$$\begin{cases} |\bar{\alpha}_i\times\bar{\alpha}_k|^2 - 4(\gamma-1)^{-2}|\bar{a}\ (\bar{\alpha}_i\times\bar{\alpha}_k)|^2 = 0 \\ |\bar{\alpha}_i| = 1 \ , \ |\bar{a}\times\bar{\alpha}_k|^2 = (\gamma-1)^2(3-\gamma)^{-1} \end{cases} \qquad (1.3)$$

$$x_i = (\gamma-1)^{-1}\frac{\partial\Pi}{\partial u_i} + u_i + t\left((\gamma-1)^{-1}\frac{\partial c^2}{\partial u_i} + u_i\right), \ i=1,2,3. \qquad (1.4)$$

Here c is the sound velocity, $\bar{u}$ -velocity vector, γ -adiabatic index, $\bar{a}, \bar{\alpha}_i$ -vector constants satisfying conditions (1.1) (1.3), $T_i(\xi_i)$ -arbitrary functions; equations (1.4) surve to define function u_i in physical space x_1, x_2, x_3, t.

From (1.1)-(1.4) one may as well obtain classes of exact solutions of plane two-wave type with independent variables x_1, x_2, t.

Class of solutions (1.1)-(1.4) allows [2-5] to construct precise solutions of the following problems. Let at the initial moment of time $t = 0$ homogeneous gas be at rest within some three-sided infinite angle formed by planes P_i orthogonal to vectors $\bar{\alpha}_2 \times \bar{\alpha}_3$ $\bar{\alpha}_1 \times \bar{\alpha}_3$, $\bar{\alpha}_1 \times \bar{\alpha}_2$ respectively. Planes P_i playing the role of inpervious pistons start to move in gas in parallels to themselves with the velocities $V_i(t)$. It is evident that away from the vertex near three-edged angle ribs there lie regions of interaction of two Riemann waves, where the flow may be constructed in the class of plane double-waves described by the system (1.1)-(1.4) which takes into account that the dimension of independent variables space reduces by unity. At the angle vertex within two-waves interaction region the solution in constructed in the triple-wave class (1.1)-(1.4), functions $T_i(\xi_i)$ being readily determined from (1.4) according to set velocities $V_i(t)$. Class of solutions (1.1)-(1.4) allows either to construct the solution of the problem set as a whole (e.g., when $V_1(t) = V_2(t) = 0$, $V_3(t) = \infty$ and gas discharges into vacuum along some two-sided angle), or to obtain a part of three-dimensional unsteady flow region, which can be locked by plunging a mobile flexible piston of variable shape.

B. Class of plane double waves in which $c = c(r)$ and distribution function $\chi(r,\varphi) = b\cos 2\varphi f(r)$, $r = \sqrt{u_1^2 + u_2^2}$, $\varphi = \operatorname{arctg} u_2 u_1^{-1}$, $b = const$ [6] allows to construct two-dimensional unsteady flow with a curvilinear shock wave of constant intensity propogating over homogeneous background.

Functions $\theta(r) = 2(\gamma-1)^{-1} c(r)$ and $f(r)$ are defined from the system of two ordinary differential equations with $r > a$

$$(\gamma-1)\theta(\theta' + r\theta'' - \theta'^3) + r(\gamma-3)\theta'^2 + 4r = 0$$

$$r^2 f'' + r(1-\theta'^2) f' - 4(1-\theta'^2) f = 0$$

with the following initial data

$$\theta(a) = \frac{2}{\gamma-1}\sqrt{\gamma \frac{P}{\rho}}, \quad \theta'(a) = \frac{2(\mathcal{D}-a)}{(\gamma-1)\theta(a)}, \quad a = \frac{2(\mathcal{D}^2-\gamma)}{(\gamma+1)\mathcal{D}},$$

$$P = \frac{2}{\gamma+1}\mathcal{D}^2 - \frac{\gamma-1}{\gamma+1}, \quad \rho = \frac{(\gamma+1)\mathcal{D}^2}{(\gamma-1)\mathcal{D}^2 + 2\gamma}, \quad f(a) = a, \quad f'(a) = 0,$$

where $\mathcal{D}$ is the shock wave velocity, the propagation law of which is determined at $r = a$ by equations:

$$x_1 = (r + \tfrac{\gamma-1}{2}\theta\theta')\cos\varphi t + \chi_r \cos\varphi - \chi_\varphi r^{-1} \sin\varphi$$

$$x_2 = (r + \tfrac{\gamma-1}{2}\theta\theta')\sin\varphi t + \chi_r \sin\varphi + \chi_\varphi r^{-1} \cos\varphi$$

2. Two- and Three-dimensional Automodel non-Isentropic Flows.

Unsteady flows of type 7,8 (-density, -entropy)

$$u_1 = \frac{x_1}{t} f_1\left(\frac{x_1}{x_2}\right), \; u_2 = \frac{x_1}{t} f_2\left(\frac{x_1}{x_2}\right), \; \rho = \left(\frac{x_1}{t}\right)^{\alpha} f_3\left(\frac{x_1}{x_2}\right), \; S = \left(\frac{x_1}{t}\right)^{\beta} f_4\left(\frac{x_1}{x_2}\right) \quad (2.1)$$

$$(\gamma - 1)\alpha + \beta - 2 = 0 \quad (2.2)$$

and steady flows with

$$u_1 = \frac{x_1}{x_3} f_1\left(\frac{x_2}{x_3}\right), \; u_2 = \frac{x_1}{x_3} f_2\left(\frac{x_2}{x_3}\right), \; u_3 = f_3\left(\frac{x_2}{x_3}\right), \; \rho = \left(\frac{x_1}{x_3}\right)^{-2} f_4\left(\frac{x_2}{x_3}\right), \; P = f_5\left(\frac{x_2}{x_3}\right) \quad (2.3)$$

(P -pressure) are described by the system of ordinary differential equations for function f_i .

This system may be readily obtained by substituting representations (2.1), (2.3) into equations of gas dynamics. It follows from (2.1) that stream lines for this class of motion are steady.

Setting over both sides of the planes $\xi = x_1 x_2^{-1} = a = const$ or $\eta = x_2 x_3^{-1} = b = const$ initial data for function f_i determined from Hugoniot's conditions and integrating the systems of equations for f_i to the direction of increasing and decreasing ξ and η we shall obtain gas flow in plane and space tubes with fixed walls which at $\xi = a$ or $\eta = b$ break up so that plane shock waves $\xi = a$ and $\eta = b$ are generated, on the fronts of which all Hugoniot's conditions are precisely satisfied.

3. Exact Solutions with Sub and Super-sonic Zones.

Let us describe two types of precise solutions corresponding to potential plane parallel gas flows in the gravity field [9]. Sound velocity c in the hodograph plane for this type of flows has the form:

$$c^2 = -0.5\gamma(\gamma-1)^{-1}\left[\gamma(u_1^2 + u_2^2) + 2(2\gamma - 1)u_1 u_2\right], \quad (3.1)$$

$$c^2 = 0.5(\gamma-1)^2(3-\gamma)^{-1}(u_1 - u_2)^2, \qquad \gamma < 3 \quad (3.2)$$

Transition into physical plane x_1, x_2 is realized by the formulae

$$x_i = \alpha(2\gamma - 1)(\gamma - 1)^{-2} f_i(u_1, u_2) \quad (3.3)$$

$$f_1(u_1, u_2) = f_2(u_2, u_1) - 0.5(2-\gamma)u_1^2 + \gamma u_1 u_2 + 0.5\gamma u_2^2$$

for flows (3.1) ($\alpha = const$ specifies gravity field) and

$$x_i = \alpha(3-\gamma)^{-1} g_i(u_1, u_2)$$
$$g_1(u_1, u_2) = g_2(u_2, u_1) = -0.5(3+\gamma)u_1^2 + (\gamma-1)u_1u_2 + 0.5(\gamma-1)u_2^2 \qquad (3.4)$$

for the case (3.2).

With $\gamma > 2$ solution (3.1), (3.3) describes super-sonic flow rising from infinity through flat tube against gravity which slows down and having passed the plane sound line, becomes sub-sonic and then after the tube curvature the flow speeds up, again passes the plane sound line and becomes super-sonic. Sound velocity in such flow on infinity falls to zero.

For the case (3.2) when $2 < \gamma < 3$ analogous picture is observed but the sound velocity on infinity is increasing unrestrictedly.

Formulae (3.1), (3.3) define, in principle, the gas flow being in the gravity field over the entire plane $|x_k| < \infty$ which contains the zone of flow T and vacuum zone W, where $c = 0$. Then, with the help of (3.1), (3.3) exact solution of the next unsteady two-dimensional gas dynamics problem is being constructed. Let the obtained steady flow in the gravity field set initial data of Cauchy problem in the plane x_1, x_2 for unsteady equations of gas dynamics provided mass forces are not present. Solution of such problem corresponds to that on gas discharge into vacuum from region T when at the moment of time $t = 0$ gravity field instantly vanishes. Unsteady flow is obtained by substituting equations $\xi_k = x_k - 0.5\alpha t^2$ into (3.3) instead of x_k. The front of discharge into vacuum from region T will be formed by two travelling planes

$$x_2 = l_\pm x_1 + \frac{\alpha}{2\gamma}\sqrt{3\gamma-1}\left(\sqrt{3\gamma-1} \mp \sqrt{\gamma-1}\right)t^2,$$
$$l_\pm = \gamma^{-1}\left(1-2\gamma \pm \sqrt{(\gamma-1)(3\gamma-1)}\right),$$

intersecting on bisectrix of the first square. At $t \to \infty$ the velocity of discharge into vacuum is unrestrictedly growing.

4. Application of Characteristic Series for Describing the Flows in the Vicinity of Weak Discontinuities.

Let at the moment of $t = 0$ a space piston of arbitrary rather smooth shape with zero normal initial velocity and non-zero initial acceleration $W(\lambda,\mu)$ (parameters λ and μ specify the surface S_0 of the piston initial position) start to move into a rest gas. The surface of weak discontinuity R_t, on which $z=0$ ($z=|\vec{u}|$) will propagate into a rest gas with the sound velocity. In [9, 10] representations of the solution for velocities potential $\Phi(u_1,u_2,u_3,t)$ in the vicinity of R_t have been constructed in hodograph space u_1,u_2,u_3 in the form of a rapidly converging characteristic series

$$\Phi = \sum_{K=0}^{\infty} a_K(\varphi,\theta,t)\, z^K ,$$

where $z=0$ corresponds to surface R_t, φ and θ are spherical coordinates in the hodograph space, coefficients $a_K(\varphi,\theta,t)$ a determined explicitly. The above series give good approximation of compression zones with large gradients of values. It appeared that the moment t^* and the locality where the potential flow starts to discontinue and a shock wave is formed may be accurately estimated.

$$t^* = \min_{\lambda,\mu}\left\{ R_1(\lambda,\mu),\ R_2(\lambda,\mu),\ Q(\lambda,\mu)\right\}$$

$$Q(\lambda,\mu) = \frac{2}{K^2}\,\mathrm{sh}\,\frac{K}{(\gamma+1)W}\left[K\,\mathrm{ch}\,\frac{K}{(\gamma+1)W} - H\,\mathrm{sh}\,\frac{K}{(\gamma+1)W}\right]$$

where R_1, R_2 are curvature radii of main normal sections of surface S_0, $K(\lambda,\mu)$ and $H(\lambda,\mu)$ -are Gauss and average curvatures of surface S_0 respectively.

5. Vortical Gas Flows, Linear for some Part of Space Variables.

Let us consider flat and space steady vortical motions of the type [12]

$$u_1 = g_1(x_1),\ u_2 = f_2(x_1)x_2,\ \rho^{\gamma-1} = g(x_1),\ S = const \qquad (5.1)$$

in the plane case

$$u_1 = g_1(x_1),\quad u_2 = \ell_2(x_1)x_2 + f_2(x_1)x_3,$$
$$u_3 = \ell_3(x_1)x_2 + f_3(x_1)x_3,\quad \rho^{\gamma-1} = g(x_1) \qquad (5.2)$$

in the space case. From gas dynamics equations one can easily obtain the system of ordinary equations for functions g, g_1, f_K, ℓ_K. It turns out to be integrated till the end in quadratures, one-(5.1) or two-(5.2) velocity vector components maintain constant value along the stream lines. Function g_1 is defined for the case (5.2) by means of the integral:

$$x_1 = \int g_1^{-\frac{1}{2}} (\beta - g_1^2)^{\frac{1-2\gamma}{2(\gamma-1)}} \left(g_1^2(\gamma+1) - \beta(\gamma-1)\right)\left[C_1 + C_2 g_1 (\beta - g_1^2)^{\frac{1}{\gamma-1}}\right]^{-\frac{1}{2}} dg_1$$

where $C_K = const$, $\beta = const$.

Similary to point 2 solution (5.1) specifies flow in infinite plane tube with the walls break up at $x_1 = const$ and with a shock wave. Solution (5.2) describes, in particular, vortical rotational flows of gas.

References

1. A.F.Sidorov, On exact solutions of equations of gas dynamics of triple-wave type, Dokl.AN USSR, t.194, № 4, 782-785(1970)
2. Ju.Ja.Pogodin, V.A.Suchkov, N.N.Janenko, On running wave equations of gas dynamics, Prikl.mat. i meh., t.22, № 2, 188-196(1958)
3. V.A.Suchkov, Discharge in vacuum over oblique wall, Prikl.mat. i meh., t.27, № 4, 739-740(1963).
4. A.F.Sidorov, O.B.Khairullina, On precise solutions of some edge problems of gas dynamics in double- and triple-wave classes. In: Metody Reshenia Zadach Meh. Sploshnoi Sredy, Trudy IMM USC AN USSR, № 25, 52-66(1978).
5. A.F.Sidorov, Two exact solutions of hydrodynamic equations of triple-wave type, Prikl.mat. i meh., t.28, № 6, 1139-1142(1964).
6. A.F.Sidorov, Some exact solutions of non-stationary two-dimensional gas dynamics, Prikl.mat. i meh., t.26, № 2, 380-386(1962).
7. A.F.Sidorov, On one class of exact solutions of hydrodynamics, Izv. SO AN USSR, tehnich. ser., № 13, t.3, 26-29(1967).
8. S.N.Martushov, A.F.Sidorov, On one class of steady conic non-isentropic flows of inviscid gas. In: Metody Reshenia Zadach Meh. Sploshnoi Sredy, Trudy IMM USC AN USSR, № 25, 43-46(1978).
9. A.F.Sidorov, On some gas flows in the field of gravity, Prikl. mat. i meh., t.42, № 1, 96-104(1978).
10. A.F.Sidorov, On breaking the gas potential flows adjacent to the region of rest, Prikl.mat. i meh., t.35, № 3, 482-491(1971).
11. E.N.Zubov, A.F.Sidorov, On solution of one edge problem for unsteady space flow of gas and propagation of weak spherical shock waves, Chisl. Metody Meh. Sploshnoi Sredy (Novosibirsk), t.3, № 3, 32-50(1972).
12. A.F.Sidorov, On two classes of solutions of gas dynamics equations, Jurn.prikl.mat. i tehn.fis., t.5, 16-24(1980).

Discrete Models of the Kinetic Boltzman Equation

U.M.Sultangazin

Institute of Mathematics and Mechanics
of the Kazakh Academy of Sciences
Alma-Ata, Vinogradova 34, 480100/ USSR

We will discuss for simple models the connection between the kinetic equations and the equations of gas dynamics. It is very important for a proper understanding of the thermodynamical structure of the equations of mechanics and of the problems in the the theory of quasilinear equations of gas dynamics. By examining below a simple model of Carleman type of the kinetic theory of gases we explain how the "doubly divergent" form of the corresponding equations of a continuous medium is a consequence of the H-theorem. We indicate the meaning for the kinetic model of energy integral of the difference of the two solutions of a symmetric non-linear system describing a model of a continuous medium. The structure of shock waves is studied in the kinetic model and in the corresponding model of the Navier-Stokes equations.

We now pass on to a description of the kinetic model with a finite set of velocities $\vec{\Omega}_1, \vec{\Omega}_2, \ldots, \vec{\Omega}_N$. We denote by $f_1, f_2, \ldots, f_N$ the densities corresponding to these velocities. We assume the laws of collisions are such that as a result of collisions the molecules assume only velocities from the chosen set. Then the number of particles dispersed in unit time in the direction of $\vec{\Omega}_i$ as a result of the collision $\vec{\Omega}_k \leftrightarrow \vec{\Omega}_e$ of molecules from groups with velocities $\vec{\Omega}_k$, $\vec{\Omega}_e$ is equal to $\sigma_{ke}^{ij} f_k f_e$ with $\sigma_{ke}^{ij} \geq 0$. Here $\sigma_{ke}^{ij} = S\,\delta_{ke}^{ij}\,|\vec{\Omega}_k - \vec{\Omega}_e|$, where S is the effective collision cross-section and where δ_{ke}^{ij} is the probability the pair $\vec{\Omega}_k$, $\vec{\Omega}_e$ gives the pair $\vec{\Omega}_i$, $\vec{\Omega}_j$ after collision. Suppose that these coefficients are equal for a collision and its opposite, that is $\sigma_{ke}^{ij} = \sigma_{ij}^{ke}$. This equation is a model for the so-called law of detailed equilibrium, which usually holds for real systems. Then the Boltzmann equation for this model can be written as follows:

$$(1)\quad \frac{\partial f_i}{\partial t} + \Omega_{ix}\frac{\partial f_i}{\partial x} + \Omega_{iy}\frac{\partial f_i}{\partial y} + \Omega_{iz}\frac{\partial f_i}{\partial z} = \sum_{k,e,j} \sigma_{ke}^{ij}\,(f_k f_e - f_i f_j)$$

We define the equilibrium distribution function f_i^0 as the minimum of the $H = \sum_{i=1}^{N} f_i \ln f_i$ function for a given mass, momentum, energy densities $\sum_{i=1}^{N} f_i = \rho$, $\sum_{i=1}^{N} \vec{\Omega}_i f_i = \rho \vec{U}$, $\sum_{i=1}^{N} \vec{\Omega}_i^2 f_i = \rho (E + \frac{U^2}{2})$

We study the mathematical theory of the equations describing the flow of a model gas in the simplest one-dimensional model. The model studied below is described by the system

$$
\begin{aligned}
&\frac{\partial f_1}{\partial t} + \frac{\partial f_1}{\partial x} = \frac{f_2^2 - f_1 f_3}{\varepsilon} , \\
(2) \qquad &\frac{\partial f_2}{\partial t} = -2 \frac{f_2^2 - f_1 f_3}{\varepsilon} , \\
&\frac{\partial f_3}{\partial t} - \frac{\partial f_3}{\partial x} = \frac{f_2^2 - f_1 f_3}{\varepsilon} .
\end{aligned}
$$

Multiplying the equations of (2) by $\ln f_1$, $\ln f_2$, $\ln f_3$, respectively, and adding we arrive at an identity, which is an analogue to the H-theorem for this model:

$$
(3) \qquad \frac{\partial \sum f_i \ln f_i}{\partial t} + \frac{\partial (f_1 \ln f_1 - f_3 \ln f_3)}{\partial x} = -\frac{1}{\varepsilon}(f_2^2 - f_1 f_3) \ln \frac{f_2^2}{f_1 f_3}
$$

We construct local functions of distribution f_1^0, f_2^0, f_3^0 guaranteeing a minimum of the H-function. Then $f_1^0 = e^{q_1+q_2-1}$ $f_2^0 = e^{q_1-1}$, $f_3 = e^{q_1-q_2-1}$. The values of the Lagrange multipliers q_1, q_2 are selected so as to satisfy the conditions

$$
f_1 + f_2 + f_3 = e^{q_1+q_2-1} + e^{q_1-1} + e^{q_1-q_2-1} = \rho
$$

$$
f_1 - f_3 = e^{q_1+q_2-1} - e^{q_1-q_2-1} = \rho u
$$

The equations of hydrodynamics corresponding to our kinetic model is to be written in the form

$$
\begin{aligned}
&\frac{\partial \rho}{\partial t} + \frac{\partial \rho u}{\partial x} = 0 , \\
(4) \qquad &\frac{\partial \rho u}{\partial t} + \frac{\partial}{\partial x}\left[\frac{2}{\varepsilon} \rho \left(2 - \sqrt{1 + \frac{3}{4} u^2}\right)\right] = 0
\end{aligned}
$$

It is of interest to rewrite the "hydrodynamical" system (4) in variables q_1, q_2.

$$\frac{\partial(e^{q_1+q_2-1}+e^{q_1-q_2-1})}{\partial t}+\frac{\partial(e^{q_1+q_2-1}-e^{q_1-q_2-1})}{\partial x}=0$$

(5)
$$\frac{\partial(e^{q_1+q_2-1}-e^{q_1-q_2-1})}{\partial t}+\frac{\partial(e^{q_1+q_2-1}+e^{q_1-q_2-1})}{\partial x}=0$$

It is not difficult to check that these equations have the form

$$\frac{\partial L_{q_1}}{\partial t}+\frac{\partial L'_{q_1}}{\partial x}=0, \qquad \frac{\partial L_{q_2}}{\partial t}+\frac{\partial L'_{q_2}}{\partial x}=0$$

where we have put

$$L=e^{q_1+q_2-1}+e^{q_1-1} \quad L'=e^{q_1+q_2-1}-e^{q_1-q_2-1}$$

The integral

(6)
$$\frac{\partial(q_1L_{q_1}+q_2L_{q_2}-L)}{\partial t}+\frac{\partial(q_1L'_{q_1}+L'_{q_2}-L')}{\partial x}=0$$

for smooth solution of this system. It is also not difficult to make sure that

$$q_1L_{q_1}+q_2L_{q_2}-L=f_1\ln f_1+f_2\ln f_2+f_3\ln f_3$$
$$q_1L'_{q_1}+q_2L'_{q_2}-L'=f_1\ln f_1-f_3\ln f_3$$

Then the equation (6) can be rewritten as

$$\frac{\partial(f_1\ln f_1+f_2\ln f_2+f_3\ln f_3)}{\partial t}+\frac{\partial(f_1\ln f_1-f_3\ln f_3)}{\partial x}=0$$

This assertion can be interpreted as follows: the right-hand side in the H-theorem almost vanishes in almost equilibrium solutions. Hence, if we take into account that the right-hand side at (3) is negative, it becomes clear that when we consider discontinuous general solutions of the system

$$\frac{\partial L_{q_i}}{\partial t}+\frac{\partial L'_{q_i}}{\partial x}=0$$

then for integrals around contours containing the discontinuities we have

$$\oint\Big(\sum_i q_iL_{q_i}-L\Big)dx-\Big(\sum q_iL'_{q_i}-L'\Big)dt\geq 0$$

In narrow zones of non-equilibrium, which in the hydrodynamical model simulate the discontinuities, the H-integral is dissipative.

As a measure of the distance between the points $f^{(1)}=(f_1^{(1)}, f_2^{(1)}, f_3^{(1)})$ and $f^{(2)}=(f_1^{(2)}, f_2^{(2)}, f_3^{(2)})$ we take the quantity

$$R(f^{(1)}, f^{(2)}) = \sum_{i=1}^{3} (f_i^{(1)} - f_i^{(2)}) \ln \frac{f_i^{(1)}}{f_i^{(2)}} .$$

Then for solutions "almost in equilibrium" for the equations (2) we obtain $R(f^{(1)}, f^{(2)}) \leq R(f^{(1)}(0,x), f^{(2)}(0,x))\, e^{\kappa t}$

This kind can be obtained for smooth solutions of the limiting equations ($\varepsilon \to 0$) of a continuous medium.

Theorem 1. There exist a unique bounded solution of the equations (1) with initial vector-function $\vec{f}(0,\bar{z}) \in W_2^3$ for the time interval

$$t_0 \leq \frac{1}{\varepsilon \sigma \| \vec{f}(0,\bar{z}) \|_{W_2^3}} + \varepsilon \tau , \quad \sigma = const, \quad \tau = const$$

Theorem 2. The kinetic model

$$\frac{\partial f_1}{\partial t} + \frac{\partial f_1}{\partial x} = \frac{1}{\varepsilon} \frac{f_2^2 - f_1 f_3}{\sqrt{1+\delta^2 (f_1+f_2+f_3)^2}} ,$$

$$\frac{\partial f_2}{\partial t} = -\frac{2}{3} \frac{f_2^2 - f_1 f_3}{\sqrt{1+\delta^2 (f_1+f_2+f_3)^2}} ,$$

$$\frac{\partial f_3}{\partial t} - \frac{\partial f_3}{\partial x} = \frac{1}{\varepsilon} \frac{f_2^2 - f_1 f_3}{\sqrt{1+\delta^2 (f_1+f_2+f_3)^2}} .$$

yields an approximate general solution of the "hydrodynamical" system (5) or (4) for small ε.

If in the expansion of the solution of the equation by powers of the small parameters ε we restrict ourselves to two terms, we obtain

$$\frac{\partial (e^{q_1+q_2-1} + e^{q_1-1} + e^{q_1-q_2-1})}{\partial t} + \frac{\partial (e^{q_1+q_2-1} - e^{q_1-q_2-1})}{\partial x} = 0 ,$$

$$\frac{\partial (e^{q_1+q_2-1} - e^{q_1-q_2-1})}{\partial t} + \frac{\partial (e^{q_1+q_2-1} - e^{q_1-q_2-1})}{\partial x} = \frac{\partial}{\partial x} \left[\frac{4 \varepsilon e^{q_2}}{(e^{q_2}+4+e^{-q_2})^2} \frac{\partial q_2}{\partial x} \right] .$$

These equations are analogues for a continuous medium of the Navier-

Stokes equations.

The figures below illustrate the results of the numerical integration of the structure equations for a shock wave for both our models.

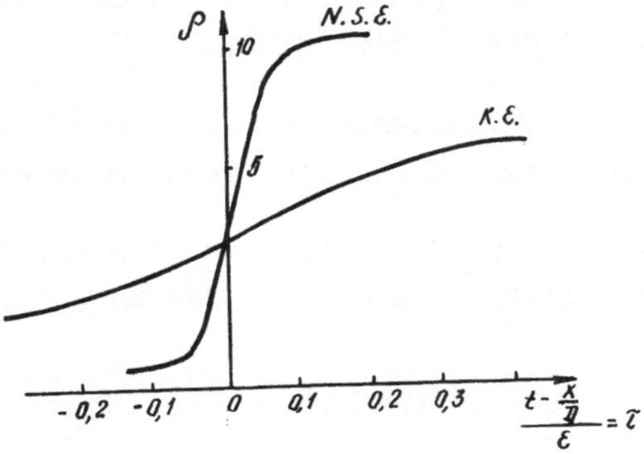

It is interesting that not only the fine structure of the wave, but also its thickness turns out to be substantially different.

In the final part of the report a conservative differential scheme for discrete kinetic equations on the basis of splitting method is considered [4].

$$\frac{u_n^{k+\frac{1}{2}} - u_n^k}{\tau} + \frac{u_n^k - u_{n-1}^k}{h} = 0,$$

$$\frac{v_n^{k+\frac{1}{2}} - v_n^k}{\tau} = 0,$$

$$\frac{w_n^{k+\frac{1}{2}} - w_n^k}{\tau} - \frac{w_{n+1}^k - w_n^k}{h} = 0.$$

$$\frac{u_n^{k+1} - u_n^{k+\frac{1}{2}}}{\tau} = \frac{1}{\varepsilon}\left(v_n^{2\,k+1} - u_n^{k+1} w_n^{k+1}\right),$$

$$\frac{v_n^{k+1} - v_n^{k+\frac{1}{2}}}{\tau} = -\frac{2}{\varepsilon}\left(v_n^{2\,k+1} - u_n^{k+1} w_n^{k+1}\right),$$

$$\frac{w_n^{k+1} - w_n^{k+\frac{1}{2}}}{\tau} = \frac{1}{\varepsilon}\left(v_n^{2\,k+1} - u_n^{k+1} w_n^{k+1}\right).$$

References

1. T.Carleman, Problems mathematiques dans la theorie cinetique des gaz, Almqvist ans Nilsells, Uppsala, 1957.

2. J.E.Brodwell, Shock structure in a Simple Discrete Velocity Gas, Phys. Fluids 7 (1964), 1243 - 1247.

3. S.K.Godunov and U.M.Sultangazin, On discrete models of the kinetic Boltzmann equation, Uspehi Mathematic Nauk, 26, 1971.

4. U.M.Sultangazin, The spherical harmonics and discrete methods of the transport equation, Alma-Ata, "Nauka", 1979.

APPLICATION OF A POLYNOMIAL SPLINE IN HIGHER-ORDER ACCURATE VISCOUS-FLOW COMPUTATIONS

M.G. Turner,* K.N. Ghia, U. Ghia and J.S. Keith*

Department of Aerospace Engineering and Applied Mechanics
University of Cincinnati, Cincinnati, Ohio 45221, U.S.A.

*General Electric Company, Cincinnati, Ohio 45215, U.S.A.

INTRODUCTION

Large amounts of computer time and storage are generally required for the solution of viscous-flow problems of engineering interest. These demanding requirements, which are larger when the Navier-Stokes equations rather than the boundary-layer equations are employed, can be significantly reduced with the use of higher-order accurate discretization procedures. In the past few years, a number of higher-order accurate schemes have appeared in the literature, e.g., [1-8]. Many of these are either Hermite or spline collocation procedures. These latter techniques yield fourth-order accurate solutions on uniform meshes with a 2x2 block-tridiagonal form of the governing matrix system and lead to smaller truncation errors than those of five-point finite-difference formulas with a pentadiagonal matrix system. In general, spline collocation techniques provide second- and higher-order accurate solutions with minimum truncation errors with uniform as well as nonuniform meshes.

However, for certain flow problems with localized high-gradient regions, it has been observed [6, 9] that, although the resulting solutions themselves may be smooth, their first- and/or second-order derivatives are frequently not smooth. This motivated the present study wherein a new spline technique is developed by employing a quartic spline polynomial S(4,2) of deficiency two, i.e., with two continuous derivatives. The 1-D spline S(4,2) has five unknowns at each node. The integrated form of the governing equation, which is generally used in finite-volume techniques, is used here to complete the equation set and comprises the key feature in the success of this new scheme. The scheme has been studied via application to four model problems in order to achieve the following objectives: (i) to provide the numerical analysis for 1-D spline S(4,2) technique for viscous-flow problems and demonstrate the advantages of this new spline over the existing splines S(5,3) and S(3,1); (ii) to assess the accuracy of spline S(4,2) by comparing with other existing fourth-order schemes; and, finally, (iii) to formulate the bi-quartic spline S(4,2) and demonstrate its feasibility by application to a 2-D model problem and show its potential for use in the solution of the Navier-Stokes equations.

FORMULATION OF QUARTIC SPLINE S(4,2)

The 1-D quartic spline polynomial S(4,2) for a function F(x) on the interval [j-1,j] with $(x_j - x_{j-1}) = h_j$, is given as

$$F(x) \equiv S(x;\, 4,2) = \sum_{i=0}^{4} A_i t^i \quad \text{where} \quad t = (x - x_{j-1})/h_j \tag{1}$$

and, at the grid points j-1 and j, the spline function is denoted as

$$S(x_{j-1};\ 4,2) = F_{j-1} \quad ; \quad S(x_j;\ 4,2) = F_j \quad . \tag{2}$$

Also, the spline derivative approximations of F'(x) and F"(x) are designated as m_j and M_j, respectively, so that

$$S'(x_{j-1};\ 4,2) = m_{j-1} \ , \qquad S'(x_j;\ 4,2) = m_j$$

$$\text{and} \quad S''(x_{j-1};\ 4,2) = M_{j-1} \ , \qquad S''(x_j;\ 4,2) = M_j \quad . \tag{3a,b}$$

As given by Eq. (1), the spline S(4,2) is a fourth-order polynomial with five unknowns per interval. For N nodes, there are a total of 5x(N-1) unknowns for which 4x(N-1) pieces of information consist of N differential equations together with 2 boundary conditions (assuming a second-order differential equation) and 3x(N-2) matching conditions. In order to close this system, (N-1) additional conditions are specified by applying the integrated form of the governing equation over each of the (N-1) intervals. The additional use of these integrated equations enables the present scheme to better represent the physics of a given flow problem. With some careful algebra, and by alternatively expressing the unknown A_i's in terms of the solution function F_j and its derivatives, the resulting (5x5) block-tridiagonal system can be reduced to a (2x2) block-tridiagonal matrix system.

BI-QUARTIC SPLINE S(4,2)

The discretization procedure is analogous to that for the 1-D S(4,2). For a given function $\phi(x,y)$, the bi-quartic polynomial in the region consisting of the cell encompassing the intervals [i-1,i] and [j-1,j] is given as

$$\phi(x,y) \equiv S[(x;\ 4,2),\ (y;\ 4,2)] = \sum_{m=0}^{4} \sum_{n=0}^{4} B_{mn}\, s^m t^n \tag{4}$$

where $s = (x-x_{i-1})/(x_i-x_{i-1})$ and $t = (y-y_{j-1})/(y_j-y_{j-1})$.

Thus, there are 25 unknowns per rectangular grid cell referred to as a patch. For a basic 4-patch region, there will be a total of 100 unknowns. Use of the differential equation, boundary conditions, matching conditions on the function and its first- and second-order derivatives with respect to x as well as y, together with the implementation of appropriately integrated equations, helps to form a complete system. The coefficients B_{mn} are alternatively expressed in terms of the function and its derivatives; the overall algebra of this process is quite involved and was, therefore, performed using an algebraic manipulation program ALGE.

APPLICATION TO MODEL PROBLEMS

The quartic spline S(4,2) developed in this study is implemented in the discretization and numerical solution of various model problems; the bi-quartic spline

S(4,2) is also implemented in the solution of Laplace's equation. The resulting discretized equations form a coupled set which is solved using Crout's direct banded-matrix solver. For purposes of comparison, solutions have been also generated for the spline-discretized equations obtained by implementing splines S(3,1) and S(5,3). These equations were solved using coupled solvers of the form given by U. Ghia et al. [10]. An analogous form of the coupled solver was also implemented by K. Ghia et al. [6] in their study of higher-order techniques using spline discretization. The spline S(4,2) results to be presented are generated using the VAX 780 computer, whereas all other results are obtained using the AMDAHL 470 V/7 computer. At present, no effort has been made to optimize any of the numerical methods, since extremely small CPU times are involved, e.g., with 25 η-intervals, a ξ-step of the flow past a parabola requires 0.0016 second of AMDAHL CPU time.

I. SOLUTION OF BURGERS' EQUATION

The spline technique developed is first applied to the 1-D nonlinear Burgers' equation

$$u_t + (u-U)\, u_\eta = \nu\, u_{\eta\eta} \quad , \qquad \eta = x - Ut \quad , \tag{5}$$

with the boundary conditions $u(-\infty) = 2U$ and $u(\infty) = 0$ and the exact solution

$$u = u[1 - \tanh\{U\eta/(2\nu)\}] \quad . \tag{6}$$

For $\nu \ll 1$ in Burgers' equation (1), the solution exhibits a thin region with large gradients, representative of a strong shock wave, highly concentrated shear layers and thin boundary layers. Unless handled carefully, this can lead to large truncation errors in the solution. Results have been obtained for a range of values of ν, but only representative results are presented here, in Table 1, for $\nu = 1/24$ using a nonuniform grid and three different spline techniques. With 18 intervals, the function u and its first and second derivatives as computed by spline S(4,2) are in much better agreement with the exact results as compared to both S(3,1) and S(5,3). All the results were obtained using double-precision arithmetic. The spurious oscillations which appear in the results with splines S(3,1) and S(5,3) do not appear in the results with spline S(4,2). For finer meshes, spline S(5,3) does lead to improvement over the results of spline S(3,1). For the case of $\nu = 1/8$, Hirsh [2] has provided fourth-order accurate results using compact differencing. Again, although not shown here, spline S(4,2) results are in slightly better agreement with the exact results than are those of Hirsh [2].

IIa. SELF-SIMILAR BOUNDARY-LAYER FLOWS

The laminar incompressible boundary-layer equations in Görtler variables (ξ,η) are given [Ref. 11] as

Momentum Equation: $$F_{\eta\eta} - VF_\eta + \beta(1-F^2) - 2\xi\, FF_\xi = 0 \quad , \tag{7}$$

Equation for V: $V_\eta + F + 2\xi F_\xi = 0$, (8)

where β is the pressure gradient parameter. For self-similar flows, the last term in Eqs. (7) and (8) vanishes to yield the similarity form of the equations. The reduced equations were solved for β = -0.1, 0.0 and 1.0. Table 2 shows the results obtained with η_{max} = 24.254 and 10 intervals. The results of Wornom [7] obtained by a fourth-order accurate box scheme (B4S) are also included. Although the results of splines S(3,1) and S(5,3) show some oscillations and overshoots for $\eta > 2.0$, those of spline S(4,2) on this coarse grid show excellent agreement with the exact results obtained using 640 intervals. The overall accuracy of the spline S(4,2) results is at least similar to, if not better than, that of Wornom [7]. It should be stated that the grid used by Wornom [7] is not optimum for the present scheme. The S(4,2) results obtained using a modified distribution of the given 11 grid points do show better agreement for large η; these results are also shown in Table 2. Fourth-order accuracy of the procedure for both F and V is demonstrated in Fig. 1 for the self-similar stagnation-point flow solution for which β = 1.

IIb. NONSIMILAR BOUNDARY-LAYER FLOW - FLOW PAST A PARABOLA

For symmetric flow past a parabola, the nonsimilar form of the boundary-layer equations (7) and (8) are solved in terms of the Görtler variables (ξ,η). To maintain second-order accuracy in the streamwise direction ξ, the quasilinearized coupled set of equations was solved using a Crank-Nicolson scheme. This led to oscillations in the function V for each of the η-derivative discretization schemes used, including the conventional second-order accurate finite-difference procedure. Mild oscillations in the function V persisted even when the grid was refined first in ξ-direction alone, and then in both the ξ and η directions. As Fig. 2 shows, these oscillations vanish with the use of a 3-point second-order accurate backward difference for the ξ-direction and, as such, all of the results presented here are obtained using this scheme. The present second-order accurate finite-difference results are in excellent agreement with those of Werle and Davis [11] obtained by the Crank-Nicolson scheme. Results for V were not presented in [11]. The present S(4,2) results for flow past a parabola are compared with the second-order accurate finite-difference results as well as with the fourth-order accurate spline S(5,3) solutions for the function F in Fig. 3, and for the skin friction and displacement thickness in Fig. 4. All of the spline results were obtained using 1/4 of the grid points used for the finite-difference technique. For spline S(4,2), still coarser grids lead to almost the same results.

III. SOLUTION OF LAPLACE'S EQUATION

The bi-quartic spline S(4,2) formulated in two-dimensions is implemented in the numerical solution procedure. The attendant mathematical difficulties of this procedure are quite severe. Therefore, as a model problem, Laplace's equation in a square region, with a sinusoidal function distribution along one boundary has been solved. Results of the accuracy study are shown in Fig. 5 where fourth-order accuracy is again realized with a rather coarse grid.

CONCLUSION

The quartic spline S(4,2) developed in this study overcomes some of the difficulties experienced by the authors in using spline S(5,3) or S(3,1) and provides fourth-order accurate results with relatively few grid points. In the course of this study, many more detailed numerical results have been generated than could be included in the present paper with its space limitations. These results are presently being assembled [12] and should provide detailed quantitative assessment of the various schemes examined for the model problems considered here. The accuracy of spline S(4,2) is comparable to, if not better than, that of the fourth-order box scheme and compact differencing scheme. This study suggests the use of spline S(4,2) as a potential means for fourth-order accurate solutions of Navier-Stokes equations.

ACKNOWLEDGEMENT

This research was supported, in part, by NASA-Lewis Grant No. NSG-3267, with Dr. Hue Kao as Technical Monitor. The authors are thankful to Mr. C.T. Shin for providing some of the results used in the comparative study.

REFERENCES

1. Krause, E., Hirschel, E.H. and Kordulla, W. (1976), Computers and Fluids, Vol. 4, pp. 77-92.
2. Hirsh, R. (1975), J. Computational Physics, Vol. 19, pp. 90-109.
3. Adam, Y. (1975), Comp. Math. Appl., Vol. 1, pp. 393-406.
4. Rubin, S.G. and Khosla, P.K. (1977), J. Computational Physics, Vol. 24, No. 3, pp. 217-244.
5. Rubin, S.G. and Khosla, P.K. (1978), Lecture Notes in Physics, edited by H. Cabannes, M. Holt and V. Rusanov, Vol. 90, pp. 468-476.
6. Ghia, K.N., Shin, C.T. and Ghia, U. (1979), AIAA CP-799, pp. 284-291.
7. Wornom, S.F. (1977), AIAA Paper No. 77-637, pp. 61-71.
8. Weinberg, B.C., Leventhal, S.H. and Ciment, M. (1977), AIAA Paper No. 77-638, pp. 72-81.
9. Shin, C.T., Ghia, K.N. and Ghia, U. (1981), Aerospace Engineering and Applied Mechanics Report No. AFL 81-9-60, University of Cincinnati, Ohio.
10. Ghia, U., Ghia, K.N. and Studerus, C. (1976), Lecture Notes in Physics, edited by A.I. van de Vooren and P.J. Zandbergen,vol. 59, pp. 197-204.
11. Werle, M.J. and Davis, R.T. (1972), J. of Applied Mechanics, Vol. 39, Series E, No. 1, pp. 7-12.
12. Ghia, K.N. Ghia, U. and Shin, C.T. (1982), Aerospace Engineering and Applied Mechanics Report under preparation, University of Cincinnati, Ohio.

TABLE 1. SOLUTION OF BURGERS' EQUATION: $\nu = 1/24$, $h_j/h_{j-1} = 1.4$, $N = 18$, $U = 0.5$.

η	u_{Exact}	S(4,2)	S(3,1)	S(5,3)	u_η - Exact	u_η - S(4,2)	u_η - S(3,1)	u_η - S(5,3)
-5.0000	1.0000	1.0000	1.0000	1.0000	-1.0508×10^{-25}	-3.7207×10^{-6}	7.2092×10^{-4}	5.7132×10^{-3}
-3.4988	1.0000	1.0000	1.0035	1.0218	-7.0021×10^{-18}	-7.2416×10^{-6}	-8.9389×10^{-4}	-2.8817×10^{-3}
-2.4265	1.0000	1.0000	1.0013	1.0098	-2.7139×10^{-12}	-1.8382×10^{-5}	1.2267×10^{-3}	3.4852×10^{-3}
-1.6605	1.0000	1.0000	1.0028	1.0173	-2.6625×10^{-8}	-6.7437×10^{-5}	-1.8992×10^{-3}	-4.3811×10^{-3}
-1.1134	1.0000	1.0000	1.0016	1.0123	-1.8902×10^{-5}	-4.0567×10^{-4}	3.5619×10^{-3}	5.8592×10^{-3}
-0.7226	0.9998	0.9996	1.0025	1.0152	-2.0556×10^{-3}	-4.3137×10^{-3}	-8.8027×10^{-3}	-1.4433×10^{-2}
-0.4435	0.9951	0,9949	1.0027	1.0052	-5.8016×10^{-2}	-6.0874×10^{-2}	3.4537×10^{-2}	-4.6322×10^{-2}
-0.2441	0.9493	0.9500	0.9668	0.9626	-0.5776	-0.5697	-0.6512	-0.5488
-0.1017	0.7722	0.7719	0.7753	0.7884	-2.1110	-2.1127	-2.2112	-2.2132
0	0.5000	0.5000	0.5000	0.5000	-3.0000	-3.0000	-2.9543	-3.1456

η	$u_{\eta\eta}$ - Exact	$u_{\eta\eta}$ - S(4,2)	$u_{\eta\eta}$ - S(3,1)	$u_{\eta\eta}$ - S(5,3)
-5.0000	-1.2609×10^{-24}	-4.4648×10^{-5}	8.6510×10^{-3}	6.8558×10^{-2}
-3.4988	-8.4026×10^{-17}	-8.6899×10^{-5}	-1.0802×10^{-2}	-3.6087×10^{-2}
-2.4265	-3.2567×10^{-11}	-2.2058×10^{-4}	1.4758×10^{-2}	4.2645×10^{-2}
-1.6605	-3.1950×10^{-7}	-8.0924×10^{-4}	-2.2920×10^{-2}	-5.4397×10^{-2}
-1.1134	-2.2682×10^{-4}	-4.8678×10^{-3}	4.2884×10^{-2}	7.2039×10^{-2}
-0.7226	-2.4658×10^{-2}	-5.1727×10^{-2}	-0.1062	-0.1785
-0.4435	-0.6894	-0.7230	0.4167	-0.5616
-0.2441	-6.2286	-6.1532	-7.2959	-6.0924
-0.1017	-13.7899	-13.7878	-14.6113	-15.3176
0.0000	0	1.0893×10^{-13}	4.5632×10^{-7}	3.6267×10^{-13}

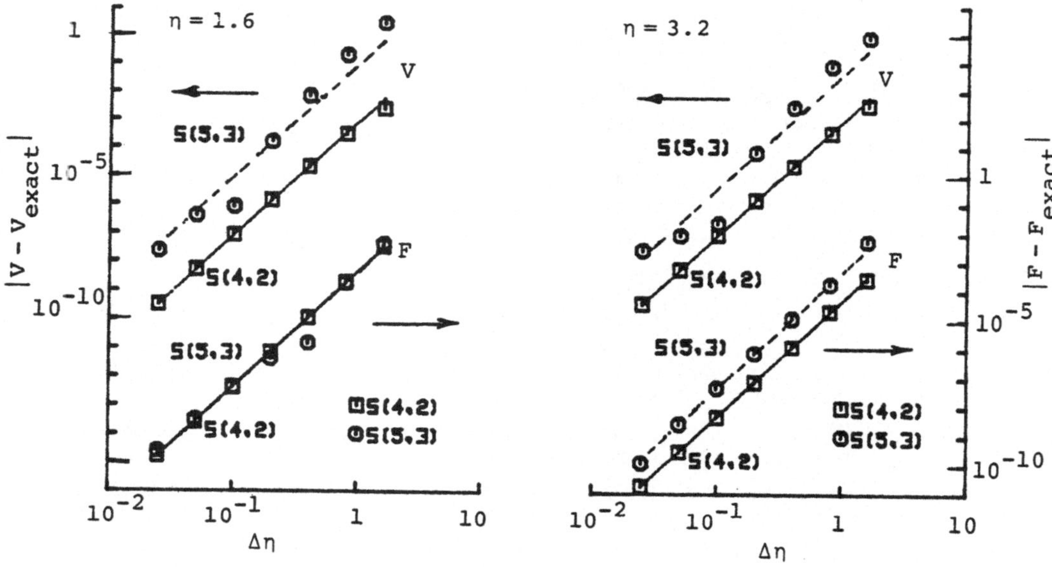

FIG. 1. ACCURACY STUDY FOR 1-D SPLINE S(4,2) FOR STAGNATION-POINT FLOW SOLUTION; $\beta = 1$, $\eta_{max} = 6.4$.

TABLE 2. COMPARATIVE STUDY FOR HIGHER-ORDER SOLUTIONS FOR SELF-SIMILAR BOUNDARY-LAYER FLOWS.

$\beta = 1.0$

η	EXACT	B4S	S(5,3)	S(4,2)	S(4,2)*
0.00000E+00	0.00000E+00	0.00000E+00	0.00000E+00	0.00000E+00	0.00000E+00
0.49975E-01	0.60350E-01	0.60359E-01	0.60262E-01	0.60359E-01	0.60260E-01
0.14205E+00	0.16503E+00	0.16506E+00	0.16478E+00	0.16505E+00	0.16482E+00
0.30761E+00	0.33236E+00	0.33242E+00	0.33181E+00	0.33242E+00	0.33223E+00
0.60231E+00	0.56784E+00	0.56795E+00	0.56667E+00	0.56796E+00	0.56845E+00
0.11266E+01	0.82381E+00	0.82393E+00	0.82133E+00	0.82410E+00	0.82443E+00
0.20651E+01	0.97721E+00	0.97683E+00	0.97107E+00	0.97803E+00	0.97741E+00
0.37657E+01	0.99989E+00	0.99904E+00	0.99162E+00	0.10003E+01	0.99989E+00
0.69004E+01	0.10000E+01	0.99972E+00	0.10064E+01	0.10012E+01	0.10000E+01
0.12809E+02	0.10000E+01	0.99984E+00	0.95336E+00	0.10018E+01	0.10000E+01
0.24254E+02	0.10000E+01	0.10000E+01	0.10000E+01	0.10000E+01	0.10000E+01

$\beta = 0.0$

η	EXACT	B4S	S(5,3)	S(4,2)	S(4,2)*
0.00000E+00	0.00000E+00	0.00000E+00	0.00000E+00	0.00000E+00	0.00000E+00
0.49975E-01	0.23468E-01	0.23480E-01	0.21918E-01	0.23736E-01	0.23483E-01
0.14205E+00	0.66705E-01	0.66738E-01	0.62300E-01	0.67466E-01	0.66744E-01
0.30761E+00	0.14437E+00	0.14444E+00	0.13485E+00	0.14602E+00	0.14443E+00
0.60231E+00	0.28164E+00	0.28178E+00	0.26315E+00	0.28485E+00	0.28167E+00
0.11266E+01	0.51474E+00	0.51498E+00	0.48167E+00	0.52044E+00	0.51482E+00
0.20651E+01	0.83286E+00	0.83293E+00	0.78480E+00	0.84078E+00	0.83266E+00
0.37657E+01	0.99552E+00	0.99616E+00	0.93983E+00	0.10054E+01	0.99571E+00
0.69004E+01	0.10000E+01	0.99947E+00	0.96372E+00	0.10075E+01	0.10000E+01
0.12809E+02	0.10000E+01	0.99983E+00	0.95143E+00	0.10057E+01	0.10000E+01
0.24254E+02	0.10000E+01	0.10000E+01	0.10000E+01	0.10000E+01	0.10000E+01

$\beta = -0.1$

η	EXACT	B4S	S(5,3)	S(4,2)	S(4,2)*
0.00000E+00	0.00000E+00	0.00000E+00	0.00000E+00	0.00000E+00	0.00000E+00
0.49975E-01	0.16080E-01	0.16092E-01	0.19978E-01	0.16934E-01	0.16099E-01
0.14205E+00	0.46360E-01	0.46394E-01	0.57438E-01	0.48787E-01	0.46417E-01
0.30761E+00	0.10289E+00	0.10297E+00	0.12686E+00	0.10814E+00	0.10303E+00
0.60231E+00	0.20967E+00	0.20981E+00	0.25630E+00	0.21989E+00	0.20997E+00
0.11266E+01	0.41286E+00	0.41312E+00	0.49611E+00	0.43113E+00	0.41336E+00
0.20651E+01	0.75159E+00	0.75173E+00	0.86601E+00	0.77676E+00	0.75160E+00
0.37657E+01	0.98964E+00	0.99005E+00	0.10587E+01	0.10103E+01	0.98993E+00
0.69004E+01	0.10000E+01	0.99900E+00	0.10666E+01	0.10132E+01	0.10001E+01
0.12809E+02	0.10000E+01	0.99971E+00	0.10751E+01	0.10089E+01	0.10000E+01
0.24254E+02	0.10000E+01	0.10000E+01	0.10000E+01	0.10000E+01	0.10000E+01

*WITH 11-POINT GRID ALTERED TO BE COARSER FOR $\eta < 2$, BUT FINER FOR $\eta > 2$.

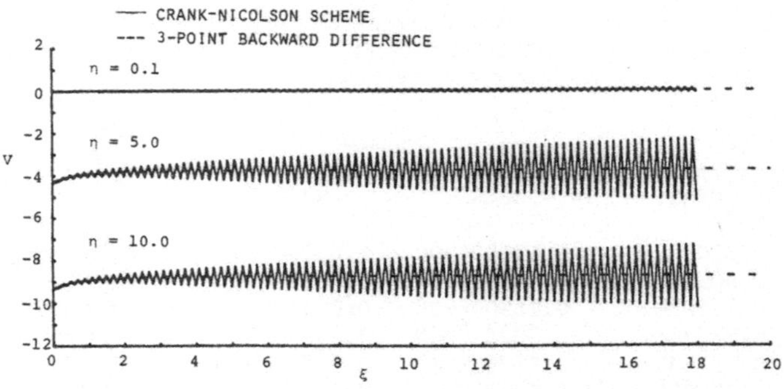

FIG. 2. EFFECT OF CRANK NICOLSON AND 3-POINT BACKWARD DIFFERENCE SCHEMES ON STREAMWISE VARIATION OF NORMAL VELOCITY V FOR PARABOLA.

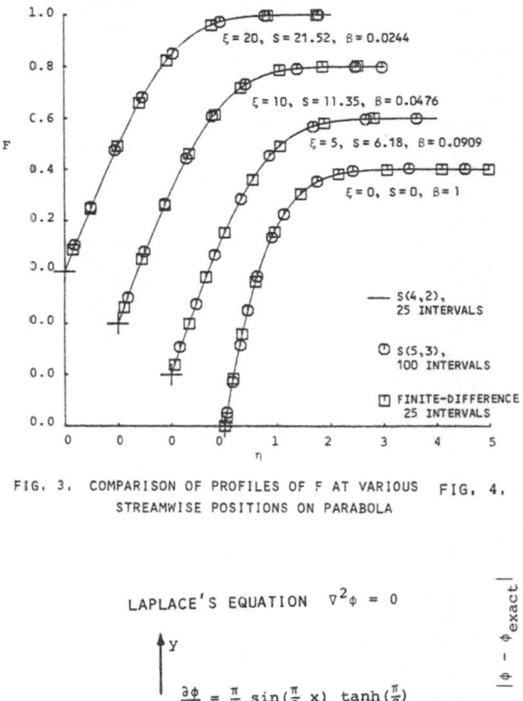

FIG. 3. COMPARISON OF PROFILES OF F AT VARIOUS STREAMWISE POSITIONS ON PARABOLA

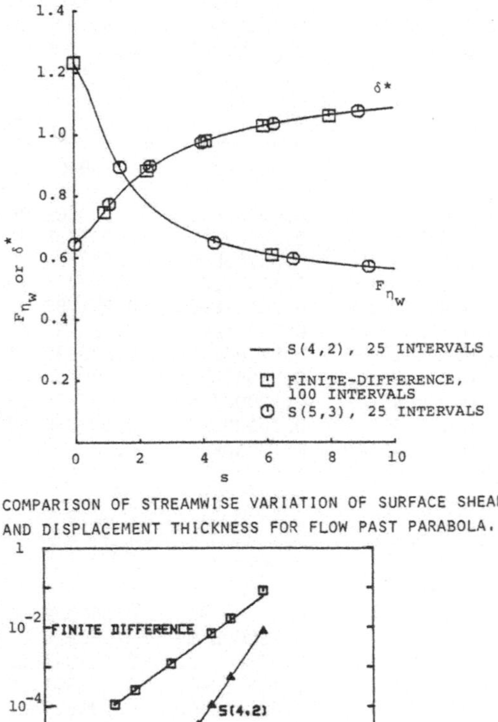

FIG. 4. COMPARISON OF STREAMWISE VARIATION OF SURFACE SHEAR AND DISPLACEMENT THICKNESS FOR FLOW PAST PARABOLA.

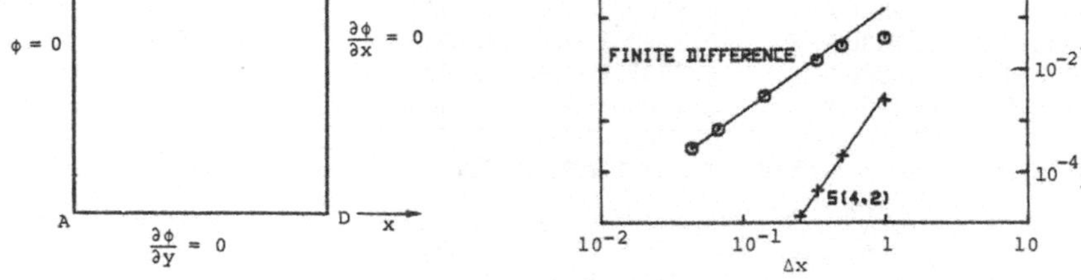

FIG. 5. S(4,2) SOLUTION OF LAPLACE'S EQUATION: ACCURACY STUDY.

FLUX-VECTOR SPLITTING FOR THE EULER EQUATIONS

Bram van Leer
Leiden University Observatory
P.O. Box 9513, 2300 RA Leiden, The Netherlands

Introduction

When approximating a hyperbolic system of conservation laws $w_t + \{f(w)\}_x = 0$ with so-called upwind differences, we must, in the first place, establish which way the wind blows. More precisely, we must determine in which direction each of a variety of signals moves through the computational grid. For this purpose, a physical model of the interaction between computational cells is needed; at present two such models are in use.

In one model, neighboring cells interact through discrete, finite-amplitude waves. The nature, propagation speed and amplitude of these waves are found by solving, exactly or approximately, Riemann's initial-value problem for the discontinuity at the cell interface. We may call this the Riemann approach (Fig. 1a). The numerical technique of distinguishing between the influence of the forward- and the backward-moving waves is called flux-difference splitting; examples are the methods of Roe [1] and of Osher [2].

In the other model, the interaction of neighboring cells is accomplished through mixing of pseudo-particles that move in and out of each cell according to a given velocity distribution. We may call this the Boltzmann approach (Fig. 1b). The numerical technique of distinguishing between the influence of the forward- and the backward-moving particles is called flux-vector splitting or simply flux-splitting; an example is the "beam scheme" of Prendergast [3], rediscovered by Steger and Warming [4].

Both kinds of splitting are discussed by Harten, Lax and Van Leer [5].

The present paper is restricted to flux-vector splitting for the Euler equations of compressible flow, with the ideal-gas law used as equation of state.

Goal

We wish to split the flux $f(w)$ in a forward flux $f^+(w)$ and a backward flux $f^-(w)$, that is,

(1) $f(w) = f^+(w) + f^-(w)$,

(2) df^+/dw must have all eigenvalues ≥ 0,
df^-/dw must have all eigenvalues ≤ 0,

under the following restrictions:

(3) $f^\pm(w)$ must be continuous, with
$f^+(w) \equiv f(w)$ for Mach-numbers $M \geq 1$,
$f^-(w) \equiv f(w)$ for $M \leq -1$,

(4) the components of f^+ and f^- together must mimic the symmetry of f with respect to M (all other state quantities held constant), that is,
$f_k^+(M) = \pm f_k^-(-M)$ if $f_k(M) = \pm f_k(-M)$,

(5) $df^\pm/dw$ must be continuous,

(6) $df^\pm/dw$ must have one eigenvalue vanish for $|M| < 1$,

(7) $f^\pm(M)$, like $f(M)$, must be a polynomial in M, and of the lowest possible degree.

Restriction (3) ensures that in supersonic regions flux-vector splitting leads to standard upwind differencing. Restrictions (4) and (5) are self-evident, although (5) was not satisfied in [4], with negative consequences for the smoothness of numerical results. Restriction (6) is crucial, greatly narrowing down the choice of functions. The degeneracy of $f^\pm(w)$ in subsonic cells makes it possible to build stationary shock structures with no more than two interior zones, just as does the flux-difference splitting technique in [2]. Finally, requirement (7) makes the splitting unique.

Research was supported under NASA Contract No. NAS1-15810 while the author was in residence at ICASE, NASA Langley Research Center, Hampton, VA 23665.

Derivation

Consider the one-dimensional Euler equations. We shall regard the full, forward, and backward fluxes as functions of density ρ, sound speed c and Mach number M. The full flux reads

$$(8) \quad f(\rho,c,M) = \begin{pmatrix} \rho c M \\ \rho c^2(M^2 + \frac{1}{\gamma}) \\ \rho c^3 M(\frac{1}{2}M^2 + \frac{1}{\gamma-1}) \end{pmatrix},$$

where γ is the specific-heat ratio.

From conditions (3) and (5) it follows that $f^+(\rho,c,M)$ as well as $\partial f^+(\rho,c,M)/\partial M$ must vanish for $M \downarrow -1$, while $f^-(\rho,c,M)$ and $\partial f^-(\rho,c,M)/\partial M$ must vanish for $M \uparrow 1$. Condition (7) then leads to the restriction that f^+ includes a factor $(M+1)^2$ and f^- a factor $(-M+1)^2$, for $|M| < 1$. Without introducing further factors depending on M we can now achieve the splitting of the mass flux:

$$(9) \quad \rho c M \equiv \rho c\{\tfrac{1}{2}(M+1)\}^2 - \rho c\{\tfrac{1}{2}(-M+1)^2\}, \qquad |M| < 1,$$

satisfying (1), (3), (4), (5) and (7). In order to split the momentum flux in agreement with (3) and (5) we need cubic polynomials in M:

$$(10) \quad \rho c^2(M^2 + \frac{1}{\gamma}) \equiv \rho c^2\{\tfrac{1}{2}(M+1)\}^2(\frac{\gamma-1}{\gamma}M + \frac{2}{\gamma}) + \rho c^2\{\tfrac{1}{2}(-M+1)\}^2(-\frac{\gamma-1}{\gamma}M + \frac{2}{\gamma}), \quad |M| < 1;$$

again, (1), (3), (4), (5) and (7) are satisfied.

The splitting of the energy flux can now be achieved by combining the split mass- and momentum-fluxes:

$$(11) \quad f^{\pm}_{energy} = \frac{\gamma^2}{2(\gamma^2-1)} (f^{\pm}_{momentum})^2/f^{\pm}_{mass}, \quad |M| < 1.$$

The scale factor is needed to satisfy (3). The relation (11) between the components of $f^{\pm}$ makes these fluxes degenerate: $df^{\pm}(w)/dw$ has a zero eigenvalue for $|M| < 1$. Thus condition (6) is fulfilled. Moreover, since the vanishing eigenvalue is continuous for $M = \pm 1$, we have sufficient information about the smoothness of $f^{\pm}$ to conclude that condition (5) is fully satisfied. Testing the fulfilment of (7), (4) and (1) by (11) is trivial.

We still must determine if condition (2) is satisfied by the splitting (9), (10), (11). There is a good reason to believe this is indeed the case: the eigenvalue $\mu_1^{\pm}$, of $df^{\pm}/dw$ that is most likely to violate condition (2), i.e. to have the wrong sign, has been forced to vanish for $|M| < 1$. We find that the non-zero eigenvalues $\mu^+_{2,3}$ of df^+/dw are the roots of the quadratic equation:

$$(12) \quad (\mu^+)^2 - c\mu^+\cdot\frac{3}{2}(1+M)\left[1-\frac{\gamma-1}{12\gamma(\gamma+1)}(M-1)\{\gamma(M-1)^2 + 2\gamma(M-1) - 2(\gamma+3)\}\right]$$
$$+ c^2\cdot\frac{1}{4}(1+M)^3\left[1-\frac{M-1}{8\gamma(\gamma+1)}\{4\gamma(\gamma-1)(M-1) + (\gamma+1)(3-\gamma)\}\right] = 0, \quad |M| < 1.$$

In the relevant range $1 \le \gamma \le 3$, the roots of (12) are positive. The negativity of $\mu^-_{2,3}$ follows by symmetry.

This completes the derivation of the split fluxes for the one-dimensional Euler equations. The formulas for the three-dimensional equations are given in Table I (full equations) and Table II (constant enthalpy assumed).

Stability

The stability analysis for the first-order upwind scheme based on the above split fluxes is complicated by the fact that df^+/dw and df^-/dw commute neither with each other nor with df/dw, for $|M| < 1$. This leads to a reduction of the CFL limit; in the worst case, M=0, we find for the shortest waves:

$$(13) \quad \frac{\Delta t}{\Delta x} c \le 2\gamma/(\gamma+3).$$

A practical local stability criterion is

$$(14) \quad \frac{\Delta t}{\Delta x}(|u|+c) \le \{2\gamma + |M|(3-\gamma)\}/(\gamma+3), \quad |M| < 1.$$

Steady discontinuities

The degeneracy of $f^{\pm}$ for $|M| < 1$ makes it possible to build numerical shock structures that satisfy the steady upwind-difference equations using only two interior zones.

Consider the one-dimensional Euler equations. We denote the supersonic pre-shock state by L, the subsonic post-shock state by R and the interior states by P and Q, as in Fig. 2. To require stationarity means to require constancy of net flux:

$$(15)\quad \begin{aligned} f_R &= f_L \\ &= f_L^+ + f_P^- \\ &= f_P^+ + f_Q^- \\ &= f_Q^+ + f_R^- . \end{aligned}$$

Assume that the first equality, i.e. the jump condition across the full shock structure, holds. The second equality is automatically satisfied if zone P, like L, is supersonic, with $f_P^- = 0$, $f_L^+ = f_L$. The fourth equality boils down to

$$(16)\quad f_R^+ = f_Q^+ ,$$

which, if zone Q is subsonic, implies only 2 independent equations. The third equality stands for 3 equations, so that we end up with 5 equations for 6 unknowns, the components of w_P and w_Q. The solutions form a one-parameter family of steady shock profiles, the parameter relating to the sub-grid shock position. There exists one profile with only one interior zone that is precisely sonic. In contrast, Godunov's and Roe's schemes yield steady shocks with one or no interior zone.

Another scheme that yields steady shock structures with two interior states is Osher's flux-difference scheme [2]. For a scalar conservation law it boils down to a flux-splitting method.

Unlike Osher's scheme, the present flux-vector splitting can not preserve a stationary contact discontinuity; in fact, no flux-vector splitting scheme can (see [5]). This is readily understood from the underlying physical model: cell-interactions are achieved through mixing, a diffusive process. Present research is aimed at the development of a computationally simple "collision term" that could prevent the diffusion across a steady contact discontinuity.

A comparison

For one-dimensional isothermal flow ($\gamma=1$, constant c) the flux-vector splitting of Steger and Warming, alias the beam scheme, can be derived from the assumption that there are two kinds of gas particles, equally abundant but moving at different rates: the beams. Per unit volume, $\frac{1}{2}\rho$ moves with velocity u-c and $\frac{1}{2}\rho$ with velocity u+c, yielding the correct average flow speed u. For $|M| < 1$ there is a forward and a backward beam, with associated fluxes satisfying (1), (3) and (4):

$$(17)\quad f^{\pm}_{B/SW} = \begin{pmatrix} \frac{1}{2}\rho(u\pm c) \\ \frac{1}{2}\rho(u\pm c)^2 \end{pmatrix}, \quad |u| < c.$$

In Fig. 3a the dependence of the components of $f_{B/SW}$ on M is shown. The momentum flux is continuously differentiable at $M = \pm 1$; the mass flux is not. The splitting given by Eqs. (9) and (10) for $\gamma = 1$ reads

$$(18)\quad f^{\pm}_{VL} = \begin{pmatrix} \pm\frac{1}{4}\rho(u\pm c)^2/c \\ \frac{1}{4}\rho(u\pm c)^2 \end{pmatrix}, \quad |u| < c,$$

with the same momentum-flux splitting but improved mass-flux splitting. The dependence on M is shown in Fig. 3b. The figs. 4a and 4b are graphs of the eigenvalues of $df^{\pm}_{B/SW}/dw$ and $df^{\pm}_{VL}/dw$. Their values are given by

$$(19)\quad \begin{aligned} (\mu^{+}_{1,2})_{B/SW} &= \tfrac{1}{4}c\{3+2M\mp\surd(5+4M)\}, \\ (\mu^{+}_{1})_{VL} &= 0, \qquad\qquad |M| < 1, \\ (\mu^{+}_{2})_{VL} &= \tfrac{1}{4}c(5-M)(1+M), \end{aligned}$$

and the symmetry condition (4). The eigenvalues shown in Fig. 4a, while satisfying (2), have a discontinuity at M = +1 or -1. In a stationary numerical solution this causes a discontinuity in the gradient at the sonic point, as seen from Fig.5, curve B/SW. Correcting the mass-flux splitting performs a small miracle, as seen from curve VL. The gradient-discontinuity disappears, and the numerical diffusion in the subsonic region is substantially reduced. This reduction corresponds to the reduction of the eigenvalues $\mu_1^{\pm}$ to zero.

Conclusions

For the full or isenthalpic Euler equations combined with the ideal-gas law, the flux-vector splitting presented here is, by a great margin, the simplest means to implement upwind differencing. For a polytropic gas law, with $\gamma > 1$, closed formulas have not yet been derived.

The scheme produces steady shock profiles with two interior zones. There is evidence [10] that, among implicit versions of upwind methods, those with a two-zone steady-shock representation give faster convergence to a steady solution than those with a one-zone representation.

A disadvantage in using any flux-vector splitting is that it leads to numerical diffusion of a contact discontinuity at rest. This diffusion can be removed; present research is aimed at achieving this with minimal computational effort.

Numerical solutions by first- and second-order schemes including the above split fluxes can be found in Refs. [6], [7] (one-dimensional) and [8], [9] (two-dimensional).

References

1. P.L. Roe, J. Computational Phys. 43 (1981), 357.
2. S. Osher, "Numerical solution of singular perturbation problems and hyperbolic systems of conservation laws", North-Holland Mathematical Studies 47 (1981), 179.
3. R.H. Sanders and K.H. Prendergast, Astrophys. J. 188 (1974), 489.
4. J.L. Steger and R.F. Warming, J. Computational Phys. 40 (1981), 263.
5. A. Harten, P.D. Lax and B. van Leer, "Upstream differencing and Godunov-type schemes", ICASE Report No. 82-5; to appear in SIAM Review.
6. J.L. Steger, "A preliminary study of relaxation methods for the inviscid conservative gasdynamics equations using flux-vector splitting", Report No. 80-4, August 1980, Flow Simulations, Inc.
7. G.D. van Albada, B. van Leer and W.W. Roberts, Jr., "A comparative study of computational methods in cosmic gas dynamics", Astron. Astrophys. 108 (1982), 76.
8. G.D. van Albada and W.W. Roberts, Jr., Ap. J. 246 (1981), 740.
9. M.D. Salas, "Recent developments in transonic flow over a circular cylinder", NASA Technical Memorandum 83282 (April 1982).
10. P.M. Goorjian and R. van Buskirk, "Implicit calculations of transonic flow using monotone methods", AIAA Paper 81-331 (1981).

TABLE I. Flux-splitting for the Euler equations. $M = u/c$.

conserved quantity	x-flux f	forward x-flux f^+, $\lvert M\rvert < 1$
mass	ρu	$\rho c\{\frac{1}{2}(M+1)\}^2$
x-momentum	$\rho(u^2+c^2/\gamma)$	$f^+_{mass}\cdot\{(\gamma-1)u+2c\}/\gamma$
y-momentum	ρuv	$f^+_{mass}\cdot v$
z-momentum	ρuw	$f^+_{mass}\cdot w$
total energy	$\rho u\{\frac{1}{2}(u^2+v^2+w^2) + c^2/(\gamma-1)\}$	$f^+_{mass}\cdot[\{(\gamma-1)u+2c\}^2/\{2(\gamma^2-1)\} + \frac{1}{2}(v^2+w^2)]$

TABLE II. Flux-splitting for the isenthalpic Euler equations. $H \equiv$ enthalpy, $c_* = \sqrt{[\{2(\gamma-1)/(\gamma+1)\}\{H - \frac{1}{2}(v^2+w^2)\}]}$, $M_* = u/c_*$.

conserved quantity	x-flux f	forward x-flux f^+, $\lvert M_*\rvert < 1$
mass	ρu	$\rho c_*\{\frac{1}{2}(M_*+1)\}^2$
x-momentum	$\rho\{(\gamma+1)/(2\gamma)\}(u^2+c_*^2)$	$f^+_{mass}\cdot\{(\gamma+1)/\gamma\}c_*$
y-momentum	ρuv	$f^+_{mass}\cdot v$
z-momentum	ρuw	$f^+_{mass}\cdot w$

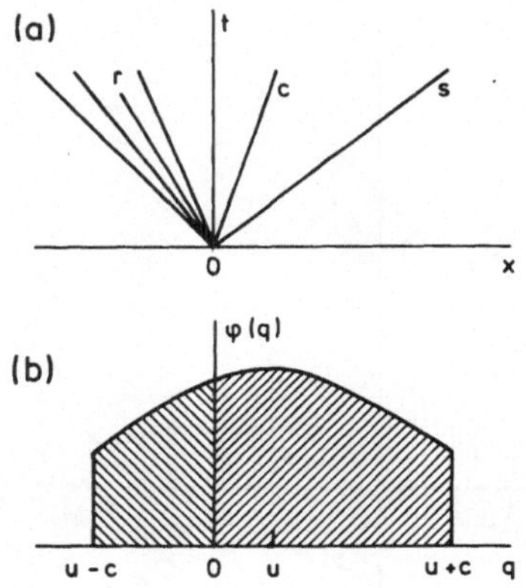

FIGURE 1. Two physical models for upwind differencing. (a) Space-time diagram showing waves in the solution of a Riemann problem: forward shock **s** and contact discontinuity **c**, backward rarefaction **r**. (b) Velocity-distribution function used in the Boltzmann model. The densities of the forward-moving and backward-moving particles are represented by the differently shaded areas.

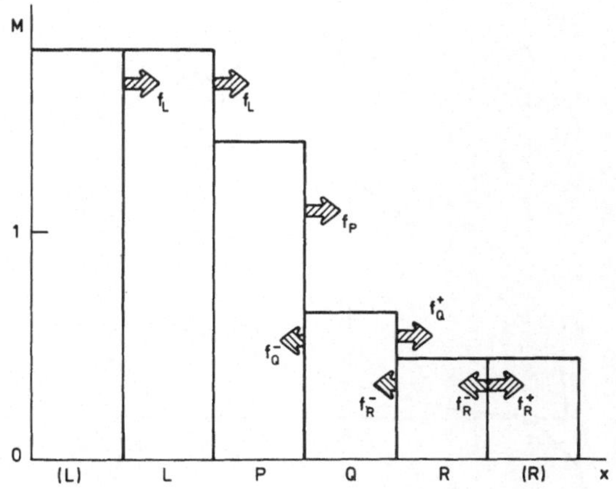

FIGURE 2. Steady shock profile from a flux-split upwind scheme.

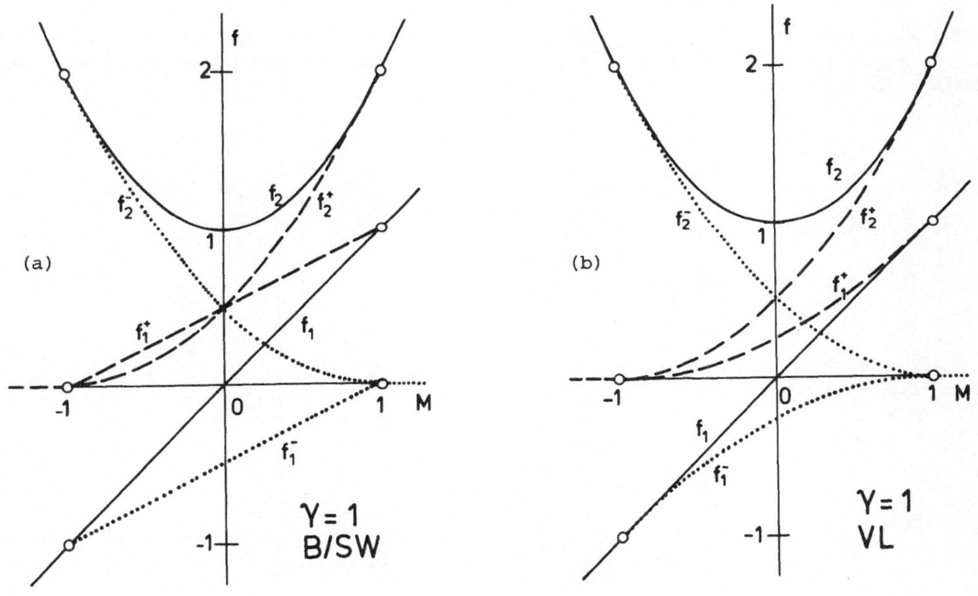

FIGURE 3. Flux-splitting for the isothermal Euler equations ($\gamma=1$). Plotted against Mach number are mass fluxes (subscript 1) and momentum fluxes (subscript 2), normalized by ρc and ρc^2, respectively. (a) Beam scheme/Steger and Warming; (b) Van Leer.

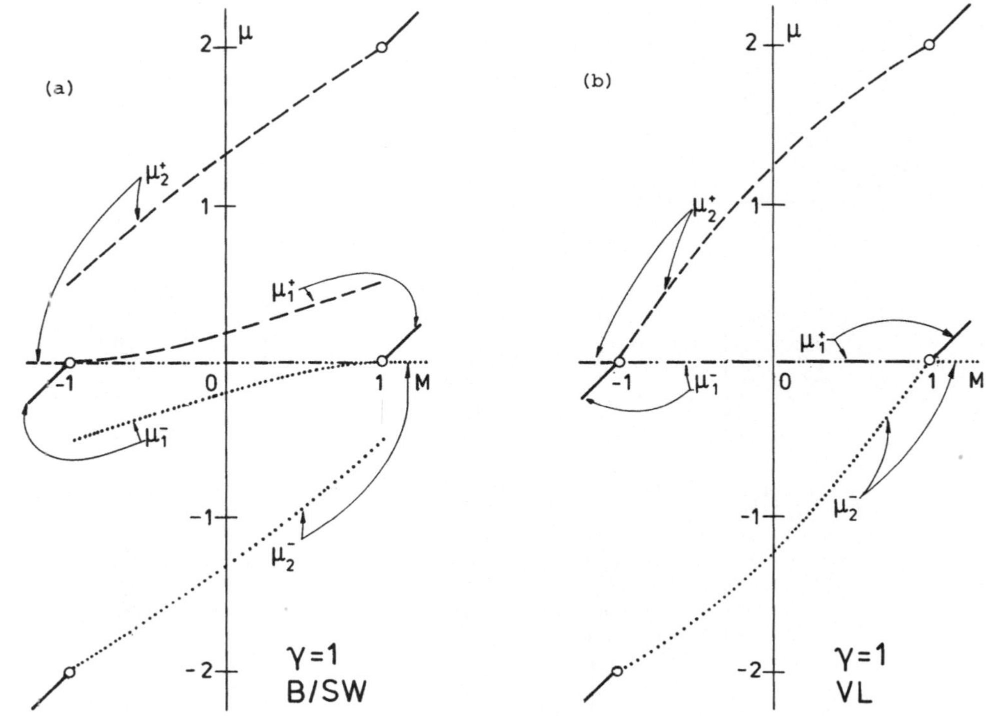

FIGURE 4. Eigenvalues of split-flux derivatives (γ=1). Plotted against Mach number are the eigenvalues of df^+/dw and df^-/dw, normalized by c. (a) Beam scheme/Steger and Warming; (b) Van Leer.

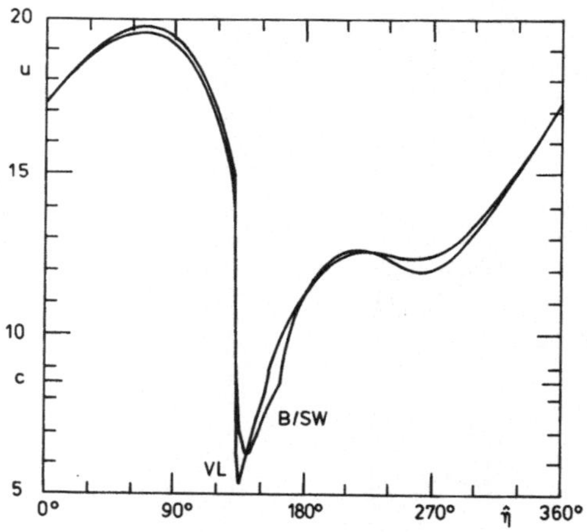

FIGURE 5. Numerical solutions of the periodic, nozzle-type, cosmic flow problem (γ=1) from [7] by flux-split upwind schemes, on a 128-zone grid. Plotted is velocity against coordinate angle. B/SW ≡ steady solution by Beam scheme/Steger and Warming; VL ≡ steady solution by Van Leer.

NUMERICAL SOLUTION OF THE PROBLEM OF SUPERSONIC FLOW PAST WINGS OF ARBITRARY FORM WITH A DETACHED SHOCK WAVE

G.P. Voskresensky
Keldysh Institute of Applied Mathematics
USSR Academy of Sciences
Moscow, USSR

Abstract

A numerical method has been developed to calculate supersonic inviscous gas flow around wings of arbitrary form with a shock wave detached from the leading edge. The solution domain is divided into three subdomains. A total algorithm is formed from the specific ones of each subdomains, depending on the wing configuration. For the front part of the wing where the flow is subsonic and supersonic the problem is solved by using the time-dependent stationary principle. The remaining parts of the wing including its tip side are treated as three-dimensional stationary cases. A number of test cases are computed for the flow around the wings of various configurations.

Introduction

The supersonic inviscous gas flow around the aircraft wings is considered. The planform and the airfoil of the wings must be feasible to construct but otherwise arbitrary. Such wings may have a shock wave attached to or detached from the leading edge. When the airfoil is blunt-nosed the shock wave is always detached. In the sharp-nosed case it separates at the leading edge only for some specific combination of the number $M_{n\infty}$ measured along the normal to the leading edge, of the airfoil tip angle and the angle of attack. As there is always a slight technological bluntness at the leading edge the detached shock wave should be considered as a more general case.

One of the major difficulties in solving the flow problem around the arbitrary form wings is to design the coordinate system in which the wing surface can well be described. Such an unified coordinate system covering the entire solution domain cannot be constructed for all wings actually used in aircraft. In some cases it is expedient to divide the solution domain into a number of subdomains, each with its own but compatible coordinate system. Such division is carried out in the proposed method. For each subdomain an algorithm is developed and a total algorithm is formed from the specific ones depending on the wing planform and its airfoils. For example, in the delta-wing case the algorithm is comprised of two parts described within the unified coordinate system, and for trapezoidal low sweep wings it consists of three parts described in different coordinate systems.

The elliptical wing case proves to be most complex. In this method it is chosen as basic. There is a detached shock wave separating from the blunt nose of such a wing. Behind it a subsonic flow zone is located. The latter is bounded downstream and at the tip side of the wing by the sonic surface. Next is a supersonic flow zone. Thus if the boundary of the solution domain is situated in the supersonic zone the flow on the front part of the wing is merged, i.e. consisting of subsonic and supersonic flows. Such a flow can be described by differential equations of elliptic-hyperbolical type. The formulation of the problem is very complex, however, it may be treated as a nonstationary time-hyperbolic case. A simpler problem with initial data is considered to be correct here. Introducing the fourth variable (time) into the equations is a penalty for this simplification.

Beyond the boundary of the front part the flow downstream and at the tip side is essentially supersonic.

I.The above considerations define a general statement of the problem. The entire solution domain is given in an initial Cartesian coordinate system, consisting of three subdomains (Fig. I). The first subdomains refers to the front part of the wing. It is bounded by the wing and shock wave surfaces, by the surfaces Π downstream and by the plane L along the tip side. Π and L are lying entirely in the supersonic flow zone, i.e. being of space type. The second subdomain corresponds to the tip side part of the wing and the third subdomain - to its central part. In these two subdomains the shock wave is an outer boundary while the wing surface is an inner boundary. The subdomains are separated by the planes N whose position is determined below.

In all three subdomains curvilinear body-surface-fixed coordinate systems are introduced so that complex physical solution domains are mapped onto simple orthogonal computational subdomains. New coordinates are introduced by using the following transformation of variables (Fig. 2).

In the first subdomain $(t,x,y,z) \to (\tau,\xi,\eta,\zeta)$,

$$x = \zeta_1(\tau,\zeta) + [\zeta(\eta,\zeta) - A\cos\eta]\cos\varphi, \qquad A = G + \xi(F-G),$$

$$y = \zeta_2(\tau,\eta,\zeta) + A\sin\eta,$$

$$z = \zeta_3(\tau,\zeta) - [\zeta(\eta,\zeta) - A\cos\eta]\sin\varphi,$$

$G=G(\tau,\eta,\zeta)$, $F=F(\tau,\eta,\zeta)$ are the equations of the wing and shock wave surfaces; $\varphi,\zeta,\zeta_1,\zeta_2,\zeta_3$ are the parameters of the computational grid.

In the second domain $(x,y,z) \to (\tau,\xi,\eta)$, (here τ is the space coord.)

$$x = \zeta_1(\tau) - [\zeta(\tau,\eta) + A\cos\eta]\cos\varphi$$

$$y = \zeta_2(\tau,\eta) + A\sin\eta$$

$$z = \zeta_3(\tau) + [\zeta(\tau,\eta) + A\cos\eta]\sin\varphi$$

$G=G(\tau,\eta)$, $F=F(\tau,\eta)$ are the equations of the wing and shock wave surfaces; $\varphi,\zeta,\zeta_1,\zeta_2,\zeta_3$ are the parameters with the same meaning as in the first domain.

In the third subdomain $(x,y,z) \to (\tau,\xi,\zeta)$,

$$x = \zeta_1(\tau,\zeta)$$

$$y = \zeta_2(\tau,\zeta) + A$$

$$z = \zeta_3(\tau,\zeta)$$

$G=\pm G(\tau,\zeta)$, $F=\pm F(\tau,\zeta)$ are the equations of the wing and shock wave surfaces; the sign "+" stands for the upper subdomain (above the wing) and the sign "-" for the lower one. The coordinates of subdomain II are transformed to the ones of subdomain III when $\eta = \pm\pi/2$ and $\zeta = 0$ at the boundary between the subdomains N $(\zeta=1)$. Local clustering of the computational grid can be achieved with the help of some intermediate coordinates: $(\alpha,\beta,\gamma,\tau)$. The relations $\xi=\xi(\alpha,\beta,\gamma,\tau)$, $\eta=\eta(\alpha,\beta,\gamma,\tau)$, $\zeta=\zeta(\alpha,\beta,\gamma,\tau)$ for the first subdomain or ones $\tau=\alpha$ $\xi=\xi(\alpha,\beta,\gamma)$, $\eta=\eta(\alpha,\beta,\gamma)$ for the second and third subdomains are determined specifically by their purposes. In this case the grid irregular in ξ, η, ζ will be regular in α,β,γ

II. The inviscous gas flow is described by the gasdynamic equations:

$$\frac{\partial X}{\partial t} + \bar{A}\frac{\partial X}{\partial x} + \bar{B}\frac{\partial X}{\partial y} + \bar{C}\frac{\partial X}{\partial z} = 0, \quad X = \{\rho, u, v, w, p\}$$

The nonzero elements of the matrices $\bar{A}$, $\bar{B}$, $\bar{C}$ are: $a_{11}=a_{22}=a_{33}=a_{44}=a_{55}=u$, $a_{25}=b_{35}=c_{45}=1/\rho$, $a_{52}=b_{53}=c_{54}=\rho a^2$, $b_{11}=b_{22}=b_{33}=b_{44}=b_{55}=v$, $c_{11}=c_{22}=c_{33}=c_{44}=c_{55}=w$

where a is the sound velocity.

After the introduction of the new variables the system is reduced to

$$\frac{\partial X}{\partial \tau} + A\frac{\partial X}{\partial \xi} + B\frac{\partial X}{\partial \eta} + C\frac{\partial X}{\partial \zeta} = 0$$

where $A = \bar{A}\xi_x + \bar{B}\xi_y + \bar{C}\xi_z + \xi_\tau E$, $B = \bar{A}\eta_x + \bar{B}\eta_y + \bar{C}\eta_z + \eta_\tau E$, $C = \bar{A}\zeta_x + \bar{B}\zeta_y + \bar{C}\zeta_z + \zeta_\tau E$
and E is an unit matrix. In the stationary case $\frac{\partial X}{\partial \tau} = 0$.

III. In the first subdomain the problem is formulated for a 3D-unsteady hyperbolic system of equations. It is solved by using the time-dependent stationary principle. In this case for initial conditions one can use data obtained from the flow around a body with a configuration close to the wing or even a sphere. In the course of the solution the wing shape and the freestream parameters may be monotonically changed. Thus by starting with the flow around the front part of the sphere; then flattening and stretching it one can obtain any desired form of the wing for any M_∞ and angle of attack with a prescribed accuracy. The wing surface is approximated at the grid points by using local cubic splines at each time step.

In the second and third subdomains the problem is formulated for a 3D-stationary hyperbolic system of equations at the initial conditions on the surfaces π and in the plane L, obtained from solving the first problem. The boundary between the second and third subdomains are the planes N crossing the intersection line of the surfaces π and the plane L tangentially to the characteristic conoid with a larger angle of entrance comparing to neighbouring conoids. This plane is of space type.

IV. In all the subdomains an orthogonal space grid is introduced. In the first subdomains the derivatives are written down by using second-order finite differences at the points of the cliche in Fig. 3a. The cliche corresponds to a half cell at some t, the other half corresponds to the time $t+\Delta t$. In the second and third subdomains the derivatives are written by using the cliche in Fig. 3b. In both cases the system of initial differential equations is approximated by implicit second-order finite-difference schemes. Computational algorithms are based on the layer-to-layer advancement by one step in time for the non-steady case and one step in the coordinate x for the stationary case. The grid lines in the layer, connecting the wing and shock wave surfaces, are called rays. The system of difference equations for them is given by

$$aX_{m+1} + bX_m = \mathcal{F} .$$

The right-hand side of this equation involves data from the points of the previous layer (n) as well as the two neighbouring rays of this layer (n+1). For this system a boundary value problem is formulated with one boundary condition on the body and four boundary conditions in the shock wave. The boundary value problem is solved successively on all the rays by the sweep method, known in Russian literature as the "progonka" method. Since the boundary value problems on the rays are connected with each other through the right-hand sides the iterations on the layer are needed.

V. Using the algorithms developed a number of test cases were computed for the flow around the elliptical planform wing with a thick blunted airfoil as well as around a delta wing with a sweep $\chi = 70^\circ$ and a blunted airfoil with a relative thickness of 6%. Note that if $\chi > 45^\circ$ only one algorithm of the second subdomain can be used. The initial data for it can be obtained from the calculation of the flow around an elliptic cone used as a substitution for a small vicinity of the wing tip. Some information about the flow field for the elliptic wing at $M_\infty = 2$ and the angle of attack $\alpha = 5^\circ$ is shown. The position of the shock wave is given in Fig. 4 and the pressure distribution over the upper and lower parts of the longitudinal cross-sections of the wing in Fig. 5. The pressure distribution in the cross-sections of the delta wing at $M_\infty = 1.5$ and $\alpha = 0^\circ$ is shown in Fig. 6.

References

1. G.P. Voskresensky. Numerical solution of the problem of unsteady supersonic flow around the front part of the wings with a detached shock wave. J.Comp. Math. in Appl. Mech. and Eng. N I9, I979, 257-275.
2. G.P. Voskresensky, M.G. Orlova, and V.A. Stebunov. A supersonic inviscous flow around wings with a detached shock wave. Preprint IPM AN SSSR, N I52, I98I.

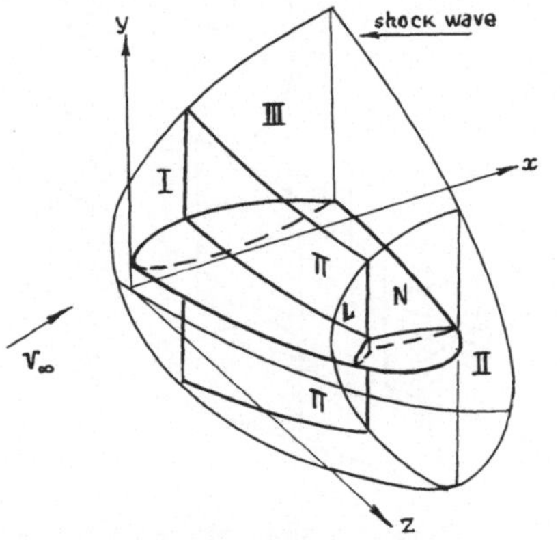

Fig.I Coordinate system. Subdomains I,II,III.

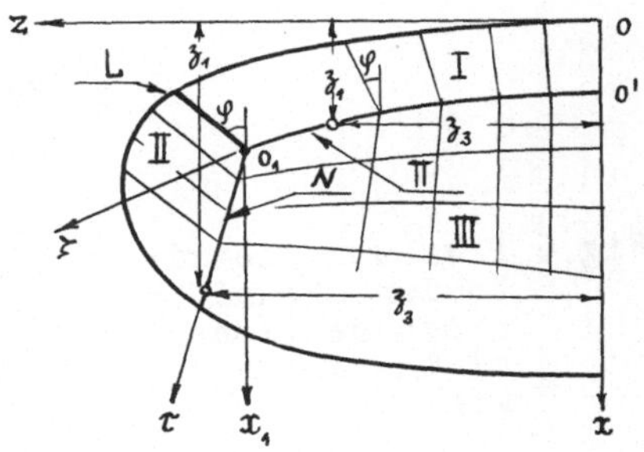

Fig.2 Subdomains of the solution and the parameters of the computational mesh

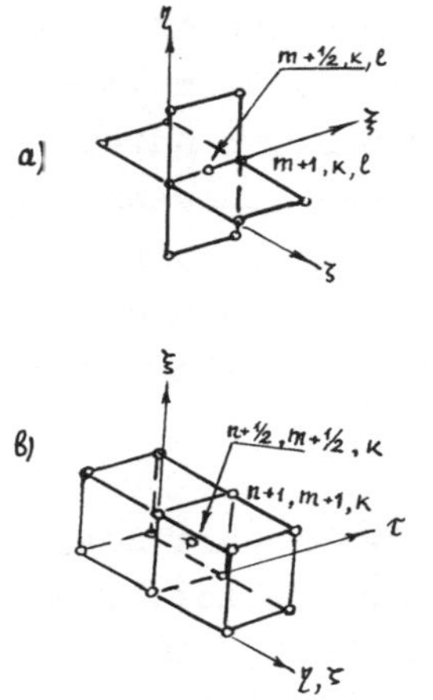

Fig.3 Cells of the computational mesh

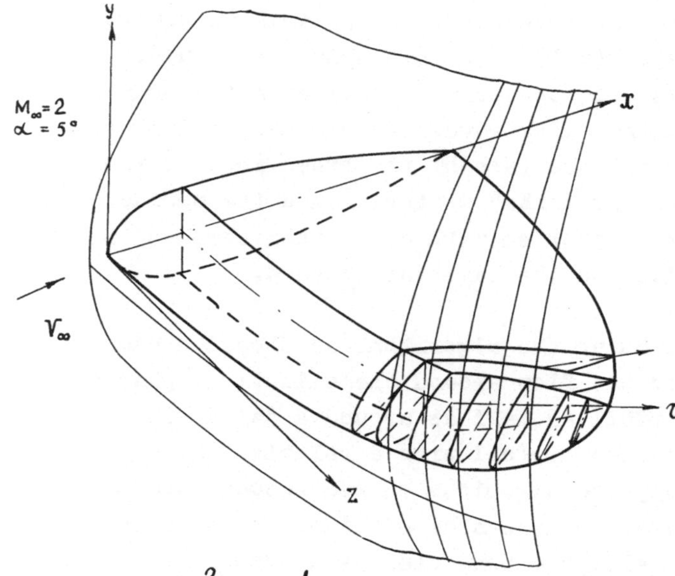

Fig.4 Flow around elliptical wing. Surfaces of the wing and the shock wave

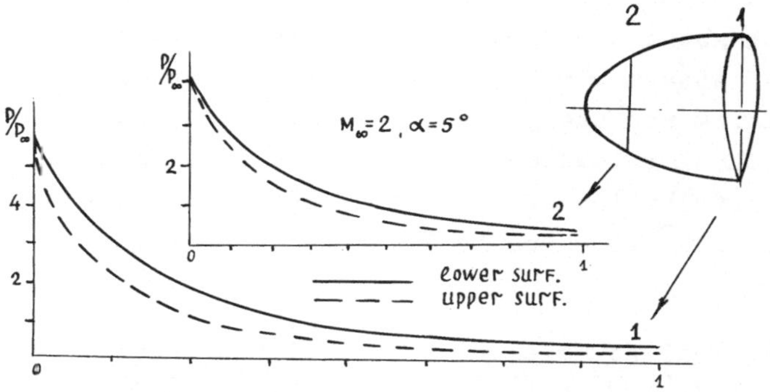

Fig.5 Pressure distribution around airfoils of the wing

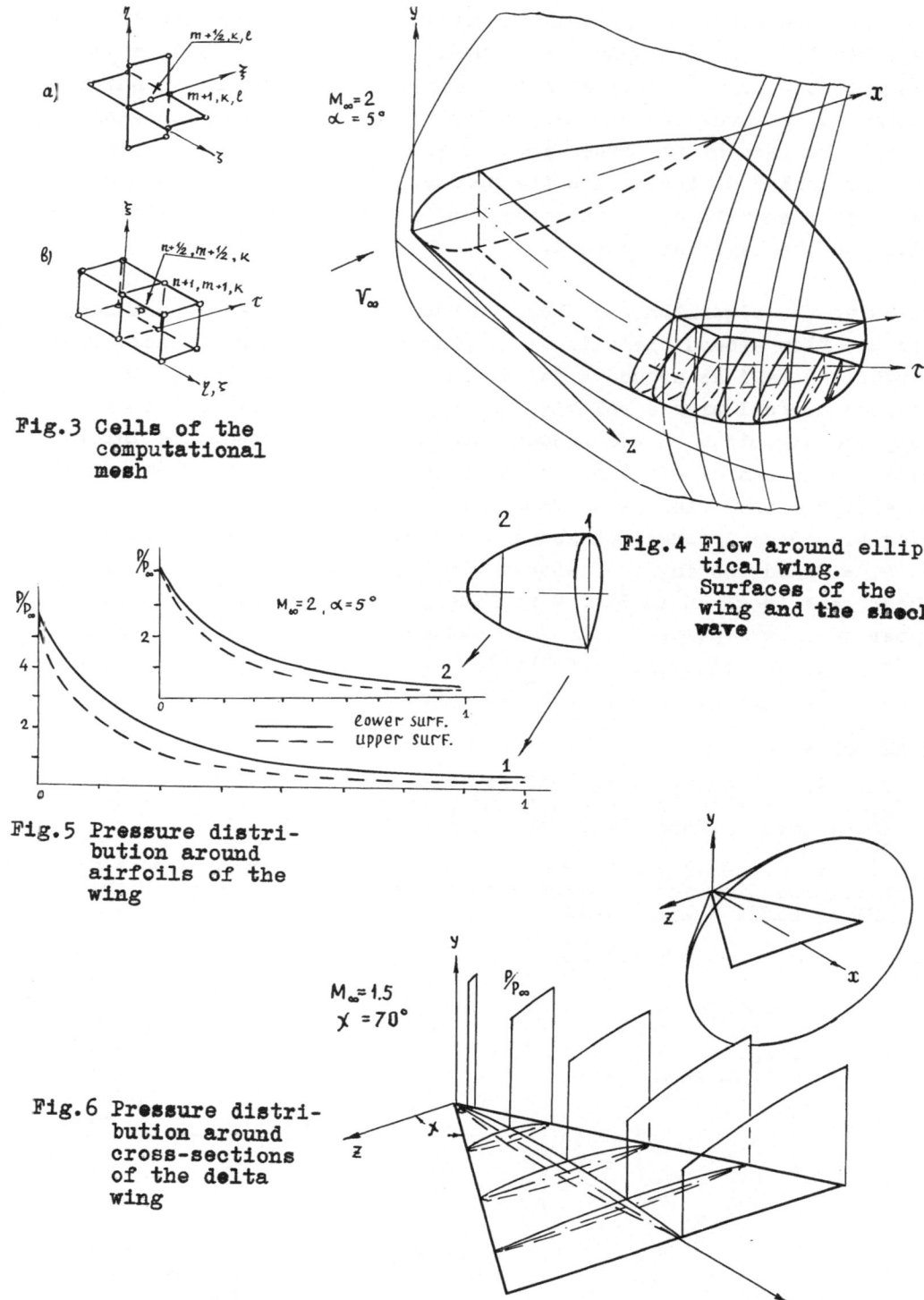

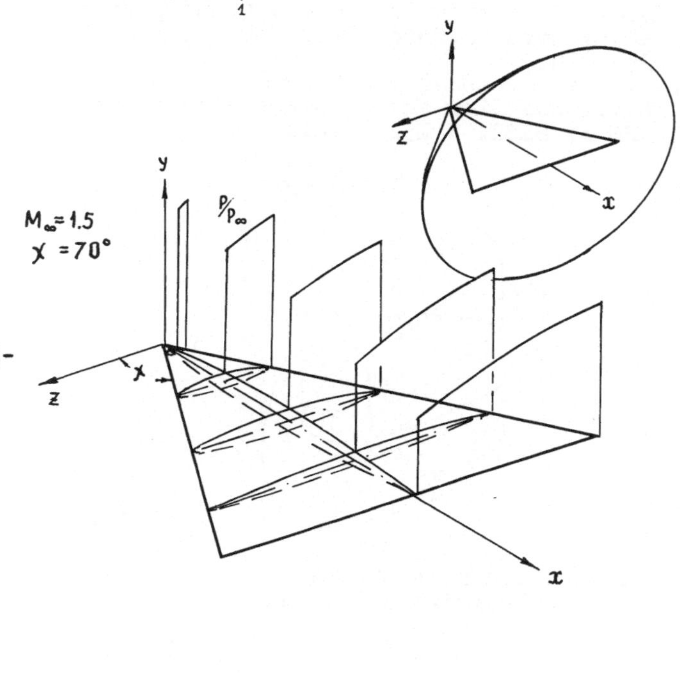

Fig.6 Pressure distribution around cross-sections of the delta wing

SOLUTION OF THREE-DIMENSIONAL TIME-DEPENDENT VISCOUS FLOWS

by

B. C. Weinberg and H. McDonald
Scientific Research Associates, Inc.
P.O. Box 498, Glastonbury, CT 06033
USA

ABSTRACT

A procedure for solving three-dimensional, time-dependent turbulent flows is presented. The consistently split Linearized Block Implicit (LBI) scheme is used in conjunction with the QR Operator scheme to solve an approximate form of the Navier-Stokes equations in generalized nonorthogonal coordinates employing physical velocity components. Results of computations for both second order finite differences and the fourth order generalized Operator Compact Implicit (OCI) schemes are presented.

INTRODUCTION

The behavior of boundary layers on wings and bodies in both steady and unsteady flows has long been of interest to aerodynamicists. When the flow is steady, boundary layer prediction schemes have reached a high level of sophistication and predictive accuracy, even in three space dimensions. In unsteady flows, such as are commonly encountered in rotary winged aircraft, some progress has been made in two space dimensions but little to date has appeared on unsteady three-dimensional boundary layers, especially for cases where negative cross flows are encountered, i.e. the spanwise component of velocity changes sign. To be of practical value, time-dependent three-dimensional boundary layer prediction schemes require high computational efficiency and transient accuracy coupled to the ability to treat arbitrary cross flow profiles. Conventional boundary layer integration schemes have developed by forward marching the streamwise velocity u in the streamwise x direction and simultaneously marching out along the span in the z position direction. However, in cases where there is "reverse cross flow", one must resort to specialized differencing (Ref. 1) to allow one to march through such a region. Recently, (cf. Ref. 2) the benefits of performing an implicit spanwise construction to remove the restriction of only positive cross flow has been noted. As a consequence of these observations a spanwise implicit formulation (retaining spanwise diffusion) seems mandatory for the rotary wing applications and at least desirable for fixed wing applications, especially as it can be had for a very modest increase in code computational labor. Since the solution is being time marched, the opportunity to take a streamwise implicit sweep at roughly the same cost as the explicit sweep (forward march) does arise. If an implicit streamwise structure is adopted, then full time linearization can be utilized. That is the linearization of the nonlinear terms is performed about the known time level, rather than a known spatial marching level. As is pointed out in Ref. 3, it is easier to obtain a consistent spatial-temporal order accurate linearization by marching in time than in space (in time the nonlinear marching derivatives have the form u_i whereas in space marching they have the form $u_i u_j$). Further by structuring implicitly in the space marching direction, regions of axial reverse flow would be permitted. As a result of these combined benefits a streamwise implicit structure in all three spatial directions has been implemented.

The governing equations that are considered here are the Navier-Stokes equations, continuity, energy and the equation of state which are written in generalized tensor form for a body oriented nonorthogonal coordinate system (boundary layer coordinates). In accordance with the boundary layer assumptions, the normal momentum equation is eliminated and the pressure is specified throughout the viscous layer in its stead. For the energy equation constant stagnation temperature T_0 is assumed. This assumption is a good approximation for the flow fields considered, and is thus included here only for purposes of simplification. In the analysis that follows, the full energy equation could equally well have been used. The problem can be further reduced to one involving only the three velocity components, u, w, and v and three equations, the streamwise and spanwise momentum equations and the continuity equation. A block-three system rather than a block-four system is considered which leads to a significant reduction in computer time. If the full energy equation were to be considered, a block-four system would result due to the inclusion of the temperature as an additional unknown.

GOVERNING EQUATIONS

In the following, the governing equations are nondimensionalized as follows, x^i with respect to the characteristic length L, the velocity with respect to U_∞, density, pressure and temperature with respect to ρ_∞, $\rho_\infty U_\infty^2$ and U_∞^2/c_p respectively and time with respect to L/U_∞. Viscosity is nondimensionalized with respect to μ_∞.

Continuity Equation

$$\frac{\partial \rho}{\partial t} + \frac{1}{J}(J\rho u^k)_{,k} = 0 \tag{1}$$

where J is the Jacobian, ρ the density, and u^k is the kth contravariant velocity component

Momentum

$$\rho\left[\frac{\partial u^i}{\partial t} + u^k u^i\big|_k\right] = -g^{ik}\left(p + \frac{2}{3}\frac{\mu}{Re}\Delta\right)_{,k} + g^{mk}\left[\frac{\mu}{Re}u^i\big|_m\right]\Big|_k + g^{mi}\left[\frac{\mu}{Re}u^k\big|_m\right]\Big|_k \tag{2}$$

Energy Equation

$$T_0 = \text{CONSTANT} \tag{3}$$

Equation of State

$$P = \frac{\gamma - 1}{\gamma}\rho T \tag{4}$$

A complete derivation is given in Ref. 4. In order to employ the QR Operator formulation for the spatial differencing the equations are cast into quasi-linear form (cf. Ref. 5).

SPATIAL DIFFERENCE APPROXIMATIONS

In this section tridiagonal finite difference approximations to the spatial differential operator are discussed.

Consider a uniform mesh $x_j = jh$ $(j = 0, 1, \ldots, J)$ where J is a positive integer and $h = 1/J$, then the OCI approximation of

$$L(u) = u_{xx} + b(x)u_x + c(x)u = f(x) \tag{5}$$

which has an implicit tridiagonal form is

$$r_j^- u_{j-1} + r_j^c u_j + r_j^+ u_{j+1} = h^2(q_j^- f_{j-1} + q_j^c f_j + q_j^+ f_{j+1}) \qquad (6)$$

where u_j, $(Lu)_j$ are the approximates of $u(x_j)$ and $Lu(x_j)$, respectively. For each $j = 1, \ldots, J-1$, the r's and q's depend only on h, b_{j-1}, b_j and b_{j+1}. Note that for the standard second order finite approximation $q_j^- = q_j^+ = 0$, $q_j^c = 1$ and r_j^-, r_j^c, r_j^+ are a linear combination of the first and second central difference formulas. Hence, by appropriately defining the q_j and r_j coefficients, a variety of schemes including the fourth order OCI scheme as well as second order methods can be implemented in an efficient manner (cf. Ref. 6).

The tridiagonal difference operators Q and R can now be defined

$$\begin{aligned} R[u_j] &= r_j^- u_{j-1} + r_j^c u_j + r_j^+ u_{j+1} \\ Q[f_j] &= q_j^- f_{j-1} + q_j^c f_j + q_j^+ f_{j+1} \end{aligned} \qquad (7)$$

Equation [7] can thus be rewritten in more compact form as

$$R[u_j] = h^2 Q[L(u)_j] = h^2 Q[f_j] \qquad (8)$$

Alternatively by employing the inverse operator Q^{-1} an expression for $(Lu)_j$ can be obtained

$$L(u)_j = \frac{1}{h^2} Q^{-1} R u_j \qquad (9)$$

For standard central finite differences $Q = Q^{-1} = I$, the identity matrix and the spatial operator is given explicitly in terms of u_{j-1}, u_j and u_{j+1}. However, in general, for higher order methods Q is tridiagonal and Q^{-1} is a full matrix. Thus, Equation [7] expresses the spatial operator as a function of the u_j's for a wider class of difference approximations. This formalism is termed the QR Operator technique.

In multidimensions where coupling of the equations and variables is essential in obtaining the desired accuracy and convergence rates, an OCI type scheme becomes even more attractive. The block size of the resulting matrix is held to a minimum, while other higher order techniques may lead to blocks two to three times as large, (Ref. 7 and Ref. 8).

As with many numerical methods, e.g., standard finite differences and splines, the standard OCI scheme has a cell Reynolds number stability condition which if exceeded admits oscillatory solutions in some instances. For certain viscous flow applications such a limitation could be overly restrictive. In an attempt to eliminate such undesirable behavior, a family of "generalized" OCI schemes were constructed (Ref. 9). These schemes were derived with the condition that the r's and q's possess certain desirable properties such that the discrete system satisfies a maximum principle and there is no cell Reynolds number stability condition per se. The "generalized" OCI schemes are implemented identically as the standard OCI scheme, but do not exhibit the behavior associated with the violation of a cell Reynolds number condition. Further details and a listing of the q and r coefficients is given in Ref. 9. This scheme has been used to solve the two-dimensional laminar boundary equations (Ref. 10).

LINEARIZED BLOCK IMPLICIT SCHEME

Consider a system of nonlinear partial differential equations

$$A\vec{\Phi}_t = \mathcal{D}\vec{\Phi} + \vec{\Psi} \tag{10}$$

where $\vec{\phi}$ is a vector of unknowns and $\vec{\psi}$ is a source term vector which is a function of x^1, x^2, x^3 and t. Extension to source terms which are functions of $\vec{\phi}$ are discussed in Ref. (3). $\mathcal{D}$ is a three-dimensional nonlinear differential operator and of the matrix A acting on the momentum equations is equal to ρI where ρ is the density and I the unity matrix.

Equation (10) can be linearized by Taylor series expansion in time about the nth time level by the procedure described in Ref. (3) to give a second order linearization. After some manipulation and centering the equation at $n + \beta$ $(0 \leq \beta \leq 1)$ a two-level hybrid implicit explicit scheme is obtained.

$$A^n\left[\Phi^{n+1}-\Phi^n\right]/\Delta t = \beta\left[\mathcal{L}_1^n+\mathcal{L}_2^n+\mathcal{L}_3^n\right]\left[\Phi^{n+1}-\Phi^n\right]+\left[\mathcal{L}_1^n+\mathcal{L}_2^n+\mathcal{L}_3^n\right]\Phi^n+S^{n+\beta} \tag{11}$$

where the linearized differential operator has been expressed as a sum of suboperators. In the usual ADI framework these suboperators are associated with a specific direction. Furthermore, all explicit terms have been absorbed into one source term $S^{n+\beta}$.

To solve this system efficiently it is split into a sequence of easily invertible operations following a generalization of the procedure of Douglas and Gunn (Ref. 11) in its natural extension to systems of partial differential equations. The Douglas-Gunn splitting of Eq. (11) can be written in increment or delta form as

$$\begin{aligned}\left[A^n-\Delta t\beta\mathcal{L}_1^n\right]\left[\Phi^*-\Phi^n\right] &= \Delta t\left[\mathcal{L}_1^n+\mathcal{L}_2^n+\mathcal{L}_3^n\right]\Phi^n+\Delta t S^{n+\beta}\\ \left[A^n-\Delta t\beta\mathcal{L}_2^n\right]\left[\Phi^{**}-\Phi^n\right] &= A^n\left[\Phi^*-\Phi^n\right]\\ \left[A^n-\Delta t\beta\mathcal{L}_3^n\right]\left[\Phi^{***}-\Phi^n\right] &= A^n\left[\Phi^{**}-\Phi^n\right]\end{aligned} \tag{12}$$

Due to the use of a boundary layer coordinate system, the normal velocity appears only in conjunction with terms associated with the normal "3" direction. Hence, in the first two sweeps, for the streamwise and spanwise momentum equations one is required to solve only for the two corresponding velocity components without the need of considering the continuity equation. However, on the third sweep where all 3 velocity components appear, one must solve all 3 equations. This strategy reduces the solution procedure to the inversion of two 2 x 2 block matrices and one 3 x 3 block matrix rather than three 3 x 3 block matrices which leads to a reduction in computation time.

The discretization of the continuity equation is now considered. Since the continuity equation is only solved on the third sweep special attention must be given. Employing the Douglas-Gun procedure to approximate the continuity equation as a third sweep equation, a consistent approximation is obtained, i.e., the x^1 derivative term is evaluated at the * level and the x^2 derivative term is evaluated at the ** level. The values of the intermediate derivative terms are obtained after the solution of the first two sweeps of the two momentum equations. As mentioned earlier, these terms do not contain the normal velocity. Since the only term involving v is the x^3 derivative term, the equation can be integrated directly with respect to x^3, i.e.

$$\int_{x^3} \left[A^n \Delta u^{n+1} + B^n \Delta w^{n+1} \right] dx^3 + \Delta t \frac{\beta \Delta t}{J} \left[v^n A^n \Delta u^{n+1} + v^n B^n \Delta w^{n+1} + \rho^n \Delta v^{n+1} \right]$$

$$\int_{x^3} \left\{ S^n - \frac{\beta \Delta t}{J} \frac{\partial}{\partial x^1} \left[(\rho^n + u^n A^n) \Delta u^* + (u^n B^n) \Delta w^* \right] \right. \tag{13}$$

$$\left. - \frac{\beta \Delta t}{J} \frac{\partial}{\partial x^2} \left[(\rho^n + w^n B^n) \Delta w^{**} + (w^n A^n) \Delta u^{**} \right] \right\} dx^3$$

where A^n and B^n are linearization coefficients (Ref. 4). Note that conceptually the continuity equation in integrated form is treated on each sweep of the Douglas-Gunn splitting, although in actuality this can be viewed as having the same form as each sweep and the integration operator can be incorporated into the $\mathcal{L}$ and $\mathcal{D}$ difference operators, and as a result the stability and consistency of the original splitting is retained.

The resulting matrix that is obtained from the discretized equations is in block tridiagonal form (Q and R are tridiagonal operators), and it is inverted by standard LU decomposition.

NUMERICAL RESULTS

The numerical results that are presented have been chosen to demonstrate the capabilities of the solution algorithm in treating three-dimensional, turbulent and time-dependent flows.

The first case considered is the three-dimensional viscous flow over a flat plate skewed at a 45° angle to the flow direction (cf. Fig. 1). An equally spaced mesh of 11 points in each of the three directions is employed. Calculations were obtained with the full equations (all diffusion terms retained), and the boundary layer form of the equations with diffusion retained only in the normal direction. The boundary conditions are no-slip at the surface, velocity specified at the outer edge, velocity profiles prescribed at the inflow boundaries 1 and 2 and extrapolation conditions at the outflow boundaries 3 and 4.

Both cases compared well with one another and to the Blasius solution. This is shown in Fig. 2 where the displacement thickness δ^*, and the momentum thickness θ along the diagonal of the computational domain are compared to the theoretical values.

The second case considered is the turbulent flow over a flat plate at a Re = 661,000/ft. corresponding to the data of Weighardt (Ref. 12). A 36 x 36 grid is employed with a uniform mesh in the streamwise direction, and a Roberts type transformed mesh in the normal direction. At the inflow boundary, a Coles velocity profile (Ref. 12) is specified. McDonald's mixing length model (Ref. 13) with Van Driest damping is used to obtain the results shown in Figs. 3 and 4. The skin friction distribution shown in Fig. 3 for both the second order and fourth order G/OCI calculations compare well with the data. Good agreement with the data is also obtained for the velocity profile (cf. Fig. 4).

The final case is the unsteady laminar flow on a flat plate where the external velocity oscillates according to

$$u = u_o(1 + A\cos\omega t)$$

where u_o = 10, B = .125 and $\omega = 5\pi/2$. Of interest is the phase angle between the wall shear and the external velocity distribution. In Fig. 5 the phase angle distribution as a function of reduced frequency, $\bar{\omega} = \omega x/u_o$, is presented and compared to the calculations of Ref. 14 and Lighthill's low and high frequency limits. The computations were obtained noniteratively on a 36 x 36 grid with $\beta = 1$ and a time step corresponding to $\omega = 10°$. For $\bar{\omega} \geq .8$ the fourth order solution compares well with the results of Cebeci and Carr (Ref. 14). The discrepencies for small $\bar{\omega}$ are due to imposition of the inflow boundary at a small finite value of x rather than at x = 0. Future efforts are aimed at alleviating this difficulty by incorporating a y/δ transformation.

CONCLUDING REMARKS

In this paper a numerical procedure is described that can be applied to the solution of an approximate form of the three-dimensional, time-dependent Navier-Stokes equations. The governing equations are more general than the conventional boundary layer equations, notably in the inclusion of streamwise and spanwise diffusion terms, although the pressure is still imposed by the external flow, as in conventional boundary layer theory. The method incorporates the split LBI scheme in conjunction with the QR operator scheme that permits a variety of spatial difference schemes. Results of several problems are presented which demonstrate the capabilities of the method. Future efforts will aim at solving three-dimensional unsteady flows with application to the helicopter rotor problem.

ACKNOWLEDGEMENT

This work was supported by NASA-Ames Research Center under Contract NAS2-10016.

REFERENCES

1. Blottner, F.G.: Computer Methods in Applied Mechanics and Engineering, Vol. 6, 1975, pp. 1-30.

2. Lin, T.C. and Rubin, S.G.: J. Fluid Mech., Vol. 59, pt. 3, 1973.

3. McDonald, H. and Briley, W.R.: J. of Comp. Physics, Vol. 19, 1975.

4. Weinberg, B.C. and McDonald, H.: Final Contractor's Report (NAS2-10016), 1980.

5. Weinberg, B.C. and McDonald, H.: Final Contractor's Report (NAS2-10016), 1979.

6. Ciment, M., Leventhal, S.H. and Weinberg, B.C.: J. of Comp. Physics, Vol. 23, 1978.

7. Hirsh, R.S.: J. of Comp. Physics, Vol. 19, 1975.

8. Rubin, S.G. and Khosla, P.K.: J. of Comp. Physics, Vol. 24, 1977.

9. Berger, A.E., Solomon, J.M., Ciment, M., Leventhal, S.H. and Weinberg, B.C.: Math. Comp., Vol. 35, 1980.

10. Weinberg, B.C.: AIAA Paper 79-1468, 1979.

11. Douglas, J. and Gunn, J.E.: Numerische Math., Vol. 6, 1964.

12. <u>Proceedings 1968 AFOSR-IFP Stanford Conference</u>, Vol. II, 1969.

13. McDonald, H. and Camaratta, F.J.: <u>Proceedings 1968 AFOSR-IFP Stanford Conference</u>, Vol. I, 1969.

14. Cebeci, T. and Carr, L.W.: NASA TM-78470, 1978.

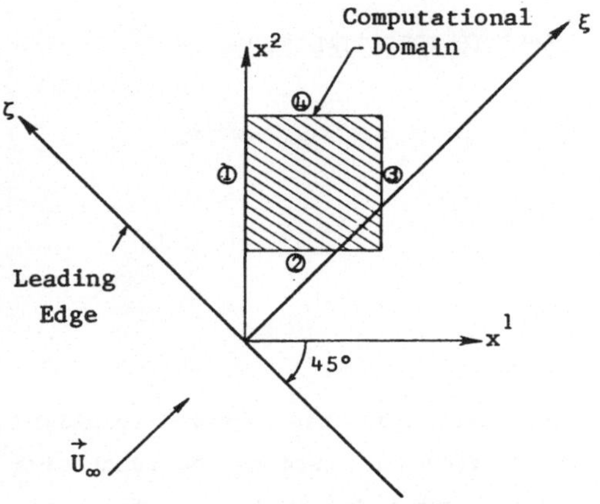

Fig. 1 - Computational Domain for Three-Dimensional Blasius Flow Problem.

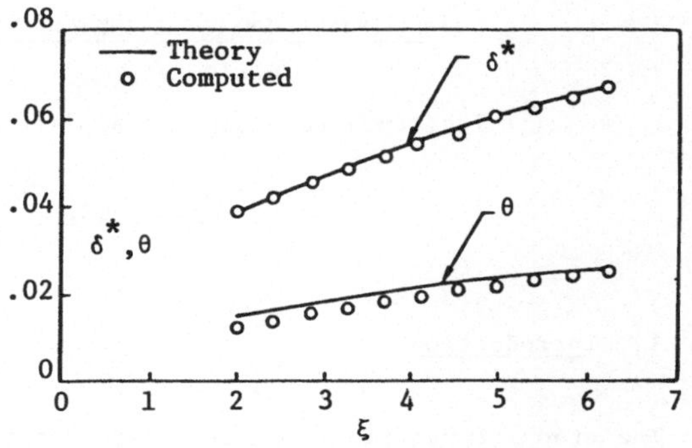

Fig. 2 - Displacement Thickness and Momentum Thickness Distribution Along Diagonal.

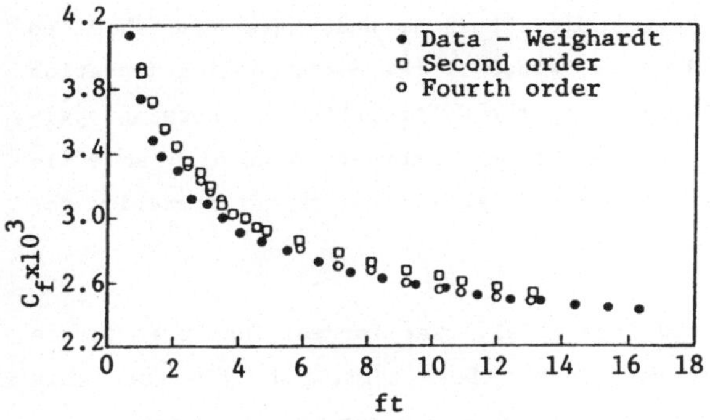

Fig. 3 - Skin Friction Distribution As a Function of Distance Along Plate.

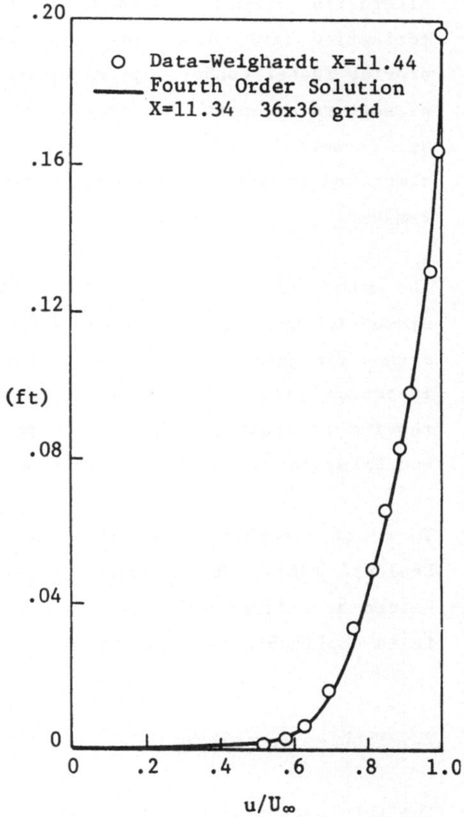

Fig. 4 - Comparison of Velocity Profile.

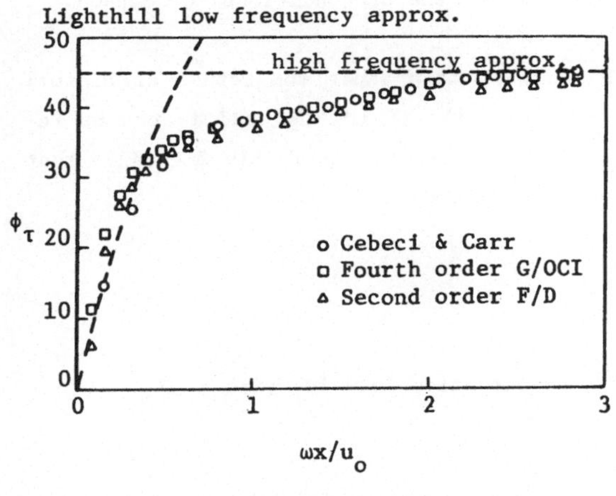

Fig. 5 - Phase Angle Between Edge Velocity and Wall Shear Stress for an Oscillating Flow Over a Flat Plate.

A MINIMAL RESIDUAL METHOD FOR TRANSONIC POTENTIAL FLOWS

Y. S. Wong*
Institute for Computer Applications in Science and Engineering

M. M. Hafez
George Washington University

1. Introduction

The study of numerical solution for transonic potential flows has received considerable attention in the recent past. The standard procedure based on the successive line over-relaxation (SLOR) method is reliable but rather inefficient in the sense that even with an optimal relaxation parameter it suffers from slow convergence. Alternative numerical methods such as the ZEBRA scheme [7] and the Approximate Factorization (AF) method [4] have been suggested. These methods have been shown to provide faster convergence rates than the SLOR method if the corresponding iteration parameters are properly determined. Motivated by the difficulties in choosing optimal parameters a priori, it is, therefore, of strong interest to develop an efficient and reliable method which does not require an estimation of any iteration parameter.

The method of Conjugate Gradient (CG) has been widely used for solving large sparse symmetric and positive definite linear equations. Note that, not only does this method provide a fast rate of convergence, it also does not require any knowledge of iteration parameters. Khosla and Rubin [5] had adapted this method and presented results for subsonic flows over an airfoil. For transonic flow calculations, Wong and Hafez [8] suggest a combination of the SLOR and the preconditioned CG method.

To avoid a combined iteration, we shall study in this paper the method of Minimal Residual (MR). The method is closely related to the CG method, and it can be regarded as a first order gradient method. The main advantage of this method is that it is applicable to symmetric and nonsymmetric matrices.

2. Problem Formulation

The full potential equation expressed in a conservation form is written as

---

*This work was supported by the National Aeronautics and Space Administration under Contracts NAS1-15810 and NAS1-16394 while the author was in residence at ICASE, NASA Langley Research Center, Hampton, VA 23665.

$$(\rho\phi_x)_x + (\rho\phi_y)_y = 0$$

where (1)

$$\rho = [1 - \frac{\gamma-1}{\gamma+1}(\phi_x^2 + \phi_y^2)]^{\frac{1}{\gamma-1}},$$

where ϕ is the velocity potential, ρ, the density and γ, the ratio of specific heats. A grid generation is applied to transform the cartesian coordinates into the computational domain in a rectangle

$$(\rho U/J)_\xi + (\rho V/J)_\eta = 0, \tag{2}$$

where J is the Jacobian of the transformation, U and V are the contravarient velocity components along the transformed coordinates ξ and η. The fluid density is modified in such a way

$$\tilde{\rho} = \rho - \mu\rho_\xi\Delta\xi,$$

so that an artificial viscosity is introduced. Here μ is a switching function, which is zero in subsonic regions and non-zero in supersonic regions. ρ_ξ is the density gradient in the streamwise direction.

Equation (2) is a nonlinear mixed elliptic-hyperbolic partial differential equation. However, assuming ρ is given, a central finite difference approximation to eqn. (2) gives

$$L\phi = 0, \tag{3}$$

where L is a large sparse matrix operator.

3. Iterative Schemes

AF Scheme [2,4]: The AF scheme is given by:

$$\text{Step 1:} \quad (\alpha - \overleftarrow{\delta}_\eta B_j)f^n_{i,j} = \alpha\omega L\phi^n_{i,j}$$

$$\text{Step 2:} \quad (\alpha\overrightarrow{\delta}_\eta \mp \alpha\beta\overleftrightarrow{\delta}_\xi - \overleftarrow{\delta}_\xi B_i \overrightarrow{\delta}_\xi)\delta\phi^n_{i,j} = f^n_{i,j}, \tag{4}$$

where B_i and B_j are the metric quantities. α is an acceleration parameter, $f_{i,j}$ is an intermediate result, $\delta\phi_{i,j}$ is the correction and $L\phi_{i,j}$ is the residual. Note that, the difference approximation is split between the two steps in the η direction. This generates a $\phi_{\eta t}$ - type term. A $\phi_{\xi t}$ - type term has been added in step 2. The parameter β is updated depending on the changes in the residual.

ZEBRA Scheme: The successive line overrelaxation scheme is given by

$$\sigma_u B_{i-1}\delta\phi_{i-1,j} + \sigma_\ell B_i \delta\phi_{i+1,j} + B_{j-1}\delta\phi_{i,j-1}$$

$$- [(B_{i-1}+B_i) + (B_{j-1}+B_j)]\delta\phi_{i,j} + B_j\delta\phi_{i,j+1} = - L\phi^n_{i,j}$$

$$\phi^{n+1}_{i,j} = \phi^n_{i,j} + \omega\delta\phi_{i,j},$$

where ω is the relaxation parameter, on the upper surface $\sigma_u = 1.0$, $\sigma_\ell = 0.0$, and on the lower surface $\sigma_u = 0.0$, $\sigma_\ell = 1.0$.

A tridiagonal solver is used for each line (ξ = const.). Notice that marching is always with the flow direction. Alternatively, successive ring can be overrelaxed where a periodic tridiagonal solver is needed. The ZEBRA scheme is essentially based on the successive line (ring) overrelaxation with red and black ordering.

MR Scheme: The solution of eqn. (2) can be obtained by the following iterative scheme

$$M\delta\phi^n = - L\phi^n, \tag{5}$$

where $\delta\phi^n = \phi^{n+1} - \phi^n$, $L\phi^n$ is the residual vector at the nth iteration, and M is a matrix operator. Here M is chosen from the operator L by ignoring the skewness effect due to the grid transformation. Thus M is an approximation to L. Note that if M is the Jacobian of L, then (5) is a Newton iterative method. Thus the procedure considered here can be regarded as a Newton-like iterative scheme. However, it will converge only for subsonic flows. For mixed subsonic-supersonic flow calculations M must be modified so that a $\phi_{\xi t}$ term is included.

$$\tilde{M} = M \mp \mu\beta\vec{\delta}_\xi,$$

where μ is the switching function defined as before, β determines the amount of $\phi_{\xi t}$ introduced.

The MR method is then applied to solve eqn. (5), and a preconditioning matrix operator C is introduced to accelerate the rate of convergence. The preconditioned MR algorithm is as follows:

Set $r_0 = L\phi_0 - M\delta\phi_0$, solve $Cp_0 = r_0$, then for $k = 0, 1, 2,\ldots,K$

$$\delta\phi_{k+1} = \delta\phi_k + \alpha_k p_k$$

$$r_{k+1} = r_k + \alpha_k M p_k$$

$$\alpha_k = (r_k, Mp_k)/(Mp_k, Mp_k)$$

$$Cp_{k+1} = r_{k+1}.$$

The main computational works required are one matrix-vector multiplication and one solution of $Cp = r$ for each k. The preconditioning matrix C is based on an incomplete LU factorization of M, which can be factorized into a product of sparse lower and upper triangular matrix. Thus the solution of $Cp = r$ is obtained very efficiently by simply backward and forward substitution. Different preconditioning matrices are considered. The detail regarding the determination of C and the implementation of the algorithm can be found in [9].

4. Numerical Results

Transonic potential flows around NACA 0012 airfoil are calculated for different Mach numbers and angles of attack. Figures 1 - 2 compare the convergence histories of the present method with other methods given by [1], the grid size is 65×33. Figures 3 - 6 compare the convergence rates of the methods based on TAIR [2] codes, and the grid size is 149×30.

The TAIR code based on AF scheme is reliable and very fast. There are three parameters in the calculations (α_ℓ, α_H and ω) and the rate of convergence is not really very sensitive to small variations of these parameters. All reported results are obtained using default values. It is obvious that the performance of TAIR is very satisfactory. There are, however, some questions about the approximate factorization involved, and it is not clear how the error terms effect the convergence of transonic calculations. Another point of concern is the treatment of the boundary conditions for the intermediate variable. It is handled in the code in an empirical way. (The parameter α which is cyclically varied between α_ℓ and α_H is restricted to never fall below AFAC $\cdot\ \alpha_\ell$ in the neighborhood of the airfoil boundary, where AFAC's default value is 14.0).

The ZEBRA scheme is simple and reliable. The convergence, however, depends on a relaxation parameter ω, which is chosen by numerical experiments. The CPU time of

ZEBRA iteration and AF iteration (using the same grid) are almost the same. An improved version of this scheme can be found in [6].

The main advantage of the MR algorithm is that no relaxation parameters ω or α's as in other schemes are required. The convergence histories for MR method are smooth compared to AF and ZEBRA tested here. It should be pointed out that the present code is not efficient because the CPU time per iteration requires about three times that required for AF or ZEBRA schemes. However, it can certainly be improved. A new version of the MR method is currently being investigated, and a detailed comparison with the AF scheme will be reported later [9].

5. Concluding Remarks

Preliminary results are presented using a preconditioned minimum residual scheme. No parameter estimation is needed and convergence is fast for subsonic flows. For transonic calculations, the present method is not competitive with the AF method. Improvement may be achieved with better preconditioning techniques. Other modifications of MR method are in progress. Recently, a new version of the computer code requiring only 60% of the CPU time has been achieved [9].

Acknowledgement

The authors would like to thank Terry Holst of NASA Ames, Jerry South of NASA Langley, and Mike Doria of Valparaiso University, for providing them with their codes to test the MR and ZEBRA algorithms.

References

[1] Doria, M. L. and South, J. C.: Transonic Potential Flow and Coordinate Generation for Bodies in a Wind Tunnel, presented at the AIAA Aerospace Sciences Meeting, Orlando, FL, January 1982.

[2] Dougherty, F. C., Holst, T. L., Gundy, K. L. and Thomas, S. D., TAIR - A Transonic Airfoil Analysis Computer Code, NASA Technical Memorandum 81296.

[3] Hafez, M. M. and South, J. C., Vectorization of Relaxation Methods for Solving Transonic Full Potential Equation, GAMM Workshop on Numerical Methods for the Computation of Inviscid Transonic Flow with Shock Waves, FFA Stockholm, Sweden, September 1979.

[4] Holst, T. L. and Ballhaus, W. F.: Fast Conservative Schemes for the Full Potential Equation Applied to Transonic Flow, AIAA J., Vol. 17, 1978, pp. 1038-1045.

[5] Khosla, P. K. and Rubin, S. G.: A Conjugate Gradient Iterative Method, Lecture Notes in Physics, 141, Springer-Verlag, 1981, pp. 248-253.

[6] Hafez, M. M. and Lovell, D.: Improved Relaxation Schemes for Transonic Potential Calculations, submitted to AIAA Aerospace Sciences Meeting, Reno, NV, January 1983.

[7] South, J. C., Keller, J. D., and Hafez, M. M.: Vector Processor Algorithms for Transonic Flow Calculations, AIAA J., Vol. 18, 1980, pp. 786-792.

[8] Wong, Y. S. and Hafez, M. M.: Application of Conjugate Gradient Methods to Transonic Finite Difference and Finite Element Calculations, Proceedings of the AIAA 5th Computational Fluid Dynamics Conference, Palo Alto, CA, June, 1981, pp. 272 - 283.

[9] Wong, Y. S.: Comparison Between Newton-Like Method and Approximate Factorization Scheme for Transonic Flow Calculations, to be submitted to AIAA J., 1982.

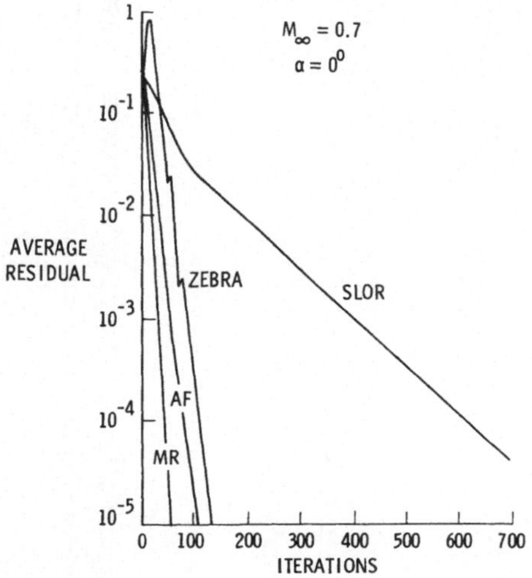

Figure 1.

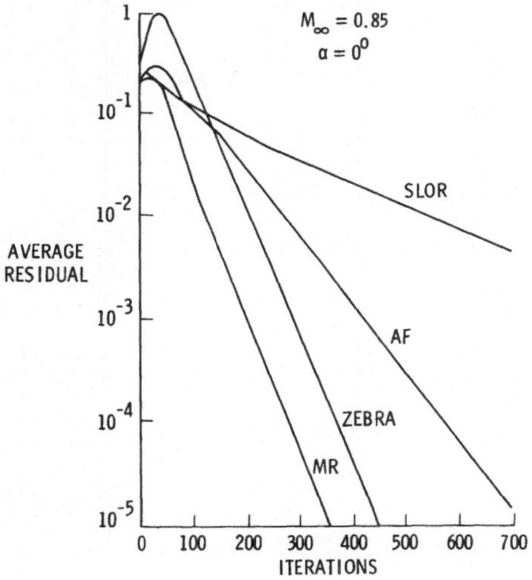

Figure 2.

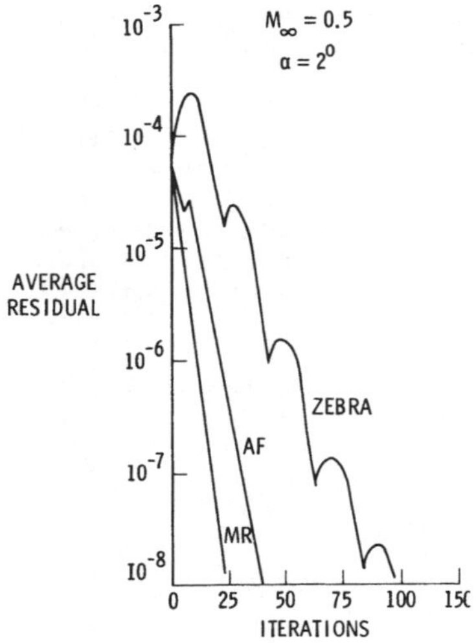

Figure 3.

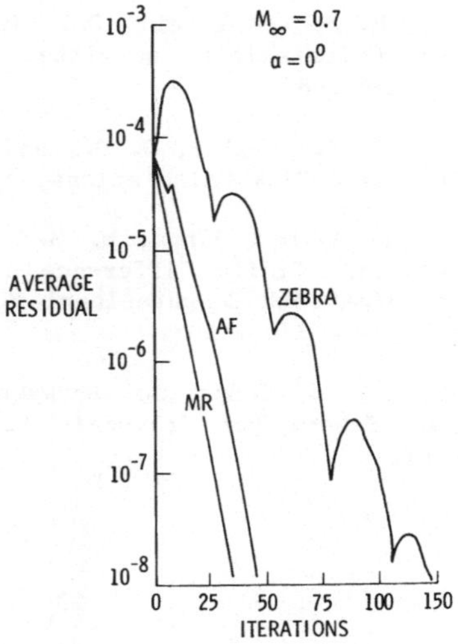

Figure 4.

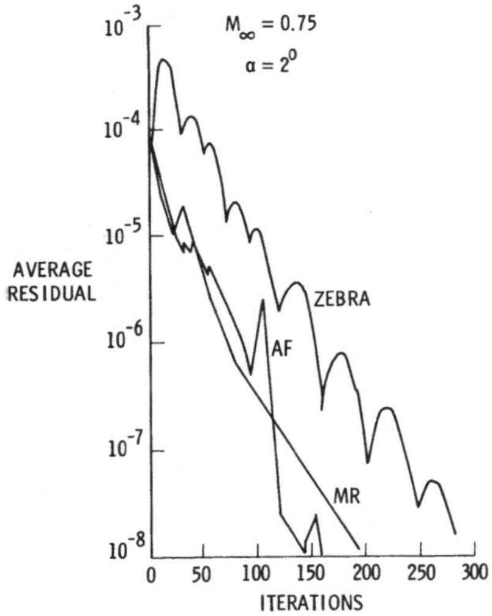

Figure 5.

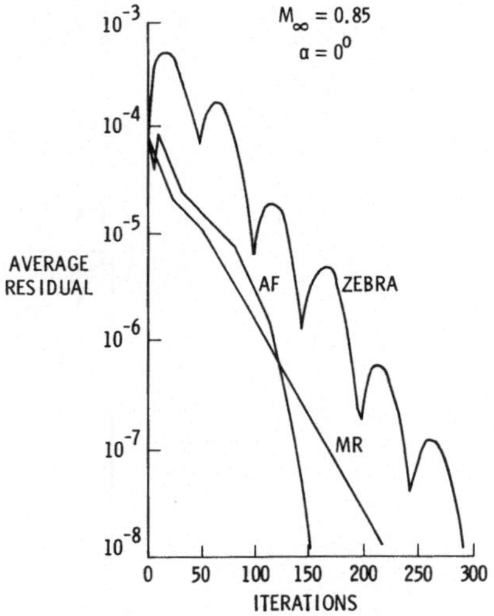

Figure 6.

APPLICATION OF TWO-POINT DIFFERENCE SCHEMES TO THE CONSERVATIVE EULER EQUATIONS FOR ONE-DIMENSIONAL FLOWS

Stephen F. Wornom
NASA LANGLEY RESEARCH CENTER
HAMPTON, VIRGINIA 23665 U.S.A.

SUMMARY

An implicit finite-difference method is presented for obtaining steady-state solutions to the time-dependent, conservative Euler equations for flows containing shocks. The method uses a two-point central difference scheme with dissipation added at supersonic points via the retarded density concept. Application of the method to the one-dimensional nozzle flow equations for various combinations of subsonic and supersonic boundary conditions show the method to be very efficient. Residuals are typically reduced to machine zero in approximately 35 time steps for 50 mesh points. It is shown that the scheme offers certain advantages over the more widely-used three-point schemes, especially in regard to application of boundary conditions.

INTRODUCTION -

Numerical solutions to the Euler equations appear to be sensitive to the boundary conditions imposed. This is seen in the numerical results of [1] which examined the effect of different boundary approximations on stability of three-point implicit schemes. That study was motivated by the extra numerical boundary conditions which are required by the three-point formulation to close the system of difference equations. This point can be illustrated with the one-dimensional Euler equations (conservation of mass, momentum, and energy) where the characteristic speeds are u, $u + c$, $u - c$. For subsonic inflow, the characteristics associated with u and $u+c$ will be coming from upstream and can be replaced by two boundary conditions. For subsonic outflow the $u-c$ characteristic can be replaced by an outflow boundary condition. Thus three physical boundary conditions exist for subsonic inflow and outflow. The three-point difference equations of [1] require 6 boundary conditions -thus 3 extra boundary conditions are required in addition to the 3 physical ones.

To avoid the need for extra boundary conditions, the idea of representing first derivatives by a two-point approximation is examined. This concept has been applied to solve the boundary-layer equations [2].

The two-point differencing idea appears to be very attractive for solving natural first-order systems, such as the Euler equations, for several reasons. First, since the system contains only first derivatives, there is no need to introduce new dependent variables as was necessary for the boundary-layer equations. Second, when applied to the mass, momentum and energy equations only three boundary conditions are needed to close the system. Thus, for the previously mentioned case of subsonic inflow and outflow, the difficulties introduced at boundaries with a three-point method do not occur.

GOVERNING EQUATIONS

The governing equations are the quasi-one-dimensional, time-dependent Euler equations which define flow through a nozzle. These are written in conservation form as :

$$\rho_t + (\rho u)_x = 0 \qquad \text{(mass)} \qquad (1)$$

$$(\rho u)_t + (P + \rho u^2)_x - \frac{P}{A} A_x = 0 \qquad \text{(momentum)} \qquad (2)$$

$$P = \frac{(\gamma - 1)}{\gamma} \rho \left(T_{total} - \frac{1}{2} u^2\right) \qquad (3)$$

where T_{total} is the nondimensional reservoir stagnation temperature (or enthalpy). ρ and P are the physical density and pressure times the nozzle cross-sectional area A(x). More details of all phases of this paper can be found in [3].

In this paper only the steady-state solution is of interest, thus the pressure equation replaces the energy equation and assumes the total enthalpy to be constant.

FINITE-DIFFERENCE EQUATIONS

The partial-differential equations (1) and (2) are replaced with the following two-point central-difference approximation :

$$\frac{1}{2}\left[\bar{\rho}_t\big|_i + \bar{\rho}_t\big|_{i-1}\right] + \left[(\bar{\rho}u)_i - (\bar{\rho}u)_{i-1}\right]\Big/\Delta x = 0 \qquad (4)$$

$$\frac{1}{2}\left[(\rho u)_t\big|_i + (\rho u)_t\big|_{i-1}\right] + \left[(P + \rho u^2)_i - (P + \rho u^2)_{i-1}\right]\Big/\Delta x$$

$$- \frac{1}{2}\left[\left(\frac{P}{A}\right)_i + \left(\frac{P}{A}\right)_{i-1}\right](A_i - A_{i-1})/\Delta x = 0 \qquad (5)$$

$\bar{\rho}$ is a retarded density used to introduce dissipation at supersonic points. This idea was developed [4] for obtaining solutions to the potential equation and is given by :

$$\bar{\rho} = \rho - \mu\rho_x\Delta x \qquad \bar{\rho}_i = \rho_i - \mu_i(\rho_i - \rho_{i-1})$$

$$\mu_i = \max\left[0, 1 - (M_c/M_{i-\frac{1}{2}})^2\right]$$

where M_c is a cutoff Mach number and $M_{i-1/2}$ the mid-cell value.

The retarded density method was chosen because : 1) the solution algorithm can be written to implicitly include the extra mesh point introduced at supersonic points and it does not alter the algorithm ; and 2) dissipation is only added at supersonic or near supersonic points.

SOLUTION ALGORITHM

Equations (4) and (5) are linearized by Newton's method and have the following form :

$$B_i\,(\Delta\rho,\Delta u)^T_i + A_i\,(\Delta\rho,\Delta u)^T_{i-1} + C_i\Delta\rho_{i-2} = R_i$$

where the right-hand-side are the steady-state terms in equations (4) and (5). These are inverted with the following algorithm (right-to-left sweep) :

$$\Delta\rho_i = d_i^{(1)} + e_i^{(1)}\Delta u_i$$

$$i = I,\ I-1,\ \ldots\ 2$$

$$\Delta u_{i-1} = d_i^{(2)} + e_i^{(2)}\Delta u_i$$

The coefficients d_i and e_i are determined from a left-to-right sweep. The values of $d_1^{(1)}$, $e_1^{(1)}$ and Δu_I are determined from the boundary conditions.

BOUNDARY CONDITIONS

Equations (1) and (2) have two characteristics, which determines how many boundary conditions may be applied at each boundary.

Table I summarizes the number of physical boundary conditions available for four classes of flow as well as the number of boundary conditions required by the present two-point method. It can be seen from Table I that the two-point scheme is ideally suited for class 1 and 3 flows since the number of physical boundary conditions available equals the number of boundary conditions required to close the two-point system and these can be applied at the appropriate boundaries. For flows where the outflow is supersonic (classes 2 and 4), the two-point method requires one more boundary condition to close the system than there are physical boundary conditions available. The additional boundary condition needed for closure when the outflow is supersonic is obtained by extrapolation.

The boundary conditions used are : 1) subsonic inflow, $P/\rho^\gamma = P_o/\rho_o^\gamma$, 2) subsonic outflow $P = P_{exit}$, 3) supersonic inflow, $P/\rho^\gamma = P_o/\rho_o^\gamma$, ρ_o and M_o, 4) supersonic outflow, extrapolate Δu_I. ρ and P here are the physical values.

RESULTS -

Table II summarizes the five flow conditions investigated. These are the same cases investigated in [1] using a three-point method. The initial conditions were obtained by linear interpolation between the exact entrance and exit conditions. Typical results are shown here.

a) Subsonic Flow-No Shock (Case I) :

Results for the symmetric case are shown in figure 1. This same case-when computed with a three-point method [1] was the most sensitive (of all five cases) to the type of extra boundary conditions required by the three-point method. Two types of extra numerical boundary conditions imposed produced converged solution with shocks when a CFL number > 1 was used. A $CFL_{max} = 1{,}000$ calculation was reported with a profile given at 500 time steps. The results shown here were obtained with a $CFL_{max} = 10^8$. Accurate profiles were reached in 10 time steps.

b) Subsonic Flow--With Shock (Case II) :

Results for this case are given in figure 2. The value of M_c for case I was 1.0 and .9 for all other cases. A value < 1 was necessary for transonic flows to prevent an expansion shock from occuring at the sonic point. According to [1] the maximum CFL number for this case using a three-point method was 20. ; a solution was given for CFL = 5 at 1500 time steps. The present calculations were computed with a CFL_{max} of 10^8. On this grid (65 points), the residuals were reduced to machine zero in approximately 45 time steps with plotting accuracy achieved in 10 time steps.

c) Supersonic Inflow--Subsonic Outflow (Case V) :

Results for this case are shown in figure 3. In order to retard the density at the inflow boundary, the density and Mach number at the previous station must be specified. These were obtained from the exact solution. The maximum CFL number for the three-point method was 10. Results were shown for a CFL = 5 at 150 time steps. The experimental maximum CFL number for the present calculations was approximately 250. Plotting accuracy was achieved in 10 time steps and the residuals reduced to machine zero in approximately 35 time steps.

d) Overall Comparison With Three-Point Method :

Table III shows a comparison between the experimental maximum stable CFL numbers for all five test cases for both the three-point [1] and the present two-point method. The maximum CFL shown for the three-point method is the maximum value reported for all possible choices for the extra numerical boundary conditions investigated for the three-point scheme. As shown in Table III, the present method permitted significantly higher CFL values to be used for three of the five cases.

CONCLUSIONS :

For subsonic inflow and outflow, with or without shocks, the present two-point scheme is preferred to a three-point scheme since it requires no extra numerical boundary conditions and these flows can be computed at CFL numbers on the order of 10^3 larger than the three-point method. For three of the five test cases, the present two-point method permitted calculations with much larger CFL numbers than the three-point method, thus further research towards possible extension to two-dimensional flows is merited.

REFERENCES

1. Yee, H.C. ; Beam, R.M. ; and Warming R.F. : Proceedings of AIAA 5th Computational Fluid Dynamics Conference, June 1981, pp. 125-135.
2. Keller, Herbert B. ; and Cebeci, Tuncer : Proceedings of the Second International Conference on Numerical Methods in Fluid Dynamics. Volume 8 of Lecture Notes in Physics, Maurice Holt, ed., Springer Verlag, 1971, pp. 92-100.
3. Wornom, Stephen F. : NASA TM 83262, May 1982.
4. Hafez, M. ; South, J.; and Murman, E. : AIAA J., vol. 17, n° 8, Aug. 1979, pp. 838-844.

TABLE I.- PHYSICAL AND NUMERICAL BOUNDARY CONDITIONS FOR VARIOUS FLOWS

Class	Case[a]	Boundary Conditions			
		Inflow		Outflow	
		P	N	P	N
1	I	1	1	1	1
2	III	1	1	0	1
3	V	2	2	1	1
4	IV	2	2	0	1

P = PHYSICAL N = NUMERICAL

[a] see TABLE II

RESULTS

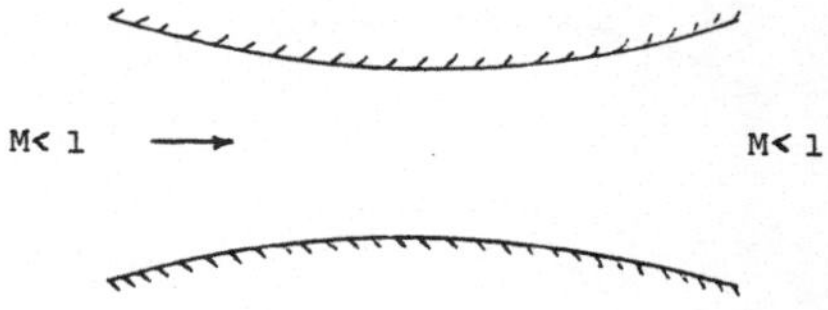

a) Geometry
CASE I

Figure 1 :Subsonic inflow, subsonic outflow no shock ($CFL_{max} = 10^8$, 65-point grid).

TABLE II.- SUMMARY OF FLOW CONDITIONS

Case	Nozzle Type	Description	
		Inflow	Outflow
I[a]	C-D	Subsonic	subsonic
II[b]	C-D	Subsonic	subsonic
III	C-D	Subsonic	supersonic
IV	D	Supersonic	supersonic
V	D	Supersonic	subsonic

C-D = CONVERGENT-DIVERGENT

D = DIVERGENT

[a] NO SHOCK [b] WITH SHOCK

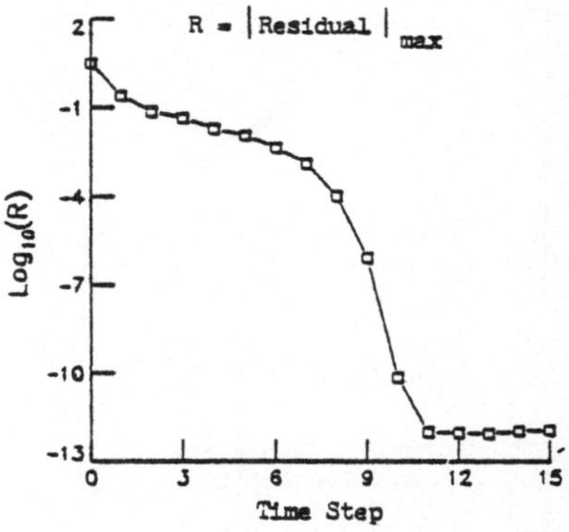

(b) Convergence history

TABLE III.- EXPERIMENTAL MAXIMUM STABLE CFL NUMBER

Case	Two-Point Method	Three-Point Method
I	10^6	10^3
II	10^6	20
III	10^6	10^6
IV	10^6	10^6
V	250^a	10

[a]Unlimited CFL number can be used with "gradual start." see [3].

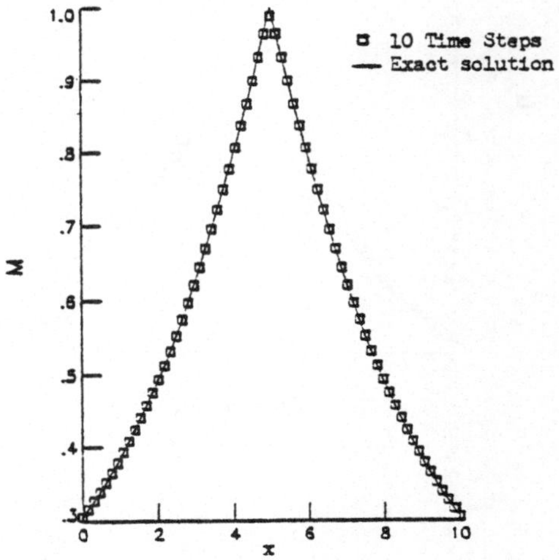

(c) Mach Number

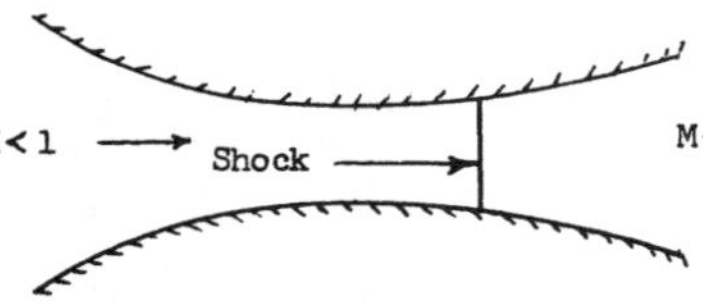

a) Geometry
CASE II

Figure 2 : Subsonic inflow, subsonic outflow with shock ($CFL_{max} = 10^8$, 65-point grid).

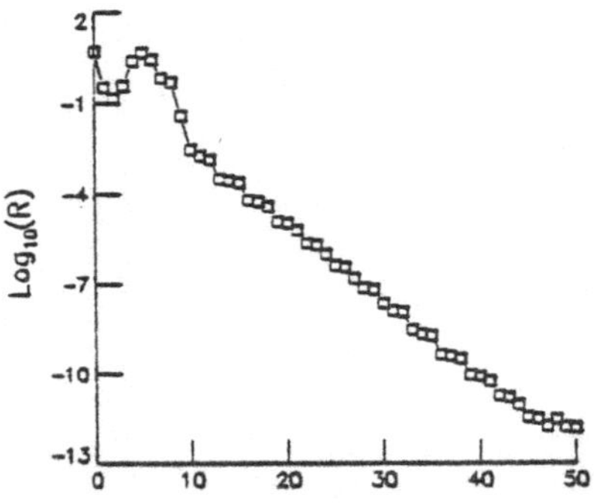

(b) Convergence history

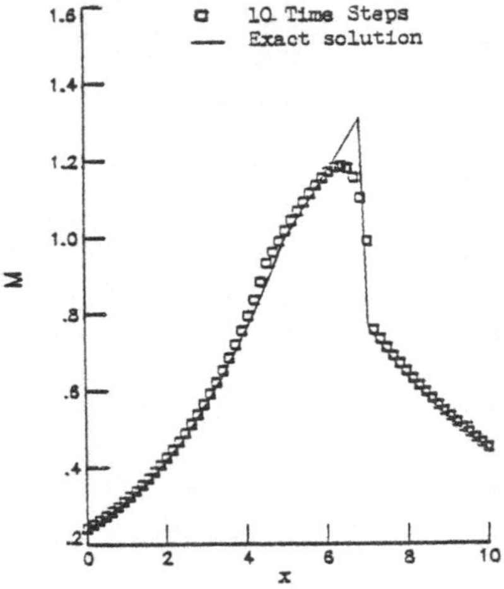

(c) Mach Number

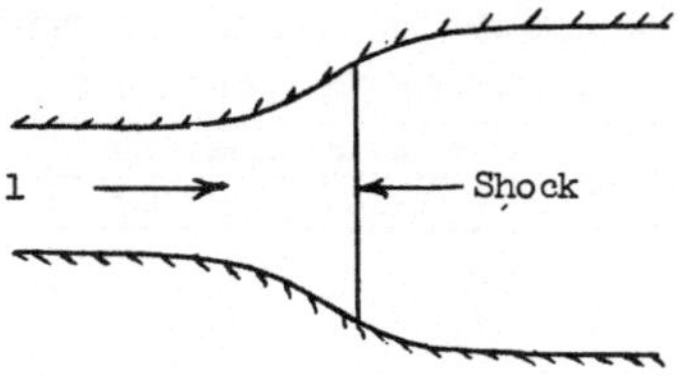

a) Geometry
CASE V

Figure 3 : Supersonic inflow, subsonic outflow ($CFL_{max} = 250$, 65-point grid).

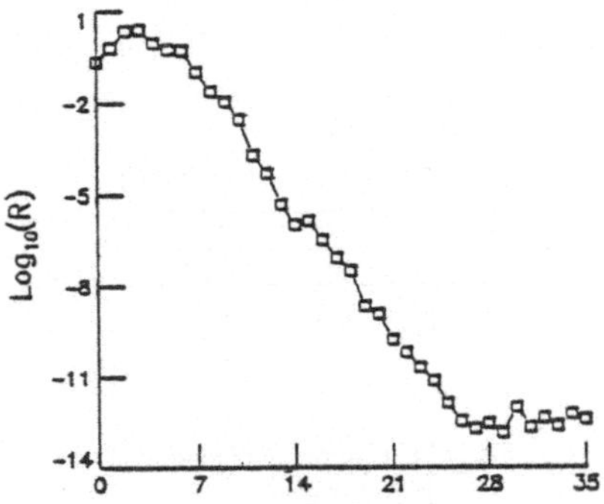

(b) Convergence history

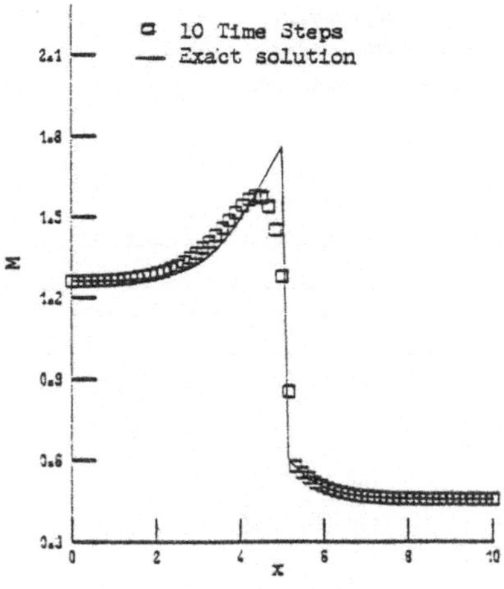

(c) Mach Number

NUMERICAL SOLUTION OF THE EULER EQUATION FOR A COMPRESSIBLE FLOW PROBLEM

S. M. Yen and S. H. Lee
University of Illinois
Urbana, Illinois 61801

Introduction

The objective of the present work is to find the numerical solution of the Euler equation for a triconic body in compressible flow. We used the implicit factored scheme which was developed by Beam and Warming [1] to solve the Navier-Stokes equations. Computations were made using the version of the AIR3D Navier-Stokes Computer Program (developed at NASA/AMES) adapted by Nietubicz [2] for an azimuthally invariant thin layer. Our major effort was in developing a mesh system adaptable to the triconic body as well as to the flow features encountered in the compressible flow regime. Also, the implementation of boundary conditions was studied very carefully.

Mesh System

The mesh system in the near-field for the subsonic case is shown in Fig. 1. The guidelines in designing the mesh systems for all cases are as follows:

1. The surface geometry is accomodated exactly. The shape of the outer boundary is similar to that of the surface.
2. The downstream boundary is perpendicular to the cylindrical surface and is set at a large distance from the last shoulder.
3. The other boundary is the symmetry line.
4. The grid spacing is controlled according to the local gradient of the flow variables. More nodes should, therefore, be placed around the shoulders, near the surface, near the nose and near a shock wave. Also, the spacing should be changed gradually in order to minimize the error due to the numerical transformations.

The mesh systems for transonic and supersonic flows, both of which have dissimilar features, are different from that shown in Fig. 1.

Fig. 2 shows schematically the transformation of the mesh system in the physical domain to a uniform mesh system in the computational domain. The mesh system in the physical domain is divided into six distinct regions. The first one is over the spherical cap, the next three over the conical surfaces and the last two over the cylindrical surface. Regions II, III, IV and V are trapezoids with common boundaries at each of the three shoulders. In all cases, the interface is chosen as the bisector of the angle formed by the two normals to the two adjacent cone surfaces. The transformation to the coordinate system (ζ_1, ξ_1) consists of transforming region I to a rectangle and rotating regions II, III and IV so that the inner and outer boundaries are horizontal. In the next step, the trapezoids are transformed to rectangles. The final step is to transform from the mesh system in coordinates (ζ', ξ') to a uniform system in coordinates (ζ, ξ).

We shall describe our method to control the grid spacings in the physical domain. Since the mesh system in coordinates (ζ', ξ') has orthogonal grid lines, separate transformations can be used for the ζ' and ξ' directions. We thus used two transformations to control the grid spacings in these two directions. Let $g(\zeta) = d\zeta'/d\zeta$ = the transformation function in the ζ direction. We control the spacing $\Delta\zeta$ and $\Delta\zeta'$ using the following expression.

$$\zeta'_{i+1} - \zeta'_i = \int_{\zeta_i}^{\zeta_{i+1}} g(\zeta) d\zeta \tag{1}$$

This means that the spacing $\Delta\zeta'_{i+1} = \zeta'_{i+1} - \zeta'_i$ is equal to the area under $g(\zeta)$ between ζ_{i+1} and ζ_i. We chose $g(\zeta)$ to be the error function which gives good control of grid points where needed. This method can easily be extended to more complex distributions of grid points by using a sum of error functions. The spacing $\Delta\zeta'$ corresponding to a given uniform spacing $\Delta\zeta$ is controlled by changing appropriate parameters in the error function.

In the longitudinal direction, the distribution of spacing in the physical domain is chosen and is, therefore, known a priori. The transformation function from ξ' to ξ is determined numerically.

We treat the interface between two regions in the following way. No nodes are placed along the interface but the flow variables on the next nodal line in the neighboring region are assumed to exist along the interface. This treatment is suitable for a relatively small change in cone angles and also for a large concentration of points near the corner.

Implementation of Boundary Conditions

The boundary condition at the solid surface is usually implemented in the following way: (1) the normal component of the velocity is set to zero, (2) the tangential component of the velocity is obtained by extrapolation, (3) the pressure is calculated from the normal momentum equation, and (4) the surface density needed in the normal momentum equation for calculating pressure is calculated either directly by extrapolation of density or indirectly by setting the density or the temperature gradient to zero.. We tested all three methods for calculating the density and found that the adiabatic wall condition gives a more accurate surface entropy distribution for the entire range of Mach numbers considered. As shown in Fig. 3, the fluctuations of the function P/ρ^{γ} near the shoulders are much smaller for the case of the adiabatic wall condition. Also, this wall condition yields correct entropy increases for transonic and supersonic calculations. The outflow boundary conditions are implemented by setting the longitudinal derivatives to zero. However, for supersonic flow, the free stream conditions are set at the outflow boundary.

Results

Numerical solutions were obtained for Mach numbers in the range 0.5 to 2.75. Two hundred time steps with $\Delta t = 0.025$ are needed to obtain numerical solutions for M_∞ between 0.5 and 2. For $M_\infty > 2$, a smaller Δt has to be used, since the maximum Courant number exceeds the allowable value of 30. In order to speed up convergence, we initialized the solution for a given Mach number by using the results linearly extrapolated from the solution for a lower Mach number. Solutions were obtained for Mach number intervals of 0.1 for $M_\infty < 1.2$ and 0.2 for $M_\infty > 1.5$. About one hundred time steps are needed to obtain solutions for $M_\infty > 1.2$. For transonic and supersonic flows, the grid systems have to be regenerated by adjusting the parameters of the transformation in order to capture the shock wave and to obtain enough resolution near the nose region. Fig. 4 shows the surface pressure distribution for $M_\infty = 0.5$, 1.2, 0.95 and 2.75. These results are compared with calculations by Gustafson [3], who used a panel method for $M_\infty = 0.5$, a relaxation method for transonic flows for $M_\infty = 0.95$ and 1.2 and a finite difference method to solve the Euler equation for the shock layer for $M_\infty = 2.75$. The local Mach number contour lines in the physical domain for three Mach numbers are shown in Fig. 5.

Acknowledgments

We wish to thank Mr. T. Gustafson for the use of his unpublished calculations; Mr. C. J. Nietubicz for making his computer code available to us; Mr. C. J. Nietubicz and D. T. H. Pulliam for discussions on using the code.

References

1. Beam, R. and Warming, R. F., AIAA Paper 77-645, June 1977.
2. Nietubicz, C. J., Pulliam, T. H. and Steger, J. L., AIAA Paper 79-0010, January 1979.
3. Gustafson, T., private communications.

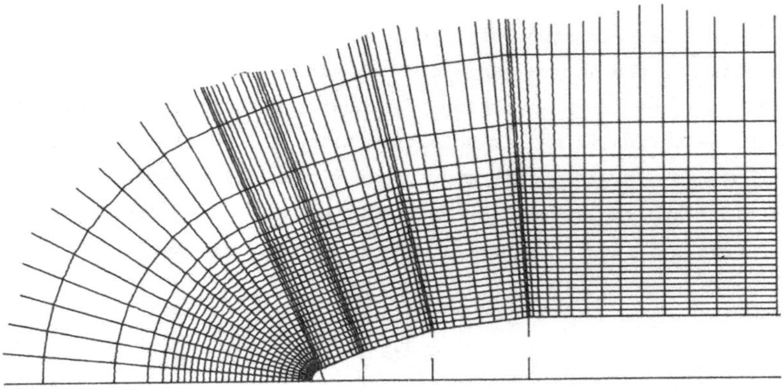

Fig. 1 Near-field mesh system for subsonic flows.

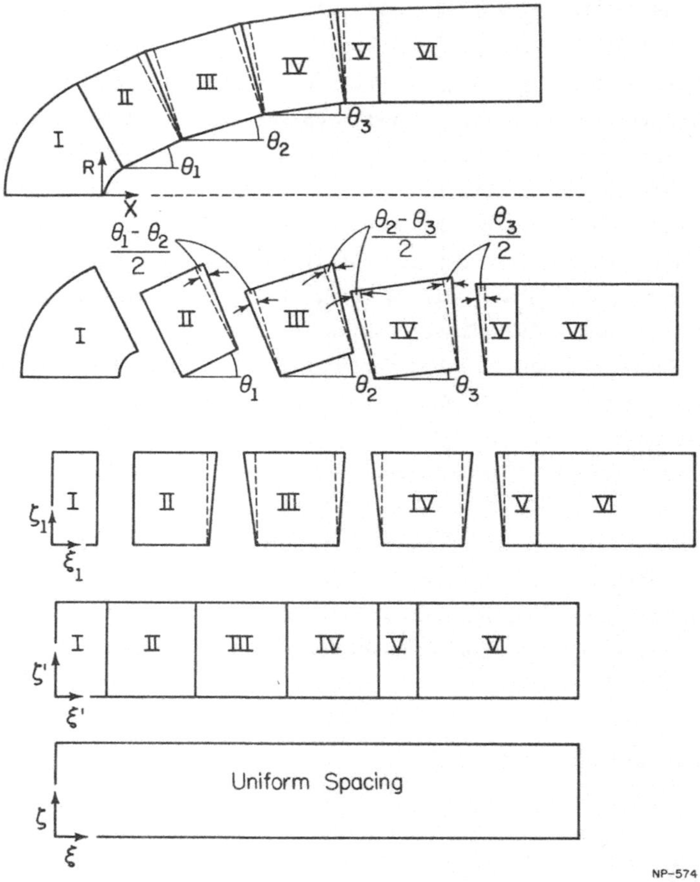

Fig. 2 Schematic representation of the transformations from the physical to the computational domain.

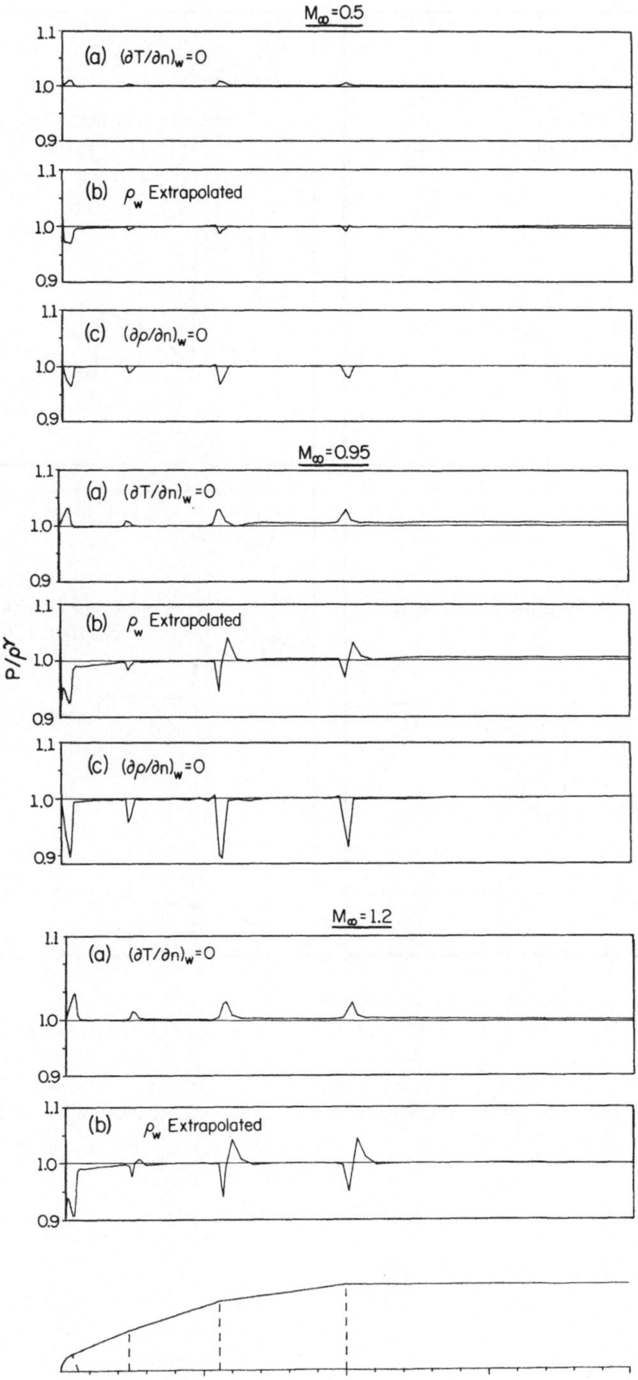

Fig. 3 Comparison of the function P/ρ^{γ} over the surface obtained from different implementations of the surface boundary condition. M_{∞} = 0.5, 0.95 and 1.2.

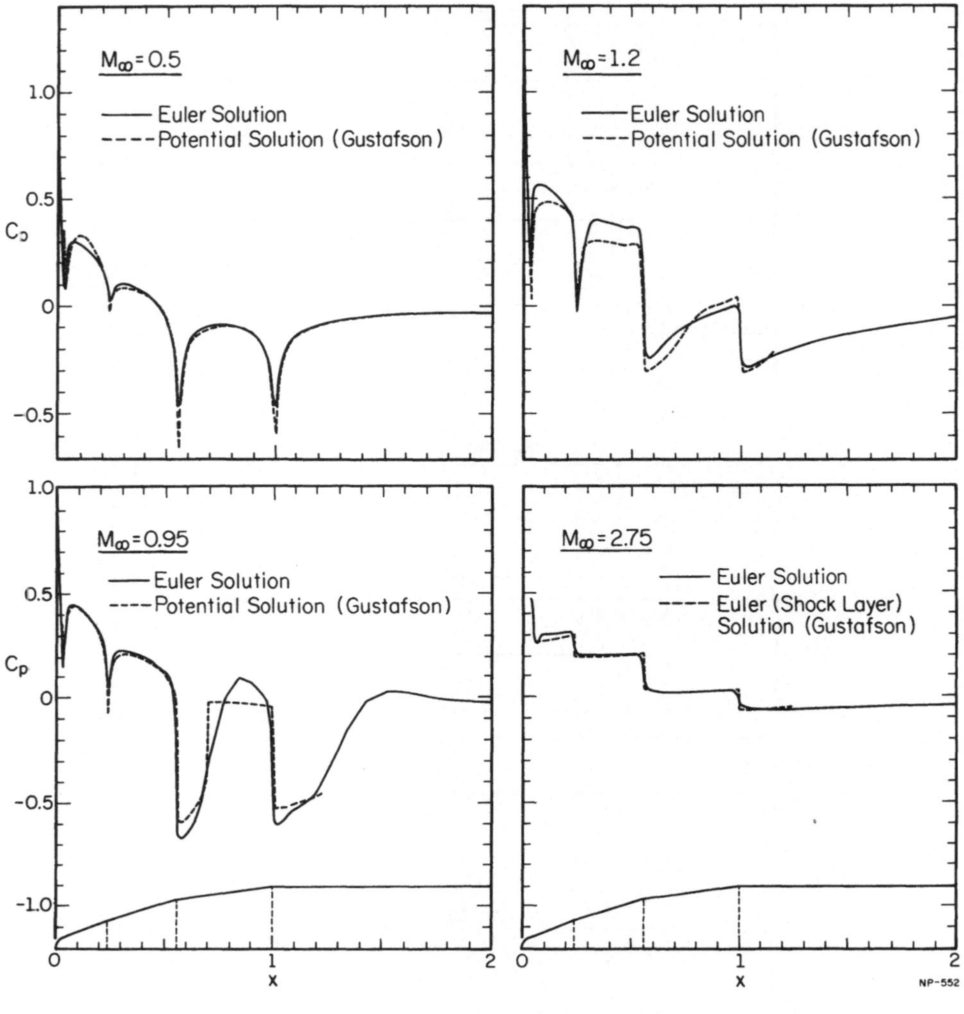

Fig. 4 Surface pressure distribution, C_p vs x, for M_∞ = 0.5, 0.95, 1.2 and .275.

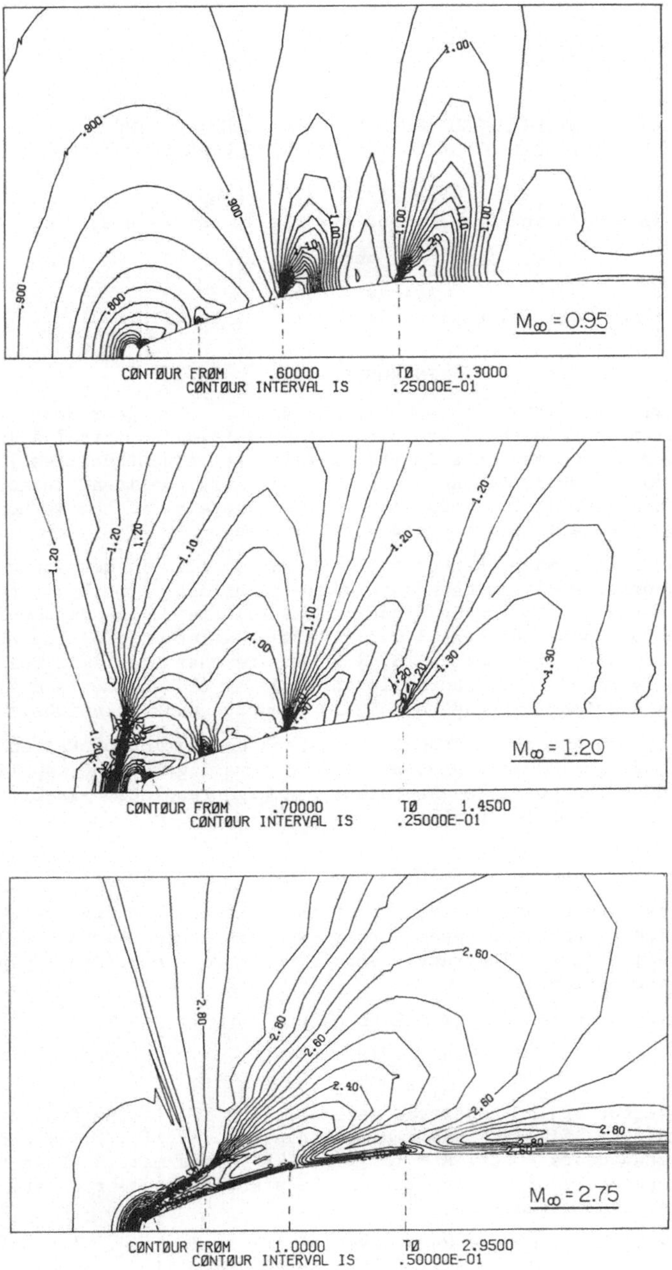

Fig. 5 Mach number contour lines. $M_\infty = 0.95$, 1.2 and 2.75.

A HIGH-RESOLUTION NUMERICAL TECHNIQUE FOR INVISCID GAS-DYNAMIC PROBLEMS WITH WEAK SOLUTIONS

H. C. Yee and R. F. Warming
NASA Ames Research Center, Moffett Field, California, U.S.A.

and

Amiram Harten
New York University, New York, New York, U.S.A.

INTRODUCTION

In problems with shocks where conventional second- or higher-order accurate central spatial difference methods are used, the resulting numerical solution exhibits overshoots and undershoots in the vicinity of discontinuities [1]. The oscillations not only degrade the accuracy but can cause nonlinear instabilities. One remedy is to add numerical dissipation. Unfortunately, ad hoc methods of adding dissipation generally smear the discontinuities.

This work was motivated by Harten's [2] recent success in developing a high-resolution second-order explicit method for one-dimensional hyperbolic conservation laws. This method has the following properties: (a) the scheme is developed in conservation form to ensure that the limit is a weak solution, (b) the scheme satisfies a proper entropy inequality [3,4] to ensure that the limit solution will have only physically relevant discontinuities, and (c) the scheme is designed such that the numerical dissipation produces highly accurate weak solutions.

The goal of this paper is to examine the shock resolution of Harten's method for a two-dimensional gas-dynamic problem, and to investigate other applications of his method, including the possible extension to a high resolution implicit method for both one- and two-dimension problems.

HARTEN'S SCHEME

To aid the development and motivation of the next section, we briefly describe a particular version of Harten's second-order explicit scheme for a one-dimensional system of conservation laws. The reader should refer to the original paper [2] for a more detailed description.

Consider a system of conservation laws of the form

$$\frac{\partial U}{\partial t} + \frac{\partial F(U)}{\partial x} = 0 \quad , \tag{1}$$

where U is the vector of m conserved variables and F is the flux vector. The Jacobian matrix $A(U) = \partial F(U)/\partial U$ has real eigenvalues $(\lambda^1,\lambda^2,\ldots,\lambda^m)$ and a complete set of (right) eigenvectors. Let $R = (R^1,R^2,\ldots,R^m)$ be a matrix whose columns are the right eigenvectors of A and let L be a matrix whose rows are the left eigenvectors of A normalized such that $LR = I$.

Let $U_{j+\frac{1}{2}} = V(U_j,U_{j+1})$ denote an average of U_j and U_{j+1}. A simple example is the arithmetic average $U_{j+\frac{1}{2}} = (U_{j+1} + U_j)/2$ (see Roe [5] for a more sophisticated formula for $U_{j+\frac{1}{2}}$ for inviscid gas-dynamic problems). Also let $L_{j+\frac{1}{2}}$ be the matrix L evaluated at $U_{j+\frac{1}{2}}$, i.e.,

$$L_{j+\frac{1}{2}} = L(U_{j+\frac{1}{2}})$$

and

$$\Delta_{j+\frac{1}{2}}U = U_{j+1} - U_j \tag{2}$$

$$\alpha_{j+\frac{1}{2}} = \begin{pmatrix} \alpha^1_{j+\frac{1}{2}} \\ \alpha^2_{j+\frac{1}{2}} \\ \vdots \\ \alpha^m_{j+\frac{1}{2}} \end{pmatrix} = L_{j+\frac{1}{2}} \cdot (\Delta_{j+\frac{1}{2}} U) \quad . \tag{3}$$

Then Harten's explicit conservative difference scheme can be written as follows:

$$U^{n+1}_j - U^n_j = -\frac{\Delta t}{\Delta x}\left(\hat{F}^n_{j+\frac{1}{2}} - \hat{F}^n_{j-\frac{1}{2}}\right) \tag{4a}$$

$$\hat{F}_{j+\frac{1}{2}} = \frac{1}{2}\,[F(U_j) + F(U_{j+1})] + \frac{\Delta x}{2\Delta t}\sum_{\ell=1}^{m}\left\{\beta^\ell_{j+\frac{1}{2}}\left(g^\ell_j + g^\ell_{j+1}\right) - Q^\ell\left(\nu^\ell_{j+\frac{1}{2}} + \beta^\ell_{j+\frac{1}{2}}\gamma^\ell_{j+\frac{1}{2}}\right)\alpha^\ell_{j+\frac{1}{2}}\right\} R^\ell_{j+\frac{1}{2}} \tag{4b}$$

where Δt is the time increment, Δx is the spatial increment, with

$$\beta^\ell_{j+\frac{1}{2}} = \left(1 + 2\theta^\ell_{j+\frac{1}{2}}\right) \tag{4c}$$

and

$$\theta^\ell_{j+\frac{1}{2}} = \max\left(\theta^\ell_j, \theta^\ell_{j+1}\right) \tag{4d}$$

$$\theta^\ell_j = \left(\left|\alpha^\ell_{j+\frac{1}{2}}\right| - \left|\alpha^\ell_{j-\frac{1}{2}}\right|\right)\Big/\left(\left|\alpha^\ell_{j+\frac{1}{2}}\right| + \left|\alpha^\ell_{j-\frac{1}{2}}\right|\right)$$

$$\nu^\ell_{j+\frac{1}{2}} = \frac{\Delta t}{\Delta x}\,\lambda^\ell(U_{j+\frac{1}{2}}) \tag{4e}$$

$$g^\ell_j = S^\ell_{j+\frac{1}{2}} \cdot \max\left[0, \min\left(\left|\tilde{g}^\ell_{j+\frac{1}{2}}\right|, \tilde{g}^\ell_{j+\frac{1}{2}} \cdot S^\ell_{j+\frac{1}{2}}\right)\right] \tag{4f}$$

$$S^\ell_{j+\frac{1}{2}} = \mathrm{Sign}\left(\tilde{g}^\ell_{j+\frac{1}{2}}\right) \tag{4g}$$

$$\tilde{g}^\ell_{j+\frac{1}{2}} = \frac{1}{2}\left[Q^\ell\left(\nu^\ell_{j+\frac{1}{2}}\right) - \left(\nu^\ell_{j+\frac{1}{2}}\right)^2\right]\alpha^\ell_{j+\frac{1}{2}} \tag{4h}$$

where $Q^\ell\left(\nu^\ell_{j+\frac{1}{2}}\right)$ is a function of $\nu^\ell_{j+\frac{1}{2}}$

$$\gamma^\ell_{j+\frac{1}{2}} = \begin{cases} \left(g^\ell_{j+1} - g^\ell_j\right)\Big/\alpha^\ell_{j+\frac{1}{2}} & \text{when } \alpha^\ell_{j+\frac{1}{2}} \neq 0 \\ 0 & \text{when } \alpha^\ell_{j+\frac{1}{2}} = 0 \end{cases} \tag{4i}$$

For a discussion of the choice and properties of the function $Q(\nu)$, the reader should refer to Harten's original paper. Two choices of Q that we have investigated are:

$$\text{(a)} \qquad Q(\nu) = \nu^2 + 1/4 \tag{4j}$$

$$\text{(b)} \qquad Q(\nu) = \begin{cases} \dfrac{\nu^2}{4\varepsilon} + \varepsilon & \nu < 2\varepsilon \\ |\nu| & |\nu| \geq 2\varepsilon \\ & \varepsilon \text{ a fixed constant} \end{cases} \tag{4k}$$

The last term in equation (4b) is designed in such a way that the approximation satisfies properties a, b and c in the introduction, i.e., this term ensures that a true physical weak solution can be obtained with high accuracy.

We can view this conservative differencing as a form of the second-order-accurate Lax-Wendroff scheme plus an additional term. The global accuracy of this scheme is second-order and it is a nonlinear differencing scheme even if the Jacobian A is a constant matrix. In addition, Harten's method, like that of Lax-Wendroff, has a steady-state dependence on Δt.

EXTENSIONS

It is well-known that the time step of conventional explicit schemes are limited by a CFL number of order one and result in long computation times for "stiff" problems or for steady-state calculations. For stiff problems we would like to retain Harten's high resolution feature but modify the scheme to be implicit. For steady-state calculations, we are considering the following ways of incorporating high resolution schemes:

(a) Compute an approximate steady-state solution by some efficient scheme (explicit, implicit or a mixture of the two) and then apply an explicit high-resolution scheme like Harten's as a postprocessor.

(b) Modify an implicit scheme in delta form [6], where the iteration matrix on the left-hand side is a simple modification of the original implicit operator, while the right-hand side is an appropriate representation of the explicit high-resolution scheme.

One-Dimensional Problem:

Here we describe technique (b) for the one-dimensional problems. For simplicity, consider the class of implicit, linear, one-step time differencing methods in a noniterative delta-form [7] for the system of conservative laws (1)

$$\left(I + \theta\Delta t \frac{\partial}{\partial x} A^n\right)\Delta U^n = -\Delta t \left(\frac{\partial F}{\partial x}\right)^n , \tag{5}$$

where $\Delta U^n = U^{n+1} - U^n$ and should not be confused with $\Delta_{j+\frac{1}{2}}U$ defined in equation (2). If $\theta = 1/2$, the time-differencing method is the trapezoidal formula and if $\theta = 1$, the time-differencing method is backward Euler. There is no connection between θ in equation (5) and θ_j in equation (4d). For the spatial differencing, three-point central differencing or upstream differencing is commonly used. The simplest implicit extension of Harten's method is to replace the right-hand side of (5) by the right-hand side of equation (4a). Some modification of the left-hand side of (5) can also be made. For comparison, this differencing of the spatial derivative can be viewed as three-point central differencing, plus some "appropriate" numerical dissipation, implicit on the left-hand side, explicit on the right-hand side. This method is called the "modified implicit method."

For a steady-state calculation, the right-hand side differencing determines the solution. We can keep the left-hand side the way it is or (to improve convergence) add a scalar dissipation term [8] to the left-hand side; for example,

$$\frac{\partial(A^n\Delta U^n)}{\partial x} = \frac{A^n_{j+1}\Delta U^n_{j+1} - A^n_{j-1}\Delta U^n_{j-1}}{2\Delta x} + \frac{\xi^n_{j+1}\Delta U^n_{j+1} - 2\xi^n_j\Delta U^n_j + \xi^n_{j-1}\Delta U^n_{j-1}}{2\Delta x} , \tag{6}$$

with $\xi^n_j = \max\left\{\left|\lambda^1_j\right|, \left|\lambda^2_j\right|, \ldots, \left|\lambda^m_j\right|\right\}$.

A constant version of the scalar dissipation term of equation (6) is obtained by setting ξ^n_j = constant for all j and n. This form of dissipation is sometimes called second-order implicit numerical dissipation.

Harten's Method for Two Space Dimensions:

Consider a two-dimensional system of conservation laws

$$\frac{\partial U}{\partial t} + \frac{\partial F(U)}{\partial x} + \frac{\partial G(U)}{\partial y} = 0 \quad , \tag{7}$$

where $F(U)$ and $G(U)$ are the flux vectors. Let the grid spacing be denoted by Δx and Δy such that $x = j\Delta x$ and $y = k\Delta y$.

Although Harten's scheme was developed for one-dimensional problems, the scheme can be conveniently implemented in two spatial dimensions by the method of fractional steps as follows:

$$U^*_{j,k} = U^n_{j,k} - \frac{h}{2\Delta x}\left(\hat{F}^n_{j+\frac{1}{2},k} - \hat{F}^n_{j-\frac{1}{2},k}\right) = \mathfrak{L}_x^{h/2} U^n_{j,k} \quad , \tag{8}$$

$$U^{n+1}_{j,k} = U^*_{j,k} - \frac{h}{2\Delta y}\left(\hat{G}^*_{j,k+\frac{1}{2}} - \hat{G}^*_{j,k-\frac{1}{2}}\right) = \mathfrak{L}_y^{h/2} U^*_{j,k} \quad , \tag{9}$$

that is,

$$U^{n+1}_{j,k} = \mathfrak{L}_y^{h/2}\,\mathfrak{L}_x^{h/2} U^n_{j,k} \quad ,$$

where $h = \Delta t, \hat{F}_{j+\frac{1}{2},k}$ is the appropriate analog of (4) evaluated at $y = k\Delta y$ for the flux $F(U)$ and $\hat{G}_{j,k+\frac{1}{2}}$ is the appropriate analog of (4) evaluated at $x = j\Delta x$ for the flux $G(U)$.

In order to retain the original second-order time accuracy of the method, we use a Strang type sequence of fractional step operators [9], namely

$$U^{n+2}_{j,k} = \mathfrak{L}_x^{h/2}\mathfrak{L}_y^{h/2}\mathfrak{L}_y^{h/2}\mathfrak{L}_x^{h/2}U^n_{j,k} = \left(\mathfrak{L}_x^{h/2}\mathfrak{L}_y^{h/2}\mathfrak{L}_y^{h/2}\mathfrak{L}_x^{h/2}\right)\left(\mathfrak{L}_x^{h/2}\mathfrak{L}_y^{h/2}\mathfrak{L}_y^{h/2}\mathfrak{L}_x^{h/2}\right) \cdots \left(\mathfrak{L}_x^{h/2}\mathfrak{L}_y^{h/2}\mathfrak{L}_y^{h/2}\mathfrak{L}_x^{h/2}\right)U^o_{j,k} \tag{10}$$

For steady-state calculation, the intermediate steps of equation (10) are just a part of the iteration procedure. Therefore, if we handle the boundary data for the intermediate steps correctly, we can combine the adjacent $\mathfrak{L}_y^{h/2}\mathfrak{L}_y^{h/2}$ operators into $\mathfrak{L}_y^h$ and the adjacent $\mathfrak{L}_x^{h/2}\mathfrak{L}_x^{h/2}$ operators into $\mathfrak{L}_x^h$. The half-step operators need only be applied at the first and last iterations, that is,

$$U^{n+2}_{j,k} = \mathfrak{L}_x^{h/2}\mathfrak{L}_y^h\mathfrak{L}_x^h\mathfrak{L}_y^h \cdots \mathfrak{L}_y^h\mathfrak{L}_x^{h/2}U^o_{j,k} \quad . \tag{11}$$

Modified Implicit Method for Two Space Dimensions:

The extension of technique (b) in two dimensions is not straightforward. It is well known that fractional step methods have the property that the steady-state (if one exists) depends on Δt even if the original one-dimension version of the scheme does not depend on Δt. For explicit methods, a fractional step procedure does not introduce a serious error. However, if one could take a large time-step by making the method implicit, then the steady-state accuracy will be degraded. In general, the steady-state dependence on Δt for implicit methods can be avoided by using an alternating direction implicit method. Recall that Harten's one-dimensional scheme has an inherent dependence on Δt in the steady-state. Consequently, technique (b) will also have a steady-state dependence on Δt even if we use the alternating direction implicit method. This dilemma is the subject of further investigation. For the purpose of this paper, we use a preliminary version of technique (b) in two space dimensions.

The two-dimensional alternating direction implicit version of (5) is

$$\left(I + \theta\Delta t\frac{\partial}{\partial x}A^n\right)\Delta U^* = -\Delta t\left(\frac{\partial F^n}{\partial x} + \frac{\partial G^n}{\partial y}\right) \tag{12}$$

$$\left(I + \theta\Delta t\frac{\partial}{\partial y}B^n\right)\Delta U^n = \Delta U^* \tag{13}$$

$$U^{n+1} = U^n + \Delta U^n \tag{14}$$

where B is the Jacobian matrix $\partial G/\partial U$. Then the modified implicit method is obtained by replacing $F_x^n + G_y^n$ on the right-hand side of (12) by

$$\frac{1}{\Delta x}\left(\hat{F}^n_{j+\frac{1}{2},k} - \hat{F}^n_{j-\frac{1}{2},k}\right) + \frac{1}{\Delta y}\left(\hat{G}^n_{j,k+\frac{1}{2}} - \hat{G}^n_{j,k-\frac{1}{2}}\right) \tag{15}$$

where the left-hand sides of (12) and (13) are replaced by the appropriate analogs of (6).

NUMERICAL RESULTS

A detailed implementation of Harten's method for the one-dimensional, inviscid, compressible equations of gas dynamics can be found in Harten's original paper. A description of the extension of Harten's method for two-dimensional problems can be found in [10]. For the numerical experiments, we choose the quasi-one-dimensional nozzle problem with two nozzle shapes (divergent and convergent-divergent) and the two-dimensional shock reflection problem. The exact steady-state solutions of these sample problems are used in evaluating the quality of the numerical solutions. For the numerical computations in this paper, we used Roe's averaging method [5]. A more detailed description and comparison of results for these test problems are presented in reference [10].

Comparison:

We choose three methods for comparison with the Harten's method and the modified implicit methods: i) a second-order flux-vector splitting method [8], ii) a conventional implicit scheme [7] with three-point central difference in space and backward Euler in time, and iii) the explicit-implicit MacCormack scheme [11].

Discussion of Numerical Results:

(a) Quasi-one-dimensional problems:

Figure 1a,b shows the shock resolution of Harten's method with two mesh spacings on the domain $0 \le x \le 10$. Figure 1c,d compares the conventional implicit method with the modified implicit method. Figure 2 shows the usefulness of Harten's scheme as a postprocessor on a steady-state solution computed by a conventional implicit method.

(b) Two-dimensional shock reflection problem:

In all of the numerical experiments, the computational domain is $0 \le x \le 4.1$, and $0 \le y \le 1$. The incident shock angle ψ is 29° and the freestream Mach number M_∞ is 2.9 (see fig. 3). The grid size is 61×21. The exact minimum pressure corresponding to $\psi = 29°$ and $M_\infty = 2.9$ is 0.714286; the exact maximum pressure is 2.93398. Forty-one pressure contour levels between the values of 0 and 4 with uniform increment 0.1 are used for each plot. The pressure coefficient is defined as $c_p = 2(p/p_\infty - 1)/\gamma M_\infty^2$; p_∞ is the freestream pressure and γ is the ratio of specific heat.

Figure 3 shows the exact pressure solution of the shock reflection problem. Figure 4 compares the pressure contours and the pressure coefficient at $y = 0.5$ obtained from Harten's method, the modified implicit method, the flux-vector splitting method, the conventional implicit method, and the explicit-implicit MacCormack's method. The results show a definite improvement in shock resolution by the high-resolution methods. The results for the MacCormack's method are furnished by the courtesy of W. Kordulla of the NASA Ames Research Center. All methods required 150 to 600 iterations to converge with CFL ranging from 0.5 to 0.8. Convergence rate and computation efficiency will be investigated and reported elsewhere.

CONCLUSIONS

The application of Harten's method to quasi-one-dimensional nozzle problems and to a two-dimensional shock-reflection problem resulted in high shock resolution steady-state numerical solutions. Applications of the postprocessor method and the modified implicit method for steady-state calculations show encouraging results for one-dimensional problems; however, testing in two dimensions is not complete and further investigation is needed for efficient implementation of the implicit method.

REFERENCES

[1] V.V. Rusanov, *Lecture Notes in Physics*, Vol. 141, 1981.
[2] A. Harten, *NYU Report*, March 1982.
[3] P.D. Lax, *SIAM Conference Board of the Mathematical Sciences*, 1972.
[4] A. Harten, J.M. Hyman and P.D. Lax, *Comm. Pure Appl. Math.*, Vol. 29, 1976.
[5] P.L. Roe, *J. Comp. Phys.*, Vol. 43, 1981.
[6] R.F. Warming and R.M. Beam, *SIAM-AMS Proceedings*, Vol. 11, 1978.
[7] R. M. Beam and R.F. Warming, AIAA paper 79-1446, 1979.
[8] J. Steger and R.F. Warming, *Comp. Phys.*, Vol. 40, No. 2, April 1981.
[9] G. Strang, *SIAM J. Numer. Anal.* Vol. 5, 1968, pp. 506-517.
[10] H.C. Yee, R.F. Warming and A. Harten, NASA TM-84256, June 1982.
[11] R.W. MacCormack, AIAA paper 81-0110, 1981.

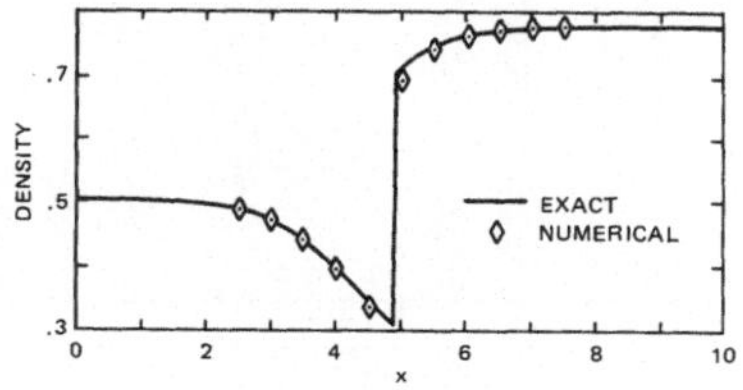

(a) Harten method (explicit).

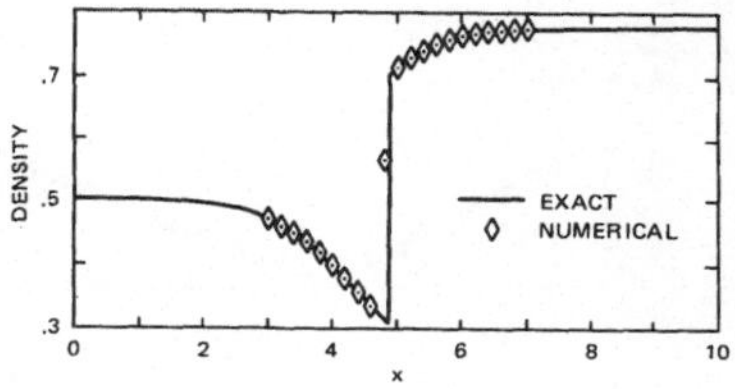

(b) Harten method (explicit).

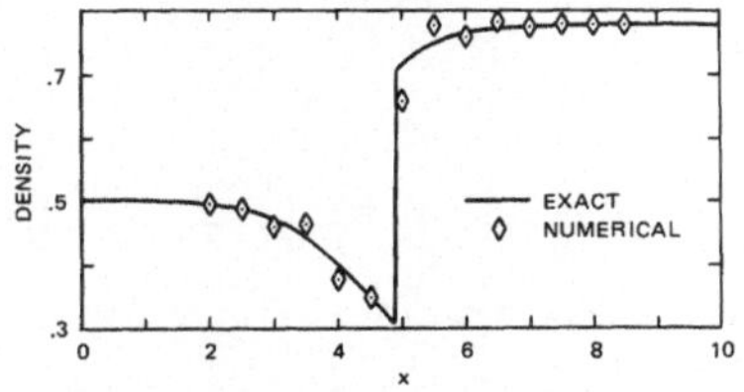

(c) Conventional implicit method.

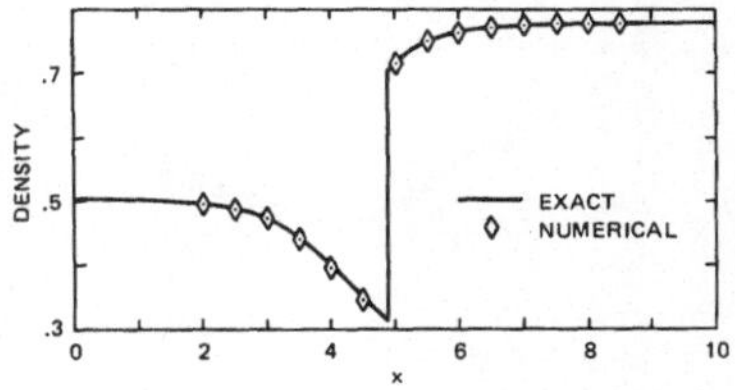

(d) Modified implicit method.

Fig. 1. Density distribution: supersonic inflow, subsonic outflow.

Nozzle shape:  ; 20 spatial intervals for figures a, c and d; 50 spatial intervals for figure b.

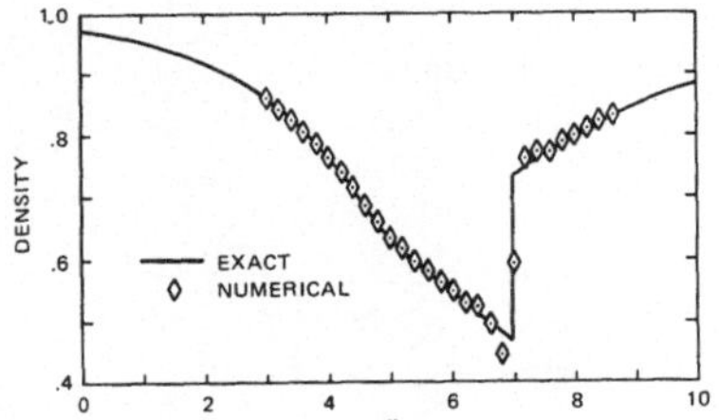

(a) Conventional implicit method.

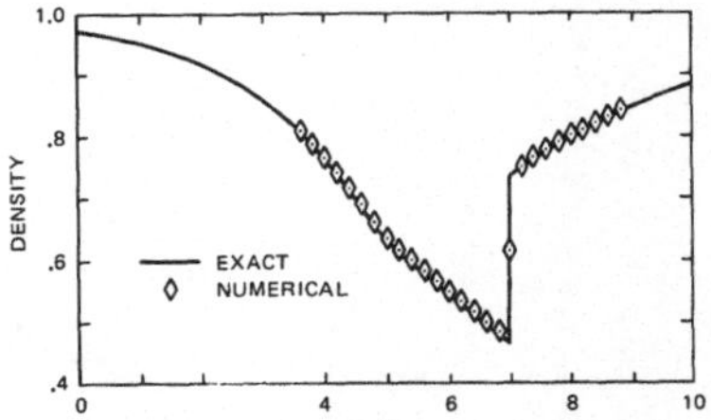

(b) Harten as postprocessor.

Fig. 2. Density distribution: subsonic inflow, subsonic outflow.

Nozzle shape:  ; 50 spatial intervals.

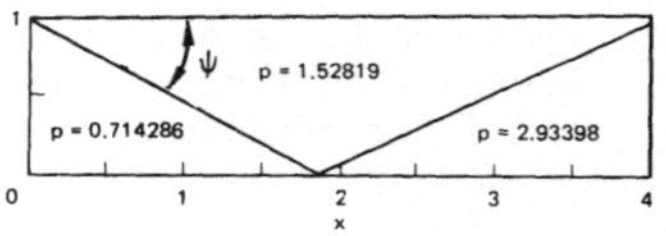

Fig. 3. Exact pressure solution for the shock reflection problem.

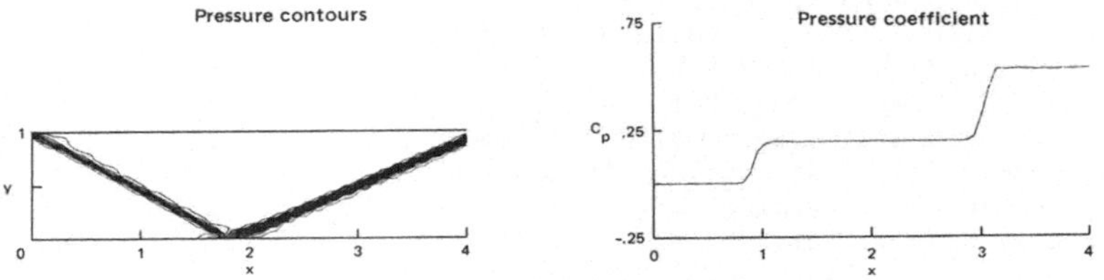

(a) Harten's method (explicit).

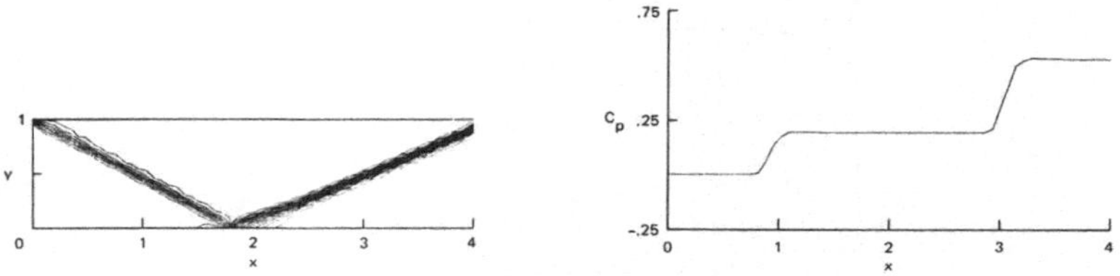

(b) Modified implicit method.

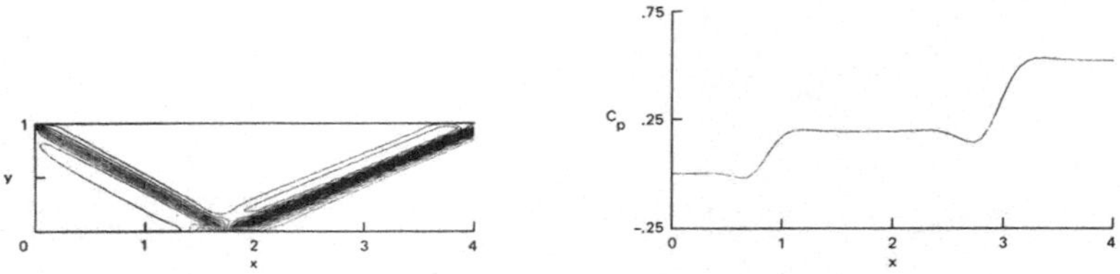

(c) Flux-vector splitting (implicit).

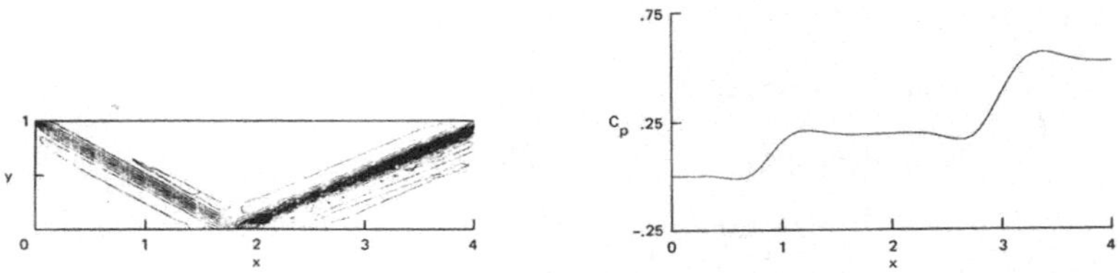

(d) Conventional implicit method.

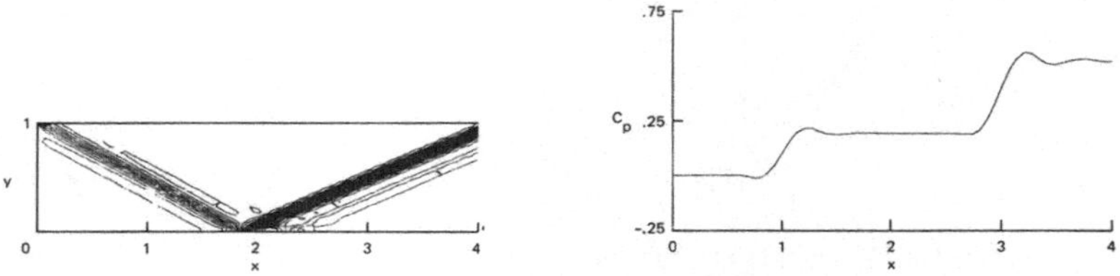

(e) Explicit-implicit MacCormack's method.

Fig. 4. Pressure contours for a two-dimensional shock-reflection problem.

SOME NEW DEVELOPMENTS OF THE SINGULARITY-SEPARATING DIFFERENCE METHOD

Y.-l. Zhu, B.-m. Chen, X.-h. Wu and Q.-s. Xu
The Computing Center of Academia Sinica, Beijing, China

The problem of interaction between three dimensional discontinuities in steady supersonic flow has been successfully solved by the singularity-separating difference method in Zhu and Chen (1981) and Zhu et al. (1980). This method was recently applied to the unsteady multimaterial flow in Wu(1982a), the hyperbolic equation with a non-convex equation of state in Wu (1982b), the heat conduction problem with phase-change in Xu (1981), the steady supersonic flow around bodies with a bent nose in Chen and Zhu (1981) and the steady supersonic flow of air in chemical equilibrium around combined bodies in Chen and Zhu (1982) with success again.

When one calculates the unsteady multimaterial flow, the difficulty encountered is how to deal with the boundaries between different materials--the contact discontinuities, and the interaction between a shock and a contact discontinuity. The accuracy of results obtained by using existing numerical methods is usually not satisfying.

We have applied the singularity-separating method to this problem. For example, we have calculated the following flow : at the beginning, there are a shock and a contact discontinuity on whose two sides the materials are different. Then the shock intersects the contact discontinuity, and a shock (or a centered wave) is reflected. Thus there are finally two shocks and a contact discontinuity (or a shock, a centered wave and a contact discontinuity) in the flow field. In order to obtain accurate solutions of this problem, we must have a method which can accurately deal with shocks, contact discontinuities and centered waves. Our singularity-separating difference method possesses these abilities. In this method, shocks, contact discontinuities, and boundaries of centered waves are taken as internal boundaries, and the problem is solved by using the second order mixed scheme for initial-boundary-value problems developed by the authors.

Several results for this problem have been obtained. In Figure 1 a density curve for a certain flow field is given, where there are two shocks and a contact discontinuity. The ratios of specific heats of the gases on the left and right sides of the contact discontinuity are 5 and 1.5 respectively. Therefore this is a multimaterial flow. From that figure, we can see that neither shocks nor contact discontinuities are smeared and there is no oscillation.

In order to compare our method with other methods, we have calculated an unsteady flow with a shock, a contact discontinuity and a centered wave by using the L-W scheme and our scheme. In Figure 2 we present three curves which are obtained respectively by the L-W scheme with 100 mesh points, the L-W scheme with 2000 mesh points, and our method with 30 mesh points. From this figure we can see the following: when our method is used, we obtain quite good results taking only tens of mesh points; when the L-W scheme is used, the results are not good if hundreds of mesh points are taken, and thousands of mesh points are needed in order to get an acceptable result.

When the equation of state is nonconvex, several new physical phenomena appear. One of them is that one shock can split up into two discontinuities, for example, into one right contact discontinuity and one left contact discontinuity, or two shocks. As is well-known, this phenomenon never appears if the equation of state is convex. Moreover, the appearance of

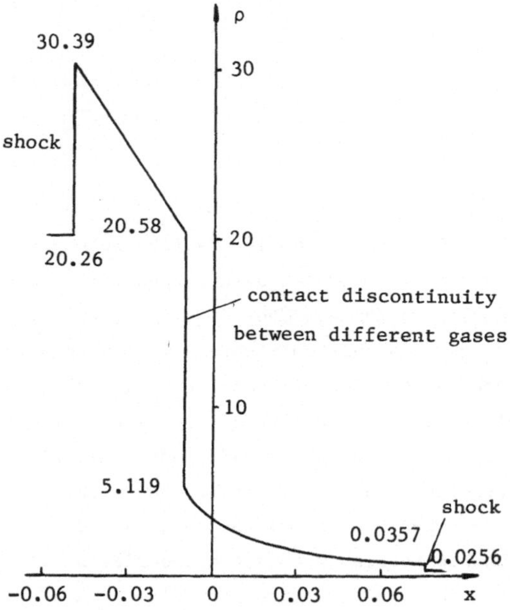

Fig. 1

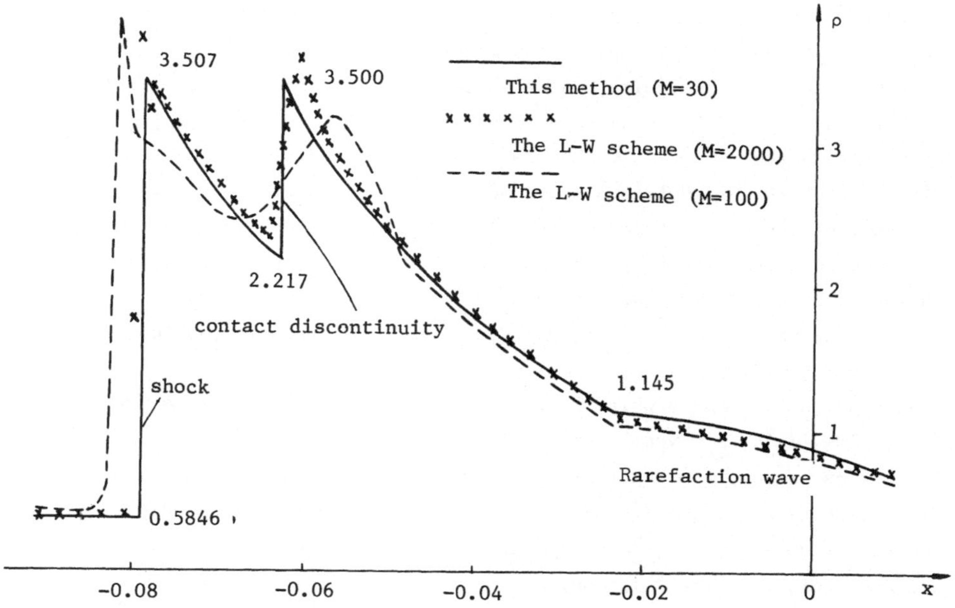

Fig. 2

non-physically relevant solutions in numerical calculation is much more common in the case of the non-convex equation of state than in the case of the convex equation of state.

This problem has also been accurately solved by the singularity-separating difference method. Since the equation of state is non-convex, certain modifications are made: the stability condition (E) for discontinuities is used in order for the numerical solutions to be always physically relevant solutions; and the algorithm of determining the quantities at "shock points" is modified because "shocks" now include normal shocks, right contact discontinuities, left contact discontinuities and double contact discontinuities.

we have solved the following equation

$$\begin{cases} u_t + f(u)_x = 0, \\ u(x,0) = \begin{cases} u_p(x), & x > 0, \\ u_n(x), & x < 0, \end{cases} \end{cases}$$

where

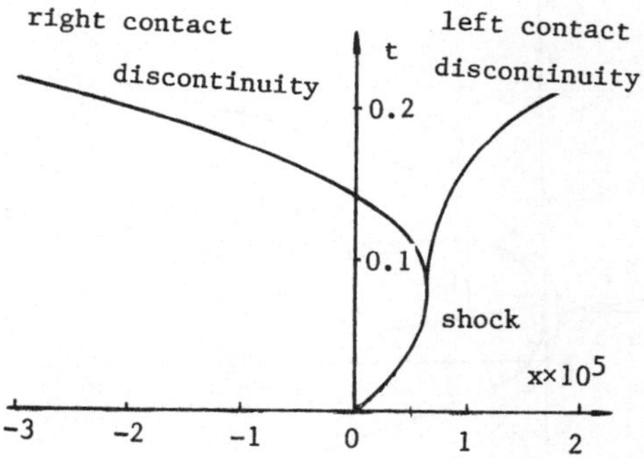

Fig.3

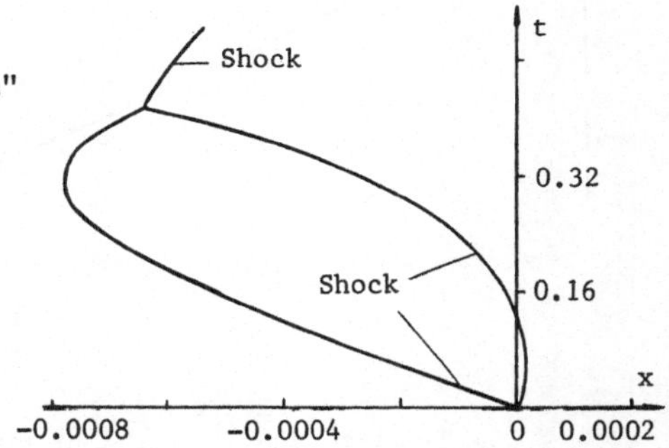

Fig.4

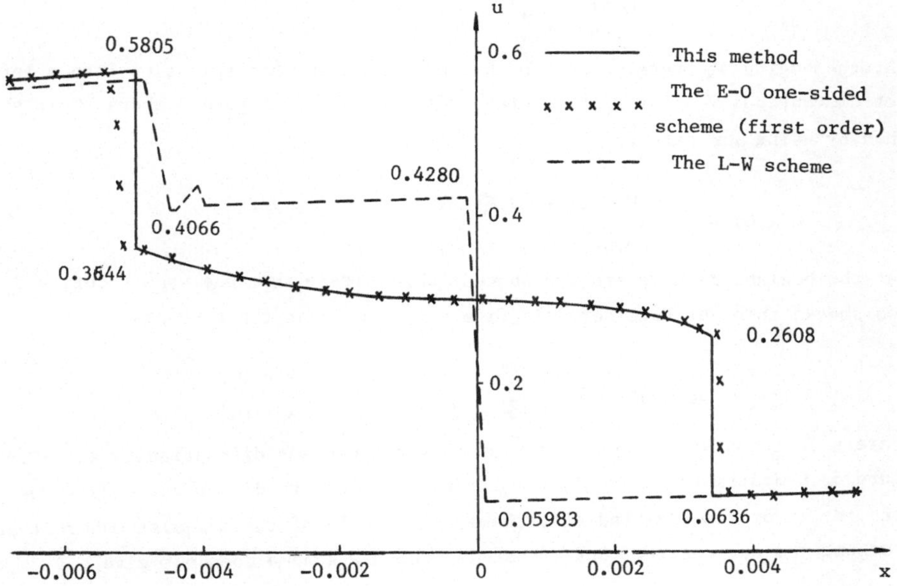

Fig.5

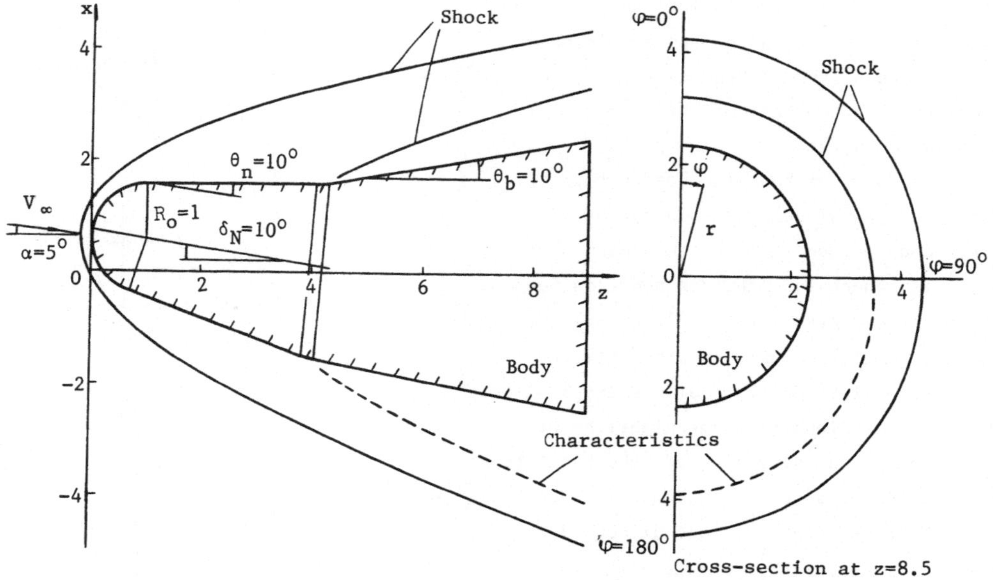

Fig.6

$$f(u)=u^4/2-1.9u^3/3 + 0.25u^2 - 0.033u.$$

Clearly the structures of the solutions might be different if the initial functions are different.

If

$$u(x,0) = \begin{cases} 0.01 + 4x, & x > 0, \\ 0.66 + 5x, & x < 0, \end{cases}$$

then at the beginning there is one shock, and the shock then splits up into a left contact discontinuity and a right contact discontinuity. Figure 3 shows their shapes obtained by using our method.

If

$$u(x,0) = \begin{cases} 0.02-4x, & x> 0, \\ 0.6-5x, & x< 0, \end{cases}$$

then at the beginning there are two shocks, which issue from the same point, and the two shocks then get together. In Figure 4 we present their shapes.

If

$$u(x,0) = \begin{cases} (10-\sqrt{91})/30 + 4x, & x > 0, \\ (10+\sqrt{91})/30 + 5x, & x < 0, \end{cases}$$

there are a left contact discontinuity and a right contact discontinuity at t=0.98. In Figure 5, u-distribution is given. In order to compare different methods, we give the result of the Lax-Wendroff scheme, the result of the Engquist-Osher first order scheme and our result. Unfortunately, the L-W scheme does not give a

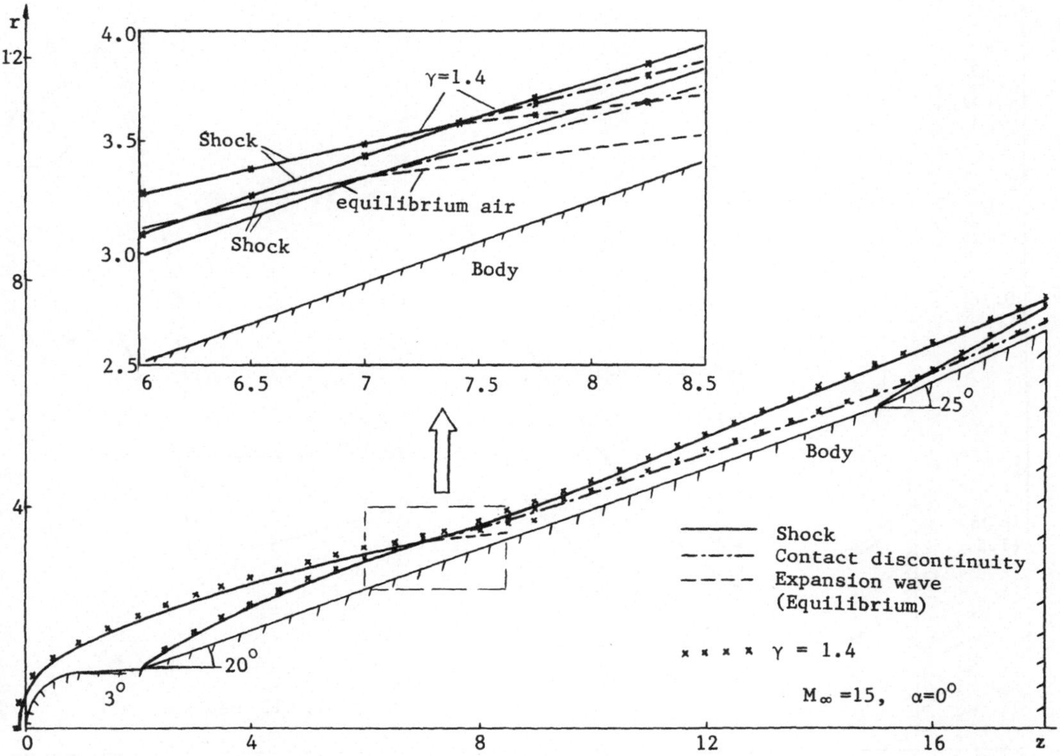

Fig.7

physically relevant solution. Both the E-O scheme and our scheme give physically relevant solutions. However, in order to obtain a result with a fixed accuracy, the E-O scheme needs much more mesh points than the singularity-separating difference method. Our result in Figure 5 is obtained taking 40 mesh points, and the result of the E-O scheme is obtained taking 800 mesh points. If we decrease the number of mesh points for the E-O scheme, then the difference between the E-O scheme's result and our scheme's result will get quite large.

The heat conduction problem with phase-change is another tough problem. The main difficulty is how to deal with the heat of transformation when the material changes from one phase to another. We have accurately calculated this problem by the singularity-separating difference method. In our calculation, the boundary between the material of the solid phase and the material of the liquid phase is taken as an internal boundary, i.e., its position is determined as one part of the solution. Thus the absorbed heat and the released heat can be accurately considered, and fine results can be obtained. In order to compare our method with other methods, we calculate a certain problem by using our method and the method in Bonacina and Comini (1973). When our method is used, we take $\Delta t=0.864$, $\Delta x=0.1$ and the maximum error is 0.004; and when the method in Bonacina and Comini (1973) is used, we take

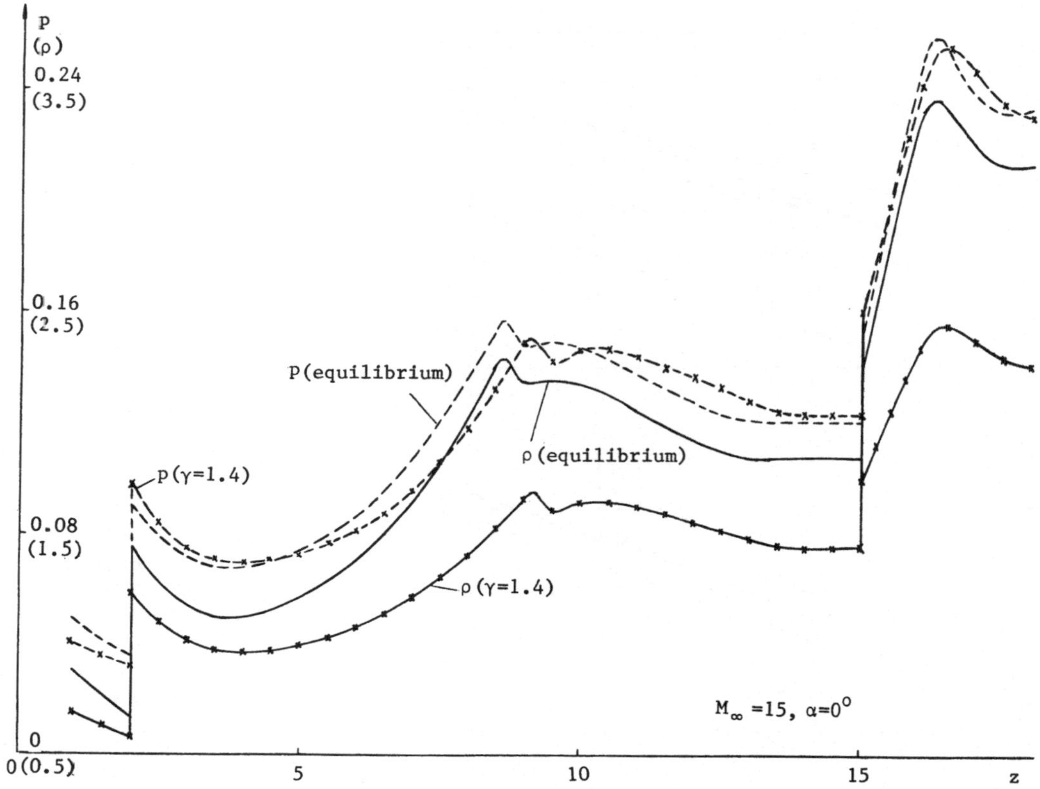

Fig.8

Δt=0.04, Δx=0.025, and the maximum error is 0.317. This means that our method can provide much more accurate results than the method in Bonacina and Comini (1973) although the mesh size for our method is larger than that for the method in Bonacina and Comini (1973).

The steady supersonic flow around bodies with a bent nose and the steady supersonic flow of air in chemical equilibrium around combined bodies are also calculated. In the flow field around bodies with a bent nose, there is an "unclosed", embedded shock, which is like the internal shock in the flow field around a space-shuttle. In our calculation, the "unclosed" shock is extended to a "closed" shock whose extended part is a wave-characteristic surface and a wave-characteristic surface is regarded as a weak shock. Because of this treatment, we can solve the flow around bodies with a bent nose by the direct use of our method. The shapes of the main shock and the "unclosed" internal shock are given in Figure 6.

The method for the perfect gas flow can be still used for the flow of air in chemical equilibrium. However, because of the complicated equation of state, the algorithm for determining the quantities at shock points needs to be slightly modified. We have calculated supersonic flow of equilibrium air around the combined bodies where there are the intersection of two shocks and the intersection of a shock with a contact discontinuity. In Figure 7, such a combined body and the shapes of discontinuities in the flow field around the body are drawn. In Figure 7 we also give the result of the perfect gas with $\gamma=1.4$. From this figure we see that the difference between the positions of the shocks in the equilibrium air flow and in the perfect gas flow is quite large if $M_\infty=15$. In Figure 8 we show the pressure distributions and density distributions on the body both for the equilibrium air flow and for the perfect gas flow, which shows that the difference between two density distributions is much larger than that between two pressure distributions.

References

Bonacina, O. and Comini, G. (1973) : Numerical solution of phasechange problems, Int. J.Heat Mass Transfer, 16, 1825-1832.

Chen, B.-m. and Zhu, Y.-l. (1981) : Numerical calculation of the flow around bodies with a bent nose, Report of the Computing Center of Academia Sinica, Beijing, China, 1981.

Chen, B.-m. and Zhu, Y.-l. (1982) : Numerical calculation of the flow of air in chemical equilibrium around combined bodies, Report of the Computing Center of Academia Sinica, Beijing, China, 1982.

Wu, X.-h. (1982a) : Numerical calculation of the multimaterial flow with shocks, Report of the Computing Center of Academia Sinica, Beijing, China, 1982.

Wu, X.-h. (1982b) : Numerical calculation of the hyperbolic equation with a non-convex equation of state, Report of the Computing Center of Academia Sinica, Beijing, China, 1982.

Xu, Q.-s. (1981) : A new numerical method for the heat conduction problem with phase-change, Report of the Computing Center of Academia Sinica, Beijing, China, 1981.

Zhu, Y.-l. and Chen, B.-m. (1981) : An accurate method for calculating the interactions between discontinuities in three dimensional flow, in "Lecture Notes in Physics", Vol. 141, Springer-Verlag, Berlin etc., 1981.

Zhu, Y.-l., Zhong, X.-c., Chen, B.-m. and Zhang, Z.-m. (1980) : Difference methods for initial-boundary-value problems and flow around bodies, Science Press, Beijing, China, 1980.

LIST OF PARTICIPANTS

HELGE ANDERSSON
INST.F.MEKAN.,UNIV. TRONDHEIM

7034 TRONDHEIM
NORWAY

PROF. PAUL ARMINJON
DEPT. OF MATHEMATICS
UNIVERSITY OF MONTREAL
MONTREAL, QUEBEC, H3C3J7
CANADA

ASS. PROF. DR. H. M. BADR
UNIVERSITY OF PETROLEUM & MINERALS
UPM BOX 322
DHARAN
SAUDI ARABIA

DR. ALLEN J. BAKER
UNIVERSITY OF TENNESSEE
317 PERKINS HALL
KNOXVILLE, TN 37996
U.S.A.

J. C. LE BALLEUR
ONERA
29, AVENUE DE LA DIVISION LECLERC
F-92320 CHATILLON
FRANCE

DR. WILLIAM F. BALLHAUS
APPL. COMP. AERO. BR., MS 202A-14
NASA AMES RESEARCH CENTER
MOFFETT FIELD, CA 94035
U.S.A.

DR. HOWARD R. BAUM
CENTER FOR FIRE RESEARCH
NATIONAL BUREAU OF STANDARDS
WASHINGTON, DC 20234
U.S.A.

YU. A. BEREZIN
INST. OF THEOR. + APPL. MECHANICS
USSR ACADEMY OF SCIENCES
SU-630090 NOVOSIBIRSK 90
U.S.S.R.

PROF. SUNE BERNDT

SVEAVAEGEN 7 B
S-182 62 DJURSHOLM
SWEDEN

DR. ISHWAR C. BHATELEY
GENERAL DYNAMICS CORP., FORT WORTH
P.O. BOX 748
FORTH WORTH, TX 76101
U.S.A.

DAVID L. BOOK, PH. D.
LAB. FOR COMPUTATIONAL PHYSICS
NAVAL RESEARCH LABORATORIES
WASHINGTON, DC 20375
U.S.A.

DR. EUGEN BOTTA
MATHEMATISCH INSTITUUT
POSTBUS 800
NL-9700 AV GRONINGEN
THE NETHERLANDS

DR. J. S. BRAMLEY
DEP. OF MATHS.,UNIV. STRATHCLYDE
LIVINGSTONE TOWER, 26 RICHMOND STR
GLASGOW G1 1XH
ENGLAND

KURT BRAND
GMD - IMA
POSTFACH 1240
D-5205 ST. AUGUSTIN 1
WEST GERMANY

DR. CHARLES HENRI BRUNEAU
ONERA
29 AV. DE LA DIVISION LECLERC
F-92320 CHATILLON
FRANCE

PROF. HENRI CABANNES
UNIVERSITE PIERRE ET MARIE CURIE
23 ALLEE DE TREVISE
92330 SCEAUX
FRANCE

DR. JAMES E. CARTER
UNITED TECHNOLOGIES RES. CENTER
SILVER LANE
EAST HARTFORD, CT 06108
U.S.A.

A. U. CHATWANI
INSTITUT FUER ANTRIEBSTECHNIK
DFVLR
D-5000 KOELN-PORZ
WEST GERMANY

DIPL. MATH. JUERGEN CLEMENS

BLUMENTHAL 30
D-2167 BURWEG
WEST GERMANY

ENG. TUAN DANG-TRONG
AEROSPATIALE/ SUBDIV. SYSTEMES
B.P. 96
F-78130 LES MUREAUX
FRANCE

DR. A. DAVEY
MATHEMATICS DEPT.
UNIV. OF NEWCASTLE UPON TYNE
NEWCASTLE UPON TYNE, NE1 7RU
ENGLAND

IR. G. J. DE BRUIN
T.H.T. , ROOM WB/W-250
P.O. BOX 217
NL-7500 AE ENSCHEDE
THE NETHERLANDS

RES. ASS. HERMAN DECONINCK
DEPT.FL.MECH.,VRIJE UNIV. BRUSSEL
PLEINLAAN 2
B-1050 BRUESSEL
BELGIUM

PROF. STANLEY C. R. DENNIS
DEPT. OF APPLIED MATHEMATICS
UNIVERSITY OF WESTERN ONTARIO
LODON, ONTARIO, N6A 5B9
CANADA

A. DI CARLO
INST. MECCANICA APPL., UNIVERSITA
VIA EUDOSSIANA 18
I-00184 ROMA
ITALY

DR. ING. ERIK DICK
DEPT. OF MACH.,ST. UNIV. OF GENT
SINT PIETERSNIEUWSTRAAT 41
B-9000 GENT
BELGIUM

DR. D. DIJKSTRA
DEPT. MATH.,TWENTE UNIV. OF TECHN.
T.H.T., P.O.B. 217
7500 AE ENSCHEDE
THE NETHERLANDS

H. EICKHOFF
DFVLR
INST. F. ANTRIEBSTECHNIK
D-5000 KOELN
WEST GERMANY

DIPL. MATH. KARL HEINZ ERKENS
CONTROL DATA GMBH
STESEMANNALLEE 30
D-6000 FRANKFURT/MAIN 70
WEST GERMANY

DR. J. A. ESSERS
VON KARMAN INST. F. FLUID DYNAMICS
CHAUSSEE DE WATERLOO 72
B-1640 RHODE-SAINT-GENESE
BELGIUM

CHERCHEUR MARIE FARGE
LAB. METEOR.DYN.,ECOLE NORM.SUPER.
24 RUE LHOMOND
F-75231 PARIS CEDEX 05
FRANCE

DR. HERMANN FASEL
INST. F. MECHANIK, UNIV. STUTTGART
PFAFFENWALDRING 9
D-7000 STUTTGART 80
WEST GERMANY

PROF. DR. ING. RAINER FRIEDRICH
LEHRST. F. STR.MECH.,TU MUENCHEN
ARCISSTR. 21
D-8000 MUENCHEN
WEST GERMANY

DR. H. H. FRUEHAUF
INST. F. RAUMFAHRTANTR.,UNI S
PFAFFENWALDRING 31
D-7000 STUTTGART 80
WEST GERMANY

DR. MARK FRY
SCIENCE APPLICATIONS, INC. -T-4
1710 GOODRIDGE DRIVE
MCLEAN, VA 22102
U.S.A.

HONG-YUAN FU
COMP. CENTER ACADEMIA SINICA
ZHONGGUANCUN
BEIJING
CHINA

DR. LASZLO FUCHS
DEPT. OF GASDYNAMICS
THE ROYAL INSTITUTE OF TECHNOLOGY
S-100 44 STOCKHOLM
SWEDEN

MR. SHOJI FUKUDA
CENTURY RESEARCH CENTER CORP.
2,3-CHOME,HONCHO,NIHONBASHI,CHUOKU
TOKYO 103
JAPAN

PROF. DR. ING. H. E. GALLUS
INST. F. STRAHLANTR. U. TURBOMASCH
RWTH AACHEN, TEMPLERGRABEN 55
D-5100 AACHEN
WEST GERMANY

PROF. DR. ING. K. GERSTEN
INST. F. THERMO- U. FLUIDDYNAMIK
BUSCHEYSTR. 10, RUHRUNIVERSITAET
D-4630 BOCHUM-QUERENBURG
WEST GERMANY

MRS. DR. URMILA GHIA
AS.E. & A.M., UNIV. OF CINCINNATI
MAIL LOCATION 70
CINCINNATI, OH 45221
U.S.A.

DR. A. F. GHONIEM
DEPT. OF MECHAN. ENG.
UNIVERSITY OF CALIFORNIA
BERKELEY, CA 94720
U.S.A.

DIPL. ING. ULRICH GIESE
AERODYNAMISCHES INST., RWTH AACHEN
WUELLNERSTR. 55
D-5100 AACHEN
WEST GERMANY

DR. HARLAND M. GLAZ
NAVAL SURFACE WEAPONS CENTER
WHITE OAK
SILVER SPRING, MD 20910
U.S.A.

DIPL. ING. K. GOERNER
UNI STUTTGART, IVD
PFAFFENWALDRING 23
D-7000 STUTTGART 80
WEST GERMANY

JEAN PIERRE GREGOIRE
DIR. D. ET.& RECH. ELECTR. FRANCE
1 AVENUE DU GENERAL DE GAULLE
F-92 CLAMART
FRANCE

DR. PHILIP M. GRESHO
LAWRENCE LIVERMORE NAT. LAB.
P. O. BOX 808
LIVERMORE, CA 94550
U.S.A.

DIPL. ING. KARL GRUBER
INST. A MECH.,UNIV. STUTTGART
PFAFFENWALDRING 9
D-7000 STUTTGART 80
WEST GERMANY

PROF. DR. W. HACKBUSCH
INST. F. MATH., RUHR-UNI BOCHUM
POSTFACH 102148
D-4630 BOCHUM 1
WEST GERMANY

E. W. HADDON
SCHOOL OF COMPUTING STUDIES
UNIVERSITY OF EAST ANGLIA
NORWICH, NR4 7TJ
ENGLAND

PROF. AMIRAM HARTEN
SCHOOL OF MATHEMATICAL SCIENCES
TELAVIV UNIVERSITY, RAMAT-AVIV
TELAVIV 69978
ISRAEL

DR. RUDI HEISER
E.-MACH-INST., ABT. BALLISTIK
HAUPTSTR. 18
D-7858 WEIL AM RHEIN
WEST GERMANY

DR. DIETER HENSEL
DEUTSCH-FRANZ. F.-INST.,ST-LOUIS
TALSTR. 13
D-7851 SCHALLBACH
WEST GERMANY

PROF. DR. FUMIO HIGASHINO
STOSSWELLENLAB., INST. L.+ RAUMF.
RWTH, TEMPLERGRABEN 55
D-5100 AACHEN
WEST GERMANY

PROF. DR. CH. HIRSCH
FLUID MECH.,VRIJE UNIV. BRUSSEL
PLEINLAAN 2
B-1050 BRUESSEL
BELGIUM

PROF. DR. ERNST H. HIRSCHEL
MBB UFE 122
POSTFACH 80 11 60
D-8000 MUENCHEN 80
WEST GERMANY

DR. RICHARD S. HIRSH
APPL. PHYS. LAB.,J. HOPKINS UNIV.
JOHNS HOPKINS ROAD
LAUREL, MD 20707
U.S.A.

MARC HITTINGER
CISI, COMP. INTERN. DE SERVICE EN
INFORMAT., B.P. NO 24
F-91190 GIF-SUR-YVETTE
FRANCE

PROF. MAURICE HOLT
UNIVERSITY OF CALIFORNIA
MECHANICAL ENGINEERING
BERKELEY, CA 94720
U.S.A.

ZHENGYONG HUANG
INST. NUMER. MATH., TU WIEN
GUSSHAUSSTR. 27-29
A-1040 WIEN
AUSTRIA

RES. ENG. RISTO HUHTANEN
TECHN. RES. CENTRE,NUCL. ENG. LAB.
P.O.B. 169
SF-00181 HELSINKI 18
FINLAND

D. A. HUMPHREYS
FFA
BOX 11021
S-16111 BROMMA
SWEDEN

DR. M. YOUSUFF HUSSAINI
ICASE, MAIL STOP 132C
NASA LANGLEY RESEARCH CENTER
HAMPTON, VA 23665
U.S.A.

ASS. PROF. MOSHE ISRAELI
COMPUTER SCIENCE DEPT.
TECHNION CITY
HAIFA 32000
ISRAEL

A. JAMESON
PRINCETON UNIVERSITY

PRINCETON, NJ 08544
U.S.A.

DR. GARY JOHNSON
MAIL STOP 5-9
NASA LEWIS RESEARCH CENTER
CLEVELAND, OH 44135
U.S.A.

PROF. REIJO KARVINEN
TAMPERE UNIVERSITY OF TECHNOLOGY
P.O. BOX 527
SF-33101 TAMPERE 10
FINLAND

DR. M. J. KASCIC, JR.
MATH. RES., CONTROL DATA CORP.
8100 34TH AVENUE SOUTH, BOX O
MINNEAPOLIS, MN 55440
U.S.A.

DR. GEORGE V. KELLY
UNIVERSITY COLLEGE

CORK
IRELAND

DR. G. DAVID KERLICK
NIELSEN ENG. & RES., INC.
510 CLYDE AVENUE
MOUNTAIN VIEW, CA 94043
U.S.A.

PROF. PREM K. KHOSLA
DEPT. OF AEROSP. ENG.+ APPL. MECH.
UNIVERSITY OF CINCINNATI
CINCINNATI, OH 45221
U.S.A.

DIPL. ING. P. KILGENSTEIN
INST. F. MECHANIK, TH DARMSTADT
HOCHSCHULSTRASSE 1
D-6100 DARMSTADT
WEST GERMANY

PROF. JOHN KIM
DEPT. OF MECHANICAL ENGINEERING
STANFORD UNIVERSITY
STANFORD, CA 94305
U.S.A.

DR. LEONHARD KLEISER
INST. F. REAKTORENTW.,KERNF.ZENTR.
POSTFACH 3640
D-7500 KARLSRUHE
WEST GERMANY

DR. MARIA KLONOWSKA

4, RUE DU REPOS
F-25000 BESANCON
FRANCE

DR. WILLI KORDULLA
DFVLR SM-TS
BUNSENSTR. 10
D-3400 GOETTINGEN
WEST GERMANY

V. M. KOVENYA
INST. OF THEOR. + APPL. MECHANICS
USSR ACADEMY OF SCIENCES
SU-630090 NOVOSIBIRSK 90
U.S.S.R.

DR. K. KOZEL
STROJNI FAKULTA CVUT
SUCHBATAROWA 4
166 07 PRAHA 6
CZECHOSLOVAKIA

PROF. EGON KRAUSE
AERODYNAM. INSTITUT, RWTH AACHEN
WUELLNERSTR. ZW. 5-7
D-5100 AACHEN
WEST GERMANY

DR. ALLEN KUHL
R & D ASSOCIATES

MARINA DEL REY, CA
U.S.A.

PROF. KUNIO KUWAHARA
INST. OF SPACE + ASTRONAUT. SC.
KOMABA, MEGURO-KU
TOKYO
JAPAN

DR. WILLIAM E. LANGLOIS

2340 STRATFORD DRIVE
SAN JOSE, CA 95124
U.S.A.

PROF. ERNST VON LAVANTE, PH. D.
AEROSP. ENG. DEPT.
TEXAS A&M UNIVERSITY
COLLEGE STATION TEXAS 77843-3141
U.S.A.

T. H. LE
ONERA
29, AVENUE DE LA DIVISION LECLERC
F-92320 CHATILLON
FRANCE

ING. JEAN-PIERRE LEGOFF
D.R.E.T.
26 BOULEVARD VICTOR
F-75015 PARIS
FRANCE

S. LEICHER
THEOR. AEROD. DEPT., DORNIER GMBH
POSTFACH 1420
D-7990 FRIEDRICHSHAFEN
WEST GERMANY

DR. ANTHONY LEONARD

1450 MARLBAROUGH CT.
LOS ALTOS, CA 94022
U.S.A.

DR. ALAIN LERAT
ECOLE NAT. SUP. D ARTS METIERS
151 BLD. DE L HOPITAL
75640 PARIS CEDEX 13
FRANCE

LESOVOY
INST. OF THEOR.+APPL. MECHANICS
U.S.S.R. ACADEMY OF SCIENCES
NOVOSIBIRSK 630090
U.S.S.R.

DR. EDDIE LIU
COMP. FLUID DYN. GROUP
NASA LANGLEY RESEARCH CENTER
HAMPTON, VA 23665
U.S.A.

DR. CHARLES K. LOMBARD
PEDA CORPORATION
1150 FIFE STREET
PALO ALTO, CA 94301
U.S.A.

CHIEF ENGINEER TAKEO MAEDA
CENTURY RESEARCH CENTER CORP.
2,3-CHOME,HONCHO,NIHONBASHI,CHUOKU
TOKYO T 103
JAPAN

ASS. PROF. PHILIP MARCUS
2-337 DEPT. OF MATHEMATICS
MASSACHUSETTS INST. OF TECHNOLOGY
CAMBRIDGE, MA 02139
U.S.A.

PROF. WALTER MELNIK
DAVID W. TAYLOR NAVAL SHIP RES. +
DEVELOPMENT CENTER
BETHESDA, MD 20742
U.S.A.

DIPL. ING. A. MERTEN
URANIT-GMBH

D-5170 JUELICH
WEST GERMANY

DR. RALPH W. METCALFE
FLOW RESEARCH CO.
21414 68 TH AVENUE
SOUTH KENT, WA 98031
U.S.A.

DR. N. K. MITRA
INST. F. THERMO- U. FLUID DYNAMIK
RUHR-UNI BOCHUM, POSTFACH 2148
D-4630 BOCHUM-QUERENBERG
WEST GERMANY

DR. PARVIZ MOIN
NASA AMES RESEARCH CENTER
MAIL STOP 202 A-1
MOFFETT FIELD, CA 94035
U.S.A.

MR. MOL
AERODYN. DEPT. FOKKER
POSTBUS 7600
SHIPOOL-OOST
THE NETHERLANDS

YVES MORCHOISNE
ONERA
29 AV. DE LA DIVISION LECLERC
F-92320 CHATILLON
FRANCE

MR. MORICE

ONERA
92320 CHATILLON
FRANCE

PROF. K. W. MORTON
DEPT. OF MATHEMATICS
UNIVERSITY OF READING
WHITEKNIGHTS, READING RG6 2AX
ENGLAND

DIPL. MATH. CLAUS-DIETER MUNZ
MATH. INST. II DER UNIV. KARLSRUHE
KAISERSTR. 12
D-7500 KARLSRUHE 1
WEST GERMANY

KAZUHIRO NAKAHASHI, PH. D.
NAT. AEROSP. LAB., KAKUDA BRANCH
P.O. BOX 7
OHGAWARA, MIYAGI 989-12
JAPAN

PROF. SHOICHIRO NAKAMURA
OHIO STATE UNIVERSITY
206 W. 18TH AVE.
COLUMBUS, OH 43210
U.S.A.

PROF. MICHELE NAPOLITANO PH. D.
INST. DI MACCHINE,UNIV. DI BARI
VIA RE DAVID, 200
I-70125 BARI
ITALY

PROF. A. NASTASE
RWTH AACHEN
TEMPLERGRABEN 55
D-5100 AACHEN
WEST GERMANY

PROF. HELMER L. NIELSEN
SAN JOSE STATE UNIVERSITY
WASHINGTON SQUARE
SAN JOSE, CA 95192
U.S.A.

PROF. DR. HELGE NORSTRUD
DIVISION OF AERO- AND GAS DYNAMICS
TECHNICAL UNIVERSITY OF NORWAY
N-7034 TRONDHEIM-NTH
NORWAY

PROF. DR. H. OERTEL, JR.
INST. F. THEOR. STR.MECH. DFVLR
BUNSENSTR. 10
D-3400 GOETTINGEN
WEST GERMANY

PROF. KOICHI OSHIMA
ISAS, UNIVERSITY OF TOKYO
4-6-1 KOMABA, MEGURO-KU
TOKYO 153
JAPAN

DR. YUKO OSHIMA
DEPT. PHYS., OCHANOMIZU UNIVERSITY
OHTSUKA, BUNKYO-KU
TOKYO 112
JAPAN

PROF. DR. E. OUTA
DEPT. MECH. ENG.,WASEDA UNIV.
OKUBO 3-4-1, SHINJUKU
TOKYO
JAPAN

DR. MICHAEL PAGE
SCHOOL OF MATH.+PHYS.
UNIVERSITY OF EAST ANGLIA
NORWICH, NR4 7TJ
ENGLAND

PROF. MAURIZIO PANDOLFI
INSTITUT DE THERMIQUE APPLIQUEE
ECOLE POLYTECHNIQUE FEDERALE
LAUSANNE
SWITZERLAND

J. PERIAUX
AVIONS M. DASSAULT/BREGUET AVIAT.
B. P. 300
F-92214 ST CLOUD
FRANCE

PROF. N. PETERS
INST. F. ALLG. MECH.,RWTH AACHEN
TEMPLERGRABEN 55
D-5100 AACHEN
WEST GERMANY

PROF. ROGER PEYRET
UNIVERSITE DE NICE
IMSP-DEPT. DE MATH., PARC VALROSE
06034 NICE CEDEX
FRANCE

J. PIQUET
UNIVERSITE DE NANTES, E.N.S.M.
1, RUE DE LA NOE
44072 NANTES CEDEX
FRANCE

PROF. R. PIVA
INST. MECCANICA APPL., UNIVERSITA
VIA EUDOSSIANA 18
I-00184 ROMA
ITALY

PROF. DR. W. J. PROSNAK
POLSKIEJ AKADEMII NAUK
UL. MARUSARZOWNY 7/12
PL-80-287 GDANSK
POLEN

PROF. LUIGI QUARTAPELLE
INST. FISICA,POLIT. DI MILANO
PIAZA LEONARDO DA VINCI, 32
I-20133 MILANO
ITALY

MR. JOHN RAKICH
NASA AMES RESEARCH CENTER

MOFFETT FIELD, CA 95014
U.S.A.

DR. RONALD REHM
ADMIN. BLDG., RM A305
NATIONAL BUREAU OF STANDARDS
WASHINGTON, DC 20234
U.S.A.

PROF. N. RILEY
UNIVERSITY OF EAST ANGLIA
SCHOOL OF MATH. AND PHYSICS
NORWICH, NR4 7TJ
ENGLAND

DR. YERMIYAHU RIMON
NEVE SHAANAN
12 KALISHER ST.
HAIFA 32713
ISRAEL

DR. ARTHUR W. RIZZI
FFA, AERON. RES. INST. OF SWEDEN
BOX 11021
S-16111 BROMMA
SWEDEN

P. L. ROE
RAE, FLUID MECHANICS DIVISION

BEDFORD
ENGLAND

DR. HEINZ ROEHE
VOLKSWAGENWERK AG
FO-ANTRIEBSTECHNIK
D-3180 WOLFSBURG 1
WEST GERMANY

MOSHE ROSENFELD
DEPT. OF AERON. ENGINEERING
TECHNION, ISRAEL INST. OF TECHNOL.
HAIFA
ISRAEL

PH. ROY
ONERA
29, AVENUE DE LA DIVISION LECLERC
F-92320 CHATILLON
FRANCE

PROF. STANLEY G. RUBIN
DEPT. OF AERO. ENG. + APPL. MECH.
UNIVERSITY OF CINCINNATI
CINCINATTI, OH 45221
U.S.A.

PROF. DR. D. RUES
DFVLR SM-TS
BUNSENSTR. 10
D-3400 GOETTINGEN
WEST GERMANY

PROF. VIKTOR V. RUSANOV
KELDYSH INST. OF APPL. MATHEMATICS
MIUSSKAYA PL. 4
MOSCOW A-47
U.S.S.R.

DIPL. ENGNG. PENTTI SAARENRINNE
TAMPERE UNIVERSITY OF TECHNOLOGY
P.O. BOX 527
SF-33101 TAMPERE 10
FINLAND

MANUEL D. SALAS
NASA LANGLEY RESEARCH CENTER
MAIL STOP 360
HAMPTON, VA 23665
U.S.A.

PROF. ABDELAZIZ SALEM
LABORATOIRE DE MECANIQUE DES FLUID
USTHB
DAR EL BEIDA
ALGERIEN

ASS. PROF. NOBUYUKI SATOFUKA
DEPT. MECH. ENG.,KYOTO TECHN. UNIV
MATSUGASAKI, SAKYO-KU
KYOTO 606
JAPAN

HERBERT SCHAEPERTOENS
VW AG, FORSCHUNG AGGREGATETECHNIK

D-3180 WOLFSBURG
WEST GERMANY

DR. WOLFGANG H. P. SCHMIDT
DORNIER GMBH, DEPT. BF 30
POSTFACH 1420
D-7990 FRIEDRICHSHAFEN
WEST GERMANY

DIPL. ING. LUDWIG SCHMITT
LEHRST. F. STR.MECH., TU MUENCHEN
POSTFACH 202420
D-8000 MUENCHEN 2
WEST GERMANY

DIPL. MATH. GEZA SCHRAUF
INST. F. ANG. MATH., UNI BONN
BERINGSTR. 4
D-5300 BONN 1
WEST GERMANY

DR. ULRICH SCHUMANN
KERNFORSCHUNGSZENTRUM KARLSRUHE
POSTFACH 3640
D-7500 KARLSRUHE 1
WEST GERMANY

DR. ING. DIETER SCHWAMBORN
DFVLR INSTITUT SM-TS
BUNSENSTR. 10
D-3400 GOETTINGEN
WEST GERMANY

DR. ZVI SHIMONI
RAKAH INSTITUTE OF PHYSICS
HEBREW UNIVERSITY JERUSALEM
JERUSALEM
ISRAEL

DR. YURII SHOKIN
INST. FOR PURE A. APPL. MECHANICS

SU-630090 NOVOSIBIRSK 90
U.S.S.R.

JACQUES SIDES
ONERA
29 AV. DIVISION LECLERC
F-92320 CHATILLON
FRANCE

DR. SOUBBARAMAYER
CENTRE D'ETUDES NUCLEAIRS-SACLAY
BOITE POSTALE NO. 2
91191 GIF-SUR-YEVETTE CEDEX
FRANCE

DR. JERRY SOUTH
COMP.FL.DYN.GR. AT SUBS. AERO. BR.
NASA LANGLEY RESEARCH CENTER
HAMPTON, VA 23665
U.S.A.

DR. BERNARD STEVERDING
U.S. ARMY RES. + STANDARDIZATION
223 OLD MARYLEBONE RD.
LONDON W1
ENGLAND

M. D. SU
LEHRST. F. STR.MECH.,TU MUENCHEN
ARCISSTR. 21
D-8000 MUENCHEN
WEST GERMANY

U. M. SULTANGAZIN
INST. OF THEOR. + APPL. MECHANICS
USSR ACADEMY OF SCIENCES
SU-630090 NOVOSIBIRSK 90
U.S.S.R.

DR. KLAUS TAUBERT
INST. F. ANG. MATH., UNI HAMBURG
BUNDESSTR. 55
D-2000 HAMBURG 13
WEST GERMANY

DR. THOMAS D. TAYLOR
APPL. MECH. LAB.,J. HOPKINS UNIV.
JOHNS HOPKINS ROAD
LAUREL, MD 20810
U.S.A.

PROF. LU TING
COURANT INST., NEW YORK UNIV.
251 MERCER STREET
NEW YORK, NY 10012
U.S.A.

PROF. P. I. TSHUSHKIN
COMP. CENTRE, ACAD. OF SCIENCE
VAVILOVA 40
MOSCOW B-333
U.S.S.R.

PROF. ELI TURKEL
DEP. OF MATH., TEL-AVIV UNIVERSITY
RAMAT AVIV
TEL AVIV
ISRAEL

DR. FRITS VAN BECKUM
DEPT. MATH.,TWENTE UNIV. TECHNOL.
P.O.B. 217
NL-7500 AE ENSCHEDE
THE NETHERLANDS

PROF. DR. ADRIAAN VAN DE VOOREN
MATHEMATICAL INSTITUTE
P.O. BOX 800
NL-9700 AV GRONINGEN
THE NETHERLANDS

DR. BRAM VAN LEER
HUYGENS LAB., STERREWACHT LEIDEN
POSTBUS 9513
NL-2300 RA LEIDEN
THE NETHERLANDS

DR. ARTHUR E. P. VELDMAN
NATIONAL AEROSPACE LABORATORY NLR
P.O. BOX 90502
NL-1006 BM AMSTERDAM
THE NETHERLANDS

ING. MARCEL VELLUET
D.R.E.T.
26 BOULEVARD VICTOR
F-75015 PARIS
FRANCE

YVON VIGNERON
A/DET/EG/AERO,S.N.IND. AEROSPAT.
316 ROUTE DE RAYONNE
F-31053 TOULOUSE
FRANCE

DR. HENRI VIVIAND

20 RUE DE PROVENCE
F-78000 VERSAILLES
FRANCE

DR. CLAUS WEILAND
MBB UF
POSTFACH 80 11 60
D-8000 MUENCHEN 80
WEST GERMANY

DR. BERNARD C. WEINBERG
SCIENTIFIC RESEARCH ASS., INC.
P. O. BOX 498
GLASTONBURY, CT 06033
U.S.A.

PROF. PIETER WESSELING
DEPT. MATH.+INF.,DELFT UNIV.TECHN.
JULIANALAAN 132
NL-2628 BL DELFT
THE NETHERLANDS

PROF. DAVID WHITFIELD
DEP. OF AEROSP. ENGR., MISS. STATE
DRAWER A
MISSISSIPPI STATE, MS 39762
U.S.A.

PROF. DR. KRISTIAN WITSCH
MATHEMAT. INST., UNI DUESSELDORF
UNIVERSITAETSSTR. 1
D-4000 DUESSELDORF 1
WEST GERMANY

DR. M. WOLFSHSTEIN
AERONAUTICAL ENG. DEPT.
TECHNION CITY
HAIFA
ISRAEL

DR. YAU SHU WONG
NASA LANGLEY RESEARCH CENTER
ICASE, MAIL STOP 132C
HAMPTON, VA 23665
U.S.A.

DR. STEPHEN F. WORNOM
O N E R A
3, ALLEE RAVEL
F-92320 CHATILLON
FRANCE

GUO-RONG XU
COMP. CENTER ACADEMIA SINICA
ZHONGGUANCUN
BEIJING
CHINA

PROF. NICKOLAY N. YANENKO
INST. OF THEOR.+APPLIED MECHANICS
U.S.S.R. ACADEMY OF SCIENCES
NOVOSIBIRSK 630090
U.S.S.R.

DR. HELEN M. C. YEE
NASA AMES RES. C.,COMP. FL.DYN.BR.
MS 202A-1
MOFFETT FIELD, CA 94035
U.S.A.

PROF. SHEE-MANG YEN
DEPT. OF AERONAUT. + ASTRON. ENG.
UNIVERSITY OF ILLINOIS
URBANA, IL 61801
U.S.A.

PROF. PIETER J. ZANDBERGEN
DEPT. OF APPLIED MATHEMATICS
TECHNISCHE HOGESCHOOL TWENTE
POSTBUS 217, ENSCHEDE 7500 AE
THE NETHERLANDS

THOMAS A. ZANG
NASA LANGLEY RESEARCH CENTER
MS 163
HAMPTON, VA 73665
U.S.A.

DR. LUCA ZANNETTI
IST. DI MACCHINE E MOT. P. AEROM.
POLITECNICO DI TORINO
TORINO 10129
ITALY

ENG. PEIYE ZHU
INST. A MECHAN.,UNIV. STUTTGART
PFAFFENWALDRING 9
D-7000 STUTTGART 80
WEST GERMANY

ASS. PROF. YOU-LAN ZHU
COMP. CENTER OF ACAD. SINICA
ZHONGGUANCUN
BEIJING
CHINA

Lecture Notes in Physics

Vol. 136: The Role of Coherent Structures in Modelling Turbulence and Mixing. Proceedings 1980. Edited by J. Jimenez. XIII, 393 pages. 1981.

Vol. 137: From Collective States to Quarks in Nuclei. Edited by H. Arenhövel and A. M. Saruis. VII, 414 pages. 1981.

Vol. 138: The Many-Body Problem. Proceedings 1980. Edited by R. Guardiola and J. Ros. V, 374 pages. 1981.

Vol. 139: H. D. Doebner, Differential Geometric Methods in Mathematical Physics. Proceedings 1981. VII, 329 pages. 1981.

Vol. 140: P. Kramer, M. Saraceno, Geometry of the Time-Dependent Variational Principle in Quantum Mechanics. IV, 98 pages. 1981.

Vol. 141: Seventh International Conference on Numerical Methods in Fluid Dynamics. Proceedings. Edited by W. C. Reynolds and R. W. MacCormack. VIII, 485 pages. 1981.

Vol. 142: Recent Progress in Many-Body Theories. Proceedings. Edited by J. G. Zabolitzky, M. de Llano, M. Fortes and J. W. Clark. VIII, 479 pages. 1981.

Vol. 143: Present Status and Aims of Quantum Electrodynamics. Proceedings, 1980. Edited by G. Gräff, E. Klempt and G. Werth. VI, 302 pages. 1981.

Vol. 144: Topics in Nuclear Physics I. A Comprehensive Review of Recent Developments. Edited by T.T.S. Kuo and S.S.M. Wong. XX, 567 pages. 1981.

Vol. 145: Topics in Nuclear Physics II. A Comprehensive Review of Recent Developments. Proceedings 1980/81. Edited by T. T. S. Kuo and S. S. M. Wong. VIII, 571-1.082 pages. 1981.

Vol. 146: B. J. West, On the Simpler Aspects of Nonlinear Fluctuating. Deep Gravity Waves. VI, 341 pages. 1981.

Vol. 147: J. Messer, Temperature Dependent Thomas-Fermi Theory. IX, 131 pages. 1981.

Vol. 148: Advances in Fluid Mechanics. Proceedings, 1980. Edited by E. Krause. VII, 361 pages. 1981.

Vol. 149: Disordered Systems and Localization. Proceedings, 1981. Edited by C. Castellani, C. Castro, and L. Peliti. XII, 308 pages. 1981.

Vol. 150: N. Straumann, Allgemeine Relativitätstheorie und relativistische Astrophysik. VII, 418 Seiten. 1981.

Vol. 151: Integrable Quantum Field Theory. Proceedings, 1981. Edited by J. Hietarinta and C. Montonen. V, 251 pages. 1982.

Vol. 152: Physics of Narrow Gap Semiconductors. Proceedings, 1981. Edited by E. Gornik, H. Heinrich and L. Palmetshofer. XIII, 485 pages. 1982.

Vol. 153: Mathematical Problems in Theoretical Physics. Proceedings, 1981. Edited by R. Schrader, R. Seiler, and D.A. Uhlenbrock. XII, 429 pages. 1982.

Vol. 154: Macroscopic Properties of Disordered Media. Proceedings, 1981. Edited by R. Burridge, S. Childress, and G. Papanicolaou. VII, 307 pages. 1982.

Vol. 155: Quantum Optics. Proceedings, 1981. Edited by C.A. Engelbrecht. VIII, 329 pages. 1982.

Vol. 156: Resonances in Heavy Ion Reactions. Proceedings, 1981. Edited by K.A. Eberhard. XII, 448 pages. 1982.

Vol. 157: P. Niyogi, Integral Equation Method in Transonic Flow. XI, 189 pages. 1982.

Vol. 158: Dynamics of Nuclear Fission and Related Collective Phenomena. Proceedings, 1981. Edited by P. David, T. Mayer-Kuckuk, and A. van der Woude. X, 462 pages. 1982.

Vol. 159: E. Seiler, Gauge Theories as a Problem of Constructive Quantum Field Theory and Statistical Mechanics. V, 192 pages. 1982.

Vol. 160: Unified Theories of Elementary Particles. Critical Assessment and Prospects. Proceedings, 1981. Edited by P. Breitenlohner and H.P. Dürr. VI, 217 pages. 1982.

Vol. 161: Interacting Bosons in Nuclei. Proceedings, 1981. Edited by J.S. Dehesa, J.M.G. Gomez, and J. Ros. V, 209 pages. 1982.

Vol. 162: Relativistic Action at a Distance: Classical and Quantum Aspects. Proceedings, 1981. Edited by J. Llosa. X, 263 pages. 1982.

Vol. 163: J. S. Darrozes, C. Francois, Mécanique des Fluides Incompressibles. XIX, 459 pages. 1982.

Vol. 164: Stability of Thermodynamic Systems. Proceedings, 1981. Edited by J. Casas-Vázquez and G. Lebon. VII, 321 pages. 1982.

Vol. 165: N. Mukunda, H. van Dam, L.C. Biedenharn, Relativistic Models of Extended Hadrons Obeying a Mass-Spin Trajectory Constraint. Edited by A. Böhm and J.D. Dollard. VI, 163 pages. 1982.

Vol. 166: Computer Simulation of Solids. Edited by C.R.A. Catlow and W.C. Mackrodt. XII, 320 pages. 1982.

Vol. 167: G. Fieck, Symmetry of Polycentric Systems. VI, 137 pages, 1982.

Vol. 168: Heavy-Ion Collisions. Proceedings, 1982. Edited by G. Madurga and M. Lozano. VI, 429 pages. 1982.

Vol. 169: K. Sundermeyer, Constrained Dynamics. IV, 318 pages. 1982.

Vol. 170: Eighth International Conference on Numerical Methods in Fluid Dynamics. Proceedings, 1982. Edited by E. Krause. X, 569 pages. 1982.

Selected Issues from

Lecture Notes in Mathematics

Vol. 809: I. Gumowski and Ch. Mira, Recurrences and Discrete Dynamic Systems. VI, 272 pages. 1980.

Vol. 810: Geometrical Approaches to Differential Equations. Proceedings 1979. Edited by R. Martini. VII, 339 pages. 1980.

Vol. 816: L. Stoica, Local Operators and Markov Processes. VIII, 104 pages. 1980.

Vol. 819: Global Theory of Dynamical Systems. Proceedings, 1979. Edited by Z. Nitecki and C. Robinson. IX, 499 pages. 1980.

Vol. 821: Statistique non Paramétrique Asymptotique. Proceedings, 1979. Edited by J.-P. Raoult. VII, 175 pages. 1980.

Vol. 826: Ph. G. Ciarlet et P. Rabier, Les Equations de von Kármán. VI, 181 pages. 1980.

Vol. 827: Ordinary and Partial Differential Equations. Proceedings, 1978. Edited by W. N. Everitt. XVI, 271 pages. 1980.

Vol. 831: Representation Theory I. Proceedings, 1979. Edited by V. Dlab and P. Gabriel. XIV, 373 pages. 1980.

Vol. 832: Representation Theory II. Proceedings, 1979. Edited by V. Dlab and P. Gabriel. XIV, 673 pages. 1980.

Vol. 836: Differential Geometrical Methods in Mathematical Physics. Proceedings, 1979. Edited by P. L. García, A. Pérez-Rendón, and J. M. Souriau. XII, 538 pages. 1980.

Vol. 837: J. Meixner, F. W. Schäfke and G. Wolf, Mathieu Functions and Spheroidal Functions and their Mathematical Foundations Further Studies. VII, 126 pages. 1980.

Vol. 838: Global Differential Geometry and Global Analysis. Proceedings 1979. Edited by D. Ferus et al. XI, 299 pages. 1981.

Vol. 840: D. Henry, Geometric Theory of Semilinear Parabolic Equations. IV, 348 pages. 1981.

Vol. 841: A. Haraux, Nonlinear Evolution Equations- Global Behaviour of Solutions. XII, 313 pages. 1981.

Vol. 842: Séminaire Bourbaki vol. 1979/80. Exposés 543–560. IV, 317 pages. 1981.

Vol. 843: Functional Analysis, Holomorphy, and Approximation Theory. Proceedings. Edited by S. Machado. VI, 636 pages. 1981.

Vol. 845: A. Tannenbaum, Invariance and System Theory: Algebraic and Geometric Aspects. X, 161 pages. 1981.

Vol. 849: P. Major, Multiple Wiener-Itô Integrals. VII, 127 pages. 1981.

Vol. 851: Stochastic Integrals. Proceedings, 1980. Edited by D. Williams. IX, 540 pages. 1981.

Vol. 852: L. Schwartz, Geometry and Probability in Banach Spaces. X, 101 pages. 1981.

Vol. 856: R. Lascar, Propagation des Singularités des Solutions d'Equations Pseudo-Différentielles à Caractéristiques de Multiplicités Variables. VIII, 237 pages. 1981.

Vol. 858: E. A. Coddington, H. S. V. de Snoo: Regular Boundary Value Problems Associated with Pairs of Ordinary Differential Expressions. V, 225 pages. 1981.

Vol. 861: Analytical Methods in Probability Theory. Proceedings 1980. Edited by D. Dugué, E. Lukacs, V. K. Rohatgi. X, 183 pages. 1981.

Vol. 866: J.-M. Bismut, Mécanique Aléatoire. XVI, 563 pages. 1981.

Vol. 878: Numerical Solution of Nonlinear Equations. Proceedings, 1980. Edited by E. L. Allgower, K. Glashoff, and H.-O. Peitgen. XIV, 440 pages. 1981.

Vol. 881: R. Lutz, M. Goze, Nonstandard Analysis. XIV, 261 pages. 1981.

Vol. 888: Padé Approximation and its Applications. Proceedings, 1980. Edited by M. G. de Bruin and H. van Rossum. VI, 383 pages. 1981.

Vol. 898: Dynamical Systems and Turbulence, Warwick, 1980. Proceedings. Edited by D. Rand and L.-S. Young. VI, 390 pages. 1981.

Vol. 901: Séminaire Bourbaki vol. 1980/81 Exposés 561–578. III, 299 pages. 1981.

Vol. 904: K. Donner, Extension of Positive Operators and Korovkin Theorems. XII, 182 pages. 1982.

Vol. 905: Differential Geometric Methods in Mathematical Physics. Proceedings, 1980. Edited by S.J. Andersson, H.-D. Doebner, and. H.R. Petry. VI, 309 pages. 1982.

Vol. 909: Numerical Analysis. Proceedings, 1981. Edited by J.P. Hennart. VII, 247 pages. 1982.

Vol. 912: Numerical Analysis. Proceedings, 1981. Edited by G. A. Watson. XIII, 245 pages. 1982.

Vol. 920: Séminaire de Probabilités XVI, 1980/81. Proceedings. Edité par J. Azéma et M. Yor. V, 622 pages. 1982.

Vol. 921: Séminaire de Probabilités XVI, 1980–81 Supplément: Géométrie Différentielle Stochastique. Proceedings. Edité par J. Azéma et M. Yor. III, 285 pages. 1982.

Vol. 922: B. Dacorogna, Weak Continuity and Weak Lower Semicontinuity of Non-Linear Functionals. V, 120 pages. 1982.

Vol. 923: Functional Analysis in Markov Processes. Proceedings, 1981. Edited by M. Fukushima. V, 307 pages. 1982.

Vol. 926: Geometric Techniques in Gauge Theories. Proceedings, 1981. Edited by R. Martini and E.M.de Jager. IX 219 pages. 1982.

Vol. 927: Y. Z. Flicker, The Trace Formula and Base Change for GL (3). XII, 204 pages. 1982.

Vol. 928: Probability Measures on Groups. Proceedings 1981. Edited by H. Heyer. X, 477 pages. 1982.

Vol. 929: Ecole d'Eté de Probabilités de Saint-Flour X – 1980. Proceedings, 1980. Edited by P.L. Hennequin. X, 313 pages. 1982.

Vol. 930: P. Berthelot, L. Breen, et W. Messing, Théorie de Dieudonné Cristalline II. XI, 261 pages. 1982.

Vol. 931: D.M. Arnold, Finite Rank Torsion Free Abelian Groups and Rings. VII, 191 pages. 1982.

Vol. 932: Analytic Theory of Continued Fractions. Proceedings, 1981. Edited by W.B. Jones, W.J. Thron, and H. Waadeland. VI, 240 pages. 1982.

Vol. 934: M. Sakai, Quadrature Domains. IV, 133 pages. 1982.

Vol. 935: R. Sot, Simple Morphisms in Algebraic Geometry. IV, 146 pages. 1982.

Vol. 936: S.M. Khaleelulla, Counterexamples in Topological Vector Spaces. XXI, 179 pages. 1982.

Vol. 937: E. Combet, Intégrales Exponentielles. VIII, 114 pages. 1982.

Vol. 938: Number Theory. Proceedings, 1981. Edited by K. Alladi. IX, 177 pages. 1982.

Vol. 942: Theory and Applications of Singular Perturbations. Proceedings, 1981. Edited by W. Eckhaus and E.M. de Jager. V, 363 pages. 1982.

Vol. 953: Iterative Solution of Nonlinear Systems of Equations. Proceedings, 1982. Edited by R. Ansorge, Th. Meis, and W. Törnig. VII, 202 pages. 1982.

Lecture Notes in Mathematics